T0281680

NONEQUILIBRIUM QUANTUM FIELD THEORY

Research into the nonequilibrium processes of quantum many body systems, and the statistical mechanical properties of interacting quantum fields, constitutes a fast developing and widely applicable area of theoretical physics.

Bringing together the key ideas from nonequilibrium statistical mechanics and powerful methodology from quantum field theory, this book captures the essence of nonequilibrium quantum field theory.

Beginning with the foundational aspects of the theory, the book presents important concepts and useful techniques, discusses issues of basic interest such as decoherence and entropy generation, and shows how thermal field, linear response, kinetic theories and hydrodynamics emerge. It also illustrates how these concepts and methodology are applied to current research topics such as nonequilibrium phase transitions, thermalization in relativistic heavy ion collisions, the nonequilibrium dynamics of Bose-Einstein condensation, and the generation of structures from quantum fluctuations in the early Universe.

The book is divided into five parts, with each part addressing a particular stage in the conceptual and technical development of the subject. Full derivations or detailed plausibility arguments are presented throughout. This self-contained book is a valuable reference for graduate students and researchers in particle physics, gravitation, cosmology, atomic-optical and condensed matter physics.

This title, first published in 2009, has been reissued as an Open Access publication on Cambridge Core.

ESTEBAN A. CALZETTA is a Professor in the Departamento de Física at the Universidad de Buenos Aires and Researcher at CONICET, Argentina.

BEI-LOK B. HU is a Professor in the Department of Physics and a Fellow of the Joint Quantum Institute at the University of Maryland, College Park.

CAMBRIDGE MONOGRAPHS ON MATHEMATICAL PHYSICS

General editors: P. V. Landshoff, D. R. Nelson, S. Weinberg

J. Polchinski *String Theory Volume 2: Superstring Theory and Beyond*
V. N. Popov *Functional Integrals and Collective Excitations*[†]
R. J. Rivers *Path Integral Methods in Quantum Field Theory*[†]
R. G. Roberts *The Structure of the Proton: Deep Inelastic Scattering*[†]
C. Rovelli *Quantum Gravity*[†]
W. C. Saslaw *Gravitational Physics of Stellar and Galactic Systems*[†]
H. Stephani, D. Kramer, M. MacCallum, C. Hoenselaers and E. Herlt *Exact Solutions of Einstein's Field Equations, 2nd edition*
J. Stewart *Advanced General Relativity*[†]
T. Thiemann *Modern Canonical Quantum General Relativity*
D. J. Toms *The Schwinger Action Principle and Effective Action*
A. Vilenkin and E. P. S. Shellard *Cosmic Strings and Other Topological Defects*[†]
R. S. Ward and R. O. Wells Jr *Twistor Geometry and Field Theory*[†]
J. R. Wilson and G. J. Mathews *Relativistic Numerical Hydrodynamics*

[†]Issued as a paperback

Nonequilibrium
Quantum Field Theory

ESTEBAN A. CALZETTA
University of Buenos Aires and CONICET

BEI-LOK B. HU
University of Maryland

CAMBRIDGE
UNIVERSITY PRESS

CAMBRIDGE
UNIVERSITY PRESS

Shaftesbury Road, Cambridge CB2 8EA, United Kingdom

One Liberty Plaza, 20th Floor, New York, NY 10006, USA

477 Williamstown Road, Port Melbourne, VIC 3207, Australia

314–321, 3rd Floor, Plot 3, Splendor Forum, Jasola District Centre, New Delhi – 110025, India

103 Penang Road, #05–06/07, Visioncrest Commercial, Singapore 238467

Cambridge University Press is part of Cambridge University Press & Assessment, a department of the University of Cambridge.

We share the University's mission to contribute to society through the pursuit of education, learning and research at the highest international levels of excellence.

www.cambridge.org
Information on this title: www.cambridge.org/9781009289986

DOI: 10.1017/9781009290036

First published 2009
Reissued as OA 2022

A catalogue record for this publication is available from the British Library.

ISBN 978-1-009-28998-6 Hardback
ISBN 978-1-009-29002-9 Paperback

Dedicated to—
María Isabel and Francisco from E.C.
Tung-Hui, Tung-Fei and Chun-Chu from B-L.

Contents

III GAUGE INVARIANCE, DISSIPATION, ENTROPY, NOISE AND DECOHERENCE

Preface

In the last decade or two we see increasing research activities in areas where quantum field processes of nonequilibrium many-body systems prevail. This includes nuclear particle physics in the relativistic heavy ion collision (RHIC) experiments, early universe cosmology in the wake of high-precision observations (such as WMAP), cold atom (such as Bose–Einstein) condensation (BEC) physics in highly controllable environments, quantum mesoscopic processes and collective phenomena in condensed matter systems. There is a demand for a new set of tools and concepts from quantum field theory to treat the nonequilibrium dynamics of relativistic many-particle systems and for the understanding of basic issues like dissipation, entropy, fluctuations, noise and decoherence in these systems.

The subject matter of this book is at the intersection of nonequilibrium statistical mechanics (NEqSM) and quantum field theory (QFT). It deals with the nonequilibrium quantum processes of relativistic many-body systems with techniques from quantum field theory. To a lesser extent it also touches on the nonequilibrium statistical mechanical aspects of interacting quantum field theory itself. This subject matter is a natural extension of thermal field theory from equilibrium (finite temperature) to nonequilibrium systems. One major technical challenge is that the usual Euclidean or imaginary time quantum field theoretical methods applicable to stationary quantum systems are no longer valid (except for linear response in near-equilibrium conditions) and real-time formulations are required.

The book has five parts: The first part comprising Chapters 1–3 deals with the basics. After an introductory chapter on basic notions and issues in NEqSM, two chapters are devoted to the basic ideas and techniques of nonequilibrium systems. The second part comprising Chapters 4–6 begins with Chapter 4 on quantum field processes in dynamical backgrounds. Chapters 5–6 form the backbone of the book, in establishing the real-time quantum field theory framework based on the so-called closed time path (CTP or Schwinger–Keldysh) effective action and the influence functional (IF, or Feynman–Vernon) formalisms. This is followed by three chapters in Part III to illustrate the use of these formalisms for addressing issues like gauge invariance, dissipation, entropy, noise and decoherence. From these formalisms we proceed to Part IV, including Chapters 10–12 in the development of thermal, kinetic and hydrodynamics theories for interacting quantum fields, with linear response and thermal field theory as the near-equilibrium limits. Part V of this book shows how to apply this body of knowledge with examples drawn from three areas: Bose–Einstein condensates (BEC), relativistic heavy ion collisions (RHIC) and early universe cosmology discussed in Chapters 13–15 respectively. The range of application is much broader than that represented by these chosen examples.

We assume the reader has a good knowledge of quantum field theory as taught in a typical first or second year graduate course, with standard textbooks, but we do not require the reader to have much more knowledge of nonequilibrium statistical mechanics beyond those discussed (regrettably light) in a standard first/second year statistical mechanics course. Prior exposure to nonrelativistic many-body theory is helpful but not required, so is thermal field theory, as they will be developed as subcases of the fully relativistic nonequilibrium quantum field theory, the main theme of this book.

Below is a quick guide to the use of this book for readers with different backgrounds. Readers with some good understanding of NEqSM may go directly to Part II while readers familiar with the CTP-IF formalism may start with Part I and go to Part III. Readers more interested in the structure of the kinetic and hydrodynamic theories can delve into Part IV after Part II, while readers more interested in statistical mechanical issues manifested in quantum field theory may want to focus on Chapters 1, 8, 9. Recognizing that readers may come from different disciplines with solid knowledge of their own field who want to learn nonequilibrium quantum field theory for applications to their own problems, we suggest the following streams:

(1) Atomic-optical and condensed matter physics: Chapters 2, 3, 5, 6, 8, 10, 11, 13
(2) Nuclear-particle physics: Chapters 2, 3, 4, 5, 6, 7, 8, 9, 10, 11, 12, 13, 14
(3) Gravitation and cosmology: Chapters 1, 2, 3, 4, 5, 6, 8, 9, 10, 11, 12, 15

A book on this relatively new but fast developing subject can be of some substance or usefulness only if the authors make a serious attempt to capture or represent the collective research effort by their colleagues working in this field. With this belief we sent out each chapter before it saw its final form to experts on that particular topic, and were blessed with many careful comments and insightful suggestions. For this we are indebted to Alejandra Kandus, Gabor Kunstatter, Da-Shin Lee, Daniel Litim, Fernando Lombardo, Sabino Matarrese, Diego Mazzitelli, Stanislav Mrowczynski, Marcello Musso, Kin-Wang Ng, Juan Pablo Paz, Robert Pisarski, Ana Maria Rey, Antonio Riotto, Ray Rivers, Albert Roura, Dam Son, Rafael Sorkin, Enric Verdaguer, Alexander Vilenkin, Serge Winitzki and Laurence Yaffe. We received able help from A. Eftekharzadeh and Taihung Wu in the preparation of the bibliography and from Chad Galley in checking the consistency of some notations and conventions.

Many of our colleagues and co-workers have contributed to our understanding and development of nonequilibrium quantum field theory over the years. We would like to thank Mario Castagnino, the late Bryce DeWitt, J. Robert

Dorfman, Michael E. Fisher, James B. Hartle, Werner Israel, Leonard E. Parker, Zhao-Bin Su, John A. Wheeler, Lu Yu and Robert Zwanzig, as well as our former postdocs and students who worked with us on this subject.

We also wish to thank Simon Capelin and Rufus Neel, senior editors of Cambridge University Press for their sustained interest and patience, and Lindsay Barnes for advice in the production of the book.

The writing of this book as an intellectual challenge was a joy for the authors, but it also meant significant sacrifice at the personal level for a long duration, taking our time and attention away from our families and friends; without their understanding and forbearance it would not have been possible.

EC dedicates this book to María Isabel and Francisco. To be able to dedicate it to them was one of the main reasons to write it in the first place.

BLH would like to take this special opportunity to express his sincere appreciation to his wife of 35 years, Chun-Chu Yee, for her companionship in meeting the challenges of life, inward and outward; and her understanding and forbearance at both the ethereal and mundane levels of coexistence and communion. He wants to tell his beloved son Tung-Hui and daughter Tung-Fei that they are the best that have ever happened to him and the most precious in his life: Just being with you or simply hearing from you gives me the greatest pleasure. I value you each for being a fiercely independent individual with such keen or even painful senses of being, and yet able to reach out and connect with the closest and farthest reaches of humanity. He wants to express his love and appreciation to his brothers Bambi Hu and Shiu-Lok Hu, and to his cousin Kwen-Wai Lau for their constant encouragement and unfailing support. Finally he expresses his deep gratitude and indebtedness to his late parents Mr. and Mrs. I-Ping and Pie Wang Hu who raised and cared for their sons with love and dedication, inspiring within them the virtues of kindness and generosity; and to his aunt Sheng-Shuen Hu for her selfless devotion to our family. He is grateful to his many loving friends and the many lovely persons who share with him the laughter and tears, the daring and the follies of life.

Part I

Fundamentals of nonequilibrium
statistical mechanics

1

Basic issues in nonequilibrium statistical mechanics

Perhaps due to its technical complexity, oftentimes one sees in research papers on nonequilibrium quantum field theory (NEqQFT) more emphasis placed on the field-theoretical formalisms than the ideas these sophisticated techniques attempt to capture, or the issues such problems embody. All the more so, we need some basic understanding of the important issues and concepts in nonequilibrium statistical mechanics (NEqSM), and how they are manifested in the context of quantum field theory. Many important advances in this field came from asking such questions and finding out how to answer them in the language of quantum field theory. Because of this somewhat skewed existing emphasis in NEqQFT, and since we do not assume the reader to have had a formal course on NEqSM before, we shall give a brief summary of the basic concepts of NEqSM relevant to the field-theoretical processes discussed in this book. Many fine monographs and reviews written on this subject take a more formal mathematical approach. Since our purpose here is to familiarize readers with these issues and their subtleties, rather than training them to work in the rich field of NEqSM (which includes in addition to the traditional subject matter such as the projection operator formalism and open system concepts, also current topics at their foundation, such as dynamical systems and quantum chaos), we choose to approach these topics in a more intuitive and physical way, sacrificing by necessity rigor and completeness.

We first examine some commonly encountered physical processes and try to bring out in each a different key concept in NEqSM. To have a concrete bearing and a common ground, let us focus on just one such issue which is of paramount importance and poses a constant challenge to theoretical physicists: How does apparent irreversibility in the macroscopic world arise from the time-reversal invariant laws of microphysics [Leb93, HaPeZu94, Mac92, Sch97]?

We begin with an analysis of the nature and origin of irreversibility in well-known physical processes such as dispersion (referring in the specific context here to the divergence of neighboring trajectories in configuration or phase space due to dynamical instability), diffusion, dissipation and mixing. We will seek the microdynamical basis of these processes and clarify the distinction between processes whose irreversibility arises from the stipulation of special initial conditions, and those arising from the system's interaction with a coarse-grained environment. It is beneficial to keep in mind these processes and the issues they embody when we begin our study of quantum field processes so that they will not be marred by the technical complexity of quantum field theory. We can ask questions such as (1) "What is the entropy generation from particle creation in an external field or a dynamical spacetime, as in cosmology?"; (2) "How could an interacting field thermalize?"; (3) "Is there irreversibility associated with quantum fluctuations in field theoretical processes like particle creation?" Or, more boldly, "Can the 'birth of the Universe' be viewed as a large fluctuation?" "Might it not happen at all – a 'still' birth – due to the powerful dissipative effects of particle creation which suppresses the tunneling rate?" (4) "Can one use thermodynamic relations to characterize certain quantum field processes?"

These questions reveal how deeply one can probe into the NEqSM features of quantum field theory and how quantum field processes can lend themselves to statistical mechanical and thermodynamic depiction or characterization. Asking question (1) reveals the differences resulting from many levels of coarse graining between a quantum field understanding of particle creation processes (no entropy production because the vacuum is a pure state) and a thermodynamic description (yes, entropy is proportional to the number of particles produced). Asking question (2) forces us to reckon with the intricate NEqSM features of an interacting quantum field such as how a correlation entropy can be defined from the Schwinger–Dyson hierarchy. These aspects are not usually discussed in quantum field theory textbooks. The first part of question (3) brings out the often used yet poorly understood aspects of noise – beginning with quantum noise associated with vacuum fluctuations, properties of multiplicative colored noise, and nonlocal dissipation and their effects on the dynamical processes. The second part of question (3) is the so-called "back-reaction" effect of quantum fields on a background field or background spacetime. Question (4) asks if this effect can have a thermodynamic interpretation. To the degree that thermodynamics is the long-wavelength, heavily coarse-grained limit of microphysics and quantum field theory is a theory of microphysics, we certainly expect such relations to exist and their discovery will reveal the relation between micro–macro and quantum-to-classical transitions. A well-known relation is the black hole thermodynamics of Bekenstein [Bek73] and the quantum Hawking radiation [Haw75]. Sciama [Sci79] suggested that this can be understood from the viewpoint of quantum dissipative systems. This view also applies to dissipation of anisotropy in the early universe due to particle creation from the vacuum. We will find out later that both for the black hole and the early universe these processes can indeed be understood as manifestations of a fluctuation–dissipation relation, relating fluctuations of quantum fields to dissipation in the dynamics of the background field or spacetime.

1.1 Macroscopic description of physical processes

Let us begin by examining a few examples of irreversible processes to illustrate their different natures and origins. Consider the following processes:

Dispersion
Diffusion
Dissipation
Relaxation
Mixing
Recurrence
Decoherence
Recoherence

They contain different aspects of irreversibility. The usage of these terms appearing in general-purpose books could be rather loose or even confusing. For example, diffusion, relaxation and dissipation are often seen used interchangeably. Even the same word could mean different things in different contexts. For example, classical diffusion is often viewed as a form of dissipation, while quantum diffusion refers to phase dispersion, usually occurs at a much faster time-scale and is more closely related to decoherence than dissipation. We will discuss quantum phenomena in Chapter 3. Here we will focus on the first six such processes listed above and aim at providing some microdynamics basis to these processes in order to give them a more precise meaning. In so doing we hope to elucidate some basic notions and issues of NEqSM through examples.

We first highlight the distinction between dissipative processes (which are always irreversible) and irreversible or "apparently" irreversible processes (which are not necessarily dissipative). For example, in elastic scattering, neighboring trajectories diverge exponentially fast. This is characteristic of mixing systems, which are reversible. Relaxation and diffusion referring to dissipative systems are irreversible. They are mixing systems with some type of coarse graining introduced. As we shall see, not any type of coarse graining leads to irreversibility. Many factors enter, such as the large size of the system, the particular initial conditions chosen, or the time-scales at work. This is where it calls for special caution in doing the analysis. Better understanding of the chaotic behavior in classical molecular dynamics has provided a firmer microscopic basis for nonequilibrium statistical mechanics. Such studies for quantum systems are less developed and for this reason we shall refrain from describing them. In Chapter 3 we shall have occasion to discuss quantum decoherence and dissipation where the interplay of quantum and thermal fluctuations in the environment and their effects on the system will be discussed. We shall also revisit these issues of irreversibility and approach to equilibrium in Chapter 12.

A. Dispersion

Consider a system of dilute gas made up of interacting particles modeled as hard spheres with diameter d. For simplicity, let us work in two dimensions with hard disks. (Our illustration here follows [Gas98]; see also [Ma85] which contains excellent conceptual discussions.) Assume the particles move with constant velocity v

and traverse a distance given by the mean free path $\ell \gg d$ before colliding with another particle elastically. The trajectory of any particle governed by the laws of mechanics is of course reversible in time. However, upon just a few collisions two neighboring trajectories will deviate from each other very rapidly if the scattering surface is convex, as a sphere is. To see this, let's set our stop watch time zero $(t = 0)$ right after the first collision (call this collision the $n = 0$ one) and follow the particle's trajectory for n subsequent collisions. Call the scattering angle of the first collision $\theta(0)$ and the uncertainty associated with it $\delta\theta(0)$ and likewise for the scattering angle after an additional n collisions $\theta(t)$ and its uncertainty $\delta\theta(t)$. For each additional collision the uncertainty in the scattering angle increases by a factor of ℓ/d deduced from the simple trigonometry of incident and scattered trajectories. So after n collisions then

$$|\delta\theta(t)| \sim |\delta\theta(0)|^n \equiv |\delta\theta(0)|e^{\lambda t} \qquad (1.1)$$

The second equivalence relation above defines the parameter λ, which is called the Lyapunov exponent (actually its maximal value enters into this expression). The time for n successive collisions is given by $t = n\tau$ where τ is the time between collisions related to the mean free path ℓ by $v = \ell/\tau$. Thus the (maximal) Lyapunov exponent is given by

$$\lambda \sim \frac{1}{\tau}\ln\frac{\ell}{d} \qquad (1.2)$$

This simple way of estimating the maximum Lyapunov exponent first given by Krylov [Kry44, Kry79] remains very useful in illustrating the elemental process of divergence of neighboring trajectories due to dynamical instability, referred to here as "dispersion" for short. For hard sphere collisions we see that after a sufficiently long time $|\delta\theta(t)| \approx 1$, the exit direction becomes completely indeterminate due to the accumulated error.

The asymmetry in the initial and final conditions of the collection of trajectories (congruence) comes from the accumulation and magnification of the *uncertainty in the initial conditions* due to the collisions, even though the dynamical law governing each trajectory is time-symmetric. To trace a particular trajectory backwards in time after a large number of collisions requires an exponentially high degree of precision in the specification of the initial condition. This ultra-sensitivity of dynamics to initial conditions is characteristic of chaotic systems. Note that the divergence of neighboring trajectories in phase space or parameter space is an intrinsic property of the nonlinear Hamiltonian of the system, not a result of coarse graining by the truncation of the Bogoliubov–Born–Green–Kirkwood–Yvon (BBGKY) series *and* the causal factorizability of the two-particle correlation function as in Boltzmann's molecular chaos hypothesis. (Initially uncorrelated particles become correlated after collisions, thus giving rise to time-asymmetry in the dissipative dynamics of Boltzmann's equation.) The evolution of an ensemble of such systems at some finite time from the initial moment often appears to be unrelated to their initial conditions, not because the individual systems are insensitive to the initial conditions but because they are overly-sensitive to them, thus making it difficult to provide an accurate prediction of each system's state in the future. It is in this sense that these systems manifest irreversibility. In contrast, for an integrable system the trajectories stay close to each other because the regions in phase space for its dynamics are limited by the constants of motion. Such trajectories in integrable systems are referred to as "stable" while those in chaotic systems are "unstable" as they become dispersive in the sense defined above owing to their dynamical instability. We will return in a later section to irreversibility and nonequilibrium thermodynamics considered from the framework of Hamiltonian dynamics.

B. Diffusion

Let us look at some simple examples in kinetic theory: gas expansion, ice melting and an ink drop in water. These are irreversible processes because the initial states of 10^{23} molecules on one side of the chamber and a piece of ice or ink drop immersed in a bath of water are *highly improbable configurations* out of all possible arrangements. These initial conditions are states of very low entropy. The only reason why they are special is because we arrange them to be so. For these problems, we also know that the system–environment separation and interaction make a difference in the outcome. In the case of an expanding ideal gas, for example, for free expansion the change of entropy is $\delta S_{\text{system}} > 0$, $\delta S_{\text{environ}} = 0$, $\delta S_{\text{total}} > 0$. For isothermal *quasistatic* expansion: $\delta S_{\text{system}} = -\delta S_{\text{environ}} > 0$, $\delta S_{\text{total}} = 0$ instead (see, e.g. [Rei65]).

Another important factor in determining whether a process is irreversible is the *time-scale* of observation compared to the dynamic time-scale of the process. We are familiar with the irreversible process of an ink drop dispersing in water which happens in a matter of seconds, but the same dye suspension put in glycerine takes days to diffuse, and for a short duration after the initial mixing (say, by cranking the column of glycerine with a vertical stripe of dye one way) one can easily "unmix" them (by reversing the direction of cranking [UMDdemo]). We will discuss in the next section under what conditions and in what sense a "mixing" system, though time-reversible, can be viewed as capable of approaching equilibrium. Diffusion, when used in the sense of dissipation, is nevertheless an irreversible process.

C. Dissipation

There are two basic models of dissipation in nonequilibrium statistical mechanics: the Boltzmann kinetic theory of dilute gas, and the Langevin theory of Brownian motion. Each invokes a different set of concepts, and even their relation is illustrative. In kinetic theory, the equations governing the n-particle distribution functions (the BBGKY hierarchy) preserve the full information of an

n-particle system. It is (1) ignoring (more often restricted by the precision of one's observation than by choice) the information contained in the higher-order correlations (truncation of the BBGKY hierarchy), *and* (2) the imposition of causal factorization conditions, like the molecular chaos assumption, that brings about dissipation and irreversibility in the dynamics of the lower-order correlations [Zwa01, Bal75].

In the lowest order truncation of the BBGKY hierarchy valid for the description of dilute gases, the Liouvillian operator L acting on the one-particle distribution function $f_1(r_1, p_1, t)$ is driven by a collision integral involving a two-particle distribution function $f_2(r_1, p_1, r_2, p_2, t)$ (cf. Chapters 2 and 11). Boltzmann's molecular chaos ansatz (MCA) assumes an initial uncorrelated state between two particles: $f_2(1, 2) = f_1(1)f_1(2)$, i.e. that the probability of finding particle 1 at (r_1, p_1, t) and particle 2 at (r_2, p_2, t) at the same time t is equal to the product of the single-particle probabilities (a factorizable condition). Note that this condition is assumed to hold only initially, but not finally. A short-range interaction in a collision process will almost certainly generate dynamical correlations between the two collision partners. The truncated BBGKY hierarchy (with MCA) is an example of what we call an effectively open system (see Section 1.5 of this chapter). Boltzmann's explanation of dissipation in macroscopic dynamics is one of the crowning achievements of theoretical physics.

Dissipation in an open system described by the Langevin dynamics has similarities with and differences from that of an effectively open system (as exemplified by the Boltzmann system). The open system can be one distinguished oscillator, the Brownian particle (with mass M), interacting with many oscillators (with mass m) serving as its environment (see Chapter 2). Dissipation in the dynamics of the open system arises from ignoring details of the environmental variables and only keeping their averaged effect on the system (this also brings about a renormalization of the mass and the natural frequency of the Brownian particle). Usually one assumes $M \gg m$ and weak coupling between the system and the environment to simplify calculations. The effect of the environment on a particular system can be summarized by its spectral density function, but other environments can produce equivalent effects. In both of these models, as well as in more general cases, the following conditions are essential for the appearance of dissipation (see, e.g. [Hu89]):

(a) *System–environment separation.* This split depends on what one is interested in, which defines the system: it could be the slow variables, the low modes, the low order correlations, the mean fields; or what one is restricted to: the local domain, the late history, the low energy, the asymptotic region, outside the event horizon, inside the particle horizon, etc.

(b) *Coupling.* The environment must have many degrees of freedom to share with and spread the information from the system; its coupling with the system must be effective in the transfer of information (e.g. nonadiabatic) and the response of the coarse-grained environment must be sufficiently nonsystematic in that it will only react to the system in an incoherent and retarded way. (An example of almost the opposite condition is a dressed atom, i.e. an atom in a high finesse electromagnetic cavity where the quantum coherence of the system can be preserved to a high degree [CoPaPe95].)

(c) *Coarse graining.* One must ignore or down-grade the full information in the environmental variables to see dissipation appearing in the dynamics of the open system. (The time of observation enters also, in that it has to be greater than the interaction time of the constituents but shorter than the recurrence time in the environment.) Coarse graining can be the causal truncation of a correlation hierarchy, the averaging of the higher modes, the "integrating out" of the fluctuation fields, or the tracing of a density matrix (discarding phase information).

(d) *Initial conditions.* Whereas a dissipative system is generally less sensitive to the initial conditions in that for a wide range of initial states dissipation can drive the system to the same final (equilibrium) state, the process is nevertheless possible only if the initial state is off-equilibrium. The process manifests irreversibility also because the initial time is singled out as a special temporal reference point when the system is prepared in that particular initial state. Thus in this weaker sense, dissipation is also a consequence of specially prescribed initial conditions.[1]

While the dynamics of the combined system made up of a subsystem and its environment is unitarity, and its entropy remains constant in time, when certain coarse graining is introduced in the environment, the subsystem turns into an open system, and the entropy of this open system (constructed from the reduced density matrix by tracing out the environmental variables) increases in time. In this open system dynamics, the effect of the coarse-grained environment on the subsystem leads to dissipation and irreversibility in its dynamics.

In our prior discussion of dynamical instability or "dispersion" with the example of hard-disk scattering we were introduced to irreversible but nondissipative processes. Irreversibility there refers to the ultra-sensitivity of the dynamics to the initial conditions. It is extremely difficult to trace back in time a highly divergent congruence of trajectories. The source of irreversibility

---

[1] Note the distinction between these cases: If one defines t_0 as the time when a dissipative dynamics begins and t_1 as when it ends, then the dynamics from t_0 to $-t$ is exactly the same as from t_0 to t, i.e. the system variable at $-t_1$ is the same as at t_1. This is expected because of the special role assigned to t_0 in the dynamics with respect to which there is time-reversal invariance, but it is not what is usually meant by irreversibility in a dissipative dynamics. The arrow of time there is defined as the direction of increase of entropy and irreversibility refers to the inequivalence of the results obtained by reversing t_0 and t_1 (or, for that matter reversing t_0 and $-t_1$), but not between t_1 and $-t_1$. The time-reversal invariance of the H-theorem has the same meaning.

there is by nature fundamentally different from that found in open systems discussed here. The former dynamics is irreversible but nondissipative, while the latter is both dissipative and irreversible. Both types of processes depend on the stipulation of initial conditions. The difference is that the former depends sensitively so, the latter less sensitively. Thus dissipative processes must involve some measure of coarse graining, but coarse graining alone need not lead to dissipation. We will have a subsection later on the issue of coarse graining.

D. Phase mixing

Two well-known effects fall under this category: Landau damping and spin echo (e.g. [Bal75, Ma85]). Let us examine the first example. If one considers long-ranged forces such as the Coulomb force in a dilute plasma gas where close encounters and collisions are rare, the factorizable condition can be assumed to hold throughout, before and after each collision (thus there is no causal condition like the molecular chaos assumption imposed). Under these conditions the Boltzmann kinetic equation becomes a Vlasov (or collisionless Boltzmann) equation (see, e.g. [Bal75, Kre81]). This problem will be discussed in Chapters 10 and 11. The dependence on the one-particle distribution function $f_1(\mathbf{r}, \mathbf{p}, \mathbf{t})$ makes the Vlasov equation nonlinear, and it has to be solved in a self-consistent way. (This aspect is analogous to the Hartree approximation in many-body theory.) Note that the Vlasov equation which has a form depicting free streaming is time-reversal invariant: the Vlasov term representing the effect of the averaged field does not cause dissipation. This mean-field approximation in kinetic theory, which yields a unitary evolution of reversible dynamics, is, however, only valid for times short compared to the relaxation time of the system in its approach to equilibrium. This relaxation time is associated with the collision-induced dissipation process.

Landau damping in the collective local charge oscillations, being a solution of the Vlasov equation, is intrinsically a reversible process. The appearance of apparent "irreversibility" is a consequence of some specially stipulated initial conditions. One may even be able to find a function which is monotonically increasing and refer to it as representing entropy generation. However, upon the choice of some other condition, this feature can disappear and the entropy function can decrease. (An example in Chapter 4 is the entropy function defined in the particle number basis.) Landau "damping" is a mixing process, illustrated here by the Vlasov dynamics. It is fundamentally different from the dissipation process, in that the latter has an intrinsic damping time-scale but not the former, and that while dissipation depends only weakly on the initial conditions, mixing is very sensitive to the initial conditions. Spin echo is another well-known example of phase mixing [Bal75]. For quantum plasma, one needs to coarse grain the phase information in the wavefunctions *and* consider special initial conditions to see this apparent "damping" effect (more in Chapter 4).

From the array of examples above we see that irreversibility and dissipation involve very different causes. The effect of interaction, the role of coarse graining, the choice of time-scales, and the specification of initial conditions in any process can give rise to very different results. We will expand on these physical conditions later, after we have had a chance to look at the microscopic characterization of these macroscopic processes, i.e. their molecular dynamics basis.

1.2 Microscopic characterization from dynamical systems behavior

From a sampling of these macroscopic processes we see a variety of physical behavior. The underlying causes should all be traceable to the microscopic molecular dynamics, to which we now turn our attention. Let us start with a deceptively simple question: An isolated mechanical system is time-reversible. Under what conditions and in what sense does a large isolated system reach equilibrium?

1.2.1 Ergodicity describes a system in equilibrium

An isolated system of N molecules in a volume V has a constant total energy E under the Hamiltonian $H(\mathbf{r}, \mathbf{p})$, where $\mathbf{r}, \mathbf{p}$ each is a $3N$-dimensional vector denoting the position and momenta of all the particles in a $6N$-dimensional phase space Γ. The density function $\rho(\gamma)$ is defined such that the probability of finding a member γ of the ensemble in a differential volume $d\Gamma \equiv d\mathbf{r}_1 \cdots d\mathbf{r}_N d\mathbf{p}_1 \cdots d\mathbf{p}_N$ is equal to $\rho(\gamma)d\Gamma$. Its dynamics is described by the **flow** of each member of the ensemble restricted to the constant energy surface or manifold $\mathcal{E}$ in Γ. Since the number of members flowing in and out of a region in phase space should be equal for all times we have ρ satisfying the Liouville equation,

$$\frac{d\rho}{dt} \equiv \frac{\partial\rho}{\partial t} + \sum_1^N \left(\dot{\mathbf{r}}_i \cdot \frac{\partial}{\partial\mathbf{r}_i} + \dot{\mathbf{p}}_i \cdot \frac{\partial}{\partial\mathbf{p}_i} \right) \rho = 0 \qquad (1.3)$$

where an overdot denotes derivative with respect to time.

In statistical mechanics the microcanonical **ensemble** describes such an isolated system. The number of states is represented by the area of the energy surface E in phase space:

$$\Omega(E) = \int_{H=E} d\mu \equiv \int_\Gamma \delta(H - E)d\mu, \qquad (1.4)$$

where μ is the invariant measure on Γ. The entropy is defined as $S = k_B \ln \Omega(E)$. The ensemble average of a phase space function F over the energy surface E is given by

$$\langle F \rangle_\mu \equiv \frac{\int_{H=E} d\mu F(\gamma)}{\int_{H=E} d\mu} = \frac{\int_\Gamma F(\gamma)\delta(H - E)d\mu}{\int_\Gamma \delta(H - E)d\mu} \qquad (1.5)$$

We also learned that a system in *equilibrium* (either by itself, as in a micro-canonical ensemble, or in contact with a heat bath, as in a canonical ensemble) will have *equal a priori probability* to occupy any of its accessible microstates. How do these concepts: ensemble average, flows in phase space and equilibrium state, connect with each other? Equivalence between the kinetic theory and statistical mechanics description implies that there must be a relation between the way the system points in phase space move (the Liouville flow) and what makes up a typical copy of the system (ensemble average). Equilibrium suggests that the system is stationary. Thus a typical system point must spend an equal amount of time in regions of phase space of equal measure on the energy surface. This is the gist of Boltzmann's **ergodic hypothesis**. If we define the time average of a phase space function $F(\gamma)$ on the energy surface E as

$$\langle F \rangle_t \equiv \lim_{t \to \infty} \frac{1}{T} \int_0^T F(\gamma_t) dt \tag{1.6}$$

where γ_t denotes the point in Γ space after evolving a time t, then the ergodic hypothesis states that

$$\langle F \rangle_t = \langle F \rangle_\mu . \tag{1.7}$$

This says that an arbitrary snapshot (time) of the system provides a typical copy (ensemble) of the system in equilibrium, or, loosely, that time average is equivalent to ensemble average.

Examples of an ergodic system include a one-dimensional harmonic oscillator, an automorphism on a 2-torus in phase space such as the baker's transform or the Arnold cat map. For a quantum system to be ergodic it has to have a nondegenerate energy spectrum. Many simple yet important systems in statistical mechanics are nonergodic. Examples are an ideal gas and multiple harmonic oscillators. For nonergodic systems the energy manifold is metrically decomposable, i.e. E can be partitioned into two or more invariant submanifolds each of which is invariant under the flow in phase space Γ. An equilibrium condition is described by an ensemble density which is constant on each submanifold, but not necessarily on the entire energy manifold.

Note that ergodicity is a microdynamics condition depicting a system in equilibrium but the property is irrelevant to whether a system can approach equilibrium [Far64].

1.2.2 Mixing system is time-reversible; weak sense
of approach to equilibrium

One would think that if the flow in a system is chaotic enough such that the initial probability distribution spreads sufficiently evenly throughout the phase space then there may be a chance for the energy surface to be uniformly occupied. The first condition constitutes what is known as a **mixing system**. The second condition is close to, but still insufficient to define a state of equilibrium. Let $\mu(A)$ be the measure[2] on a set A on the energy surface (the complete energy surface is denoted by $\mathcal{E}$) in phase space. Denote by A_t the same set at time t. It is obvious that $\mu(A) = \mu(A_t)$. A system is mixing if for all sets B on the energy surface the following holds:

$$\lim_{t \to \infty} \frac{\mu(A_t \cap B)}{\mu(B)} = \frac{\mu(A)}{\mu(\mathcal{E})} \tag{1.8}$$

In practice the infinite time limit can be just the laboratory or observation time-scale. What this means is that in a mixing system the stretching of the original set will enable it to intersect with almost any region in the entire energy surface. This requires two conditions. First, there must exist in the system trajectories which spread out rapidly in certain directions of the phase space on the energy surface. Second, that the flow can traverse the whole energy surface $\mathcal{E}$, so it has to be metrically nondecomposable (i.e. that it cannot be subdivided into two or more regions of nonzero measure such that a trajectory starting in one region will never leave it). Common examples of mixing systems are the baker's transform and the Arnold cat map. Since a mixing system has flows which are nearly uniform in the phase space it can be understood to imply ergodicity. However the converse is not true. Both ergodic and mixing systems are time-reversible.

To see irreversible behavior and the approach to equilibrium one needs to introduce some measure of coarse graining, such as considering only the slow variable associated with the unstable direction of the flows, or imposing certain assumptions on the initial conditions in the distributions.

1.2.3 Dissipative system: coarse-grained mixing permits approach to equilibrium

For a system whose unstable trajectories stretch out any initial distribution into very "long and narrow" filaments on the energy surface in the course of time they can produce a uniform spread in phase space. We refer to systems having these properties as satisfying the "chaotic hypothesis" of Cohen and Gallavoti [GalCoh95], or chaotic systems (strictly speaking, the system needs to satisfy the set of criteria which define a hyperbolic or an Anosov system, which is much stronger than mixing; to delve into this topic will go beyond the scope of our book, and we refer interested readers to nice monographs such as [Dor99, Gas98]). In such systems neighboring trajectories diverge from each other exponentially fast – with positive Liapunov exponents, in an unstable direction. The chaotic hypothesis is to dissipative systems as ergodicity is to equilibrium

---

[2] For hyperbolic systems, to capture the smoothness in the unstable directions and the fractal nature in the stable directions, one needs to use the Sinai–Ruelle–Bowen measure [RueEck85, Sinai72, BowRue75, Rue76].

systems. Thereupon one can speculate that averages taken with the distribution function defined on this extended set be equal to the average taken with a smooth equilibrium distribution. (To fulfill the exact criterion of equilibrium one also needs to consider whether the equal spreading in any region of phase space on the energy surface is also uniform in time.) Gibbs correctly observed that a mixing system will not reach the uniform phase space density $\bar{\rho}(\gamma)$ in the fine-grained sense, i.e. $\lim_{t \to \infty} \rho_t(\gamma) = \bar{\rho}(\gamma)$ for each phase space point γ in Γ. But it is likely to do so in a coarse-grained sense, i.e. that the average of $\rho_t(\gamma)$ over each fixed region of phase space will become uniform. $\bar{\rho}$ is called the weak limit of the family of functions ρ_t [Pen70].

It is only in this weak sense that a statement like "a mixing system approaches equilibrium" becomes valid. Bear in mind that a mixing system is time-reversal invariant without coarse graining. Even coarse graining does not automatically turn a mixing system into a dissipative one. We will expand on this point in the next section. Some additional conditions need to be introduced to turn a mixing system into one which shows irreversibility and approaches equilibrium.

Averaging and tracing. Since irreversibility appears in macroscopic systems one may attempt to scale up the system and hope that averaging over a larger phase space may lead to irreversibility. Coarse graining in this way does lead to entropy increase, but it does so in both time directions, so the system remains time-reversible. On the other hand, projection to a lower dimension, or "tracing" (in a sense defined by e.g. [Mac74]) does allow for increase in one time direction.

Molecular chaos assumption. Boltzmann's *Stosszahlansatz*, or molecular chaos assumption, is a causal condition, i.e. before each collision the two molecules are uncorrelated but afterwards they are: This *is* where the irreversibility enters. The dynamical origin of Boltzmann's *Stosszahlansatz* is not clear. Note the choice of time-scales involved here: the shorter time-scales describing higher order correlation functions play a less important role in the long-time behavior which is dominated by the time-scale associated with the behavior of the one-particle distribution function.

Unstable versus stable direction. On the microscopic level we notice already some coarse graining is introduced when we focus on the divergent trajectories in defining the unstable directions of flow in phase space. In the simplest example ([Dor99]) of a Boltzmann equation derived from the baker's map it is the distribution function projected in the slow variables associated with the unstable direction which has a chance to approach the equilibrium distribution. This happens at a time-scale which is much shorter than the time it takes for any small region of phase space to get mixed in the full space. Again this presumes two conditions: that we are dealing with a projected distribution function by paying attention only to the trajectories with positive Lyapunov exponents, and that the initial conditions are conducive to starting the chaotic flow. Notice the time factor involved. Here we see some similarity between our description of the molecular (micro) dynamics of a few degrees of freedom and the gas (macro) dynamics in terms of invoking projection (or reduction), imposing initial conditions, and choosing time-scales. Their connection certainly depends on the role of very large numbers, multi-time-scales or interaction strength. We now turn our attention to these issues.

1.2.4 Nonequilibrium thermodynamics and chaotic dynamics[3]

It is often said that information loss is the source of irreversibility and the approach to equilibrium. From the molecular dynamics description this is not always the case. The baker model gives a good example here. It is essential that, for chaotic systems at least (we really don't understand much about nonchaotic systems, paradoxically) the projection catches at least a piece of the unstable directions in phase space, so the stretching mechanism can smooth out an irregular distribution function. It is also important that the distribution functions be smooth enough not to be concentrated on special orbits which are insensitive to the projection. The application of the molecular chaos assumption in the Boltzmann theory is interesting in the sense that if one takes the correlations to be destroyed by collisions, then one gets an anti-Boltzmann equation, with funny properties. So to get irreversibility one needs both the projection onto a space that has some stretching mechanism in phase space as well as some special conditions imposed on an initial state distribution function.

The other form of coarse graining is more subtle and connected to chaos. Even if no projections onto lower dimensional phase spaces are made, chaotic dynamics, when present, forces a distribution function to become closer and closer to a fractal with structure on arbitrarily fine scales. In the limit of large times, distribution functions do not have nice mathematical properties. They are not differentiable, for example. Thus some coarse graining is required to go from an SRB fractal measure to a distribution function that can be used to calculate averages. This necessitates a loss of information and is a source of entropy increase.

The source of irreversibility in Landau damping is also connected with the construction of a fractal structure.[4] Mathematically the distribution function becomes a Schwartz distribution and lives in a space where the usual theorems

---

[3] The authors are grateful to Professor Robert Dorfman for sharing with them the latest view on these issues in a correspondence from which some of the description in this subsection is adopted.

[4] In the opinion of some leading statistical physicists, Dorfman being one of them, who conveyed this to the authors, the earliest notion of fractals was introduced in the physics literature by van Kampen in his discussion of Landau damping. He called the distribution function "corrugated" for lack of a better word.

about the spectra of differential operators no longer apply, and decays can appear in unexpected ways. The notion of a "Gelfand triple" is useful here for its description [LaCaId99, ACGI00]. To delve further into these directions is beyond the scope and intent of this book, but interested readers should consult the excellent books of Dorfman [Dor99], Gaspard [Gas98] on the micro–macro relations. Pierre Gaspard has pioneered this approach to irreversibility and the Second Law by showing explicitly the deep connection between Kolmogorov–Sinai (KS) entropy and thermodynamic entropy production as well as other distinct properties of nonequilibrium thermodynamics from the theories of dynamical systems (see, e.g. the book of Nicolis [Nic95]). Some salient features are mentioned below (see Gaspard's 2006 summer school lectures [Gas06]).

The aim is to understand the statistical behavior of a collection of particles such as relaxation, diffusion, dissipation, viscosity from the microdynamics of the particles and the divergence properties of their trajectories (congruence) in time. The starting point is the familiar Liouville equation for Hamiltonian systems. One can extract the instrinsic relaxation rates from this equation under certain assumptions on the dynamics. Two important quantities characterizing the microdynamics of the particle congruence are the Lyapunov exponents and the KS entropy per unit time. (The Lyapunov exponents characterize the sensitivity to initial conditions of the underlying microscopic dynamics while the KS entropy per unit time measures the degree of dynamical randomness developed by the trajectories of the system during their time evolution.) The new focus in this recent work is on the large deviations or large fluctuations that the dynamical properties of a system develop in time. In the escape-rate formalism, these large-deviation relationships relate these microscopic quantities to the transport functions in the macroscopic dynamics of the collective particles (see, e.g. [Gas98, Dor99]). These large-deviation relationships are also the basis for the formulation of new fluctuation theorems [EvCoMo93, EvaSea94, GalCoh95, Jar97]. The concepts and techniques in these interfaces have also proven to be invaluable in treating new mesoscopic physical phenomena at the nanoscale [BuLiRi05].

1.3 Physical conditions

Let us be reminded that in addition to examining the microscopic basis of nonequilibrium statistical mechanics via abstract dynamical systems we also need to consider the fact that we are dealing with a large system. It is well-known that the thermodynamic limit is obtained by taking N, V to infinity while keeping the ratio of these two quantities finite. What is the effect of a large system on fluctuations and irreversibility? How does the imposition of some specific initial condition alter the macroscopic dynamics? How can we understand the fundamental difference between microscopic and macroscopic behavior in terms of time-scales or interaction strength? How do the averaging or coarse-graining procedures affect the outcome of a macroscopic observation? One needs to seek answers to these questions in order to address the fundamental issue of how the macroscopic features arise from its microscopic dynamics. We shall now combine the micro and macro descriptions in exploring these important physical factors. The following items also make up a useful checklist to examine whenever we encounter a new quantum field process and attempt to understand its basic statistical mechanical meaning.

1.3.1 Large systems: Fluctuations, Poincaré recurrence and thermodynamic limit

We are familiar with the advantage of taking the large number N and large volume V limit. *Thermodynamics* obtained in this limit while keeping $n = N/V$ constant is a simple yet powerful theory which captures the essential features of macroscopic phenomena. From microdynamics, a chaotic system (one which satisfies the "chaotic hypothesis") approaches equilibrium in a coarse-grained sense ([Dor99]). For systems whose microdynamics has the right properties the average $\langle F \rangle_t$ of a dynamical variable F taken over the appropriate ensembles approaches an equilibrium value $\langle F \rangle_{Eq}$. To infer that in any of the individual systems in the ensemble F is close to $\langle F \rangle_{Eq}$ one needs to ensure that the fluctuations of F are small, and for this one needs to invoke the large size of the system as well as its mixing or chaotic properties ([Pen79]).

On the relation of fluctuations, the size of the system, and the time-scales involved, it is instructive to bring up the Poincaré recurrence and Zermelo's (1896) objection to Boltzmann's theory. Poincaré (1892) stated that any isolated, finite, conservative system will in a finite time come arbitrarily close to its initial configuration. Boltzmann's $H_B(t)$ function cannot decrease monotonically but must eventually increase to reach its original value $H_B(0)$ in a finite time. Thus, Zermelo argued that Poincaré recurrence would undermine Boltzmann's theory of approach to equilibrium.

Boltzmann's answer to this paradox invokes **fluctuations** and **probabilistic** arguments. We know from statistics that if N were just a few particles the fluctuations are comparable to the mean. The Poincaré recurrence time T_P is short and there is no discernible trend of irreversibility. This case is not addressed by Boltzmann's theory. The larger the system the smaller the fluctuations become, and the longer the Poincaré recurrence time. For example, Mazur and Montroll [MazMon70] considered a linear chain of N classical point masses m harmonically (with natural frequency ω_0) coupled to each other's nearest neighbors. For $N = 10$ and $\omega_0 = 10/\text{sec}$ they found that $T_P = 10^{10}$ years, about the age of the universe. Only with a long Poincaré recurrence time will the distribution function for the macroscopic variables become sharp and the tenets of statistical mechanics apply. In addition to the size of the system the recurrence time depends sensitively and irregularly on the initial state. Because of random fluctuations individual sample occurrences cannot be used for reliable prediction of the robust behavior of the overall physical systems, which can only be made in a probabilistic sense. Taking the thermodynamic limit permits one to construct a simpler, asymptotic, statistical theory for large systems. In this limit Poincaré's recurrence is probabilistically suppressed.

From kinetic theory considerations, the **dynamical correlations** established between particles after collisions will become less significant, at least for a dilute gas, when a larger system is considered by observers interested in the long-time behavior of the system. This enables one to focus on those physical quantities of most interest in the long-time limit, such as the expected value of the one-particle distribution function. It is in this same context where Boltzmann proposed his truly original and remarkable theory in depicting the dynamical behavior of the macroscopic world.

1.3.2 Initial conditions: Specific, randomized, dynamical correlations

For a mixing system any set of nonzero measure will be spread out in time uniformly on the energy surface. This suggests that the trajectories must be very sensitive to the initial conditions. Indeed it is so.

Boltzmann assumes that the molecular chaos assumption holds for each collision. Lanford in 1975 [Lan75], using a Lorenz gas of hard spheres of radius a, showed that in the (Grad) limit: $a \to 0$ while $n = N/V \to \infty$ in such a way that the mean free path $\lambda = (n\pi a^2)^{-1}$ remains constant (it thus applies for all values of the mean free path), and Boltzmann's *Stosszahlansatz* can be replaced by the assumption that the particles are uncorrelated initially: $\rho(z_1 \ldots z_N; 0) = \prod_{i=1}^N f_i(z_i, 0)$ [$z_i = (\mathbf{r}_i, \mathbf{p}_i)$ denotes the coordinates and momenta of the i-th particle] since in this limit the r-particle distribution converges almost everywhere to products of one-particle distribution functions at all times, i.e. $\lim f_r(z_1, \cdots z_r, t) = \prod_{i=1}^r f_1(z_1, t)$. (Note that the Grad limit is different from the thermodynamic limit in that the volume is kept constant.)

Time-reversal of Boltzmann's dynamics can be exact but any small uncertainty or error can wipe out reversibility. Thus random initial conditions ensure that we can extract the system's generic and not specific behavior (by design, such as putting all particles on one side of a partition). Let us see how the **Loschmidt paradox** (1876) can illuminate the role played by the initial conditions on irreversibility. An isolated system is time-reversal invariant. If a system evolves towards equilibrium there must be an equally acceptable evolution which takes the system away from equilibrium which is not seen in nature. At the dawn of the computer age, one of the first computer simulations of molecular dynamics was performed by Orban and Bellemans in 1967 [OrbBel67, Ald73] who numerically integrated the equations of motion for a two-dimensional dilute gas (at a density of 0.04 of close packing) of 100 hard disks in a square box colliding

with each other and the box. They let the system evolve for a definite number
of collisions up to time t_1 short compared to the equilibration time t_{Eq}, and
then reversed all the velocities. Since this is a reversible microdynamics one may
expect to recover the initial state after a time $2t_1$. They found that the accuracy with which the original state is restored at time $2t_1$ falls off rapidly as t_1 is
increased, due to the rounding errors in the numerical integration. This can be
understood in light of the divergence of trajectories ("dispersion," or dynamical
instability) discussed in Section 1.1. It is also a good illustration of the important
role played by initial conditions: In the numerical simulation, Orban and Bellemans chose as *initial condition* $t = 0$ the molecules being placed at the vertices of
a square mesh in the box with equal speed but *random direction*. The gas reaches
equilibrium after about 200 collisions as the distribution approaches a Maxwell–
Boltzmann form and the Boltzmann H-function of the velocities $H_B(t)$ reaches a
minimum. In contrast the initial condition at t_1 for the time-reversed evolution
is a very special one [Pri73], because the correlations established amongst the
particles (hard disks) are very particular to that instant in the entire history
of the system. If we consider the condition of the system close to equilibrium
($t \approx t_{Eq}$) as natural (highest probability of occurrence) then the condition of
the system at t_1 is highly unnatural (very low probability) with respect to the
equilibrium state. Indeed it shows anti-kinetic (contrary to Boltzmann's predictions) behavior when $H_B(t)$ increases over a period of time. The result of these
numerical simulations in spin-echo experiments was obtained by Rhim, Pines
and Waugh in 1971 [RhPiWa71]. For experimental realization of the Loschmidt
echo see [PLURH00].

Note also that the anti-kinetic behavior (Boltzmann's H-function $H_B(t)$
increasing) cannot be obtained from solutions of the Boltzmann equation,
because it is predicated upon a molecular chaos assumption. To do the velocity
or time reversal one must solve for the correlations in time from the complete
BBGKY hierarchy of the N-body system, which is difficult but possible numerically or experimentally, but almost impossible analytically.

Thus the resolution of the Loschmidt paradox is that Boltzmann's equation is
only an approximation to the exact equations of motion which describe systems
with random initial states and no dynamical correlations. This is a very different situation from the time-reversed evolution, where the initial condition at t_1
registers information of strong **dynamical correlations**.

1.3.3 Time-scales and interaction

We have already seen how one can characterize the condition for a system to
approach equilibrium by the discrepancy between characteristic time-scales. In
Bogoliubov's explanation of the kinetic conditions, Boltzmann's equation governing the one-particle distribution function measures the time between collisions
which is the slow variable (the relevant variable), while the fast variables giving

the time during collisions are ignored (the irrelevant variables). We saw a similar division of time-scales in the microdynamics of the chaotic systems. In such systems, the unstable direction defines a slow variable while a stable direction defines a fast variable. One can construct a Boltzmann equation (in the form of a gain–loss equation) which permits the approach to equilibrium (see [Dor99]). Such a Markovian equation shows irreversibility.

We also saw the relation of long time-scale (Poincaré recurrence) and large systems in relation to the formulation of thermodynamic and kinetic theories in the depiction of physical reality, likewise the dependence on the initial conditions.

It is often remarked that interaction (e.g. collisions amongst gas molecules) is needed for a system to equilibrate and to show irreversibility. Interaction is necessary for equilibration but interaction does not generate irreversibility. Mean field dynamics such as that described by the Vlasov equation has interaction but the dynamics is reversible. Equilibration (or thermalization when we refer to energy specifically but not particle numbers or chemical species) shows irreversibility, but irreversibility does not imply equilibration. We already saw at the molecular dynamics level that divergent trajectories show irreversibility, but it takes more to show equilibration (e.g. Anosov systems under coarse graining).

1.3.4　Coarse graining

Coarse graining in the most general sense refers to some information lost, removed, or degraded from a system. It could come about because this information is *inaccessible to us*, due to the limited accuracy in our observation or measurement. A drastic example is Planck-scale physics, the details of which are mostly lost (hard to retrieve) because the world we live in today is an ultra-low-energy construct. For this one needs to invoke ideas like effective field theory [Wei95]. Even when information is fully accessible to us in principle, in practice one may only be interested in some aspects of the system. *We choose to ignore* certain variables such as ignoring the higher order correlations in Boltzmann's kinetic theory, or ignoring the phase information in a quantum system by imposing a random phase approximation. We do this by "integrating over" or "projecting out" these "irrelevant" variables.[5]

Let us see some examples of coarse graining in action. We start with the familiar Boltzmann theory: implementation of the molecular chaos assumption (i.e.

---

[5] Quotation marks are put here to emphasize the colloquial usage and the warning that operations bearing the same name could bring forth different results depending on the assumptions introduced. For example, the projection operator formalism of Zwanzig and Mori *et al.* ([Nak58] [Zwa60] [Zwa61] [Mor65] [WilPic74] [Gra82] [Kam85] [GoKaZi04] [GorKar04] [Bal75]) applied to a closed system will turn the differential equations of motion for each subsystem into an integro-differential equation for a particular subsystem. Without casting away some information somewhere in the system this equation is just another way to express the interaction of the subsystems. It contains no more or no less information as the original equations describing the total system.

the two-particle distribution function $f_2 = f_1 f_1$ can be expressed schematically as a product of two one-particle distribution functions f_1) entails performing a coarse graining in the collision integral of space over the range of interaction and of time over the duration of a collision.

Another example concerns particle creation from an external background field or changing spacetime. Since particle pairs originate from the vacuum which is a pure state, there should not be any entropy generation. On the other hand, in a thermodynamics description, the entropy S is related to the number N of particles present. We may wonder whether to trust either or both of these statements. The key lies in understanding that the thermodynamic description has undergone several levels of coarse graining from the fundamental quantum field theory description. Indeed it is a very educational intellectual exercise to see what coarse-graining measures are introduced and what concepts are at work as we move from a microscopic (quantum field theory) to a macroscopic (nonequilibrium thermodynamics) description of this same system, but with different degrees of precision.

Note also that coarse graining is a necessary but not sufficient condition for entropy generation. It does not always produce a dissipative system. Truncation of the BBGKY hierarchy leads to a closed subsystem composed of n-particle correlation functions whose dynamical equations are unitary. (An example mentioned before is the Vlasov equation describing particle interaction via long-range forces.) In quantum field theory equations derived from a finite-loop effective action are also unitary – at one loop the effect of the quantum field on the particles manifests through the renormalized masses and charges (to be exact, the equations of motion derived from a finite-loop effective action are unitary if none of the relevant correlation functions are "slaved" – see Chapter 6 Section 6.3 and Chapter 9 Section 9.2.3 for a discussion of this concept; for ℓ loops, one must keep the first $(\ell + 1)$th-order correlations, otherwise dissipation in the sense defined above sets in – dissipation is absent only in very specific situations, such as a free theory or equilibrium initial conditions). That is perhaps why (if one limits one's attention to loop expansions) statistical mechanical concepts rarely came to the fore, until one starts asking questions of a distinct nature, such as how dissipative dynamics appears in an otherwise unitary system, and the origin and nature of noise in quantum field theory. A causal condition needs to be introduced to render the dynamics of the subsystem irreversible. This opens up another important theme in this book: effective field theory viewed in the open system framework, which will be developed in later chapters.

1.4 Coarse graining and persistent structure in the physical world

We have seen from the above discussions that the appearance of irreversibility is often traced to the initial condition being special in some sense. The dynamics of the system and how it interacts with its environment also enter in determining whether the system exhibits mixing or dissipative behavior. For the sake of highlighting the contrast we could broadly divide the processes into two classes depending on how sensitive they are to the initial conditions versus the dynamics.[6] One can say that the first class is *a priori* determined by the initial conditions, the other is *a posteriori* rather insensitive to the initial conditions. Of the examples we have seen, the first group includes divergent trajectories in molecular (micro) dynamics, Landau damping, vacuum particle creation, and the second class includes gas (macro) or fluid dynamics (see the discussions at the end of Section 1.2.3), diffusion, particle creation with interaction, decoherence. Appearance of dissipation is accompanied by a degradation of information via coarse graining (such as the molecular chaos assumption in kinetic theory, restriction to one-particle distribution in particle creation with interaction, "integrating out" some class of histories in decoherence). An arrow of time appears because of some special prearranged conditions; how it manifests in the system also depends on the system dynamics and the coarse graining introduced to the system. The issues we have touched on involve the transformation of a closed to an open system, the relation between the microscopic and the macroscopic world, and the transition from quantum multiplicities to classical realities. Many perceived phenomena in the observable physical world, including the phenomenon of time-asymmetry, can be understood in the open-system viewpoint via the approximations introduced to the objective microscopic world by a macroscopic observer [GKJKSZ96, Omn94, Per93].

Thus, time asymmetry in these processes is influenced by many factors: the way one stipulates the boundary conditions and initial states, the time-scale of observation in comparison with the dynamical time-scale, how one decides what the relevant variables are and how they are separated from the irrelevant ones, how the irrelevant variables are coarse grained, and what assumptions one makes and what limits one takes in shaping a macroscopic picture from one's imperfect knowledge of the underlying microscopic structure and dynamics.

We have discussed the procedures which can bring about these results. However, a set of more important and challenging issues remain largely unexplored, i.e. under what conditions the outcomes become less subjective and less sensitive to these procedures, such as the system–environment split and the coarse graining of the environment. These procedures provide one with a viable prescription to get certain general qualitative results, but are still not specific and robust enough to explain how and why the variety of observed phenomena in the physical world arise and stay in their particular ways. To address these issues one should ask a different set of questions:

---

[6] As discussed earlier, dissipation also requires the stipulation of a somewhat special initial condition, i.e. that the system is not in an equilibrium state; but, in the words of R. Sorkin, "not more special than it needs to be".

(1) By what criteria are the system variables chosen? Collectivity and hierarchy of structure and interactions

In a model problem, one picks out the system variables – be it the Brownian particle or the mini-superspace variables – by fiat. One defines one's system in a particular way because one wants to calculate the properties of that particular system. But in the real world, certain variables distinguish themselves from others because they possess a relatively well-defined, stable, and meaningful set of properties for which the observer can carry out measurements and derive meaningful results. Its meaningfulness is defined by the range of validity or degree of precision or the level of relevance to what the observer chooses to extract information from. In this sense, it clearly carries a certain degree of subjectivity – not in the sense of arbitrariness in the exercise of free will of the observer, but in the specification of the parameters of observation and measurement. For example, the thermodynamic and hydrodynamic variables are only good for systems close to equilibrium; in other regimes one needs to describe the system in terms of kinetic-theoretical or statistical-mechanical variables.

The soundness in the choice of a system in this example thus depends on the time-scale of measurement compared to the relaxation time. As another example, contrast the variables used in the nuclear collective model and the independent nucleon models. One can use the rotational–vibrational degrees of freedom to depict some macroscopic properties of the motion of the nucleus, and one can carry out meaningful calculations of the dissipation of the collective trajectories (in the phase space of the nucleons) due to stochastic forces. In such cases, the noncollective degrees of freedom can be taken as the noise source. However, if one is interested in how the independent nucleons contribute to the properties of the nucleus, such as the shell structure, one's system variable should, barring some simple cases, not be the elements of the $SO(3)$ group, or the $SU(6)$ group. At a still higher energy where the attributes of the quarks and the gluons become apparent, the system variables for the calculation of, say, the stability of the quark–gluon plasma should change accordingly. The level of relevance which defines one's system changes with the level of structure of matter and the relative importance of the forces at work at that level. The improvement of the Weinberg–Salam model with W, Z intermediate bosons over the Fermi model of four-point interactions is what is needed in probing a deeper level of interaction and structure which puts the electromagnetic and weak forces on the same footing. Therefore, one needs to explore the rules for the formation of such relatively distinct and stable levels, before one can sensibly define one's system (and the environment) to carry out meaningful inquiries of a statistical nature.

What is interesting here is that these levels of structures and interactions come in approximate hierarchical order (so one doesn't need QCD to calculate the rotational spectrum of a nucleus, and the Einstein spacetime manifold picture will hopefully provide most of what we need in the post-Planckian era). One needs both some knowledge of the hierarchy of interactions and the way effective theories emerge from "integrating out" variables at very different energy scales in the hierarchical structure (e.g. ordinary gravity plus particle theory regarded as a low-energy effective higher-dimension or Kaluza–Klein theory). The first part involves fundamental constituents and interactions and the second part the application of statistical methods. One should also keep in mind that what is viewed as fundamental at one level can be a composite or statistical mixture at a finer level. There are system–environment separation schemes which are designed to accommodate or reflect these more intricate structures, from the mean-field–fluctuation-field split to the multiple source or nPI formalism (see Chapter 6) for the description of the dynamics of correlations and fluctuations. The validity of these approximations depends quite sensitively on where exactly one wants to probe in between any two levels of structure. Statistical properties of the system such as the appearance of dissipative effects and the associated irreversibility character of the dynamics in an open system certainly depend on this separation.

(2) How does the behavior of the subsystem depend on coarse graining? Sensitivity and variability of coarse graining, stability and robustness of emergent structure

Does there exist a common asymptotic regime as the result of including successively higher order iterations in the same coarse-graining routine? This measures the sensitivity of the end result to a particular kind of coarse graining. How well can different kinds of coarse-graining measure produce and preserve the same result? This is measured by its variability. Based on these properties of coarse graining, one can discuss the relative stability of the behavior of the resultant open system after a sequence of coarse grainings within the same routine, and its robustness with respect to changes to slightly different coarse-graining routines.

Let us illustrate this point with some simple examples. When we present a microscopic derivation of the transport coefficients (viscosity, heat conductivity, etc.) in kinetic theory via the system–environment separation scheme, we usually get the same correct answer independent of the way the environment is chosen or coarse grained. Why? It turns out that this is likely only if we operate in the linear-response regime (see [FeyVer63]). The linear coupling between the system and the environment makes this dependence simple. This is something we usually take for granted, but has some deeper meaning. For nonlinear coupling, the above problem becomes nontrivial. Another aspect of this problem can be brought out [BalVen87, Spo91] by comparing these two levels of structure and interaction, e.g. the hydrodynamic regime and the kinetic regime. Construct the relevant entropy from the one-particle classical distribution function f_1, that gives us the kinetic theory entropy S_{kt} which is simply $-kH_B$, where H_B is Boltzmann's H-function. Now comparing it with the hydrodynamic entropy function S_{hd} given in terms of the hydrodynamic variables (in this case, the number and energy density), one sees that $S_{\text{hd}} > S_{\text{kt}}$. A simple physical argument for this result is that the information contained in the correlations amongst the particles is not included in the hydrodynamic approximation. Even within the kinetic theory regime there exist intermediate stages described by suitably chosen variables [Spo91]. The entropy functions constructed therefrom will reflect how much fine-grained information is lost. In this sense S_{hd} is a maximum in the sequence of different coarse-graining procedures. In the terminology we introduced above, by comparison with the other regimes, the hydrodynamic regime is more robust in its structure and interactions with respect to varying levels of coarse graining. One way to account for this is that, as we know, the hydrodynamic variables enter in the description of systems in equilibrium and they obey conservation laws [HaLaMa95, Bru96, Hal98]. Further coarse graining on these systems is expected to produce the same results, i.e. the hydrodynamic regime is a limit point of sorts after the action from a sequence of coarse grainings. Therefore, a kind of "maximal entropy principle" with respect to variability of coarse graining is one way where thermodynamically robust systems can be located.

While including successively higher orders of the same coarse-graining measure usually gives rise to quantitative differences (if there is a convergent result, that is, but this condition is not guaranteed, especially if a phase transition intervenes), coarse graining of a different nature will in general result in very different behavior in the dynamics of the open system. Let us look further at the relation of variability of coarse graining and robustness of structure.

Sometimes the stability of a system with respect to the variability of coarse graining is an implicit criterion behind the proper identification of a system. For example, Boltzmann's equation governing the one-particle distribution function which gives a very adequate depiction of the physical world is, as we have seen, only the lowest order equation in an infinite (BBGKY) hierarchy. If coarse graining is by the order of the hierarchy – e.g. if the second and higher order correlations are ignored – then one can calculate without ambiguity the error introduced by such a truncation. The dynamics of the open system which includes dissipation effects and irreversible behavior will not change drastically if one uses a different (say more fine-grained) procedure, such as retaining the fourth-order correlations (if the series converges, which is a nontrivial issue, see, e.g. [Dor81]). Consider now a different approximation: For a binary gas of large mass discrepancy, if one considers the system as the heavy mass particles, ignore their mutual interactions and coarse grain the effect of the light molecules on the heavy ones, the system now behaves like a Brownian particle motion described by a Fokker–Planck equation. We get a qualitatively very different result in the behavior of the system.

In general the variability of different coarse grainings in producing a qualitatively similar result is higher (more variations allowed) when the system one works with is closer to a stable level in the interaction range or in the hierarchical order of structure of matter. The result is more sensitive to different coarse-graining measures if it is far away from a stable structure, usually falling in between two stable levels.

One tentative analogy may help to fix these concepts. Robust systems are like the stable fixed points in a parameter space in the renormalization group theory description of critical phenomena: the points in a trajectory are the results of performing successive orders of the same coarse-graining routine on the system (e.g. the Kadanoff–Migdal scaling [Kad76, Kad77, WilKog74, Fis74, Fis83]), a trajectory will form if the coarse-graining routine is stable. An unstable routine will produce in the most radical situations a random set of points. Different trajectories arise from different coarse-graining routines. Neighboring trajectories will converge if the system is robust, diverge if not. Therefore the existence of a stable fixed point where trajectories converge to is an indication that the system is robust. Only robust systems survive in nature and carry definite meaning in terms of their persistent structure and systematic evolution. This is where the relation of coarse graining and persistent structures enters.

So far we have only discussed the activity around one level of robust structure. To investigate the domain lying in-between two levels of structures (e.g. between nucleons and quark–gluons) one needs to first know the basic constituents and interactions of the two levels. This brings back our consideration of levels of structures above. Studies in the properties of coarse graining can provide a useful guide to venture into the often nebulous and elusive area between the two levels and extract meaningful results pertaining to the collective behavior of the underlying structure. But one probably cannot gain new information about the fine structure and the new interactions from the old just by these statistical measures (cf. the old bootstrapping idea in particle physics versus the quark model).[7]

1.5 Physical systems: Closed, open, effectively closed and effectively open

1.5.1 Open systems: Coarse graining and back-reaction

In treating physical systems containing many degrees of freedom one often attempts to select out a small set of variables to render the problem technically tractable while preserving its physical essence. Familiar examples abound: e.g. thermodynamics from statistical mechanics, hydrodynamic limit of kinetic theory, collective dynamics in condensed matter and nuclear physics [Wil82].

---

[7] In this sense, one should not expect to gain new fundamental information about quantum gravity just by extrapolating what we know about the semiclassical theory, although studying the way the semiclassical theory takes shape (viewed as an effective theory) from possible more basic theories is useful. It may also be sufficient for what we can understand or care about in this later stage of the universe we now live in.

When one starts from the microscopic picture, one distinguishes the variables which depict the system of interest from those which can affect the system but whose detail is otherwise of lesser interest or importance. Making a sensible distinction involves recognizing and devising a set of criteria to separate the relevant from the irrelevant variables. This procedure is simplified when the two sets of variables possess very different characteristic time or length or energy scales or interaction strengths. An example is the separation of slow–fast variables as in the Born–Oppenheimer approximation in molecular physics where the nuclear variables are assumed to enter adiabatically as parameters in the electronic wavefunction. Similar separation is possible in quantum cosmology between the "heavy" gravitational sector characterized by the Planck mass and the "light" matter sector. In statistical physics this separation can be made formally with projection operator techniques. This usually results in a nonlinear integro-differential equation for the relevant variables, which contains the causal and correlational information from their interaction with the irrelevant variables.

Apart from finding some way of *separating* the overall closed system into a "relevant" part of primary interest (the open system) and an "irrelevant" part of secondary interest (the environment) in order to render calculations possible, one also needs to devise some *averaging* scheme to reduce or reconstitute the detailed information of the environment such that its effect on the system can be represented by some macroscopic functions, such as the transport functions. This involves introducing certain *coarse-graining* measures. It is usually by the imposition of such measures that an environment is turned into a bath, and certain macroscopic characteristics such as temperature and chemical potential can be introduced to simplify its description. A coarse-grained description of the effect of the environment on the system (in terms of, say, thermodynamic or hydrodynamic variables and their associated response functions) is qualitatively very different from the detailed description (in terms of the underlying microscopic variables and dynamics). A familiar example in many-body theory used for simplifying the effect of the environment is by assuming that each independent particle interacts with an effective potential depicting the averaged effect of all other particles. Vlasov dynamics in a plasma is of such a nature, so is the mean field approximation in quantum field theory (where the effect of quantum fluctuations of fields is described at this level of approximation in terms of a renormalized interaction potential and couplings to the system).

How good an *effective theory* is in its depiction of physical phenomena at a particular scale is usually determined by the appropriateness in the choice of the collective variables, the correctness and extent of coarse graining in relation to the probing scale and the precision of measurement. How the environment affects the open system is determined by the **back-reaction** effects. By referring to an effect as a back-reaction, it is implicitly assumed that a system of interest is preferentially identified, that one cares much less about the details of the other sector (the "irrelevant" variables in the "environment"). The back-reaction can be significant, but should not be too overpowering, so as to invalidate the separation scheme. To what extent one views the interplay of the two sectors as *interaction* (between two subsystems of approximately equal weight) or as *back-reaction* (of a less relevant environment on the more important system) is reflexive of and determined by the degree to which one decides to keep or discard the information in one subsystem versus the other. It also depends on their interaction strength. Through reaction and back-reaction the behavior of each sector is linked to the other in an inseparable way, i.e. by their interplay.

Self-consistency is thus a necessary requirement in back-reaction considerations. This condition can manifest itself as the fluctuation–dissipation theorem (FDT). When the environment is a bath, for systems near equilibrium, their response can be depicted by linear response theory. Even though such relations are usually presented in such a context, its existence in a more general form can be shown to cover nonequilibrium systems. Indeed as long as back-reaction is included, such a relation can be understood as a corollary of the self-consistency requirement, which ultimately can be traced to the unitary condition of the original closed system.

A familiar example of a self-consistent back-reaction process is the time-dependent Hartree–Fock approximation in atomic physics or nuclear physics, where the system could be described by the wavefunction of the electrons obeying the Schrödinger equations with a potential determined by the charge density of the electrons themselves via the Poisson equation. In a cosmological back-reaction problem, one can view [Hu89] the system as a classical spacetime, whose dynamics is determined by Einstein's equations with sources given by particles produced by the vacuum excited by the dynamics of spacetime and depicted by the appropriate wave equations in this particular curved spacetime [BirDav82].

Much of the physics of open systems is concerned with the appropriateness in the devising and the implementation of these procedures. They are: (1) the identification and separation of the physically interesting variables which make up the open system – one needs to first come up with the appropriate collective variables; (2) the "averaging" away of the environment or irrelevant variables – how different coarse-graining measures affect the final result is important (as discussed in the last section); and (3) the evaluation of the averaged effect of the environment on the system of interest. We will refer to these procedures as *separation, coarse graining* and *back-reaction* for short.

These considerations surrounding an open system are common and essential not only to well-posed and well-studied examples of many-body systems like molecular, nuclear and condensed-matter physics, they also bear on some basic issues at the foundations of quantum mechanics and statistical mechanics, such as decoherence and the existence of the classical limit [HarGel93], with profound implications on the emergence of time and spacetime [Har92], or quantum mechanics itself [Adl04].

1.5.2 From closed to effectively open systems

There are many systems in nature which are apparently closed (to the observer), in that there is no obvious way to define a system which is so much different from an environment. These systems do not possess a parameter which can enable the observer to distinguish possible heavy–light sectors, high–low frequency behavior or slow–fast dynamics. Boltzmann's theory of molecular gas is a simple good example: All molecules in the gas are on an equal footing, in that no one can claim to be more special than the others. Because of the lack of parameters which marks the discrepancy of one component from the other, these systems do not lend themselves to an obvious or explicit separation from their environment (like open systems would), and appear like closed systems. However, usually in their effective description a separation is introduced implicitly or operationally because of their restricted appearance or due to the imprecision in one's measurement. These are called effectively open systems.

In this example, on the microscopic level (of molecular dynamics) all molecular movements are time-reversal invariant, but on the macroscopic level (of our observation), dissipation and violation of time-reversal invariance obviously exist. To reconcile this difference and understand the origin of dissipation in nature was of course the great challenge Boltzmann posed for himself and which he so ingeneously resolved. Boltzmann came up with the idea that if only one-particle distribution functions were observed, and the molecular chaos assumption was imposed (for any collision process), there is an explanation for the origin of dissipation in macroscopic phenomena. Using the correlation functions (the lowest order being the one-particle distribution function) as a way to systemize the information in the gas, one would get the BBGKY hierarchy, which contains the full information of the gas. It is only upon the truncation of the hierarchy and the re-expression of the higher correlation functions in terms of the lower ones, e.g. the causal factorization condition (assuming that colliding partners are uncorrelated initially, what we shall call "slaving," to be discussed in detail later), that the otherwise closed system expressed by the full hierarchy is rendered open, and dissipation appears. It is in this sense that it is called an effectively open system.

1.5.3 Two major paradigms of nonequilibrium statistical mechanics

We can highlight the distinction between open and effectively open systems by comparing the two primary models which characterize these two major paradigms of nonequilibrium statistical mechanics (see, e.g. [AkhPel81, Pri62, ToKuSa92]): the Boltzmann–BBGKY theory of molecular kinetics, and the Langevin (Einstein–Smoluchowski) theory of Brownian motions. The differences between the two are of both formal and conceptual nature.

To begin with, the *setup* of the problem is different: As we remarked above, in kinetic theory one studies the overall dynamics of a system of gas molecules, treating each molecule in the system on the same footing, while in Brownian motion one (Brownian) particle which defines the system is distinct, the rest are relegated as the environment. The terminology of "relevant" versus "irrelevant" variables not so subtly reflects the discrepancy.

The *object* of interest in kinetic theory is the (one-particle) distribution function (or the nth-order correlation function), while in Brownian motion it is the reduced density matrix. The emphasis in the former is the behavior of the gas as a whole (e.g. dissipative dynamics) taking into account the correlations amongst the particles, while in the latter it is the motion of the Brownian particle under the influence of the environment.

The nature of *coarse graining* is also very different: in kinetic theory coarse graining resides in confining one's attention to one-particle distribution functions, a factorization condition for the two-point functions and the adoption of the molecular chaos assumption. This corresponds formally to a truncation of the BBGKY hierarchy and introducing a causal slaving condition, while in Brownian motion it is in the integration over the environmental variables. The part that is truncated or "ignored" is where the noise comes from, while its main physical effect on the "system" is to render its dynamics dissipative. Thus the fluctuation–dissipation relation and other features.

Finally the *philosophies* behind these two paradigms are quite different: In Brownian motion problems, the separation of the system from the environment is prescribed: it is usually determined by some clear disparity between the two systems. These models represent "autocratic systems," where some degrees of freedom are more relevant than others. By contrast, molecular gas models subscribing to the effectively open systems represent "democratic systems" where all particles in a gas are equally relevant. In the lack of any clear discrepancy in scales, making a separation "by hand" is *ad hoc*, contrived, and often leading to wrong description. Coarse graining in Boltzmann's kinetic theory is also very different from that of Brownian motion. The latter is explicit while the former appears implicit (having its own systematics). However, as we shall see later, the coarse graining in the Boltzmann theory lies in the truncation and slaving procedures, where information attached to higher correlation orders is not kept in full. Now just what correlation order is sufficient for the physics under study is an objectively definable and verifiable fact, which ultimately is determined by the degree of precision in a measurement and judged by how well it depicts the relevant physics.

In Chapter 2 we will provide a physical discussion of the Boltzmann and Langevin dynamics, two prime examples of these two major paradigms. To see the mathematical origin of these stochastic equations it is best to acquire some knowledge of stochastic processes. A brief summary of this subject is given in Appendix A, which starts with rudimentary probability theory and ends with a derivation of the Chapman–Komogorov/Einstein–Smolochousky equation and the Kramers–Moyal/Fokker–Planck equation.

1.6 Appendix A: Stochastic processes and equations in a (tiny) nutshell

We give here a brief summary of the theory of stochastic processes, leading to the derivation of the Chapman–Komogorov/Einstein–Smolochousky equation and the Kramers–Moyal/Fokker–Planck equation. We will convey the necessary yet minimal set of information to enable the reader without prior knowledge of this subject to follow the development of later chapters leading to its application to quantum field theory of nonequilibrium processes. The emphasis here is more on physical ideas than mathematical rigor. One can find nice discussions of these topics in standard books such as those by van Kampen [Kam81], Papooulis [Pap84], Gardiner [Gar90], Gardiner and Zoller [GarZol00b], Carmichael [Car93], and Reichl [Rei98]. More succinct and accessible summaries can be found in, for example, Weissbluth [Wei88] and Mandel and Wolf [Man95]. Here we follow mainly the discussions by van Kampen.

1.6.1 Probability, random variables and stochastic processes

Probability

We start with some basic concepts and definitions in *probability theory*. A probability space, or simply, an experiment, consists of the triplet (Ω, F, P) where Ω is the sample space containing all possible outcomes of the experiment. An event A is a subset of Ω, and F is a collection of subsets of Ω. P is the probability of finding A in such an experiment. Example: In a single throw of a dice (our experiment) what is the probability of finding an even number facing up? Then $\Omega = \{1, 2, 3, 4, 5, 6\}, A = \{2, 4, 6\}, P = \frac{1}{2}$. Set theory is usefully applied to probability theory starting with $P(\Omega) = 1, P(\emptyset) = 0$, where $\emptyset$ denotes the empty set. Two events are said to be **mutually exclusive**, or disjoint, if $P(A \cup B) = P(A) + P(B)$ or $P(A \cap B) = 0$. Two events A, B are **independent** iff $P(A \cap B) = P(A)P(B)$. Note that independent events are not mutually exclusive events because for mutually exclusive events $P(A \cap B) = 0$.

Let $A_1, \ldots, A_n$ be a finite collection of events. They are called mutually independent if for any $1 \leq i_1 < i_2 < \cdots < i_k \leq n$

$$P(A_{i_1} \cap A_{i_2} \cap \cdots \cap A_{i_k}) = P(A_{i_1})P(A_{i_2}) \cdots P(A_{i_k}) \tag{1.9}$$

The events are called pairwise independent if for any $1 \leq i_1 < i_2 \leq n$

$$P(A_{i_1} \cap A_{i_2}) = P(A_{i_1})P(A_{i_2}) \tag{1.10}$$

Obviously pairwise independence is a much weaker condition than mutual independence.

Finally we define the conditional probability $P(A|B)$ as the probability that event A will occur if B occurs, or simply, A given B. Obviously the probability that both events A, B will occur is $P(A \cap B) = P(A|B)P(B)$. From this it is easy to derive Bayes' rule

$$P(A|B)P(B) = P(B|A)P(A) \tag{1.11}$$

Random variables

A **random variable** X defined on a sample space Ω is a function which maps Ω into the set of real numbers. It assigns a real number to each sample point. In the example of throwing a dice, winning a certain amount of money x_i (out of the whole range $X(\Omega) = \{x_i\}$) when some number in the set $i = \{1 \cdots 6\}$ faces up is one such mapping. One can define a probability distribution $P(x_i) = f(x_i)$. For continuous variables the probability P that an event occurs resulting in X taking on values in an interval $a \leq X \leq b$ is given by

$$P(a \leq X \leq b) = \int_a^b f_X(x)dx \tag{1.12}$$

$f_X(x)$ is called the probability density for such an occurrence. The cumulative probability distribution function (cdf) $F_X(x) \equiv P(X \leq x)$ is obtained by integrating f_X up to the value x, i.e.

$$F_X(x) = \int_{-\infty}^x f_X(x')dx' \tag{1.13}$$

Now consider two stochastic variables X, Y on the same sample space Ω, $X(\Omega) = \{x_i\}, Y(\Omega) = \{y_j\}$. We define the **joint probability distribution** $f(X, Y)$ of X and Y as the probability of an ordered pair occurring, $P(x_i \epsilon X, y_j \epsilon Y) = f(x_i, y_j)$. For continuous variables $f(x, y) \geq 0$ is normalized $\int \int dx dy f(x, y) = 1$. The single-variable distribution f_X is obtained if one disregards (integrates over) the value of Y in $f(x, y)$, i.e.

$$f_X(x) = \int dy f(x, y), \quad f_Y(y) = \int dx f(x, y) \tag{1.14}$$

We now generalize the number of stochastic variables to r and define an r-dimensional vector $\mathbf{X} = (X_1, X_2, \dots, X_r)$. We can think of this as the vector denoting the space and momenta of N particles in the phase space Γ, in which case $r = 6N$. (Note we used i, j earlier to denote the (discrete) sample space variables, while r, s here denote the dimension of the space of stochastic variables.) The probability density $P_r(\mathbf{X})$ is the joint probability density of the r variables $(X_1, X_2, \dots, X_r)$. For a projection of $\mathbf{X}$ into a smaller space of dimension $s, s < r$, the joint probability density of a subset $s < r$ of variables $P_s(X_1, X_2, \dots, X_s)$ regardless of the remaining variables $X_{s+1}, \dots, X_r$ is obtained from integrating over these variables, i.e.

$$P_s(X_1, X_2, \dots, X_s) = \int P_r(X_1, \dots, X_s, X_{s+1}, \dots, X_r)dX_{s+1} \cdots dX_r \tag{1.15}$$

In probability theory this is called the **marginal distribution** of subset r. In statistical mechanics this is called the **reduced (probability density) distribution** function.

One can define the **nth moments** of a stochastic variable X by

$$\langle X^n \rangle = \sum_i x_i^n f_X(x_i) = \int dx \, x^n f_X(x) \tag{1.16}$$

where the first is for discrete and the second for continuous variables. The first two moments are familiar: For $n = 1, \langle X \rangle$ is called the **mean**; for $n = 2, \sigma_X^2 = \langle X^2 \rangle - \langle X \rangle^2$ is called the **variance**.

One can also define the **characteristic function** Φ as the Fourier transform of the probability density

$$\Phi_X(k) = \langle e^{ikX} \rangle = \int_\infty^{+\infty} dx \, e^{ikx} f_X(x) = \sum_{n=0}^\infty \frac{(ik)^n}{n!} \langle X^n \rangle \tag{1.17}$$

with inverse transform

$$f_X(x) = \frac{1}{2\pi} \int dk \, e^{-ikx} \Phi_X(k) \tag{1.18}$$

Equivalently, in terms of Laplace transforms we can define the **moment generating function**

$$M_X(s) = \langle e^{sX} \rangle \tag{1.19}$$

This name becomes obvious when we rewrite the **moments** as

$$\langle X^n \rangle = \frac{d^n M_X(s)}{ds^n} \Big|_{s=0} \tag{1.20}$$

Likewise one can define a **cumulant** expansion by the relation

$$\ln \Phi_X(k) = \sum_{n=1}^\infty \frac{(ik)^n}{n!} C_n(X) \tag{1.21}$$

The relations between cumulants and moments are as follows:

$$
\begin{aligned}
C_1(X) &= \langle X \rangle \quad C_2(X) = \langle X^2 \rangle - \langle X \rangle^2 \\
C_3(X) &= \langle X^3 \rangle - 3 \langle X^2 \rangle \langle X \rangle + 2 \langle X \rangle^3 \\
C_4(X) &= \langle X^4 \rangle - 4 \langle X^3 \rangle \langle X \rangle - 3 \langle X^2 \rangle^2 \\
&\quad + 12 \langle X^2 \rangle \langle X \rangle^2 - 6 \langle X^4 \rangle
\end{aligned}
\tag{1.22}
$$

Note again that the first two cumulants are the mean and the variance. The **covariance** and **correlation** of two different stochastic variables X, Y are defined respectively as

$$\text{Cov}(X, Y) \equiv \int \int dx dy \, (x - \langle X \rangle)(y - \langle Y \rangle) f(x, y)$$

$$= \langle XY \rangle - \langle X \rangle \langle Y \rangle$$

$$\text{Cor}(X, Y) \equiv \frac{\text{Cov}(X, Y)}{\sigma_X \sigma_Y} \tag{1.23}$$

Stochastic processes

Given a stochastic variable X, one can define a **stochastic function** Ξ obtained from X by some mapping

$$\Xi_X(t) = g(X, t) \tag{1.24}$$

where t is some smooth variable. If t denotes time, $\Xi(t)$ is called a **stochastic process**. When X takes on the value x, $\Xi_x(t) = g(x, t)$ becomes a sample function or a realization of the process.

The probability density for a stochastic function $\Xi_x(t)$ to take on value ξ at time t is given by

$$P_1(\xi, t) = \int \delta(\xi - \Xi_x(t)) f_X(x) dx \tag{1.25}$$

We recognize that $f_X(x)$ is the probability density for the stochastic variable X. The subscript 1 denotes a function of one stochastic variable. The probability over all values of ξ_1 at any particular time t_1 should be unity, thus the normalization condition is $\int P_1(\xi, t_1) d\xi = 1$. Generalizing to n we can define the **joint probability density** as

$$P_n(\xi_1, t_1, \cdots \xi_n, t_n) \equiv \int \delta(\xi_1 - \Xi_x(t_1)) \cdots \delta(\xi_n - \Xi_x(t_n)) f_X(x) dx \tag{1.26}$$

When one ignores one stochastic function ξ_n one obtains the **reduced joint probability density**

$$\int P_n(\xi_1, t_1, \cdots \xi_n, t_n) d\xi_n = P_{n-1}(\xi_1, t_1, \cdots \xi_{n-1}, t_{n-1}) \tag{1.27}$$

The correlation between values of Ξ at different times is measured by the time-dependent moments

$$\langle \xi_1(t_1) \xi_2(t_2) \cdots \xi_n(t_n) \rangle = \int \cdots \int d\xi_1 d\xi_2 \cdots d\xi_n \, \xi_1 \xi_2 \cdots \xi_n \, P_n(\xi_1, t_1, \cdots \xi_n, t_n) \tag{1.28}$$

For **stationary processes**

$$P_n(\xi_1, t_1, \cdots \xi_n, t_n) = P_n(\xi_1, t_1 + \tau, \cdots \xi_n, t_n + \tau) \tag{1.29}$$

for all n, t_j and τ.

The conditional probability density $P_{1|1}(\xi_2, t_2 | \xi_1, t_1)$ for Ξ to take on values ξ_2 at t_2 given that it took on values ξ_1 at t_1 is defined by the joint probability density

$$P_{1|1}(\xi_2, t_2|\xi_1, t_1)P_1(\xi_1, t_1) = P_2(\xi_1, t_1, \xi_2, t_2) \tag{1.30}$$

In physics language this is often referred to as the **transition probability** between state 1 and state 2. Generalizing this, the conditional probability density

$$P_{m|k}(\xi_{k+1}, t_{k+1}, \cdots \xi_{k+m}, t_{k+m}|\xi_1, t_1, \cdots \xi_k, t_k) \tag{1.31}$$

for Ξ to take on the value ξ_{k+1} at $t_{k+1} \cdots \xi_{k+m}$ at t_{k+m} given that it took on the value ξ_1 at $t_1, \cdots \xi_k$ at t_k is defined by

$$P_{m|k}(\xi_{k+1}, t_{k+1}, \cdots \xi_{k+m}, t_{k+m}|\xi_1, t_1, \cdots \xi_k, t_k) \equiv \frac{P_{k+m}(\xi_1, t_1, \cdots \xi_{k+m}, t_{k+m})}{P_k(\xi_1, t_1, \cdots \xi_k, t_k)} \tag{1.32}$$

where P_{k+m} is the joint probability density.

1.6.2 Markov processes

A **Markov process** is a stochastic process where the random variable has memory only of its immediate past, i.e.

$$P_{1|n-1}(\xi_n, t_n|\xi_1, t_1, \cdots \xi_{n-1}, t_{n-1}) = P_{1|1}(\xi_n, t_n|\xi_{n-1}, t_{n-1}) \tag{1.33}$$

A Markov process is entirely determined by $P_1(\xi_1, t_1)$ and $P_{1|1}(\xi_2, t_2|\xi_1, t_1)$. It is easy to show by using the Bayes rule that for Markov processes

$$P_{1|1}(\xi_3, t_3|\xi_1, t_1) = \int d\xi_2 P_{1|1}(\xi_3, t_3|\xi_2, t_2)P_{1|1}(\xi_2, t_2|\xi_1, t_1) \tag{1.34}$$

i.e. the two steps are statistically independent. This is the **Chapman–Komogorov** (CK) or **Einstein–Smolochousky** (ES) equation. For stationary Markov processes, if we define

$$P_{1|1}(\xi_2, t_2|\xi_1, t_1) \equiv P_\tau(\xi_2|\xi_1) \tag{1.35}$$

since they depend only on $\tau = t_2 - t_1$, the CK or ES equation can be written schematically as

$$P_{\tau'+\tau} = P_{\tau'}P_\tau \tag{1.36}$$

in the sense of integral kernels.[8] We now derive a differential form of the CK equation which is known as the (Markovian) Pauli master equation. Consider a small increment τ in time from t_1 and expand $P_{1|1}(\xi_2, t_1 + \tau|\xi_1, t_1)$ in a Taylor

---

[8] In probability theory understandably the symbol P is used profusely. Here a single subscript τ denotes the conditional probability density $P_{1|1}$ in a stationary process. Notice this equation which we see quite commonly in physics actually presupposes the Markovian property. When it involves probability concepts, as in quantum mechanics, interpreting physics equations in the stochastic process sense may reveal a deeper layer of meaning for these common objects.

series around t_1 making sure that the normalization condition is preserved to all orders in τ; we have

$$P_{1|1}(\xi_2, t_1 + \tau | \xi_1, t_1) = P_{1|1}(\xi_2, t_1 | \xi_1, t_1) + \tau \partial P_{1|1} / \partial \tau + \dots \qquad (1.37)$$

Extract the singular part from $\partial P_{1|1} / \partial \tau$

$$\frac{\partial P_{1|1}}{\partial \tau} \equiv -a_0(\xi_1)\delta(\xi_2 - \xi_1) + W(\xi_2 | \xi_1) \qquad (1.38)$$

We say $W(\xi_2 | \xi_1)$ is the **transition probability per unit time**. In physics language this is called the transition rate. $a_0(\xi_1)$ is determined by the condition that the normalization condition $\int P_{1|1}(\xi_2, t_2 | \xi_1, t_1) d\xi_2 = 1$ is satisfied to all orders of τ. To first order in τ, the condition yields

$$a_0(\xi_1) = \int W(\xi_2 | \xi_1) d\xi_2 \qquad (1.39)$$

Using this we have

$$P_\tau(\xi_2 | \xi_1) = (1 - a_0(\xi_1)\tau)\delta(\xi_2 - \xi_1) + \tau W(\xi_2 | \xi_1) \qquad (1.40)$$

Writing down a copy of this equation for $P_{\tau'}(\xi_3 | \xi_2)$

$$P_{\tau'}(\xi_3 | \xi_2) = (1 - a_0(\xi_2)\tau')\delta(\xi_3 - \xi_2) + \tau' W(\xi_3 | \xi_2) \qquad (1.41)$$

and putting them back into the CK equation we obtain

$$P_{\tau+\tau'}(\xi_3 | \xi_1) = \int (1 - a_0(\xi_2)\tau')\delta(\xi_3 - \xi_2) P_\tau(\xi_2 | \xi_1) d\xi_2$$
$$+ \int \tau' W(\xi_3 | \xi_2) P_\tau(\xi_2 | \xi_1) d\xi_2 \qquad (1.42)$$

Performing the integral in the first term, we obtain upon dividing by τ' on both sides and letting $\tau' \to 0$

$$\frac{\partial P_\tau(\xi_3 | \xi_1)}{\partial \tau} = \int d\xi_2 [-W(\xi_2 | \xi_3) P_\tau(\xi_3 | \xi_1) + W(\xi_3 | \xi_2) P_\tau(\xi_2 | \xi_1)] \qquad (1.43)$$

where we have used the expression for a_0 above. This is the CK equation for stationary Markov process which include the familiar gain–loss, birth–death processes.

To cast this in a more familiar form we can eliminate ξ_1 by introducing the two conditional probability densities

$$P_\tau(\xi_3 | \xi_1) = P_1(\xi_3, t) \to P_n(t) \qquad (1.44)$$
$$P_\tau(\xi_2 | \xi_1) = P_1(\xi_2, t) \to P_{n'}(t) \qquad (1.45)$$

The right arrow indicates transforming to a notation for processes via discrete variables Ξ, as in quantum states. We get the familiar Pauli master equation

$$\frac{dP_n}{dt} = \sum_{n'} [-W_{n \to n'} P_n(t) + W_{n' \to n} P_{n'}(t)] \qquad (1.46)$$

In conventional (less rigorous) physics language we call $P_n(t)$ the probability to find the system in state n, and $W_{n \to n'} P_n(t)$ the transition probability from state n to state n' in time t. Thus the first term measures the "loss" of system in state n (depletion), and the second term its "gain" (increase).[9]

1.6.3 Kramers–Moyal and Fokker–Planck equations

For linear systems and in the limit where the jumps in a Markov process are small, this equation takes a special form known as the Fokker–Planck equation. Define $\eta = \xi - \xi'$ as the jump size. The transition probability $W(\xi|\xi') = W(\xi'; \eta)$ is assumed to vary slowly with ξ, ξ', and is a sharply peaked function of η. From the CK equation,

$$\frac{\partial P_1(\xi, t)}{\partial t} = \int d\xi' [W(\xi|\xi') P_1(\xi', t) - W(\xi'|\xi) P_1(\xi, t)]$$

$$= \int d\eta [W(\xi - \eta; \eta) P_1(\xi - \eta, t) - P_1(\xi, t) \int d\eta W(\xi - \eta; -\eta) \quad (1.47)$$

Taylor expanding P_1 around ξ in the integrand of the first term on the right-hand side, i.e.

$$P_1(\xi - \eta, t) = P_1(\xi) - \eta \frac{\partial P_1}{\partial \eta} + \frac{\eta^2}{2} \frac{\partial^2 P_1}{\partial \eta^2} + \cdots \quad (1.48)$$

we have

$$\frac{\partial P_1(\xi, t)}{\partial t} = \sum_{\nu=1}^{\infty} \frac{(-1)^\nu}{\nu!} \frac{\partial^\nu}{\partial \xi^\nu} [a_\nu(\xi) P_1(\xi, t)] \quad (1.49)$$

where

$$a_\nu(\xi) = \int_{\infty}^{\infty} \eta^\nu W(\xi; \eta) d\eta \quad (1.50)$$

This is called the Kramers–Moyal expansion of the Markovian master equation. Keeping only the first two terms and dropping the subscript 1 on P_1 (to convert

---

[9] The Pauli equation could be the first instance we learn about the master equation, usually in the context of quantum mechanics (e.g. Chapter 15 of Reif [Rei67]), but it is not tied to any quantum notion whatsoever. (The only relevant concept from quantum physics is the discrete state, but we know there is a corresponding version of the CK equation for continuous variables.) To begin with, it deals with probabilities, not amplitudes, so there is no phase information, and thus is useless in dealing with issues like quantum decoherence, which probes into how the quantum phase information gets lost as a system's classical behavior emerges. More importantly it describes only Markovian stationary process – we will see that it is far from the most general conditions. For example these are the conditions behind the Fermi Golden rule, or the Wigner-Weisskopf lineshape, which are built upon time-dependent perturbation theory. What this tells us is that it is always helpful to ask a few questions about the tacit assumptions behind any physical law, no matter how familiar they appear.

to physics notation) yields the Fokker–Planck (FP) equation

$$\frac{\partial P(\xi,t)}{\partial t} = -\frac{\partial}{\partial \xi}[a_1(\xi)P(\xi,t)] + \frac{1}{2}\frac{\partial^2}{\partial \xi^2}[a_2(\xi)P(\xi,t)] \tag{1.51}$$

For small changes in time we can write the coefficients a_1, a_2 as follows:

$$a_1(\xi) = \int \eta W(\xi,\eta)d\eta \simeq \frac{\langle \Delta\xi \rangle}{\Delta t} = \langle v_\xi \rangle \tag{1.52}$$

$$a_2(\xi) = \int \eta^2 W(\xi,\eta)d\eta \simeq \frac{\langle (\Delta\xi)^2 \rangle}{\Delta t} \tag{1.53}$$

If there is no external force $a_1 = 0$, the FP equation is in the form of a diffusion equation

$$\frac{\partial P(\xi,t)}{\partial t} = D_\xi \frac{\partial^2 P(\xi,t)}{\partial \xi^2} \tag{1.54}$$

with diffusion coefficient

$$D_\xi = \frac{a_2}{2} = \frac{\langle (\Delta\xi)^2 \rangle}{2\Delta t} \tag{1.55}$$

This is known as the first Einstein relation.

In Chapter 2 we shall use intuitive physical reasoning to give a derivation of the Boltzmann and Langevin equation, and their quantum version in Chapter 3. These equations, together with the general (not just the Pauli) master equation, will be the starting point for our expedition into nonequilibrium quantum field processes.

2

Relaxation, dissipation, noise and fluctuations

2.1 A simple model of Brownian motion

In this chapter we shall continue the study of relaxation, dissipation, noise and fluctuations by analyzing how they appear in simple models extracted from classical physics. We shall also introduce some specific concepts, such as the fluctuation–dissipation relation, which will be central to the development of our subject matter.

Possibly the simplest manifestation of the relaxation process is the damping of a pendulum swinging in open air. The simplest model of a pendulum is the harmonic oscillator

$$\ddot{x} + \Omega^2 x = 0 \tag{2.1}$$

At this level of description, it belongs to the realm of mechanics rather than thermodynamics [LanLif69]; it obeys the conservation of phase space volume theorem, it generates no entropy, and it does not relax. To see relaxation, we must introduce damping. Let us proceed phenomenologically by adding a "damping constant" γ to our oscillator equation (2.1), which becomes

$$\ddot{x} + 2\gamma \dot{x} + \Omega^2 x = 0 \tag{2.2}$$

Later we will probe into the microscopic origin of dissipation.

Introduce an angle φ such that $\gamma = \Omega \sin \varphi$ and write $\Omega_1 = \Omega \cos \varphi$. The solution to equation (2.2) is

$$x(t) = e^{-\gamma t} \left\{ x(0) \frac{\cos [\Omega_1 t - \varphi]}{\cos \varphi} + \frac{p(0)}{M \Omega_1} \sin \Omega_1 t \right\} \tag{2.3}$$

where M is the mass of the oscillator. Although this system does relax, it is a little boring: the only possible equilibrium is at the bottom of the potential. But we know that a classical pendulum at finite temperature has nonzero average kinetic and potential energies, obeying the energy equipartition theorem. So something is missing. Let us call $\xi(t)$ the missing term, so that the system (2.2) becomes

$$\ddot{x} + 2\gamma \dot{x} + \Omega^2 x = \frac{\xi(t)}{M} \tag{2.4}$$

A solution is in the form $x = x_h + x_p$, where x_h is the homogeneous solution [given by equation (2.3)], and x_p is the particular solution

$$x_p(t) = \int_0^t dt' \, e^{-\gamma(t-t')} \frac{\sin \Omega_1 (t - t')}{M \Omega_1} \xi(t') \tag{2.5}$$

Let us consider the source $\xi(t)$ as some kind of "noise" or stochastic forcing term, and assume that the expectation value at any time is zero $\langle \xi(t) \rangle = 0$, where $\langle \rangle$ stands for the average over realizations of the noise. Then $\langle x_p \rangle \equiv 0$, so that $\langle x \rangle \to 0$ as $t \to \infty$. As for $\langle x^2 \rangle$, we know that x_h will eventually die away, so for long times $\langle x^2(t) \rangle \sim \langle x_p^2(t) \rangle$, given by

$$\langle x_p^2(t) \rangle = \frac{1}{M^2\Omega_1^2} \int_0^t dt'dt'' \, e^{-\gamma(2t-t'-t'')} \sin\Omega_1(t-t') \sin\Omega_1(t-t'') \langle \xi(t')\xi(t'') \rangle$$

$$(2.6)$$

To proceed we must say something about the noise correlator. If in our intuitive picture $\xi(t)$ represents the stochastic bombardment of the ball of the pendulum by its surrounding air molecules, then the simplest property is that the noise is stationary and statistically independent at macroscopically distinguishable times, hence $\langle \xi(t)\xi(t') \rangle = \sigma^2\delta(t-t')$. Discarding exponentially decaying and other small terms, we obtain $\langle x^2(t) \rangle \sim \sigma^2/4M^2\Omega^2\gamma$. Comparing with the equipartition theorem $\langle x^2(t) \rangle = k_BT/M\Omega^2$, where k_B is Boltzmann's constant, this suggests that the system is equilibrating at a temperature given by the Einstein relation [Ein05]

$$\sigma^2 = 4\gamma M k_B T \qquad (2.7)$$

We have succeeded (our model successfully describes relaxation) where we ought to have failed (we violated the time reversibility of the original model equation (2.1)). Let us take our model apart, and try to understand the secret of its working.

Observe that the system–environment interaction goes both ways: while the γ term steadily dumps system energy into the environment, whereby the information on initial conditions is lost, the noise term works in the opposite direction, feeding the right amount of fluctuations into the system and compensating its tendency to drop to the bottom of the potential. Neither alone would do the job, as clearly shown by equation (2.7), which, when seen in this light, goes under the name of a *fluctuation–dissipation theorem* [Nyq28, CalWel51].

In this view of the fluctuation–dissipation theorem, *if* we wish the system to relax at a certain temperature T, we'd better throw in white noise with the proper amplitude. But it could be that Nature does not care about relaxation, and therefore that it does not need a fluctuation–dissipation theorem. Well, as we know from everyday experience, it does, and there is a deeper reason for equation (2.7). In the final analysis, the Einstein relation is an expression of the unitarity of the dynamics of the system–environment complex. To understand how this comes about, we shall backtrack a little, and offer a simple mechanical model of how the environment works.

2.1.1 The linear oscillator model

The simplest possible mechanical model of the environment is to consider it as a large set of linear harmonic oscillators with displacement q_α, proper frequency ω_α

and mass m_α coupled to the system through a time-dependent coupling constant $c_\alpha(t)$ (see below) [Rub60, Rub61, FoKaMa65]. This is a very poor model of an environment; in a certain sense, it is no environment at all, as we may and will easily integrate the full dynamics, so there is little to be gained in regarding the q_α as different or "irrelevant," as the word "environment" may imply. In the real world, environments are huge nonlinear systems, and the information dumped in them is lost for all practical purposes as far as the observer is concerned. However, this modest ansatz for an environment will be adequate for our purpose here, which is why the above model actually works.

The full dynamics is given by

$$\ddot{x}(t) + \Omega^2 x(t) + \sum_\alpha \frac{c_\alpha(t)}{M} q_\alpha(t) = 0$$

$$\ddot{q}_\alpha(t) + \omega_\alpha^2 q_\alpha(t) + \frac{c_\alpha(t)}{m_\alpha} x(t) = 0 \qquad (2.8)$$

The second set of equations is easily solved as $q_\alpha(t) = q_{\alpha p}(t) + q_{\alpha h}(t)$, where

$$q_{\alpha h}(t) = \left[q_\alpha(0) + \frac{c_\alpha(0)}{m_\alpha \omega_\alpha^2} x(0) \right] \cos \omega_\alpha t + \frac{p_\alpha(0)}{m_\alpha \omega_\alpha} \sin \omega_\alpha t$$

$$q_{\alpha p}(t) = \frac{-1}{m_\alpha \omega_\alpha} \int_0^t dt' \, \sin \omega_\alpha (t - t') c_\alpha(t') x(t') - \frac{c_\alpha(0)}{m_\alpha \omega_\alpha^2} x(0) \cos \omega_\alpha t$$

$$= -\frac{c_\alpha(t)}{m_\alpha \omega_\alpha^2} x(t) + \frac{1}{m_\alpha \omega_\alpha^2} \int_0^t dt' \, \cos \omega_\alpha (t - t') \frac{d}{dt'} (c_\alpha x) \qquad (2.9)$$

We have kept this level of detail just to show that the evolution of the environment is not indifferent to the way the interaction is switched on. The simplest assumption is that the interaction is introduced adiabatically, but quickly settles to a constant value. In this scheme, we have $c_\alpha(0) = 0$ but $\dot{c}_\alpha = 0$ at any macroscopically positive time. Introducing this into the equation for the system, we obtain

$$\ddot{x}(t) + \int_0^t dt' \, \gamma(t - t') \dot{x}(t') + \Omega_r^2 x(t) = \frac{\xi(t)}{M} \qquad (2.10)$$

where

$$\Omega_r^2 = \Omega^2 - \frac{1}{M} \sum_\alpha \frac{c_\alpha^2}{m_\alpha \omega_\alpha^2} \qquad (2.11)$$

$$\gamma(t - t') = \frac{1}{M} \sum_\alpha \frac{c_\alpha^2}{m_\alpha \omega_\alpha^2} \cos \omega_\alpha (t - t') \qquad (2.12)$$

$$\xi(t) = -\sum_\alpha c_\alpha q_{\alpha h}(t) \qquad (2.13)$$

There are three differences between equations (2.10) and (2.4). First, the frequency of the system has been renormalized. The second difference is that γ now has a finite memory, reducing to the simple ohmic case $\gamma(t - t') = 4\gamma \delta(t - t')$

only for a rather special (and unphysical) choice of the bath; this is unimportant for our present purpose. The real difference is that $\xi(t)$ is *not* a stochastic variable: it is a complex function of the bath's initial conditions. For this reason, equation (2.10) does *not* describe relaxation. It is simply the unitary dynamics of the system–bath complex, written in a different set of variables. So, what is missing?

Could it be that we forgot to record the actual initial conditions for the environment? If so, we may consider that these initial conditions are taken at random. To make it even simpler, we may assume that the initial conditions are taken independently for each oscillator, and that they sample each classical orbit homogeneously. Under these conditions, we have, from the classical virial theorem

$$\frac{1}{m_\alpha} \langle p_\alpha(0) p_{\alpha'}(0) \rangle = m_\alpha \omega_\alpha^2 \langle q_\alpha(0) q_{\alpha'}(0) \rangle = \delta_{\alpha\alpha'} \langle \varepsilon_\alpha \rangle \tag{2.14}$$

$$\langle p_\alpha(0) q_{\alpha'}(0) \rangle = 0 \tag{2.15}$$

where $\langle \varepsilon_\alpha \rangle$ is the expectation value of the energy of the αth oscillator at $t = 0$. Now $\xi(t)$ is a bona fide stochastic variable, and

$$\langle \xi(t) \xi(t') \rangle = \sum_\alpha \frac{c_\alpha^2}{m_\alpha \omega_\alpha^2} \langle \varepsilon_\alpha \rangle \cos \omega_\alpha (t - t') \tag{2.16}$$

If the bath itself is at equilibrium, then $\langle \varepsilon_\alpha \rangle = k_B T$, and $\langle \xi(t) \xi(t') \rangle = M k_B T \gamma (t - t')$. This is Einstein's relation for the non-ohmic case, reducing to the case above in the ohmic limit.

Somewhere between equations (2.13) and (2.16) the environment oscillators lose their role as dynamical variables. The "ordered" part of the system–environment energy transfer is replaced by the γ term in equation (2.10), which refers to the system alone (we say that the bath variables have been *slaved to the system*); the "disordered" part is replaced by a generic stochastic force, whose effect is to compensate the dissipation and thus to make a nontrivial steady equilibrium possible. Time-reversal invariance becomes devoid of operational meaning, because the choice of a random initial condition for the bath forfeits one's ability to reverse the initial velocities of each oscillator in the bath. This introduces an arrow of time in the macrodynamics.

Of course, the actual time development of $\xi(t)$ as given in equation (2.13) looks a lot like a realization of the stochastic process defined by equation (2.16) for any finite period. But as time goes by, correlations build up between the system and its environment which are not contained in the stochastic model. Because these correlations are neglected, the stochastic model describes a nonunitary evolution; therein lies the true reason for Boltzmann's H-theorem – if all correlations were kept, unitarity would be restored.

This basic framework for irreversibility will be the backdrop for our future discussions. Of course, the Brownian motion paradigm which we discussed here is an example of an *autocratic* system: the ball is the king, the relevant party, the center of attention, and the molecules in its environment are subservient, slaved and "irrelevant." Irreversibility also obtains in *democratic* systems, such as a Boltzmann gas: all molecules are born equal and treated equally. However, limitation of observational precision introduces coarse graining of a different sort. In particular, we shall see below how irreversibility in the Boltzmann gas is actually a consequence of the *slaving* of irrelevant, many-particle correlations to the one-particle distribution function, which is of special interest as the coarsest yet most accessible level of description.

2.1.2 Fluctuation–dissipation theorem

Let us discuss the fluctuation–dissipation theorem (FDT) in a still simple but more general framework. This formulation of the FDT will be relevant when we come to discuss fluctuations in the Boltzmann equation later on. This presentation follows closely that given by Landau and Lifshitz [LaLiPi80a].

The simplest setting for the FDT is a homogeneous system described by variables X^i. Equilibria are located at the maxima of a thermodynamic potential $S(X^i)$. For an isolated system, S is the entropy, for an isothermal system, $S = -F/k_B T$, where F is the free energy, etc.

The thermodynamic forces are the components of the gradient of S, $L_i = -S_{,i}$ (a comma denotes a derivative). We chose coordinates so that thermodynamic equilibrium lies at $X^i = 0$. Then L_i also vanish at the origin, and for small deviations, we get a linear relationship $L_i = C_{ij} X^j$, where the matrix C is nonnegative.

For example, we could consider an isolated system made of a system proper and an environment. Let us choose as coordinates the energy, volume and particle number of the system $X^i = (E, V, N)$. The function S is the total entropy, and from the first law

$$dS = \left(\frac{1}{T_s} - \frac{1}{T_e} \right) dE + \left(\frac{p_s}{T_s} - \frac{p_e}{T_e} \right) dV - \left(\frac{\mu_s}{T_s} - \frac{\mu_e}{T_e} \right) dN \qquad (2.17)$$

where T, p and μ stand for temperature, pressure and chemical potential, and the subscripts "s" and "e" denote system and environment, respectively. The coefficients in this differential form are (minus) the forces, and we see that they indeed vanish at equilibrium. The matrix elements of C are the specific heat and compressibility functions, etc. (for example, $C_{EE} = 1/T^2 C_V$), and the condition of C being nonnegative engenders a set of thermodynamic inequalities such as positivity of the specific heat.

We wish to motivate a dynamics for this system, under the basic requirement that it should describe regression to equilibrium. This suggests writing $\dot{X}^i = -\Gamma^{ij} L_j$, where Γ is nonnegative; then $\dot{S} = \Gamma^{ij} S_{,i} S_{,j} \geq 0$, and we obtain an H-theorem of sorts. But this dynamics is too efficient, because we know that in true equilibrium the system is not just sitting at $X = 0$, but fluctuating around it. Following Einstein, we identify the probability of a fluctuation carrying the system from 0 to X as $\exp S\,[X]$, whereby (in equilibrium) $\langle X^i L_j \rangle = \delta^i_j$. To obtain these fluctuations, we must modify our ansatz to

$$\dot{X}^i = -\Gamma^{ij} L_j + \Xi^i. \tag{2.18}$$

The first term describes the mean regression of the system towards a local entropy maximum, Γ^{ij} being the dissipative coefficient or function, and the second term describes the random microscopic fluctuations induced by its interaction with an environment. To simplify, let us assume that Ξ^i is a Gaussian white noise, namely $\langle \Xi^i (t) \Xi^j (t') \rangle = \sigma^{ij} \delta (t - t')$, where the matrix σ is, of course, symmetric and nonnegative. The FDT will allow us to relate the matrices σ and Γ.

In equilibrium, correlation functions are stationary. In particular

$$\frac{d}{dt} \left\langle X^i (t)\, X^j (t) \right\rangle = \left\langle \dot{X}^i (t)\, X^j (t) + X^i (t)\, \dot{X}^j (t) \right\rangle = 0 \tag{2.19}$$

Therefore

$$\left\langle \Xi^i (t)\, X^j (t) + X^i (t)\, \Xi^j (t) \right\rangle = \Gamma^{ij} + \Gamma^{ji} \tag{2.20}$$

If the noise is Gaussian, we have the Novikov identity [Nov65]

$$\left\langle X^i (t)\, \Xi^j (t') \right\rangle = \int dt'' \, \frac{\delta X^i (t)}{\delta \Xi^k (t'')} \left\langle \Xi^k (t'')\, \Xi^j (t') \right\rangle \tag{2.21}$$

which for our chosen autocorrelation becomes

$$\left\langle X^i (t)\, \Xi^j (t') \right\rangle = \sigma^{kj} \frac{\delta X^i (t)}{\delta \Xi^k (t')} \tag{2.22}$$

Since the dynamics is linear, we may write

$$X^i (t) = X^i_h (t) + \int_{}^{t} dt' \, G^i_k (t - t')\, \Xi^k (t') \tag{2.23}$$

where the homogeneous solution $X^i_h (t)$ is independent of the noise, and the propagator G satisfies $G^i_k (0) = \delta^i_k$. In the coincidence limit $t' = t$ we find

$$\frac{\delta X^i (t)}{\delta \Xi^k (t)} = \int_{}^{t} dt' \, G^i_k (t - t')\, \delta (t - t') = \frac{1}{2} \delta^i_k \tag{2.24}$$

From equations (2.20), (2.22) and (2.24), we get

$$\sigma^{ik} = \Gamma^{ik} + \Gamma^{ki} \tag{2.25}$$

which is the FDT in a simple classical formulation.

In the case of a one-dimensional system, the above argument can be simplified even further because there is only one variable X, and Γ, C, σ are simply constants. In equilibrium, we have $\langle X^2 \rangle = C^{-1}$. On the other hand, the late time solution of the equations of motion reads

$$X (t) = \int_{}^{t} du \, e^{-\Gamma C (t - u)} \Xi (u) \tag{2.26}$$

which implies $\langle X^2 \rangle = \sigma / 2\Gamma C$. Thus $\sigma = 2\Gamma$, in agreement with equation (2.25).

As an example of this view of the FDT, let us return to the problem of the dissipative pendulum. The system is described by two degrees of freedom x and $p = M\dot{x}$. Since we are interested in the pendulum coming to equilibrium at a given temperature, the relevant thermodynamic potential is $S = -F/k_{\mathrm{B}}T$. We identify the free energy associated with a phase space point (x, p) as the work necessary to bring the pendulum from rest to (x, p), in a reversible way and at constant temperature. This work is, of course, the mechanical energy, so

$$S = -\frac{p^2}{2Mk_{\mathrm{B}}T} - \frac{M\Omega^2 x^2}{2k_{\mathrm{B}}T} \tag{2.27}$$

The forces are then $L_x = M\Omega^2 x / k_{\mathrm{B}}T$ and $L_p = p/Mk_{\mathrm{B}}T$. In these terms, Hamilton's equations become

$$\dot{x} = \frac{p}{M} = k_{\mathrm{B}}TL_p; \qquad \dot{p} = -M\Omega^2 x = -k_{\mathrm{B}}TL_x \tag{2.28}$$

This corresponds to an antisymmetric Γ matrix, and therefore the potential S is conserved. We get no H-theorem, as expected.

In order to obtain regression to equilibrium, we must include dissipation. As is stressed by Landau and Lifshitz, it makes no sense to modify the first of equation (2.28), since this represents the definition of p rather than a true dynamical law. Thus our only possibility is to modify the second equation

$$\dot{p} = -M\Omega^2 x - 2\gamma p = -k_{\mathrm{B}}T\left(L_x + 2M\gamma L_p\right) \tag{2.29}$$

The new understanding is that this modification must be *necessarily* followed by the inclusion of noise $\Xi^i = \left(\tilde{\xi}, \xi\right)$

$$\dot{x} = \frac{p}{M} + \tilde{\xi}; \qquad \dot{p} = -m\omega^2 x - 2\gamma p + \xi \tag{2.30}$$

and that we have no freedom in choosing the noise autocorrelation, as this is given by the FDT. In our case, discarding the antisymmetric part of γ^{ij}, we get $\sigma_{xx} = \sigma_{xp} = 0$, $\sigma_{pp} = 4\gamma Mk_{\mathrm{B}}T$, which of course reproduces the result from the last section.

2.2 The Fokker–Planck and Kramers–Moyal equations

Let us now consider a single variable $X(t)$ evolving according to the Langevin equation [Cha43, Kam81]

$$\frac{dX}{dt}(t) + \Gamma(t) X(t) = \Xi(t) \tag{2.31}$$

(that is, in comparison with equation (2.18), we now take the entropy as simply $S = (-1/2) X^2$, thus $L = X$, and allow Γ to depend on time), where Ξ is a Gaussian *colored* noise

$$\langle \Xi(t) \Xi(t') \rangle = s^2(t, t') \tag{2.32}$$

Under the influence of noise the variable X will show a complicated behavior, even if its initial value is accurately known. It becomes uninteresting to try and follow the evolution of X in all its detail; just knowing the probability density $f(x,t)$ for actually finding X in a neighborhood of x at times t is enough. Formally

$$f(x,t) = \langle \delta (X(t) - x)\rangle \tag{2.33}$$

where the average is over realizations of the noise and also over all possible initial conditions $X(0)$. For simplicity, we assume the noise acts independently of the initial condition.

The probability density f evolves according to the so-called Fokker–Planck equation [Ris89, Gar90]. To derive this equation, observe that [SanMig89]

$$\frac{\partial}{\partial t} f(x,t) = \left\langle \frac{dX}{dt}(t) \frac{\partial}{\partial X(t)} \delta(X(t) - x)\right\rangle$$

$$= -\frac{\partial}{\partial x}\left\langle \frac{dX}{dt}(t)\, \delta(X(t) - x)\right\rangle$$

$$= \frac{\partial}{\partial x}\left[\Gamma(t)\, x f(x,t)\right] - \frac{\partial}{\partial x}\left\langle \Xi(t)\, \delta(X(t) - x)\right\rangle \tag{2.34}$$

To compute the last expectation value, we appeal to the Novikov identity (2.21)

$$\langle \Xi(t)\, \delta(X(t) - x)\rangle = \int_0^t dt'\, s^2(t,t')\left\langle \frac{\delta}{\delta\Xi(t')} \delta(X(t) - x)\right\rangle$$

$$= -\frac{\partial}{\partial x}\left[\sigma(t)\, f(x,t)\right] \tag{2.35}$$

where

$$\sigma(t) = \int_0^t dt'\, s^2(t,t') \frac{\delta X(t)}{\delta \Xi(t')} \tag{2.36}$$

which in this simple case can be computed almost explicitly. The final result takes the form of a continuity equation

$$\frac{\partial}{\partial t} f(x,t) = \frac{\partial}{\partial x}\left\{\left[\Gamma(t)\, x + \frac{\partial}{\partial x}\sigma(t)\right] f(x,t)\right\} \tag{2.37}$$

One remarkable feature of this equation is that it is *local* in time, in spite of the noise being colored. Moreover, it does not seem possible to reconstruct $s^2(t,t')$ from $\sigma(t)$ in general, unless some further hypothesis is added (for example, that the noise is actually white). In this sense, the original Langevin description contains more information about the system than the Fokker–Planck one [CaRoVe03].

Equation (2.31) may be generalized to nonlinear dynamics [BixZwa71, Zwa73]

$$\frac{dX}{dt}(t) + \Gamma[X(t),t] = \xi(t) \tag{2.38}$$

Repeating our earlier steps, we find

$$\frac{\partial}{\partial t} f(x,t) = \frac{\partial}{\partial x} \left[\Gamma[x,t] f(x,t) \right] + \frac{\partial^2}{\partial x^2} \int_0^t dt' \, s^2(t,t') \left\langle \frac{\delta X(t)}{\delta \xi(t')} \delta(X(t) - x) \right\rangle$$

(2.39)

where

$$\frac{\partial}{\partial t} \frac{\delta X(t)}{\delta \xi(t')} + \frac{\partial \Gamma[X(t),t]}{\partial X(t)} \frac{\delta X(t)}{\delta \xi(t')} = \delta(t - t')$$

(2.40)

In general, this will be a complicated function of the base trajectory $X(t)$. However, if the noise is white

$$s^2(t,t') = \sigma^2(t) \delta(t - t')$$

(2.41)

then the Fokker–Planck equation simplifies to

$$\frac{\partial}{\partial t} f(x,t) = \frac{\partial}{\partial x} \left\{ \left[\Gamma[x,t] + \frac{\sigma^2(t)}{2} \frac{\partial}{\partial x} \right] f(x,t) \right\}$$

(2.42)

An important particular case of the above is when the Langevin dynamics follows from adding local dissipation and white noise to an otherwise Hamiltonian system. We then have two variables X and P, with

$$\frac{dX}{dt} = \frac{\partial H}{\partial P}$$

(2.43)

$$\frac{dP}{dt} = -\frac{\partial H}{\partial X} - 2\gamma P + \xi$$

(2.44)

$$H = \frac{P^2}{2M} + V(X)$$

(2.45)

where H is the Hamiltonian (following Landau, we only add noise to the second equation). Then

$$f(x,p,t) = \langle \delta(X(t) - x) \delta(P(t) - p) \rangle$$

(2.46)

and

$$\frac{\partial}{\partial t} f(x,t) = -\{H, f\} + \frac{\partial}{\partial p} \left[\left(2\gamma p + \frac{\sigma^2(t)}{2} \frac{\partial}{\partial p} \right) f \right]$$

(2.47)

where

$$\{H, f\} = \frac{\partial H}{\partial p} \frac{\partial f}{\partial x} - \frac{\partial H}{\partial x} \frac{\partial f}{\partial p}$$

(2.48)

is the Poisson bracket. This is the so-called Kramers–Moyal equation [Kra40, Moy49].

In the derivation of the Kramers–Moyal equation we have used the fact that a change in the external force changes the acceleration, but neither the position nor the velocity, instantaneously, so

$$\frac{\delta X\left(t\right)}{\delta \xi\left(t\right)} = 0 \tag{2.49}$$

The resulting Kramers–Moyal equation contains only second-order p-derivatives. This is the so-called *normal* diffusion. For colored noise there are both normal and anomalous diffusion (we shall see an example in Chapter 3).

For a thermodynamic system in contact with a heat bath any spontaneous transformation decreases the free energy $F = U - TS$. For a system described by the Kramers–Moyal equation, if both γ and σ^2 are time-independent, there is an analog to this statement. We replace the internal energy U by the average value of the Hamiltonian, the entropy S by the Boltzmann H_B function

$$H_B = -k_B \int dX dP\, f \ln\left[f\right] \tag{2.50}$$

and the temperature T by $\sigma^2/4M\gamma k_B$ [cf. equation (2.7)]. Thus we obtain Kramers' nonequilibrium free energy [Kur98, Kur05]

$$F_K = \int dX dP\, f \left\{ H + \frac{\sigma^2}{4M\gamma} \ln\left[f\right] \right\} \tag{2.51}$$

and an H-theorem of sorts

$$\frac{dF_K}{dt} = -\frac{2\gamma}{M} \int dX dP\, f \left[P + \frac{\sigma^2\left(t\right)}{4\gamma f} \frac{\partial f}{\partial P} \right]^2 \tag{2.52}$$

This also shows that there is only one stationary solution

$$f_{\text{eq}} \propto e^{-\left(4\gamma M/\sigma^2\right) H} \tag{2.53}$$

so we are led to the identify $\sigma^2 = 4\gamma M k_B T$, as expected.

If γ and σ^2 go to zero, the Kramers–Moyal equation reduces to the Liouville equation

$$\frac{\partial}{\partial t} f\left(x, t\right) = -\left\{ H, f \right\} \tag{2.54}$$

In the opposite limit, it reduces to a Fokker–Planck equation. For very large damping, we have

$$P \sim \frac{1}{2\gamma} \left[-V' + \xi \right] \tag{2.55}$$

$$\frac{dX}{dt} = -\frac{V'\left(X\right)}{2\gamma M} + \Xi \sim -\frac{V''\left(0\right)}{2\gamma M} X + \Xi \tag{2.56}$$

where $\Xi = \xi/2\gamma M$. This is the kind of dynamics we studied at the beginning of this section. Since the kinetic energy is negligible compared to the potential energy, we have $S = -V\left(X\right)/k_B T$, $C = V''\left(0\right)/k_B T$ and $\Gamma = k_B T/2\gamma M$.

The fluctuation–dissipation relation appropriate to the Fokker–Planck equation $\langle \Xi(t)\,\Xi(t')\rangle = 2\Gamma\delta(t-t')$ leads us back to $\langle \xi(t)\,\xi(t')\rangle = 4\gamma M k_{\mathrm{B}} T \delta(t-t')$.

2.3 The Boltzmann equation

We shall now examine the other major paradigm of irreversible behavior in classical physics, namely, Boltzmann's theory of dilute gases [Bol64, ChaCow39, LifPit81]. As we already mentioned, the Brownian motion paradigm we examined in the last section corresponds to an *autocratic* system where an environment is subservient to our system of interest. The Boltzmann model of a gas, on the other hand, seems to be *democratic* in that it embraces all molecules on equal terms. In this sense, the Boltzmann gas appears as a truly closed system. However, this system will be shown to be an effectively open system in the space of correlation functions. Specifically, our relevant system shall be the one-particle distribution function, and its environment consists of the higher correlations. When seen in this light, we shall see that irreversibility in the Boltzmann equation follows a similar pattern as in the Brownian motion problem.

This view of the Boltzmann theory as describing an effectively open system shows how nontrivial it may be to identify the right degrees of freedom to describe a given system. We may say that the genius of Boltzmann has been to realize that, while the characteristic time for the dynamics of individual molecules is the collision time, the characteristic time for the dynamics of the one-particle distribution function is the relaxation time, which is much longer. Thus, the one-particle distribution function is the collective degrees of freedom in whose terms the dynamics becomes slow and simple. The very first step in treating the nonequilibrium dynamics of a system, i.e. identifying the right collective degrees of freedom in a given situation, may turn out to be the most important, and at times the most difficult, task.

Consider a gas of N identical molecules interacting through a binary central potential $V(r)$; we shall assume the forces are short range and the gas is dilute, $Na^3/V \ll 1$ where a is the range of the potential. We shall consider no external forces. The Hamiltonian

$$H = \sum_{i=1}^{N} \frac{\mathbf{p}_i^2}{2m} + \frac{1}{2}\sum_{i\neq j} V_{ij}; \qquad V_{ij} = V(|\mathbf{x}_i - \mathbf{x}_j|) \tag{2.57}$$

(we assume no self-energies: $V_{ii} = 0$) leads to the Hamilton equations

$$\frac{d\mathbf{x}^i}{dt} = \frac{\partial H}{\partial \mathbf{p}_i} = \frac{\mathbf{p}^i}{m}; \qquad \frac{d\mathbf{p}^i}{dt} = -\frac{\partial H}{\partial \mathbf{x}_i} = -\sum_{i\neq j}\frac{\partial V_{ij}}{\partial \mathbf{x}_i} \tag{2.58}$$

Equivalently we may describe the state of the system through a $6N$-dimensional distribution function $\rho = \rho((\mathbf{x}_1,\mathbf{p}_1),\ldots,(\mathbf{x}_N,\mathbf{p}_N),t)$, which satisfies the

Liouville equation

$$\frac{\partial \rho}{\partial t} = -\{H, \rho\} \qquad (2.59)$$

where we introduced the Poisson bracket (generalizing (2.48))

$$\{f, g\} = \sum_{i=1}^{N} \left[\frac{\partial f}{\partial \mathbf{p}_i} \frac{\partial g}{\partial \mathbf{x}_i} - \frac{\partial g}{\partial \mathbf{p}_i} \frac{\partial f}{\partial \mathbf{x}_i} \right] \qquad (2.60)$$

ρ integrates to 1 over the whole phase space. We shall assume that ρ is totally symmetric, which in the quantum case yields Bose–Einstein statistics.

Given a (one-particle) phase space point $(\mathbf{x}, \mathbf{p})$, we may define the density at that point

$$\mathcal{F}(\mathbf{x}, \mathbf{p}) = \sum_{i=1}^{N} \delta(\mathbf{x}_i - \mathbf{x})\, \delta(\mathbf{p}_i - \mathbf{p}) \qquad (2.61)$$

The one-particle distribution function f_1 is the expectation value of the density

$$f_1(\mathbf{x}, \mathbf{p}) = \sum_{i=1}^{N} \langle \delta(\mathbf{x}_i - \mathbf{x})\, \delta(\mathbf{p}_i - \mathbf{p}) \rangle \qquad (2.62)$$

$$\langle \delta(\mathbf{x}_i - \mathbf{x})\, \delta(\mathbf{p}_i - \mathbf{p}) \rangle$$
$$= \int \prod_j d^3\mathbf{x}_j d^3\mathbf{p}_j\, \rho\left((\mathbf{x}_1, \mathbf{p}_1), \ldots, (\mathbf{x}_N, \mathbf{p}_N), t\right) \delta(\mathbf{x}_i - \mathbf{x})\, \delta(\mathbf{p}_i - \mathbf{p}) \quad (2.63)$$

which from symmetry becomes

$$\langle \delta(\mathbf{x}_i - \mathbf{x})\, \delta(\mathbf{p}_i - \mathbf{p}) \rangle = \int \prod_{j=2}^{N} d^3\mathbf{x}_j d^3\mathbf{p}_j\, \rho((\mathbf{x}, \mathbf{p}), (\mathbf{x}_2, \mathbf{p}_2), \ldots, (\mathbf{x}_N, \mathbf{p}_N), t)$$

$$\qquad (2.64)$$

and is independent of i. Therefore

$$f_1(\mathbf{x}, \mathbf{p}) = N \int \prod_{j=2}^{N} d^3\mathbf{x}_j d^3\mathbf{p}_j\, \rho((\mathbf{x}, \mathbf{p}), (\mathbf{x}_2, \mathbf{p}_2), \ldots, (\mathbf{x}_N, \mathbf{p}_N), t) \qquad (2.65)$$

For later use, we shall introduce also the s-particle distribution function

$$f_s\left((\mathbf{x}_1, \mathbf{p}_1), \ldots, (\mathbf{x}_s, \mathbf{p}_s)\right)$$
$$= \frac{N!}{(N-s)!} \int \prod_{j=s+1}^{N} d^3\mathbf{x}_j d^3\mathbf{p}_j\, \rho((\mathbf{x}_1, \mathbf{p}_1), \ldots, (\mathbf{x}_s, \mathbf{p}_s), (\mathbf{x}_{s+1}, \mathbf{p}_{s+1}) \ldots) \quad (2.66)$$

We obtain the dynamics of f_1 integrating side by side in Liouville's equation

$$\frac{\partial f_1}{\partial t} = -N \int \prod_{j=2}^{N} d^3\mathbf{x}_j d^3\mathbf{p}_j\, \{H, \rho\} \qquad (2.67)$$

Developing the Poisson bracket, we observe that all terms involving derivatives with respect to $\mathbf{x}_j$ or $\mathbf{p}_j$, $j \neq 1$, may be reduced to total derivatives and discarded (under suitable boundary conditions at infinity). The only surviving terms yield

$$\frac{\partial f_1}{\partial t}(\mathbf{x}_1, \mathbf{p}_1) = -\frac{\mathbf{p}_1}{m} \frac{\partial f_1}{\partial \mathbf{x}_1} + \frac{\partial}{\partial \mathbf{p}_1} \int d^3\mathbf{x}_2 d^3\mathbf{p}_2 \left[\frac{\partial}{\partial \mathbf{x}_1} V(|\mathbf{x}_1 - \mathbf{x}_2|) \right]$$
$$\times f_2((\mathbf{x}_1, \mathbf{p}_1), (\mathbf{x}_2, \mathbf{p}_2)) \tag{2.68}$$

To obtain the dynamics for f_1 we need the dynamics for f_2. This is obtained in an analogous way

$$\frac{\partial f_2}{\partial t} = -N(N-1) \int \prod_{j=3}^{N} d^3\mathbf{x}_j d^3\mathbf{p}_j \; \{H, \rho\} \tag{2.69}$$

Repeating the above argument, we get

$$\frac{\partial f_2}{\partial t} = -\{H_2, f_2\} + \int d^3\mathbf{x}_3 d^3\mathbf{p}_3 \; K f_3 \tag{2.70}$$

where H_2 is the two-particle Hamiltonian

$$H = \frac{\mathbf{p}_1^2}{2m} + \frac{\mathbf{p}_2^2}{2m} + V(|\mathbf{x}_1 - \mathbf{x}_2|) \tag{2.71}$$

The precise form of the kernel K in equation (2.70) is unimportant. What matters is that, if the dynamics of f_1 depends on f_2, it will depend on f_3, which in turn depends on f_4, etc. Thus we obtain an infinite hierarchy of equations, commonly known as the Bogoliubov–Born–Green–Kirkwood–Yvon (BBGKY) hierarchy.

We face a situation which is different from our oversimplified Brownian motion model. In the linearly coupled harmonic oscillators problem the dynamics is so simple that one is seriously tempted to just solve it, without ever mobilizing all the Langevin equation machinery. In the BBGKY case, a solution of the infinite hierarchy is close to impossible. So we need to find ways to reduce the problem to a simpler one. Usually the first step in this simplification is to reduce the infinite hierarchy to a finite system by just discarding an infinite set of distribution functions. We shall call this brute force reduction a *truncation* of the hierarchy.

For example, we may argue that, since the integral over $\mathbf{x}_2$ is effectively reduced to a sphere of radius a around $\mathbf{x}_1$, the collision term in equation (2.68) is smaller than the first term by a factor Na^3/V, which is $\ll 1$ by assumption. In turn, the collision term in equation (2.70) will be smaller than the other terms in this equation by about the same factor. For a dilute gas with short-range interactions, we would be dealing with small corrections to ever smaller terms, and at some point they may become negligible. For simplicity, we shall assume that we are interested in a situation where the first nontrivial truncation works, namely, we put $K = 0$ in equation (2.70).

We stress that this strategy is by no means guaranteed to work. If there were long-range interactions (like Coulomb forces), something drastically different may be required, such as a Vlasov scheme where all far away particles are replaced by an effective continuous charge distribution supporting an average potential. This is another example of why finding the right collective degrees of freedom may constitute the hardest part of the work, as we already mentioned.

2.3.1 Slaving of higher correlations in the Boltzmann equation

Our goal is to solve equation (2.70) for f_2 (with $K = 0$) and to substitute the solution in equation (2.68) for f_1. At first sight it may look like these equations are decoupled, but, as we shall see, they couple through the boundary conditions, as the behavior of f_2 for large separations will be determined by f_1, through the so-called *molecular chaos* hypothesis [AkhPel81].

Equation (2.70) expresses the conservation of probability as the particles move along the classical orbits generated by the Hamiltonian H_2. These trajectories are easiest to study if we decompose the motion in center of mass and relative variables

$$\mathbf{X} = \frac{1}{2}(\mathbf{x}_1 + \mathbf{x}_2); \qquad \mathbf{u} = \mathbf{x}_1 - \mathbf{x}_2 \tag{2.72}$$

Introducing the conjugate momenta

$$\mathbf{P} = \mathbf{p}_1 + \mathbf{p}_2; \qquad 2\mathbf{p} = \mathbf{p}_1 - \mathbf{p}_2 \tag{2.73}$$

we get the Hamiltonian

$$H_2 = \frac{\mathbf{P}^2}{2M} + \frac{\mathbf{p}^2}{2\mu} + V(u); \qquad M = 2m, \ \mu = \frac{m}{2} \tag{2.74}$$

The center of mass motion represents a particle of mass M moving with uniform speed, while the relative motion represents a particle of mass μ scattering off a fixed center of force at the origin.

Let us observe that the integral in equation (2.68) is effectively restricted to the range $|\mathbf{x}_1 - \mathbf{x}_2| \sim a$, and so the center of mass variable changes little. Thus we may ignore the dependence of f_2 on X (on a more formal level, we are computing the first term in a development of the collision integral in derivatives with respect to X). Also an initial domain of initial conditions will move along the classical orbits and be distorted. Since relative motion is very fast with respect to macroscopic time-scales, we may assume that on the time-scales relevant to our observations, the initial domain has been elongated and fills the classical trajectory uniformly (this effect is known as phase diffusion, or the running men effect: a line of runners with differential velocities will elongate and eventually go uniformly round the track). Under the twin hypothesis of center of mass independence and phase diffused relative motion, we get $f_{2,t} = f_{2,\mathbf{X}} = 0$, and the equation for f_2 becomes

$$\frac{\mathbf{p}}{\mu} \cdot \nabla_{\mathbf{u}} f_2 - (\nabla_{\mathbf{u}} V) \cdot \nabla_{\mathbf{p}} f_2 = 0 \tag{2.75}$$

We may add a term

$$[\nabla_{\mathbf{x}_2} V(|\mathbf{x}_1 - \mathbf{x}_2|)] \nabla_{\mathbf{p}_2} f_2((\mathbf{x}_1, \mathbf{p}_1), (\mathbf{x}_2, \mathbf{p}_2)) \tag{2.76}$$

under the integral in equation (2.68), since it integrates to zero anyway. Now observe that

$$\nabla_{\mathbf{x}_1} V(|\mathbf{x}_1 - \mathbf{x}_2|) = -\nabla_{\mathbf{x}_2} V(|\mathbf{x}_1 - \mathbf{x}_2|) = \nabla_{\mathbf{u}} V \tag{2.77}$$

$$\nabla_{\mathbf{p}_1} f_2 - \nabla_{\mathbf{p}_2} f_2 = \nabla_{\mathbf{p}} f_2 \tag{2.78}$$

Changing variables from $\mathbf{x}_2$ to $\mathbf{u}$, we get

$$\frac{\partial f_1}{\partial t}(\mathbf{x}_1, \mathbf{p}_1) = -\frac{\mathbf{p}_1}{m} \nabla_{\mathbf{x}_1} f_1 + \int d^3 p_2 \int d\mathbf{u} \, \frac{\mathbf{P}}{\mu} \nabla_{\mathbf{u}} f_2 \tag{2.79}$$

For a given $\mathbf{p}$, we may choose adapted cylindrical coordinates (u, b, φ). Then this simplifies to

$$\frac{\partial f_1}{\partial t} + \frac{\mathbf{p}_1}{m} \frac{\partial f_1}{\partial \mathbf{x}_1} = \int d^3 p_2 \int 2\pi b db \, \frac{|\mathbf{p}|}{\mu} [f_2(\mathbf{p}_1, \mathbf{p}_2, b, u = \infty) $$
$$- f_2(\mathbf{p}_1, \mathbf{p}_2, b, u = -\infty)] \tag{2.80}$$

where by $u = \pm\infty$ we mean a relative coordinate which is large enough to take the particles out of interaction range, but still small in macroscopic terms.

It is at this point that the crucial step is taken. At $u = -\infty$, the two particles have not yet interacted. Here we impose the molecular chaos condition, namely, that there are no correlations among them initially, and thus

$$f_2(\mathbf{p}_1, \mathbf{p}_2, b, u = -\infty) \sim f_1(\mathbf{x}_1, \mathbf{p}_1) f_1(\mathbf{x}_1, \mathbf{p}_2) \tag{2.81}$$

At $u = \infty$ the particles have interacted and are correlated. However, since f_2 is constant along the trajectories, we have

$$f_2(\mathbf{p}_1, \mathbf{p}_2, b, u = \infty) = f_2(\mathbf{p}_1', \mathbf{p}_2', b, u = -\infty) \sim f_1(\mathbf{x}_1, \mathbf{p}_1') f_1(\mathbf{x}_1, \mathbf{p}_2') \tag{2.82}$$

where $\mathbf{p}_1', \mathbf{p}_2'$ are the momenta which evolve into $\mathbf{p}_1, \mathbf{p}_2$ after a collision with impact parameter b. Equations (2.81) and (2.82) implement the *slaving* of the two-particle correlation to the one-particle distribution. After this, no trace of f_2 is left, but only functionals of f_1.

To make the content of these equations even clearer, let us write

$$f_2(\mathbf{p}_1, \mathbf{p}_2, b, u = \infty) = \int d^3 p_3 dp_4 \, \delta(\mathbf{p}_3 - \mathbf{p}_1') \delta(\mathbf{p}_4 - \mathbf{p}_2') f_1(\mathbf{x}_1, \mathbf{p}_3) f_1(\mathbf{x}_1, \mathbf{p}_4) \tag{2.83}$$

and also the trivial identity

$$f_2(\mathbf{p}_1, \mathbf{p}_2, b, u = -\infty) = \int d^3 p_3 dp_4 \, \delta(\mathbf{p}_3 - \mathbf{p}_1') \delta(\mathbf{p}_4 - \mathbf{p}_2') f_1(\mathbf{x}_1, \mathbf{p}_1) f_1(\mathbf{x}_1, \mathbf{p}_2) \tag{2.84}$$

The final result is

$$\frac{\partial f_1}{\partial t} + \frac{\mathbf{p}_1}{m}\frac{\partial f_1}{\partial \mathbf{x}_1} = \int d^3\mathbf{p}_2 d^3\mathbf{p}_3 d^3\mathbf{p}_4 \, \mathbf{T}\left(\mathbf{p}_1, \mathbf{p}_2, \mathbf{p}_3, \mathbf{p}_4\right)$$

$$\times \left\{ f_1\left(\mathbf{p}_3\right) f_1\left(\mathbf{p}_4\right) - f_1\left(\mathbf{p}_1\right) f_1\left(\mathbf{p}_2\right) \right\} \qquad (2.85)$$

where $\mathbf{T}$ is the transition probability

$$\mathbf{T} = \int 2\pi b db \, \frac{|\mathbf{p}|}{\mu} \delta\left(\mathbf{p}_3 - \mathbf{p}_1'\right) \delta\left(\mathbf{p}_4 - \mathbf{p}_2'\right) \qquad (2.86)$$

$\mathbf{T}$ is zero unless $\mathbf{p}_3, \mathbf{p}_4$ do evolve into $\mathbf{p}_1, \mathbf{p}_2$ for *some* impact parameter. Equation (2.85) is the Boltzmann equation, and it is dissipative. We observe that the source of dissipation is the slaving of f_2 to f_1, similar in philosophy as in our Brownian motion toy model. As in Brownian motion, in equilibrium there will be density fluctuations. Thus equation (2.85) is incomplete: there must also be a stochastic term, which is determined by the fluctuation–dissipation theorem. We shall derive this term, but first let us consider the changes in equation (2.85) brought by relativity and quantum statistics.

2.3.2 Corrections from quantum statistics

The Boltzmann equation has the structure of a balance equation where changes in the particle number within a given cell in phase space are attributed (other than transport along classical one-particle trajectories) either to gain or loss processes. Gain obtains when one of two particles with momenta p_3, p_4 are injected into the cell through a collision, and loss when a particle within the cell is scattered off by collision with another particle of momentum p_2. If the particles obey quantum statistics, we must take into account the effect of stimulated emission for Bose–Einstein (BE) statistics and Pauli blocking for Fermi–Dirac (FD) statistics [Lib98]. The kinetic equation is then changed into

$$\frac{\partial f_1}{\partial t} + \frac{\mathbf{p}_1}{m}\frac{\partial f_1}{\partial \mathbf{x}_1} = \int d^3\mathbf{p}_2 d^3\mathbf{p}_3 d^3\mathbf{p}_4 \, \mathbf{T} \, \mathbf{I} \qquad (2.87)$$

where $\mathbf{T}$ is a suitable transition probability, not necessarily identical to (2.86), and

$$\mathbf{I} = (1 \pm f_1)(1 \pm f_2) f_3 f_4 - (1 \pm f_3)(1 \pm f_4) f_1 f_2 \qquad (2.88)$$

Hereafter we drop the subindex 1 in f (we shall not consider higher correlation functions) and use the shorthand $f_i = f(\mathbf{x}, \mathbf{p}_i)$. The upper sign holds for BE, and the lower sign for FD.

In equilibrium the collision integral must vanish, and therefore $\ln\left[f/(1 \pm f)\right]$ must be an additive constant of motion [Hua87]. If the gas is globally at rest, we may discard a term proportional to $\mathbf{p}$, which would conflict with rotational invariance, to get

$$\ln\frac{f}{1 \pm f} = -\beta\left(\varepsilon - \mu\right) \qquad (2.89)$$

where ε is the one-particle energy, and β and μ are constants. Therefore

$$f_{eq} = \frac{1}{e^{\beta(\varepsilon-\mu)} \mp 1} \tag{2.90}$$

where again the signs correspond to BE (upper) or FD (lower) statistics. We recognize that $\beta = 1/k_B T$, and μ is the chemical potential. From now on we shall assume BE statistics.

The lesson from equation (2.90) is that to specify an equilibrium state we need five numbers: the three components of the velocity of the rest frame, and the temperature and chemical potential in that frame. In other words, equilibrium states are astonishingly simpler than the generic states of the theory, which live in a $6N$-dimensional parameter space. This essential simplicity is the ultimate reason why we can describe real physical systems so elegantly by thermodynamics and statistical mechanics.

2.3.3 Relativistic kinetic theory

Let us now add the demands of relativity [Isr72, Isr88]. We consider our particles as living in a four-dimensional spacetime with coordinates x^μ ($x^0 = ct$, $x^i = x, y, z$), endowed with a metric tensor $g_{\mu\nu}$, which in Minkowski space is just $\eta_{\mu\nu} = \text{diag}(-1, 1, 1, 1)$ (we use Misner–Thorne–Wheeler conventions (MTW) [MiThWh72]). The system is described by the one-particle distribution function $f(x^\mu, p_\mu)$, where x is a position variable, and p is a momentum variable. Momentum is assumed to lie on a mass shell $p^2 + M^2 = 0$ and have positive energy $p^0 > 0$.

We assume there is a conserved charge which allows us to define a meaningful conserved particle number. Given a spatial element $d\Sigma^\mu = n^\mu d\Sigma$ and a momentum space element $d^4 p$, the number of particles with momentum p lying within that phase space volume element is

$$dn = -4\pi f(x,p)\, \theta(p^0)\, \delta(p^2 + M^2)\, p^\mu n_\mu\, d\Sigma \frac{d^4 p}{(2\pi)^4 \sqrt{-g}} \tag{2.91}$$

where the normalization will be useful later on. In this formula, $g = \det g_{\mu\nu}$; of course, $-g = 1$ in Minkowski space, which we shall assume from now on. Observe that this definition is covariant. The particle number *density* is defined as (minus) the flux of the particle number *current*

$$N^\mu(x) = 2 \int Dp\, p^\mu f(x,p) \tag{2.92}$$

where we introduced the momentum space volume element

$$Dp = \theta(p^0)\, \delta(p^2 + M^2)\, \frac{d^4 p}{(2\pi)^3 \sqrt{-g}} \tag{2.93}$$

If we are only concerned with the particle number flux across equal time surfaces, we may decompose the particle current into $N^\mu = (c\rho, \mathbf{J})$, where ρ is the ordinary density and $\mathbf{J}$ the ordinary particle flux.

The energy–momentum density is defined in terms of the energy–momentum tensor

$$dP^\mu = -T^{\mu\nu} d\Sigma_\nu; \qquad T^{\mu\nu} = 2 \int Dp\, p^\mu p^\nu f(x, p) \tag{2.94}$$

$T^{00} = cE$, where E is the ordinary energy density, $T^{0i} = \mathbf{E}$ are the energy flux, $T^{i0} = c\mathbf{P}$ are the momentum density, and T^{ij} are the components of the momentum flux. Since $T^{\mu\nu}$ is symmetric, we get $\mathbf{P} = \mathbf{E}/c$.

The dynamics of the distribution function is given by the Boltzmann equation

$$p_1^\mu \frac{\partial}{\partial x^\mu} f = I_{\text{col}} \tag{2.95}$$

$$I_{\text{col}} = \int \left[\prod_{i=2}^4 Dp_i \right] \left[(2\pi)^4 \delta(p_1 + p_2 - p_3 - p_4) \right] \mathbf{T}\,\mathbf{I} \tag{2.96}$$

where once again $\mathbf{T}$ is a suitable transition probability, not necessarily identical to (2.86), and $\mathbf{I}$ is given in equation (2.88). We have made explicit the momentum conservation delta function, and assume that the transition probability $\mathbf{T}$ is symmetric under particle exchange and time reversal. These symmetry conditions lead directly to the conservation laws for particle number and energy–momentum $N^\mu_{;\mu} = T^{\mu\nu}_{;\nu} = 0$, which hold for *any* distribution function. In equilibrium, we have the stronger result $\mathbf{I} = 0$, leading to

$$f_{\text{eq}} = \frac{1}{e^{-\beta_\mu p^\mu - \alpha} - 1}, \tag{2.97}$$

where $\beta^\mu = u^\mu / k_B T$, u^μ is the macroscopic four-velocity of the gas ($u^2 = -1$) and $\alpha = \mu / k_B T$. The number of parameters which identify an equilibrium state remains at five.

Besides the conserved currents N^μ and $T^{\mu\nu}$, we may define the entropy current

$$S^\mu(x) = \int Dp\, p^\mu \left\{ [1 + f(p)] \ln [1 + f(p)] - f(p) \ln f(p) \right\} \tag{2.98}$$

Unlike the other currents, entropy is not conserved: $S^\mu_{;\mu} \geq 0$ is the relativistic H-theorem.

Consider a small deviation from the equilibrium distribution $f = f_{\text{eq}} + \delta f$ corresponding to the same particle and energy fluxes

$$\int Dp\, p^\mu \delta f(p) = \int Dp\, p^\mu p^0 \delta f(p) = 0 \tag{2.99}$$

The variation in entropy becomes

$$\delta S^0 = -\frac{1}{2} \int Dp\, p^0 \frac{(\delta f)^2}{[1 + f_{\text{eq}}(p)] f_{\text{eq}}(p)} \tag{2.100}$$

showing that entropy is indeed a maximum at equilibrium.

In the classical theory, the distribution function is concentrated on the positive frequency mass shell. Therefore, it is convenient to label momenta just by their spatial components $\mathbf{p}$, the temporal component being necessarily $\omega_p = \sqrt{M^2 + \mathbf{p}^2} > 0$. In the same way, it is simplest to regard the distribution function as a function of the three momentum $\mathbf{p}$ alone, according to the rule

$$f^{(3)}(x, \mathbf{p}) = f[x, (\omega_p, \mathbf{p})] \tag{2.101}$$

where f represents the distribution function as a function on four-dimensional momentum space, and $f^{(3)}$ its restriction to the three-dimensional mass shell. With this understood, we shall henceforth drop the superscript, using the same symbol f for both functions, since only the distribution function on mass shell enters into our discussion. The variation of the total entropy now reads

$$\delta S = -\frac{1}{2} \int d^3x \int \frac{d^3\mathbf{p}}{(2\pi)^3} \frac{(\delta f)^2}{[1 + f_{eq}(p)] f_{eq}(p)} \tag{2.102}$$

This formula shall be relevant to our discussion of fluctuations in the Boltzmann equation.

2.3.4 Fluctuations in the Boltzmann equation

We have seen that the Boltzmann equation has a dissipative character: by virtue of the H-theorem, *any* initial condition is eventually transformed into the equilibrium solution. On the other hand, we have seen that there is a well-defined entropy decrease associated with fluctuations in the distribution function. If we believe in Einstein's formula for the probability of a fluctuation, we must conclude that in equilibrium the number of particles in a phase space cell must not have a definite value, but rather be a Gaussian stochastic variable with mean deviation

$$\langle \delta f(\mathbf{x}, \mathbf{p}) \delta f(\mathbf{x}', \mathbf{p}') \rangle \sim (2\pi)^3 \delta(\mathbf{x} - \mathbf{x}') \delta(\mathbf{p} - \mathbf{p}') f_{eq}[1 + f_{eq}] \tag{2.103}$$

It is not hard to derive this result. The formula for the equilibrium distribution function is equivalent to considering the gas in a grand canonical ensemble, and therefore there must be number fluctuations

$$\left\langle (\delta N)^2 \right\rangle = \beta \frac{\partial \langle N \rangle}{\partial \mu} = \int d^3x \frac{d^3\mathbf{p}}{(2\pi)^3} f_{eq}[1 + f_{eq}] \tag{2.104}$$

On the other hand

$$N = \int d^3x \frac{d^3\mathbf{p}}{(2\pi)^3} f \tag{2.105}$$

so

$$\left\langle (\delta N)^2 \right\rangle = \int d^3x \frac{d^3\mathbf{p}}{(2\pi)^3} d^3x' \frac{d^3\mathbf{p}'}{(2\pi)^3} \langle \delta f(\mathbf{x}, \mathbf{p}) \delta f(\mathbf{x}', \mathbf{p}') \rangle \tag{2.106}$$

taking us back to equation (2.103). This means that if at $t = 0$ we actually measure the number of particles f_0 in each phase space cell, we will rarely obtain those given by f_{eq} (although we will get numbers that will remain statistically close to it). However, if we adopt f_0 as the initial condition and solve the Boltzmann equation, after a long enough time the solution converges to f_{eq} in each and every cell. To obtain these occupation numbers from an actual measurement would be highly unlikely for a system in equilibrium under a grand canonical distribution.

This outrage against Gibbsian common sense means that the Boltzmann equation is not telling the whole story. There is another term besides the collision integral, which sustains the right amount of deviations from the equilibrium state. We could trace back to the derivation of the Boltzmann equation to see where the relevant information was disregarded (and for this reason, we unfolded that derivation in some detail). However, in practice, we know this extra term represents fluctuations which may be quantified by a noise distribution, whose statistics is determined from fluctuation–dissipation considerations; for some classic implementations of this insight see [LanLif57, LanLif59, FoxUhl70a, FoxUhl70b, BixZwa69, KacLog76, KacLog79]. The two points we wish to stress are (1) the incompleteness of the Boltzmann equation which only accounts for dissipation, and (2) the possibility of using fluctuation–dissipation relation considerations to add fluctuations to the Boltzmann equation, valid *for all practical purposes.*

Let us consider the regression of a small deviation δf from the equilibrium distribution f_{eq}. In order to apply the fluctuation–dissipation theorem we must obtain an expression for the time derivative of δf in terms of the thermodynamic force

$$F(x, \mathbf{p}) = -\frac{\delta S}{\delta(\delta f)} = \frac{1}{(2\pi)^3} \frac{\delta f(x, \mathbf{p})}{[1 + f_{eq}(p)] f_{eq}(p)} \tag{2.107}$$

Writing the linearized equation as

$$\frac{\partial f}{\partial t} + \frac{\mathbf{p}}{\omega_p} \nabla f = \frac{1}{\omega_p} I_{col} + \xi(X, \mathbf{p}) \tag{2.108}$$

the Γ matrix has an asymmetric part (coming from the spatial gradients term) and a symmetric part (coming from the linearization of the collision integral). Only the latter contributes to the noise autocorrelation, and so we obtain

$$\langle \xi(X, \mathbf{p}) \xi(Y, \mathbf{q}) \rangle = -\left\{ \frac{1}{\omega_p} \frac{\delta I_{col}(X, \mathbf{p})}{\delta F(Y, \mathbf{q})} + \frac{1}{\omega_q} \frac{\delta I_{col}(Y, \mathbf{q})}{\delta F(X, \mathbf{p})} \right\} \tag{2.109}$$

To obtain a crude idea of what is going on, we may keep only those terms in I_{col} which are proportional to $F(\mathbf{p})$, as is usually done in deriving the "collision time approximation" to the Boltzmann equation (also related to the Krook–Bhatnager–Gross kinetic equation [Lib98, Cer69]), thus we write

$$\delta I_{col}(\mathbf{p}) \sim -\omega_p \nu^2(\mathbf{p}) F(x, \mathbf{p}) \tag{2.110}$$

where

$$\nu^2(X, \mathbf{p}) = \frac{(2\pi)^3}{\omega_p} \int \left[\prod_{i=2}^{4} Dp_i \right] \left[(2\pi)^4 \delta (p_1 + p_2 - p_3 - p_4) \right] \mathbf{T} I_+ \qquad (2.111)$$

$$I_+ = [1 + f_{eq}(p_1)] [1 + f_{eq}(p_2)] f_{eq}(p_3) f_{eq}(p_4) \qquad (2.112)$$

Under this approximation we find the noise autocorrelation

$$\langle \xi(y, \mathbf{k}) \xi(x, \mathbf{p}) \rangle = 2\delta^{(4)}(x - y) \delta(\mathbf{k} - \mathbf{p}) \nu^2(x, \mathbf{p}) \qquad (2.113)$$

Equations (2.108) and (2.113) are the solution to our problem, that is, they describe the fluctuations in the Boltzmann equation, required by consistency with the FDT. Observe that, unlike equation (2.103), the mean square value of the stochastic force vanishes for a free gas. This does not mean that there are no fluctuations (equation (2.103) does not vanish) but that in the collisionless case it is enough to include the fluctuations in the initial conditions, since they are preserved by the dynamics. It is only in the dissipative case that an explicit noise term is necessary to keep fluctuations at the required level [CalHu00].

3

Quantum open systems

Before we develop the nonequilibrium aspects, we want to go over some basics of ordinary quantum mechanics [Dir58, LanLif76, Bes04]. Our goal is to review some formal manipulations which will be used later in the statistical physics contexts, and along the way establish some common notations. We shall develop the theory of quantum open systems from the point of view of the so-called Feynman–Vernon influence functional [FeyVer63, FeyHib65, CalLeg83a, GrScIn88, Kle90, HuPaZh92, HuPaZh93a, Wei93]. This approach and its closely related Schwinger–Keldysh or closed time path method will underlie the analysis of nonequilibrium quantum fields in the rest of the book. We refer the reader to the literature for alternative approaches to quantum open systems [GarZol00b, Car93, Per98].

3.1 A quick review of quantum mechanics

Let us consider a quantum mechanical system described by a single degree of freedom x. The states $|\alpha\rangle$ of the system live in a Hilbert space $\mathcal{H}$ and observables A are represented by Hermitian linear operators $\hat{A}$ in this space. We have different "pictures" of the dynamics, of which the most useful are the Schrödinger and Heisenberg ones. In the former, observables are time-independent, while states evolve in time according to the Schrödinger equation

$$i\hbar \frac{\partial}{\partial t} |\alpha\rangle = \hat{H} |\alpha\rangle \tag{3.1}$$

where the Hamiltonian operator $\hat{H}$ is associated with the observable "energy." This equation may be integrated

$$|\alpha(t)\rangle = U(t, t_0) |\alpha(t_0)\rangle, \tag{3.2}$$

with the evolution operator

$$U = T \left[\exp\left(-\frac{i}{\hbar} \int_{t_0}^{t} dt' \, \hat{H}(t') \right) \right] \tag{3.3}$$

where T stands for temporal order. We are mostly interested in cases where the Hamiltonian is time-independent, whereby $U(t, t_0) = \exp\left(-i\hat{H}(t - t_0)/\hbar \right)$. In the Heisenberg picture states do not evolve, but observables do, according to the rule

$$\hat{A}(t) = U^\dagger(t) \hat{A} U(t) \tag{3.4}$$

The rationale for this rule is that we get consistent values for expectation values of observables in either picture: $\left\langle \hat{A}(t) \right\rangle = \left\langle \alpha(t) \left| \hat{A}(0) \right| \alpha(t) \right\rangle_{\text{Sch}} = \left\langle \alpha(0) \left| \hat{A}(t) \right| \alpha(0) \right\rangle_{\text{Hei}}$. The evolution of operators in the Heisenberg picture is summarized by the Heisenberg equation

$$\frac{d\hat{A}}{dt} = \frac{i}{\hbar} \left[\hat{H}, \hat{A} \right] \tag{3.5}$$

As shown by Einstein, Podolsky and Rosen (EPR) [EiPoRo35], this quantum mechanical description of physical reality cannot be considered complete, since there are states of the system which are not described by kets in the Hilbert space. They occur when we know that the state of the system belongs with certainty to a given class of states $|\alpha_i\rangle$, but our knowledge does not allow us to go beyond assigning a probability of occurrence ρ_i to each member of this class. These situations are depicted by density matrices $\rho = \sum_i \rho_i |\alpha_i\rangle \langle \alpha_i|$, where we assume that the $|\alpha_i\rangle$ states are orthonormal. We always have Tr $\rho = 1$. Kets in the Hilbert space are particular cases of density matrices with Tr $\rho^2 = 1$, the general case being Tr $\rho^2 \leq 1$. In the Schrödinger picture, ρ is time-dependent, and obeys the Liouville–von Neumann equation

$$\frac{d\rho}{dt} = -\frac{i}{\hbar} \left[\hat{H}, \rho \right] \tag{3.6}$$

Observe that this is *not* the Heisenberg equation for the ρ matrix.

Let us now assume that the variable X is continuous and unbounded, and that the states $|x\rangle$ where this variable is well defined form a basis. We have the translation operators Π_a given by $\langle x| \Pi_a |\alpha\rangle = \langle x + a \mid \alpha\rangle$, which are unitary, and given the semigroup structure of these operators, we must have a Hermitian generator $\hat{P}$ such that $\Pi_a = \exp\left(ia\hat{P}/\hbar\right)$. The action of the generator is

$$\langle x| \hat{P} |\alpha\rangle = -i\hbar \frac{\partial}{\partial x} \langle x \mid \alpha \rangle \tag{3.7}$$

$\hat{P}$ has eigenstates $|p\rangle$ such that

$$\langle x \mid p \rangle = \frac{e^{ipx/\hbar}}{\sqrt{2\pi\hbar}} \tag{3.8}$$

The momentum observable $\hat{P}$ and the position observable $\hat{X}$ do not commute, but rather $\left[\hat{P}, \hat{X} \right] = -i\hbar\mathbf{1}$.

Consider a Hamiltonian of the form $\hat{H} = K(\hat{P}) + V(\hat{X})$, $K = \hat{P}^2/2M$. Since tK and tV do not commute, we cannot factor out the evolution operator as a product of a function of $\hat{P}$ times a function of $\hat{X}$. But since the commutator is of order t^2, factorization becomes a good approximation when t is small enough. This gives rise to the Trotter formula [Sch81]

$$e^{-it\hat{H}/\hbar} = \left[e^{-i\tau K/\hbar} e^{-i\tau V/\hbar} \right]^{N+1}, \qquad (N+1)\tau = t, \quad N \to \infty \tag{3.9}$$

and thereby to the path integral representation of the evolution operator [FeyHib65, Sch81], since

$$\langle x_{N+1}| U(t) |x_0\rangle = \langle x_{N+1}| \left[e^{-i\tau K/\hbar} e^{-i\tau V/\hbar} \right]^{N+1} |x_0\rangle$$

$$= \int \left[\prod_{i=1}^{N} dx_i \right] \left\{ \prod_{j=0}^{N} \langle x_{j+1}| e^{-i\tau K/\hbar} |x_j\rangle\, e^{-i\tau V(x_j)/\hbar} \right\}$$

$$= \int \left[\prod_{i=1}^{N} dx_i \right] \left[\prod_{i=1}^{N+1} \frac{dp_i}{2\pi\hbar} \right]$$

$$\times \left\{ \prod_{j=0}^{N} e^{ip_{j+1}(x_{j+1}-x_j)/\hbar} e^{-i\tau p_{j+1}^2/2M\hbar} e^{-i\tau V(x_j)/\hbar} \right\}$$

$$= \int \left[\prod_{i=1}^{N} \sqrt{\frac{-iM}{2\pi\hbar\tau}} dx_i \right] \left\{ \prod_{j=0}^{N} e^{iM(x_{j+1}-x_j)^2/2\tau\hbar} e^{-i\tau V(x_j)/\hbar} \right\}$$

$$(3.10)$$

which as $N \to \infty$ yields

$$\langle x_t| U(t) |x_0\rangle = \int_{x(t)=x_t, x(0)=x_0} Dx\, e^{iS/\hbar} \qquad (3.11)$$

The converse is also true, namely, if we take equation (3.11) as the definition of the evolution operator, we may derive the Schrödinger equation. We have

$$\langle x_t| U(t+\tau) |x_0\rangle = \int \sqrt{\frac{-iM}{2\pi\hbar\tau}} dx'\, e^{iM(x_t-x')^2/2\tau\hbar} e^{-i\tau V(x')/\hbar} \langle x'| U(t) |x_0\rangle$$

$$(3.12)$$

The Gaussian factor makes sure that only values $y \approx x_t$ contribute, so we may expand everything else in powers of $(y - x_t)$ and integrate term by term, whereby

$$\langle x_t| U(t+\tau) |x_0\rangle = \left[1 - \frac{i\tau}{\hbar} V(x_t) \right] \langle x_t| U(t) |x_0\rangle$$

$$+ \frac{i\hbar\tau}{2M} \frac{\partial^2}{\partial x_t^2} \langle x_t| U(t) |x_0\rangle + O\left(\tau^2\right) \qquad (3.13)$$

QED

3.1.1 Wigner functions

So far, we have described states of a quantum system in terms of kets $|\alpha\rangle$ in a Hilbert space. Considering the position and momentum states $|x\rangle$ and $|p\rangle$, we may introduce the wavefunctions in position and momentum representations $\psi(x) = \langle x \mid \alpha\rangle$ and $\psi(p) = \langle p \mid \alpha\rangle$, which are related to each other through a Fourier transform

$$\psi(p) = \int \frac{dx}{\sqrt{2\pi\hbar}}\, e^{-ipx/\hbar} \psi(x) \qquad (3.14)$$

$|\psi(x)|^2$ and $|\psi(p)|^2$ represent the probability distribution functions for position and momentum, respectively. The question arises on whether these distributions may be obtained as marginal distributions from a joint probability for position and momentum. The answer is of course not, at least in general, since the existence of such a joint probability density would be almost conjured as saying that position and momentum may be simultaneously well defined. Nevertheless, in 1932 Wigner found an object which comes remarkably close [Wig32, HOSW84]. This object is the Wigner function

$$f^W(x,p) = \int \frac{du}{2\pi\hbar} e^{-ipu/\hbar} \psi^*\left(x - \frac{u}{2}\right)\psi\left(x + \frac{u}{2}\right) \tag{3.15}$$

Indeed, if we integrate over p we get the probability distribution for x

$$\int dp\, f^W(x,p) = |\psi(x)|^2 \tag{3.16}$$

while integrating over x and switching variables to $x \pm u/2$ we get

$$\int dx\, f^W(x,p) = |\psi(p)|^2 \tag{3.17}$$

The reason why f^W cannot be directly identified as a probability distribution function is that f^W, although real, is not necessarily nonnegative. We shall see examples below.

The dynamics of the Wigner function is also quite remarkable. If the wavefunction obeys the Schrödinger equation (equation (3.1) in the coordinate representation)

$$i\hbar\frac{\partial\psi}{\partial t} = -\frac{\hbar^2}{2M}\frac{\partial^2\psi}{\partial x^2} + V(x)\psi(x) \tag{3.18}$$

then

$$\frac{\partial f^W}{\partial t} = \frac{1}{i\hbar}\int \frac{du}{2\pi\hbar} e^{-ipu/\hbar}$$
$$\times \left\{\left(-\frac{\hbar^2}{2M}\right)\left[\psi^*\left(x - \frac{u}{2}\right)\frac{\partial^2\psi}{\partial x^2}\left(x + \frac{u}{2}\right) - \psi\left(x + \frac{u}{2}\right)\frac{\partial^2\psi^*}{\partial x^2}\left(x - \frac{u}{2}\right)\right]\right.$$
$$\left. + \left[V\left(x + \frac{u}{2}\right) - V\left(x - \frac{u}{2}\right)\right]\psi^*\left(x - \frac{u}{2}\right)\psi\left(x + \frac{u}{2}\right)\right\} \tag{3.19}$$

In the first line, we observe that

$$\psi^*\left(x - \frac{u}{2}\right)\frac{\partial^2\psi}{\partial x^2}\left(x + \frac{u}{2}\right) - \psi\left(x + \frac{u}{2}\right)\frac{\partial^2\psi^*}{\partial x^2}\left(x - \frac{u}{2}\right)$$
$$= 2\frac{\partial^2}{\partial u\partial x}\left[\psi^*\left(x - \frac{u}{2}\right)\psi\left(x + \frac{u}{2}\right)\right]$$

After integration by parts, this term contributes

$$\frac{1}{i\hbar}\left(-\frac{\hbar^2}{2M}\right)\left(\frac{2ip}{\hbar}\right)\frac{\partial f^W}{\partial x} \equiv \frac{-p}{M}\frac{\partial f^W}{\partial x} \tag{3.20}$$

The second term is much harder to handle. If the potential is smooth, one can try a Kramers–Moyal expansion [Kra40, Moy49, Kam81]

$$V\left(x + \frac{u}{2}\right) - V\left(x - \frac{u}{2}\right) = 2\sum_{k=0}^{\infty} \frac{V^{(2k+1)}(x)}{(2k+1)!}\left(\frac{u}{2}\right)^{2k+1} \tag{3.21}$$

Commuting the integral and the sum, we obtain the second term as

$$\frac{2}{i\hbar}\sum_{k=0}^{\infty}\frac{V^{(2k+1)}(x)}{(2k+1)!}\left[\frac{i\hbar}{2}\frac{\partial}{\partial p}\right]^{2k+1} f^{W} \tag{3.22}$$

In terms of the classical Hamiltonian $H = p^2/2m + V$, our result reads

$$\frac{\partial f^{W}}{\partial t} = -\{H, f^{W}\} + O\left(\hbar^2\right) \tag{3.23}$$

where the Poisson bracket $\{H, f^{W}\}$ was introduced in Chapter 2, equation (2.48). In other words, the dynamics of the Wigner function follows remarkably closely the classical transport equation with external potential $V(x)$. If V is harmonic, there are no higher order terms, and the dynamics followed by the Wigner function is exactly the classical dynamics of a distribution function [Hab04, CDHR98]. However, as we have already remarked, that does not mean that f is classical, as it may be negative in some regions of phase space.

It is clear that we may compute the Wigner function f^{W} associated with any wavefunction ψ, but the converse is not true: it is easy to imagine phase space functions f^{W} which cannot be obtained as Wigner functions from *any* ψ. Indeed, it is enough to imagine a distribution function violating Heisenberg's uncertainty principle to exclude such an identification. To the best of our knowledge, there is no simple sufficient condition to see whether a given f^{W} is a Wigner function, although there are many necessary conditions (such as positivity of the marginal distributions).

To summarize, although f^{W} itself cannot be understood as a probability density, conveniently smeared versions of f^{W} are nonnegative and may be used to assign probabilities to different events. This restricted interpretation of the Wigner function will be enough for our requirements below.

Some examples

The simplest possible example of a Wigner function is a momentum state

$$\psi(x) = \frac{e^{i\mathbf{p}x/\hbar}}{\sqrt{2\pi\hbar}} \tag{3.24}$$

Then

$$f^{W} = \frac{1}{2\pi\hbar}\delta(p - \mathbf{p}) \tag{3.25}$$

Now consider a stationary wave

$$\psi(x) = \frac{1}{\sqrt{\pi\hbar}}\cos\left(\frac{\mathbf{p}x}{\hbar}\right) \tag{3.26}$$

representing a coherent superposition of two states of opposite momentum. Then

$$f^W(x,p) = \frac{1}{4\pi\hbar}\left[\delta(p-\mathbf{p}) + \delta(p+\mathbf{p})\right] + \cos\left(\frac{2\mathbf{p}x}{\hbar}\right)\delta(p) \qquad (3.27)$$

We see that f^W is not nonnegative. The oscillatory terms are related to the interference between the two components of the wave packet [PaHaZu93].

As a second example, let us consider a Gaussian wave packet

$$\psi(x) = \frac{e^{-x^2/4\sigma^2}}{(2\pi\sigma^2)^{1/4}} \qquad (3.28)$$

Then

$$f^W(x,p) = \frac{1}{\pi\hbar}e^{-x^2/2\sigma^2}e^{-2\sigma^2(p/\hbar)^2} \qquad (3.29)$$

In this case f^W is positive definite, and the dispersions in x and p are what may be expected for a minimum uncertainty state.

In particular, suppose our state is the ground state for a harmonic oscillator. Then $\sigma^2 = \hbar/2M\Omega$, and

$$f^W(x,p) = \frac{1}{\pi\hbar}\exp\left\{-\frac{E}{\varepsilon}\right\}; \qquad \varepsilon = \frac{1}{2}\hbar\Omega, \ E = \frac{p^2}{2M} + \frac{M\Omega^2x^2}{2} \qquad (3.30)$$

As a final example, let us consider a superposition of two Gaussian wave packets

$$\psi(x) = \frac{1}{(2\pi\sigma^2)^{1/4}}\left\{Ae^{-(x-a)^2/4\sigma^2} + Be^{-(x+a)^2/4\sigma^2}\right\} \qquad (3.31)$$

leading to

$$f^W(x,p) = \frac{e^{-2\sigma^2(p/\hbar)^2}}{\pi\hbar}\left\{|A|^2 e^{-(x-a)^2/2\sigma^2} + |B|^2 e^{-(x+a)^2/2\sigma^2} \right.$$
$$\left. + e^{-x^2/2\sigma^2}\left[AB^*e^{-2ipa/\hbar} + A^*Be^{2ipa/\hbar}\right]\right\} \qquad (3.32)$$

Again, we see nonpositive terms arising from the interference between the different components. If A and B had random phases, f^W would be nonnegative.

Wigner functions and probabilities

We know that if the system is in the state $\psi(x)$, the probability of observing it in the state $\phi(x)$ is

$$P = \left|\int dx\, \phi^*(x)\psi(x)\right|^2 \qquad (3.33)$$

If we call f_ψ^W and f_ϕ^W the corresponding Wigner functions, and call

$$Q = 2\pi\hbar\int dx dp\, f_\psi^W(x,p) f_\phi^W(x,p) \qquad (3.34)$$

then $P = Q$. Indeed

$$Q = \int dx dp \int \frac{du du'}{2\pi\hbar} \, e^{-ip(u+u')/\hbar} \psi^* \left(x - \frac{u}{2}\right)$$

$$\times \psi \left(x + \frac{u}{2}\right) \phi^* \left(x - \frac{u'}{2}\right) \phi \left(x + \frac{u'}{2}\right)$$

$$= \int dx du \, \psi^* \left(x - \frac{u}{2}\right) \psi \left(x + \frac{u}{2}\right) \phi^* \left(x + \frac{u}{2}\right) \phi \left(x - \frac{u}{2}\right) = P \quad (3.35)$$

This implies in particular that the inner product (3.34) of two Wigner functions must be positive. Since Gaussian distributions consistent with Heisenberg's principle are allowed Wigner functions, this implies that Gaussian smearings of a Wigner function are positive definite.

3.1.2 Closed time path (CTP) integrals

Recall that states evolve according to equation (3.2). Using the matrix elements (3.11) for the evolution operator, we obtain

$$\psi(x,t) = \int dx(0) \, U(x, x(0), t) \, \psi(x(0), 0) = \int_{x(t)=x} Dx \, e^{iS/\hbar} \psi(x(0), 0)$$

$$(3.36)$$

in the coordinate representation, where $U(x, x(0), t) = \langle x| U(t) |x(0)\rangle$. By linearity, we infer that the density matrix evolves according to

$$\rho(x, x', t) = \langle x| U(t) \rho U^\dagger(t) |x'\rangle$$

$$= \int_{x(t)=x, x'(t)=x'} Dx Dx' \, e^{i(S[x]-S[x'])/\hbar} \rho(x(0), x'(0), 0) \quad (3.37)$$

The possibility of cyclic permutations under a trace shows that $\mathrm{Tr}\, \rho(t) = \mathrm{Tr}\, \rho(0) = 1$, as it should.

We see that the path integral representation involves *two* histories, rather than a single history of the system as in equation (3.11). This observation is the departure point of the so-called closed time path formalism, which we shall develop at length in this book, especially in Chapters 5 and 6; for source references see [Sch60, Sch61, BakMah63, Kel64, ChoSuHa80, CSHY85, SCYC88, DeW86, Jor86, CalHu87, CalHu88, CalHu89]. To investigate further the meaning of these two-time-path integrals, let us consider the expression

$$G^{11}(\tau, \tau') = \int_{x(t)=x'(t)} Dx Dx' \, e^{i(S[x]-S[x'])/\hbar} \rho(x(0), x'(0), 0) \, x(\tau) x(\tau')$$

$$(3.38)$$

The upper limit is free, provided it is the same for both histories. We may describe this as an integral over single histories defined on a *closed time path* (CTP). This time path has a first branch from 0 to t, where the history takes the values $x(t)$, and a second branch from t back to 0, where the history takes the values $x'(t)$. The CTP boundary condition $x(t) = x'(t)$ says that the history is continuous as a function on the time path.

To understand why we are describing the second branch as going backwards in time, let us translate $G^{11}(\tau, \tau')$ to canonical language. To this end, let us assume $\tau > \tau'$, and make explicit the value of the histories at these two preferred times, namely

$$G^{11}(\tau, \tau') = \int dx(0)\, dx'(0)\ dx(\tau')\, dx(\tau)\, dx(t)$$

$$\times \left[\int_{0 \leq t \leq \tau'} Dx\ e^{iS[x]/\hbar}\right] x(\tau') \left[\int_{\tau' \leq t \leq \tau} Dx\ e^{iS[x]/\hbar}\right] x(\tau)$$

$$\times \left[\int_{\tau \leq t \leq t} Dx\ e^{iS[x]/\hbar}\right] \left[\int_{x'(t)=x(t)} Dx'\ e^{-iS[x']/\hbar}\right]$$

$$\times \rho(x(0), x'(0), 0) \tag{3.39}$$

Identifying each bracket as a matrix element for some evolution operator, we get

$$G^{11}(\tau, \tau') = \int dx(0)\, dx'(0)\ dx(\tau')\, dx(\tau)\, dx(t)$$

$$\times \langle x(t)| U(t, \tau) |x(\tau)\rangle\ x(\tau)\ \langle x(\tau)| U(\tau, \tau') |x(\tau')\rangle\ x(\tau')$$

$$\times \langle x(\tau')| U(\tau', 0) |x(0)\rangle\ \langle x(0)| \rho |x'(0)\rangle\ \langle x'(0)| U(0, t) |x(t)\rangle$$

$$= Tr \left\{ U(t, \tau)\, \hat{X}\, U(\tau, \tau')\, \hat{X}\, U(\tau', 0)\, \rho(0)\, U(0, t) \right\} \tag{3.40}$$

in the Schrödinger representation, or equivalently

$$G^{11}(\tau, \tau') = Tr \left\{ \hat{X}(\tau)\, \hat{X}(\tau')\, \rho \right\} \equiv \left\langle \hat{X}(\tau)\, \hat{X}(\tau') \right\rangle \tag{3.41}$$

in the Heisenberg representation. Observe that if we had not specified the relationship between τ and τ', then the path integral would have automatically set the largest time to the left. This expresses the "time ordering" of the two Heisenberg operators, so that we may generalize the result to $G^{11}(\tau, \tau') \equiv \left\langle T\left[\hat{X}(\tau)\, \hat{X}(\tau')\right]\right\rangle$, where T stands for temporal ordering.

Now consider instead

$$G^{12}(\tau, \tau') = \int_{x(t)=x'(t)} Dx\, Dx'\ e^{i(S[x]-S[x'])/\hbar} \rho(x(0), x'(0), 0)\ x(\tau)\, x'(\tau')$$

$$\tag{3.42}$$

The corresponding Schrödinger picture canonical expression is

$$G^{12}(\tau, \tau') = Tr \left\{ U(0, \tau')\, \hat{X}\, U(\tau', t)\, U(t, \tau)\, \hat{X}\, U(\tau, 0)\, \rho(0) \right\} \tag{3.43}$$

or, in Heisenberg's representation, $G^{12}(\tau, \tau') \equiv \left\langle \hat{X}(\tau')\, \hat{X}(\tau) \right\rangle$. In this case, the primed Heisenberg operator comes out to the left, whichever time is greatest. We may think of this as a *path*, rather than a time, ordering. Finally, with the same argument we see that

$$G^{22}\left(\tau,\tau'\right) = \int_{x(t)=x'(t)} Dx\,Dx'\; e^{i\left(S[x]-S[x']\right)/\hbar}\rho\left(x\left(0\right),x'\left(0\right),0\right)\left\{x'\left(\tau\right)x'\left(\tau'\right)\right\}$$

$$\equiv \left\langle \tilde{T}\left[\hat{X}\left(\tau\right)\hat{X}\left(\tau'\right)\right]\right\rangle \tag{3.44}$$

where $\tilde{T}$ stands for *anti*-time ordering (that is, the latest time to the right). This anti-time ordering property justifies regarding the second branch as going backwards with respect to the first branch.

If necessary, more involved time paths may be considered. For example, it may be that the initial density matrix corresponds to a thermal state $\rho\left(0\right) = e^{-\beta H}/Z$, which can be regarded as an evolution operator in Euclidean time $\tau_\beta = -i\hbar\beta$. Then its matrix elements admit a path integral representation on a time branch going from 0 to τ_β, which appears as a third branch in the path integral representation for average values [Mil69, McL72a, McL72b]. We will have a lot more to say on thermal states in Chapter 10.

3.2 Influence functional

We wish to use the above to study the dynamics of a quantum open system. The set-up is the usual one: a system S described by a variable x interacts with an environment E described by variable(s) $q = \{q_n\}$. The classical action takes the form $S\left[x,q\right] = S_S\left[x\right] + S_E\left[q\right] + S_{\text{int}}\left[x,q\right]$. The Hamiltonian $\hat{H} = \hat{H}_S + \hat{H}_E + \hat{H}_{\text{int}}$, where

$$\hat{H}_s = \frac{1}{2}p^2 + V\left(x\right); \qquad \hat{H}_{\text{int}} = V_{\text{int}}\left(x,q\right) \tag{3.45}$$

The quantum state of the total system is described by the density matrix $\rho\left(xq,x'q',t\right)$ depending on both system and environment variables. It evolves unitarily under $\hat{H}$ from an initial density matrix $\rho(0)$ at $t = 0$ to $\rho(t) = e^{-it\hat{H}/\hbar}\rho\left(0\right)e^{it\hat{H}/\hbar}$ at finite time t. Explicitly, using completeness conditions in a path integral representation:

$$\rho(x\,q,\,x'\,q,\,t) = \langle x\,q,\,t|\rho|x'\,q,\,t\rangle$$

$$= \int dx_i\,dq_i \int dx_i'\,dq_i'\; \langle x\,q,\,t|x_i\,q_i,\,0\rangle\langle x_i\,q_i,\,0|\rho|x_i'\,q_i',0\rangle\langle x_i'\,q_i',\,0|x'\,q,t\rangle$$

$$= \int dx_i\,dq_i \int dx_i'\,dq_i' \int_{x_i}^{x} Dx \int_{q_i}^{q} Dq\; e^{iS[x,q]/\hbar}\rho(x_i\,q_i,\,x_i'\,q_i',\,0)$$

$$\times \int_{x_i'}^{x'} Dx' \int_{q_i'}^{q} Dq'\; e^{-iS[x',q']/\hbar}$$

$$\equiv \int dx_i\,dq_i \int dx_i'\,dq_i'\; \mathcal{J}(x\,q,\,x'\,q,\,t|x_i\,q_i,\,x_i'\,q_i',\,0)\,\rho(x_i\,q_i,\,x_i'\,q_i',\,0) \tag{3.46}$$

where $\mathcal{J}$ is seen to be an evolution operator for the system plus environment.

Since we care more about the system's behavior than the environment, we need not keep track of the details of the environment in the specifics of its Hamiltonian. In particular, we are mostly interested in computing the expectation values of system observables. Considered as operators on the whole Hilbert space for the system, these take the form $\hat{A} \otimes \mathbf{1}$, where $\hat{A}$ is an operator in the system Hilbert space, and $\mathbf{1}$ is the unit operator on the environment Hilbert space. The expectation value of such observables may be computed with the *reduced* density matrix ρ_r. This is obtained from the total density matrix as a partial (Landau's) trace over the environment variables, namely $\rho_r = \text{Tr}_q \, \rho$. Explicitly,

$$\rho_r(x\,x',\,t) = \int_{-\infty}^{\infty} dq \, \rho(x\,q,\,x'\,q,\,t) \tag{3.47}$$

Let us further assume that at $t = 0$ the system and environment (variables with subscript i) are uncorrelated,

$$\rho(x_i\,q_i,\,x_i'\,q_i',\,0) = \rho_S(x_i\,x_i',\,0)\,\rho_E(q_i\,q_i',\,0) \tag{3.48}$$

(Thus we are bringing the system and its environment together with all due care to avoid the complications associated with the sudden switching on and off of interactions. For the general case, see [HakAmb85, MorCal87, DavPaz97].) As such, we are able to rearrange the order of integration to write the reduced density matrix in the following way:

$$\rho_r(x\,x',\,t) = \int dx_i \, dx_i' \, \mathcal{J}_r(x\,x',\,t|x_i\,x_i',\,0)\,\rho_S(x_i\,x_i',\,0) \tag{3.49}$$

where the evolution operator for the reduced density matrix is defined by

$$\mathcal{J}_r(x\,x',\,t|x_i\,x_i',\,0) \equiv \int_{x_i}^{x} Dx \int_{x_i'}^{x'} Dx' \, e^{i\hbar^{-1}\left(S[x]-S[x']\right)} \, \mathcal{F}[x,x'] \tag{3.50}$$

$\mathcal{F}[x,x']$ is the so-called Feynman–Vernon influence functional [FeyVer63, FeyHib65, Wei93]:

$$
\begin{aligned}
\mathcal{F}[x, x'] &\equiv e^{iS_{\text{IF}}[x,x',t]/\hbar} \\
&= \int dq \, dq_i \, dq_i' \, \rho_E(q_i\,q_i'\,0) \int_{q_i}^{q} Dq \, e^{i\hbar^{-1}(S_E[q]+S_{int}[x,q])} \\
&\quad \times \int_{q_i'}^{q} Dq' \, e^{-i\hbar^{-1}(S_E[q']+S_{int}[x',q'])}
\end{aligned}
\tag{3.51}
$$

Here, S_{IF} is called the influence action. Equation (3.49) looks like the evolution of a density matrix for a closed system, but it contains a nonlocal term S_{IF}, which induces an explicit interaction between the two histories in the CTP. All the influence of the environment on the system is encoded into the influence action S_{IF}.

We can also write the influence functional in a basis-independent form as follows. In terms of the propagators $U(t), U'(t)$ for $S_E[q] + S_{int}[x, q]$ and

$S_E[q] + S_{\text{int}}[x', q]$, respectively, the path integrals can be expressed as

$$\mathcal{F}[x, x'] = \int dq \, dq_i \, dq'_i \, \rho_E(q_i \, q'_i, \, 0) \, \langle q|U(t)|q_i \rangle \, \langle q'_i|U'^{\dagger}(t)|q \rangle \qquad (3.52)$$

Then upon integrating over q, q_i and writing the remaining integral as a trace, we obtain:

$$\mathcal{F}[x, x'] = \text{Tr} \, U(t) \, \rho_E(0) \, U'^{\dagger}(t) \qquad (3.53)$$

3.2.1 *Some properties of the influence action*

Let us explore the main properties of the influence action. From equation (3.53)

$$e^{i S_{\text{IF}}[x, x', t]/\hbar} = \text{Tr} \, \{U_{x'}(0, t) \, U_x(t, 0) \, \rho_E(0)\} \qquad (3.54)$$

The U's represent evolution operators with respect to a dynamics where the system variable x plays the role of an external, time-dependent parameter. For two different histories $x(t)$ and $x'(t)$ the U's do not cancel each other. But when $x = x'$, they do, and we get $S_{\text{IF}}[x, x, t] \equiv 0$. Even in the presence of an explicit time dependence, the evolution operator U_x is unitary, whereby $S_{\text{IF}}[x', x, t] \equiv -S_{\text{IF}}[x, x', t]^*$. This means that, in a functional Taylor expansion in terms of the difference variable $u = x - x'$ and the "center of mass" variable $X = (x + x')/2$,

$$S_{\text{IF}}[X, u, t] = \sum_{k=1} \frac{1}{k!} \int dt_1 \ldots dt_k \, S^{(k)}[X(\tau), t_1, \ldots t_k, t] \, u(t_1) \ldots u(t_k) \quad (3.55)$$

all the odd terms are real, and all the even terms are imaginary. Taking a variation along the diagonal we get the additional property $S_{\text{IF},x}|_{x=x'} = - \, S_{\text{IF},x'}|_{x=x'}$.

At this point, it is convenient to introduce a notation that will stay with us for the rest of the book. Let us call $x(t) = x^1(t)$, $x'(t) = x^2(t)$. We shall think of x^a, $a = 1, 2$, as a single field doublet defined on a conventional (single branch) time path. Moreover, as in a σ model, we define a metric tensor $c_{ab} = \text{diag}(1, -1)$ in target space. The metric tensor, together with its contravariant $(c^{ab} = (c^{-1})^{ab} = \text{diag}(1, -1))$ and mixed $(c^a_b = c^{ad}c_{db} = \delta^a_b)$ forms may be used to raise and/or lower indices, as in $x_1 = c_{1a}x^a = x^1 = x$, $x_2 = c_{2a}x^a = -x^2 = -x'$. From now on, the Einstein convention of summation over repeated indices will be assumed; for example, the kinetic terms in the system action will be written as

$$\frac{1}{2} \int dt \, c_{ab} \dot{x}^a \dot{x}^b = \frac{1}{2} \int dt \, \dot{x}_a \dot{x}^a = \frac{1}{2} \int dt \, [\dot{x}^2 - \dot{x}'^2] \qquad (3.56)$$

and we shall refer to the *CTP action* $S[x^a] \equiv S[x] - S[x']$ without discriminating the contributions from either branch.

3.2.2 The linear bath model

As an example, let us assume that the environment action is quadratic in the (many) q variable(s), the initial environment density matrix is Gaussian, and the interaction term is bilinear $S_{\text{int}} = \int dt\, x^a(t)\, Q_a[q(t)]$ (in CTP notation!), where the Q's are linear combinations of the q's [CalLeg83a, CalLeg83b, GrScIn88]. Under all these assumptions, the influence action must also be quadratic in x and x' (equation (3.51) is a functional Fourier transform of an elaborate Gaussian functional of histories $Q(t)$ and $Q'(t)$, and the Fourier transform of a Gaussian is another Gaussian). Therefore we write $S_{\text{IF}} = (1/2) \int dt dt'\, x^a(t)\, \mathbf{M}_{ab}(t,t')\, x^b(t')$, where

$$\mathbf{M}_{ab}(t,t') = -i\hbar \left. \frac{\delta^2}{\delta x^a(t)\, \delta x^b(t')} e^{iS_{\text{IF}}[x^a,T]/\hbar} \right|_{x^a=0} \tag{3.57}$$

A direct variation from equation (3.51) yields

$$\left. \frac{\delta^2 e^{iS_{\text{IF}}[x^a,T]/\hbar}}{\delta x^a(t)\, \delta x^b(t')} \right|_{x^a=0} = \frac{-1}{\hbar^2} \int_{q^1(T)=q^2(T)} Dq^a\, e^{iS_E[q^a]/\hbar} Q_a(t)$$
$$\times Q_b(t')\, \rho_e\left(q^1(0), q^2(0), 0\right) \tag{3.58}$$

As per the earlier discussion, we obtain

$$\mathbf{M}_{ab}(t,t') = \frac{i}{\hbar} \begin{pmatrix} \langle T[Q(t)Q(t')]\rangle & -\langle Q(t')Q(t)\rangle \\ -\langle Q(t)Q(t')\rangle & \langle \tilde{T}[Q(t)Q(t')]\rangle \end{pmatrix} \tag{3.59}$$

(where the expectation values are computed *disregarding* the interaction with the system), or, in terms of the original variables

$$S_{\text{IF}} = \frac{i}{2\hbar} \int dt dt' \left\{ \langle T[Q(t)Q(t')]\rangle\, x(t)\, x(t') - \langle Q(t')Q(t)\rangle\, x(t)\, x'(t') \right.$$
$$\left. - \langle Q(t)Q(t')\rangle\, x'(t)\, x(t') + \left\langle \tilde{T}[Q(t)Q(t')]\right\rangle x'(t)\, x'(t') \right\} \tag{3.60}$$

If we now write $x = X + u/2$, $x' = X - u/2$, we get the equivalent expression

$$S_{\text{IF}} = \int dt dt' \left\{ u(t)\, \mathbf{D}(t,t')\, X(t') + \frac{i}{2} u(t)\, \mathbf{N}(t,t')\, u(t') \right\} \tag{3.61}$$

where we encounter for the first time the *dissipation* $\mathbf{D}$ and *noise* $\mathbf{N}$ kernels

$$\mathbf{D}(t,t') = \frac{i}{\hbar} \langle [Q(t), Q(t')]\rangle\, \theta(t-t')\,; \qquad \mathbf{N}(t,t') = \frac{1}{2\hbar} \langle \{Q(t), Q(t')\}\rangle \tag{3.62}$$

Square and curly brackets stand for commutator and anticommutator, respectively. They are both real, as expected, and $\mathbf{D}$ is also causal.

Unraveling the physical meaning of these kernels and applying them to different situations will be a major theme for the rest of the book.

3.3 The master equation

When the influence functional is quadratic, and hence may be written as in
equation (3.61), it is possible to derive a dynamical equation for the evolution of
the reduced density matrix [Zha90, HuPaZh92, HuPaZh93a, PaHaZu93, Paz94,
HalYu96].

First use equation (3.61), plus the observation that $\mathbf{D}$ is causal and vanishes
on the diagonal (cf. equation (3.62)), to obtain an explicit representation for
$S_{\mathrm{IF}}\left[t + dt\right]$

$$S_{\mathrm{IF}}\left[x^a, t + dt\right] = S_{\mathrm{IF}}\left[x^a, t\right] + dt\, u\left(t\right) \int_0^t dt'\, \left\{\mathbf{D}\left(t, t'\right) X\left(t'\right) + i\mathbf{N}\left(t, t'\right) u\left(t'\right)\right\}$$
(3.63)

Now use this in equation (3.49) to obtain

$$\frac{\partial}{\partial t}\rho_r\left(x, x', t\right) = -\frac{i}{\hbar}\left[\hat{H}_s, \rho_r\left(t\right)\right]_{x, x'}$$

$$- \frac{1}{\hbar}\left(x - x'\right)\int_0^t dt'\, \left\{\mathbf{N}\left(t, t'\right)\left[\mathbf{X} - \mathbf{X}'\right]\left(x, x', t'\right)\right.$$

$$\left. - \frac{i}{2}\mathbf{D}\left(t, t'\right)\left[\mathbf{X} + \mathbf{X}'\right]\left(x, x', t'\right)\right\}$$
(3.64)

The first term is just the Liouville–von Neuman equation (3.6) for the closed
system. In the second term

$$\mathbf{X}\left(x, x', t'\right) = \int_{x(t)=x, x'(t)=x'} Dx Dx'\, e^{i\left(S_S\left[x\right] - S_S\left[x'\right] + S_{\mathrm{IF}}\left[x, x', t\right]\right)/\hbar}$$

$$\times \rho_s\left(x\left(0\right), x'\left(0\right), 0\right) x\left(t'\right)$$
(3.65)

with a similar expression for $\mathbf{X}'\left(x, x', t'\right)$, replacing the last factor $x\left(t'\right)$ by $x'\left(t'\right)$.

In general, $\mathbf{X}\left(x, x', t'\right)$ and $\mathbf{X}'\left(x, x', t'\right)$ are complicated functions of x, x' and t.
However, since in general $\mathbf{N}$ and $\mathbf{D}$ are of second order in the system–bath inter-
action (cf. equation (3.62)), S_{IF} may be neglected within the integral in equation
(3.65) to third order in this interaction, and $\mathbf{X}\left(x, x', t'\right)$ and $\mathbf{X}'\left(x, x', t'\right)$ may
be expressed in terms of quantities belonging to the system alone. Concretely,
$\mathbf{X}\left(x, x', t'\right)$ is the $\left(x, x'\right)$ matrix element of the operator

$$\mathbf{X}\left(t'\right) = e^{-i\hat{H}_s\left(t - t'\right)/\hbar}\hat{X}e^{-i\hat{H}_s t'/\hbar}\rho_s\left(0\right)e^{i\hat{H}_s t/\hbar}$$
(3.66)

where $\hat{X}$ is the position operator in the Schrödinger representation. Introducing
the Heisenberg operator $\hat{X}\left(t\right) = e^{i\hat{H}_s t/\hbar}\hat{X}e^{-i\hat{H}_s t/\hbar}$ and writing $t' = t - \tau$, we get,
to second order in the system–bath coupling

$$\mathbf{X}\left(t'\right) = \hat{X}\left(-\tau\right)\rho_r\left(t\right)$$
(3.67)

Similarly, $\mathbf{X}'\left(x, x', t'\right)$ is the $\left(x, x'\right)$ matrix element of the operator

$$\mathbf{X}'\left(t'\right) = \rho_r\left(t\right)\hat{X}\left(-\tau\right)$$
(3.68)

whereby we get the so-called master equation

$$\hbar \frac{\partial}{\partial t} \rho_r(t) = -i\left[H_s, \rho_r(t)\right]$$

$$-\int_0^t d\tau \left\{ \mathbf{N}(t, t-\tau) \left[\hat{X}, \left[\hat{X}(-\tau), \rho_r(t)\right]\right] \right.$$

$$\left. -\frac{i}{2}\mathbf{D}(t, t-\tau) \left[\hat{X}, \left\{\hat{X}(-\tau), \rho_r(t)\right\}\right] \right\} \qquad (3.69)$$

3.3.1 The linear system model

When the system is also linear we can give an explicit formula for the Heisenberg operators

$$\hat{X}(-\tau) = \cos\left[\Omega\tau\right] \hat{X} - \frac{\sin\left[\Omega\tau\right]}{M\Omega} \hat{P} \qquad (3.70)$$

and we can write the master equation in a way which is explicitly local in time

$$\hbar \frac{\partial}{\partial t} \rho_r(x, x', t) = \left\{ \frac{i\hbar^2}{2M} \left[\frac{\partial^2}{\partial x^2} - \frac{\partial^2}{\partial x'^2}\right] - \frac{iM\left[\Omega^2 + \delta\Omega^2(t)\right]}{2}(x^2 - x'^2) \right.$$

$$-\frac{\sigma^2(t)}{2\hbar}(x - x')^2 - i\Delta_{ad}(t)(x - x')\left[\frac{\partial}{\partial x} + \frac{\partial}{\partial x'}\right]$$

$$\left. -\hbar\Gamma(t)(x - x')\left[\frac{\partial}{\partial x} - \frac{\partial}{\partial x'}\right] \right\} \rho_r(x, x', t) \qquad (3.71)$$

where

$$\frac{\sigma^2(t)}{2} = \int_0^t d\tau \, \hbar\mathbf{N}(t, t-\tau) \cos\left[\Omega\tau\right] \qquad (3.72)$$

$$\Delta_{ad}(t) = \frac{1}{M\Omega}\int_0^t d\tau \, \hbar\mathbf{N}(t, t-\tau) \sin\left[\Omega\tau\right] \qquad (3.73)$$

$$\Gamma(t) = \frac{1}{2}\int_0^t d\tau \, \gamma(\tau) \cos\left[\Omega\tau\right] \qquad (3.74)$$

$$\delta\Omega^2(t) = \Omega \int_0^t d\tau \, \gamma(\tau) \sin\left[\Omega\tau\right] - \gamma(0) \qquad (3.75)$$

and we have written $\mathbf{D}(t, t-\tau) = -M(d\gamma(\tau)/d\tau)$, with the convention that $\gamma(t) = 0$. Observe that besides the effects of noise and dissipation, the σ^2 term clearly acts to suppress the off-diagonal elements of the density matrix. Therefore we must add decoherence to the list of effects of the environment on the system, together with dissipation, diffusion and renormalization.

3.4 The Langevin equation

We now present two ways to derive the Langevin equation: first, formally from the influence action using the Feynman–Vernon identity [FeyVer63, FeyHib65] to reduce the part containing the noise kernel to an integral over a new classical stochastic forcing term, and second, through the time evolution of the reduced Wigner function. When the influence functional has the form (3.61), either rigorously or as a result of approximations, it is possible to read the Langevin equation directly off the path integral representation for the reduced density matrix, without explicit reference to the Wigner function. The idea is to substitute the Gaussian identity (3.76) into the path integral representation (3.51). We then commute the integrals, and perform the x and x' integrations by the method of stationary phase. The coupled equations for the stationary paths admit solutions where $x = x'$, and the Langevin equation (3.93) is just the stationarity condition for these solutions. The final integration over ξ is, of course, necessary to compute physical observables.

Later, in Chapter 5, when we treat open systems of quantum fields, we shall use this method as an efficient way to derive the functional Langevin equation. In Chapter 9 we will discuss in greater detail a class of problems where the fluctuations predicted by this Langevin equation have a direct physical meaning.

From the influence action via a noise average

For linear coupling to a linear bath, the influence functional has the form (3.61). In this case Feynman and Vernon showed that the noise kernel part of the influence functional can be written as a classical stochastic force ξ acting on the system. The following is an identity of the Gaussian functional integral:

$$\exp\left\{\frac{-1}{2\hbar}\int dt dt'\, u(t)\,\mathbf{N}(t,t')\,u(t')\right\} = \int D\xi\, P\,[\xi]\,\exp\left[\frac{i}{\hbar}\int_0^\infty dt\,\xi u\right] \quad (3.76)$$

where $P\,[\xi]$ is a Gaussian measure such that

$$\langle\xi\rangle = 0, \quad \langle\xi(t)\,\xi(t')\rangle = \hbar\mathbf{N}(t,t') \quad (3.77)$$

The stochastic force ξ has zero mean and correlation function given by $\mathbf{N}(t,t')$ the noise kernel, thus its name. We observe that $P\,[\xi]$ does not depend on t. The probability density functional is a functional of $X(s)$ if we allow the statistical properties of ξ to depend on the system history. This functional defines a stochastic average $\langle\ \rangle_\xi$ as a functional integral over $\xi(s)$ multiplied by a normalized Gaussian probability density functional $\mathcal{P}[\xi(s); X(s)]$.

One can then write the total influence functional (3.51) as

$$\mathcal{F}[X,u] = \left\langle \exp\left[\frac{i}{\hbar}\int_{t_i}^{t_f}\xi_{\text{full}}(s)u(s)ds\right]\right\rangle \quad (3.78)$$

$$\xi_{\text{full}}(s) = \int_{t_i}^s ds'\mathbf{D}(s,s')X(s') + \xi(s) \quad (3.79)$$

The equation of motion generated by the influence action is

$$\frac{\partial L}{\partial x} - \frac{d}{dt}\frac{\partial L}{\partial \dot{x}} + \int_{t_i}^s ds'\,\mathbf{D}(s,s')X(s') = -\xi(t) \quad (3.80)$$

whereby we obtain the Langevin equation. In general $\mathbf{D}$ generates nonlocal dissipation while ξ represents a colored noise source.

The reduced Wigner function

We have seen that any wavefunction is associated with a function in phase space, the so-called Wigner function (cf. equation (3.15)). Suppose the system is described by a density matrix rather than a single wavefunction. Decomposing the density matrix in terms of its own eigenfunctions

$$\rho\left(x, x', t\right) = \sum_\alpha \rho_\alpha\left(t\right) \psi_\alpha\left(x, t\right) \psi_\alpha^*\left(x', t\right) \tag{3.81}$$

$$\int dx \, \psi_\alpha^*\left(x, t\right) \psi_\beta\left(x, t\right) = \delta_{\alpha\beta} \tag{3.82}$$

we see that $\rho_\alpha(t)$ is the probability of finding the system in one of the ψ_α states. Let us associate each ψ_α state with its corresponding Wigner function f_α^W, and compute the expectation value

$$
\begin{aligned}
f^W\left(x, p, t\right) &= \sum_\alpha \rho_\alpha\left(t\right) f_\alpha^W\left(x, p, t\right) \\
&= \int \frac{du}{2\pi\hbar} \, e^{-ipu/\hbar} \sum_\alpha \rho_\alpha\left(t\right) \psi_\alpha^*\left(x - \frac{u}{2}, t\right) \psi_\alpha\left(x + \frac{u}{2}, t\right) \\
&= \int \frac{du}{2\pi\hbar} \, e^{-ipu/\hbar} \, \rho\left(x + \frac{u}{2}, x - \frac{u}{2}, t\right)
\end{aligned} \tag{3.83}
$$

The Wigner function is directly given as the partial Fourier transform of the density matrix, without any explicit reference to the latter eigenstates.

For a quantum open system, we define the reduced Wigner function as the partial Fourier transform of the reduced density matrix

$$f_r^W\left(X, P, t\right) = \int \frac{du}{2\pi\hbar} \, e^{-iPu/\hbar} \, \rho_r\left(X + \frac{u}{2}, X - \frac{u}{2}, t\right) \tag{3.84}$$

From the path integral representation of the reduced density matrix containing its dynamics one can derive how the reduced Wigner function evolves in time [CaRoVe01, Rou02, CaRoVe03].

Let us replace ρ_r in equation (3.84) by its path integral representation, with the initial reduced density matrix given in terms of the initial reduced Wigner function

$$
\begin{aligned}
f_r^W\left(X_f, P_f, t\right) = \int \frac{du_f}{2\pi\hbar} \, e^{-iP_f u_f/\hbar} \int_{x(t)=X_f+u_f/2, x'(t)=X_f-u_f/2} & \mathcal{D}x \mathcal{D}x' \\
\times \exp\left\{\frac{i}{\hbar}\left[S_S\left[x\right] - S_S\left[x'\right] + S_{\mathrm{IF}}\left[x, x', t\right]\right]\right\} & \\
\times \int dP_i \, \exp\left[\frac{i}{\hbar} P_i\left(x\left(0\right) - x'\left(0\right)\right)\right] f_r^W\left(\frac{x\left(0\right) + x'\left(0\right)}{2}, P_i, 0\right) &
\end{aligned} \tag{3.85}
$$

Insert the momentum variables by means of the identity

$$e^{iS_S[x]/\hbar} = \exp\left[-\frac{i}{\hbar}\int_0^t dt\, V\left[x\left(t\right)\right]\right] \int Dp\, \exp\left\{\frac{i}{\hbar}\int_0^t dt\,\left[p\dot{x} - \frac{p^2}{2M}\right]\right\} \quad (3.86)$$

and introduce new variables

$$x, x' = X \pm \frac{u}{2} \quad (3.87)$$

$$p, p' = P \pm \frac{\pi}{2} \quad (3.88)$$

We assume a linear bath so the influence functional has the form (3.61). We then use the Gaussian identity (3.76) to introduce the stochastic variable ξ and write

$$V\left(x\right) - V\left(x'\right) = uV'\left(X\right) + \mathbf{V}\left(X, u\right), \qquad \mathbf{V} \sim O\left(u^3\right) \quad (3.89)$$

We may now formally integrate over the u and π variables, to get

$$f_r^W\left(X_f, P_f, t\right) = \int_{X(t)=X_f, P(t)=P_f} DX DP\, f_r^W\left(X\left(0\right), P\left(0\right), 0\right)$$

$$\times \int D\xi\, P_Q\left[\xi, t\right]\, \delta\left[\dot{P} + V'\left(X\right) + D\left(t\right) - \xi\right]\delta\left[\dot{X} - \frac{P}{M}\right]$$

$$(3.90)$$

where

$$D\left(t\right) = -\int_0^t dt'\, \mathbf{D}\left(t, t'\right) X\left(t'\right) \quad (3.91)$$

$$P_Q\left[\xi, t\right] = \exp\left[\frac{-i}{\hbar}\int_0^t dt\, \mathbf{V}\left(X, i\hbar\frac{\delta}{\delta\xi\left(t\right)}\right)\right] P\left[\xi\right] \quad (3.92)$$

In other words, the Wigner function evolves as if it described an ensemble of particles following trajectories which obey the equations

$$\dot{X}\left(t\right) = \frac{P\left(t\right)}{M}, \qquad \dot{P}\left(t\right) = -V'\left(X\left(t\right)\right) - D\left(t\right) + \xi\left(t\right) \quad (3.93)$$

with random initial conditions weighted by the initial Wigner function and noise autocorrelation given by equation (3.77). These are the Hamilton equations of the system but now acquiring two extra terms, D and ξ, describing the influence of the environment. D is a deterministic, memory-dependent term, while ξ plays the role of "noise" with a "probability" distribution P_Q. Observe that this is an exact relation; in particular, the system retains fully its quantum coherence, which is encoded in P_Q. This means that we can use averages over the "noise" and initial conditions to compute exact quantum expectation values of system variables. In this sense, the Langevin equation gives the most detailed description of the quantum open system we shall see in this chapter.

Incidentally, observe that since **V** involves cubic derivatives or higher, the noise autocorrelation is given by equation (3.77), independent of the self-interaction potential V.

The path integral representation (3.90) may be simplified greatly if the X dependence of **V** can be ignored (as it happens when the potential is cubic), or at least X may be replaced by the solution $\bar{X}[X(0), P(0), 0; t]$ to the classical equations of motion with Cauchy data $(X(0), P(0))$ at time 0. In this case the path integral over X and P may be performed, and we get

$$f_r^W(X_f, P_f, t) = \langle \delta(X(t) - X_f)\, \delta(P(t) - P_f)\rangle \qquad (3.94)$$

where $X(t)$ and $P(t)$ are the solutions of the Langevin equation (3.93) and the average is over the initial conditions and noise realizations. This average is more involved than the one we considered in Chapter 2, because of the more complex noise distribution function.

3.4.1 The linear bath model

It is interesting to compare equations (3.93) to the simple linear bath model we discussed in Chapter 2. To this end, we shall use expressions (3.62) for the dissipation and noise kernels. Let us write $q = \{q_\alpha\}$, $Q = \sum c_\alpha q_\alpha$. Recall that the expectation values in equation (3.62) are computed at $X = 0$, and that for a linear system the commutator of two field operators, being a c-number, is state independent. Thus, we may write

$$[Q(t), Q(t')] = \sum_\alpha c_\alpha^2\, [q_\alpha(t), q_\alpha(t')] \qquad (3.95)$$

For a linear system we may solve Heisenberg's equations

$$q_\alpha(t) = q_\alpha(0)\cos\omega_\alpha t + p_\alpha(0)\,\frac{\sin\omega_\alpha t}{m_\alpha \omega_\alpha} \qquad (3.96)$$

$$[q_\alpha(t), q_\alpha(t')] = \frac{\hbar}{i}\,\frac{\sin\omega_\alpha(t - t')}{m_\alpha \omega_\alpha} \qquad (3.97)$$

To compare with the Brownian motion model, we write

$$\mathbf{D}\left(t, t'\right) = -M\frac{\partial}{\partial t}\gamma(t - t') \qquad (3.98)$$

After an integration by parts, and discarding the term from the lower limit because we assume the interaction is switched on smoothly, we get

$$D(t) = M\delta\Omega_0^2 X(t) + \int_0^t dt'\,\gamma(t - t')\,P(t') \qquad (3.99)$$

$$\gamma(t - t') = \frac{1}{M}\sum_\alpha \frac{c_\alpha^2}{m_\alpha \omega_\alpha^2}\cos\omega_\alpha(t - t')$$

$$\delta\Omega_0^2 = -\gamma(0)$$

as in Chapter 2, equation (2.12). We now see the origin of the name *dissipation kernel* for $\mathbf{D}(t, t')$.

In reference to the noise, we observe first of all that

$$\frac{1}{2}\langle\{Q(t), Q(t')\}\rangle = \langle Q(t)\rangle\langle Q(t')\rangle + \frac{1}{2}\langle\{Q(t) - \langle Q(t)\rangle, Q(t') - \langle Q(t')\rangle\}\rangle$$

(3.100)

If, for example, the initial state for the environment is thermal, then $\langle Q(t)\rangle = 0$, and we recover the result from Chapter 2, equation (2.16)

$$\langle \xi(t)\xi(t')\rangle = \hbar\mathbf{N}(t, t') = \sum_\alpha \frac{c_\alpha^2}{m_\alpha\omega_\alpha^2}\langle\varepsilon_\alpha\rangle\cos\omega_\alpha(t - t') \qquad (3.101)$$

only now we must use the quantum energy expectation value

$$\langle\varepsilon_\alpha\rangle = \hbar\omega_\alpha\left[\frac{1}{2} + \frac{1}{e^{\hbar\omega_\alpha/k_\mathrm{B}T} - 1}\right] \qquad (3.102)$$

whereby we recover the quantum form of the fluctuation–dissipation theorem. Of course, the noise is truly Gaussian only if $\mathbf{V}$ is zero, which means the system itself is linear.

If the bath frequencies span a continuum, we should replace

$$\sum_\alpha \rightarrow \int_0^\infty d\omega\,\rho(\omega) \qquad (3.103)$$

where $\rho(\omega)\,d\omega$ is the number of oscillators with frequencies between ω and $\omega + d\omega$. We say the bath is ohmic if

$$\rho(\omega) = \frac{4\gamma M}{\pi}\frac{m_\omega\omega^2}{c(\omega)^2} \qquad (3.104)$$

for some constant γ. Observe that for an ohmic bath $\gamma(t - t') = 4\gamma\delta(t - t')$ and $D(t) = 2\gamma P(t)$, so the Langevin equation is local in time. No physical bath can be exactly ohmic, because it would require either an infinite number of oscillators or else arbitrarily strong coupling to the bath, but many physical systems exhibit ohmic dissipation (for example, a biased Josephson junction) and may be modeled as *if* they were in contact with an ohmic bath.

Let us investigate the noise autocorrelation for an ohmic bath in equilibrium. We have

$$\langle\xi(t)\xi(t')\rangle = \hbar\mathbf{N}(t, t') = \frac{4\hbar M\gamma}{\pi}\int_0^\infty d\omega\,\omega\left[\frac{1}{2} + \frac{1}{e^{\hbar\omega/k_\mathrm{B}T} - 1}\right]\cos\omega(t - t')$$

(3.105)

For high temperature and $t - t' \gg \hbar/k_\mathrm{B}T$, we may argue that the integral is dominated by low frequencies, whereby the noise is white and we recover the classical fluctuation–dissipation theorem

$$\hbar\mathbf{N}(t, t')|_{T\to\infty} = 4M\gamma k_\mathrm{B}T\delta(t - t') \qquad (3.106)$$

As $T \to 0$, however, the integral becomes singular. Let us define

$$\mathrm{Pf}\left[\frac{1}{t^2}\right] = -\int_0^\infty d\omega \, \omega \cos \omega t, \qquad (3.107)$$

where Pf stands for the Hadamard finite part prescription. For example, if we regularize the integral by including a convergence factor $e^{-\omega/\Lambda}$, then

$$\mathrm{Pf}\left[\frac{1}{t^2}\right] = \lim_{\Lambda \to \infty} \frac{t^2 - \Lambda^{-2}}{(t^2 + \Lambda^{-2})^2} \qquad (3.108)$$

With this definition, the noise correlation at $T = 0$ becomes

$$\hbar \mathbf{N}\,(t,t')|_{T=0} = -\frac{2\hbar M\gamma}{\pi}\,\mathrm{Pf}\left[\frac{1}{(t-t')^2}\right] \qquad (3.109)$$

Observe that the decay of the noise correlation obeys a power law, which implies a very strongly colored noise.

3.5 The Kramers–Moyal equation

As in Chapter 2, the Langevin equation for the "trajectories" of the quantum open system may be turned into a Kramers–Moyal equation for the reduced Wigner function. To obtain this equation, we simply take the time derivative of the path integral representation (3.90). Observe that we get a new term coming from the explicit time dependence of P_Q. Indeed, write

$$P_Q\,[\xi,t] = P_Q\,[\xi,t^*] - \frac{i}{\hbar}\,(t-t^*)\,\mathbf{V}\left(X_f, i\hbar\frac{\delta}{\delta\xi\,(t^*)}\right) P_Q\,[\xi,t^*] \qquad (3.110)$$

where the reference time $t^* < t$ is taken to t after computing the derivatives. Then the noise averages may be split in two, and

$$\frac{\partial}{\partial t} f_r^W\,(X_f, P_f, t) = -\,\{H, f_r^W\} + M\delta\Omega_0^2 X_f \frac{\partial f_r^W}{\partial P_f} + \frac{\partial}{\partial P_f}\,[\mathbf{A} + \mathbf{B}] + \mathbf{C} \quad (3.111)$$

where the first term contains the Poisson brackets. The new terms are

$$\mathbf{A} = \int_0^t dt' \,\gamma\,(t-t')\,\langle P\,(t')\,\delta\,(X\,(t) - X_f)\,\delta\,(P\,(t) - P_f)\rangle \qquad (3.112)$$

$$\mathbf{B} = -\,\langle \xi\,(t)\,\delta\,(X\,(t) - X_f)\,\delta\,(P\,(t) - P_f)\rangle \qquad (3.113)$$

$$\mathbf{C} = \left(\frac{-i}{\hbar}\right)\left\langle \mathbf{V}\left(X_f, -i\hbar\frac{\delta}{\delta\xi\,(t^*)}\right)\,\delta\,(X\,(t) - X_f)\,\delta\,(P\,(t) - P_f)\right\rangle \qquad (3.114)$$

We may use certain approximations to extract the leading behavior of these expressions. To simplify the $\mathbf{A}$ term, for example, we replace $P\,(t')$ by the solution $\bar{P}\,[X_f, P_f, t; t']$ to the classical equations of motion with Cauchy data $[X_f, P_f]$ at time t. Observe that even for a strong system–bath interaction this approximation is justified if the kernel γ decays fast enough; it is exact for an ohmic bath. So we approximate

$$\mathbf{A} \sim \mathbf{\Gamma}\left(X_f, P_f, t\right) f_r^W\left(X_f, P_f, t\right) \tag{3.115}$$

$$\mathbf{\Gamma}\left(X_f, P_f, t\right) = \int_0^t dt'\, \gamma\left(t - t'\right) \bar{P}\left[X_f, P_f, t; t'\right] \tag{3.116}$$

To simplify the $\mathbf{B}$ term, let us first neglect the X dependence in $\mathbf{V}$. This approximation is actually exact for a cubic potential. Also recall that since the t dependence of $\mathbf{V}$ is explicitly considered through the $\mathbf{C}$ term, the time-integral in P_Q in the $\mathbf{B}$ term is truncated at t^-. Then we have

$$-\xi\left(t\right) P_Q\left[\xi, t\right] = -\exp\left[\frac{-i}{\hbar}\int_0^{t^-} dt'\, \mathbf{V}\left(X, i\hbar\frac{\delta}{\delta\xi\left(t\right)}\right)\right]\xi\left(t\right) P\left[\xi\right]$$

$$= \int_0^t dt'\, \hbar\mathbf{N}\left(t - t'\right)\frac{\delta}{\delta\xi\left(t'\right)}P_Q\left[\xi, t\right] \tag{3.117}$$

and after a further integration by parts

$$\mathbf{B} = \int_0^t dt'\, \hbar\mathbf{N}\left(t - t'\right)\left\{\left\langle\frac{\delta X\left(t\right)}{\delta\xi\left(t'\right)}\frac{\partial}{\partial X_f}\delta\left(X\left(t\right) - X_f\right)\delta\left(P\left(t\right) - P_f\right)\right\rangle \right.$$
$$\left. + \left\langle\frac{\delta P\left(t\right)}{\delta\xi\left(t'\right)}\delta\left(X\left(t\right) - X_f\right)\frac{\partial}{\partial P_f}\delta\left(P\left(t\right) - P_f\right)\right\rangle\right\} \tag{3.118}$$

To compute the variations with respect to the noise, recall the identities

$$\frac{\delta X\left(t'\right)}{\delta\xi\left(t'\right)} = 0; \qquad \frac{\delta P\left(t'^+\right)}{\delta\xi\left(t'\right)} = 1 \tag{3.119}$$

and use the chain rule

$$0 = \frac{\delta X\left(t'\right)}{\delta X\left(t\right)}\frac{\delta X\left(t\right)}{\delta\xi\left(t'\right)} + \frac{\delta X\left(t'\right)}{\delta P\left(t\right)}\frac{\delta P\left(t\right)}{\delta\xi\left(t'\right)} \tag{3.120}$$

$$1 = \frac{\delta P\left(t'\right)}{\delta X\left(t\right)}\frac{\delta X\left(t\right)}{\delta\xi\left(t'\right)} + \frac{\delta P\left(t'\right)}{\delta P\left(t\right)}\frac{\delta P\left(t\right)}{\delta\xi\left(t'\right)} \tag{3.121}$$

Now assume that $(X, P)\left(t\right)$ and $(X, P)\left(t'\right)$ are linked through the classical equations of motion. The determinant of the system is 1 from Liouville's theorem, and so

$$\frac{\delta X\left(t\right)}{\delta\xi\left(t'\right)} = -\frac{\delta\bar{X}\left[X_f, P_f, t; t'\right]}{\delta P_f}; \qquad \frac{\delta P\left(t\right)}{\delta\xi\left(t'\right)} = \frac{\delta\bar{X}\left[X_f, P_f, t; t'\right]}{\delta X_f} \tag{3.122}$$

The final result is

$$\mathbf{B} = -\left\{\Phi, f_r^W\right\}, \tag{3.123}$$

$$\Phi = \int_0^t dt'\, \hbar\mathbf{N}\left(t - t'\right)\bar{X}\left[X_f, P_f, t; t'\right] \tag{3.124}$$

Finally, to compute $\mathbf{C}$ we use the identities (3.119) at time t to get

$$\mathbf{C} = \left(\frac{-i}{\hbar}\right) \mathbf{V}\left(X_f, i\hbar\frac{\delta}{\delta P_f}\right) f_r^W\left(X_f, P_f, t\right) \tag{3.125}$$

To summarize, the quantum Kramers–Moyal equation reads

$$\frac{\partial}{\partial t} f_r^W\left(X_f, P_f, t\right)$$

$$= -\{H, f_r^W\} - \left(\frac{i}{\hbar}\right)\mathbf{V}\left(X_f, i\hbar\frac{\delta}{\delta P_f}\right) f_r^W\left(X_f, P_f, t\right)$$

$$+ M\delta\Omega_0^2 X_f \frac{\partial f_r^W}{\partial P_f}$$

$$+ \frac{\partial}{\partial P_f}\left[\mathbf{\Gamma}\left(X_f, P_f, t\right) f_r^W\left(X_f, P_f, t\right) - \{\Phi, f_r^W\}\right] \tag{3.126}$$

The first line gives the evolution of the Wigner function without interaction with the environment, while the second and third lines describe the renormalization, dissipation, diffusion and decoherence effects.

3.5.1 The linear system model

If the system itself is linear, we can obtain simple analytic expressions for $\bar{X}[X_f, P_f, t; t')$ and $\bar{P}[X_f, P_f, t; t')$ and thus derive an explicit result. We have (cf. equation (3.70))

$$\bar{X}[X_f, P_f, t; t') = X_f \cos\Omega\left(t - t'\right) - \frac{P_f}{M\Omega} \sin\Omega\left(t - t'\right) \tag{3.127}$$

$$\bar{P}[X_f, P_f, t; t') = P_f \cos\Omega\left(t - t'\right) + M\Omega X_f \sin\Omega\left(t - t'\right) \tag{3.128}$$

The Kramers–Moyal equation now reads (for a linear system, $\mathbf{V} = 0$)

$$\frac{\partial}{\partial t} f_r^W\left(X_f, P_f, t\right) = -\{H, f_r^W\} + M\delta\Omega^2\left(t\right) X_f \frac{\partial f_r^W}{\partial P_f}$$

$$+ \frac{\partial}{\partial P_f}\left[2\Gamma\left(t\right) P_f + \frac{\sigma^2\left(t\right)}{2}\frac{\partial}{\partial P_f} + \Delta_{ad}\left(t\right)\frac{\partial}{\partial X_f}\right] f_r^W \tag{3.129}$$

where the coefficients $\sigma^2\left(t\right)$, $\Delta\left(t\right)$, $\Gamma\left(t\right)$ and $\delta\Omega^2\left(t\right)$ were defined above, from equations (3.72)–(3.75). The identity of the coefficients to those in the master equation (3.71) is not surprising, since for linear systems the Kramers–Moyal equation (3.129) and the master equation (3.71) are equivalent. For nonlinear systems, they are still closely related, but the approximations which go into one or the other are not exactly the same.

The form (3.129) of the Kramers–Moyal equation makes it clear that the coefficient $\gamma\left(t\right)$ is associated with dissipation and $\sigma^2\left(t\right)$ with "normal" diffusion. We call $\Delta_{ad}\left(t\right)$ the "anomalous" diffusion constant.

$\sigma^2(t)$ also pertains to decoherence. To see this, consider the pseudo-entropy

$$\tilde{S} = 1 - \operatorname{tr} \rho_r^2$$

$$= 1 - (2\pi\hbar) \int dX dP \, (f_r^W)^2 \, (X, P) \tag{3.130}$$

Then

$$\frac{d\tilde{S}}{dt} = 2\Gamma(1 - \tilde{S}) + (4\pi\hbar) \int dX dP \left\{ \frac{\sigma^2(t)}{2} \left(\frac{\partial f_r^W}{\partial P_f} \right)^2 + \Delta_{ad}(t) \frac{\partial f_r^W}{\partial X_f} \frac{\partial f_r^W}{\partial P_f} \right\} \tag{3.131}$$

The first term represents heat loss to the environment and the second induces decoherence. The third does not have a definite sign.

To conclude, let us evaluate these coefficients for an ohmic bath. At high temperature, we get the expected relations $\Gamma(t) = \gamma$, $\sigma^2(t) = 4\gamma M k_B T$, $\Delta_{ad}(t) = \delta\Omega^2(t) = 0$. At $T = 0$, though, the naive expressions diverge. Suppose we use an exponential cut-off to regularize them, as in (3.108). Then as the cut-off is removed, we get $\Gamma(t) = \gamma$ and $\delta\Omega^2(t) = 0$. For the expressions involving the noise kernel (3.109), we get that Δ_{ad} diverges logarithmically, while σ^2 diverges linearly in the cut-off Λ. This result suggests that at late times the system perceives the environment as a heat bath at a temperature $k_B T_{\text{eff}} \approx \hbar\Lambda$ [ALMV06].

3.6 Derivation of the propagator and the master equation

For the influence functional path integral treatment of quantum Brownian motion (QBM) the formal expression of the evolutionary operator for the reduced density matrix was derived by Grabert, Schramm and Ingold [GrScIn88] and an exact master equation for QBM in a general (non-ohmic) environment at an arbitrary temperature was derived by Hu, Paz and Zhang [HuPaZh92, HuPaZh93a]. In this section we give a discussion of this problem based on their work. This is useful not only as a model example of this important method, but also because in some problems such as the calculation of entropy generation (to be discussed in Chapter 9) in quantum open systems we need some of these details.

Let us consider the general case of a quantum harmonic oscillator with time-dependent mass, cross-term and natural frequency undergoing Brownian motion through its interaction with an environment made up of n harmonic oscillators with the same time-dependent parameters. The total Lagrangian of the system is given by

$$S[x, \mathbf{q}] = S[x] + S_E[\mathbf{q}] + S_{\text{int}}[x, \mathbf{q}]$$

$$= \int_{t_i}^t ds \left\{ \frac{1}{2} M(s) \left[\dot{x}^2 + 2\mathcal{E}(s) x\dot{x} - \Omega^2(s) x^2 \right] \right.$$

$$\left. + \sum_n \left[\frac{1}{2} m_n(s) \left[\dot{q}_n^2 + 2\varepsilon_n(s) q_n \dot{q}_n - \omega_n^2(s) q_n^2 \right] \right] + \sum_n \left[-c(s) x q_n \right] \right\} \tag{3.133}$$

where the particle and the bath oscillators have coordinates x and q_n, respectively; we may also let the system variable interact with the environment variable through a more general $f(x)$ functional form. This Hamiltonian is considered in detail by Hu and Matacz [HuMat94] as an example of a squeezed quantum open system. We will discuss this in the last section of Chapter 4.

3.6.1 Evolution of the reduced density matrix

Given some initial system density matrix $\rho_S(x_i\, x_i'\, 0)$ we want to evolve it in time using (3.49). The formal expression for $\mathcal{J}_r$ was derived by Grabert *et al.* [GrScIn88] using path integral methods, and calculated explicitly in [HuPaZh92, HuPaZh93a, HuMat94] for a general (non-ohmic) environment.

In terms of the sum and difference variables the classical paths followed by the system, $X_{\rm cl}, u_{\rm cl}$, can be written in terms of more elementary functions u, v:

$$X_{\rm cl}(s) = X_{\rm cl}(t_i)u_1(s) + X_{\rm cl}(t)u_2(s)$$
$$u_{\rm cl}(s) = u_{\rm cl}(t_i)v_1(s) + u_{\rm cl}(t)v_2(s) \tag{3.134}$$

Then it can be shown [HuMat94] that the evolutionary operator $\mathcal{J}_r$ is equal to

$$\mathcal{J}_r(x, x', t | x_i, x_i', t_i) = \frac{|b_2|}{2\pi\hbar} \exp\left[\frac{i}{\hbar}(b_1 Xu - b_2 Xu_i + b_3 X_i u - b_4 X_i u_i)\right.$$
$$\left. - \frac{1}{\hbar}\left(a_{11}u_i^2 + a_{12}u_i u + a_{22}u^2\right)\right] \tag{3.135}$$

The functions $b_1 \to b_4$ can be expressed as

$$b_1(t, t_i) = M(t)\dot{u}_2(t) + M(t)\mathcal{E}(t)$$
$$b_2(t, t_i) = M(t_i)\dot{u}_2(t_i)$$
$$b_3(t, t_i) = M(t)\dot{u}_1(t)$$
$$b_4(t, t_i) = M(t_i)\dot{u}_1(t_i) + M(t_i)\mathcal{E}(t_i) \tag{3.136}$$

while the functions a_{ij} are defined by

$$a_{ij}(t, t_i) = \frac{1}{1 + \delta_{ij}} \int_{t_i}^{t} ds \int_{t_i}^{t} ds'\, v_i(s)\, \mathbf{N}(s, s')\, v_j(s') \tag{3.137}$$

The functions $u_1 \to v_2$ are solutions to the following equations (dropping subscripts on u, v):[1]

$$\ddot{u}(s) + \frac{\dot{M}}{M}\dot{u} + \left(\Omega^2 + \dot{\mathcal{E}} + \frac{\dot{M}}{M}\mathcal{E}\right)u - \frac{1}{M(s)}\int_{t_i}^{s} ds'\, \mathbf{D}(s, s')\, u(s') = 0 \tag{3.138}$$

$$\ddot{v}(s) + \frac{\dot{M}}{M}\dot{v} + \left(\Omega^2 + \dot{\mathcal{E}} + \frac{\dot{M}}{M}\mathcal{E}\right)v - \frac{1}{M(s)}\int_{s}^{t} ds'\, v(s')\mathbf{D}(s', s) = 0 \tag{3.139}$$

---

[1] Do not confuse u here with $\mathbf{u} \equiv \mathbf{x}_1 - \mathbf{x}_2$ in Chapter 2 or $u \equiv x - x'$ in Chapter 3.

subject to the boundary conditions

$$u_1(t_i) = v_1(t_i) = 1, \quad u_1(t) = v_1(t) = 0 \tag{3.140}$$

$$u_2(t_i) = v_2(t_i) = 0, \quad u_2(t) = v_2(t) = 1 \tag{3.141}$$

To proceed further we need explicit expressions for $a_{11} \to b_4$. These are expressed in terms of $u_1 \to v_2$, which in turn come from solving equations (3.138) and (3.139). To solve these equations we need to know the dissipation $\mathbf{D}$ kernel of the environment, which is determined by the coupling and the spectral density function of the environment. We consider an ohmic bath and assume an unsqueezed (coherent) thermal bath made up of unit mass static (time-independent frequency) oscillators so the dissipation and noise kernels simplify to the form

$$\mathbf{D}(s, s') = -4\gamma_0\, c(s)c(s')\, \delta'(s - s')$$

$$\mathbf{N}(s, s') = \frac{2\gamma_0}{\pi}\, c(s)c(s') \int_0^\infty \omega \coth \frac{\hbar\omega}{2k_\mathrm{B}T} \cos\omega(s - s')\, d\omega \tag{3.142}$$

If $c(s) = c =$ constant, we may identify $\gamma_0 c^2 = M\gamma$. In this case, in the high-temperature limit the noise becomes white, that is, $\mathbf{N}$ tends toward a delta function.

3.6.2 Master equation

We now proceed with the derivation of the master equation from the evolution operator using the simplified method of Paz [Paz94]. We first take the time derivative of both sides of equation (3.135), multiply both sides by $\rho_r(X_i, u_i, t_i)$ and integrate over X_i, u_i to obtain

$$\dot{\rho}_r(X_f, u_f, t) = \left[\frac{\dot{b}_2}{b_2} + \frac{i}{\hbar} b_1 X_f u_f - \dot{a}_{22} \frac{u_f^2}{\hbar} \right] \rho_r(X_f, u_f, t)$$

$$+ \frac{i}{\hbar} u_f b_3 \int du_i dX_i\, X_i \mathcal{J}_r \rho_r(X_i, u_i, t_i)$$

$$- \frac{1}{\hbar} (i \dot{b}_2 X_f + \dot{a}_{12} u_f) \int du_i dX_i\, u_i \mathcal{J}_r \rho_r(X_i, u_i, t_i) \tag{3.143}$$

$$- \frac{i}{\hbar} b_4 \int du_i dX_i\, X_i u_i \mathcal{J}_r \rho_r(X_i, u_i, t_i)$$

$$- \frac{\dot{a}_{11}}{\hbar} \int du_i dX_i\, u_i^2 \mathcal{J}_r \rho_r(X_i, u_i, t_i)$$

Here the dot denotes the derivative with respect to t. We can perform the integrals in (3.143) by using

$$u_i \mathcal{J}_r = \frac{i\hbar}{b_2} \frac{\partial \mathcal{J}_r}{\partial X_f} + \frac{b_1 u_f}{b_2} \mathcal{J}_r \tag{3.144}$$

$$X_i \mathcal{J}_r = -\frac{i}{b_3}\left[\hbar\frac{\partial \mathcal{J}_r}{\partial u_f} + (u_i a_{12} + 2u_f a_{22})\mathcal{J}_r)\right] - \frac{b_1}{b_3} X_f \mathcal{J}_r \tag{3.145}$$

$$X_i u_i \mathcal{J}_r = -\left(\frac{i\hbar}{b_2}\frac{\partial}{\partial X_f} + \frac{b_1 u_f}{b_2}\right)$$
$$\times \left(\frac{i\hbar}{b_3}\frac{\partial}{\partial u_f} + \frac{i}{b_3}[u_i a_{12} + 2u_f a_{22}] + \frac{b_1}{b_3}X_f\right)\mathcal{J}_r \tag{3.146}$$

The u_i functions obey mixed boundary conditions. It is convenient to express them in terms of functions w_i obeying initial conditions only. We write

$$u_1(s) = w_1(s) - w_2(s)\frac{w_1(t)}{w_2(t)}, \quad u_2(s) = \frac{w_2(s)}{w_2(t)} \tag{3.147}$$

In order to satisfy the boundary conditions (3.140) we require

$$w_1(t_i) = \dot{w}_2(t_i) = 1, \quad w_2(t_i) = \dot{w}_1(t_i) = 0 \tag{3.148}$$

In this representation we can show that

$$\frac{\dot{b}_4}{b_2 b_3} = -\frac{1}{M(t)}, \quad b_1 = -M(t)\frac{\dot{b}_2}{b_2} + M(t)\mathcal{E}, \quad \dot{a}_{11} = -\dot{v}_1(t)a_{12} \tag{3.149}$$

With these relations the master equation is the same as equation (3.71) with two additional terms

$$i\hbar\frac{\partial}{\partial t}\rho_r(x, x', t) = \left\{-\frac{\hbar^2}{2M(t)}\left(\frac{\partial^2}{\partial x^2} - \frac{\partial^2}{\partial x'^2}\right) + i\hbar\mathcal{E}\left(x\frac{\partial}{\partial x} + x'\frac{\partial}{\partial x'}\right)\right.$$
$$+ \frac{M(t)}{2}\left[\Omega^2 + \delta\Omega^2(t)\right](x^2 - x'^2) + i\hbar\mathcal{E}\bigg\}\rho_r(x, x', t)$$
$$- i\hbar\Gamma(t, t_i)(x - x')\left(\frac{\partial}{\partial x} - \frac{\partial}{\partial x'}\right)\rho_r(x, x', t)$$
$$- i\frac{\sigma^2(t)}{2\hbar}(x - x')^2 \rho_r(x, x', t)$$
$$+ \Delta_{ad}(t)(x - x')\left(\frac{\partial}{\partial x} + \frac{\partial}{\partial x'}\right)\rho_r(x, x', t)$$
$$- i\hbar^2 D_{xx}(t, t_i)\left(\frac{\partial}{\partial x} + \frac{\partial}{\partial x'}\right)^2 \rho_r(x, x', t) \tag{3.150}$$

where we identify

$$\left[\Omega^2 + \delta\Omega^2\left(t\right)\right] = \frac{b_1\dot{b}_3}{M(t)b_3} - \frac{\dot{b}_1}{M(t)} + \mathcal{E}^2 - \frac{\dot{b}_2}{b_2}\mathcal{E} \tag{3.151}$$

$$\Gamma(t,t_i) = -\frac{1}{2}\left(\frac{\dot{b}_3}{b_3} - \frac{\dot{b}_2}{b_2}\right) \tag{3.152}$$

$$-\frac{\sigma^2\left(t\right)}{2\hbar} = \frac{b_1^2}{b_2}\left(\frac{a_{12}}{M(t)} - \frac{\dot{a}_{11}}{b_2}\right) + \frac{2b_1}{M(t)}a_{22} - \dot{a}_{22} + 2\frac{\dot{b}_3}{b_3}a_{22} + a_{12}\frac{b_1\dot{b}_3}{b_2b_3} - \dot{a}_{12}\frac{b_1}{b_2} \tag{3.153}$$

$$\Delta_{ad}\left(t\right) = \hbar\left[\frac{\dot{a}_{12}}{b_2} - 2\frac{a_{22}}{M(t)} - \frac{\dot{b}_3a_{12}}{b_3b_2} - \frac{2b_1}{b_2}\left(\frac{a_{12}}{M(t)} - \frac{\dot{a}_{11}}{b_2}\right)\right] \tag{3.154}$$

$$D_{xx}(t,t_i) = \frac{1}{b_2}\left(\frac{a_{12}}{M(t)} - \frac{\dot{a}_{11}}{b_2}\right) \tag{3.155}$$

The dot in these equations denotes taking the derivative with respect to t.

The factor $a_{12}/M(t) - \dot{a}_{11}/b_2$ vanishes only when the dissipation kernel is stationary (i.e. a function of $s - s'$) and the system is a time-independent harmonic oscillator. When this happens $v_1(s) = u_2(t - s)$ and we have $\dot{v}_1(t) = -b_2/M(t)$. We see from equation (3.149) that the factor $a_{12}/M(t) - \dot{a}_{11}/b_2$ is zero in this case. All the diffusion coefficients contain this factor and D_{xx} depends solely on it.

3.7 Consistent histories and decoherence functional

The question which remains unanswered is whether individual solutions of the Langevin equation are actually observable. This question contains two aspects, namely, (a) whether the evolution of the quantum open system may be analyzed in terms of trajectories, and (b) whether these trajectories describe any recognizable dynamics. As we shall see, the answer is not straightforward, because it involves a new component, namely, the accuracy of our observations. Out of quantum common sense, we expect that if we follow the trajectories too closely, we would be feeding noise into the system (Heisenberg's principle), eventually masking the system–environment interaction. Still the question remains whether there is *any* range where the Langevin equation is a satisfactory description of the observed evolution of the system.

To analyze this question we shall adopt the consistent histories approach to quantum mechanics, in the version advanced by Gell-Mann and Hartle (see [Gri84, Gri93, Omn88, Omn90, Omn92, KoEzMuNo90, Har92, Har93, GelHar90, HarGel93, Bru93, GelHar06]). The idea is to define a history by a set of projectors P_α acting at times t_i. In canonical terms, a history is given by an evolution of the state vector such that at every time t_i, it belongs to the proper space of

$P_\alpha(t_i)$. In path integral terms, the projectors are represented by window functions $w_\alpha[x(t_i)]$, which take on unit value if the instantaneous configuration x satisfies the requirements of the history α, and vanish otherwise. The limiting case of a *fine-grained history*, namely, when $x(t)$ is specified for all times, is assigned an amplitude $\exp iS/\hbar$, as usual in the Feynman path integral formulation. The amplitude for a *coarse-grained* history defined by window functions $w_\alpha[x(t_i)]$ is defined by the superposition

$$A[\alpha] = \int Dx \, e^{iS/\hbar} \psi[x(0)] \left\{ \prod_i w_\alpha[x(t_i)] \right\} \tag{3.156}$$

The probability is naturally expressed in terms of a closed time path integral

$$\mathbf{P}[\alpha] = |A[\alpha]|^2 = \int Dx Dx' \, e^{i[S-S']/\hbar} \rho[x(0), x'(0)]$$

$$\times \left\{ \prod_i w_\alpha[x(t_i)] \right\} \left\{ \prod_i w_\alpha[x'(t_i)] \right\} \tag{3.157}$$

In this way we may assign a probability to any coarse-grained history, but these probability assignments are not generally *consistent*, namely, the probabilities of two mutually exclusive histories do not generally add up. Indeed, let us define the *decoherence functional* of two histories α and β

$$\mathcal{D}[\alpha, \beta] = \int Dx Dx' \, e^{i[S-S']/\hbar} \rho[x(0), x'(0)] \left\{ \prod_i w_\alpha[x(t_i)] \right\} \left\{ \prod_j w_\beta[x'(t_j)] \right\} \tag{3.158}$$

$\mathbf{P}[\alpha] = \mathcal{D}[\alpha, \alpha]$ but $\mathbf{P}[\alpha \vee \beta] = \mathcal{D}[\alpha, \alpha] + \mathcal{D}[\beta, \beta] + 2\mathrm{Re}\mathcal{D}[\alpha, \beta] \neq \mathbf{P}[\alpha] + \mathbf{P}[\beta]$. The probability sum rule $\mathbf{P}[\alpha \vee \beta] = \mathbf{P}[\alpha] + \mathbf{P}[\beta]$ only applies when the third term vanishes, and in particular when there is *strong decoherence*, $\mathcal{D}[\alpha, \beta] = 0$ for $\alpha \neq \beta$. As physicists, who deal with reality, we shall be satisfied that a set of mutually exclusive histories is consistent when $|\mathcal{D}[\alpha, \beta]| \ll \mathcal{D}[\alpha, \alpha], \mathcal{D}[\beta, \beta]$ whenever $\alpha \neq \beta$.

A simple set of consistent histories refers to the values of conserved quantities [HaLaMa95]. First observe that the path integral expression (3.158) translates into the canonical expression

$$\mathcal{D}[\alpha, \beta] = \mathrm{Tr} \left\{ \tilde{T}\left[\prod_j P_\beta(t_j) \right] T\left[\prod_i P_\alpha(t_i) \right] \rho(0) \right\} \tag{3.159}$$

The projectors at different times are related in the usual way $P_\alpha(t) = U(t) P_\alpha(0) U^\dagger(t)$. If a projector commutes with the Hamiltonian, then it is time-independent, and expression (3.159) collapses unless all projectors are indeed identical. The only histories with nonzero probabilities are those defined by ranges of conserved quantities in the initial state, and they are automatically consistent if these ranges do not overlap.

For open quantum systems we are interested in histories where the system variable X is specified to follow a trajectory $\chi(t)$ with a given accuracy $\sigma(t)$, while the environment variable q is left unspecified. For technical reasons, it is convenient to use Gaussian, rather than sharp, windows. We also make a Gaussian ansatz for the initial state, which we assume to be pure. Therefore, we replace

$$\rho\left[x\left(0\right), x'\left(0\right)\right]\left\{\prod_i w_\alpha\left[x\left(t_i\right)\right]\right\}\left\{\prod_j w_\beta\left[x'\left(t_j\right)\right]\right\} \tag{3.160}$$

by

$$\exp\left[-\int\frac{dt}{2\sigma^2\left(t\right)}\left\{\left(x-\chi\right)^2+\left(x'-\chi'\right)^2\right\}\right] \tag{3.161}$$

The unconstrained integration over environment variables yields the action functional, which has the structure we already know. Adopting a shorthand notation

$$\mathcal{D}\left[\chi, \chi'\right] = \int Dx Dx'\ \exp\left(\frac{-1}{2}\right)$$
$$\times\left\{-2iuLX + \mathbf{N}u^2 + \frac{1}{\sigma^2}\left[\left(x-\chi\right)^2+\left(x'-\chi'\right)^2\right]\right\} \tag{3.162}$$

where $X = x + x'/2$, $u = x - x'$, the symbols L, $\mathbf{N}$ and $1/\sigma^2$ denote operators (which we shall handle as if they were c-numbers) and we have applied Einstein's convention to time integrals. Write $\chi, \chi' = \Upsilon \pm y/2$ and develop the last term to get

$$\mathcal{D}\left[\chi, \chi'\right] = \exp\left(\frac{-1}{2\sigma^2}\right)\left[2\Upsilon^2 + \frac{y^2}{2}\right]$$
$$\times\int DX Du\ \exp\left(\frac{-1}{2}\right)\left\{-2iuLX + \left[\mathbf{N} + \frac{1}{2\sigma^2}\right]u^2 + \frac{2}{\sigma^2}X^2\right\}$$
$$\times\exp\frac{1}{2\sigma^2}\left[uy + 4X\Upsilon\right] \tag{3.163}$$

Now consider the matrix

$$M = \begin{pmatrix} 2\sigma^{-2} & \left(-i\right)L \\ \left(-i\right)L & \mathbf{N} + \left(2\sigma^2\right)^{-1} \end{pmatrix} \tag{3.164}$$

Already from the fact that the noise kernel appears in the combination $\mathbf{N} + \left(2\sigma^2\right)^{-1}$ we see that there must be a limit where the "Langevin noise" is drowned in the "Heisenberg noise." The determinant of this matrix is $\mathrm{Det}\left(M\right) = \left(\mathbf{N} + \left(2\sigma^2\right)^{-1}\right)2\sigma^{-2} + L^2$, and the inverse is (we assume all operators commute)

$$M^{-1} = \left[\mathrm{Det}\left(M\right)\right]^{-1}\begin{pmatrix} \mathbf{N} + \left(2\sigma^2\right)^{-1} & iL \\ iL & 2\sigma^{-2} \end{pmatrix} \tag{3.165}$$

Therefore

$$|\mathcal{D}\left[\chi, \chi'\right]| \sim \exp\left\{-\left[1 - \frac{2\left[Det\left(M\right)\right]^{-1}}{\sigma^2}\left(\mathbf{N} + \left(2\sigma^2\right)^{-1}\right)\right]\frac{\Upsilon^2}{\sigma^2}\right.$$
$$\left. - \left[1 - \frac{\left[Det\left(M\right)\right]^{-1}}{\sigma^4}\right]\frac{y^2}{4\sigma^2}\right\} \tag{3.166}$$

We see that the dynamics and the decoherence aspects are clearly separated. To obtain a simpler expression, we shall assume that the L operator is "small," so we can expand in powers of L. Keeping only the first nonzero contributions, we get

$$|\mathcal{D}\left[\chi, \chi'\right]| \sim \exp\left(\frac{-1}{2}\right)\left\{\left(\mathbf{N} + \left(2\sigma^2\right)^{-1}\right)^{-1}\left(L\Upsilon\right)^2 + \mathbf{N}\left(2\sigma^2\mathbf{N} + 1\right)^{-1}y^2\right\} \tag{3.167}$$

To find the probability of a given history, we must set $y = 0$. We see that the most likely histories are those which satisfy the "classical" equations of motion $L\chi = 0$; these are the equations of motion for the expectation value of the system variable, and include the dissipative terms, but not the noise. The magnitude of the expected deviations from the deterministic behavior is given by $\mathbf{N} + \left(2\sigma^2\right)^{-1}$. The noise kernel provides a lower bound for the "noisiness" of the dynamics, but we can say that the deviations from the classical motion are well described by the Langevin equation only in the limit of "fuzzy" observations, $\left(2\sigma^2\right)^{-1} \ll \mathbf{N}$. In the opposite limit, the dominant effect is the Heisenberg noise.

To study consistency, we must follow the decoherence functional as y increases. We see that our histories tend to decohere, and they become approximately consistent whenever $y^2 \geq \left(2\sigma^2 + \mathbf{N}^{-1}\right)$. The relevant question is whether any two histories which may be resolved by our apparatus are automatically consistent. The limit of resolution is $y^2 \sim \sigma^2$; therefore, consistency is obtained only asymptotically for strong noise $\sigma^2\mathbf{N} \gg 1$.

In conclusion, a picture of the system evolution based on actual nearly classical trajectories may only result from a compromise whereby the accuracy of observations is adjusted to the noise level, $\sigma^2 \sim \mathbf{N}^{-1}$. Larger noise for a given σ means more decoherence but less predictability; for a weaker noise, predictability is only limited by the Heisenberg bounds, but individual trajectories will not decohere. If we are satisfied with predictability within the limits imposed by the Langevin equation, then in the strong noise limit we may consider individual trajectories as actually depicting physical reality.

For a critique of the consistent history approach to quantum mechanics, see [DowKen96, BasGhi99].

Part II

Basics of nonequilibrium quantum field theory

4

Quantum fields on time-dependent backgrounds: Particle creation

Beginning with this chapter we will introduce quantum field theory (QFT) and develop the necessary ideas and methods which form the basis of nonequilibrium (NEq) QFT. We focus on quantum field systems in external fields or in a time-varying background spacetime. The latter is included here because many basic concepts and techniques in QFT in external fields were developed historically in the area of QFT in curved spacetimes, especially in time-dependent backgrounds used in relativistic cosmology. Cosmology is also the arena where some of the basic tenets of NEqQFT were established and tested out.

In a dynamical background some basic concepts of QFT need to be reexamined. We point out the problem in straightforwardly extending the methodology of Minkowski spacetime QFT, such as the definition of particles by way of instantaneous diagonalization of the Hamiltonian. The vacuum state defined this way is nonviable since particles are being created as the system evolves. We introduce the Bogoliubov transformation between two sets of mode functions of the field, and discuss how two different particle models defined at different times are related to each other. Particle creation is a nonadiabatic process. We introduce the nth order adiabatic vacuum and number state as the proper way to construct a QFT in dynamical backgrounds. We derive expressions for spontaneous particle production as parametric amplification of vacuum fluctuations, and stimulated production as amplification of particles already present in the quantum or thermal state.

Following this we give two examples for the problem of charged particle motion in an external field. The first one is for a uniform electric field. We show how to use the adiabatic number state and the Bogoliubov transformation to obtain the famous result of Schwinger. In the second problem we study periodically driven fields based on the Floquet theory of parametric resonance. For charged particles in an external field we derive a quantum Vlasov equation for the rate of particle creation and show that particle creation is a non-Markovian (history dependent) process. We point out the intrinsic relation between number and phase of a quantum system, and under what conditions particle number may increase and others when it may decrease.

We then turn to a discussion of the second class of problems, that of quantum fields in dynamic background spacetimes. These are useful for the study of quantum processes in the early universe. We introduce the wave equation in curved spacetime, and discuss the conditions where one can construct a physically meaningful particle model, including the conformal vacuum for conformal fields in conformally flat spacetimes, which are relevant to the standard model in cosmology. We use a simple observation to show why gravitons are not produced in a radiation-dominated universe, and a simple model to illustrate how thermal particle creation arises. We then demonstrate how one can identify and remove the ultraviolet divergences in the stress–energy tensor of the quantum field by the method of adiabatic regularization. Obtaining a physically reasonable regularized stress–energy tensor is an essential step in approaching the so-called "back-reaction problem," i.e. finding a self-consistent solution of the quantum particle–EM field or quantum field–background spacetime system.

This is followed by a self-contained description of particle creation in the squeezed state language which can better elucidate the relation between number and phase representations. We first give the result of spontaneous and stimulated production, discuss the difference between bosons and fermions, and their dependence on the initial state. We then introduce the statistical mechanics of particle creation and relate entropy generation to the specification of the initial state and the choice of representations, such as the number state, the coherent and the squeezed state. Finally we present results for the fluctuations in particle number as it is relevant to defining noise in quantum fields and the vacuum susceptibility of spacetime. In the last section we give a description of squeezed quantum open systems. These discussions bring out some basic issues in the statistical mechanics of quantum fields and prepare the ground for investigating the statistical, kinetic, and stochastic features of quantum processes such as back-reaction and dissipation, entropy generation, fluctuations, correlations, noise and decoherence, which will be elaborated in later chapters.

4.1 Basic field theory

4.1.1 Classical fields

A field theory is concerned with extended physical systems, whose configurations are defined by giving some set of numbers at each spacetime point associated with an event, with coordinates denoted by a 4-vector $x^\mu = (t, \mathbf{x})$ containing the time and space components respectively. The simplest field theories have only one (real) number assigned to each event (or, attached to each spacetime point) and this number is prescribed to be the same for all observers. These are the so-called scalar field theories. For example, if we imagine spacetime as a continuous fluid, we may define a temperature (scalar) field $T(x)$ whose field configuration is given by the temperature T reading (a number) at each spatial point $\mathbf{x}$ at a given time t as measured by an observer at rest with respect to the fluid. Another familiar example of a scalar field is the magnetization density $\mu(x)$ in a ferromagnetic material, again in the continuous spacetime approximation.

For pedagogical reasons we shall be using the scalar field theory to illustrate new ideas and methods in this book. Extensions to vector (e.g. electromagnetic), tensor (e.g. gravitational) and spinor (e.g. electron) fields can be made with proper treatment of their specific tensor characters. In most parts of this book, except Chapters 9 and 15, we shall work in flat spacetime, endowed with the Minkowski metric $\eta^{\mu\nu} = \text{diag}(-1, 1, 1, 1)$, with time being the zeroth coordinate.

A scalar field theory describes the field variable $\phi(x)$, namely, the single real number to be prescribed at every event. Its dynamics is given by the action $S[\phi]$ of the theory; for example

$$S[\phi] = \int d^4x \left\{ -\frac{1}{2} (\nabla\phi)^2 - V[\phi(x)] \right\} \tag{4.1}$$

where $(\nabla\phi)^2$ in Minkowski space is equal to $\partial_\mu\phi\partial^\mu\phi = (\partial\phi)^2$ and the potential $V(\phi)$ is a real functional of the field variable ϕ. In this chapter we choose units such that the speed of light $c = 1$. A common example for massive interacting fields is the ϕ^4 potential

$$V(\phi) = \frac{1}{2}m^2\phi^2 + \frac{\lambda}{4!}\phi^4 \tag{4.2}$$

where m is the *mass* of the field (also known as the inverse correlation length) and λ is the coupling constant. The equations of motion are given by the variational principle $\delta S/\delta\phi = S_{,\phi} = 0$. In our case they read

$$\nabla^2\phi - V'(\phi) = 0 \tag{4.3}$$

where $\nabla^2 = \partial_\mu\partial^\mu$ in Minkowski space and $V' = dV/d\phi$. We can define the field momentum as $\pi = \phi_{,t}$ $\left(\phi_{,t} \equiv \phi_{,0} \text{ or } \dot{\phi}\right)$. A particular solution of the equations of motion is identified by its *Cauchy data* ϕ, π on a constant time surface. (There are more general surfaces one can use, the so-called *Cauchy surfaces*, but we won't go into that here.) The dynamics inherits the symmetries of the action, which in Minkowski spacetime possesses Poincaré invariance, and, for an even potential such as in equation (4.2), $\phi \to -\phi$ symmetry.

The second-order equation (4.3) can also be written as a first-order equation for π, namely

$$\frac{\partial\pi}{\partial t} = \nabla^2\phi - V'(\phi) \tag{4.4}$$

The definition of π and (4.4) together have the structure of canonical equations derivable from a Hamiltonian

$$H = \int d^3\mathbf{x} \left[\frac{1}{2}\pi^2 + \frac{1}{2}(\nabla_i\phi)^2 + V(\phi) \right] \tag{4.5}$$

Observe that the integral extends over space variables only. In other words, the nondenumerable set $\{\phi(t, \mathbf{x}), \mathbf{x}\epsilon R^3\}$ defines the canonical coordinates at time t, and the π's are their conjugate momenta. These canonical variables obey the equal-time Poisson brackets

$$\{\phi(t, \mathbf{x}), \phi(t, \mathbf{x}')\} = \{\pi(t, \mathbf{x}), \pi(t, \mathbf{x}')\} = 0; \quad \{\pi(t, \mathbf{x}), \phi(t, \mathbf{x}')\} = \delta(\mathbf{x} - \mathbf{x}') \tag{4.6}$$

This formulation is called the canonical formalism of field theory.

4.1.2 Quantum fields

The theory is quantized by replacing the field variable ϕ by an operator-valued distribution Φ. In the *Heisenberg picture*, for each event x there is an operator $\Phi(x)$ acting on some Hilbert space $\mathcal{H}$ of states. The conjugate momentum π goes over to the momentum operator Π, and the Poisson brackets equation (4.6) become the *equal-time canonical commutation relations* (ETCCRs)

$$[\Phi(t,\mathbf{x}),\Phi(t,\mathbf{x}')] = [\Pi(t,\mathbf{x}),\Pi(t,\mathbf{x}')] = 0; \quad [\Pi(t,\mathbf{x}),\Phi(t,\mathbf{x}')] = -i\hbar\delta(\mathbf{x}-\mathbf{x}')$$
$$(4.7)$$

The field operator moreover obeys the equation

$$\nabla^2\Phi - V'(\Phi) = 0 \qquad (4.8)$$

which is equivalent to the first-order system

$$\dot{\Phi} = \frac{i}{\hbar}[H,\Phi]; \quad \dot{\Pi} = \frac{i}{\hbar}[H,\Pi] \qquad (4.9)$$

leading to the rule

$$\Phi(t,\mathbf{x}) = U^\dagger(t,t')\,\Phi(t',\mathbf{x})\,U(t,t') \qquad (4.10)$$

where U is the *evolution operator*

$$U(t,t') = e^{-i(t-t')H/\hbar} \qquad (4.11)$$

More generally, we may introduce the generators P^μ of translations. The P^μ operators commute among themselves, as dictated by the algebra of the Poincaré group, and equation (4.11) is a particular case of the transformation rule

$$\Phi(x) = e^{-iPx/\hbar}\Phi(0)\,e^{iPx/\hbar} \qquad (4.12)$$

after identifying the Hamiltonian $H = P^0$.

4.1.3 Free fields

A free field corresponds to a quadratic potential $V(\Phi)$. A generic example is a free massive scalar field with $V(\Phi) = (1/2)m^2\Phi^2$. The Heisenberg equation of motion for this field becomes the *Klein–Gordon equation* $\nabla^2\Phi(x) - m^2\Phi(x) = 0$.

Assuming that the field lives in a finite large volume V and expanding the scalar field operator in (spatial) Fourier modes, we have

$$\Phi(t,\mathbf{x}) = \frac{1}{\sqrt{V}}\sum_{\mathbf{k}}\varphi_{\mathbf{k}}(t)u_{\mathbf{k}}(\mathbf{x}) \qquad (4.13)$$

where $\mathbf{k} = 2\pi\mathbf{n}/L$, and $\mathbf{n} = (n_1,n_2,n_3)$ in general consists of a triplet of integers. In Minkowski space the spatial mode functions are simply $u_{\mathbf{k}} = e^{i\mathbf{k}\cdot\mathbf{x}}$. In the infinite volume continuum limit this becomes

$$\Phi(t, \mathbf{x}) = \int \frac{d^3\mathbf{k}}{(2\pi)^{3/2}} e^{i\mathbf{kx}} \varphi_{\mathbf{k}}(t) \tag{4.14}$$

The (operator-valued) amplitude function $\varphi_{\mathbf{k}}(t)$ for each mode $\mathbf{k}$ obeys a harmonic oscillator equation

$$\frac{d^2\varphi_{\mathbf{k}}}{dt^2} + \omega_{\mathbf{k}}^2 \varphi_{\mathbf{k}} = 0 \tag{4.15}$$

where $\omega_{\mathbf{k}}^2 = |\mathbf{k}|^2 + m^2$ in Minkowski space.

Given two complex independent solutions $f_{\mathbf{k}}, f_{\mathbf{k}}^*$ of equation (4.15), we may write

$$\varphi_{\mathbf{k}}(t) = f_{\mathbf{k}}(t) a_{\mathbf{k}} + f_{\mathbf{k}}^*(t) a_{-\mathbf{k}}^\dagger \tag{4.16}$$

Let us introduce the Wronskian $(f, g) = f\dot{g} - g\dot{f}$, which is conserved by equation (4.15), and impose the normalization

$$(f_k, f_k^*) = i\hbar \tag{4.17}$$

The ETCCRs are equivalent to

$$[a_{\mathbf{k}}, a_{\mathbf{k}'}] = \left[a_{\mathbf{k}}^\dagger, a_{\mathbf{k}'}^\dagger\right] = 0; \quad \left[a_{\mathbf{k}}, a_{\mathbf{k}'}^\dagger\right] = \delta(\mathbf{k} - \mathbf{k}') \tag{4.18}$$

These operators may be interpreted as particle destruction and creation operators. We say that each choice of the basis functions $f_{\mathbf{k}}$ constitutes a *particle model*, where $f_{\mathbf{k}}$ is the *positive frequency* component and $f_{\mathbf{k}}^*$ is the *negative frequency* component of the $\mathbf{k}$th mode; the state which is destroyed by all the $a_{\mathbf{k}}$'s is the vacuum of the particle model. The vacua of different particle models are in general inequivalent. This situation becomes more challenging for quantum fields in a dynamical background field or spacetime, which is the central theme of this chapter.

In terms of the creation and destruction operators, the Hamiltonian is

$$H = \int \frac{d^3\mathbf{k}}{(2\pi)^3} \left\{ \mathcal{A}\hbar\omega_{\mathbf{k}} \left(\hat{N}_{\mathbf{k}} + \frac{1}{2} \right) + F_{\mathbf{k}} a_{\mathbf{k}} a_{-\mathbf{k}} + F_{\mathbf{k}}^* a_{\mathbf{k}}^\dagger a_{-\mathbf{k}}^\dagger \right\} \tag{4.19}$$

Here,

$$\hat{N}_{\mathbf{k}} = a_{\mathbf{k}}^\dagger a_{\mathbf{k}}; \quad \mathcal{A}\hbar\omega_{\mathbf{k}} \equiv \left(\left| \dot{f}_{\mathbf{k}} \right|^2 + \omega_{\mathbf{k}}^2 |f_{\mathbf{k}}|^2 \right); \quad F_{\mathbf{k}} \equiv \dot{f}_{\mathbf{k}}^2 + \omega_{\mathbf{k}}^2 f_{\mathbf{k}}^2 \tag{4.20}$$

We may diagonalize the Hamiltonian at any time $t = 0$ by imposing the condition $\dot{f}_{\mathbf{k}}(0) = -i\omega_{\mathbf{k}} f_{\mathbf{k}}(0)$, making $F_{\mathbf{k}}(0) = 0$. In Minkowski space, and with the natural time coordinate, the Hamiltonian stays diagonal at all times. The corresponding particle model in *Minkowski* space is given by

$$f_{\mathbf{k}}(t) = \sqrt{\frac{\hbar}{2\omega_{\mathbf{k}}}} e^{-i\omega_{\mathbf{k}} t}; \quad \mathcal{A} = 1 \tag{4.21}$$

which possesses a well-defined meaning of particles at all times. This is the framework of (flat space) quantum field theory implicitly assumed in textbooks.

4.1.4 Particle creation

We now consider quantum fields propagating on dynamic backgrounds. When a mode decomposition is available the (c-number) amplitude function of the kth mode obeys, from equation (4.15), the wave equation

$$\frac{d^2 f_{\mathbf{k}}}{dt^2} + \omega_{\mathbf{k}}^2(t) f_{\mathbf{k}}(t) = 0 \tag{4.22}$$

where the natural frequency $\omega_{\mathbf{k}}$ now acquires an explicit time dependence.

In Minkowski space QFT we are accustomed to the notion that positive energy solutions to the wave equation for every normal mode correspond to particles while negative energy solutions correspond to antiparticles. One can diagonalize the Hamiltonian to select a preferred particle model, e.g. the Minkowski modes (4.21). However for a time-dependent background field this notion becomes meaningless and the criterion of instantaneous diagonalization of the Hamiltonian is inviable as a particle model. This is because the mode equation (4.22) generally possesses time-dependent solutions which have no clear a priori physical meaning in terms of particles or antiparticles. The energy of individual particle/antiparticle modes is not conserved, and a consistent separation into positive and negative energy solutions of the wave equation is not always possible. This is just a reflection of the fact that physical particle number does not correspond to an operator which commutes with the Hamiltonian. We can see this point more clearly by way of the Bogoliubov transformation.

The transformation between any two Fock space bases $a_{\mathbf{k}}$ and $\tilde{a}_{\mathbf{k}}$ is known as the Bogoliubov transformation. Let the first basis $a_{\mathbf{k}}$ be associated with modes $(f_{\mathbf{k}}, f_{\mathbf{k}}^*)$, the second basis $\tilde{a}_{\mathbf{k}}$ with modes $\left(\tilde{f}_{\mathbf{k}}, \tilde{f}_{\mathbf{k}}^*\right)$. We may expand the field operators in either base, leading to equation (4.16) in the first case, and to

$$\varphi_{\mathbf{k}}(t) = \tilde{f}_{\mathbf{k}}(t)\,\tilde{a}_{\mathbf{k}} + \tilde{f}_{\mathbf{k}}^*(t)\,\tilde{a}_{-\mathbf{k}}^\dagger \tag{4.23}$$

in the second. Since both sets of solutions of the mode equations are complete, we must have

$$f_{\mathbf{k}}(t) = \alpha_{\mathbf{k}} \tilde{f}_{\mathbf{k}}(t) + \beta_{\mathbf{k}} \tilde{f}_{\mathbf{k}}^*(t) \tag{4.24}$$

and its inverse

$$\tilde{f}_{\mathbf{k}}(t) = \alpha_{\mathbf{k}}^* f_{\mathbf{k}}(t) - \beta_{\mathbf{k}} f_{\mathbf{k}}^*(t) \tag{4.25}$$

The Wronskian condition $(f_{\mathbf{k}}, f_{\mathbf{k}}^*) = \left(\tilde{f}_{\mathbf{k}}, \tilde{f}_{\mathbf{k}}^*\right) = i\hbar$ imposes a condition on the Bogoliubov coefficients

$$|\alpha_{\mathbf{k}}|^2 - |\beta_{\mathbf{k}}|^2 = 1 \tag{4.26}$$

for each $\mathbf{k}$. We can thus write

$$|\alpha_{\mathbf{k}}(t)| = \cosh r_{\mathbf{k}}(t)$$
$$|\beta_{\mathbf{k}}(t)| = \sinh r_{\mathbf{k}}(t) \tag{4.27}$$

where $r_{\mathbf{k}}(t)$ is called the squeeze parameter for mode $\mathbf{k}$, a terminology adopted from quantum optics. In Section 4.7 we will give a description of particle creation in the squeezed state language.

The linear relationship between the $\tilde{f}$'s and f's induces a corresponding transformation between $a, \tilde{a}$

$$\tilde{a}_{\mathbf{k}} = \alpha_{\mathbf{k}} a_{\mathbf{k}} + \beta_{\mathbf{k}}^* a_{-\mathbf{k}}^\dagger \tag{4.28}$$

with inverse

$$a_{\mathbf{k}} = \alpha_k^* \tilde{a}_{\mathbf{k}} - \beta_{\mathbf{k}} \tilde{a}_{-\mathbf{k}}^\dagger \tag{4.29}$$

Each particle model is associated with a particular vacuum state, in this case, $|0\rangle$ and $|\tilde{0}\rangle$, defined by

$$a_{\mathbf{k}}|0\rangle = 0 \quad \text{and} \quad \tilde{a}_{\mathbf{k}}|\tilde{0}\rangle = 0 \tag{4.30}$$

separately for all $\mathbf{k}$. Fock spaces can be constructed from the vacuum states by the action of the creation operators. One can easily see that generally $\tilde{a}_{\mathbf{k}}|0\rangle \neq 0$ because the two vacua are different by the coefficients α, β. Introducing the particle number operator $\left(\tilde{N}_{\mathbf{k}}\right)^\wedge \equiv \tilde{a}_{\mathbf{k}}^\dagger \tilde{a}_{\mathbf{k}}$ of the second particle model, we see that its expectation value with respect to the vacuum of the first model is nonzero, but equal to

$$\tilde{N}_{\mathbf{k}} = \langle 0 | \left(\tilde{N}_{\mathbf{k}}\right)^\wedge | 0 \rangle = V \, |\beta_{\mathbf{k}}|^2 \tag{4.31}$$

where V is the "volume" of space. An observer of the second particle model would say that $\tilde{N}_{\mathbf{k}}$ particles have been created from the first vacuum (from now on, we shall disregard factors of V, assuming that particle counts are always referred to a unit volume). When we think of the second particle model as defined at a time t while the first particle model is defined at the initial time t_0, we may write the particle numbers at these two times as $\langle \hat{N}_{\mathbf{k}}(t) \rangle_t$, $\langle \hat{N}_{\mathbf{k}}(t_0) \rangle_0$ respectively, i.e. $\hat{N}$ denotes a generic number operator which takes on eigenvalues N and $\tilde{N}$ in the two Fock spaces respectively. (We may at times use the notation n and $\mathcal{N}$ for these two values also.) In an S-matrix formulation of quantum field theory in a dynamical background (field or spacetime), where one assumes an asymptotic region where the background field is constant or the spacetime is static (so the modes obtained by the diagonalization of the Hamiltonian in those regions give a preferred particle model), the states of the first ($a_{\mathbf{k}}$) and second ($\tilde{a}_{\mathbf{k}}$) particle models are conventionally called the *in* and the *out* states respectively. We will use these nomenclatures interchangeably.

It is interesting to give a closed expression for the amplitude for finding $\tilde{n}$ pairs in the $|0\rangle$ state in terms of the Bogoliubov coefficients. We have

$$\langle \tilde{n}_k, \tilde{n}_{-k} | 0 \rangle = \frac{1}{\alpha_k^*} \left[\frac{\beta_k^*}{\alpha_k^*} \right]^{\tilde{n}_k} \qquad (4.32)$$

4.1.5 Adiabatic vacua

The transformation of the Fock space operators described by the Bogoliubov transformation (4.28), despite its appearance, is only a formal expression. The creation and annihilation operators do not give particle creation unless the vacuum state is well defined. We will discuss below situations where there are preferred particle models asymptotically, such as constant background fields or stationary spacetimes at $t = \pm\infty$, or conformally-invariant fields in conformally-static spacetimes without asymptotic conditions. Then the Fock spaces are well defined and one can calculate the amplitude for particle creation in a S-matrix sense. Under general conditions the particle number at any one time during the evolution is not well-defined. A straightforward intuitive generalization from flat space field theory – the so-called method of instantaneous diagonalization of the Hamiltonian – leads to severe problems; see e.g. [Ful89]. One has to appeal to other methods. If the external field (or background spacetime) does not change too rapidly (to be quantified below by the nonadiabaticity parameter) there is a conceptually clear and technically simple method which has proven to be useful in problems involving time-dependent fields (as in the external field problem) and spacetimes (as in cosmological particle creation). It is the nth order *adiabatic vacuum or number state*, and, when applied to the removal of ultraviolet divergences in the current or energy–momentum tensor, it is called *adiabatic regularization*. A selection of influential papers on this subject is [Park66, Park69, ParFul74, FulPar74, Park76, Park77].

Both the time-dependence of the Fock space operators and the evolution of the amplitude functions are dictated by the wave equations for the normal modes of the quantum scalar field with time-dependent natural frequency ω_k as in equation (4.22). To single out a solution, we need to specify initial data for $f_{\mathbf{k}}$ and $df_{\mathbf{k}}/dt$ at some time t_0. When ω_k is constant one can use the same Fock space representation of the field theory as it remains the same as originally defined at t_0. Staticity means that the dynamics is invariant in time, and implies the existence of a Killing vector in time ∂_t, which enforces the positive and negative frequency components to remain separated. This means, in second quantized language, that the particles and antiparticles are separately well-defined and their number remains a constant. Therefore the possibility of defining a positive frequency component in a field theory is the precondition for a vacuum state to exist. We learned that maintaining such a condition in the evolution is not always possible.

If the external field or background spacetime changes gradually one can extend this idea and define an adiabatic vacuum or number state. Recall from elementary wave or quantum theory that a WKB solution can give a reasonable approximation to the wave equation when the system changes gradually enough. Successively higher order WKB (or adiabatic) solutions can encompass more rapid changes in the background field as they show up in the natural frequency function. This is the lead idea behind the adiabatic method.

The sequence of successively higher order WKB solutions to this wave equation has been explored quite extensively by researchers working on wave propagation in inhomogeneous media. There, the reflection of waves due to successively higher order derivatives in the dielectric media can be treated with successively higher order WKB solutions. Translating the variation in spatial homogeneity to time dependence is a physically intuitive way to understanding the adiabatic vacuum. This route explored by Hu [Hu72, Hu74] gives the same result as that established first by Parker and Fulling. It was also shown to be equivalent to the result obtained by Zeldovich and Starobinsky [ZelSta71, FuPaHu74] in their "n-wave regularization."

Consider the wave equation (4.22) in t time for the amplitude function of the kth mode. (We shall omit the **k** subscript, as only one mode is being considered.) The idea is to use a transformation of both time t and dependent variable f to reduce this equation to one we can solve.

Define a new time variable $t_1 = t_1(t)$, and write equation (4.22) as

$$\left(\frac{dt_1}{dt}\right)^2 \frac{d^2 f}{dt_1^2} + \left(\frac{d^2 t_1}{dt^2}\right)\frac{df}{dt_1} + \omega^2 f = 0 \qquad (4.33)$$

The equation is simplified by choosing

$$\frac{dt_1}{dt} = \omega(t) \qquad (4.34)$$

whereby

$$\frac{d^2 f}{dt_1^2} + \frac{1}{\omega}\left(\frac{d\omega}{dt_1}\right)\frac{df}{dt_1} + f = 0 \qquad (4.35)$$

The first-order term is eliminated by writing

$$f = \omega^{-1/2} f_1 \qquad (4.36)$$

obtaining

$$\frac{d^2}{dt_1^2} f_1 + w_1^2 f_1 = 0 \qquad (4.37)$$

where

$$w_1^2 = 1 + \epsilon_2, \quad \epsilon_2 = -\frac{1}{\omega^{1/2}}\frac{d^2}{dt_1^2}(\omega^{1/2}) \tag{4.38}$$

Observe that equation (4.37) has the same structure as the original equation (4.22). If ω varies sufficiently slowly, we can neglect ϵ_2, and it becomes trivial.

Higher order WKB approximations to the wave equation are obtained by iterating this procedure. Define (note r here is an adiabatic order parameter, not the squeeze parameter introduced earlier)

$$dt_r \equiv w_{r-1}dt_{r-1} \equiv W_r dt \quad (w_0 \equiv \omega, t_0 \equiv t) \tag{4.39}$$

$$f_r \equiv w_{r-1}^{1/2}f_{r-1} = W_r^{1/2}f \tag{4.40}$$

$$W_r \equiv w_0 w_1 \cdots w_{r-1} \tag{4.41}$$

$$\Theta_r \equiv \int W_r dt \tag{4.42}$$

The $n(= 2r)$th-order WKB equation is given by $(r = 1, 2, \ldots)$

$$\frac{d^2}{dt_r^2}f_r + w_r^2 f_r = 0 \tag{4.43}$$

where, for $r = 1, 2, 3 \ldots$,

$$w_r^2 = 1 + \epsilon_{2r}, \quad \epsilon_{2r} = -\frac{1}{w_{r-1}^{1/2}}\frac{d^2}{dt_r^2}(w_{r-1}^{1/2}) \tag{4.44}$$

The quantities ϵ_{2r} are called the *adiabatic frequency corrections* [FuPaHu74]. If $|\epsilon_{2r}| \ll 1$, the solution of the wave equation correct up to the $n(= 2r)$th order of derivatives of the natural frequency $w^2(t)$ with respect to t_r is given by

$$f_{(n)}(t) = \frac{\hbar^{1/2}}{(2W_r)^{1/2}}\left[Ae^{-i\int W_r dt} + Be^{i\int W_r dt}\right] \tag{4.45}$$

where A, B are complex functions. The subscript (n) on f indicates that a solution to the full wave equation is sought *including up to* the nth adiabatic order. In contradistinction, we define a $n(= 2r)$th-order adiabatic solution as the solution with ϵ_{2r} set equal to zero.

The *nth-order adiabatic vacuum* is defined such that there is no negative frequency component in the nth-order WKB solution. What this means is that, at the nth adiabatic order approximation the nth-order adiabatic number state is obtained by assuming that the wavefunction $f(t)$ is given only by the positive frequency nth-order WKB solution

$$f(t) \simeq f_{(n)}^+(t) = \frac{Ae^{-i\int^t W_{n/2}dt}}{\sqrt{2W_{n/2}}} \tag{4.46}$$

So intrinsically this is a quasi-local (in time) expansion counting time derivative orders, which can be translated to frequency ranges. In terms of what adiabatic order will encompass what range of frequencies we shall see how this method becomes useful for identifying and isolating ultraviolet divergences in quantum field theory in dynamical spacetimes, as in cosmology. This method, known as adiabatic regularization, will be discussed in a later section.

4.1.6 Hamiltonian mean field dynamics and general Gaussian ansatz

Let us broaden our scope somewhat to introduce an important class of approximations in quantum field theory which shares the same dynamics as the problem under discussion so far. This is the mean field (or Gaussian) approximation. Mean field methods have a long history in such diverse areas as atomic physics (Born–Oppenheimer), nuclear physics (Hartree–Fock), condensed matter (BCS) and statistical physics (Landau–Ginzburg), quantum optics (coherent/squeezed states), and semiclassical gravity. Because no higher than second moments of the fluctuations are incorporated, the mean field approximation is related to a Gaussian variational ansatz for the wavefunction of the system.

For the mixed state density matrix ρ Habib *et al.* [HKMP96] have shown that the time-dependent mean field approximation is equivalent to the general Gaussian ansatz. It is instructive to follow the exposition of this feature.

As a matter of principle, the Hamiltonian nature of the evolution makes it clear from the outset that the mean field approximation does *not* introduce dissipation or time irreversibility at a fundamental level. Any such behavior must come from some assumption in coarse graining some information of this closed system away. We shall remark on this aspect at the end of this section and further in Chapter 9 on entropy generation.

Consider again the one-dimensional harmonic oscillator with Hamiltonian

$$H_{osc}(q, p; t) = \frac{1}{2} \left(p^2 + \omega^2(t) \, q^2 \right) \tag{4.47}$$

where $\omega(t)$ is the natural frequency. The most general Gaussian ansatz for the mixed state normalized density matrix is

$$\langle x' | \rho | x \rangle = (2\pi\xi^2)^{-\frac{1}{2}} \exp \left\{ i \frac{\bar{p}}{\hbar} (x' - x) - \frac{\zeta^2 + 1}{8\xi^2} \left[(x' - \bar{q})^2 + (x - \bar{q})^2 \right] \right.$$
$$\left. + i \frac{\eta}{2\hbar\xi} \left[(x' - \bar{q})^2 - (x - \bar{q})^2 \right] + \frac{\zeta^2 - 1}{4\xi^2} (x' - \bar{q})(x - \bar{q}) \right\} \tag{4.48}$$

in the coordinate representation. The five parameters $(\bar{q}, \bar{p}, \xi, \eta, \zeta)$ of this Gaussian may be identified with the two mean values, $\bar{q} = \langle q \rangle \equiv \text{Tr}(\rho q)$, $\bar{p} = \langle p \rangle \equiv \text{Tr}(\rho p)$, and the three symmetrized variances via

$$\langle (q - \bar{q})^2 \rangle = \xi^2, \qquad \langle (pq + qp - 2\bar{q}\bar{p}) \rangle = 2\xi\eta$$
$$\langle (p - \bar{p})^2 \rangle = \eta^2 + \frac{\hbar^2 \zeta^2}{4\xi^2} \tag{4.49}$$

The one antisymmetrized variance is fixed by the commutation relation, $[q, p] = i\hbar$. The parameter ζ measures the degree to which the state is mixed: $\mathrm{Tr}\,\rho^2 = \zeta^{-1} \leq 1$, with unity for pure states. If the state is pure, $\rho = |\psi\rangle\langle\psi|$, and only two of the three symmetrized variances in (4.49) are independent.

The Gaussian ansatz for the density matrix is preserved under time evolution. In the Schrödinger picture ρ evolves according to the Liouville equation, $\dot{\rho} = -i[H, \rho]$. Substitution of the Gaussian form (4.48) into this equation with Hamiltonian (4.47) and equating coefficients of x, x', x^2, x'^2 and xx' gives five evolution equations for the five parameters specifying the Gaussian,

$$\begin{aligned}
\bar{q}_{,t} &= \bar{p} \; ; & \bar{p}_{,t} &= -\omega^2(t)\bar{q} \\
\xi_{,t} &= \eta \; ; & \eta_{,t} &= -\omega^2(t)\xi + \tfrac{\hbar^2\zeta^2}{4\xi^3}
\end{aligned} \tag{4.50}$$

and $\dot{\zeta} = 0$. Since ζ is a constant and the von Neumann entropy $-\mathrm{Tr}\,\rho\ln\rho$ of the state (4.48) is a (monotonic) function of ζ alone, this quantity is also a constant of the motion. This establishes the equivalence between mean field methods and Gaussian density matrices for all evolutions of the form of equations (4.50).

An essential property of the evolution equations (4.50) is that they are Hamilton's equations (hence, time reversible) for an effective classical Hamiltonian [RajMar82], with η playing the role of the momentum conjugate to ξ,

$$H_{\text{eff}}(\bar{q}, \bar{p}; \xi, \eta) = \mathrm{Tr}(\rho H) = \frac{1}{2}\left(\bar{p}^2 + \eta^2\right) + V_{\text{eff}} \tag{4.51}$$

and $V_{\text{eff}}(\bar{q}, \xi)$ depending on the particular form of $\omega^2(\bar{q}(t), \xi(t); t)$.

The unitary time evolution operator $U(t)$ for the density matrix (4.48),

$$\rho(t) = U(t)\rho(0)U^\dagger(t) \,, \quad U(t) = \exp\left(-i\hbar^{-1}\int_0^t H\,dt\right) \tag{4.52}$$

is given explicitly in the coordinate basis by

$$\langle x'|U(t)|x\rangle = (2\pi i\hbar v(t))^{-\frac{1}{2}}\exp\left\{\frac{i}{2\hbar v(t)}\left(u(t)x^2 + \dot{v}(t)x'^2 - 2xx'\right)\right\} \tag{4.53}$$

in terms of the two linearly independent solutions to the classical evolution equation,

$$\left(\frac{d^2}{dt^2} + \omega^2(t)\right)\begin{pmatrix} u \\ v \end{pmatrix} = 0 \; ; \qquad \begin{aligned} u(0) &= \dot{v}(0) = 1 \\ \dot{u}(0) &= v(0) = 0 \end{aligned} \tag{4.54}$$

The Gaussian dynamics may be expressed as well by means of a Fock representation of the time-dependent Heisenberg operators,

$$\begin{aligned}
q(t) &= U^\dagger(t)\,q(0)\,U(t) = \bar{q}(t) + af(t) + a^\dagger f^*(t) \\
p(t) &= U^\dagger(t)\,p(0)\,U(t) = \bar{p}(t) + a\dot{f}(t) + a^\dagger \dot{f}^*(t)
\end{aligned} \tag{4.55}$$

where $[a, a^\dagger] = 1$. The complex mode functions f satisfy the evolution equation (4.54) and the Wronskian condition (4.17). This shows that Gaussian time evolution is essentially classical, with $\hbar$ appearing only in the time-independent condition (4.17) enforcing the quantum uncertainty relation.

Time-dependent basis

One can choose a basis in which all expectation values vanish, except

$$\langle a^\dagger a \rangle = \langle aa^\dagger \rangle - 1 \equiv N \geq 0 \tag{4.56}$$

The Gaussian density matrix is diagonal in the corresponding $a^\dagger a$ time-independent number basis,

$$\langle n'|\rho|n \rangle = \frac{2\delta_{n'n}}{\zeta + 1} \left(\frac{\zeta - 1}{\zeta + 1} \right)^n \tag{4.57}$$

with $\zeta = 2N + 1 = \mathcal{A}$ (the parametric amplification factor as introduced in equation (4.20)) and $\xi^2(t) = \zeta|f(t)|^2$. Upon identifying $\zeta = \coth(\hbar\omega/2k_\mathrm{B}T)$, the diagonal form (4.57) will be recognized as a thermal density matrix at temperature T. The pure state Gaussian wavefunction ($\zeta = 1$) corresponds therefore to a coherent, squeezed zero temperature vacuum state. The smoothness of the finite temperature classical limit $\hbar\zeta \to 2k_\mathrm{B}T/\omega$ as $\hbar \to 0$, $\zeta \to \infty$ shows that quantum and thermal fluctuations are treated by the mean field approximation in a unified way.

Instantaneous diagonalization

It is always possible to diagonalize (4.47) at any given time, bringing the quadratic Hamiltonian into the standard harmonic oscillator form, $H_\mathrm{osc} = \frac{\hbar\omega}{2}\left(\tilde{a}\tilde{a}^\dagger + \tilde{a}^\dagger\tilde{a}\right)$ with $\tilde{a}$ time dependent. This time-dependent basis is defined by the relations,

$$q(t) = \tilde{a}\tilde{f} + \tilde{a}^\dagger \tilde{f}^*, \qquad p(t) = -i\omega\tilde{a}\tilde{f} + i\omega\tilde{a}^\dagger \tilde{f}^*$$

$$\tilde{f}(t) = \sqrt{\frac{\hbar}{2\omega(t)}} \exp\left(-i\int_0^t dt'\,\omega(t')\right) \tag{4.58}$$

in place of (4.55). In the $\tilde{a}^\dagger\tilde{a}$ number basis, ρ is no longer diagonal, $\langle \tilde{a} \rangle$, $\langle \tilde{a}\tilde{a} \rangle$, etc. are nonvanishing, and $\tilde{N} \equiv \langle \tilde{a}^\dagger\tilde{a} \rangle \neq N$ in general, becoming equal only in the static case of constant ω. As cautioned by Fulling [Ful89] this is the incorrect way to establish a quantum field theory in dynamical backgrounds.

Adiabatic basis

If $\omega(t)$ varies slowly in time, an adiabatic invariant may be constructed from the Hamilton–Jacobi equation corresponding to the effective classical Hamiltonian (4.51). By a simple quadrature we find the adiabatic invariant,

$$\frac{W}{2\pi\hbar} = \frac{\langle H \rangle}{\hbar\omega} - \frac{\zeta}{2} = \tilde{N}(t) - N \tag{4.59}$$

Since N is time independent, $\tilde{N}(t)$ is an adiabatic invariant of the evolution. On the other hand, the phase angle conjugate to the action variable W varies rapidly in time. Since the diagonal matrix elements of ρ in the $\tilde{N}$ basis are independent of this phase angle, they are slowly varying, whereas the *off-diagonal* matrix elements of ρ in this basis (which depend on the phase angle) are *rapidly* varying functions of time. If we are interested only in the effects of the fluctuations on the more slowly varying mean fields it is natural to define an *effective* density matrix $\rho_{\text{eff}}(t)$ by *time-averaging* the density matrix (4.48), thereby truncating ρ to its diagonal elements only, in the adiabatic $\tilde{N}$ basis [HuPav86, Kan88a, Kan88b]. Clearly, for this truncation to be justified there must be very efficient phase cancellation, *i.e. dephasing*, either by averaging the fluctuations over time or by summing over many independent fluctuating degrees of freedom at a fixed time. This is perhaps the most direct way to understand the decoherence of the mean field. We shall discuss this issue in Chapter 8.

4.2 Particle production in external fields

After the above simple introduction we can begin to explore two classes of problems involving quantum fields in dynamical backgrounds. In this section we study the production of charged scalar particles in an external field, relevant to problems of collective excitations in QED plasma (and by extension to QCD quark–gluon processes). There are good introductions to this topic in standard texts, such as [ItzZub80]. In the next section we study a neutral scalar field in a dynamical spacetime, applicable to cosmological problems, such as vacuum particle creation at the Planck time or reheating after GUT (Grand-Unified Theory) scale inflationary expansion. Both problems have been studied extensively; the former began with the works of Klein [Kle29], Sauter [Sau31, Sau32], Heisenberg and Euler [HeiEul36], Schwinger [Sch51], and others [Greiner, GrMaMo88, FrGiSh91, Ginz87, Ginz95]; the latter by Parker, Sexl and Urbantke, Zel'dovich and Starobinsky, Fulling, Hu, and many others. For later and current developments, see [DeW75, BirDav82, Bordag]. For the first part in this section our treatment follows the work of Kluger, Mottola and Eisenberg [KlMoEi98]. For the second part in the next section, we follow the approach of Zel'dovich, Starobinsky [ZelSta71] and Hu [Hu72, Hu74, FuPaHu74].

Assuming that the electric field is spatially homogeneous, in the Coulomb gauge, we can express the vector potential as

$$\mathbf{A} = A(t)\hat{\mathbf{z}}, \ A_0 = 0 \tag{4.60}$$

and the electric field as

$$\mathbf{E} = -\dot{A}\hat{\mathbf{z}} = E\hat{\mathbf{z}} \tag{4.61}$$

Assuming also the field lives in a finite large volume V we can expand the charged scalar field operator in Fock space in Fourier modes. Since particles are physically distinct from antiparticles, we need two independent sets of destruction operators

$$\Phi(\mathbf{x}, t) = \frac{1}{\sqrt{V}} \sum_{\mathbf{k}} e^{i\mathbf{k}\cdot\mathbf{x}} \varphi_{\mathbf{k}}(t) = \frac{1}{\sqrt{V}} \sum_{\mathbf{k}} \left\{ e^{i\mathbf{k}\cdot\mathbf{x}} f_{\mathbf{k}}(t) a_{\mathbf{k}} + e^{-i\mathbf{k}\cdot\mathbf{x}} f^*_{-\mathbf{k}}(t) b^{\dagger}_{\mathbf{k}} \right\}$$

(4.62)

Denote the time-independent annihilation operator of a particle in mode $\mathbf{k}$ by $a_{\mathbf{k}}$ and the creation of an antiparticle in mode $-\mathbf{k}$ by $b^{\dagger}_{\mathbf{k}}$. They obey the commutation relations

$$[a_{\mathbf{k}}, a^{\dagger}_{\mathbf{k}'}] = [b_{\mathbf{k}}, b^{\dagger}_{\mathbf{k}'}] = \delta_{\mathbf{k}\mathbf{k}'}$$

(4.63)

Therefore

$$N_+(\mathbf{k}) \equiv \langle a^{\dagger}_{\mathbf{k}} a_{\mathbf{k}} \rangle$$
$$N_-(\mathbf{k}) \equiv \langle b^{\dagger}_{\mathbf{k}} b_{\mathbf{k}} \rangle$$

(4.64)

are the mean numbers of particles and antiparticles respectively. Without loss of generality we can make use of the freedom in defining the initial phases of the mode functions to set the correlation densities $\langle a_{\mathbf{k}} a_{\mathbf{k}} \rangle = \langle b_{\mathbf{k}} b_{\mathbf{k}} \rangle = 0$. In a Hamiltonian description we can take for each mode $\mathbf{k}$

$$\varphi_{\mathbf{k}}(t) \equiv f_{\mathbf{k}}(t) a_{\mathbf{k}} + f^*_{\mathbf{k}}(t) b^{\dagger}_{-\mathbf{k}}$$

(4.65)

as the (complex) generalized coordinates of the field Φ and

$$\pi_{\mathbf{k}}(t) = \dot{\varphi}^{\dagger}_{\mathbf{k}}(t) = \dot{f}^*_{\mathbf{k}}(t) a^{\dagger}_{\mathbf{k}} + \dot{f}_{\mathbf{k}}(t) b_{-\mathbf{k}}$$

(4.66)

as the momentum canonically conjugate to it. By virtue of the commutation relation (4.63) they obey the canonical commutation relation,

$$[\varphi_{\mathbf{k}}, \pi_{\mathbf{k}'}] = i\hbar \delta_{\mathbf{k}\mathbf{k}'}$$

(4.67)

provided that the mode functions satisfy the Wronskian condition (4.17).

The complex amplitude function $f_{\mathbf{k}}(t)$ of the kth mode satisfies the equations of motion (4.22), where the time-dependent frequency $\omega^2_{\mathbf{k}}(t)$ is given by

$$\omega^2_{\mathbf{k}}(t) = (\mathbf{k} - e\mathbf{A})^2 + m^2 = (k_z - eA(t))^2 + k^2_{\perp} + m^2$$

(4.68)

where k_z is the constant canonical momentum in the $\hat{\mathbf{z}}$ direction while the physical (gauge-invariant) kinetic momentum is given by

$$p_z(t) = k_z - eA(t); \dot{p}_z = -e\dot{A} = eE$$

(4.69)

(In the directions transverse to the electric field the kinetic and canonical momenta are the same: $p_{\perp} = k_{\perp}$.) Any function of the kinetic momenta contains these two components, e.g. $\omega(p_z, p_{\perp}) = \sqrt{p^2_z + p^2_{\perp} + m^2}$.

Since the definition of particle number becomes very different from that conceived in QFT in Minkowski space, especially in arbitrarily strong and rapidly time-varying fields, it is often easier to deal with the conserved physical currents like $j(t)$ in an external field problem (or the stress–energy tensor $T_{\mu\nu}(x)$ in curved spacetimes). For a spatially homogeneous electric field (i.e. $\nabla \cdot \mathbf{E} = 0$), by Gauss' law, the mean charge density must vanish,

$$j^0(t) = e \int d^3\mathbf{k}\, [N_+(\mathbf{k}) - N_-(-\mathbf{k})] = 0 \qquad (4.70)$$

The mean current in the $\hat{\mathbf{z}}$ direction is

$$j(t) = 2e \int d^3\mathbf{k}\, [k_z - eA(t)]|f_{\mathbf{k}}(t)|^2 (1 + N_+(\mathbf{k}) + N_-(-\mathbf{k})) \qquad (4.71)$$

One can further restrict to the subspace of states for which

$$N_+(\mathbf{k}) = N_-(-\mathbf{k}) \equiv N_{\mathbf{k}} \qquad (4.72)$$

Clearly the vacuum $N_+(\mathbf{k}) = N_-(-\mathbf{k}) = 0$ (as well as a thermal state) belongs to this class of states.

Particle pairs will be produced in a strong background field, and in turn, affect the strength and evolution of this background field. At the first level of sophistication (simplification), one can assume the background field (electric field or spacetime) is fixed in what is called a "test field" approximation (language also used in QFT in curved spacetime). At the second level, one looks for a self-consistent solution of the mean electric field $\mathbf{E}(t)$ (or the classical background spacetime) coupled to the expectation value of the current $j(t)$ of the quantum charged scalar field (or, in the case of cosmology, the energy–momentum tensor of the quantized matter field). This is known as the dynamical back-reaction problem. For the creation of charged particles in a homogeneous electric field, the back-reaction problem involves solving for the current $j(t)$ from the charge field $\varphi_{\mathbf{k}}(t)$, and using it as source in the Maxwell equation for the vector potential $\mathbf{A}$. In a spatially homogeneous electric field, the only nontrivial Maxwell equation is simply

$$-\dot{E}(t) = \ddot{A}(t) = j(t) \qquad (4.73)$$

where the current is given by (4.71). Since the charged scalar field depends on the vector potential A to begin with, $f_{\mathbf{k}}(t)$ and $A(t)$ need to be solved self-consistently from equations (4.22) with (4.68) and (4.73).

4.2.1 Particle creation in a constant electric field

As a concrete example of particle creation in strong fields, let us review the well-known case of a uniform time-independent electric field worked out by Schwinger [Sch51]. There is a very detailed treatment of this problem in [KlMoEi98]. We may take E to be along the z direction with $A(t) = -Et$. The wave equation for the amplitude function of the kth mode can be written in terms of a new time τ (we omit the mode subscripts, since only one mode is considered):

$$\frac{d^2 f}{d\tau^2} + \omega^2(\tau)f = 0 , \quad \omega^2(\tau) = \nu_0^2 + \nu_1^4 \tau^2 \qquad (4.74)$$

where

$$\tau = t + \frac{k_z}{eE}, \quad \nu_0^2 = k_\perp^2 + m^2, \quad \nu_1^4 = e^2 E^2 \tag{4.75}$$

We are interested in the strong-field case $\nu_1^2 \geq \nu_0^2$. It is obvious that the natural frequency is never constant. However, the second-order adiabatic frequency correction (4.38)

$$\epsilon_2 = -\frac{1}{\omega^{\frac{1}{2}}} \frac{d^2}{d\tau_1^2} (\omega^{\frac{1}{2}}) = \frac{\nu_1^4}{[\nu_0^2 + \nu_1^4 \tau^2]^2} \left\{ \frac{3}{4} - \frac{5}{4} \frac{\nu_0^2}{[\nu_0^2 + \nu_1^4 \tau^2]} \right\} \tag{4.76}$$

is small provided $|\tau| \gg \nu_1^{-1} \geq \nu_0/\nu_1^2$. Therefore, for this problem, the zeroth-order adiabatic vacua already can provide a consistent particle definition both in the distant past and future.

Let us then consider the nth adiabatic order positive frequency solution $f_{(n)}^+$ in equation (4.46) with $n = 0$ and $W_0 = \omega$. To be precise, we adopt the convention that the WKB exponent (adiabatic phase) takes on the values

$$\Theta(\tau) \equiv \int_0^\tau \omega(\tau') \, d\tau' \ (\tau \geq 0) \quad \text{and} \quad \Theta(\tau) \equiv -\Theta(-\tau) \ (\tau < 0) \tag{4.77}$$

Computing the integral, we get the parametric form

$$\Theta_0 = \frac{\nu_0^2}{2\nu_1^2} [u + \sinh u \cosh u], \quad \tau = \frac{\nu_0}{\nu_1^2} \sinh u \tag{4.78}$$

For large τ,

$$u \sim \ln \left[\frac{2\nu_1^2 \tau}{\nu_0} \right] + O(\tau^{-2}) \tag{4.79}$$

$$\Theta = \frac{\nu_1^2}{2} \tau^2 + \frac{\nu_0^2}{2\nu_1^2} \ln \left[\frac{2\nu_1^2 \tau}{\nu_0} \right] + \frac{\nu_0^2}{4\nu_1^2} + O(\tau^{-2}) \tag{4.80}$$

$$\omega \sim \nu_1^2 \tau + O(\tau^{-1}) \tag{4.81}$$

Using equation (4.77) we obtain the corresponding form for $\tau \to -\infty$,

$$\Theta(\tau) \sim \frac{-\nu_1^2}{2} \tau^2 - \frac{\nu_0^2}{2\nu_1^2} \ln \left[\frac{2\nu_1^2 |\tau|}{\nu_0} \right] - \frac{\nu_0^2}{4\nu_1^2} + O(\tau^{-2}) \tag{4.82}$$

$$\omega \sim \nu_1^2 |\tau| + O(\tau^{-1}) \tag{4.83}$$

The asymptotic behavior of the WKB-approximate positive frequency mode function f^+ is

$$f^+(\tau) \sim \frac{\hbar^{1/2}}{\nu_1} \frac{1}{\sqrt{2|\tau|}} \left[\frac{2\nu_1^2 |\tau|}{\nu_0} \right]^{i\nu_0^2/2\nu_1^2} \exp\left\{ \frac{i\nu_1^2}{2} \tau^2 + \frac{i\nu_0^2}{4\nu_1^2} \right\} \ (\tau \to -\infty) \tag{4.84}$$

We define the positive frequency mode associated to the *in* vacuum as the exact solution f_{in} of equation (4.74) which matches this behavior in the distant past.

Similarly, for $\tau \to \infty$,

$$f^+ (\tau) \sim \frac{\hbar^{1/2}}{\nu_1} \frac{1}{\sqrt{2\tau}} \left[\frac{2\nu_1^2 \tau}{\nu_0} \right]^{-i\nu_0^2/2\nu_1^2} \exp\left\{ \frac{-i\nu_1^2}{2}\tau^2 - \frac{i\nu_0^2}{4\nu_1^2} \right\} \quad (\tau \to \infty) \quad (4.85)$$

and we define the positive frequency mode associated with the *out* vacuum as the exact solution f_{out} of equation (4.74) which matches this behavior in the distant future. The whole point of the analysis is that $f_{in} \neq f_{out}$.

A basis of solutions of equation (4.74) is given by the parabolic cylinder functions $D_p(z)$ and its conjugate $D_{p^*}(z^*)$, where $z = (-1 + i)\nu_1\tau$ and $p = \left(i(\nu_0/\nu_1)^2 - 1 \right)/2$. When $\tau \to -\infty$, $z \sim \sqrt{2}\nu_1 |\tau| e^{-i\pi/4}$, and

$$D_p(z) \sim z^p e^{-z^2/4} = \left(\sqrt{2}\nu_1 |\tau| \right)^{-1/2} \left(\sqrt{2}\nu_1 |\tau| \right)^{i\nu_0^2/2\nu_1^2}$$

$$\times e^{i\pi/8} e^{\pi(\nu_0/\nu_1)^2/8} \exp\left\{ \frac{i}{2} (\nu_1\tau)^2 \right\} \quad (4.86)$$

$(\tau \to -\infty)$. Comparing with the corresponding expansion of f^+, equation (4.84), we find that the normalized mode function associated with the *in* vacuum is

$$f_{in} = \frac{\hbar^{1/2}}{\sqrt{\sqrt{2}\nu_1}} \left[\frac{\sqrt{2}\nu_1}{\nu_0} \right]^{i\nu_0^2/2\nu_1^2} e^{-i\pi/8} e^{-\pi(\nu_0/\nu_1)^2/8} \exp\left\{ \frac{i\nu_0^2}{4\nu_1^2} \right\} D_p(z) \quad (4.87)$$

When $\tau \to \infty$, $z \sim \sqrt{2}\nu_1\tau\, e^{3i\pi/4}$, and

$$D_p(z) \sim z^p e^{-z^2/4} - \frac{\sqrt{2\pi}}{\Gamma[-p]} e^{i\pi p} z^{-p-1} e^{z^2/4}$$

$$= \left(\sqrt{2}\nu_1\tau \right)^{-1/2} e^{-3i\pi/8} \left\{ \left(\sqrt{2}\nu_1\tau \right)^{i\nu_0^2/2\nu_1^2} e^{-3\pi(\nu_0/\nu_1)^2/8} \exp\left\{ \frac{i}{2}(\nu_1\tau)^2 \right\} \right.$$

$$\left. - \frac{\sqrt{2\pi}}{\Gamma[-p]} e^{-i\pi/2} \left(\sqrt{2}\nu_1\tau \right)^{-i\nu_0^2/2\nu_1^2} e^{-\pi(\nu_0/\nu_1)^2/8} \exp\left\{ -\frac{i}{2}(\nu_1\tau)^2 \right\} \right\}$$

$$(4.88)$$

$(\tau \to \infty)$. Substituting this into equation (4.87) and comparing with the development equation (4.85) we find that f_{in} and f_{out} are related in a way given exactly by the Bogoliubov transformation

$$f_{in} = \alpha\, f_{out} + \beta\, f_{out}^* = \alpha\, f^+ + \beta\, f^- \quad (\tau \to \infty) \quad (4.89)$$

where f^- is the corresponding negative frequency solution of the same adiabatic order [the term with coefficient B in (4.45)]. Note that the first identity actually holds everywhere. Hence we can identify the Bogoliubov coefficients as

$$\alpha = \left[\frac{\sqrt{2}\nu_1}{\nu_0} \right]^{i\nu_0^2/\nu_1^2} \exp\left\{ \frac{i\nu_0^2}{2\nu_1^2} \right\} \left(\frac{\sqrt{2\pi}}{\Gamma[-p]} \right) e^{-\pi(\nu_0/\nu_1)^2/4} \quad (4.90)$$

and

$$\beta = e^{-i\pi/2} e^{-\pi(\nu_0/\nu_1)^2/2} \quad (4.91)$$

As a check, observe that

$$|\alpha|^2 = 2\cosh\left[\frac{\pi}{2}\left(\frac{\nu_0}{\nu_1}\right)^2\right]\exp\left[\frac{-\pi}{2}\left(\frac{\nu_0}{\nu_1}\right)^2\right] \tag{4.92}$$

$$|\beta|^2 = \exp\left[-\pi\left(\frac{\nu_0}{\nu_1}\right)^2\right] \tag{4.93}$$

obeys the Wronskian condition $|\alpha|^2 - |\beta|^2 = 1$.

It is clear that if we set up the quantum state to be the *in* vacuum, when we arrive at the *out* region we find

$$N = |\beta|^2 = \exp\left\{-\pi\left(\frac{k_\perp^2 + m^2}{eE}\right)\right\} \tag{4.94}$$

particles in each mode. This is Schwinger's celebrated result [Sch51].

4.3 Spontaneous and stimulated production

So far we have focused on how to define a physically meaningful vacuum state and the number of particles produced in a changing external field or dynamical space-time. We learn how to define adiabatic vacuum states in a dynamical setting, via the adiabatic expansion. The nth order adiabatic number state is well-defined to the nth adiabatic order. In this section we will show how to derive the energy density of these particles produced in adiabatic orders. A related problem is the identification and subtraction of ultraviolet divergences in the stress–energy tensor of quantum fields in a dynamical background. Here we will explain how to apply the adiabatic method introduced above in what is called the adiabatic regularization scheme.

We begin with a formal rendition to the parametric oscillator equation (4.22) describing the amplitude function of the kth normal mode. We want an expression of $s_k \equiv |\beta_k|^2$ in terms of $|f_k|$ and $|\dot{f}_k|$. Here following [ZelSta71, Hu74] we seek a solution in the form:

$$f_k(t) = \sqrt{\frac{\hbar}{2\omega_k}}\left\{\alpha_k e_k^- + \beta_k e_k^+\right\}; \quad e_k^\pm \equiv \exp\left\{\pm i\int\omega_k dt\right\} \tag{4.95}$$

The two functions α_k, β_k are the positive and negative frequency components of a formal solution f_k, but without a well-defined vacuum they do not convey the meaning of particles and antiparticles, as we forewarned with regard to the Bogoliubov coefficients. Since the single equation (4.95) does not determine the coefficients α_k and β_k uniquely, we need another condition, which is chosen so that the Wronskian condition equation (4.17) is satisfied. The auxiliary condition imposed on $\dot{f}_k$ is

$$\dot{f}_k(t) = -i\sqrt{\frac{\hbar\omega_k}{2}}\left(\alpha_k e_k^- - \beta_k e_k^+\right) \tag{4.96}$$

Inverting these two equations we can express the complex function $\beta_{\mathbf{k}}$ in terms of $|f_{\mathbf{k}}|^2$, $|\dot{f}_{\mathbf{k}}|^2$ as follows:

$$\alpha_{\mathbf{k}} = \sqrt{\frac{\omega_{\mathbf{k}}}{2\hbar}} \left(f_{\mathbf{k}} + \frac{i}{\omega_{\mathbf{k}}} \dot{f}_{\mathbf{k}} \right) e_{\mathbf{k}}^+, \qquad \beta_{\mathbf{k}} = \sqrt{\frac{\omega_{\mathbf{k}}}{2\hbar}} \left(f_{\mathbf{k}} - \frac{i}{\omega_{\mathbf{k}}} \dot{f}_{\mathbf{k}} \right) e_{\mathbf{k}}^- \qquad (4.97)$$

Making use of the Wronskian condition we obtain

$$s_{\mathbf{k}} \equiv |\beta_{\mathbf{k}}|^2 = \frac{1}{2\hbar\omega_{\mathbf{k}}} \left(\left| \dot{f}_{\mathbf{k}} \right|^2 + \omega_{\mathbf{k}}^2 |f_{\mathbf{k}}|^2 \right) - \frac{1}{2} \qquad (4.98)$$

It is tempting to regard $s_{\mathbf{k}} = |\beta_{\mathbf{k}}|^2$ as the amount of particle production. However, we need to be careful that the vacuum state is well-defined to make sense of particles. To which adiabatic order one needs to carry out the expansion is determined by the physical conditions (foremost how rapidly the natural frequency changes) of the system and by the accuracy demanded in its description. For slowly varying fields if one is interested in problems concerning the adiabatic particle number or mean current distribution as used in quantum kinetic theory [GrLeWe80] (low particle creation rate and minimal phase information) the adiabatic number state of [KlMoEi98] to be introduced in a later section, which is in the lowest adiabatic order, will suffice.

4.3.1 Spontaneous production

The energy–momentum tensor of a massive scalar field in flat space is

$$T_{\mu\nu}^{\text{Mink}} = \nabla_\mu \phi \nabla_\nu \phi - \frac{1}{2}\eta_{\mu\nu}\nabla^\rho \phi \nabla_\rho \phi - \frac{1}{2}\eta_{\mu\nu}m^2\phi^2 \qquad (4.99)$$

The energy density associated with these particles is given by the expectation value of the 00 component of $T_{\mu\nu}$ with respect to the Minkowski vacuum, i.e.

$$\rho_0^{\text{Mink}} \equiv \langle 0 \mid T_{00} \mid 0 \rangle = \int \frac{d^3\mathbf{k}}{2(2\pi)^3}(|\dot{f}_{\mathbf{k}}|^2 + \omega_{\mathbf{k}}^2 | f_{\mathbf{k}} |^2) = \int \frac{d^3\mathbf{k}}{(2\pi)^3}(2s_{\mathbf{k}} + 1)\frac{\hbar\omega_{\mathbf{k}}}{2} \qquad (4.100)$$

In a Hamiltonian description of the dynamics of a finite system of parametric oscillators, the Hamiltonian is simply

$$H^{\text{Mink}}(t) = \frac{1}{2}\sum_k(\pi_{\mathbf{k}}^2 + \omega_{\mathbf{k}}^2 q_{\mathbf{k}}^2) = \sum_k \left(N_{\mathbf{k}} + \frac{1}{2} \right)\hbar\omega_{\mathbf{k}} \qquad (4.101)$$

Comparing this with (4.100) one can identify $| f_{\mathbf{k}} |^2$ and $| \dot{f}_{\mathbf{k}} |^2$ with the canonical coordinates $q_{\mathbf{k}}^2$ and moment $\pi_{\mathbf{k}}^2$, the eigenvalue of H_0 being the energy $E_{\mathbf{k}} = (N_{\mathbf{k}} + \frac{1}{2})\hbar\omega_{\mathbf{k}}$. The analogy of particle creation with parametric amplification is formally clear: equation (4.98) defines the number operator

$$N_{\mathbf{k}}(t) = \frac{1}{2\hbar\omega_{\mathbf{k}}}(\pi_{\mathbf{k}}^2 + \omega_{\mathbf{k}}^2 q_{\mathbf{k}}^2) - \frac{1}{2} = s_{\mathbf{k}} \qquad (4.102)$$

and equation (4.100) says that the energy density of vacuum particle creation comes from the amplification of vacuum fluctuations $\hbar\omega_{\mathbf{k}}/2$ by the factor $\mathcal{A}_{\mathbf{k}} = 2s_{\mathbf{k}} + 1$. Now it is easy to recognize that the Minkowski result in equation (4.21) corresponds to $\mathcal{A}_{\mathbf{k}} = 1$, no particle creation or zero amplification.

In general there are ultraviolet divergences appearing in the integral (4.100) which requires a subtraction scheme. The adiabatic method comes in handy for such a task, because as we have explained before, the lowest few orders of the WKB solutions encompass particle production from the high-frequency range downwards in the spectrum. This is just what one needs for the subtraction of ultraviolet divergences. For renormalization of the energy–momentum tensor of quantum fields in curved spacetimes the zeroth, second and fourth adiabatic order expressions give the quartic, quadratic and logarithmic divergences. We will discuss this method in the context of cosmological particle creation in Section 4.6. To facilitate adoption of the formula there for flat space field theory in a dynamical background field, just replace χ by ϕ, η by t (thus primes by overdots) and set $a = 1$.

A quantity which enters in the expressions for the adiabatic expansion of the energy–momentum tensor of quantum fields is the *nonadiabaticity parameter* defined as (for the $\mathbf{k}$th mode with natural frequency $\omega_{\mathbf{k}}$ in t time) $\bar{\omega}_{\mathbf{k}} \equiv \dot{\omega}_{\mathbf{k}}/\omega_{\mathbf{k}}^2$. Particle production is more pronounced in modes which evolve nonadiabatically, i.e. $\bar{\omega}_{\mathbf{k}}(t) \simeq 1$ (or $\bar{\omega}_{\mathbf{k}}(\eta) \simeq 1$ in the conformal wave equation of Section 4.6). Thus particle production is a nonadiabatic process. We will learn soon that it is also a non-Markovian process (nonlocal in time, memory, or history, dependent).

4.3.2 Stimulated production

Equation (4.98) gives the vacuum energy density of particles produced from an initial vacuum, a pure state. If the initial state at t_0 is a statistical mixture of pure states, each of which contains a definite number of particles, then an additional mechanism of particle creation enters. This is known as induced or stimulated creation. In particular, if the statistical density matrix μ is diagonal in the representation whose basis consists of the eigenstates of the number operators $a_{\mathbf{k}}^\dagger a_{\mathbf{k}}$ at time t_0, then for bosons this process increases the average number of particles (in mode $\mathbf{k}$ in a unit volume) at a later time t over and above the initial amount present. From (4.28) we have

$$\tilde{N} \equiv \langle N_{\mathbf{k}}(t)\rangle_t = \text{Tr}[\mu \tilde{a}_{\mathbf{k}}^\dagger(t)\tilde{a}_{\mathbf{k}}(t)] = \langle N_{\mathbf{k}}(t_0)\rangle + \mid \beta_{\mathbf{k}}(t)\mid^2 [1 + 2\langle N_{\mathbf{k}}(t_0)\rangle] \quad (4.103)$$

where angular brackets without a subscript t refers to that taken at the initial time t_0, $\langle N_{\mathbf{k}}(t_0)\rangle = \text{Tr}[\mu a_{\mathbf{k}}^\dagger a_{\mathbf{k}}]$, if the system is in a pure state at t_0. For fermions induced (or stimulated) particle creation decreases the initial number.

The above result can be understood in the parametric oscillator description as the sum of two parts: First, the amount $s_{\mathbf{k}} = |\beta_{\mathbf{k}}(t)|^2$ from spontaneous production of particles from the amplification of vacuum fluctuations by the factor

$\mathcal{A}_{\mathbf{k}} = 2s_{\mathbf{k}} + 1$. Second, an amplification by the same factor $\mathcal{A}_{\mathbf{k}}$, of the particles already present $N_{\mathbf{k}}(t_0)$, i.e.

$$\langle N_{\mathbf{k}}(t)\rangle_t = \mid \beta_{\mathbf{k}}(t)\mid^2 + \mathcal{A}_{\mathbf{k}}\langle N_{\mathbf{k}}(t_0)\rangle \tag{4.104}$$

where $s_{\mathbf{k}} = |\beta_{\mathbf{k}}(t)|^2$. The second part is called stimulated production. It yields an energy density ρ_n with respect to the n-particle state at t_0 given by

$$\rho_n^{\text{Mink}} = \langle n \mid T_{00} \mid n \rangle = \int \frac{d^3\mathbf{k}}{(2\pi)^3} (\mid \dot{f}_{\mathbf{k}} \mid^2 + \omega_{\mathbf{k}}^2 \mid f_{\mathbf{k}} \mid^2)\langle a_{\mathbf{k}}^\dagger a_{\mathbf{k}}\rangle$$

$$= \int \frac{d^3\mathbf{k}}{(2\pi)^3}(2s_{\mathbf{k}} + 1)\hbar\omega_{\mathbf{k}}\langle N_{\mathbf{k}}(t_0)\rangle \tag{4.105}$$

Combining (4.100) and (4.105), for a density matrix diagonal in the number state, the total energy density of particles created from the vacuum and from those already present in the n-particle state is given by

$$\rho^{\text{Mink}} = \rho_0^{\text{Mink}} + \rho_n^{\text{Mink}} = \int \frac{d^3\mathbf{k}}{(2\pi)^3}(\mid \dot{f}_{\mathbf{k}} \mid^2 + \omega_{\mathbf{k}}^2 \mid f_{\mathbf{k}} \mid^2)\left(\frac{1}{2} + \langle a_{\mathbf{k}}^\dagger a_{\mathbf{k}}\rangle\right)$$

$$= \int \frac{d^3\mathbf{k}}{(2\pi)^3}\mathcal{A}_{\mathbf{k}}\hbar\omega_{\mathbf{k}}\left(\frac{1}{2} + \langle N_{\mathbf{k}}(t_0)\rangle\right) \tag{4.106}$$

This can be understood as the result of parametric amplification by the factor $\mathcal{A}_{\mathbf{k}}$ of the energy density of vacuum fluctuations $\hbar\omega_{\mathbf{k}}/2$ plus that of the particles originally present in the kth mode at t_0, i.e. $\langle N_{\mathbf{k}}(t_0)\rangle\hbar\omega_{\mathbf{k}}$.

4.4 Quantum Vlasov equation

Having familiarized ourselves with the general scheme of adiabatic vacuum and number states, we now return to the problem of charged particle production in an external electromagnetic field. We continue to follow the treatment given by Kluger, Mottola and Eisenberg [KlMoEi98] for pedagogical advantage.

4.4.1 Adiabatic number state

An adiabatic number state $\tilde{f}_{\mathbf{k}(0)}^+(t)$ was suggested by [KlMoEi98] for the description of a kinetic theory of charged particles moving in an electromagnetic field. That corresponds to the $n = 0$ adiabatic state defined in equation (4.46)

$$\tilde{f}_{\mathbf{k}}^{(0)}(t)\left(= f_{\mathbf{k}(0)}^+ \text{ equation (4.46)}\right) \equiv \sqrt{\frac{\hbar}{2\omega_{\mathbf{k}}(t)}} \exp\left(-i\Theta_{\mathbf{k}(n=0)}\right) \tag{4.107}$$

where $\Theta_{\mathbf{k}(n=0)} \equiv \int^t \omega_{\mathbf{k}}(t')dt'$ is the $(n = 0)$th-order adiabatic phase. At this level of accuracy one measures particle numbers at all times with respect to the initial vacuum state at time t_0. This definition of a number state makes use of the fact that under adiabatic evolution, particle number is an adiabatic invariant. This restricts its validity from the start to weak or slowing varying background fields.

The adiabatic particle number is defined to be [KlMoEi98]

$$
\begin{aligned}
\tilde{N}_{\mathbf{k}}(t) &\equiv \langle \tilde{a}_{\mathbf{k}}^{\dagger}(t)\tilde{a}_{\mathbf{k}}(t)\rangle + \langle \tilde{b}_{-\mathbf{k}}^{\dagger}(t)\tilde{b}_{-\mathbf{k}}(t)\rangle = |\alpha_{\mathbf{k}}|^2 \langle a_{\mathbf{k}}^{\dagger}a_{\mathbf{k}}\rangle + |\beta_{\mathbf{k}}|^2 \langle b_{-\mathbf{k}}b_{-\mathbf{k}}^{\dagger}\rangle \\
&= \left(1 + |\beta_{\mathbf{k}}|^2\right) N_{+}(\mathbf{k}) + |\beta_{\mathbf{k}}|^2 \left(1 + N_{-}(-\mathbf{k})\right) \\
&= |\beta_{\mathbf{k}}|^2 + (1 + 2|\beta_{\mathbf{k}}|^2)N_{\mathbf{k}} = N_{\mathbf{k}} + (1 + 2N_{\mathbf{k}})\,|\beta_{\mathbf{k}}(t)|^2 \qquad (4.108)
\end{aligned}
$$

where the last line is valid only if the number of positive and negative charges are equal (cf. (4.72)). To verify that $\tilde{N}_{\mathbf{k}}$ is an adiabatic invariant we show that it is proportional to the ratio of the energy to frequency for any mode $\mathbf{k}$, $\epsilon_{\mathbf{k}}(t)/\hbar\omega_{\mathbf{k}}(t)$, which is known as such for a harmonic oscillator with time-dependent frequency. After the discussions on spontaneous and stimulated production we can actually read off this expression from (4.106): Viewing $\int d^3k/(2\pi)^3$ as $1/V$, the inverse volume, the integrand there is the energy in mode $\mathbf{k}$. Dividing by $\hbar\omega$ and multiplying it by 2 for the presence of both $\pm$ charges gives the expression we are looking for:

$$
\frac{\epsilon_{\mathbf{k}}(t)}{\hbar\omega_{\mathbf{k}}(t)} = 1 + 2\tilde{N}_{\mathbf{k}}(t) \qquad (4.109)
$$

The amount of particle production at time t in this basis is given by the expectation value of the number operator $\tilde{a}^{\dagger}\tilde{a}$ at time t with respect to the vacuum state $|\rangle_0$ defined at t_0 (not the vacuum state $|\rangle_t$ defined at t). As discussed above this is fine if the vacuum states are well defined at the initial t_0 *and* final times t, as in an asymptotically-static evolution. Otherwise one needs to specify the adiabatic order to make the vacuum well-defined: the adiabatic number state of [KlMoEi98] corresponds to the lowest adiabatic order.

From earlier discussions, we know that this level of approximation will not give a good measure for on-going particle creation, as particle creation is basically a nonadiabatic process. It is however useful for quantum kinetic theory descriptions, where a quasi-particle approximation is usually introduced which amounts to incorporating only the quantum radiative corrections to the particles but not fully field theoretical effects such as particle creation. In other words, quantum kinetic theory is usually treated at the same level of approximation described by the adiabatic number basis. We will describe quantum kinetic *field* theory in Chapter 11.

4.4.2 Number and correlation

We now proceed to derive an equation for the time rate of change of the number of particles created in each mode with respect to the time-dependent particle number basis. Differentiating (4.108), we obtain

$$
\frac{d}{dt}\tilde{N}_{\mathbf{k}} = 2\left(1 + 2N_{\mathbf{k}}\right)\operatorname{Re}\left(\beta_{\mathbf{k}}^{*}\dot{\beta}_{\mathbf{k}}\right) \qquad (4.110)
$$

We need an expression for $\dot{\beta}_{\mathbf{k}}$ in terms of α, β and $\Theta_{\mathbf{k}0}(t) \equiv \int^t \omega_{\mathbf{k}}(t')dt'$ (we will omit the subscript 0 on Θ_0 in this subsection). To do so we use equations (4.97) and (4.22) to get

$$\dot{\alpha}_{\mathbf{k}} = \frac{\dot{\omega}_{\mathbf{k}}}{2\omega_{\mathbf{k}}}\beta_{\mathbf{k}}\exp(2i\Theta_{\mathbf{k}}), \quad \dot{\beta}_{\mathbf{k}} = \frac{\dot{\omega}_{\mathbf{k}}}{2\omega_{\mathbf{k}}}\alpha_{\mathbf{k}}\exp(-2i\Theta_{\mathbf{k}}) \tag{4.111}$$

thus

$$\frac{d}{dt}\tilde{N}_{\mathbf{k}} = \frac{\dot{\omega}_{\mathbf{k}}}{\omega_{\mathbf{k}}}(1 + 2N_{\mathbf{k}})\,\mathrm{Re}\,\{\alpha_{\mathbf{k}}\beta_{\mathbf{k}}^*\exp(-2i\Theta_{\mathbf{k}})\} = \frac{\dot{\omega}_{\mathbf{k}}}{\omega_{\mathbf{k}}}\,\mathrm{Re}\,\{\mathcal{C}_{\mathbf{k}}\exp(-2i\Theta_{\mathbf{k}})\} \tag{4.112}$$

where we have defined the time-dependent pair correlation function

$$\mathcal{C}_{\mathbf{k}}(t) \equiv \langle\tilde{a}_{\mathbf{k}}(t)\tilde{b}_{-\mathbf{k}}(t)\rangle = (1 + 2N_{\mathbf{k}})\,\alpha_{\mathbf{k}}\beta_{\mathbf{k}}^* \tag{4.113}$$

The pair correlation $\mathcal{C}_{\mathbf{k}}(t)$ is a very rapidly varying function, since the time-dependent phases on the right side of (4.113) *add* rather than cancel. The phases, however, nearly cancel in the final combination of (4.112) to render $\tilde{N}_{\mathbf{k}}$ a slowly varying function. The time derivative of the pair correlation function is given by

$$\frac{d}{dt}\mathcal{C}_{\mathbf{k}} = \frac{\dot{\omega}_{\mathbf{k}}}{2\omega_{\mathbf{k}}}(1 + 2N_{\mathbf{k}})\exp(2i\Theta_{\mathbf{k}})\left(1 + 2|\beta_{\mathbf{k}}|^2\right) = \frac{\dot{\omega}_{\mathbf{k}}}{2\omega_{\mathbf{k}}}\left(1 + 2\tilde{N}_{\mathbf{k}}\right)\exp(2i\Theta_{\mathbf{k}}) \tag{4.114}$$

4.4.3 Current and energy density

To obtain the current (4.71) in terms of the particle number and its time derivative, we need to express $|f_{\mathbf{k}}(t)|^2$ in terms of the Bogoliubov coefficients. For this we use equations (4.98), (4.96) and obtain

$$j(t) = e\hbar\int d^3\mathbf{k}\frac{(k_z - eA(t))}{\omega_{\mathbf{k}}(t)}(1 + 2|\beta_{\mathbf{k}}(t)|^2 + 2\mathrm{Re}\{\alpha_{\mathbf{k}}\beta_{\mathbf{k}}^*e^{-2i\Theta_{\mathbf{k}}(t)}\})(1 + 2N_{\mathbf{k}}) \tag{4.115}$$

The vacuum term in this expression, $\int d^3\mathbf{k}\,(k_z - eA(t))\,/\omega_{\mathbf{k}}(t)$, vanishes by charge conjugation symmetry, when proper gauge invariant integration boundaries are chosen. Using the mean value of particles in the adiabatic number basis (4.108), its time derivative and the equations of motion (4.112), we can rewrite the current as

$$j(t) = 2e\hbar\int d^3\mathbf{k}\frac{(k - eA(t))}{\omega_{\mathbf{k}}(t)}\tilde{N}_{\mathbf{k}}(t) + \frac{2\hbar}{E}\int d^3\mathbf{k}\,\omega_{\mathbf{k}}(t)\frac{d\tilde{N}_{\mathbf{k}}}{dt}(t) = j_{\mathrm{cond}} + j_{\mathrm{pol}} \tag{4.116}$$

Classically, if the particle distribution $\tilde{N}_{\mathbf{k}}$ is coupled to a uniform electric field the energy density and its time derivative are given by

$$\varepsilon = \frac{E^2}{2} + 2\int d^3\mathbf{k}\,\hbar\omega_{\mathbf{k}}\tilde{N}_{\mathbf{k}} \tag{4.117a}$$

$$\dot{\varepsilon} = \dot{E}E + 2\int d^3\mathbf{k}\left(e\hbar E\frac{(k - eA)}{\omega_{\mathbf{k}}}\tilde{N}_{\mathbf{k}} + \omega_{\mathbf{k}}\hbar\frac{d\tilde{N}_{\mathbf{k}}}{dt}\right) = 0 \tag{4.117b}$$

Using the Maxwell equation $-\dot{E} = j$ this last relation is precisely the same as the mean value of the quantum current in (4.116). Hence we may identify the adiabatic particle number $\tilde{N}_{\mathbf{k}}(t)$ with the (quasi) classical single-particle distribution. This is the starting point of a quantum kinetic theory description.

4.4.4 Quantum Vlasov equation

Let us return now to the two equations for the rates of change of the particle number and the quantum correlations. Solving equation (4.114) formally for $\mathcal{C}_{\mathbf{k}}$, assuming that $\mathcal{C}_{\mathbf{k}}$ vanishes at some $t = t_0$ which could be taken to $-\infty$, and substituting into (4.112) we obtain

$$\frac{d}{dt}\tilde{N}_{\mathbf{k}} = \frac{\dot{\omega}_{\mathbf{k}}}{2\omega_{\mathbf{k}}} \int_{t_0}^{t} dt' \left\{ \frac{\dot{\omega}_{\mathbf{k}}}{\omega_{\mathbf{k}}}(t') \left(1 + 2\tilde{N}_{\mathbf{k}}(t')\right) \cos\left[2\Theta_{\mathbf{k}}(t) - 2\Theta_{\mathbf{k}}(t')\right] \right\} \quad (4.118)$$

Equation (4.118) may be called a "quantum Vlasov equation," in the sense that it gives the rate of particle creation in an arbitrary time-varying mean field. Note the appearance of the Bose enhancement factor $(1 + 2\tilde{N}_{\mathbf{k}})$ in (4.118) indicates that both spontaneous and induced particle creation are present. One important feature of equation (4.118) is that it is nonlocal in time, the particle creation rate depending on the entire previous history of the system. Thus particle creation in general is a *non-Markovian* process [BirDav82, Rau94, RauMue96, SRSBTP97]. Note that the nonlocal form of (4.118) results from solving one variable $\mathcal{C}$ in terms of the other $\tilde{N}$, each obeying a Hamiltonian equation of motion. This is a general feature of coupled subsystems.

Equation (4.118) becomes exact in the limit in which the electric field can be treated classically, i.e. the limit in which real and virtual photon emission is neglected, and there is no scattering. We will learn later that this *semiclassical limit* is obtained at the leading order of a large N approximation [CoJaPo74, Roo74].

Inclusion of scattering processes leads to collision terms on the right side of (4.118) which are also nonlocal in general. This nonlocality is essential to the quantum description in which phase information is retained for all times. The phase oscillations in the cosine term are a result of the quantum coherence between the created pairs, which must be present in principle in any unitary evolution. However, precisely because these phase oscillations are so rapid it is clear that the integral in (4.118) receives most of its contribution from t' close to t, which suggests that some local approximation to the integral should be possible, provided that we are not interested in resolving the short-time structure or measuring the phase coherence effects. The time-scale for these quantum phase coherence effects to wash out is the time-scale of several oscillations of the phase factor $\Theta_{\mathbf{k}}(t) - \Theta_{\mathbf{k}}(t')$, which is of order $\tau_{qu} = 2\pi/\omega_{\mathbf{k}} = 2\pi\hbar/\epsilon_{\mathbf{k}}$, where $\epsilon_{\mathbf{k}}$ is the single-particle energy.

We will return to this equation in Chapter 9 to construct the density matrix and discuss entropy generation in these quantum field processes.

4.5 Periodically driven fields

As another example of particle production from parametric amplification, we give in this section a brief discussion of the solutions of equation (4.22) in the important case when the natural frequency depends periodically on time, that is, $\omega^2(t+T) = \omega^2(t)$ for some period T. Again we drop the mode label **k**, as only one mode will be considered. This is a case of parametric resonance. In the mathematical literature, the corresponding problem is the subject of the so-called Floquet theory [WhiWat40, Inc56]. In physics there are many applications (e.g. [Shi65, MilWya83, MonPaz01]). One such area in cosmology which has drawn considerable attention is particle creation by parametric resonance during the preheating epoch after the universe came out of inflation, see Chapter 15. Our treatment here is influenced by the work of Kofman, Linde and Starobinsky [KoLiSt97].

The key insight is that, if $f(t)$ is a solution, then $f(T+t)$ is a solution too. If f_1 and f_2 are linearly independent solutions, then we must have

$$f_i(t+T) = A_{i1}f_1(t) + B_{i2}f_2(t) \qquad (i=1,2) \tag{4.119}$$

Thus there must exist solutions $F_1(t)$, $F_2(t)$ such that

$$F_i(t+T) = e^{\mu_i T} F_i(t) \tag{4.120}$$

or equivalently

$$F_i(t) = e^{\mu_i t} \tilde{f}_i(t) \tag{4.121}$$

where the functions $\tilde{f}_i(t)$ are periodic with period T. The eigenvalues μ_i are the so-called Floquet exponents. Sometimes Floquet energies $i\hbar\mu_i$ are introduced. As we shall see presently, the Floquet exponents may be real, leading to exponential amplification of the solution (or, in quantum language, exponential squeezing of the quantum state, see later in this chapter).

The second key insight is that if μ is a Floquet exponent, then $-\mu$ and μ^* must be exponents as well. The first follows from the fact that the Wronskian of two solutions must be a constant, and the second because the equation is real. So we only have two possibilities, either the Floquet exponents are imaginary and complex conjugate to each other, or real and opposite to each other. In the second case, we say there is parametric resonance.

To be concrete, we shall restrict ourselves to the Mathieu equation, which is obtained when

$$\omega^2(t) = \omega_0^2 + \omega_1^2[1 + \cos\gamma t] \tag{4.122}$$

where ω_0, ω_1 and γ are constants. There are two interesting regimes, namely, the so-called broad resonance when $\omega_1 \gg \omega_0, \gamma$, and the narrow resonance when the opposite obtains. Of course, we may take $\gamma = 1$ with no loss of generality.

4.5.1 Broad resonance

In the broad resonance regime, $\omega \gg \omega_0, 1$ unless $t \sim (2j + 1)\pi$, where j is an integer. The second-order adiabatic frequency correction is given by

$$\epsilon_2 = \left(-\frac{1}{4\omega_1^2}\right) \frac{1 + \left[1 + \left(\frac{\omega_0}{\omega_1}\right)^2\right]\cos t + \frac{1}{4}\sin^2 t}{\left[1 + \left(\frac{\omega_0}{\omega_1}\right)^2 + \cos t\right]^3} \tag{4.123}$$

which is much smaller than 1 unless $\cos t \sim -1$. Therefore we may describe the evolution as a series of adiabatic periods, separated by nonadiabatic transitions when $t \sim t_j = (2j + 1)\pi$. Between transitions we may use the $(n = 0)$ adiabatic function $\tilde{f}^0 = f_0^+$ of (4.107), there being no net amplification. Near t_j, we may approximate $\omega^2(t) = \omega_0^2 + \omega_1^2(t - t_j)^2/2$. We have already encountered the resulting equation in our study of pair creation by a constant electric field.

Let $t_{k-1} < t \leq t_k$, and consider the exact solution f which behaves as a positive frequency $(+)$ lowest WKB order $(n = 0)$ solution near t. For $t \geq t_k$ this solution plays the same role as the *in*-region positive frequency wave in the calculation of particle creation. Thus, for $t \geq t_k$, it assumes the form (cf. equation (4.89))

$$f = \alpha f_0^+ + \beta f_0^- \tag{4.124}$$

with α and β given in equations (4.90) and (4.91) respectively. Neglect any further evolution of the Bogoliubov coefficients, and write

$$f_0^+(t + 2\pi) = e^{-i\Theta_0} f_0^+(t) \tag{4.125}$$

Therefore

$$f(t + 2\pi) = \alpha e^{-i\Theta_0} f(t) + \beta e^{i\Theta_0} f^*(t) \tag{4.126}$$

The general solution is $F = Af + Bf^*$, and the eigenvalue condition (4.120) becomes a set of linear equations for the coefficients

$$\begin{pmatrix} \alpha e^{-i\Theta_0} & \beta^* e^{-i\Theta_0} \\ \beta e^{i\Theta_0} & \alpha^* e^{i\Theta_0} \end{pmatrix} \begin{pmatrix} A \\ B \end{pmatrix} = \lambda \begin{pmatrix} A \\ B \end{pmatrix} \tag{4.127}$$

where $\lambda = \exp(2\pi\mu)$. We see that the Floquet exponents must satisfy

$$(\alpha^* - \lambda e^{-i\Theta_0})(\alpha - \lambda e^{i\Theta_0}) - |\beta|^2 = 0 \tag{4.128}$$

The condition for μ to be real is

$$\text{Re}\left[\alpha e^{-i\Theta_0}\right] > 1 \tag{4.129}$$

We see that it is not sufficient to have $\beta \neq 0$.

4.5.2 Narrow resonance

Let us now consider the case of narrow resonance. Consider the case when at the boundary of a resonant region the Floquet exponents vanish, meaning that there are purely periodic solutions; a second family of unstable regions corresponds to antiperiodic solutions at the boundary, and can be treated in a similar way. When $\omega_1 \to 0$, we obtain periodic solutions if $\omega_0 = \ell$, where ℓ is an integer. So we expect to find an infinite sequence of resonant regions in the (ω_1, ω_0) plane, the ℓth region reducing to $\omega_0 = \ell$ when $\omega_1 = 0$. Our goal is to describe these regions and the corresponding Floquet exponents when $\omega_1 \ll 1, \omega_0$.

To this end, observe that if we write a solution as a linear combination of $\pm$ frequency solutions in the WKB form, as in equations (4.95) and (4.96) (exact), then the evolution of the α and β coefficients to a sufficiently high adiabatic order (presently $r = 0$) is dictated by equations (4.111) and (4.107), where

$$\frac{\dot{\omega}}{2\omega} \sim \left(\frac{-\omega_1^2}{4\omega_0^2} \right) \sin t \tag{4.130}$$

$$\Theta_0(t) \sim \left[1 + \frac{\omega_1^2}{2\omega_0^2} \right] \omega_0 t + \frac{\omega_1^2}{2\omega_0} \sin t \tag{4.131}$$

leading to

$$\exp \left\{ 2i\Theta_0 \right\} = \exp \left\{ 2i \left[1 + \frac{\omega_1^2}{2\omega_0^2} \right] \omega_0 t \right\} \sum_{n=-\infty}^{\infty} J_n \left[\frac{\omega_1^2}{\omega_0} \right] e^{int} \tag{4.132}$$

where the J_n are Bessel functions (recall that for integer n, $J_{-n} = (-1)^n J_n$).

Now consider the ℓ-th resonant region where $\omega_0 = \ell + \delta_\ell$. Keeping only the slowly varying terms in equation (4.111), we get

$$\dot{\alpha}_\ell = i\beta_\ell \kappa_\ell \exp \left(2i\sigma_\ell t \right) \tag{4.133}$$

$$\dot{\beta}_\ell = -i\alpha_\ell \kappa_\ell \exp \left(-2i\sigma_\ell t \right) \tag{4.134}$$

where

$$\kappa_\ell \equiv \frac{\omega_1^2 K_\ell}{8\omega_0^2}, \qquad \sigma_\ell \equiv \frac{\omega_1^2}{2\omega_0} + \delta_\ell \tag{4.135}$$

and

$$K_\ell = J_{2\ell-1} \left[\frac{\omega_1^2}{\omega_0} \right] - J_{2\ell+1} \left[\frac{\omega_1^2}{\omega_0} \right] \sim \frac{1}{(2\ell - 1)!} \left(\frac{\omega_1^2}{2\omega_0} \right)^{2\ell-1} \tag{4.136}$$

We seek a solution of the form

$$\alpha_\ell (t) = \alpha_{\ell 0} e^{\mu_\ell t} \exp \left(i\sigma_\ell t \right) \tag{4.137}$$

$$\beta_\ell (t) = \beta_{\ell 0} e^{\mu_\ell t} \exp \left(-i\sigma_\ell t \right) \tag{4.138}$$

where μ_ℓ is the Floquet exponent of the ℓth resonance band. We get

$$(\mu_\ell + i\sigma_\ell)\, \alpha_{\ell 0} = i\beta_{\ell 0}\kappa_\ell \tag{4.139}$$

$$(\mu_\ell - i\sigma_\ell)\, \beta_{\ell 0} = -i\dot{\alpha}_{\ell 0}\kappa_\ell \tag{4.140}$$

Therefore

$$\mu_\ell^2 = \kappa_\ell^2 - \sigma_\ell^2 \tag{4.141}$$

The boundaries of the resonant region are given by

$$\delta_\ell \sim -\frac{\omega_1^2}{2\ell} \pm \frac{\omega_1^2 K_\ell}{8\ell^2} \tag{4.142}$$

We see that the regions become narrower, and the Floquet indices become weaker, as we go to higher resonance bands. Particle production by parametric resonance is the principal mechanism in the pre-heating stage when the universe is warmed up after an inflationary expansion.

4.6 Particle creation in a dynamical spacetime

Another important class of problems similar to the external field model above is cosmological particle creation. There, a classical dynamical background spacetime governs the quantum field and imparts a time-dependence in the natural frequencies of its normal modes. Historically this is a major arena where nonequilibrium field theory was inculcated and constructed. It has wide ranging implications in modern cosmology since many late era phenomena have originated from quantum effects in the very early universe including inflationary cosmology. The era from the Planck to the GUT era is depicted by quantum field theory in curved spacetime (the test field description) and semiclassical gravity (including the back-reaction).

Cosmological particle creation is a physical process of basic theoretical interest in quantum field theory in curved spacetime [Park66, Park68, Park69, Park71, ZelSta71, SexUrb69, Zel70, Hu72, Hu74, FuPaHu74, Gri74, Berger74, Berger75a, Berger75b, HuPar77, HuPar78, HarHu79, DeW67, DeW75, BirDav82], and important practical interest in the quantum dynamics of the early universe. Our summary here is based on earlier work of [Hu74, ZelSta71]. We begin with the underlying physics, which is rooted in parametric amplification of classical waves [Zel70]. This effect in second quantized language manifests itself as particle creation. A modern representation of such processes is by means of the squeezed state language developed in quantum optics. It is useful for the discussion of entropy and coherence issues. We defer such a discussion to Section 4.7.

4.6.1 Wave equations in curved spacetimes

Consider a massive (m) neutral scalar field ϕ coupled arbitrarily (ξ) to a background spacetime with metric $g_{\mu\nu}$ and scalar curvature R. Its dynamics is described by the action

$$S = \int d^4x \mathcal{L}(\phi, \nabla\phi, g_{\mu\nu}) \tag{4.143}$$

where the Lagrangian density is given by

$$\mathcal{L}(\phi, \nabla\phi, g_{\mu\nu}) = -\frac{1}{2}\sqrt{-g}\left[g^{\mu\nu}(x)\nabla_\mu\phi\nabla_\nu\phi + (m^2 + \xi R)\phi^2(x)\right] \tag{4.144}$$

where $g \equiv \det g_{\mu\nu}$ and ∇ denotes taking the covariant derivative defined on the background spacetime. Here $\xi = 1/6$ and 0 denote, respectively, conformal and minimal coupling. The indices $\mu = (0, 1, 2, 3)$ denote time and spatial components. The scalar field satisfies the wave equation

$$[-\nabla^2 + m^2 + \xi R]\phi(\mathbf{x}, t) = 0 \tag{4.145}$$

where

$$\nabla^2 \equiv g^{\mu\nu}\nabla_\mu\nabla_\nu = \frac{1}{\sqrt{-g}}\frac{\partial}{\partial x^\mu}\left(g^{\mu\nu}\sqrt{-g}\frac{\partial}{\partial x^\nu}\right) \tag{4.146}$$

is the Laplace–Beltrami operator defined on the background spacetime.

In the canonical quantization approach, one assumes a foliation of spacetime into dynamically evolving, time-ordered, spacelike hypersurfaces Σ. If the three-dimensional space Σ possesses some symmetry, such as a homogeneous space with a group of motion, a separation of variables is usually possible which permits a normal mode decomposition of the field. (The spacetimes considered in this book, e.g. Friedmann–Lemaitre–Robertson–Walker (FLRW) and De Sitter (DS) all possess these properties.) One can then impose canonical commutation relations on the creation and annihilation operators corresponding to the (time-dependent) amplitude functions of each normal mode, define the vacuum and number states, and then construct the Fock space. In flat space, Poincaré invariance guarantees the existence of a unique global Killing vector ∂_t orthogonal to all constant-time spacelike hypersurfaces, an unambiguous separation of the positive- and negative-frequency modes, and a unique and well-defined vacuum. In curved spacetime, general covariance precludes any such privileged choice of time and slicing. There is no natural mode decomposition and no unique vacuum [Ful72, Ful89]. We assume the background spacetime under consideration has at least enough symmetry to allow for a normal mode decomposition of the invariant operator at any constant-time slice.

The classical field theory is quantized by replacing the field variable ϕ by the operator-valued distribution Φ. In the Heisenberg picture, Φ and its conjugate momentum $\Pi = \delta\mathcal{L}/\delta(\partial_0\Phi)$ obey the equal time commutation relation

(4.7). Note that the scalar delta function $\delta(\mathbf{x}, \mathbf{x}')$ in curved spacetime is defined by $\int \sqrt{-g} \delta(\mathbf{x}, \mathbf{x}') h(\mathbf{x}) = h(\mathbf{x}')$, where h is any test function.

Consider the field Φ in a coordinate volume $V = L^3$ with coordinate length L. We can expand the field Φ in terms of a complete set of (spatial) orthonormal modes $u_{\mathbf{k}}(\mathbf{x})$ as in equation (4.13). We use $\mathbf{x}$ as a generic notation for the spatial coordinates. (This is also applicable for spatially nonflat spacetimes, e.g. in S^3 with radius a, $V = 2\pi^2 a^3$, one can use the hyperspherical coordinates, $\mathbf{x} = (\chi, \theta, \phi)$, and the wavenumbers are then labeled by the corresponding principal quantum numbers $\mathbf{k} = (n, l, m)$. See, e.g. [Wig68].) As before, we write the operator-valued amplitude function $\varphi_{\mathbf{k}}(t)$ in terms of the time-independent annihilation operators $a_{\mathbf{k}}$ and the (c-number) amplitude functions $f_{\mathbf{k}}(t)$ as in (4.16). The canonical commutation rules on Φ then imply the conditions on $a_{\mathbf{k}}$ and $a_{\mathbf{k}'}^\dagger$ as in (4.18).

For the spatially-flat Friedmann–Lemaitre–Robertson–Walker (FLRW) spacetime [Park69], the spatial mode functions are simply $u_{\mathbf{k}} = e^{i\mathbf{k}\cdot\mathbf{x}}$ and the wave equation for the amplitude function of the $\mathbf{k}$th mode in cosmic time t becomes (because of spatial isotropy, f depends only on $k \equiv |\mathbf{k}|$)

$$\ddot{f}_k(t) + 3H \, \dot{f}_k(t) + [\omega_k^2(t) + q(t)] f_k(t) = 0 \qquad (4.147)$$

where an overdot denotes taking the derivative with respect to cosmic time, $\cdot = d/dt$. Here

$$\omega_k^2(t) = \frac{k^2}{a^2} + m^2; \quad q = \xi R \qquad (4.148)$$

$$R = 6\left[\dot{H}(t) + 2H^2(t)\right] \qquad (4.149)$$

$H(t) \equiv \frac{\dot{a}}{a}$ being the expansion (Hubble) rate of the background space. We have grouped terms containing two time derivatives of a (second derivative or first derivative squared) and call them q. As we will define below, they are of second adiabatic order while ω_k is of zero adiabatic order.

In curved space the inequivalence of Fock representation due to the lack of a global time-like Killing vector makes the constant separation of positive and negative-frequency components in general impossible. The mixing of positive- and negative-frequency components is the source of particle creation (in the second quantization description). Particle creation may arise from topological, geometrical, or dynamical causes. In cosmological spacetimes the inequivalence of vacua appears at different times of evolution, and thus cosmological particle creation is by nature a dynamically induced effect. Note that we are dealing here with a free field: particles are not produced from interactions, but rather from the excitation (parametric amplification [Zel70]) of vacuum fluctuations (or quantum noise) by the changing background gravitational field. The basic mechanism is also different from thermal particle creation in black holes [Haw75], accelerated detectors [Unr76] or moving mirrors [FulDav76, DavFul77], which involves the presence of an event horizon or the exponential red-shifting of outgoing modes [HuRav96, RaHuAn96, RaHuKo97].

4.6.2 Conformal vacuum in conformally-static spacetimes

In the class of conformally-static spacetimes where the metric is conformally related to a static spacetime by a conformal factor a there exists a global conformal Killing vector ∂_η, where $\eta = \int dt/a(t)$ is the conformal time. For example, the spatially-flat FRW spacetime with metric

$$g_{\mu\nu}(x) = a^2(\eta)\eta_{\mu\nu} \tag{4.150}$$

is conformally related to the Minkowski metric $\eta_{\mu\nu}$:

$$ds^2 = a^2(\eta)(-d\eta^2 + d\mathbf{x}^2) \tag{4.151}$$

In this case the vacuum defined by the mode decomposition with respect to ∂_η is globally well-defined, known as the conformal vacuum. For conformally-invariant fields (e.g. a massless scalar field with $\xi = 1/6$ in equation (4.145)) in conformally-static spacetimes, it is easy to see that there is no particle creation [Park69]. Thus any small deviation from these conditions, e.g. small m, $\xi - (1/6)$, can be treated perturbatively from these states.

Consider a neutral massive scalar field coupled to a spatially-flat FRW metric with constant ξ. It is convenient to define a conformal amplitude function $\chi_\mathbf{k}(\eta) \equiv a(\eta)f_\mathbf{k}(\eta)$ related to the c-number amplitude function $f_\mathbf{k}$ for the kth normal mode. It satisfies the following wave equation (cf. equation (4.147))

$$\chi_\mathbf{k}''(\eta) + [\omega_\mathbf{k}^2(\eta) + Q]\chi_\mathbf{k}(\eta) = 0 \tag{4.152}$$

where a prime denotes differentiation: $\prime \equiv d/d\eta$ and

$$\omega_k^2(\eta) \equiv \omega_k^2(t)a^2 = k^2 + m^2 a^2 \tag{4.153}$$

is the time-dependent natural frequency. For spatially flat FRW spacetime $Q = Q_\xi = (\xi - \frac{1}{6})Ra^2$. For anisotropic spatially homogeneous universe (Bianchi type-I) where the expansion rates $H_i(t) \equiv \frac{\dot{a}_i}{a_i}$ are different in the three directions $i = 1, 2, 3$ ($a^3 = a_1 a_2 a_3$), the wave equation in conformal time has in addition to Q_ξ another term $Q_\beta \equiv -\frac{1}{2}\sum_{i>j}(H_i - H_j)^2$, which, like Q_ξ, is also of second adiabatic order.

One sees that, for massless ($m = 0$) conformally coupled ($\xi = \frac{1}{6}$) fields in a spatially flat FLRW universe ($Q = 0$), the conformal wave equation admits solutions

$$\chi_\mathbf{k}(\eta) = Ae^{i\omega_\mathbf{k}\eta} + Be^{-i\omega_\mathbf{k}\eta} \tag{4.154}$$

which are of the same form as traveling waves in flat space. Since $\omega_\mathbf{k}(m = 0, \xi = \frac{1}{6}) = k =$ const., the positive- and negative-frequency components remain separated and there is no particle production.

In this connection, Grishchuk [Gri74] showed that there is no production of gravitons in a radiation-dominated FLRW universe. This is easily seen as follows:

The gravitons are quantized linear perturbations. In a FLRW universe, just as in Minkowski spacetime, there are two polarizations, each obeying an equation (the Lifshitz equation [Lif46]) which has the same form as a massless ($m = 0$) minimally coupled ($\xi = 0$) scalar field [ForPar77]. For a FLRW universe $R = 6a''/a^3$, the wave equation (4.152) reads, in conformal time,

$$\chi''_{\mathbf{k}}(\eta) + (k^2 - a''/a)\chi_{\mathbf{k}}(\eta) = 0 \tag{4.155}$$

For a radiation-dominated FLRW universe, $a \sim \sqrt{t} \sim \eta$, and thus $R = 0$. The natural frequency is a constant and there is no production of massless minimally coupled scalar particles or gravitons in the conformal vacuum.

More generally, the wave equation for each mode has a time-dependent natural frequency. The negative-frequency modes can thus be excited by the dynamics of the background through $a(\eta)$ and $R(\eta)$. In analogy with the time-dependent Schrödinger equation, one can view the $\omega_k^2 + Q$ term in (4.152) as a time-dependent potential $V(\eta)$ which can induce back-scattering of waves [Zel70, Hu74], thus mixing the positive and negative frequency components in each mode. This, as we have learned, signifies particle creation.

4.6.3 Thermal radiance

It is rather commonly known that black holes emit thermal radiation, known as the Hawking effect [Haw75]. Hawking radiation has a deep meaning and many ways to derive and understand it. One way is to view it as arising from the exponential red-shifting of outgoing modes from the black hole. This condition is responsible for thermal radiance observed in uniformly accelerated detectors, known as the Unruh effect [Unr76], and in an exponential expansion of the early universe [Park76]. We can see this from the simple theory we have presented above.

Consider a conformally coupled massive field in a spatially-flat FRW universe. One can define the conformal vacua at $\eta_{\pm}$ with $\chi^{in,out}$ in terms of the positive frequency components. The probability $P_n(\mathbf{k})$ of observing n particles in mode $\mathbf{k}$ at late time is given by the modulus of the ratio of the Bogoliubov coefficients [Park76]: $P_n(\mathbf{k}) = |\beta_k/\alpha_k|^{2n} |\alpha_k|^{-2}$. One can find the average number of particles $\langle N_{\mathbf{k}} \rangle$ created in mode $\mathbf{k}$ (in a comoving volume) at late times to be $n_k \equiv \langle N_k \rangle = \sum_{n=0}^{\infty} n P_n(\mathbf{k}) = |\beta_k|^2$.

The model studied by Bernard and Duncan [BerDun77, BirDav82] has the scale factor $a(\eta)$ evolving like $a^2(\eta) = A + B \tanh \rho\eta$ which tends to constant values $a_{\pm}^2 \equiv A \pm B$ at asymptotic times $\eta \to \pm\infty$. Here ρ measures how fast the scale factor rises, and is the relevant parameter which enters in the temperature of thermal radiance. With this form for the scale function, α_k and β_k have analytic forms in terms of products of gamma functions. One obtains

$$|\beta_k/\alpha_k|^2 = \sinh^2(\pi\omega_-/\rho)/\sinh^2(\pi\omega_+/\rho) \tag{4.156}$$

where

$$\omega_{\pm} = (1/2)(\omega^{out} \pm \omega_{in}) \tag{4.157}$$

$$\omega_{in}^{out} = \sqrt{k^2 + m^2 a_{\pm}^2} \tag{4.158}$$

For cosmological models in which $a(+\infty) \gg a(-\infty)$, the argument of sinh is very large (i.e. $(\pi/\rho)\omega_{\pm} \gg 1$). To a good approximation this has the form $|\beta_k/\alpha_k|^2 = \exp(-2\pi\omega_{in}/\rho)$. For high momentum modes, one can recognize the Planckian distribution with temperature given by $k_B T_\eta = \hbar\rho/(2\pi a_+)$ as detected by an observer (here in the conformal vacuum) at late times.

4.6.4 Conformal stress–energy tensor

The conformal vacuum in the above section is well defined at all times and is useful to describe particle creation for fields which are nearly conformal and in spacetimes which are nearly conformally flat. We shall use the conformal wave equation (4.152) for the amplitude function χ for the **k** mode in conformal time to derive the corresponding number density and energy density of conformally invariant fields from spontaneous and stimulated particle production studied before for Minkowski space in Section 4.3. and also to illustrate the adiabatic regularization method.

The appropriate energy–momentum tensor which is conformally related to the flat space counterpart is the so-called "new, improved" one, or simply the conformal energy–momentum tensor [CaCoJa70]

$$\Lambda_{\mu\nu} = \nabla_\mu\phi\nabla_\nu\phi - \frac{1}{2}g_{\mu\nu}\nabla^\rho\phi\nabla_\rho\phi - \frac{1}{2}g_{\mu\nu}m^2\phi^2$$

$$+ \xi\left(R_{\mu\nu} - \frac{1}{2}g_{\mu\nu}R\right)\phi^2 + \xi[g_{\mu\nu}\nabla^2(\phi^2) - \nabla_\mu\nabla_\nu(\phi^2)] \tag{4.159}$$

The conformal wave equation (4.152) has the same form as the generic wave equation for Minkowski space in t time because they are conformally related. So all the results for external field problems in flat space given before are identical for conformal fields in curved spacetime upon the substitution of f by χ, t by η, and $T_{\mu\nu}$ by $\Lambda_{\mu\nu}$ plus a suitable power of the scale factor a to give the correct dimensionality.

The vacuum energy density associated with these particles is given by the expectation value of the $t-t$ component of $\Lambda_{\mu\nu}$ with respect to the conformal vacuum, i.e.

$$\rho_0^{conf} \equiv \langle 0 \mid \Lambda_{00} \mid 0 \rangle = \frac{1}{a^4} \int \frac{d^3\mathbf{k}}{2(2\pi)^3}(\mid \chi'_{\mathbf{k}} \mid^2 + \omega_{\mathbf{k}}^2 \mid \chi_{\mathbf{k}} \mid^2) \tag{4.160}$$

$$= \frac{1}{a^4} \int \frac{d^3\mathbf{k}}{(2\pi)^3}(2s_{\mathbf{k}} + 1)\frac{\hbar\omega_{\mathbf{k}}}{2} \tag{4.161}$$

The energy density of particles produced from an initial n particle state by stimulated production is

$$\rho_n^{\text{conf}} \equiv \langle n \mid \Lambda_{00} \mid n \rangle = \frac{1}{a^4} \int \frac{d^3\mathbf{k}}{(2\pi)^3} (\mid \chi'_\mathbf{k} \mid^2 + \omega_\mathbf{k}^2 \mid \chi_\mathbf{k} \mid^2) \langle a_\mathbf{k}^\dagger a_\mathbf{k} \rangle \quad (4.162)$$

$$= \frac{1}{a^4} \int \frac{d^3\mathbf{k}}{(2\pi)^3} (2s_\mathbf{k} + 1)\hbar\omega_\mathbf{k} \langle N_\mathbf{k}(t_0) \rangle \quad (4.163)$$

Combining (4.160) and (4.162), for a density matrix diagonal in the number state, the total energy density of particles created from the vacuum and from those already present in the n-particle state is given by

$$\rho^{\text{conf}} = \rho_0^{\text{conf}} + \rho_n^{\text{conf}} = \frac{1}{a^4} \int \frac{d^3\mathbf{k}}{(2\pi)^3} \mathcal{A}_\mathbf{k} \hbar\omega_\mathbf{k} \left(\frac{1}{2} + \langle N_\mathbf{k}(t_0) \rangle\right).$$

For a thermal density matrix μ at temperature $T = \beta^{-1}$ the magnification of the n-particle thermal state gives the finite-temperature contribution to particle creation, with energy density

$$\rho_T^{\text{conf}} = \frac{1}{a^4} \int \frac{d^3\mathbf{k}}{(2\pi)^3} (2s_\mathbf{k} + 1)\hbar\omega_\mathbf{k}/(e^{\beta\hbar\omega_\mathbf{k}} - 1) \quad (4.164)$$

If $s_\mathbf{k} = 0$ the Stefan–Boltzmann relation holds for a massless conformal field in a FLRW universe

$$\rho_T^{\text{conf}} = \frac{\pi^2}{30\hbar^3} T^4 \quad (4.165)$$

Thus Ta is a constant throughout the evolution of the radiation-dominated FLRW universe. $N_\gamma \sim (Ta)^3$ is proportional to the number of relativistic particles present or the entropy content of the universe [HarHu79, DeW67, Hu81]. Further discussions of finite-temperature particle creation and the related entropy generation problem can be found in [Hu82, Hu84].

4.6.5 Adiabatic regularization

To apply the adiabatic method to the regularization of the stress energy tensor in an external field or dynamical spacetime, we need to carry out a fourth-order adiabatic expansion. We study a slightly more general wave equation (4.152) for $\chi_\mathbf{k}(\eta_\mathbf{k})$ with natural frequency $\sqrt{\omega_\mathbf{k}^2(\eta) + Q}$ where Q is a term of second adiabatic order. In the cosmological context Q stands for either Q_ξ for a nonconformally coupled scalar field in a FLRW universe or for Q_β for a conformally coupled scalar field in an anisotropic Bianchi I universe.

Taking $n = 6$ in (4.46), we have the fourth adiabatic order positive frequency solution (we will suppress the mode index $\mathbf{k}$ in $\chi, W, \omega, \epsilon$ below)

$$\chi_{(6)} = \hbar^{1/2} \frac{e^{-i\int W_2 dt}}{\sqrt{2W_3}} \quad (4.166)$$

where

$$W_3 = \omega(1 + \epsilon_2 + \epsilon_4)^{1/2} \tag{4.167}$$

Assuming that the solution χ is well-approximated by $\chi_{(6)}$ we have

$$|\chi|^2 = \hbar(2W_3)^{-1}, \qquad |\chi'|^2 = \hbar(2W_3)^{-1}\left[W_3^2 + \frac{1}{4}\left(\frac{d}{d\eta}\ln W_3\right)^2\right] \tag{4.168}$$

The adiabatic frequency corrections are given by

$$\epsilon_{2(2)} = \frac{Q}{\omega^2} - \frac{\bar{\omega}^2}{4} - \frac{\bar{\omega}'}{2\omega}, \qquad \epsilon'_{2(2)} = \frac{Q'}{\omega^2} - 2Q\frac{\bar{\omega}}{\omega} - \frac{\bar{\omega}''}{2\omega} \tag{4.169}$$

where the subscripts in parentheses denote the adiabatic order and we have defined the nonadiabaticity parameter (here, for frequency ω in conformal time η) as $\bar{\omega}_{\mathbf{k}} \equiv \omega'_{\mathbf{k}}/\omega_{\mathbf{k}}^2$. Substituting these into (4.102) and keeping terms of the same adiabatic order (as measured by the time derivatives) we get

$$s_{k(2)} = \frac{1}{16}\bar{\omega}^2$$

$$s_{k(4)} = \frac{1}{16}\left(-(1/2)\frac{\bar{\omega}\bar{\omega}''}{\omega^2} + \frac{1}{4}\frac{\bar{\omega}'^2}{\omega^2} + (1/2)\frac{\bar{\omega}'\bar{\omega}^2}{\omega} + \frac{3}{16}\bar{\omega}^4\right.$$

$$\left. + \frac{Q^2}{\omega^4} + Q'\frac{\bar{\omega}}{\omega^3} - Q\frac{\bar{\omega}'}{\omega^3} - 3Q\frac{\bar{\omega}^2}{\omega^2}\right) \tag{4.170}$$

The adiabatic expansion for particle production in the high-frequency range at the zeroth, second and fourth adiabatic order above matches the quartic, quadratic and logarithmic divergences in the vacuum energy density respectively. Substituting these expressions for $s_{k(div)} = s_{k(2)} + s_{k(4)}$ for each $\mathbf{k}$ mode into the vacuum energy density (4.160) we can identify the divergent vacuum energy density contributions as

$$\rho_{0(div)} = \frac{1}{a^4}\int\frac{d^3\mathbf{k}}{(2\pi)^3}(2s_{k(div)} + 1)\frac{\hbar\omega_{\mathbf{k}}}{2} \tag{4.171}$$

Subtracting these we get the regularized vacuum energy density given by $\rho_{0(reg)} = \rho_0 - \rho_{0(div)}$. These results were obtained by [ZelSta71, Hu74, FuPaHu74].

We note again that the above adiabatic expressions give the amount of particle creation only in the high-frequency modes when $\bar{\omega}_{\mathbf{k}} \leq 1$. That is why they are suitable for the identification and removal of ultraviolet divergences in the energy–momentum tensor. Adiabatic regularization has been applied to cosmological particle creation with back-reaction [ParFul73, HuPar77, HuPar78].

4.6.6 A simple model of a cosmological phase transition

As a final example of quantum field dynamics in conformally flat universes, we shall show a simple model of the development of a cosmological phase transition through spinodal decomposition. Our discussion follows [SCHR99].

Let us consider a $\lambda \Phi^4$ theory on a spatially flat, expanding Friedmann–Lemaitre–Robertson–Walker universe. We assume the field is conformally coupled ($\xi = 1/6$) but has a bare mass m_b^2, thus breaking conformal invariance. The field equation now has an extra term $a^2(\eta) \lambda_B \Phi^3 / 6$ describing the self-interaction. However, at early times we may adopt the Hartree approximation

$$\Phi^3 \sim 3 \langle \Phi^2 \rangle (\eta) \, \Phi \qquad (4.172)$$

and the wave equation becomes formally the equation for a free field with a self-consistent mass

$$m_{\text{eff}}^2 (\eta) = m_b^2 + \frac{\lambda_B}{2} \langle \Phi_{\text{HF}}^2 \rangle (\eta) \qquad (4.173)$$

We are assuming of course that the initial condition is also spatially homogeneous, so that $\langle \Phi^2 \rangle$ depends only on η. The "free" field Φ_{HF} admits a mode expansion in terms of conformal amplitudes χ_k which obey equation (4.152) with $Q = 0$ and $m^2 = m_{\text{eff}}^2(\eta)$, and boundary conditions

$$\chi_k(0) = \sqrt{\frac{\hbar}{2\omega_k(0)}} \qquad (4.174)$$

$$\chi_k'(0) = -i\sqrt{\frac{\hbar \omega_k(0)}{2}} \qquad (4.175)$$

We assume the expectation values

$$\left\langle a_{\mathbf{k}}^\dagger a_{\mathbf{k'}} \right\rangle = n_k \delta(\mathbf{k} - \mathbf{k'}); \qquad \left\langle a_{\mathbf{k}}^{\dagger 2} \right\rangle = \left\langle a_{\mathbf{k}}^2 \right\rangle = 0 \qquad (4.176)$$

whereby

$$\left\langle \Phi_{\text{HF}}^2 \right\rangle (\eta) = \frac{1}{a^2(\eta)} \int \frac{d^3\mathbf{k}}{(2\pi)^3} \, |\chi_k(\eta)|^2 \, (1 + 2n_k) \qquad (4.177)$$

As in flat spacetime, $\left\langle \Phi_{\text{HF}}^2 \right\rangle (\eta)$ diverges. The theory may be rendered finite by imposing a cut-off at some physical scale $\Lambda_{\text{phys}} = \Lambda(\eta)/a(\eta)$. However the resulting renormalized parameters are strongly cut-off dependent. To eliminate this dependence, let κ_{phys} be a second physical scale, large enough that for modes higher than $\kappa(\eta) = a(\eta) \kappa_{\text{phys}}$ the mode functions χ_k are well approximated by adiabatic modes, but still much lower than Λ. Then we write

$$\left\langle \Phi_{\text{HF}}^2 \right\rangle (\eta) = \hbar \left\{ \frac{\Lambda^2}{8\pi^2} - \frac{m_{\text{eff}}^2(\eta)}{8\pi^2} \ln\left[\frac{\Lambda}{\kappa}\right] + \mu^2(\eta) \right\} \qquad (4.178)$$

$$\mu^2(\eta) = \frac{1}{a^2(\eta)} \int^{a\Lambda} \frac{d^3\mathbf{k}}{(2\pi)^3} \left\{ \frac{|\chi_k(\eta)|^2}{\hbar} (1 + 2n_k) - \frac{1}{2k} + \frac{a^2(\eta) m_{\text{eff}}^2(\eta)}{4k^3} \theta(k - a\kappa) \right\}$$

$$(4.179)$$

The point is that $\mu^2(\eta)$ is essentially cut-off independent. The gap equation now reads

$$m_{\text{eff}}^2(\eta) = m_{\text{b}}^2 + \frac{\hbar \lambda_B}{2} \left\{ \frac{\Lambda^2}{8\pi^2} - \frac{m_{\text{eff}}^2(\eta)}{8\pi^2} \ln \left[\frac{\Lambda}{\kappa} \right] + \mu^2(\eta) \right\} \qquad (4.180)$$

The bare mass m_{b}^2 is defined by the condition that in flat space time $(a = 1)$ and at the critical temperature T_{C}, $m_{\text{eff}}^2 = 0$. Thus

$$0 = m_{\text{b}}^2 + \frac{\hbar \lambda_B}{2} \left\{ \frac{\Lambda^2}{8\pi^2} + \frac{T_{\text{C}}^2}{6} \right\} \qquad (4.181)$$

where we set the Boltzmann constant $k_{\text{B}} = 1$, for simplicity. We now have the finite gap equation

$$m_{\text{eff}}^2(\eta) = \frac{\hbar \lambda}{2} \left\{ \mu^2(\eta) - \frac{T_{\text{C}}^2}{6} \right\} \qquad (4.182)$$

where

$$\frac{1}{\lambda} = \frac{1}{\lambda_B} + \frac{\hbar}{16\pi^2} \ln \left[\frac{\Lambda}{\kappa} \right] \qquad (4.183)$$

We may now start discussing the early time evolution of the field. The central aspect of this behavior is the suppression factor $a^{-2}(\eta)$ in $\mu^2(\eta)$ (cf. equation (4.179)). Because of this factor, $m_{\text{eff}}^2(\eta)$ decreases and eventually becomes negative. Indeed, assume the initial spectrum n_k corresponds to a Planck distribution with temperature $T_0^2 \gg m_{\text{eff}}^2(0)$, T_{C}^2. Then, when $m_{\text{eff}}^2(\eta)$ is small we get

$$m_{\text{eff}}^2(\eta) \sim \frac{\hbar \lambda}{12} \left\{ \frac{T_0^2}{a^2(\eta)} - T_{\text{C}}^2 \right\} \qquad (4.184)$$

and so

$$\omega_k^2(\eta) = k^2 + \frac{\hbar \lambda}{12} \left[T_0^2 - a^2(\eta) T_{\text{C}}^2 \right] \qquad (4.185)$$

If the expansion is slow enough, we may approximate this by

$$\omega_k^2(\eta) = k^2 - \frac{1}{\tau} \frac{\hbar \lambda T_0^2}{12} (\eta - \eta_{\text{C}}) \qquad (4.186)$$

where η_{C} is the conformal time at which m_{eff}^2 vanishes for the first time, and $\tau^{-1} = (2aH)(\eta_{\text{C}})$ is the quench rate. $H = a'/a^2$ is the Hubble constant.

At $\eta = \eta_{\text{C}}$ the homogeneous mode becomes unstable. If m_{eff}^2 actually becomes negative, then other infrared modes become unstable as well, and the corresponding mode functions start to grow exponentially. The result is the formation of an infrared peak. Eventually, though, the approximation (4.186) becomes invalid.

To obtain an improved estimate, observe that once $\left| (\omega_k^2)' \right| \leq \left| \omega_k^3 \right|$ we may approximate the mode functions by WKB wave forms. This inequality translates into

$$\frac{1}{\tau}\frac{\hbar\lambda T_0^2}{12} \leq \left[\frac{1}{\tau}\frac{\hbar\lambda T_0^2}{12}\left(\eta - \eta_C\right) - k^2\right]^{3/2} \tag{4.187}$$

When this inequality holds, we may write

$$\chi_k\left(\eta\right) \sim \frac{\sqrt{\hbar}}{\sqrt{2\left|\omega_k\left(\eta\right)\right|}}e^{S_k\left(\eta\right)} \tag{4.188}$$

where

$$S_k\left(\eta\right) = \int_{\eta_k}^{\eta}\left|\omega_k\left(\eta'\right)\right|d\eta' = \frac{2}{3}\left(\frac{12\tau}{\hbar\lambda T_0^2}\right)\left[\frac{\hbar\lambda T_0^2}{12\tau}\left(\eta - \eta_C\right) - k^2\right]^{3/2} \tag{4.189}$$

$$\eta_k = \eta_C + \tau\left(\frac{12k^2}{\hbar\lambda T_0^2}\right) < \eta \tag{4.190}$$

In the infrared, we may approximate

$$S_k\left(\eta\right) = S_0\left(\eta\right) - \frac{1}{2}\sigma^2\left(\eta\right)k^2 \tag{4.191}$$

$$\sigma^2\left(\eta\right) = \left(\frac{48\tau}{\hbar\lambda T_0^2}\left(\eta - \eta_C\right)\right)^{1/2} \tag{4.192}$$

Provided $\eta_{\sigma^{-1}} < \eta$, we get

$$\mu^2\left(\eta\right) = \frac{1}{a^2\left(\eta\right)}\left[\frac{T_0^2}{6} + \frac{e^{2S_0\left(\eta\right)}T_0}{\left(2\pi\right)^2\left|\omega_0\left(\eta\right)\right|\sigma^2\left(\eta\right)}\right] \tag{4.193}$$

The infrared peak in mode space is correlated with the appearance of correlated domains in physical space, whose comoving size is $\sigma\left(\eta\right)$ and increases with time (coarse graining). We see that the exponential growth of the infrared peak counterbalances the red-shift due to the Hubble expansion. If we simply extrapolate this model, we conclude that eventually the infrared peak becomes dominant, and the effective mass is driven again to zero from below.

The actual picture is more involved. Within these domains, there is nondiagonal long-range order, and we may describe the field as a quantum field (represented by the stable modes) evolving on a nontrivial background field (which is the "square root" of the infrared peak). When the background field gets large enough, it starts to oscillate around the true equilibrium position. The quantum field then becomes a periodically driven field and, as we have seen, parametric amplification results in copious particle production from the background field.

4.7 Particle creation as squeezing

In this section we will use the language of squeezed states [CavSch85, Sch86] to treat a neutral scalar field in a dynamic background field or spacetime. This approach will shed a clearer light on two interrelated issues:

(a) Dependence of particle creation on the initial state. We consider in particular the number state, the coherent and the squeezed state.
(b) The relation of spontaneous and stimulated particle creation and their dependence on the initial state.

We also derive the result for the fluctuations in particle number in anticipation of its relevance to defining noise in quantum fields. Our presentation here follows [HuKaMa94].

Since the concept of squeezed state was introduced to quantum optics in the 1970s [CavSch85, Sch86, Gla05], there has been much progress in seeking its experimental realizations and theoretical implications. The language of squeezed states as a way to describe cosmological particle creation was introduced by Grishchuk and Sidorov [GriSid90]. Although the physics is not new (this was also pointed out by Albrecht *et al.* [AFJP94] in the inflationary cosmology context) and the results are largely known, the use of rotation and squeeze operators gives an alternative description which allows one to explore new avenues based on interesting ideas developed in quantum optics. Work on entropy generation in cosmological perturbations by Brandenberger and coworkers [BrMuPr92, BrMuPr93] and Gasperini and Giovannini [GasGio93, GasGioVen93] make use of coarse graining via a random phase approximation. Matacz [Mat94, LafMat93] has used the squeezed state formalism as a starting point for the study of decoherence of cosmological inhomogeneities in the coherent-state representation.

The issues of initial states and entropy generation have been discussed in restricted conditions, and the issue of spontaneous and stimulated production has only been touched upon before. For the sake of completeness, we will address these issues under a common framework, using the language of squeezed states, and present the results for different initial states (the number state, the coherent state and the squeezed state).

4.7.1 Evolutionary operator, squeezing and rotation

We now present a description of particle creation by means of the evolutionary operator U defined by

$$\tilde{a}_{\pm\mathbf{k}}(t) = U(t)a_{\pm\mathbf{k}}U^{\dagger}(t) \qquad (4.194)$$

where $UU^{\dagger} = 1$. The form of U was deduced by Parker [Park69] following Kamefuchi and Umezawa [KamUme64]. In the modern language of squeezed states [CavSch85, Sch86], one can write $U = RS$ as a product of two unitary operators, the **rotation operator**

$$R(\theta) = \exp[-i\theta(a_{+}^{\dagger}a_{+} + a_{-}^{\dagger}a_{-})] \qquad (4.195)$$

and the **two-mode squeeze operator**

$$S_2(r, \phi) = \exp[r(a_{+}a_{-}e^{-2i\phi} - a_{+}^{\dagger}a_{-}^{\dagger}e^{2i\phi})] \qquad (4.196)$$

where r is the squeeze parameter with range $0 \leq r < \infty$ and ϕ, θ are the rotation parameters with ranges $-\pi/2 < \phi \leq \pi/2, 0 \leq \theta < 4\pi$. (These parameters and U, R and S should all carry the label $\mathbf{k}$. The $\pm$ on a refer to the $\pm\mathbf{k}$ modes.) Note that

$$S_2^\dagger(r, \phi) = S_2^{-1}(r, \phi) = S_2(r, \phi + \pi/2) \tag{4.197}$$

The three real functions $(\theta_\mathbf{k}, \phi_\mathbf{k}, r_\mathbf{k})$ are related to the two complex functions $(\alpha_\mathbf{k}, \beta_\mathbf{k})$ by

$$\alpha_\mathbf{k} = e^{i\theta_\mathbf{k}} \cosh r_\mathbf{k}, \qquad \beta_\mathbf{k} = e^{i(\theta_\mathbf{k} - 2\phi_\mathbf{k})} \sinh r_\mathbf{k} \tag{4.198}$$

For mode decompositions in spatially homogeneous spacetimes leading to no mode couplings, the Bogoliubov transformation connecting the $a_\mathbf{k}$ and the $\tilde{a}_\mathbf{k}$ operators is given by equation (4.28) (for more general situations, see [Hu72]). We see that because of the linear dependence of $\tilde{a}_{+\mathbf{k}}$ on $a_{+\mathbf{k}}$ and $a_{-\mathbf{k}}^\dagger$ (but not $a_{+\mathbf{k}}^\dagger$) a two-mode squeeze operator is needed to describe particle pairs in states $\pm\mathbf{k}$.

The physical meaning of rotation and squeezing can be seen from the result of applying these operators for a single-mode harmonic oscillator as follows: (the kth mode label is omitted below unless needed explicitly).

The Hamiltonian is

$$H_0 = \hbar\Omega \left(a^\dagger a + \frac{1}{2} \right) \tag{4.199}$$

Under rotation,

$$R|0\rangle = |0\rangle, \qquad RaR^\dagger = e^{i\theta} a \tag{4.200}$$

Also,

$$R(\theta)R(\theta') = R(\theta + \theta') \tag{4.201}$$

This implies that

$$R\hat{x}R^\dagger = (\cos\theta)\,\hat{x} - (\sin\theta)\,\hat{p} \tag{4.202}$$
$$R\hat{p}R^\dagger = (\sin\theta)\,\hat{x} + (\cos\theta)\,\hat{p} \tag{4.203}$$

where

$$a = \frac{1}{\sqrt{2\hbar}} \left(\sqrt{M\Omega}\hat{x} + i\frac{\hat{p}}{\sqrt{M\Omega}} \right) \tag{4.204}$$

Thus the name rotation. Let $\delta a = a - \langle a\rangle$ (where $\langle\ \rangle$ denotes the expectation value with respect to any state); then the second-order noise moments of a are defined as [CavSch85, Sch86]:

$$\langle(\delta a)^2\rangle = \langle a^2\rangle - \langle a\rangle^2 = \langle(\delta a^\dagger)^2\rangle^*$$
$$= \frac{1}{2\hbar}[M\Omega\langle(\delta x)^2\rangle - (M\Omega)^{-1}\langle(\delta p)^2\rangle] + i\langle(\delta x\delta p)_\text{sym}\rangle \tag{4.205}$$
$$\langle|\delta a|^2\rangle = \frac{1}{2}\langle\delta a\delta a^\dagger + \delta a^\dagger\delta a\rangle = \frac{1}{2\hbar}[M\Omega\langle(\delta x)^2\rangle + (M\Omega)^{-1}\langle(\delta p)^2\rangle] \tag{4.206}$$

The first quantity is the variance of a, a complex second moment, while the second is the correlation, a real second moment, which, as seen in the more familiar x, p representation, measures the mean-square uncertainty (called total noise in [CavSch85, Sch86]). Rotation preserves the number operator

$$Ra^\dagger a R^\dagger = a^\dagger a \tag{4.207}$$

It rotates the moment

$$\langle R(\delta a)^2 R^\dagger \rangle = e^{2i\theta} \langle (\delta a)^2 \rangle \tag{4.208}$$

corresponding to a redistribution between $\hat{x}, \hat{p}$, but preserves the uncertainty

$$\langle R|\delta a|^2 R^\dagger \rangle = \langle |\delta a|^2 \rangle \tag{4.209}$$

One can define a **displacement operator** as

$$D(\mu) = \exp[\mu a^\dagger - \mu^* a] \tag{4.210}$$

Note that $D^{-1}(\mu) = D^\dagger(\mu) = D(-\mu)$. The coherent state can be defined as

$$|\mu\rangle = D(\mu)|0\rangle \tag{4.211}$$

Thus

$$a|\mu\rangle = \mu|\mu\rangle \tag{4.212}$$

and

$$Da^\dagger a D^\dagger = a^\dagger a - (\mu a^\dagger + \mu^* a) + |\mu|^2 \tag{4.213}$$

Under displacement,

$$D(\mu)aD^\dagger(\mu) = a - \mu \tag{4.214}$$

The displacement operation also preserves the uncertainty

$$\langle D|\delta a|^2 D^\dagger \rangle = \langle |\delta a|^2 \rangle \tag{4.215}$$

The **single-mode squeeze operator** is defined as

$$S_1(r, \phi) = \exp\left[\frac{r}{2}(a^2 e^{-2i\phi} - a^{\dagger 2} e^{2i\phi})\right] \tag{4.216}$$

If we construct a Gaussian state in the position basis, with initially the same width σ_0 as that of the ground state of such an ordinary harmonic oscillator, displaced by some arbitrary amount and with a phase proportional to x, we find this to be an eigenstate of the lowering operator, and is called a coherent state. Suppose we locate the point (x, p) in phase space and draw an ellipse about this point, the lengths of whose axes are the uncertainties $\Delta x^2, \Delta p^2$. Then as the oscillator evolves this uncertainty ellipse revolves about the origin with angular speed Ω. A squeezed state is again such a state, but with an arbitrary initial width σ. We find that as the oscillator evolves the uncertainty ellipse again revolves about the origin, but its axes change length and it can also rotate about its own center. It turns out that the squeeze parameter r is related to the width of such a state:

$$r = \ln \frac{\sigma_0}{\sigma}, \quad \sigma_0 \equiv \sqrt{\frac{\hbar}{2M\Omega}} \tag{4.217}$$

Hence a coherent state has $r = 0$, or zero squeezing. A Gaussian that initially has a width smaller than σ_0 will evolve to a squeezed state with some $r > 0$. A squeezed state is formed by squeezing a coherent state,

$$|\sigma\rangle_\mu = S_1(r, \phi)|\mu\rangle \tag{4.218}$$

or,

$$|\sigma\rangle_\mu = |r, \phi, \mu\rangle = S_1(r, \phi)D(\mu)|0\rangle \tag{4.219}$$

Call $a_{S1} = S_1 a S_1^\dagger$. Then

$$a_{S1}|\sigma\rangle = \mu|\sigma\rangle \tag{4.220}$$

and

$$a_{S1} = S_1 a S_1^\dagger = a \cosh r + e^{2i\phi} a^\dagger \sinh r \tag{4.221}$$

Thus a squeezed state in the Fock space of a becomes a coherent state in the Fock space of a_{S1} with the same eigenvalue. From this we see the result of S_1 acting on $\hat{x}$ and $\hat{p}$:

$$S_1 \hat{x} S_1^\dagger = (\cosh r + \cos 2\phi \sinh r)\hat{x} + (\sin 2\phi \sinh r)(\hat{p}/(M\Omega)) \tag{4.222}$$
$$S_1 \hat{p} S_1^\dagger = (\cosh r - \cos 2\phi \sinh r)\hat{p} + (\sin 2\phi \sinh r)(M\Omega)\hat{x} \tag{4.223}$$

For $\phi = \pi/2$, these give

$$S_1 \hat{x} S_1^\dagger = e^{-r}\hat{x}, \quad S_1 \hat{p} S_1^\dagger = e^r \hat{p} \tag{4.224}$$

Hence the name squeezing. Two successive squeezes with the same rotation parameter result in one squeeze with the squeeze parameter as the sum of the two parameters:

$$S_1(r, \phi)S_1(r', \phi) = S_1(r + r', \phi) \tag{4.225}$$

The expectation value of squeezing the number operator is

$$\langle S_1^\dagger a^\dagger a S_1 \rangle = \sinh^2 r + (1 + 2\sinh^2 r)\langle a^\dagger a \rangle + \sinh 2r \, \mathrm{Re}[e^{-2i\phi}\langle a^2 \rangle] \tag{4.226}$$

and that of the correlation is

$$\langle S_1^\dagger |\delta a|^2 S_1 \rangle = \cosh 2r \langle |\delta a|^2 \rangle + \sinh 2r \, \mathrm{Re}[e^{-2i\phi}\langle (\delta a)^2 \rangle] \tag{4.227}$$

which for the vacuum and coherent states is always greater than or equal to the original value.

The two-mode squeeze operator S_2 defined in (4.196) is more suitable for the description of cosmological particle creation. One can show that the *out* state is generated from the *in* state by including contributions from all $\mathbf{k}$ modes,

$$|out\rangle = RS|in\rangle \qquad (4.228)$$

where

$$S = \Pi_{\mathbf{k}=0}^{\infty} S_2(r_{\mathbf{k}}, \phi_{\mathbf{k}}) \qquad (4.229)$$

In general

$$\langle out|F(\tilde{a}_{\pm}, \tilde{a}_{\pm}^{\dagger})|out\rangle = \langle in|F(a_{\pm}, a_{\pm}^{\dagger})|in\rangle \qquad (4.230)$$

where F is an arbitrary analytic function. The $|in\rangle$ state can be a number state, a coherent state or a squeezed state. If the initial state is a vacuum state, $|in\rangle = |0in\rangle$, then

$$|0out\rangle = S(r, \phi - \frac{\theta}{2})|0in\rangle \qquad (4.231)$$

where

$$S(r, \phi - \theta) = \exp\{\Sigma_{\mathbf{k}} r_{\mathbf{k}}[e^{-2i(\phi_{\mathbf{k}}-\theta_{\mathbf{k}})}a_{\mathbf{k}}a_{-\mathbf{k}} - e^{2i(\phi_{\mathbf{k}}-\theta_{\mathbf{k}})}a_{\mathbf{k}}^{\dagger}a_{-\mathbf{k}}^{\dagger}]\} \qquad (4.232)$$

The squeeze parameter $\sinh^2 r_{\mathbf{k}} = |\beta_{\mathbf{k}}|^2$ measures the number of particles created. Rotation does not play a role. Thus, as observed by Grishchuk and Sidorov [GriSid90], cosmological particle creation amounts to squeezing the vacuum. The same can be said about Hawking radiation [Haw75]. See [HuKaMa94].

4.7.2 *Dynamics of the squeezing parameters*

So far we have used the language of squeezed states to describe the integrated effect of the dynamics, as in equation (4.194). We will show now that for a linear system the dynamics itself may be described in terms of the evolution of the squeezing parameters r, ϕ and θ as functions of time.

Let us begin with a general quadratic Lagrangian. This Lagrangian has time-dependent mass and frequency, and we will also allow it to have a time-dependent cross-term, denoted $2\mathcal{E}(t)$:

$$L = \frac{M(t)}{2}\left[\dot{x}^2 + 2\mathcal{E}(t)\dot{x}x - \Omega^2(t)x^2\right] \qquad (4.233)$$

We perform a Legendre transformation to obtain the Hamiltonian, and switch to creation–destruction operators

$$a = \frac{1}{\sqrt{2\hbar}}\left(\sqrt{\kappa}\hat{x} + i\frac{\hat{p}}{\sqrt{\kappa}}\right) \qquad (4.234)$$

where κ is an arbitrary positive constant related to the frequency. The result is [HuMat94]

$$\hbar^{-1}H(t) = g(t)\frac{a^2}{2} + g^*(t)\frac{a^{\dagger 2}}{2} + h(t)(a^\dagger a + 1/2) \tag{4.235}$$

$$g = \frac{1}{2}\left[\frac{M}{\kappa}(\Omega^2 + \mathcal{E}^2) - \frac{\kappa}{M} + 2i\mathcal{E}\right]$$

$$h = \frac{1}{2}\left[\frac{M}{\kappa}(\Omega^2 + \mathcal{E}^2) + \frac{\kappa}{M}\right] \tag{4.236}$$

The value of κ can be chosen so that at the initial time $g(t_i) = 0$. Thus if $\mathcal{E} = 0$ we will usually have $\kappa = M(t_i)\Omega(t_i)$.

The evolution operator $U = SR$ may be written as the product of a single-mode squeeze operator S and a rotation operator R, which in turn are parameterized in terms of a squeeze parameter r and angles θ and ϕ as in equations (4.195) and (4.216). Acting on the destruction operator, U induces a Bogoliubov transformation as in equation (4.194), with Bogoliubov coefficients given in equation (4.198). Their equations of motion are

$$\dot{\alpha} = -ih\alpha - ig^*\beta$$
$$\dot{\beta} = ig\alpha + ih\beta$$
$$\alpha(t_i) = 1, \quad \beta(t_i) = 0 \tag{4.237}$$

with g, h as defined in equation (4.236).

A quantity of much importance turns out to be the sum of the Bogoliubov coefficients, $\chi \equiv \alpha + \beta$. It follows from equations (4.237) that χ satisfies the classical equation of motion for the system:

$$\ddot{\chi} + \frac{\dot{M}}{M}\dot{\chi} + \left(\Omega^2 + \dot{\mathcal{E}} + \frac{\dot{M}\mathcal{E}}{M}\right)\chi = 0 \tag{4.238}$$

with initial conditions

$$\chi(t_i) = 1; \quad \dot{\chi}(t_i) = \frac{-i\kappa}{M(t_i)} - \mathcal{E}(t_i) \tag{4.239}$$

With this result, the usual task of finding the Bogoliubov coefficients α, β from two coupled first-order differential equations is reduced to that of solving one second-order equation for χ. We have two equations

$$\chi = \alpha + \beta$$
$$\dot{\chi} = i(g - h)\alpha + i(h - g^*)\beta \tag{4.240}$$

so, solving for α, β using equation (4.236):

$$\left\{\begin{matrix} \alpha \\ \beta \end{matrix}\right\} = \frac{1}{2}\left(1 \pm \frac{i\mathcal{E}M}{\kappa}\right)\chi \pm \frac{iM}{2\kappa}\dot{\chi} \tag{4.241}$$

Equivalently, we can follow the behavior of r, ϕ, θ by writing equation (4.237) in terms of the squeeze parameter, with $g \equiv |g|e^{i\delta}$:

$$\dot{r} = |g| \sin(2\phi + \delta)$$
$$\dot{\phi} = -h + |g| \coth 2r \cos(2\phi + \delta)$$
$$\dot{\theta} = h - |g| \tanh r \cos(2\phi + \delta) \qquad (4.242)$$

As an example, consider an inverted oscillator, where the coefficients in the Lagrangean are time-independent and $\Omega^2 < 0$. The variable χ blows up, and so does the squeeze parameter $r \rightarrow \ln(2|\alpha|)$. If we set $r \rightarrow \infty$ then the equations for ϕ, θ become

$$\dot{\theta} = -\dot{\phi} = h - |g| \cos(2\phi + \delta) \qquad (4.243)$$

or

$$t = \int_{\phi_0}^{\phi} \frac{d\varphi}{|g| \cos(2\varphi + \delta) - h} \qquad (4.244)$$

The integral may be solved analytically but we do not need the result in what follows. Simply observe that $\Omega^2 < 0$ implies $|g| > h$, and so as $t \rightarrow \infty$, ϕ must approach a zero of $|g| \cos(2\varphi + \delta) - h$, so that the integral increases without bound. This makes $\dot{\theta} \rightarrow 0$ too. Therefore for late times ϕ and θ approach constant values, while r increases.

4.7.3 Number, coherence and initial states

We will show in this section that the number of particles produced depends very much on the initial state chosen. The number operator for a particle pair in mode k is given by

$$N = a_+^\dagger a_+ + a_-^\dagger a_- \qquad (4.245)$$

Note that the subscripts $\pm$ here denote a particle pair in states $\pm\mathbf{k}$ whereas in the charged particle case $+, -$ denote particle and antiparticle states respectively. For the charged particle case since at the end we assume the number of positive and negative charged particles is the same, it gives the same expression as a neutral particle there. However, here since we count the two states as distinct, we should have twice the amount for vacuum particle production.

The expectation value of the number operator with respect to the $|out\rangle$ vacuum for a general initial state is

$$\tilde{N} = \langle N \rangle_t = \langle S_2^\dagger R^\dagger N R S_2 \rangle = 2|\beta|^2 + (1 + 2|\beta|^2)\langle N \rangle$$
$$-2|\alpha||\beta|(e^{2i\phi}\langle a_+^\dagger a_-^\dagger \rangle + e^{-2i\phi}\langle a_+ a_- \rangle) \qquad (4.246)$$

Comparing this expression with (4.103) or (4.108), the only difference of a factor of 2 for the first $|\beta|^2$ term comes from the spontaneous creation of particles in the $\pm\mathbf{k}$ modes. The net change in the particle number from the initial to the final state is

$$\delta N \equiv \langle N \rangle_t - \langle N \rangle = 2|\beta|^2[1 + \langle N \rangle] - 2|\beta||\alpha|\{e^{2i\phi}\langle a_+^\dagger a_-^\dagger \rangle + e^{-2i\phi}\langle a_+ a_- \rangle\} \tag{4.247}$$

Here, the first two terms in the square brackets are respectively the spontaneous and stimulated emissions and the last term in the curly brackets is the interference term. The difference between spontaneous and stimulated creation of particles in cosmology was explained first by Parker [Park69] and explored in more detail by Hu and Kandrup [HuKan87]. Note that since there is no θ dependence, rotation has no effect. If $r_{\mathbf{k}} \neq 0$ for some $\mathbf{k}$ both spontaneous and stimulated contributions are positive. The interference term can be negative for states which give nonzero $\langle a_+ a_- \rangle$. Only when this term is non-zero can δN be negative.

We will calculate the change in particle number for some specific initial states.

(a) **Number state**

For an initial number state $|n\rangle = |n_+, n_-\rangle$

$$\delta N = 2|\beta|^2(1 + n_+ + n_-) \tag{4.248}$$

We see that the number of particles will always increase.

(b) **Coherent state**

For an initial coherent state

$$|\mu\rangle = D(\mu_+)D(\mu_-)|0, 0\rangle \tag{4.249}$$

we find that

$$\delta N = 2|\beta|^2[1 + \langle N_+ \rangle + \langle N_- \rangle] - 4|\beta||\alpha|\sqrt{\langle N_+ \rangle \langle N_- \rangle}\cos(2\phi - \zeta_+ - \zeta_-) \tag{4.250}$$

where

$$\mu_+ = \sqrt{\langle N_+ \rangle}e^{i\zeta_+}, \quad \mu_- = \sqrt{\langle N_- \rangle}e^{i\zeta_-} \tag{4.251}$$

Note the existence of the interference term which can give a negative contribution. It depends not only on the squeeze parameters $|\beta|$ and ϕ, but also on the particles present and the phase of the initial coherent state. Conditions favorable to a decrease in δN are $\cos(2\phi - \zeta_+ - \zeta_-) = 1$ and $\langle N_+ \rangle = \langle N_- \rangle = \langle N \rangle/2$. In this case we find δN is negative if

$$\langle N \rangle > \frac{|\beta|}{|\alpha| - |\beta|} \tag{4.252}$$

(c) **Single-mode squeezed vacuum state**

For an initial one-mode squeezed state

$$|\sigma\rangle_1 = S_{1+}(r_+, \phi_+)S_{1-}(r_-, \phi_-)|0, 0\rangle \tag{4.253}$$

generated by squeezing the vacuum with $S_{1\pm}$ for the $\pm\mathbf{k}$ modes, we get

$$\delta N = 2|\beta|^2(1 + \langle N_+ \rangle + \langle N_- \rangle) \qquad (4.254)$$

Once again particle number will always increase.

(d) **Two-mode squeezed vacuum state**

For an initial two-mode squeezed vacuum

$$|\sigma\rangle_2 = S_2(r_0, \phi_0)|0, 0\rangle \qquad (4.255)$$

where S_2 is defined earlier,

$$\delta N = 2|\beta|^2[1 + \langle N \rangle] + 2|\beta||\alpha|\sqrt{\langle N \rangle(2 + \langle N \rangle)}\cos 2(\phi - \phi_0) \qquad (4.256)$$

The cosine factor shows that particle number can decrease given the right phase relations. It can be shown that for $\cos 2(\phi - \phi_0) = -1$ particle number would decrease $(\delta N \leq 0)$ if $r_0 \geq r/2$. If the phase information is randomized the cosine factor averages to zero and there is a net increase in particle number. Since a squeezed state is the end result of squeezing a vacuum via particle creation, one might naively expect to see a monotonic increase in number. Our result shows that this is true only if the phase information is lost in the squeezed state to begin with.

In summary we can make the following observations:

(a) Rotation R in the evolution operator $U = RS$ does not influence particle creation.
(b) For an initial number state or single-mode squeezed vacuum we find a net increase in the number of particles.
(c) For an initial coherent state and two-mode squeezed vacuum, particle number can increase or decrease. A net increase can nevertheless be obtained by suitable choices of $S_2(r, \phi)$ and $S_2(r_0, \phi_0)$.
(d) If random phase is assumed for the initial state the interference term can be averaged out to zero and there will be a net increase in number of particles.

Coherence can persist

A measure of the coherence of the system is given by the uncertainty (called variance in [Hu72, Mol67, BroCar79, HuPav86])

$$|\delta a|^2 = \frac{1}{2}(\delta a \delta a^\dagger + \delta a^\dagger \delta a) \qquad (4.257)$$

where $\delta a = a - \langle a \rangle$. The expectation value of the uncertainty with respect to a state $|\psi\rangle$ is thus,

$$\langle \psi || \delta a|^2 |\psi\rangle = \langle \psi|a^\dagger a|\psi\rangle - |\langle \psi|a|\psi\rangle|^2 + \frac{1}{2} \qquad (4.258)$$

The expectation value of the uncertainty with respect to a transformed state $|\psi\rangle_t \equiv RS|\psi\rangle$ is given by

$$\langle\psi||\delta a|^2|\psi\rangle_t = \cosh 2r\langle\psi||\delta a|^2|\psi\rangle - 2\sinh 2r\mathrm{Re}[e^{-2i\phi}\langle\psi|\delta a_+\delta a_-|\psi\rangle] \quad (4.259)$$

where $|\delta a|^2 \equiv |\delta a_+|^2 + |\delta a_-|^2$. For an initial number state, $|\psi\rangle = |n\rangle$,

$$\langle n||\delta a|^2|n\rangle_t = 2\left(\frac{1}{2} + |\beta|^2\right)\langle n||\delta a|^2|n\rangle \geq \langle n||\delta a|^2|n\rangle \quad (4.260)$$

For a coherent state, $|\psi\rangle = |\mu\rangle$

$$\langle\mu||\delta a|^2|\mu\rangle_t = 2\left(\frac{1}{2} + |\beta|^2\right)\langle\mu||\delta a|^2|\mu\rangle \geq \langle\mu||\delta a|^2|\mu\rangle \quad (4.261)$$

where the first term corresponds to the vacuum fluctuation and the second term (whose sum over all modes is equivalent to $\mathrm{Tr}(v_{\mathbf{k}}^\dagger v_{\mathbf{k}})$ in [Hu72, HuPav86]) measures the mixing of the positive and negative frequency components of different modes. This result was first derived in [Hu72], and discussed further in [HuPav86]. Notice that it is always greater than the original value $\langle|\delta a|^2\rangle_\mu$.

For a squeezed state, $|\psi\rangle = |\sigma\rangle = S_2(r_0, \phi_0)|\mu\rangle$

$$\langle\sigma||\delta a|^2|\sigma\rangle_t = \cosh 2r\langle\sigma||\delta a|^2|\sigma\rangle - 2\sinh 2r\mathrm{Re}[e^{-2i\phi}\langle\sigma|\delta a_+\delta a_-|\sigma\rangle] \quad (4.262)$$

which can be smaller than the initial value.

Notice that of the three states we discussed, only the squeezed state can allow for a decrease in the uncertainty, i.e. an increase in the coherence as the system evolves. In addition, even though the total number and the total uncertainty of the initial state of the two modes change with particle creation, their difference remains a constant. This is because cosmological particle creation is described by the two-mode squeezed operator which satisfies the relations: $\langle\psi|S^\dagger(a_+^\dagger a_+ - a_-^\dagger a_-)S|\psi\rangle = \langle\psi|a_+^\dagger a_+ - a_-^\dagger a_-|\psi\rangle$,

$$\langle\psi|S^\dagger(|\delta a_+|^2 - |\delta a_-|^2)S|\psi\rangle = \langle\psi|(|\delta a_+|^2 - |\delta a_-|^2)|\psi\rangle \quad (4.263)$$

4.7.4 Fluctuations in number

Spontaneous particle creation can be viewed as the parametric amplification of vacuum fluctuations (or squeezing the vacuum). Particle number is an interesting quantity as it measures the degree to which the vacuum is excited. The fluctuation in particle number is another interesting quantity, as it can be related to the noise of the quantum field and the susceptibility of the vacuum. This is similar in nature to the energy fluctuation (measured by the heat capacity at constant volume) of a system being related to the thermodynamic stability of a canonical system, or the number fluctuation (measured by the compressibility at constant pressure) of a system being related to the thermodynamic stability of a grand canonical system. In gravity, we know that the number fluctuation of a self-gravitating system can be used as a measure of its heat capacity (negative)

[LynBel77]; and those associated with particle creation from a black hole can be used in a linear-response theory description as a measure of the susceptibility of spacetime [CanSci77, Mot86]. We expect that this quantity associated with cosmological particle creation may provide some important information about quantum noise and vacuum instability.

Define $\delta_i O \equiv [\langle O^2 \rangle - \langle O \rangle^2]$ as the variance or mean-square fluctuations of the variable O with respect to the initial state $| \, \rangle$, and the corresponding quantity $\delta_f O$ as that with respect to the final state $| \,)$. Consider the difference between the final and the initial number fluctuation of both the $\pm$ kinds,

$$\delta N = (\delta_f N_+ + \delta_f N_-) - (\delta_i N_+ + \delta_i N_-) \qquad (4.264)$$

Using the expressions given above, we obtain

$$\delta N = 2|\alpha|^2|\beta|^2[\delta N_+ + \delta N_- + \delta L + \partial(N_+ N_-)]_i$$
$$- (|\alpha|^3|\beta| + |\alpha||\beta|^3)[\partial(N_+ L) + \partial(N_- L)]_i \qquad (4.265)$$

where the subscript i refers to expectation values with respect to the initial states $| \, \rangle$, the symbol ∂ denotes

$$\partial(PQ) \equiv [\langle PQ \rangle + \langle QP \rangle - 2\langle P \rangle \langle Q \rangle] \qquad (4.266)$$

and

$$L = e^{2i\phi}a_+^\dagger a_-^\dagger + e^{-2i\phi}a_- a_+ \qquad (4.267)$$

Now for an initial number state $|n\rangle = |n_+, n_-\rangle$,

$$\delta N = 2|\alpha|^2|\beta|^2(1 + n_+ + n_- + 2n_+ n_-) \qquad (4.268)$$

we see that the number fluctuations will always increase. For an initial coherent state $|\mu\rangle = D(\mu_+)D(\mu_-)|0,0\rangle$, where $\mu_\pm = \sqrt{\langle N_\pm \rangle}e^{i\zeta_\pm}$,

$$\delta N = 2|\alpha|^2|\beta|^2[1 + 2(\langle N_+ \rangle + \langle N_- \rangle)]$$
$$- 4\sqrt{\langle N_+ \rangle \langle N_- \rangle}(|\alpha|^3|\beta| + |\alpha||\beta|^3)\cos(2\phi - \zeta_+ - \zeta_-) \qquad (4.269)$$

We find that under the conditions $\cos(2\phi - \zeta_+ - \zeta_-) = 1$ and $\langle N_+ \rangle = \langle N_- \rangle = \langle N \rangle/2$

$$\langle N \rangle > \frac{|\beta||\alpha|}{|\alpha|^2 + |\beta|^2 - |\beta||\alpha|} \qquad (4.270)$$

δN can be negative. In the weak particle creation limit $|\beta| \to 0, |\alpha| \to 1$ we find that this expression is equivalent to (4.252). In the strong particle creation limit we see that (4.252) diverges but in (4.270) $\langle N \rangle \to 1$. Clearly conditions for a decrease in number fluctuations are not the same as those for a decrease in the number.

For a single-mode squeezed state $|\sigma\rangle_1 = S_{1+}(r_+, \phi_+)S_{1-}(r_-, \phi_-)|0,0\rangle$

$$\delta N = 2|\alpha|^2|\beta|^2[(1 + \langle N_+ \rangle + \langle N_- \rangle)^2 + \langle N_+ \rangle(1 + \langle N_+ \rangle) + \langle N_- \rangle(1 + \langle N_- \rangle)$$
$$- 2\sqrt{\langle N_+ \rangle(1 + \langle N_+ \rangle)\langle N_- \rangle(1 + \langle N_- \rangle)}\cos 2(2\phi - \phi_+ - \phi_-)] \qquad (4.271)$$

From this it can be shown that, like the change in number, the change in the number fluctuations will always be positive for an initial single-mode squeezed vacuum.

For a two-mode squeezed state $|\sigma\rangle_2 = S_2(r_0, \phi_0)|0, 0\rangle$

$$\delta N = |\alpha|^2 |\beta|^2 \{2(1 + \langle N \rangle)^2 + \langle N \rangle(2 + \langle N \rangle)[1 + \cos 4(\phi - \phi_0)]\} \qquad (4.272)$$

$$+ 2(|\alpha|^3 |\beta| + |\beta|^3 |\alpha|)(1 + \langle N \rangle)\sqrt{\langle N \rangle(2 + \langle N \rangle)} \cos 2(\phi - \phi_0) \qquad (4.273)$$

Note that there is no definite relation between N and δN. For large $N \gg 1$ or small $|\beta| \ll 1$, $\delta N \leq 0$. The result obtained here for particle number fluctuations is relevant to issues of noise and fluctuation of quantum fields, and in turn, the dissipation and instability of condensates, background fields and spacetimes [HuSin95, HuMat96, HuMat95].

4.8 Squeezed quantum open systems

In this last section we discuss a squeezed quantum system interacting with an environment. From the examples given in this chapter we see that this encompasses a rather broad spectrum of systems with time-dependent background fields or spacetimes.

This theory was developed in the influence functional formalism by Hu and Matacz [HuMat94] extending the work on quantum Brownian motion by Hu, Paz and Zhang [HuPaZh92, HuPaZh93a], and Caldeira and Leggett [CalLeg83a] to oscillators with time-dependent frequencies. From this oscillator model it is an easy step to extend to quantum fields, which was done in [Zha90, Hu94b]. We shall treat open systems of quantum fields in the next chapter.

Our discussion here follows Koks *et al.* [KoMaHu97] based on the work of [HuMat94] which considers a squeezed (time-dependent, parametric) quantum open system coupled to a bath at temperature T with a time-dependent coupling constant. The results here are useful for calculating the entropy and uncertainty functions as well as for fluctuations and coherence, a topic to be discussed in Chapter 9.

4.8.1 Dissipation and noise kernels

For a parametric oscillator system interacting with a bath of many parametric oscillators at temperature T described by the Lagrangian given by equation (3.133) in Chapter 3 one can calculate the dissipation and noise kernels in closed forms ((equation (2.19) of [HuMat94]) in terms of the squeezed state parameterization (r, θ, ϕ) introduced in the previous section and the Bogoliubov coefficients (α, β) representation of the mode functions. In the case of a squeezed bath when the cross-term $(\varepsilon_n = 0)$ is absent and the mass of the bath oscillator is a constant $(m_n = 1)$ these expressions are in a manageable form. Note that the functions $\chi_n(t) = \alpha_n(t) + \beta_n(t)$ obey the equations (cf. equation (4.238)):

$$\ddot{\chi}_n(t) + \omega_n^2(t)\chi_n(t) = 0 \qquad (4.274)$$

with initial conditions (compare to (4.239))

$$\chi_n(t_i) = 1; \quad \dot{\chi}_n(t_i) = -i\kappa_n \qquad (4.275)$$

The bath canonical variables then admit a simple representation

$$q_n(s) = \frac{1}{2}\left\{[\chi_n(s) + \chi_n^*(s)]\, q_n(t_i) + \frac{i}{\kappa_n}[\chi_n(s) - \chi_n^*(s)]\, \dot{q}_n(t_i)\right\} \qquad (4.276)$$

The initial state of the bath is a squeezed thermal state. It has the form

$$\hat{\rho}_b(t_i) = \prod_n \hat{S}_n(r(n), \phi(n))\hat{\rho}_{\text{th}}\hat{S}_n^\dagger(r(n), \phi(n)) \qquad (4.277)$$

where $\hat{\rho}_{\text{th}}$ is a thermal density matrix of temperature T and $\hat{S}(r, \phi)$ is a squeeze operator defined in equation (4.216).

In this still rather general class of problems, the noise and dissipation kernels can be found from equations (3.62), where the relevant expectation values are computed with the help of (4.223)

$$\mathbf{D}(s, s') = 2\int_0^\infty d\omega\, I(\omega, s, s')\text{Im}[\chi_\omega(s)\chi_\omega^*(s')] \qquad (4.278)$$

$$\mathbf{N}(s, s') = \int_0^\infty d\omega\, I(\omega, s, s') \coth\left(\frac{\hbar\omega(t_i)}{2k_B T}\right)\left\{\cosh 2r(\omega)\text{Re}[\chi_\omega(s)\chi_\omega^*(s')]\right.$$

$$\left. -\frac{1}{2}\sinh 2r(\omega)\left[e^{-2i\phi(\omega)}\chi_\omega^*(s)\chi_\omega^*(s') + e^{2i\phi(\omega)}\chi_\omega(s)\chi_\omega(s')\right]\right\} \qquad (4.279)$$

We have adopted the convention that if f_n is a quantity defined for each mode of the bath, then we call $f(\omega) = f_\omega \equiv f_n$ evaluated at the mode that satisfies $\omega = \omega_n(t_i)$. $I(\omega, s, s')$ is the spectral density defined by

$$I(\omega, s, s') = \sum_n \delta(\omega - \omega_n(t_i))\frac{c_n(s)c_n(s')}{2\kappa_n} \qquad (4.280)$$

It contains information about the environmental mode density and coupling strength as a function of frequency. Different environments are classified according to the functional form of the spectral density $I(\omega)$. On physical grounds, one expects the spectral density to go to zero for very high frequencies. Let us introduce a certain cut-off frequency Λ (a property of the environment) such that $I(\omega) \to 0$ for $\omega > \Lambda$. The environment is classified as ohmic if in the physical range of frequencies ($\omega < \Lambda$) the spectral density is such that $I(\omega) \sim \omega$, as supra-ohmic if $I(\omega) \sim \omega^n, n > 1$ or as sub-ohmic if $n < 1$. The most studied ohmic case corresponds to an environment which induces a dissipative force linear in the velocity of the system. Also, by considering the continuum limit of the coupling constant, it can be shown that this constant's independence of n also leads to an ohmic environment.

Note that the dissipation kernel is independent of the bath's initial state. More generally, the noise and dissipation kernels are built out of symmetric and antisymmetric combinations of identical Bogoliubov factors. Thus the two kernels are intimately linked. For the case when the bath is a standard harmonic oscillator this inter-relationship can be written as a generalized fluctuation–dissipation relation [HuPaZh93a].

4.8.2 $u_1 \rightarrow v_2$ functions

In these last two subsections we present the explicit forms of the u, v and a, b functions for this squeezed quantum system. Recall that these are the functions first appearing in Chapter 3 in the derivation of the propagator for the reduced density matrix which determine the coefficients of the master equation. First consider equation (3.138). We treat the integral of a delta function and its derivative in the following way: use a smooth step function (i.e. $\theta(0) \equiv 1/2$) to write $(x_1 > x_0)$[1]

$$\int_{x_0}^{x_1} f(x)\delta(x-a)\,dx \equiv f(a)\,\theta(x_1-a)\,\theta(a-x_0) \tag{4.281}$$

$$\int_{x_0}^{x_1} f(x)\delta'(x-a)\,dx \equiv -f'(a)\,\theta(x_1-a)\,\theta(a-x_0) \tag{4.282}$$

Hence equation (3.138) together with equation (3.142) becomes (with u being either u_1 or u_2)

$$\ddot{u}(s) + \left(\frac{\dot{M}}{M} + \frac{2\gamma_0 c^2}{M}\right)\dot{u} + \left(\Omega^2 + \frac{\dot{M}\mathcal{E}}{M} + \dot{\mathcal{E}} + \frac{2\gamma_0 c\dot{c}}{M}\right)u = 0 \tag{4.283}$$

Now define $\tilde{u}$ by

$$\tilde{u} \equiv u \exp\left[\gamma_0 \int_{t_i}^{s} \frac{c^2(s')}{M(s')}ds'\right] \tag{4.284}$$

in which case it follows that

$$\ddot{\tilde{u}} + \frac{\dot{M}}{M}\dot{\tilde{u}} + \left(\Omega^2 + \frac{\dot{M}\mathcal{E}}{M} + \dot{\mathcal{E}} - \frac{\gamma_0^2 c^4}{M^2}\right)\tilde{u} = 0 \tag{4.285}$$

Comparing with (4.238), we recognize this as just the equation of motion of an oscillator with mass M, cross-term $\mathcal{E}$ and an effective frequency

$$\Omega_{\text{eff}}^2 \equiv \Omega^2 - \frac{\gamma_0^2 c^4}{M^2} \tag{4.286}$$

Hence we know a solution for $\tilde{u}(s)$ – it is the sum χ of the Bogoliubov coefficients for this new system. So we write (with g_1, g_2 constants to be determined)

$$u(s) = \exp\left[-\gamma_0 \int_{t_i}^{s} \frac{c^2}{M}ds'\right][g_1\chi(s) + g_2\chi^*(s)] \tag{4.287}$$

---

[1] These relations can easily be proved by checking the five cases individually, of $a < x_0$, $a = x_0$, $x_0 < a < x_1$, etc. Note that treating the delta function in this "smoothed" way eliminates the need for the frequency renormalization in [PaHaZu93]. This smoothing essentially just defines $\int_0^\infty \delta(x)dx = 1/2$ (see, e.g. [NeuHil27] for a discussion of this).

By including the boundary conditions for u_1 and u_2 we obtain

$$u_1(s) = \exp\left[-\gamma_0 \int_{t_i}^{s} \frac{c^2}{M} ds'\right] \frac{\text{Im}[\chi(t)\chi^*(s)]}{\text{Im}\chi(t)}$$

$$u_2(s) = \exp\left[\gamma_0 \int_{s}^{t} \frac{c^2}{M} ds'\right] \frac{\text{Im}\chi(s)}{\text{Im}\chi(t)} \tag{4.288}$$

Using the propagator formalism in the language of squeezed states with the Bogoliubov coefficients will be very useful for relating the entropy of a field mode to its squeeze parameter r.

Proceeding in the same way, equation (3.139) and (3.39) becomes

$$\ddot{v}(s) + \left(\frac{\dot{M}}{M} - \frac{2\gamma_0 c^2}{M}\right)\dot{v} + \left(\Omega^2 + \frac{\dot{M}\mathcal{E}}{M} + \dot{\mathcal{E}} - \frac{2\gamma_0 c\dot{c}}{M}\right)v = 0 \tag{4.289}$$

Now write

$$\tilde{v} \equiv v \exp\left[-\gamma_0 \int_{t_i}^{s} \frac{c^2}{M} ds'\right] \tag{4.290}$$

and just as for the case of u we have

$$\ddot{\tilde{v}} + \frac{\dot{M}}{M}\dot{\tilde{v}} + \left(\Omega^2 + \frac{\dot{M}\mathcal{E}}{M} + \dot{\mathcal{E}} - \frac{\gamma_0^2 c^4}{M^2}\right)\tilde{v} = 0 \tag{4.291}$$

So now v_1 and v_2 can also be written as combinations of χ and χ^*. Including the boundary conditions we eventually obtain

$$v_1(s) = \exp\left[\gamma_0 \int_{t_i}^{s} \frac{c^2}{M} ds'\right] \frac{\text{Im}[\chi(t)\chi^*(s)]}{\text{Im}\chi(t)}$$

$$v_2(s) = \exp\left[-\gamma_0 \int_{s}^{t} \frac{c^2}{M} ds'\right] \frac{\text{Im}\chi(s)}{\text{Im}\chi(t)} \tag{4.292}$$

4.8.3 $a_{11} \rightarrow b_4$ functions

To facilitate our calculations we introduce dimensionless parameters for time

$$z \equiv \kappa t, \quad \sigma \equiv \kappa s$$
$$\chi(\tau) \equiv \chi(t), \quad \text{etc.} \tag{4.293}$$

and a carat will denote division by κ, e.g. $\hat{\gamma}_0 = \gamma_0/\kappa$. Note that t is the Lagrangian time.

Now we have all the necessary ingredients to calculate the propagator. Making use of equation (3.136) and equation (3.137) we obtain

$$a_{11}(z, z_i) = \frac{1}{2\kappa^2} \int_{z_i}^{z} d\sigma \int_{z_i}^{z} d\sigma' \, \exp\left(\hat{\gamma}_0 \int_{z_i}^{\sigma} \frac{c^2}{M} d\sigma''\right) \frac{\mathrm{Im}[\chi(z)\chi^*(\sigma)]}{\mathrm{Im}\chi(z)} N(\sigma, \sigma')$$
$$\times \exp\left(\hat{\gamma}_0 \int_{z_i}^{\sigma'} \frac{c^2}{M} d\sigma''\right) \frac{\mathrm{Im}[\chi(z)\chi^*(\sigma')]}{\mathrm{Im}\chi(z)}$$

$$a_{12} = \frac{1}{\kappa^2} \int_{z_i}^{z} d\sigma \int_{z_i}^{z} d\sigma' \, \exp\left(\hat{\gamma}_0 \int_{z_i}^{\sigma} \frac{c^2}{M} d\sigma''\right) \frac{\mathrm{Im}[\chi(z)\chi^*(\sigma)]}{\mathrm{Im}\chi(z)} N(\sigma, \sigma')$$
$$\times \exp\left(-\hat{\gamma}_0 \int_{\sigma'}^{z} \frac{c^2}{M} d\sigma''\right) \frac{\mathrm{Im}\chi(\sigma')}{\mathrm{Im}\chi(z)}$$

$$a_{22} = \frac{1}{2\kappa^2} \int_{z_i}^{z} d\sigma \int_{z_i}^{z} d\sigma' \, \exp\left(-\hat{\gamma}_0 \int_{\sigma}^{z} \frac{c^2}{M} d\sigma''\right) \frac{\mathrm{Im}\chi(\sigma)}{\mathrm{Im}\chi(z)} N(\sigma, \sigma')$$
$$\times \exp\left(-\hat{\gamma}_0 \int_{\sigma'}^{z} \frac{c^2}{M} d\sigma''\right) \frac{\mathrm{Im}\chi(\sigma')}{\mathrm{Im}\chi(z)}$$

$$b_1(z, z_i) = -\hat{\gamma}_0 \, \kappa c^2(z) + \kappa M(z) \frac{\mathrm{Im}\chi'(z)}{\mathrm{Im}\chi(z)} + M(z)\mathcal{E}(z)$$

$$b_{\{{}^2_3\}} = \frac{\mp\kappa}{\mathrm{Im}\chi(z)} \exp\left(\pm\hat{\gamma}_0 \int_{z_i}^{z} \frac{c^2}{M} d\sigma\right)$$

$$b_4 = -\hat{\gamma}_0 \, \kappa c^2(z_i) + \kappa \frac{\mathrm{Re}\chi(z)}{\mathrm{Im}\chi(z)} + M(z_i)\mathcal{E}(z_i) \qquad (4.294)$$

These coefficients will be useful for calculating the entropy generation in a squeezed open quantum system in Chapter 9.

5

Open systems of interacting quantum fields

As introduced in Chapter 1, for many problems in statistical mechanics one is interested in the detailed behavior of only a part of the overall system (call it *the* system) interacting with its surrounding (call it the environment). In field theory one can accordingly decompose the field describing the overall system $\phi = \phi_S + \phi_E$ into a sum of the system field ϕ_S and the environment field ϕ_E. This decomposition is always possible formally but only when there is a clear physical discrepancy between the two sectors will it be physically meaningful and technically implementable. The division could be made between slow and fast variables, low and high frequencies or light and heavy mass sectors. Drawing examples from cosmology, in the stochastic inflation scenario one regards the system field as containing only the lower modes and the environmental field as containing the higher modes with the division provided by the event horizon in de Sitter spacetime. A similar problem in quark–gluon plasma is to ascertain the effect of the hard thermal loops on the soft gluon modes. Another is the effect of the atoms in the noncondensate on the Bose–Einstein condensate (BEC). These cases will be discussed in later chapters.

Usually the reason for performing such a decomposition is because one is interested more in the details of the system (the "relevant" variables or the "distinguished" sector), and less in that of the environment (the "irrelevant" variables). Since the environment often contains many more degrees of freedom than the system the details of which are not of particular interest to us, introducing some way of coarse graining them and extracting their overall influence on the system is desirable. This procedure renders the original system an open system, and its behavior would then be describable by the open system conceptual and technical framework we introduced in Chapter 3. In particular, the quantity of special interest is the influence action obtained from the integration over the environment field in a CTP path integral.

We recall that when the time limits in this path integral are taken to infinity, the influence action turns into the so-called closed time path (CTP) coarse-grained effective action (CGEA). The idea behind this quantity which originated from studies in dynamical critical phenomena ("coarse-grained" free energy density) was transplanted to nonequilibrium quantum field theory by Hu and Zhang [Hu91, Zha90] first in the "in–out" (Schwinger–Dewitt) formulation and then by Sinha and Hu [SinHu91] in the "in–in" (Schwinger–Keldysh) formulation. A clear presentation of the CTP CGEA can be found in Lombardo and Mazzitelli

[LomMaz96]. (See also [CaHuMa01] for a review.) We shall restrict usage of the term effective action (EA) to the particular case in which ϕ_S is the c-number part of the field operator. The so-called background field decomposition in quantum field theory, $\Phi = \phi_c + \phi_q$, on an interacting field Φ is a special case of this open system method, where the discrepancy parameter is the Planck constant $\hbar$, separating and systemizing the quantum contributions from the classical. The familiar loop expansion (in orders of $\hbar$) of the effective action is an example of the CGEA, with the special feature that the equations of motion it yields do not contain any dissipation (unless some causal condition like the factorizable initial state similar to the Boltzmann molecular chaos assumption is introduced). We will introduce the CTP CGEA in the language of influence functionals in this chapter and introduce more formal techniques for its development in the next chapter. The IF formalism and the CTP CGEA will be our main workhorse for the rest of the book.

Our goal in this chapter is to derive the influence action and the CGEA and the stochastic equations for two simple but fundamental quantum field scenarios. We treat first the case of two interacting scalar fields, one of which is chosen as the system and the other as its environment. This case is technically easier than the second case, that of a single quantum field split into two by separating the long and short wavelength sectors (to be defined precisely below), even though the CGEA was introduced historically for the latter situation, which exemplifies a broader class of statistical mechanical problems [Hu91]. For pedagogical reasons, we will stay within the technically simplest approach in quantum field theory familiar to the reader, using a straightforward perturbative expansion in powers of the coupling constants. More powerful methods will be introduced later in the book.

5.1 Influence functional: Two interacting quantum fields

In this section we study the problem of two quantum self-interacting scalar fields (one the system field, the other the environment field) interacting with each other in Minkowski spacetime. To do so we only need to generalize to quantum field theory the results for the quantum mechanical Brownian model based on the influence functional method introduced in Chapter 3. We first derive the influence functional, from which we identify the dissipation and noise kernels. We then derive a Langevin equation for the dissipative dynamics of the system field. The nonlinear mode–mode coupling between the system field and the environment field induces a nonlinear nonlocal dissipation and a coupled multiplicative colored noise source for the system field. Finally we write down the functional quantum master equation for the system field. Our presentation in this section follows [Hu94b, Zha90].

Consider two independent self-interacting scalar fields in Minkowski spacetime: $\phi(x)$ depicting the system, and $\psi(x)$ depicting the environment. The classical actions for these two fields are given respectively by:

$$S[\phi] = \int d^4x \left(-\frac{1}{2} \partial_\nu \phi(x) \partial^\nu \phi(x) - V(\phi) \right) = S_0[\phi] + S_I[\phi] \qquad (5.1)$$

$$S[\psi] = \int d^4x \left(-\frac{1}{2} \partial_\mu \psi(x) \partial^\mu \psi(x) - V(\psi) \right) = S_0[\psi] + S_I[\psi] \qquad (5.2)$$

where $V[\phi], V[\psi]$ are the self-interaction potentials. For a ϕ^4 interaction,

$$V[\phi] = \frac{1}{2} m_\phi^2 \phi^2(x) + \frac{1}{4!} \lambda_\phi \phi^4(x), \qquad (5.3)$$

and similarly for $V[\psi]$. Here, m_ϕ and m_ψ are the bare masses and λ_ϕ and λ_ψ are the bare self-coupling constants for the $\phi(x)$ and $\psi(x)$ fields respectively. In equation (5.2) we have written $S[\psi]$ in terms of a free part S_0 and an interacting part S_I which contains λ_ψ. Assume these two scalar fields interact via a polynomial coupling of the form

$$S_{\text{int}} = \int d^4x \, V_{\phi\psi}[\phi(x)] \psi^k(x) \qquad (5.4)$$

where $V_{\phi\psi}[\phi(x)] \equiv -\lambda_{\phi\psi} f[\phi(x)]$ is the vertex function with coupling constant $\lambda_{\phi\psi}$, which we assume to be small and of the same order as λ_ϕ, λ_ψ.

The total classical action of the combined system is

$$S[\phi, \psi] = S[\phi] + S[\psi] + S_{\text{int}}[\phi, \psi] \qquad (5.5)$$

The total density matrix of the combined system plus environment field is defined by

$$\rho[\phi^1, \psi^1, \phi^2, \psi^2, t] = \langle \phi^1, \psi^1 | \, \hat{\rho}(t) \, | \phi^2, \psi^2 \rangle \qquad (5.6)$$

where the superscripts $1, 2$ are the closed time path branches to integrate over as will be described in more detail in Chapter 6, and $|\phi\rangle$ and $|\psi\rangle$ are the eigenstates of the field operators $\hat{\phi}(x)$ and $\hat{\psi}(x)$, namely,

$$\hat{\phi}(\mathbf{x})|\phi\rangle = \phi(\mathbf{x})|\phi\rangle, \qquad \hat{\psi}(\mathbf{x})|\psi\rangle = \psi(\mathbf{x})|\psi\rangle \qquad (5.7)$$

Since we are primarily interested in the behavior of the system, and of the environment only to the extent in how it influences the system, the object of interest is the reduced density matrix defined by

$$\rho_r[\phi^1, \phi^2, t] = \int d\psi \, \rho[\phi^1, \psi^1, \phi^2, \psi^1, t] \qquad (5.8)$$

For technical convenience, let us assume that the total density matrix at an initial time is factorized, i.e. that the system and environment are statistically independent,

$$\hat{\rho}(t_i) = \hat{\rho}_\phi(t_i) \times \hat{\rho}_\psi(t_i) \qquad (5.9)$$

where $\hat{\rho}_\phi(t_i)$ and $\hat{\rho}_\psi(t_i)$ are the initial density matrix operator of the ϕ and ψ field respectively, the former being equal to the reduced density matrix $\hat{\rho}_r$ at t_i

by this assumption. The reduced density matrix of the system field $\phi(x)$ evolves in time following

$$\rho_r[\phi_f^1, \phi_f^2, t] = \int d\phi_i^1 \int d\phi_i^2 \; \mathcal{J}_r\left[\phi_f^1, \phi_f^2, t \mid \phi_i^1, \phi_i^2, t_i\right] \rho_r[\phi_i^1, \phi_i^2, t_i] \qquad (5.10)$$

As in Chapter 3, the propagator $\mathcal{J}_r[\phi_f^1, \phi_f^2, t \mid \phi_i^1, \phi_i^2, t_i]$ is given by a CTP Feynman integral of the exponent of the influence action

$$\mathcal{J}_r[\phi_f^1, \phi_f^2, t \mid \phi_i^1, \phi_i^2, t_i] = \int_{\phi_i^1(\mathbf{x})}^{\phi_f^1(\mathbf{x})} D\phi^1 \int_{\phi_i^2(\mathbf{x})}^{\phi_f^2(\mathbf{x})} D\phi^2 \; \exp\frac{i}{\hbar} S_{\text{eff}}[\phi^1, \phi^2] \qquad (5.11)$$

where

$$S_{\text{eff}}[\phi^1, \phi^2] \equiv S[\phi^1] - S[\phi^2] + S_{\text{IF}}[\phi^1, \phi^2] \qquad (5.12)$$

is the full influence functional (IF) effective action and S_{IF} is the influence action. The Feynman–Vernon influence functional $\mathcal{F}[\phi^1, \phi^2]$ is defined as

$$\mathcal{F}[\phi^1, \phi^2] = e^{\frac{i}{\hbar} S_{\text{IF}}[\phi^1, \phi^2]}$$

$$= \int d\psi_f^1(\mathbf{x}) \int d\psi_i^1(\mathbf{x}) \int d\psi_i^2(\mathbf{x}) \, \rho_\psi[\psi_i^1, \psi_i^2, t_i] \int_{\psi_i^1(\mathbf{x})}^{\psi_f^1(\mathbf{x})} D\psi^1 \int_{\psi_i^2(\mathbf{x})}^{\psi_f^1(\mathbf{x})} D\psi^2$$

$$\times \exp\frac{i}{\hbar} \left\{ S[\psi^1] + S_{\text{int}}[\phi^1, \psi^1] - S[\psi^2] - S_{\text{int}}[\phi^2, \psi^2] \right\} \qquad (5.13)$$

which summarizes the averaged effect of the bath on the system. For a zero-temperature bath (i.e. the environment field ψ is in a vacuum state, $\hat{\rho}_b(t_i) = |0\rangle\langle 0|$), the influence functional $\mathcal{F}$ is formally equivalent to the CTP vacuum generating functional, and the influence action S_{IF} in equation (5.12) is the usual CTP vacuum effective action, to be discussed in the next chapter.

5.1.1 Perturbation theory

The above formal framework is nice but often difficult to tackle. To evaluate the influence action we need to develop a perturbation theory. If $\lambda_{\phi\psi}$ and λ_ψ are assumed to be small parameters, the influence functional can be calculated perturbatively by making a power expansion of $\exp\frac{i}{\hbar}[S_{\text{int}} + S_I]$. In this section, we set $\lambda_\psi = 0$ for simplicity. Up to second order in $\lambda_{\phi\psi}$, and first order in $\hbar$ (one-loop), the influence action is given by

$$\begin{aligned}
S_{\text{IF}}[\phi^1, \phi^2] = &\langle S_{\text{int}}[\phi^1, \psi^1]\rangle_0 - \langle S_{\text{int}}[\phi^2, \psi^2]\rangle_0 \\
&+ \frac{i}{2\hbar} \left\{ \langle S_{\text{int}}[\phi^1, \psi^1]^2\rangle_0 - \langle S_{\text{int}}[\phi^1, \psi^1]\rangle_0^2 \right\} \\
&- \frac{i}{\hbar} \left\{ \langle S_{\text{int}}[\phi^1, \psi^1] S_{\text{int}}[\phi^2, \psi^2]\rangle_0 - \langle S_{\text{int}}[\phi^1, \psi^1]\rangle_0 \langle S_{\text{int}}[\phi^2, \psi^2]\rangle_0 \right\} \\
&+ \frac{i}{2\hbar} \left\{ \langle S_{\text{int}}[\phi^2, \psi^2]^2\rangle_0 - \langle S_{\text{int}}[\phi^2, \psi^2]\rangle_0^2 \right\} \qquad (5.14)
\end{aligned}$$

where the quantum average of a physical variable $\mathcal{Q}[\psi^1, \psi^2]$ over the unperturbed action $S_0[\psi]$ is defined by

$$\langle \mathcal{Q}[\psi^1, \psi^2] \rangle_0 = \int d\psi_f^1(\mathbf{x}) \int d\psi_i^1(\mathbf{x}) \int d\psi_i^2(\mathbf{x}) \; \rho_\psi[\psi_i^1, \psi_i^2, t_i]$$

$$\times \int_{\psi_i^1(\mathbf{x})}^{\psi_f^1(\mathbf{x})} D\psi^1 \int_{\psi_i^2(\mathbf{x})}^{\psi_f^1(\mathbf{x})} D\psi^2 \; \exp \frac{i}{\hbar} \{ S_0[\psi^1] - S_0[\psi^2] \} \times \mathcal{Q}[\psi^1, \psi^2]$$

$$\equiv \mathcal{Q}\left[\frac{\hbar\delta}{i\delta J^1(x)}, -\frac{\hbar\delta}{i\delta J^2(x)} \right] \mathcal{F}^{(0)}[J^1, J^2] \Big|_{J^1 = J^2 = 0} \qquad (5.15)$$

Here, $\mathcal{F}^{(0)}[J^1, J^2]$ is the influence functional of the free environment field, assuming a linear coupling with external sources J^1 and J^2:

$$\mathcal{F}^{(0)}[J^1, J^2]$$

$$= \int d\psi_f^1(\mathbf{x}) \int d\psi_i^1(\mathbf{x}) \int d\psi_i^2(\mathbf{x}) \; \rho_\psi[\psi_i^1, \psi_i^2, t_i] \int_{\psi_i^1(\mathbf{x})}^{\psi_f^1(\mathbf{x})} D\psi^1 \int_{\psi_i^2(\mathbf{x})}^{\psi_f^1(\mathbf{x})} D\psi^2$$

$$\times \exp \frac{i}{\hbar} \left\{ S_0[\psi^1] + \int d^4 x J^1(x)\psi^1(x) - S_0[\psi^2] - \int d^4 x J^2(x)\psi^2(x) \right\} \qquad (5.16)$$

Let us define the following free propagators of the ψ field

$$\langle T\psi^1(x)\psi^1(y) \rangle_0 = \Delta_F(x, y) \qquad (5.17)$$

$$\langle \psi^1(x)\psi^2(y) \rangle_0 = \Delta^-(x, y) \qquad (5.18)$$

$$\langle \tilde{T}\psi^2(x)\psi^2(y) \rangle_0 = \Delta_D(x, y) \qquad (5.19)$$

As we have seen in Chapter 3, the CTP path integral time-orders fields in the first branch, anti-time-orders fields in the second branch, and puts fields on the second branch to the left of fields on the first branch. Therefore these are just the familiar Feynman, Dyson and negative-frequency Wightman propagators of a free scalar field given respectively by

$$\Delta_{F,D}(x, x') = \mp i\hbar \int \frac{d^4 k}{(2\pi)^4} \frac{e^{ik(x-x')}}{k^2 + m_\psi^2 \mp i\varepsilon} \qquad (5.20)$$

$$\Delta^-(x, x') = \int \frac{d^4 k}{(2\pi)^4} e^{ik(x-x')} \theta(-k^0) \, 2\pi \hbar \delta(k^2 + m_\psi^2) \qquad (5.21)$$

The perturbation calculation by means of Feynman diagrams for the $\lambda \phi^4$ theory in the CTP formalism has been worked out before for quantum fluctuations [CalHu87, CalHu89] and for coarse-grained fields [Hu91, SinHu91]. For biquadratic coupling,

$$S_{\text{int}}[\phi, \psi] = -\int d^4 x \, \lambda_{\phi\psi} \phi^2(x)\psi^2(x) \qquad (5.22)$$

the influence action up to the second order in λ is given by (cf. [HuPaZh93a])

$$S_{\rm IF}[\phi,\phi'] = -\int d^4x\,\lambda_{\phi\psi}\,\Delta_F(x,x)\left[(\phi^1(x))^2 - (\phi^2(x))^2\right]$$

$$+\,i\hbar^{-1}\int d^4x\int d^4y\,\lambda_{\phi\psi}^2\,(\phi^1(x))^2\,[\Delta_F(x,y)]^2\,(\phi^1(y))^2$$

$$-\,2i\hbar^{-1}\int d^4x\int d^4y\,\lambda_{\phi\psi}^2\,(\phi^1(x))^2\,[\Delta^-(x,y)]^2\,(\phi^2(y))^2$$

$$+\,i\hbar^{-1}\int d^4x\int d^4y\,\lambda_{\phi\psi}^2\,(\phi^2(x))^2\,[\Delta_D(x,y)]^2\,(\phi^2(y))^2 \quad (5.23)$$

We now evaluate each term in the perturbation expansion. It is well known that all one-loop diagrams in equation (5.23) contain ultraviolet divergences in spacetime dimension $d = 4 - \epsilon$. By dimensional regularization, one can show that the first one-loop bubble diagram for the ψ^1 field is

$$\Delta_F(x,x) = \hbar\int\frac{d^d p}{(2\pi)^d}\frac{(-i)}{p^2 + m_\psi^2 - i\varepsilon}$$

$$= -\frac{\hbar m_\psi^2}{8\pi^2}\left[\frac{1}{\epsilon} + \text{constant} - \frac{1}{2}\ln\left(\frac{m_\psi^2}{4\pi\mu^2}\right)\right] \quad (5.24)$$

where μ^2 is the renormalization energy scale. The first term on the right-hand side is a singular part and must be canceled by mass renormalization. The counter action for this singular mass term is

$$\delta S_{r1}[\phi^1] = \int d^4x\frac{\hbar}{8\pi^2\epsilon}m_\psi^2\lambda_{\phi\psi}(\phi^1(x))^2 \quad (5.25)$$

The second term on the right-hand side is the one-loop finite mass renormalization term, which can be absorbed into the definition of the physical mass of the ϕ field.

For the one-loop bubble diagram for the ψ^2 field, since

$$\langle(\psi^2(x))^2\rangle_0 = \langle(\psi^1(x))^2\rangle_0 \quad (5.26)$$

the mass renormalization counter-action for the $\phi^2(x)$ field is the same as equation (5.25), so is the finite mass renormalization.

Next, for the "fish" diagram of the ψ^1 field, it also can be shown by dimensional regularization that

$$i\hbar^{-2}\Delta_F^2(x-y) = \int\frac{d^4p}{(2\pi)^4}\,e^{ip(x-y)}\frac{1}{8\pi^2}$$

$$\times\left[\frac{1}{\epsilon} + \text{constant} - \frac{1}{2}\int_0^1 d\alpha\,\ln\left(\frac{m_\psi^2 + \alpha(1-\alpha)(p^2 - i\varepsilon)}{4\pi\mu^2}\right)\right]$$

$$= \frac{1}{8\pi^2\epsilon}\delta^4(x-y) + \frac{1}{16\pi^2}\left(2 + \psi(1) - \ln\frac{m_\psi^2}{4\pi\mu^2}\right)\delta^4(x-y)$$

$$+\frac{1}{2}U(x-y) + i\frac{1}{2}\nu(x-y) \quad (5.27)$$

with the following two real nonlocal kernels

$$U(x-y) = -\frac{2}{16\pi^2} \int \frac{d^4p}{(2\pi)^4} \, e^{ip(x-y)} \int_0^1 d\alpha \, \ln\left|1 - i\epsilon + \alpha(1-\alpha)\frac{p^2}{m_\psi^2}\right| \qquad (5.28)$$

$$\nu(x-y) = \frac{2}{16\pi^2} \int \frac{d^4p}{(2\pi)^4} \, e^{ip(x-y)} \, \pi \sqrt{1 - \frac{4m_\psi^2}{(-p^2)}} \, \theta(-p^2 - 4m_\psi^2) \qquad (5.29)$$

The first term on the right-hand side of equation (5.27) is another singular term. Its counter-action is

$$\delta S_2[\psi^1] = \int d^4x \, \lambda_{\phi\psi}^2 \frac{\hbar}{16\pi^2\epsilon}(\phi^1(x))^4 \qquad (5.30)$$

The second term on the right-hand side of equation (5.27) represents a finite coupling constant renormalization, which can be absorbed into a redefinition of the physical coupling constant of the ϕ^1 field. The contribution from the fish diagram for the ψ^2 field is obtained from the above by changing the sign of the ν kernel; it can be renormalized with a counter-action similar to equation (5.30).

For the mixed "fish" diagram, we find

$$i\hbar^{-2}(\Delta^-(x,y))^2 = -\mu(x-y) + \frac{i}{2}\nu(x-y) \qquad (5.31)$$

where Cutkowsky rules have been used. The kernel μ in equation (5.31)

$$\mu(x-y) = \frac{i}{16\pi^2} \int \frac{d^4p}{(2\pi)^4} \, e^{ip(x-y)} \, \pi \sqrt{1 - \frac{4m_\psi^2}{(-p^2)}} \, \theta(-p^2 - 4m_\psi^2) \, \text{sgn}(p_0) \qquad (5.32)$$

is real.

Substituting equations (5.24), (5.27) and (5.31) into the influence action and adding the counter-action equations (5.25) and (5.30), finally we obtain the effective action for this biquadratically coupled system–environment scalar field model as follows

$$
\begin{aligned}
S_{\text{eff}}[\phi^1,\phi^2] = {} & S_{\text{ren}}[\phi^1] + \hbar \int d^4x \int d^4y \, \frac{1}{2}\lambda_{\phi\psi}^2 \, (\phi^1(x))^2 \, V_{\phi\psi}(x-y) \, (\phi^1(y))^2 \\
& - S_{\text{ren}}[\phi^2] - \hbar \int d^4x \int d^4y \, \frac{1}{2}\lambda_{\phi\psi}^2 \, (\phi^2(x))^2 \, V_{\phi\psi}(x-y) \, (\phi^2(y))^2 \\
& - \hbar \int d^4x \int^{x^0} d^4y \, \lambda_{\phi\psi}^2 \left[(\phi^1(x))^2 - (\phi^2(x))^2\right] \\
& \times \mu(x-y) \left[(\phi^1(y))^2 + (\phi^2(y))^2\right] \\
& + \frac{i\hbar}{2} \int d^4x \int d^4y \, \lambda_{\phi\psi}^2 \left[(\phi^1(x))^2 - (\phi^2(x))^2\right] \\
& \times \nu(x-y) \left[(\phi^1(y))^2 - (\phi^2(y))^2\right]
\end{aligned}
\qquad (5.33)
$$

where $S_{\text{ren}}[\phi^{1,2}]$ is the renormalized action of the $\phi^{1,2}$ field (with physical mass $m^2_{\phi r}$ and physical coupling constant $\lambda_{\phi r}$),

$$S_{\text{ren}}[\phi^a] = \int d^4x \left(-\frac{1}{2}\partial_\mu \phi^a \partial^\mu \phi^a - \frac{1}{2}m^2_{\phi r}(\phi^a)^2 - \frac{1}{4!}\lambda_{\phi r}(\phi^a)^4 \right) \tag{5.34}$$

where $a = 1, 2$. The kernel for the nonlocal potential in equation (5.33)

$$V_{\phi\psi}(x - y) = U(x - y) - \text{sgn}(x^0 - y^0)\mu(x - y) \tag{5.35}$$

is symmetric.

For the biquadratic interaction case analyzed here, the potential renormalization is thus

$$\Delta V^{(2)}(x - y) = U^{(2)}(x - y) - \text{sgn}(x^0 - y^0)\mu^{(2)}(x - y) \tag{5.36}$$

which is symmetric; and $\mu^{(2)}, \nu^{(2)}$ and $U^{(2)}$ are real nonlocal kernels

$$\mu^{(2)}(x - y) = \frac{1}{16\pi^2} \int \frac{d^4p}{(2\pi)^4} \, e^{ip(x-y)} \, \pi\sqrt{1 - \frac{4m^2_\psi}{(-p^2)}} \, \theta(-p^2 - 4m^2_\psi) \, \text{sgn}(p_0) \tag{5.37}$$

$$\nu^{(2)}(x - y) = \frac{2}{16\pi^2} \int \frac{d^4p}{(2\pi)^4} \, e^{ip(x-y)} \, \pi \sqrt{1 - \frac{4m^2_\psi}{(-p^2)}} \, \theta(-p^2 - 4m^2_\psi) \tag{5.38}$$

$$U^{(2)}(x - y) = -\frac{2}{16\pi^2} \int \frac{d^4p}{(2\pi)^4} \, e^{ip(x-y)} \int_0^1 d\alpha \, \ln\left|1 - i\epsilon - \alpha(1-\alpha)\frac{(-p^2)}{m^2_\psi}\right| \tag{5.39}$$

For a general polynomial-type coupling with S_{int} given by equation (5.4), the renormalized full effective action has the same form as that derived above for biquadratic coupling, except that $(\phi^a(x))^2$ would be replaced by $f[\phi^a(x)]$, etc. (and the kernels would carry superscripts indicating the proper order k instead of (2)). To second order in λ the renormalized full effective action is given by [Zha90, HuPaZh93a]

$$\begin{aligned}
S_{\text{eff}}&[\phi^1, \phi^2] \\
&= S_{\text{ren}}[\phi^1] + \hbar^{k-1} \int d^4x \int d^4y \, \frac{1}{2}\lambda^2_{\phi\psi} f[\phi^1(x)]\Delta V^{(k)}(x - y)f[\phi^1(y)] \\
&\quad - S_{\text{ren}}[\phi^2] - \hbar^{k-1} \int d^4x \int d^4y \, \frac{1}{2}\lambda^2_{\phi\psi} f[\phi^2(x)]\Delta V^{(k)}(x - y)f[\phi^2(y)] \\
&\quad - \hbar^{k-1} \int d^4x \int^{x^0} d^4y \, \lambda^2_{\phi\psi} \\
&\quad \times \Big\{ \left(f[\phi^1(x)] - f[\phi^2(x)]\right) \mu^{(k)}(x - y) \left(f[\phi^1(y)] + f[\phi^2(y)]\right) \\
&\quad - i\hbar^{k-1} \left(f[\phi^1(x)] - f[\phi^2(x)]\right) \nu^{(k)}(x - y) \left(f[\phi^1(y)] - f[\phi^2(y)]\right) \Big\}
\end{aligned} \tag{5.40}$$

Renormalization of the potential which arises from the contribution of the environment appears only for even order k couplings. This is a generalization of the result obtained in [Zha90, HuPaZh93a] where it was shown that the nonlocal kernel $\mu^{(k)}(s_1 - s_2)$ is associated with the nonlocal dissipation (or the generalized viscosity) function that appears in the corresponding Langevin equation and $\nu^{(k)}(s_1 - s_2)$ is associated with the time–time autocorrelation function of the stochastic forcing (noise) term. In general ν is nonlocal, which gives rise to colored noises. Only at high temperatures would the noise kernel become a delta function, which corresponds to a white noise source. Let us examine more closely the meaning of the noise kernel.

5.1.2 Noise and fluctuations

The real part of the influence functional comes from the imaginary part of the influence action which contains the noise kernel. This term can be rewritten using a functional Gaussian identity introduced by Feynman and Vernon [FeyVer63] and discussed in Chapter 3. Thus introducing a stochastic forcing term $\xi^{(k)}$ coupled to the field:

$$-\int d^4x \; \xi^{(k)}(x) \; \{f[\phi^1(x)] - f[\phi(x^2)] \; \}/\hbar \qquad (5.41)$$

we can view $\xi^{(k)}(x)$ as a classical nonlinear noise source external to the system arising from the environment. The reduced density matrix is calculated by taking a stochastic average over the distribution $\mathcal{P}[\xi^{(k)}]$ of this source. Since the expansion of the action is to quadratic order, the associated noise is Gaussian. It is completely characterized by

$$\langle \xi^{(k)}(x) \rangle_\xi = 0 \qquad (5.42)$$

$$\langle \xi^{(k)}(x)\xi^{(k)}(y) \rangle_\xi = \hbar^k \nu^{(k)}(x - y) \qquad (5.43)$$

where $\nu^{(k)}$ is redefined by absorbing the $\lambda_{\phi\psi}^2$. We see that the nonlocal kernel $\hbar^k \nu^{(k)}(x - y)$ is just the two-point autocorrelation function of the external stochastic source $\xi^{(k)}(x)$ called colored noise.

In this framework, the expectation value of any functional operator $\mathcal{Q}[\phi]$ of the field ϕ is then given by

$$\langle \mathcal{Q}[\phi] \rangle = \int D\xi^{(k)}(x) \, \mathcal{P}[\xi^{(k)}] \int d\phi \, \rho_r(\phi, \phi, [\xi^{(k)}]) \, \mathcal{Q}[\phi] \qquad (5.44)$$

$$= \Big\langle \langle \mathcal{Q}[\phi] \rangle_{\text{quantum}} \Big\rangle_{\text{noise}} \qquad (5.45)$$

This provides the physical interpretation of $\nu^{(k)}(x - y)$ as a noise or fluctuation kernel of the quantum field.

5.1.3 Langevin equation and fluctuation–dissipation relation

We will now derive the semiclassical equation of motion generated by the influence action S_{IF}. Define a "center-of-mass" function ϕ_+ and a "relative" function
ϕ_- as follows

$$\phi_+(x) = \frac{1}{2}[\phi^1(x) + \phi^2(x)] \tag{5.46}$$

$$\phi_-(x) = \phi^1(x) - \phi^2(x) \tag{5.47}$$

The equation of motion for ϕ is derived by demanding (cf. [CalHu87])

$$\left.\frac{\delta S_{\mathrm{eff}}}{\delta \phi_-}\right|_{\phi_-=0} = 0 \tag{5.48}$$

which gives

$$-\frac{\partial L_r}{\partial \phi} + \frac{d}{dt}\frac{\partial L_r}{\partial \dot{\phi}} + 2\frac{\partial f(\phi)}{\partial \phi}\int_0^x d^4y\, \gamma^{(k)}(x-y)\frac{\partial f(\phi(y))}{\partial y^0} = F_\xi^{(k)}(x) \tag{5.49}$$

We see that this is in the form of a Langevin equation with a nonlinear stochastic
force

$$F_\xi^{(k)}(x) = \xi^{(k)}(s)\frac{\partial f(\phi)}{\partial \phi} \tag{5.50}$$

This corresponds to a multiplicative noise arising from a nonlinear field coupling
(additive if $f(\phi) = \phi$). L_r is the renormalized effective Lagrangian of the system
action S_{eff}. The nonlocal kernel $\gamma^{(k)}(t-s)$ defined by

$$\frac{\partial}{\partial(x^0 - y^0)}\gamma^{(k)}(x-y) = \hbar^{k-1}\mu^{(k)}(x-y) \tag{5.51}$$

is responsible for nonlocal dissipation. Interaction with the environment field
imparts a dissipative force in the effective dynamics of the system field given
by

$$F_\gamma^{(k)}(x) = 2\int d^4y\, \gamma^{(k)}(x-y)\frac{\partial f(\phi(y))}{\partial y^0}\frac{\partial f(\phi(x))}{\partial \phi} \tag{5.52}$$

Only in special cases like a high temperature ohmic environment will the dissipation become local.

In the biquadratic coupling example the corresponding stochastic force is

$$F_\xi^{(2)}(x) \sim \xi^{(2)}(x)\phi(x) \tag{5.53}$$

The $\gamma^{(2)}$ kernel is

$$\gamma^{(2)}(x-y) = \frac{\hbar}{16\pi^2}\int \frac{d^4p}{(2\pi)^4}\, e^{ip(x-y)}\pi\sqrt{1 - \frac{4m_\psi^2}{(-p^2)}}\,\theta(-p^2 - 4m_\psi^2)\frac{1}{|p_0|} \tag{5.54}$$

and the dissipative force is

$$F_\gamma^{(2)}(x) \sim \hbar \int d^4y \; \mu(x-y)\phi^2(y)\phi(x) \tag{5.55}$$

As discussed in [HuPaZh93a], we can show that a general fluctuation–dissipation relation exists between the dissipation and the noise kernels in the form

$$\hbar^{k-1}\nu^{(k)}(x) = \int d^4y \; K^{(k)}(x-y)\gamma^{(k)}(y) \tag{5.56}$$

Apart from a delta function $\delta^3(\mathbf{x}-\mathbf{x}')$, the fluctuation–dissipation kernel K^k for quantum fields has exactly the same form as for the quantum Brownian harmonic oscillator. In general it is a rather complicated expression [HuPaZh93a], but simplifies at high and zero temperatures. At high temperatures,

$$K^{(k)}(s) = \frac{2k_B T}{\hbar}\delta(s) \qquad (s \equiv x - y) \tag{5.57}$$

which gives back the famous Einstein relation. At zero temperature,

$$K^{(k)}(s) = \int_0^{+\infty} \frac{d\omega}{\pi}\omega\cos\omega s \tag{5.58}$$

which is the same as in the linear coupling case. Both limiting forms are independent of k. In other words, at both high and zero temperatures, the FDT is insensitive to the way the system is coupled to the environment.

Our derivation of the fluctuation–dissipation relation shows that it has a more general meaning than the more restrictive conditions where it is usually presented, e.g. in the near-equilibrium or the linear response regimes. It should be viewed as a categorical relation depicting the stochastic stimulation of the system and the averaged response of the environment.

5.2 Quantum functional master equation

We now turn to a derivation of the functional master equation for the system field with the interaction described in the last section. The full equation is quite involved because it contains nonlinear nonlocal dissipation and multiplicative colored noise plus a nonlocal potential term. Just to see the qualitative features let us first examine the simplified case under a local truncation to the dissipation kernel and the noise kernel (i.e. white noise) and omitting the nonlocal potential term. Namely, we set

$$\gamma(x-x') = \gamma_0\delta^4(x-x') \tag{5.59}$$

$$\hbar^{k-1}\nu(x-x') = \nu_0\delta^4(x-x') \tag{5.60}$$

$$V_{\phi\psi}(x-x') = 0 \tag{5.61}$$

Under this approximation the quantum master equation derived from the influence functional is much simpler. However, we need to emphasize that this approximation violates the fluctuation–dissipation relation at zero temperature. The effective action equation (5.33) simplifies to

$$
S_{\text{eff}}[\phi^1, \phi^2] = \int d^4x \Bigg\{ -\frac{1}{2}\partial_\mu\phi^1\partial^\mu\phi^1 - \frac{1}{2}m_{\phi r}^2(\phi^1)^2 - \frac{1}{4!}\lambda_{\phi r}(\phi^1)^4
$$

$$
+ \frac{1}{2}\partial_\mu\phi^2\partial^\mu\phi^2 + \frac{1}{2}m_{\phi r}^2(\phi^2)^2 + \frac{1}{4!}\lambda_{\phi r}(\phi^2)^4
$$

$$
- 2\lambda_{\phi\psi}^2\gamma_0\left((\phi^1)^2 - (\phi^2)^2\right)\left(\phi^1\frac{\partial\phi^1}{\partial s} + \phi^2\frac{\partial\phi^2}{\partial s}\right)
$$

$$
+ (i/2)\lambda_{\phi\psi}^2\nu_0\left((\phi^1)^2 - (\phi^2)^2\right)^2 \Bigg\} \tag{5.62}
$$

We can now write down a "Hamiltonian" which corresponds to equation (5.62) as

$$
\hat{H}_\rho[\phi^1, \phi^2, t]
$$

$$
= \int d^3x \Bigg\{ -\frac{\hbar^2}{2}\frac{\delta^2}{\delta(\phi^1(\mathbf{x}))^2} + \frac{1}{2}(\nabla\phi^1(\mathbf{x}))^2 + \frac{1}{2}m_{\phi r}^2(\phi^1(\mathbf{x}))^2 + \frac{1}{4!}\lambda_{\phi r}(\phi^1(\mathbf{x}))^4
$$

$$
+ \frac{\hbar^2}{2}\frac{\delta^2}{\delta(\phi^2(\mathbf{x}))^2} - \frac{1}{2}(\nabla\phi^2(\mathbf{x}))^2 - \frac{1}{2}m_{\phi r}^2(\phi^2(\mathbf{x}))^2 - \frac{1}{4!}\lambda_{\phi r}(\phi^2(\mathbf{x}))^4
$$

$$
- 2i\hbar\lambda_{\phi\psi}^2\gamma_0\left[(\phi^1(\mathbf{x}))^2 - (\phi^2(\mathbf{x}))^2\right]\left[\phi^1(\mathbf{x})\frac{\delta}{\delta\phi^1(\mathbf{x})} - \phi^2(\mathbf{x})\frac{\delta}{\delta\phi^2(\mathbf{x})}\right]
$$

$$
- (i/2)\lambda_{\phi\psi}^2\nu_0\left[(\phi^1(\mathbf{x}))^2 - (\phi^2(\mathbf{x}))^2\right]^2 \Bigg\} \tag{5.63}
$$

which also is correct up to order of $\lambda_{\phi\psi}^2$. Therefore the quantum functional master equation is given by the following functional "Schrödinger" equation

$$
i\hbar\frac{\partial}{\partial t}\,\rho_r[\phi^1, \phi^2, t] = \hat{H}_\rho[\phi^1, \phi^2, t]\,\rho_r[\phi^1, \phi^2, t] \tag{5.64}
$$

This quantum functional master equation for the system field is very similar to the quantum master equation for the anharmonic oscillator with nonlinear dissipation and nonlinear coupled noise in the Brownian particle model treated in [HuPaZh93a].

Let us define the Wigner functional for the quantum field as follows

$$W[\phi, \pi, t] = \int d\psi(\mathbf{x}) \, \exp\left\{ i\hbar^{-1} \int d^3\mathbf{x}\, \pi(\mathbf{x})\psi(\mathbf{x}) \right\} \times \rho_r \left[\phi - \frac{1}{2}\psi, \phi + \frac{1}{2}\psi, t \right]$$

(5.65)

Applying equation (5.65) to both sides of the above functional master equation, we can obtain the following Wigner functional equation

$$
\begin{aligned}
\frac{\partial}{\partial t} & W[\phi, \pi, t] \\
&= \int d^3\mathbf{x} \left\{ -\pi(\mathbf{x})\frac{\delta}{\delta\phi(\mathbf{x})} - [\nabla\phi(\mathbf{x})\cdot\nabla + m_{\phi r}^2\phi(\mathbf{x}) + \frac{1}{6}\lambda_{\phi r}\phi^3(\mathbf{x})]\frac{\delta}{\delta\pi(\mathbf{x})} \right. \\
&\quad + 4\lambda_{\phi\psi}^2\gamma_0\phi^2(\mathbf{x})\frac{\delta}{\delta\pi(\mathbf{x})}\pi(\mathbf{x}) + 2\hbar\lambda_{\phi\psi}^2\nu_0\phi^2(\mathbf{x})\frac{\delta^2}{\delta\pi^2(\mathbf{x})} \\
&\quad \left. + \hbar^2\lambda_{\phi r}\phi(\mathbf{x})\frac{\delta^3}{\delta\phi^3(\mathbf{x})} + 2\lambda_{\phi\psi}^2\hbar^2\gamma_0\phi(\mathbf{x})\frac{\delta^3}{\delta\phi(\mathbf{x})\delta\pi^2(\mathbf{x})} \right\} W[\phi, \pi, t] \quad (5.66)
\end{aligned}
$$

It is clear that the last two terms on the right-hand side of equation (5.66) which contain third-order derivatives are the quantum corrections. In the classical limit they go to zero, and equation (5.66) becomes the functional Fokker–Planck equation. We also know that the Wigner functional (5.65) becomes the classical phase space distribution functional in the classical limit.

The quantum Wigner function contains just as much information as the wavefunction so it oscillates and can assume negative values. In particular it does not exhibit a peak along the classical trajectory in phase space except at high temperature or for harmonic oscillators. Thus viewing the quantum Wigner function as possessing the equivalent traits of a classical one-particle phase space distribution function is untenable except under special conditions. This has special significance in quantum–classical correspondence issues. See discussions in [Hab90, HabLaf90].

One can also show that the following "equilibrium" state distribution

$$W[\phi, \pi] \sim \exp -\bar{\beta} \int d^3\mathbf{x} \left\{ \frac{1}{2}\pi^2(\mathbf{x}) + \frac{1}{2}(\nabla\phi(\mathbf{x}))^2 + \frac{1}{2}m_{\phi r}^2\phi^2(\mathbf{x}) + \frac{1}{4!}\lambda_{\phi r}\phi^4(\mathbf{x}) \right\}$$

(5.67)

is the asymptotic solution of the above functional Wigner equation in the classical limit, provided that $\bar{\beta}^{-1} = \hbar\nu_0/\gamma_0$. Thus the above functional Wigner equation can describe the process of relaxation to equilibrium state.

We are now ready to present the functional master equation. The calculation closely parallels that of the QBM case studied in Chapter 3 [HuPaZh92, Paz94]. The quantum functional master equation for the case of nonlocal dissipation, colored noise and nonlocal potential is given by

$$i\hbar \frac{\partial}{\partial t} \rho_r[\phi^1, \phi^2, t]$$

$$= \int d^3\mathbf{x} \Bigg\{ -\frac{\hbar^2}{2} \frac{\delta^2}{\delta(\phi^1(\mathbf{x}))^2} + \frac{1}{2}(\nabla\phi^1(\mathbf{x}))^2 + \frac{1}{2}m^2_{\phi r}(\phi^1(\mathbf{x}))^2 + \frac{1}{4!}\lambda_{\phi r}(\phi^1(\mathbf{x}))^4$$

$$+ \frac{\hbar^2}{2} \frac{\delta^2}{\delta(\phi^2(\mathbf{x}))^2} - \frac{1}{2}(\nabla\phi^2(\mathbf{x}))^2 - \frac{1}{2}m^2_{\phi r}(\phi^2(\mathbf{x}))^2 - \frac{1}{4!}\lambda_{\phi r}(\phi^2(\mathbf{x}))^4$$

$$- i\lambda^2_{\phi\psi}a_1(\mathbf{x}, s)\left[(\phi^1(\mathbf{x}))^2 + (\phi^2(\mathbf{x}))^2\right] + 2\lambda^2_{\phi\psi}a_2(\mathbf{x}, t)\left[(\phi^1(\mathbf{x}))^2 + (\phi^2(\mathbf{x}))^2\right]$$

$$- \lambda^2_{\phi\psi}(\phi^1(\mathbf{x}))^2 \{v * \hat{O}^2_+\}(\mathbf{x}, t) + \lambda^2_{\phi\psi}(\phi^2(\mathbf{x}))^2 \{v * \hat{O}^2_-\}(\mathbf{x}, t)$$

$$- \lambda^2_{\phi\psi}\left[(\phi^1(\mathbf{x}))^2 - (\phi^2(\mathbf{x}))^2\right] \{\mu * (\hat{O}^2_+ + \hat{O}^2_-)\}(\mathbf{x}, t)$$

$$- (i/2)\lambda^2_{\phi\psi}\left[(\phi^1(\mathbf{x}))^2 - (\phi^2(\mathbf{x}))^2\right] \{\nu * (\hat{O}^2_+ - \hat{O}^2_-)\}(\mathbf{x}, t) \Bigg\} \rho_r[\phi, \phi', t]$$

$$(5.68)$$

where $*$ denotes convolution, namely

$$\{v * \phi^2\}(\mathbf{x}, t) = \int_{t_0}^{t} ds \int d^3\mathbf{x}' \; v(\mathbf{x} - \mathbf{x}', t - s)\phi^2(\mathbf{x}', s) \qquad (5.69)$$

The time-dependent coefficients in equation (5.68) are as follows

$$a_1(\mathbf{x}, t) = \int_{t_0}^{t} ds \int d^3\mathbf{x}' \; v(\mathbf{x} - \mathbf{x}', t - s)Q(s) = \{v * Q\}(\mathbf{x}, t) \qquad (5.70)$$

$$a_2(\mathbf{x}, t) = \int_{t_0}^{t} ds \int d^3\mathbf{x}' \; \nu(\mathbf{x} - \mathbf{x}', t - s)Q(s) = \{\nu * Q\}(\mathbf{x}, t) \qquad (5.71)$$

where

$$Q(s) = \int \frac{d^3\mathbf{k}}{(2\pi)} \frac{\sin\omega(\mathbf{k})(s - t_0)\sin\omega(\mathbf{k})(t - s)}{\omega(\mathbf{k})\sin\omega(\mathbf{k})(t - t_0)} \qquad (5.72)$$

and the operators

$$\hat{O}_+(\mathbf{x}, s) \equiv \{\alpha(t - s) * \phi^1_f\}(\mathbf{x}) - \left\{\beta(t - s) * i\frac{\hbar\delta}{\delta\phi^1_f}\right\}(\mathbf{x}) \qquad (5.73)$$

$$\hat{O}_-(\mathbf{x}, s) \equiv \{\alpha(t - s) * \phi^2_f\}(\mathbf{x}) + \left\{\beta(t - s) * i\frac{\hbar\delta}{\delta\phi^2_f}\right\}(\mathbf{x}) \qquad (5.74)$$

with

$$\alpha(\mathbf{x}, s) = \int \frac{d^3\mathbf{k}}{(2\pi)^3} e^{i\mathbf{k}\cdot\mathbf{x}} \cos\omega(\mathbf{k})s \qquad (5.75)$$

$$\beta(\mathbf{x}, s) = \int \frac{d^3\mathbf{k}}{(2\pi)^3} e^{i\mathbf{k}\cdot\mathbf{x}} \frac{\sin\omega(\mathbf{k})s}{\omega(\mathbf{k})} \qquad (5.76)$$

This is a nonstationary quantum functional master equation. In spite of its complicated appearance (convolution products appearing in the equation), it is not difficult to see that the structure of the equation is similar to the nonstationary quantum master equation for a Brownian anharmonic oscillator with nonlinear dissipation and multiplicative colored noise. Actually, in momentum space, the convolution product becomes a direct product, so the above equation in momentum space will become the quantum master equation for one particular mode (harmonic oscillator). However, the different modes will still be coupled together in the quantum master equation because of mode–mode coupling in the system field via the nonlinear potential. We now turn to the case of one interacting field divided into two sectors.

5.3 The closed time path coarse-grained effective action

To add some physical flavor to our derivation and in anticipation of applications to problems in cosmology, we consider the action of a massless scalar field with $\lambda\phi^4$ self-interaction coupled conformally to a spatially-flat Friedmann–Lemaitre–Robertson–Walker universe. The conformal-related field χ (introduced in Chapter 4) is related to ϕ by $\chi = a(t)\phi$ and the conformal time η is related to the cosmic time t by $\eta = \int dt/a(t)$. We shall use d^4x to denote $d^3x\, d\eta$ in the remainder of this chapter. Since our purpose here is more to illustrate the coarse-graining idea in the construction of a CGEA than to discuss cosmological applications (see Chapter 15), we can just view the scale factor $a(t) = e^\alpha$ as a scaling parameter rather than a dynamical function determined from Einstein's equations. The content of this section can thus be used without reference to cosmology by treating a as a constant, e.g. setting $a = 1$ would keep us in a Minkowski spacetime with the conformal time η acting as the global time t. However we wish to tag along the scale factor a so that later we can view the inflationary cosmology in the light of scaling [Hu91] without added effort.

We begin by separating the quantum field $\chi(x,\eta)$ into two parts, $\chi = \chi_< + \chi_>$, where $\chi_<$ contains the lower k wave modes and $\chi_>$ the higher k modes. We can refer to these two sectors as the system and the environment respectively. Two useful physical examples are

Case A (critical phenomena)

$$\chi_< :\, \mid \mathbf{k} \mid < \Lambda/s, \qquad \chi_> :\, \Lambda/s < \mid \mathbf{k} \mid < \Lambda \tag{5.77}$$

where Λ is the ultraviolet cut-off and $s > 1$ is the coarse-graining parameter which gives the fraction of total k modes counted in the environment.

Case B (stochastic inflation)

$$\chi_< :\, \mid \mathbf{k} \mid < \epsilon Ha, \qquad \chi_> :\, \mid \mathbf{k} \mid > \epsilon Ha, \qquad \epsilon \approx 1 \tag{5.78}$$

where the Hubble constant $H(t) \equiv \dot{a}/a$ (the event horizon in the de Sitter universe) serves to divide the physical wavelength $\mathbf{p} \equiv \mathbf{k}/a$ into two sectors, with a window function measuring how sharp the division is. We will have more to say about this point in Chapter 15. For now, we can build up our intuition for the coarse-graining ideas using Case A as an illustrative example. The separation of χ can also be made in other manners, depending on the physical set-up of the problem and the questions one asks. The formalism we present here is quite general. Our presentation for this model follows [LomMaz96].

Explicitly, we define the system by

$$\chi_<(\mathbf{x},\eta) = \int_{|\mathbf{k}|<\Lambda_c} \frac{d^3\mathbf{k}}{(2\pi)^3} \chi(\mathbf{k},\eta) e^{i\mathbf{k}\cdot\mathbf{x}} \tag{5.79}$$

and the environment by

$$\chi_>(\mathbf{x},\eta) = \int_{|\mathbf{k}|>\Lambda_c} \frac{d^3\mathbf{k}}{(2\pi)^3} \chi(\mathbf{k},\eta) e^{i\mathbf{k}\cdot\mathbf{x}} \tag{5.80}$$

The system field contains the modes with wavelengths longer than the critical value Λ_c^{-1}, while the environment field contains wavelengths shorter than Λ_c^{-1}. Λ_c corresponds to $s^{-1}\Lambda$. After the splitting, the total action can be written as

$$S[a,\chi] = S[\chi_<] + S_0[\chi_>] + S_{\text{int}}[a,\chi_<,\chi_>] \tag{5.81}$$

where S_0 denotes the kinetic term

$$S_0[\chi] = -\frac{1}{2}\int d\eta \int \frac{d^3\mathbf{k}}{(2\pi)^3} \left\{ \chi(\mathbf{k},\eta) \left[\frac{\partial^2}{\partial\eta^2} + k^2 \right] \chi(\mathbf{k},\eta) \right\} \tag{5.82}$$

$S[\chi_<]$ is the system action,

$$S[\chi_<] = S_0[\chi_<] - \int d^4x \left\{ \frac{1}{2}M^2\chi_<^2 + \frac{\lambda}{4!}\chi_<^4 \right\} \tag{5.83}$$

and the interaction part is given by

$$S_{int}[a,\chi_<,\chi_>] = -\int d^4x \left\{ \left[\frac{1}{2}M^2 + \frac{\lambda}{4}\chi_<^2(x) \right] \chi_>^2 + \frac{\lambda}{4!}\chi_>^4 + \frac{\lambda}{6}\chi_<^3\chi_> + \frac{\lambda}{6}\chi_<\chi_>^3 \right\} \tag{5.84}$$

with

$$M^2 = \left[m^2 + \left(\xi - \frac{1}{6} \right) R \right] a^2 \tag{5.85}$$

We are interested in the influence of the environment on the evolution of the system. To this end, we seek to construct the Feynman–Vernon influence functional, following the methods of Chapter 3. This is obtained by integrating over the environment field configurations between an initial time $\eta = -\eta_i$ and a final time $\eta = \eta_f$. When η_i, η_f are larger than any other characteristic time and the environment field is initially in the vacuum state, the Feynman–Vernon influence action turns into the so-called closed time path (CTP) coarse-grained effective action (CGEA) $S_\Lambda[a^1, \chi_<^1, a^2, \chi_<^2]$, which is defined by

$$\exp\left\{i\hbar^{-1}S_{\Lambda_c}[a^1, \chi_<^1, a^2, \chi_<^2]\right\}$$

$$= \exp i\hbar^{-1}\left\{S[\chi_<^1] - S[\chi_<^2]\right\} \int d\chi_{>f} \int^{\chi_{>f}} \mathcal{D}\chi_>^1 \int^{\chi_{>f}} \mathcal{D}\chi_>^2 \exp i\hbar^{-1}$$

$$\times \left\{S_0[\chi_>^1] + S_{\text{int}}[a^1, \chi_<^1, \chi_>^1] - S_0[\chi_>^2] - S_{\text{int}}[a^2, \chi_<^2, \chi_>^2]\right\} \qquad (5.86)$$

The integration here is performed over all fields $\chi_>^1$ ($\chi_>^2$) with positive (negative) frequency modes in the remote past that coincide at the final time $\chi_>^1 = \chi_>^2 = \chi_{>f}$. More general initial conditions will be discussed in later chapters.

We now derive the CTP CGEA perturbatively in λ and M^2, up to quadratic order in both quantities. A simple calculation leads to

$$S_{\Lambda_c}[a^1, \chi_<^1, a^2, \chi_<^2] = S[\chi_<^1] - S[\chi_<^2] + \langle S_{\text{int}}[a^1, \chi_<^1, \chi_>^1]\rangle_0 - \langle S_{\text{int}}[a^2, \chi_<^2, \chi_>^2]\rangle_0$$

$$+ \frac{i}{2\hbar}\left\{\langle S_{\text{int}}[a^1, \chi_<^1, \chi_>^1]^2\rangle_0 - \langle S_{\text{int}}[a^1, \chi_<^1, \chi_>^1]\rangle_0^2\right\}$$

$$- i\hbar^{-1}\left\{\langle S_{\text{int}}[a^1, \chi_<^1, \chi_>^1] S_{\text{int}}[a^2, \chi_<^2, \chi_>^2]\rangle_0\right.$$

$$\left. - \langle S_{\text{int}}[a^1, \chi_<^1, \chi_>^1]\rangle_0 \langle S_{\text{int}}[a^2, \chi_<^2, \chi_>^2]\rangle_0\right\}$$

$$+ \frac{i}{2\hbar}\left\{\langle S_{\text{int}}[a^2, \chi_<^2, \chi_>^2]^2\rangle_0 - \langle S_{\text{int}}[a^2, \chi_<^2, \chi_>^2]\rangle_0^2\right\} \qquad (5.87)$$

where the quantum average of a functional of the fields $\mathcal{Q}$ is defined with respect to the kinetic action S_0

$$\langle \mathcal{Q}[\chi_>^1, \chi_>^2]\rangle_0 = \int d\chi_{>f} \int^{\chi_{>f}} \mathcal{D}\chi_>^1 \int^{\chi_{>f}} \mathcal{D}\chi_>^2 \exp i\hbar^{-1}\left\{S_0[\chi_>^1] - S_0[\chi_>^2]\right\}\mathcal{Q} \qquad (5.88)$$

Equation (5.87) is the in–in version of the Dyson–Feynman series.

We define the propagators of the environment field as

$$\langle T\chi_>^1(x)\chi_>^1(y)\rangle_0 = G_F^{\Lambda_c}(x - y), \qquad (5.89)$$

$$\langle \chi_>^1(x)\chi_>^2(y)\rangle_0 = G_-^{\Lambda_c}(x - y), \qquad (5.90)$$

$$\langle \tilde{T}\chi_>^2(x)\chi_>^2(y)\rangle_0 = G_D^{\Lambda_c}(x - y). \qquad (5.91)$$

where $T, \tilde{T}$ denote time- and reversed-time ordering respectively.

Despite their appearance these propagators are not the usual Feynman, negative-frequency Wightman and Dyson propagators of the scalar field since, in this case, the momentum integration is restricted by the presence of the (infrared) cut-off Λ_c. The explicit expressions are

$$G_F^{\Lambda_c}(x - y) = -i\hbar \int_{|\mathbf{p}|>\Lambda_c} \frac{d^4p}{(2\pi)^4} e^{ip(x-y)} \frac{1}{p^2 - i\varepsilon} \qquad (5.92)$$

$$G_-^{\Lambda_c}(x - y) = \int_{|\mathbf{p}|>\Lambda_c} \frac{d^4p}{(2\pi)^4} e^{ip(x-y)} 2\pi\hbar\delta(p^2)\Theta(-p^0) \qquad (5.93)$$

$$G_D^{\Lambda_c}(x - y) = i\hbar \int_{|\mathbf{p}|>\Lambda_c} \frac{d^4p}{(2\pi)^4} e^{ip(x-y)} \frac{1}{p^2 + i\varepsilon} \qquad (5.94)$$

As an example, we show the expression for the propagator $G_F^{\Lambda_c}$. The usual mass-less Feynman propagator is

$$\hbar^{-1}\Delta_F(x) = \frac{1}{8\pi^2}\frac{1}{\sigma} + \frac{i}{8\pi}\delta(\sigma) \tag{5.95}$$

while

$$\hbar^{-1}G_F^{\Lambda_c}(x) = \frac{(-1)}{8\pi^2}\left[\frac{\cos[\Lambda_c(r-x^0)]}{r(r-x^0)} + \frac{\cos[\Lambda_c(r+x^0)]}{r(r+x^0)}\right]$$
$$+ \frac{i}{8\pi^2}\left[\frac{\sin[\Lambda_c(r-x^0)]}{r(r-x^0)} - \frac{\sin[\Lambda_c(r+x^0)]}{r(r+x^0)}\right]$$

$$G_F^{\Lambda_c}(x) \equiv \Delta_F(x) - G_F^{|\mathbf{p}|<\Lambda_c}(x) \tag{5.96}$$

where $\sigma = -\frac{1}{2}x^2$ and $r = |\mathbf{x}|$.

The CTP CGEA can be computed from equations (5.87)–(5.91) using standard techniques [GreMul97]. Defining

$$\left(\tilde{M}^{1,2}\right)^2 = M^2 + \frac{\lambda}{2}\left(\chi_<^{1,2}\right)^2 \tag{5.97}$$

$$\chi_-^{(n)} = (\chi_<^1)^n - (\chi_<^2)^n, \qquad \chi_+^{(n)} = \frac{1}{2}\left[(\chi_<^1)^n + (\chi_<^2)^n\right]$$
$$\lambda Q_- = (\tilde{M}^1)^2 - (\tilde{M}^2)^2, \qquad \lambda Q_+ = \frac{1}{2}\left[(\tilde{M}^1)^2 + (\tilde{M}^2)^2\right] \tag{5.98}$$

and using simple identities for the propagators, the CTP CGEA can be written as

$$S_{\Lambda_c} = S(\chi_<^1) - S(\chi_<^2) + \frac{\lambda}{4}\int d^4x\, G_F^{\Lambda_c}(0)Q_-(x)$$
$$+ \hbar^{-1}\lambda^2\int d^4x\int d^4y\,\Theta(y^0-x^0)\left\{\frac{1}{18}\chi_+^{(3)}(x)\,\mathrm{Im}G_F^{\Lambda_c}(x-y)\,\chi_-^{(3)}(y)\right.$$
$$\left. - \frac{1}{4}Q_+(x)\,\mathrm{Im}\left[G_F^{\Lambda_c}(x-y)\right]^2 Q_-(y) - \frac{1}{3}\chi_+^{(1)}(x)\,\mathrm{Im}\left[G_F^{\Lambda_c}(x-y)\right]^3\chi_-^{(1)}(y)\right\}$$
$$+ \frac{i\lambda^2}{4\hbar}\int d^4x\int d^4y\,\Theta(y^0-x^0)\left\{\frac{1}{18}\chi_-^{(3)}(x)\,\mathrm{Re}G_F^{\Lambda_c}(x-y)\,\chi_-^{(3)}(y)\right.$$
$$+ \frac{1}{4}Q_-(x)\,\mathrm{Re}\left[G_F^{\Lambda_c}(x-y)\right]^2 Q_-(y)$$
$$\left. - \frac{1}{3}\chi_-^{(1)}(x)\,\mathrm{Re}\left[G_F^{\Lambda_c}(x-y)\right]^3\chi_-^{(1)}(y)\right\} \tag{5.99}$$

The real part of the CTP CGEA in equation (5.99) contains divergences and must be renormalized. As the propagators in equations (5.89)–(5.91) differ from the usual ones only by the presence of the infrared cut-off, the ultraviolet divergences coincide with those of the usual $\lambda\chi^4$ theory. The effective action can therefore be renormalized using the standard procedure. (For the renormalization of quantum fields in curved spacetimes, it is necessary to add to the Einstein–Hilbert term in the gravitational action a cosmological constant, and terms quadratic in the curvature tensor. See, e.g. [BirDav82].)

Consider the square of the Feynman propagator. Using dimensional regularization we find

$$[G_F^{\Lambda_c}(x)]^2 = [\Delta_F(x)]^2 + [G_F^{(|\mathbf{p}|<\Lambda_c)}(x)]^2 - 2\Delta_F(x)G_F^{(|\mathbf{p}|<\Lambda_c)}(x) \quad (5.100)$$

where

$$\hbar^{-2}\Delta_F^2(x) = \frac{1}{16\pi^2}\left[\frac{i}{n-4} + i\psi(1) - 4\pi i + \ln(4\pi\mu^2)\right]\delta^4(x) + iR_1(x) + R_2(x)$$

$$(5.101)$$

$$R_1(x) = \frac{1}{(2\pi)^4}\int d^4p\, e^{ipx}\ln|p^2|$$

$$R_2(x) = \frac{\pi}{(2\pi)^4}\int d^4p\, e^{ipx}\Theta(-p^2)$$

Note that the divergence is the usual one, i.e. proportional to $\delta^4(x-y)$ and independent of Λ_c. Consequently, the term $\mathrm{Re}[G_F^\Lambda(x-y)]^2 Q_+(x)Q_-(y)$ in equation (5.99) is divergent and renormalizes the coupling constant λ and the constants that appear in the gravitational action. The other divergences can be treated in a similar way. One can also check that the imaginary part of the effective action does not contain divergences. Of course, a successful ultraviolet renormalization does not guarantee that an approximation scheme such as RG-improved perturbation theory will be well behaved. An example is in the RG equations for $\lambda\phi^4$ fields [Hu91] where loops depend on a factor $e^{H(t-t_0)}$ which would invalidate perturbation theory. Further "infrared" H-dependent (environmentally friendly) renormalization of λ is needed [OCoSte94a, OCoSte94b, EiOCSt95, Ste98, FEOS96].

As we have seen in Chapter 3, the (nonlocal) real and imaginary parts of S_{Λ_c} can be associated with the dissipation and noise respectively, which are related by an integral equation known as the fluctuation–dissipation relation.

5.3.1 Stochastic equations

The Langevin equation

We now show how to derive a Langevin equation for the system field from the CTP CGEA. This equation takes into account the three fundamental effects of the environment on the system: renormalization, dissipation and noise.

The CTP CGEA for our model is given in equation (5.99). Since the imaginary part is quadratic in the system field, we can invoke the Gaussian identity used by Feynman and Vernon [FeyVer63], as discussed in Chapter 3. The CTP CGEA can thus be rewritten as

$$S_{\Lambda_c}[\chi_<^1, \chi_<^2] = -i\hbar\ln\int \mathcal{D}\xi_1 P[\xi_1]\int \mathcal{D}\xi_2 P[\xi_2]$$

$$\times \int \mathcal{D}\xi_3 P[\xi_3]\exp\left\{i\hbar^{-1}S_{\mathrm{eff}}[\chi_<^1, \chi_<^2, \xi_1, \xi_2, \xi_3]\right\} \quad (5.102)$$

where

$$S_{\text{eff}}[\chi^1_<, \chi^2_<, \xi_1, \xi_2, \xi_3]$$
$$= \text{Re}S_{\Lambda_c}[\chi^1_<, \chi^2_<] - \int d^4x \left[\chi^{(3)}_-(x)\xi_1(x) + Q_-(x)\xi_2(x) + \chi^{(1)}_-(x)\xi_3(x) \right] \quad (5.103)$$

and $\xi_1(x)$, $\xi_2(x)$, and $\xi_3(x)$ are Gaussian stochastic sources with zero mean and auto-correlations

$$\langle \xi_1(x)\xi_1(y) \rangle = \frac{\lambda^2}{9}\text{Re}G_F^{\Lambda_c}(x-y) \quad (5.104)$$

$$\langle \xi_2(x)\xi_2(y) \rangle = \frac{\lambda^2}{2}\text{Re}\left[G_F^{\Lambda_c}(x-y) \right]^2 \quad (5.105)$$

$$\langle \xi_3(x)\xi_3(y) \rangle = \frac{2\lambda^2}{3}\text{Re}\left[G_F^{\Lambda_c}(x-y) \right]^3 \quad (5.106)$$

From this effective action it is easy to derive the stochastic field equation for the system

$$\left. \frac{\partial S_{\text{eff}}[\chi^1_<, \chi^2_<, \xi_1, \xi_2, \xi_3]}{\partial \chi^1_<} \right|_{\chi^1_<=\chi^2_<} = 0 \quad (5.107)$$

It is given by

$$\left(\frac{\partial^2}{\partial \xi_3^2} - \nabla^2 \right)\chi_< + \left[M^2 + \frac{\lambda}{2}G_F^{\Lambda_c}(0) \right]\chi_< + \frac{\lambda}{6}\chi^3_<$$
$$+ \frac{\lambda^2}{6}\chi^2_<(x) \int d^4y\,\theta(x_0 - y_0)\text{Im}G_F^{\Lambda_c}(x,y)\chi^3_<(y)$$
$$- \frac{\lambda^2}{2}\chi_<(x) \int d^4y\,\theta(x_0 - y_0)\text{Re}[G_F^{\Lambda_c}(x,y)]^2\chi^2_<(y)$$
$$- \frac{\lambda^2}{3} \int d^4y\,\theta(x_0 - y_0)\text{Im}[G_F^{\Lambda_c}(x,y)]^3\chi_<(y)$$
$$= 3\xi_1(x)\chi^2_<(x) + 2\xi_2(x)\chi_<(x) + \xi_3(x) \quad (5.108)$$

This is the functional Langevin equation derived from the variation of the CTP CGEA. By construction, it is real and causal. We see that it contains multiplicative and additive colored noise. The nonlinear coupling between modes adds complexity to the Langevin equation. This class of equations was first derived by Sinha and Hu [SinHu91] for considerations of the validity of minisuperspace approximations in quantum cosmology, and by Lombardo and Mazzitelli in [LomMaz96], whose treatment we follow here. Greiner and Müller [GreMul97] obtained a similar stochastic equation in flat spacetime, for a thermal environment. They found explicit expressions for the momentum-dependent dissipation function in the Langevin equation using a Markovian approximation for the soft modes.

The master equation (for one-mode system)

As we have seen in Chapter 3, an equivalent depiction of the dynamics of the open system is obtained from the master equation for the reduced density matrix. A functional master equation for the long-wavelength modes may be derived along similar lines [LomMaz96], but in general it is very complicated. A tractable result can be obtained when the system field contains only one mode $\mathbf{k} = \mathbf{k}_0$. This is a sort of "minisuperspace" (the space of modes in this case) approximation. Also we keep terms only up to $O\left(\lambda^2\right)$. Under these approximations the general form of the master equation is given by

$$
i\hbar\partial_h\rho_r[\chi^1_{<f}, \chi^2_{<f}, \eta] = \langle\chi^1_{<f}|\left[\hat{H}_{\text{ren}}, \hat{\rho}_r\right]|\chi^2_{<f}\rangle
$$

$$
- i\lambda^2 \left[-\frac{[(\chi^1_{<f})^3 - (\chi^2_{<f})^3]^2 V}{1152} D_3(\mathbf{k}_0; \eta)\right.
$$

$$
+ \frac{[(\chi^1_{<f})^2 - (\chi^2_{<f})^2]^2 V}{32} D_2(\mathbf{k}_0; \eta)
$$

$$
\left. - \frac{(\chi^1_{<f} - \chi^2_{<f})^2 V}{6} D_1(\mathbf{k}_0; \eta)\right] \rho_r[\chi^1_{<f}, \chi^2_{<f}, \eta] + \dots \qquad (5.109)
$$

Due to the complexity of the equation, we only show the correction to the usual unitary evolution term coming from the noise kernels. The full expression can be found in [LomMaz96]. This equation contains three time-dependent diffusion coefficients $D_i(\eta)$. (The subscripts 3, 2, 1 refer to the order of the system field $\phi_{<f}$.) Up to one loop, only D_3 and D_2 survive and are given by

$$
D_3(\mathbf{k}_0; \eta) = \int_0^t ds\, \cos^3(\mathbf{k}_0 s)\, \text{Im}G_F^{\Lambda_c}(3\mathbf{k}_0; \eta - s)
$$

$$
\approx \frac{1}{6k_0} \int_0^t ds\, \cos^3(k_0 s)\, \cos(3k_0 s)\, \theta(3k_0 - \Lambda_c)
$$

$$
= \frac{2k_0\eta + 3\sin(2k_0\eta) + \frac{3}{2}\sin(4k_0\eta) + \frac{1}{3}\sin(6k_0\eta)}{576\, k_0^2}
$$

$$
\text{for} \quad \frac{\Lambda_c}{3} < k_0 < \Lambda_c \quad (k_0 \equiv |\mathbf{k}_0|) \qquad (5.110)
$$

$$
D_2(\mathbf{k}_0; \eta) = \int_0^h ds\, \cos^2(k_0 s) \left\{ \text{Re}[G_F^{\Lambda_c}(2\mathbf{k}_0; \eta - s)]^2 + 2\text{Re}[G_F^{\Lambda_c}(0; \eta - s)]^2 \right\}
$$

$$
\qquad (5.111)
$$

Using the expressions

$$
\text{Re}[G_F^{\Lambda_c}(2\mathbf{k}_0; \eta - s)]^2 = \frac{\pi\hbar^2}{k_0} \left\{ \int_{\Lambda_c}^{2k_0+\Lambda_c} dp \int_{\Lambda_c}^{2k_0+p} dz\, \cos[(p + z)s]\right.
$$

$$
\left. + \int_{2k_0+\Lambda_c}^{\infty} dp \int_{p-2k_0}^{p+2k_0} dz\, \cos[(p + z)s] \right\} \qquad (5.112)
$$

$$
\text{Re}[G_F^{\Lambda_c}(0; \eta - s)]^2 = \pi\hbar^2 \left\{ 2\pi\delta(s) - 2\frac{\sin(2\Lambda_c s)}{s} \right\} \qquad (5.113)
$$

the D_2 diffusion coefficient can be written as

$$\hbar^{-2} D_2(\mathbf{k}_0; \eta) = \frac{\pi}{4} \left\{ 3\pi - \left(\frac{3}{2} - \frac{\Lambda_c}{2k_0} \right) \mathrm{Si}[2\eta(\Lambda_c - k_0)] \right.$$

$$- \left(2 - \frac{\Lambda_c}{2k_0} \right) \mathrm{Si}[2\Lambda_c \eta] - \left(\frac{3}{2} + \frac{\Lambda_c}{2k_0} \right) \mathrm{Si}[2\eta(\Lambda_c + k_0)]$$

$$- \left(1 + \frac{\Lambda_c}{2k_0} \right) \mathrm{Si}[2\eta(2k_0 + \Lambda_c)]$$

$$+ \frac{1}{4k_0 \eta} \left(\cos[2\Lambda_c \eta] - \cos[2\eta(\Lambda_c + k_0)] \right.$$

$$\left. \left. + \cos[2\eta(\Lambda_c - k_0)] - \cos[2\eta(2k_0 + \Lambda_c)] \right) \right\} \qquad (5.114)$$

where $\mathrm{Si}[z]$ denotes the sine-integral function [AbrSte72].

Equation (5.109) is the field-theoretical version of the QBM master equation we were looking for, except that the system here has nonlinear coupling. Owing to the existence of three interaction terms ($\chi_<^3 \chi_>$, $\chi_<^2 \chi_>^2$, and $\chi_< \chi_>^3$) there are three diffusion coefficients in the master equation. The form of the coefficients is fixed by these couplings and by the particular choice of the quantum state of the environment.

Note that these results are valid in the single-mode approximation. In this approximation one obtains a reduced density matrix for each mode $\mathbf{k}_0$, and neglects the interaction between different system modes. Due to this interaction, ρ_r will be different from $\prod_{\mathbf{k}_0} \rho_r(\mathbf{k}_0)$ in the general case. These results will be applied to the consideration of decoherence of quantum fields in Chapter 9 and cosmological structure formation in Chapter 15.

6
Functional methods in nonequilibrium QFT

One of the major goals in the establishment of a quantum field theory for nonequilibrium systems is to study dynamical problems, following the evolution of the expectation value of a physical variable with respect to an *in* state. This is different from a scattering problem characterized by the transition amplitude between the *in* and *out* states, as is treated in every textbook on quantum field theory. This problem is usually coupled with how one could identify a relevant sector of the theory as the system (light fields vs. heavy fields, slow modes vs. fast modes, long wavelength modes vs. short ones, etc.) and determine the effect of its other sectors as the environment on this system, as we have discussed in the last chapter.

Given a classical nonequilibrium system, described, for example, by a Langevin equation, there are essentially three possible strategies to follow. One may attempt to solve it, usually numerically. In the quantum field case, this gets difficult beyond the linear case. Second, one may try to transform it into an equation for the evolution of a probability distribution function in the system's configuration space. In the quantum field case the relevant object is the reduced density matrix, and the relevant equation is the Liouville–von Neumann equation. This is also infeasible beyond the linear case, unless under restrictive approximations (such as Gaussianity) which in fact reduce this approach to the third and coarsest. The third approach is to use the Langevin equation to obtain equations of motion for the expectation value $\langle x \rangle$ of the system variable x and its correlations. In the linear case, the relevant equation of motion is Heisenberg's equation (Ehrenfest's theorem). In the nonlinear case, the equation of motion for $\langle x \rangle$ will necessarily couple to higher correlation functions $\langle x^n \rangle$, and we will have a hierarchy of equations, just as in the BBGKY–Boltzmann paradigm. Then we will have to truncate the hierarchy, slave the higher correlations, etc., according to the particular set-up of the problem to extract the information accessible to a particular class of observers. We have given an introductory discussion of these issues in the context of the Boltzmann equation in Chapters 1 and 2.

The third approach is the one we will follow in the bulk of the book. Our immediate concern is to show that the (truncated) equations of motion may be obtained from variational principles of increasing complexity. At the simplest level, where we only seek an equation of motion for the expectation value ϕ of the Heisenberg picture field operator Φ, this equation follows from the variation of the Schwinger–Keldysh (CTP) effective action (EA). At the next level of a more comprehensive approach we seek coupled equations for ϕ and the "propagators" $G \sim \langle (\Phi - \phi)^2 \rangle$. The relevant action functional is the two-particle irreducible (2PI) effective action. Each higher order truncation of the hierarchy has its proper action functional, which are particular cases of the so-called "master" effective action (MEA).

Therefore the task at hand is to develop the techniques to compute these objects, and to learn to read the physics coded into their structure. In particular, we shall see that the CTPEA has the structure of a Feynman–Vernon influence action, and we shall simply borrow the physical insight gained from the study of quantum open systems. The analogy is less straightforward for the higher order truncations, but this approach remains essentially applicable.

Since field expectation values and propagators are going to be the main subjects of our discussion, it is appropriate that we start by gaining some insight into the different two-point functions and the information they contain. Thus, let us begin our discussion with a review of some basic scalar quantum field theory.

6.1 Propagators

A good deal of our discussion will revolve around the different properties of the propagators of the theory, that is, the expectation values of binary products of field operators with respect to the initial state. Since field operators at different locations do not generally commute, we have several different propagators according to the ordering of the field operators within the expectation value. In particular, we shall consider eight different propagators, namely:

The four basic propagators

Feynman

$$G_{\mathrm{F}} \equiv \langle T \left(\Phi \left(x \right) \Phi \left(x' \right) \right) \rangle \tag{6.1}$$

Dyson

$$G_{\mathrm{D}} \equiv \langle \tilde{T} \left(\Phi \left(x \right) \Phi \left(x' \right) \right) \rangle \tag{6.2}$$

positive frequency

$$G^{+} \equiv \langle \Phi \left(x \right) \Phi \left(x' \right) \rangle \tag{6.3}$$

and negative frequency

$$G^{-} \equiv \langle \Phi \left(x' \right) \Phi \left(x \right) \rangle \tag{6.4}$$

where T stands for time ordering and $\tilde{T}$ stands for anti-time ordering:

$$T \left[F \left(t \right) G \left(t' \right) \right] = F \left(t \right) G \left(t' \right) \theta \left(t - t' \right) + G \left(t' \right) F \left(t \right) \theta \left(t' - t \right)$$
$$\tilde{T} \left[F \left(t \right) G \left(t' \right) \right] = G \left(t' \right) F \left(t \right) \theta \left(t - t' \right) + F \left(t \right) G \left(t' \right) \theta \left(t' - t \right) \tag{6.5}$$

The Feynman and Dyson propagators are even. We also have $G_{\mathrm{F}} = G_{\mathrm{D}}^*$; $G^- = G^{+*}$; $G^-(x, x') = G^+(x', x)$. Finally we have the identity $G_{\mathrm{F}} + G_{\mathrm{D}} = G^+ + G^-$, which follows from the time ordering constraints.

Hadamard and Jordan propagators

The Hadamard propagator

$$G_1 = G^+ + G^- \equiv \langle\{\Phi(x), \Phi(x')\}\rangle \tag{6.6}$$

is real and even. The Jordan propagator

$$G = G^+ - G^- \equiv \langle[\Phi(x), \Phi(x')]\rangle \tag{6.7}$$

is imaginary and odd.

Advanced and retarded propagators

The advanced and retarded propagators are the fundamental solutions for the equations of motion for linear fluctuations in the field.

Each propagator conveys some specific information. For example, for the free scalar field G, G_1, G^+ and G^- are solutions to the homogeneous Klein–Gordon equation, while G_{F}, G_{D}, G_{ret} and G_{adv} are fundamental solutions. The retarded and advanced propagators may be obtained from consideration of the dynamics alone; they have no information on the state. The same can be said of the Jordan propagator, since for linear fields the commutator is a c-number. Indeed, consideration of the respective Cauchy data shows that we must have the identities (the Cauchy data for the Jordan propagator are prescribed by the equal time canonical commutation relations)

$$G_{\mathrm{adv}}(x, x') = -\frac{i}{\hbar} G(x, x') \theta(t' - t)$$

$$G_{\mathrm{ret}}(x, x') = G_{\mathrm{adv}}(x', x) = \frac{i}{\hbar} G(x, x') \theta(t - t')$$

$$G(x, x') = (-i\hbar)[G_{\mathrm{ret}}(x, x') - G_{\mathrm{adv}}(x, x')] \tag{6.8}$$

or else

$$G_{\mathrm{ret}} = \frac{i}{\hbar}\left(G_F - G^-\right) \tag{6.9}$$

$$G_{\mathrm{adv}} = \frac{-i}{\hbar}\left(G_D - G^-\right) \tag{6.10}$$

Therefore the state information is coded primarily in the remaining propagators, most of all in Hadamard's. Knowledge of the Hadamard and Jordan propagators determines all others

$$G^\pm(x, x') = \frac{1}{2}[G_1(x, x') \pm G(x, x')] \tag{6.11}$$

$$G_{\mathrm{F,D}}(x, x') = \frac{1}{2}[G_1(x, x') \pm G(x, x')\,\mathrm{sign}\,(t - t')]$$

$$= \frac{1}{2}[G_1(x, x') \mp i\hbar\,(G_{\mathrm{ret}}(x, x') + G_{\mathrm{ret}}(x', x))] \tag{6.12}$$

As a warm up, we shall compute the propagators (that is, the expectation values of products of fields) in the Minkowski vacuum $|0\rangle$. Let us begin with the positive and negative frequency propagators. The negative frequency propagator $\Delta^{-}(x, x')$ is given in Chapter 5, equation (5.21), and

$$\Delta^{+}(x, x') = \Delta^{-}(x', x) \tag{6.13}$$

All other propagators may be found as linear combinations of these. For example, their difference gives the commutator or Jordan propagator, which for free fields is both independent of the state and of the particle model

$$\Delta(x, x') = \langle[\Phi(x), \Phi(x')]\rangle = \int \frac{d^4 k}{(2\pi)^4} e^{ik(x-x')} \operatorname{sign}\left(k^0\right) 2\pi\hbar\delta\left(k^2 + m^2\right)$$
$$\tag{6.14}$$

The sum of the positive and negative frequency propagators gives the anticommutator or Hadamard propagator

$$\Delta_1(x, x') = \langle\{\Phi(x), \Phi(x')\}\rangle = \int \frac{d^4 k}{(2\pi)^4} e^{ik(x-x')} 2\pi\hbar\delta\left(k^2 + m^2\right) \tag{6.15}$$

The four propagators introduced so far are homogeneous solutions of the Klein–Gordon equation. The retarded propagator

$$\Delta_{\text{ret}}(x, x') = \frac{i}{\hbar}\Delta(x, x')\theta(t - t') = \int \frac{d^4 k}{(2\pi)^4} \frac{e^{ik(x-x')}}{(k + i\varepsilon)^2 + m^2} \tag{6.16}$$

$$(k + i\varepsilon)^2 = -\left(k^0 + i\varepsilon\right)^2 + \mathbf{k}^2 \tag{6.17}$$

is the (only) solution to the equation

$$\left[\nabla^2 - m^2\right]\Delta_{\text{ret}}(x, x') = -\delta(x, x') \tag{6.18}$$

with causal boundary conditions. We also have the advanced propagator

$$\Delta_{\text{adv}}(x, x') = \Delta_{\text{ret}}(x', x) = -\frac{i}{\hbar}\Delta(x, x')\theta(t' - t) = \int \frac{d^4 k}{(2\pi)^4} \frac{e^{ik(x-x')}}{(k - i\varepsilon)^2 + m^2}$$
$$\tag{6.19}$$

which is the fundamental solution with advanced boundary conditions. Finally, there are the Feynman and Dyson propagators, given in Chapter 5, equation (5.20).

6.1.1 Interacting fields

The basic property of the full propagators for an interacting field, that is, the expectation values of binary products of field operators with respect to the vacuum state, is Poincaré invariance. In particular, these propagators are translation invariant, which allows us to describe them in terms of their Fourier transforms, namely, any propagator G may be represented as

$$G(x, x') = \int \frac{d^4 p}{(2\pi)^4} e^{ipu} G(p) \tag{6.20}$$

with $u = x - x'$. Some important properties of the propagators actually follow from their definition as time-ordered products of field operators. For example, since the Feynman and Dyson propagators are even and $G_F = G_D^*$, $G^- = G^{+*}$ and $G^-(x, x') = G^+(x', x)$, G_F and $G_D(p)$ are even functions of momentum, while $G^-(p) = G^+(-p)$. Moreover, G^- and $G^+(p)$ are real, and $G_D(p)^* = G_F$.

The Hadamard propagator G_1 is real and even and therefore also is $G_1(p)$. The Jordan propagator G is imaginary and odd, and so $G(p)$ is odd but real. The Jordan and retarded propagators are related through equation (6.8)

$$G(p) = (-i\hbar)[G_{\text{ret}}(p) - G_{\text{ret}}(-p)] = 2\hbar \text{Im}\, G_{\text{ret}}(p) \tag{6.21}$$

where we have used the fact that $G_{\text{adv}}(p) = G_{\text{ret}}(p)^*$. Also observe that $G_{\text{ret}}(x, x')$ is real, so $G_{\text{ret}}(-p) = G_{\text{ret}}(p)^*$.

Since the retarded propagator is causal, it satisfies the equation $G_{\text{ret}} = \theta(t - t') G_{\text{ret}}$, and the real and imaginary parts of its transform are Hilbert transforms of each other

$$G_{\text{ret}}(p) = \frac{i}{2\pi} \int \frac{d\omega}{p^0 - \omega + i\varepsilon} G_{\text{ret}}(\omega, \mathbf{p})$$

$$= \frac{1}{2} G_{\text{ret}}(p) + \frac{i}{2\pi} PV \int \frac{d\omega}{p^0 - \omega} G_{\text{ret}}(\omega, \mathbf{p}) \tag{6.22}$$

$$\text{Re}\, G_{\text{ret}}(p) = \frac{1}{\pi} PV \int \frac{d\omega}{\omega - p^0} \text{Im}\, G_{\text{ret}}(\omega, \mathbf{p}) \tag{6.23}$$

These are the so-called Kramers–Kronig relations.

For further properties of the propagators, such as their Lehmann representation, we refer the reader to the literature on QFT; some classic textbooks are listed [Rom69, LanLif72, BjoDre64, BjoDre65, ItzZub80, Ram80, Hua98, PesSch95, LeB91, Zin93, Wei95, GrReBr96].

6.2 Functional methods

Before we confront the full nonequilibrium problem, it is instructive to review the more familiar case of finding the expectation values under equilibrium conditions. We will then be in a better position to judge whether a nonequilibrium formalism is a straightforward generalization of the equilibrium one, or where some new insights are required.

So let us begin by asking what is the expectation value of the field operator at a given spacetime event. In a theory such as $\lambda\Phi^4$, which is symmetric under the inversion $\Phi \to -\Phi$, one is tempted to say that the expectation value must vanish, by symmetry. However, this is not necessarily so; the quantum state of the field may have a lesser symmetry than the Hamiltonian, supporting a nonzero expectation value or condensate. In this case we say the symmetry is spontaneously broken. This is even more so if the theory is not even invariant under inversion, for example, for a potential $V[\Phi] = g\Phi^3/6$.

The problem is enormously simplified if we still assume that Poincaré invariance will not be even spontaneously broken. In this case, the expectation value ϕ of the field operator $\Phi(x)$ will be Poincaré invariant; in particular, it will be space and time independent.

In order to find the equilibrium value of the expectation value, it is convenient to work in two stages. First we assume that the system is constrained, by some external means, to a state where the expectation value of the field operator takes a preassigned value ϕ; we then pick the value of ϕ leading to greatest stability. To solve the first half of the problem, we must find the (properly normalized) state which minimizes the energy while having the correct expectation value of the field operators. To enforce these constraints (there is one at every event) we introduce a Lagrange multiplier J. Thus the object to be minimized is

$$\langle \,| H_J + \int d^3\mathbf{x}\, J\phi\, |\,\rangle \tag{6.24}$$

where

$$H_J = H - J \int d^3\mathbf{x}\, \Phi(x) \tag{6.25}$$

Let $|J\rangle$ be the state that minimizes this operator. It will be a proper vector of the operator H_J with proper value E_J

$$H_J |J\rangle = E_J |J\rangle \tag{6.26}$$

Because the state is assumed to be homogeneous and energy is an extensive quantity, the energy E_J will be proportional to the "volume" V of space, with a finite energy density E_J/V. First-order perturbation theory shows that

$$\frac{\delta}{\delta J}\left(\frac{E_J}{V}\right) = -\frac{1}{V}\left\langle J \left| \int d^3\mathbf{x}\, \Phi(x) \right| J \right\rangle = -\phi \tag{6.27}$$

If we introduce the Legendre transform of E_J/V, the so-called "effective potential" $V[\phi]$,

$$V[\phi] = \frac{E_J}{V} + \frac{1}{V}\int d^3\mathbf{x}\, J\phi. \tag{6.28}$$

then

$$\frac{d}{d\phi}V[\phi] = J \tag{6.29}$$

This equation determines ϕ if J is known. The most stable state is the one which does not require external intervention ($J = 0$); the true equilibrium expectation values are the extrema of the effective potential.

By the way, we see what is effective in the effective potential: it is not really the value of the energy density, but rather a Legendre transform thereof. The external source J and the expectation value ϕ are analogous, respectively, to the applied magnetic field B and the magnetization M in a model for ferromagnetism. Also keep in mind that we are describing equilibria at a prescribed temperature (prescribed to be zero). We may think of the effective potential as the thermodynamic potential whose critical points are the equilibria at constant temperature and ϕ, and therefore as a free energy (as opposed to the internal energy). However, so far we have not said how we intend to compute the effective potential, so that the full approach takes on meaning. We turn now to this important issue.

6.2.1 The generating functional and the effective action

We shall begin from the observation that the state $|J\rangle$ minimizes the operator H_J. Let $|\alpha\rangle$ be any state, and let it evolve from *Euclidean* time $\tau = -t_{\mathrm{E}}$ to $\tau = t_{\mathrm{E}}$ (subscript E under t denotes Euclidean time) adopting H_J as the Hamiltonian. Then $|\alpha\rangle$ evolves into

$$|\alpha, t_{\mathrm{E}}\rangle = e^{-2\hbar^{-1}H_J t_{\mathrm{E}}}|\alpha\rangle = e^{-2\hbar^{-1}E_J t_{\mathrm{E}}}|J\rangle\langle J|\alpha\rangle + \delta|\alpha, t_{\mathrm{E}}\rangle \qquad (6.30)$$

where $\delta|\alpha, t_{\mathrm{E}}\rangle$ decays faster than $e^{-2\hbar^{-1}E_J t_{\mathrm{E}}}$. It follows that, given any other state $|\beta\rangle$, and as long as $\langle J|\alpha\rangle \neq 0$ and $\langle J|\beta\rangle \neq 0$, then

$$E_J = -\lim_{t_{\mathrm{E}}\to\infty}\frac{\hbar}{2t_{\mathrm{E}}}\ln\langle\beta|e^{-2\hbar^{-1}H_J t_{\mathrm{E}}}|\alpha\rangle \qquad (6.31)$$

Using a path integral representation for the evolution operator, we find

$$E_J = -\lim_{t_{\mathrm{E}}\to\infty}\frac{\hbar}{2t_{\mathrm{E}}}\ln\int D\varphi\, e^{\hbar^{-1}\left[-S_{\mathrm{E}}+\int d^4x\, J\varphi(x)\right]} \qquad (6.32)$$

where S_{E} stands for the Euclidean action

$$S_{\mathrm{E}} = \int d^4x\left\{\frac{1}{2}\left(\frac{\partial\varphi}{\partial\tau}\right)^2 + \frac{1}{2}(\nabla\varphi)^2 + V[\varphi]\right\} \qquad (6.33)$$

This path integral representation displays the close connection between E_J and the Euclidean generating functional for connected Green functions $W_{\mathrm{E}}[J]$

$$e^{\hbar^{-1}W_{\mathrm{E}}[J]} = \int D\varphi\, e^{\hbar^{-1}\left[-S+\int d^4x\, J(x)\varphi(x)\right]} \qquad (6.34)$$

If the source J is spacetime independent, then

$$W_{\mathrm{E}}[J] \sim -2t_{\mathrm{E}}E_J \qquad (6.35)$$

The Legendre transform of $W_{\mathrm{E}}[J]$ is the Euclidean effective action (EA) $\Gamma_{\mathrm{E}}[\phi]$. As we all know, if we Taylor expand the EA in powers of the background field ϕ, the coefficients are given by the sum of all one-particle irreducible Feynman graphs. These are the graphs that are connected, and remain so if any internal line is cut. This method is not efficient as a practical tool, but the observation that the EA could be computed this way is at the base of a much better strategy, the loop expansion, which we shall discuss below. For the time being, simply recall that

$$\Gamma_{\mathrm{E}}[\phi] = W_{\mathrm{E}}[J] - \int d^4x\, J(x)\,\phi(x) \qquad (6.36)$$

So, if the background field $\phi(x)$ is constant, then

$$\Gamma_E[\phi] = -(2t_E V) V[\phi] \tag{6.37}$$

and the relationship between mean fields and sources is

$$\frac{\delta\Gamma_E[\phi]}{\delta\phi(x)} = -J(x) \tag{6.38}$$

In particular, the true equilibria are the Poincaré invariant extrema of the effective action. For further discussion, we refer the reader to Coleman's Lectures [Col85]; see also [JacKer79].

6.2.2 Not quite beyond equilibrium

Although only Poincaré invariant extrema are meaningful, the effective action may be constructed for arbitrary field configurations. However, equation (6.38) does not provide an off-equilibrium equation of motion for the mean field. This is an important point, and we must be sure we understand it before we carry on.

There is of course the observation that equation (6.38) applies to Euclidean field configurations. However, this difficulty is easily overcome. Define the Lorentzian or in-out generating functional $W_{in\text{-}out}[J]$

$$e^{i\hbar^{-1}W_{in\text{-}out}[J]} = \int D\varphi\, e^{i\hbar^{-1}\left[S + \int d^4x\, J(x)\varphi(x)\right]} \tag{6.39}$$

where S is the physical action. Then define

$$\frac{\delta W_{in\text{-}out}[J]}{\delta J} = \tilde{\phi}(x) \tag{6.40}$$

Performing the Legendre transform

$$\Gamma_{in\text{-}out}\left[\tilde{\phi}\right] = W_{in\text{-}out}[J] - \int d^4x\, J(x)\,\tilde{\phi}(x) \tag{6.41}$$

$W_{in\text{-}out}[J]$ is the generating functional for connected graphs, and $\Gamma_{in\text{-}out}$ generates one-particle irreducible graphs. Moreover, $\tilde{\phi}$ satisfies

$$\frac{\delta\Gamma_{in\text{-}out}\left[\tilde{\phi}\right]}{\delta\tilde{\phi}(x)} = -J(x) \tag{6.42}$$

However, in a truly off-equilibrium situation it is impossible to identify $\tilde{\phi}(x)$ with the expectation value of the field operator, and in any case equation (6.42) is unsuitable as an equation of motion. This important point is best appreciated with a concrete example, to which we turn [HarHu79, DeW67].

6.2.3 Trouble in the $g\phi^3$ theory

To be concrete, let us assume the potential

$$V[\varphi] = \frac{1}{2}m^2\varphi^2 - \frac{1}{6}g\varphi^3 - h\varphi \tag{6.43}$$

The linear term is included to enforce the constraint

$$\int D\varphi \, e^{iS} \varphi(x) = 0 \tag{6.44}$$

In spite of some formal drawbacks (for example, equation (6.43) cannot really hold for all values of the field operator, as the theory would have no stable ground state if it did), this model is appealing because of its simplicity, and it is actually a useful model for unstable quantum systems (such as a strongly underdamped Josephson junction).

We shall be concerned only with the small oscillations of the mean field around $\phi = 0$, which, given equation (6.44), is a solution of equation (6.42) by design when $J = 0$. To find the linearized "equations of motion," we need the effective action to quadratic order, which requires knowledge of the quadratic part of the generating functional only. From the definition

$$W_{in\text{-}out}[J] \sim \frac{i}{2\hbar} \int d^4x d^4y \, J(x) \langle \varphi(x) \varphi(y) \rangle_c J(y) \tag{6.45}$$

where $\langle \varphi(x) \varphi(y) \rangle_c$ is the sum of all connected Feynman graphs ending in two external legs as shown. If we further expand in powers of g, keeping the constraint equation (6.44), we may appeal to Wick's theorem to write

$$\langle \varphi(x) \varphi(y) \rangle_c = \Delta(x - y) - i \int d^4z d^4z' \, \Delta(x - z) \Sigma(z - z') \Delta(z' - y) \tag{6.46}$$

where

$$i\Sigma(z - z') = \frac{g^2}{2\hbar^2} \left[\Delta(z - z')\right]^2 \tag{6.47}$$

The important issue is which propagator is exactly $\Delta(x - y)$. It is given by

$$\Delta(x - y) = \frac{\int D\varphi \, e^{i\hbar^{-1}S_f} \varphi(x) \varphi(y)}{\int D\varphi \, e^{i\hbar^{-1}S_f}} \tag{6.48}$$

where S_f means the free action, that is, the action with $g = h = 0$. Since the path integral time orders whatever is inside, Δ must correspond to the Feynman propagator Δ_F for the free theory. We then have

$$\tilde{\phi}(x) = \frac{\delta W_{in\text{-}out}[J]}{\delta J(x)}$$

$$= i\hbar^{-1} \int d^4y \, K(x - y) J(y) \tag{6.49}$$

where

$$K(x - y) = \Delta_F(x - y) - i \int d^4z d^4z' \, \Delta_F(x - z) \Sigma_F(z - z') \Delta_F(z' - y) \tag{6.50}$$

Since

$$\left[\nabla^2 - m^2\right] \Delta_F(x, x') = i\hbar\delta(x, x') \tag{6.51}$$

we get the "equation of motion"

$$\left[\nabla^2 - m^2\right]\tilde{\phi}\left(x\right) - \hbar \int d^4z \ \Sigma_{\mathrm{F}}\left(x - z\right)\tilde{\phi}\left(z\right) = -J\left(x\right) + O\left(g^4\right) \qquad (6.52)$$

To appreciate the content of this equation, let us compute the kernel Σ explicitly, using the results from Chapter 5. The infinite and constant terms may be absorbed into a redefinition of m^2 and do not concern us now. For the remainder, we may already be able to make a crucial observation: since $\Sigma_{\mathrm{F}}\left(p\right)$ is an even function of p^0, Σ_{F} will be an even function of $x^0 - z^0$. Therefore, equation (6.52) cannot possibly yield a causal dynamics: the behavior of $\tilde{\phi}$ at any given event will depend on the whole future, and not only on the past of that event. This is clear also if we seek the response of $\tilde{\phi}$ to an impulse by setting J to be a delta function in equation (6.49): far from turning on when the source does, $\tilde{\phi}$ is nonzero everywhere.

Compared to this, the fact that equation (6.52) will generally predict a complex $\tilde{\phi}$ even for real sources is a lesser sin. This follows from the fact that the argument of the logarithm in the fish may be negative if $-p^2 > 4m^2$. Thus $\Sigma_{\mathrm{F}}\left(p\right)$ develops an even imaginary part, which passes on to $\Sigma_{\mathrm{F}}\left(x - z\right)$, Chapter 5, equation (5.27). Since the field operator is Hermitian, its expectation value must be real; we conclude that $\tilde{\phi}$ cannot possibly be that expectation value.

In summary, we find that the generating functional $W_{in\text{-}out}\left[J\right]$ is not generating expectation values of observables – it is generating something else, see below. Neither is it useful as a way to derive equation (6.52) because this equation is not admissible as a true dynamical law, since it is not causal.

In order to proceed, we must understand why an approach which worked fine in equilibrium situations fares so badly off-equilibrium. The key is in the boundary conditions which are conspicuously absent from the Euclidean path integral (6.34). The reason why we do not need to introduce explicit boundary conditions in equation (6.34) is that only the vacuum-to-vacuum amplitude survives the limiting procedure of taking the time interval t_{E} to infinity. Anything else becomes negligible against the vacuum-to-vacuum contribution.

This is not true of the Lorentzian path integral, and in fact equation (6.39) is meaningless unless the boundary conditions on the path of integration are specified. We implicitly assumed that the Lorentzian path integral was defined as the analytic continuation of the Euclidean path integral. This is implemented by replacing m^2 by $m^2 - i\varepsilon$ in the classical action. Therefore the path integral came to represent a vacuum-to-vacuum transition amplitude. While in a truly Poincaré-invariant situation there is only one vacuum (up to a phase), off-equilibrium the vacuum $|0in\rangle$ in the distant past may be very different from the vacuum $|0out\rangle$ in the far future.

Let us describe the situation in canonical terms. We have a physical idea of what the vacuum is, both in the distant past and future (for example, we have a particle detector we trust, and we know it is in the vacuum if the detector does not click). There is a state $|0in\rangle$ which corresponds to the vacuum at time $t = -t_{\rm L}$ (subscript L under t denotes Lorentzian time). If we adopt the Schrödinger picture, at time $t = t_{\rm L}$ this state has evolved into $e^{-2it_{\rm L}H}|0in\rangle$, which does *not* correspond to the physical vacuum. On the other hand, there is a *different* state $|0out\rangle$ (which we may regard as a state either in the Heisenberg picture or a Schrödinger picture at time $t = -t_{\rm L}$) evolving into $|0out, t_{\rm L}\rangle \equiv e^{-2iHt_{\rm L}}|0out\rangle$, which corresponds to the vacuum in the far future. We obtain the *in–out* generating functional by forcing $|0in\rangle$ to evolve not only under its own dynamics, but also under an external time dependent source $J(\mathbf{x}, t)$, and comparing the result to $|0out, t_{\rm L}\rangle$

$$e^{i\hbar^{-1}W_{in\text{-}out}[J]} = \langle 0out, t_{\rm L}|\, T\left[e^{-i\int_{-t_{\rm L}}^{t_{\rm L}} \hbar^{-1}dt\left(H - \int d^3\mathbf{x} J\Phi_S\right)}\right]|0in\rangle$$

$$= \langle 0out|\, e^{2i\hbar^{-1}t_{\rm L}H} T\left[e^{-i\int_{-t_{\rm L}}^{t_{\rm L}} \hbar^{-1}dt\left(H - \int d^3\mathbf{x} J\Phi_S\right)}\right]|0in\rangle \quad (6.53)$$

where H, $\Phi_S(\mathbf{x})$ are Schrödinger picture operators. Therefore

$$\tilde{\phi}(x) = \left.\frac{\delta W_{in\text{-}out}[J]}{\delta J(x)}\right|_{J=0} = \frac{\langle 0out|\,\Phi(x)\,|0in\rangle}{\langle 0out|0in\rangle} \quad (6.54)$$

where $\Phi(\mathbf{x}, t)$ is the Heisenberg picture operator. This is a very different object from the true expectation value

$$\phi(x) = \langle 0in|\,\Phi(x)\,|0in\rangle \quad (6.55)$$

In particular, being a nondiagonal matrix element, $\tilde{\phi}(x)$ will be generally complex, and since it carries the information that the state must evolve into the vacuum in the far future, it is not surprising that its dynamics is acausal.

The true expectation value $\phi(x)$ must be real and evolve causally. We should therefore forget about $\tilde{\phi}$ but concentrate on finding the right equations of motion for ϕ. We want to make the same overall strategy work: we shall find the correct generating functional, and obtain an effective action as a Legendre transform of it based on the demand that the variation of this effective action will yield real and causal equations of motion. The correct generating functional and effective action will have to be different from their counterparts above. These are the necessary requirements for a consistent nonequilibrium functional formalism.

6.3 The closed time path effective action

As mentioned earlier, we shall study nonequilibrium dynamics through the evolution of expectation values of field operators and their correlation functions. To study this evolution, we shall derive equations of motion which represent successive truncations of the Schwinger–Dyson hierarchy (this being the quantum equivalent of the BBGKY hierarchy in classical statistical mechanics). The reason why we concentrate on the equations of motion rather than the propagators themselves is because it is more efficient: an approximation to the equation may be equivalent to summing an infinite set of graphs in the solution. A dramatic example of this improved efficiency is the *hard thermal loop* resummation scheme which we will discuss at length in Chapter 10.

In the simplest approach, we choose a single indicator, namely, the expectation value of the field operator. Allowing for a mixed initial state rather than the vacuum initial conditions assumed so far, we have $\phi(x) = \text{Tr}[\Phi(x)\rho]$. Introducing the Heisenberg dynamical law for the field operator, this expectation value admits a representation as a CTP path integral

$$\phi(x) = \int D\varphi^1 D\varphi^2 \, \rho\left[\varphi^1(0,\mathbf{x}), \varphi^2(0,\mathbf{x})\right] \exp\left\{(i/\hbar) S\left[\varphi^1, \varphi^2\right]\right\} \varphi^1(x) \quad (6.56)$$

where

$$S\left[\varphi^1, \varphi^2\right] = S\left[\varphi^1\right] - S\left[\varphi^2\right]^* \quad (6.57)$$

This suggests considering two mean fields ϕ^a to be derived from the *closed time path* (CTP) *generating functional* $W\left[J^1, J^2\right]$

$$e^{(i/\hbar)W[J]} = \int D\varphi^1 D\varphi^2 \, \rho\left[\varphi^1(0,\mathbf{x}), \varphi^2(0,\mathbf{x})\right]$$

$$\exp\left\{(i/\hbar)\left[S\left[\varphi^1, \varphi^2\right] + \int d^4x \left[J^1\varphi^1 - J^2\varphi^2\right]\right]\right\} \quad (6.58)$$

through the variational formula

$$\phi^a(x) = \frac{\delta}{\delta J_a(x)} W[J, J'] \quad (6.59)$$

If after the variation we set $J_a = 0$, then $\phi^1 = \phi^2 = \langle\Phi(x)\rangle$. Here $a = 1, 2$ denotes the branch within the time path. Also there is a "metric" $c_{ab} = \text{diag}(1, -1)$, so that $J_1(x) = J^1(x)$ and $J_2(x) = -J^2(x)$.

To obtain the equation of motion for these mean fields, we introduce the *CTP effective action* (EA) as the Legendre transform of the generating functional $\Gamma[\phi] = W[J] - J_A\phi^A$. The dynamical equations for the mean fields read

$$\frac{\delta\Gamma}{\delta\phi^A} = -J_A \quad (6.60)$$

The index A is (x, a), where x is a spacetime event, and $a = 1, 2$ denotes the branch within the time path. We apply a generalized Einstein convention whereby repeated indices are summed if they are discrete, or integrated if continuous. For example

$$J\phi = J_A\phi^A = \int d^4x \, J_a\phi^a = \int d^4x \left[J_1\phi^1 + J_2\phi^2\right] = \int d^4x \left[J^1\phi^1 - J^2\phi^2\right] \quad (6.61)$$

To obtain the equation of motion for the physical expectation value, we set $J = 0$ in equation (6.60), in which case, as we shall see below, the two equations (6.60) are actually equivalent.

Given the flourishing of applications of the CTPEA, it would be impossible to give a complete set of references. Some papers which were influential in the development of the subject are [Sch60, Sch61, BakMah63, Kel64, ChoSuHa80, CSHY85, SCYC88, DeW86, Jor86, CalHu87, CalHu88, CalHu89].

6.3.1 An example

Before we proceed further with the formalism it is useful to show an example. We continue with the $g\Phi^3$ theory we introduced in (6.43).

Let us assume that the initial conditions are set in the distant past, where the initial state is the *in* vacuum. This is implemented, as above, by shifting m^2 to $m^2 - i\varepsilon$. Therefore

$$e^{(i/\hbar)W[J]} = \int D\varphi^1 D\varphi^2 \exp\left\{ (i/\hbar) \left[S\left[\varphi^1\right] - S\left[\varphi^2\right]^* + \int d^4x \left[J^1\varphi^1 - J^2\varphi^2 \right] \right] \right\}$$

(6.62)

In the second branch the mass is shifted to $m^2 + i\varepsilon$. Observe that the two branch integrations are not independent, as they couple through the "CTP boundary condition" $\varphi^1(T, \mathbf{x}) = \varphi^2(T, \mathbf{x})$ for all $\mathbf{x}$ at some very large time T. In canonical terms, this expression is equivalent to

$$e^{(i/\hbar)W[J]} = \langle 0in| \, U_{J^2}(-T, T) \, U_{J^1}(T, -T) \, |0in\rangle|_{T \to \infty}$$

(6.63)

where U_J is the evolution operator for the field interacting with the external c-number source J

$$U_J(t, t') = T\left[\exp\left\{ \left(-\frac{i}{\hbar} \right) \int_{t'}^{t} dt \int d^3\mathbf{x} \, (H - J\Phi) \right\} \right]$$

(6.64)

The CTP boundary condition arises from inserting a complete set of states at time T between the two evolution operators. It is interesting to compare the CTP generating functional (6.63) to its "open path" or *in-out* counterpart (6.53). Observe that the *out* vacuum plays no role in the CTP expression.

As in our earlier example, we shall compute only the quadratic part of the generating functional. Since we enforce

$$\left. \frac{\delta W}{\delta J^{1,2}(x)} \right|_{J=0} = 0$$

(6.65)

by conveniently tuning the linear term h in equation (6.43), the quadratic part is

$$W[J] = \frac{i}{2\hbar} G^{AB} J_A J_B$$

(6.66)

From equation (6.62)

$$G^{AB} = \int D\varphi^1 D\varphi^2 \exp\left\{ (i/\hbar) \left[S\left[\varphi^1\right] - S\left[\varphi^2\right]^* \right] \right\} \varphi^A \varphi^B$$

(6.67)

From equation (6.63)

$$G^{11}(x, y) = G_F(x, y) \tag{6.68}$$
$$G^{22}(x, y) = G_D(x, y) \tag{6.69}$$
$$G^{21}(x, y) = G^+(x, y) \tag{6.70}$$
$$G^{12}(x, y) = G^-(x, y) \tag{6.71}$$

So G^{AB} are the "path-ordered" propagators: Path-ordering is equivalent to time-ordering for points on the first time branch, anti-time ordering on the second time branch, and placing points on the second branch to the left of points on the first branch.

Formally, the mean fields are given by

$$\phi^A = i\hbar^{-1} G^{AB} J_B \tag{6.72}$$

which is inverted to

$$-J_A = i\hbar \left[G^{-1}\right]_{AB} \phi^B \tag{6.73}$$

Comparing with equation (6.60) we find

$$\Gamma = \frac{i\hbar}{2} \left[G^{-1}\right]_{AB} \phi^A \phi^B \tag{6.74}$$

The actual equation of motion is obtained when $\phi^1 = \phi^2$. Thus the equation of motion is

$$i\hbar \int d^4y \left[\left[G^{-1}\right]_{11}(x, y) + \left[G^{-1}\right]_{12}(x, y) \right] \phi(y) = -J(x) \tag{6.75}$$

This equation is real and causal as we shall soon prove.

For free fields, the Klein–Gordon equations for the fundamental propagators

$$\left[\nabla^2 - m^2\right] \Delta_F(x, x') = -\left[\nabla^2 - m^2\right] \Delta_D(x, x') = i\hbar\delta(x, x') \tag{6.76}$$
$$\left[\nabla^2 - m^2\right] \Delta^+(x, x') = \left[\nabla^2 - m^2\right] \Delta^-(x, x') = 0 \tag{6.77}$$

may be summarized as

$$c_{AB}\left[\nabla^2 - m^2\right] \Delta^{BC} = i\hbar\delta_A^C \tag{6.78}$$

So

$$\left[\Delta^{-1}\right]_{AB} = (-i\hbar^{-1})\left[\nabla^2 - m^2\right] c_{AB} \tag{6.79}$$

where $c_{AB} = c_{ab}\delta(x, x')$. Never mind for now that we are claiming $\left[\Delta^{-1}\right]_{AB}$ is diagonal, while Δ^{AB} is conspicuously not. The nondiagonal elements of Δ^{AB} are retrieved by inverting $\left[\Delta^{-1}\right]_{AB}$ under the CTP constraints

$$\Delta^{11}(x, y) = \theta(x^0 - y^0) \Delta^{21}(x, y) + \theta(y^0 - x^0) \Delta^{12}(x, y) \tag{6.80}$$
$$\Delta^{22}(x, y) = \theta(x^0 - y^0) \Delta^{12}(x, y) + \theta(y^0 - x^0) \Delta^{21}(x, y) \tag{6.81}$$

Figure 6.1 The tadpole graph; there are no propagators on the external lines.

With this form for the inverse propagators, the equation of motion is just the Klein–Gordon equation, which is correct but not very illuminating.

For interacting fields, define the self-energies Σ_{AB} from

$$[G^{-1}]_{AB} = [\Delta^{-1}]_{AB} + i\Sigma_{AB} \tag{6.82}$$

leading to the perturbative development

$$G^{AB} = \Delta^{AB} - i\Delta^{AC}\Sigma_{CD}\Delta^{DB} + \dots \tag{6.83}$$

Meanwhile the expansion of the propagators in powers of the coupling constant g reads

$$G^{ab}(x,y) = \Delta^{AB} - \frac{1}{2}\left(\frac{g}{6\hbar}\right)^2 \int d^4z d^4z' \left\langle \varphi^a(x)\left[(\varphi^1)^3 - (\varphi^2)^3\right](z)\right.$$
$$\left. \times \left[(\varphi^1)^3 - (\varphi^2)^3\right](z')\varphi^b(y)\right\rangle_{\mathrm{f}} + \dots \tag{6.84}$$

where $\langle\rangle_{\mathrm{f}}$ denotes a path ordered expectation value computed for free fields, enforcing the constraint of vanishing "tadpoles." These expectation values are reduced to products of propagators applying Wick's theorem. Comparing both expansions, we conclude

$$i\Sigma_{11}(x,y) = \frac{g^2}{2\hbar^2}[\Delta_{\mathrm{F}}(x,y)]^2 \tag{6.85}$$

$$i\Sigma_{12}(x,y) = -\frac{g^2}{2\hbar^2}[\Delta^-(x,y)]^2 \tag{6.86}$$

$$i\Sigma_{21}(x,y) = -\frac{g^2}{2\hbar^2}[\Delta^+(x,y)]^2 \tag{6.87}$$

$$i\Sigma_{22}(x,y) = \frac{g^2}{2\hbar^2}[\Delta_{\mathrm{D}}(x,y)]^2 \tag{6.88}$$

We may now write the equation of motion to order g^2

$$[\nabla^2 - m^2]\phi(x) + i\frac{g^2}{2\hbar}\int d^4y \left[[\Delta_{\mathrm{F}}(x,y)]^2 - [\Delta^-(x,y)]^2\right]\phi(y) = -J(x) \tag{6.89}$$

Comparing with equation (6.52), we see that there is an extra contribution to the nonlocal part. By simple inspection, we see that this new term makes the equation causal, since $\Delta_{\mathrm{F}}(x,y) = \Delta^-(x,y)$ when $y^0 > x^0$. From the results in

Chapter 5, we see that the equation is also real, as required by the physical meaning of ϕ as the expectation value of a Hermitian operator.

6.3.2 The structure of the closed time path effective action

The example above already shows several generic features of the CTPEA. We wish now to highlight these features which are general and *exact* (i.e. not dependent on the model or the order of coupling).

In the above example, we assumed vacuum initial conditions set up in the distant past. In general, we deal with an arbitrary initial state set up at some definite time, which we may take as $t = 0$. Then the CTP generating functional admits the representation

$$e^{(i/\hbar)W[J]} = \text{Tr}\left[U_{J^2}\left(0, T\right) U_{J^1}\left(T, 0\right) \rho\right]\big|_{T \to \infty} \tag{6.90}$$

where, as before, U_J is the evolution operator for the field interacting with the external c-number source J as in equation (6.64). Variation yields, in the coincidence limit

$$\phi\left(x\right) = \frac{\delta}{\delta J\left(x\right)} W\left[J, J'\right]\bigg|_{J' \to J} = \text{Tr}\left[\Phi\left(x\right) U_J\left(t, 0\right) \rho U_J\left(0, t\right)\right] \tag{6.91}$$

ϕ is the expectation value of the field operator with respect to the state which evolves from ρ under the influence of the source J.

The first property of the CTPEA we wish to discuss is its "Hermiticity," namely, for Hermitian field operators, $\Gamma\left[\phi^2, \phi^1\right] = -\Gamma\left[\phi^{1*}, \phi^{2*}\right]^*$ (since the field operators are Hermitian, we assume they couple to real c-number sources; however, we may be sure that the mean fields ϕ^a are real in the coincidence limit $J^1 = J^2$ only). To see this "Hermiticity," observe that, provided the density matrix ρ in equation (6.90) is itself Hermitian, then a similar property holds for the CTP generating functional, namely $W\left[J^2, J^1\right] = -W\left[J^1, J^2\right]^*$. Taking variations, we get

$$\phi_1\left[J^2, J^1\right] = -\phi_2\left[J^1, J^2\right]^*; \qquad \phi_2\left[J^2, J^1\right] = -\phi_1\left[J^1, J^2\right]^* \tag{6.92}$$

In other words, if the external sources necessary to sustain the given background fields $\left(\phi^1, \phi^2\right)$ are $\left(J^1, J^2\right)$, then the sources necessary to sustain the mean fields $\left(\phi^{2*}, \phi^{1*}\right)$ are $\left(J^2, J^1\right)$ (note the position of the indices). Thus

$$\Gamma\left[\phi^{2*}, \phi^{1*}\right] = W\left[J^2, J^1\right] - \left(J^2\phi^{2*} - J^1\phi^{1*}\right) = -\Gamma\left[\phi^1, \phi^2\right]^* \tag{6.93}$$

QED. An equivalent formulation is that, if the background fields $\left(\phi^1, \phi^2\right)$ are real, then

$$\text{Re}\Gamma\left[\phi^1, \phi^2\right] = -\text{Re}\Gamma\left[\phi^2, \phi^1\right]; \qquad \text{Im}\Gamma\left[\phi^1, \phi^2\right] = \text{Im}\Gamma\left[\phi^2, \phi^1\right] \tag{6.94}$$

If ρ is trace-class $(\text{Tr}\left[\rho\right] = 1)$ *and* the evolution operator is unitary, then the CTP generating functional vanishes on the diagonal $(W\left[J, J\right] = 0)$. In the coincidence limit, therefore, there is a single mean field, since $\phi^1 = \phi^2 \equiv \phi$ (again, the position of the indices matters). Equation (6.92) shows that ϕ must be real (this can also be seen directly from equation (6.91)). We then find that the CTPEA is also trivial along the diagonal $\Gamma[\phi, \phi] \equiv 0$. This dispels the apparent mystery of having two equations for a single mean field ϕ: they are linearly dependent. The single equation reads

$$\left. \frac{\delta\Gamma}{\delta\phi^1} \right|_{\phi^2=\phi^1=\phi} = - \left. \frac{\delta\Gamma}{\delta\phi^2} \right|_{\phi^2=\phi^1=\phi} = -J \qquad (6.95)$$

where J is the common value of J^1 and J^2. Observe that although Γ is generally complex, when ϕ is real the variation of the imaginary part must vanish in the coincidence limit (this follows from equations (6.94)), and so the physical equation (6.95) is explicitly a real equation.

The other fundamental property of equation (6.95) is that it is causal (we may say that doubling the degrees of freedom is the minimum price to pay to get a causal, real equation of motion for the mean field within a variational approach). Indeed, the solution to the physical equation (6.95) is given by the formal expression (6.91), which is obviously causal.

We may disclose further properties of the CTPEA by writing it as a function of new field variables $\phi_- = \phi^1 - \phi^2$ and $\phi_+ = (\phi^1 + \phi^2)/2$. Observe that $\Gamma[\phi, \phi] \equiv 0$ implies

$$\Gamma[\phi_- = 0, \phi_+] \equiv 0 \qquad (6.96)$$

Therefore the Taylor development of Γ reads

$$\Gamma[\phi_-, \phi_+] = \int d^4x \, \phi_-(x) \, \mathbf{D}_x^{\text{full}}[\phi_+] + \frac{i}{2} \int d^4x d^4x' \, \phi_-(x) \, \mathbf{N}(x, x') \phi_-(x') + \dots \qquad (6.97)$$

To find the equations of motion, we first take its variation with respect to ϕ^1 and then set $\phi_- = 0$. Only the first term contributes, and the equations read

$$\mathbf{D}_x^{\text{full}}[\phi] = -J(x) \qquad (6.98)$$

If the theory is set up so that $\mathbf{D}_x^{\text{full}}[0] = 0$, then $\mathbf{D}_x^{\text{full}}$ will have its own Taylor development

$$\mathbf{D}_x^{\text{full}}[\phi_+] = \int d^4x' \, \mathbf{D}^{\text{full}}(x, x') \phi_+(x') + \dots \qquad (6.99)$$

So the linearized equation of motion is

$$\int d^4x' \, \mathbf{D}^{\text{full}}(x, x') \phi(x') = -J(x) \qquad (6.100)$$

The Hermiticity conditions (6.94) and the causality of the equations of motion (6.95) imply that the kernels $\mathbf{D}^{\text{full}}$ and $\mathbf{N}$ are real, and $\mathbf{D}^{\text{full}}$ is causal.

The appearance of the kernel $\mathbf{N}$ may seem redundant, since it does not contribute to the mean field equations of motion. However, equation (6.97) also suggests another way of looking at the CTPEA which discloses a surprising role for $\mathbf{N}$. Let us observe that the quadratic CTPEA has the same structure as an influence functional, and if we regard it this way, then $\mathbf{N}$ corresponds to the noise kernel. According to the theory of quantum open systems, we ought to replace the mean field equation by a Langevin equation $\mathbf{D}^{\text{full}}\phi = -J - \xi$, where J is the external source, if any, and ξ is a stochastic, c-number source with autocorrelation

$$\langle \xi(x) \xi(x') \rangle = \hbar \mathbf{N}(x, x') \tag{6.101}$$

In attention to this role for $\mathbf{N}$, we shall henceforth refer to it as the *noise kernel*. In general $\mathbf{N}(x, x')$ will be a functional of ϕ_+, leading to colored and multiplicative noise in the dynamical equations. A large part of the remainder of this book may be seen as the development of this theme.

6.4 Computing the closed time path effective action
6.4.1 The background field method

So far we have formally introduced the CTPEA and investigated some of its properties. Now we show how to actually compute it. As a start, let us observe that it is possible to give a definition of the CTPEA as the solution of a particular integral equation. To this end, we recall the definition of the generating functional (in condensed notation)

$$e^{(i/\hbar)W[J]} = \int D\Phi \, \exp\{(i/\hbar)\,[S\,[\Phi] + J\Phi]\} \tag{6.102}$$

(observe that we do not write explicitly the initial density matrix; the mystery will be revealed below). The CTPEA is introduced as the Legendre transform $\Gamma[\phi] = W[J] - J_A\phi^A$. Note that, after all, $J = -\Gamma_{,\phi}$, and so we may write

$$e^{(i/\hbar)W[\phi]} = \int D\Phi \, \exp\left\{(i/\hbar)\left[S\,[\Phi] - \frac{\delta\Gamma}{\delta\phi}\Phi\right]\right\} \tag{6.103}$$

which is the self-contained equation for the CTPEA.

Let us begin by rewriting it as

$$\Gamma[\phi^A] = (-i\hbar)\ln \int D\Phi^A \, \exp\left\{(i/\hbar)\left[S\,[\Phi^A] - \frac{\delta\Gamma}{\delta\phi^A}\left(\Phi^A - \phi^A\right)\right]\right\} \tag{6.104}$$

Shift the integration variables by the mean fields $\Phi^A = \phi^A + \varphi^A$ and expand the classical action

$$S\,[\phi^A + \varphi^A] = S\,[\phi^A] + S_{,A}\varphi^A + S_r\,[\varphi^A] \tag{6.105}$$

For example, for a $\lambda\Phi^4$ theory with classical action

$$S\,[\phi^a] = \int d^4x \left\{ \frac{c_{ab}}{2}\,[-\partial\phi^a\partial\phi^b - m^2\phi^a\phi^b] - \frac{\lambda}{4}\,c_{abcd}\phi^a\phi^b\phi^c\phi^d \right\} \tag{6.106}$$

where c_{ab} is the CTP metric tensor and c_{abcd} is 1 if $a = b = c = d = 1$, -1 if all indices are equal to 2, and vanishes otherwise, we obtain

$$S_r\left[\varphi^a\right] = \int d^4x \left\{ \frac{c_{ab}}{2} \left[-\partial\varphi^a\partial\varphi^b - m^2\varphi^a\varphi^b\right] - \frac{\lambda}{4} c_{abcd}\phi^a\phi^b\varphi^c\varphi^d \right.$$
$$\left. - \frac{\lambda}{6} c_{abcd}\phi^a\varphi^b\varphi^c\varphi^d - \frac{\lambda}{24} c_{abcd}\varphi^a\varphi^b\varphi^c\varphi^d \right\} \quad (6.107)$$

Then

$$\Gamma\left[\phi^A\right] = S\left[\phi^A\right] - i\hbar \ln \int D\varphi^A \exp\left\{(i/\hbar)\left[S_r\left[\varphi^A\right] + (S_{,A} - \Gamma_{,A})\varphi^A\right]\right\} \quad (6.108)$$

Next write

$$\Gamma\left[\phi^A\right] = S\left[\phi^A\right] + \Gamma_1\left[\phi^A\right] \quad (6.109)$$

$$\Gamma_1\left[\phi^A\right] = -i\hbar \ln \int D\varphi^A \exp\left\{(i/\hbar)\left[S_r\left[\varphi^A\right] - \Gamma_{1,A}\varphi^A\right]\right\} \quad (6.110)$$

The quantum correction Γ_1 has the form of a generating functional for a new theory, whose classical action is obtained from the original one by shifting the fields as in equation (6.105) and discarding constant and linear terms. This new generating functional must be evaluated at a particular value of the external source.

By performing the Legendre transform in reverse, we could write this generating functional in terms of the effective action for the φ field. We stress that the φ field represents a different field theory than the original one; for example, the action S_r for the φ field contains cubic interactions, which the action S for the ϕ field does not. In any case, to compute the generating functional, we must be able to compute the corresponding effective action at a mean field doublet $\bar{\varphi}$ equal to the expectation value $\langle\varphi\rangle_J$ of the Heisenberg operator $\varphi = \Phi - \phi$. Generally the new action S_r is not invariant under sign reversal (cf. equation (6.107)), and so there is no reason for this to vanish. However, the value $\Gamma_{1,A}$ is precisely the external force necessary to kill this expectation value. The conclusion is that we may ignore the external source, and compute the generating functional Γ_1 as the sum of all 1PI vacuum Feynman graphs: vacuum because we compute the effective action at $\bar{\varphi} = 0$, and 1PI because, after all, it is an effective action.

To show that $\langle\varphi\rangle = 0$, let us take the variational derivative of equation (6.110) with respect to the background field to obtain

$$\int D\varphi^A \left[S_{r,A}\left[\varphi\right] - \Gamma_{1,A}\left[\phi\right] - \Gamma_{1,AB}\varphi^B\right] \exp\left\{(i/\hbar)\left[S_r\left[\varphi\right] - \Gamma_{1,C}\varphi^C\right]\right\} \equiv 0 \quad (6.111)$$

and observe that

$$\frac{\delta S_r}{\delta\phi^A} = \frac{\delta S\left[\phi + \varphi\right]}{\delta\phi^A} - \frac{\delta S\left[\phi\right]}{\delta\phi^A} - \frac{\delta S\left[\phi\right]}{\delta\phi^A\delta\phi^B}\varphi^B = \frac{\delta S_r}{\delta\varphi^A} - \frac{\delta S\left[\phi\right]}{\delta\phi^A\delta\phi^B}\varphi^B \quad (6.112)$$

The product

$$\left[\frac{\delta S_r}{\delta \varphi^A} - \Gamma_{1,A}\right] e^{\{(i/\hbar)[S_r[\psi] - \Gamma_{1,B}\varphi^B]\}} = -i\hbar \frac{\delta}{\delta \varphi^A} e^{\{(i/\hbar)[S_r[\varphi] - \Gamma_{1,A}\varphi^A]\}} \quad (6.113)$$

integrates to zero, and so we are left with the identity $\Gamma_{,AB} \langle \varphi^B \rangle = 0$. But the Hessian operator $\Gamma_{,AB}$ must be nonsingular, since it follows from the properties of Legendre transformation that

$$\Gamma_{,AB} \frac{\delta^2 W}{\delta J_B \delta J_C} = -\delta_A^C \quad (6.114)$$

and this establishes the vanishing of $\langle \varphi^b \rangle$, as desired. To summarize, Γ, the (vacuum) CTPEA consists of the classical action S plus a quantum correction Γ_1 (thereby the label *effective action*). This quantum correction is the sum of all *one-particle irreducible* (1PI) (that is, containing no one-particle insertions) *vacuum bubbles* (that is, containing no external vertices). This recipe will be the start of all computations based on the CTPEA.

For an exposition of the background field method, read, e.g. the classic papers by Jackiw and Iliopoulos, Itzykson and Martin [Jac74, IlItMa75].

6.4.2 The loop expansion

Having reduced the problem of computing the CTPEA in the theory with classical action S to the calculation of vacuum bubbles in the theory with classical action S_r, we proceed, as in the general case, to split the new action into its free and interacting components

$$S_r = \frac{1}{2} \frac{\delta^2 S}{\delta \phi^2} \varphi^2 + S_Q \quad (6.115)$$

For example, for a $\lambda \Phi^4$ theory the free and interacting parts correspond to the first and second lines in equation (6.107), respectively.

We generate the Feynman graphs by expanding the exponential of S_Q. The different vertices shall be connected through lines, and associated with each line there is a propagator

$$\langle \varphi^A \varphi^B \rangle = \int D\varphi^C \; [\varphi^A \varphi^B] \exp \left\{ \frac{i}{2\hbar} \frac{\delta^2 S}{\delta \phi^2} \varphi^2 \right\} \equiv i\hbar \left[\frac{\delta^2 S}{\delta \phi^A \delta \phi^B}\right]^{-1} \quad (6.116)$$

To make this formula well defined, we assume the usual Gell-Mann–Low boundary condition which states that interactions are adiabatically switched off in the distant past, so the *in* vacuum for the φ field is the same as for the Φ field. Moreover, we assume this state to be properly normalized, so it is not necessary to normalize explicitly the expectation value (6.116).

The neat split of Γ into a classical and a quantum part can be continued by analyzing further its development in powers of $\hbar$. The idea is that each vertex contributes one inverse power of $\hbar$ to the amplitude of the graph, while each line contributes $\hbar$. So the overall power of $\hbar$, including the one in the beginning of equation (6.110), is $L = I - V + 1$, where I is the number of lines, and V of vertices. This is also the number of independent loops in the graph, and so the expansion of Γ in powers of $\hbar$ is equivalent to a topological classification of graphs according to the number of loops.

Note that I and V also satisfy the constraint $2I - 3V_3 - 4V_4 = 0$, where V_3 (V_4) is the number of cubic (quartic) vertices in the graph. Solving for the number of vertices of each type, we find $V_3 = 2I - 4(L - 1)$ and $V_4 = 3(L - 1) - I$. Since each of the numbers I, V_3, V_4 must be nonnegative, we conclude that for each value of L only a finite number of graphs are allowed. For example, for $L = 2$ we must have either $I = 2$, $V_3 = 0$, $V_4 = 1$ or $I = 3$, $V_3 = 2$, $V_4 = 0$, etc.

If $L = 1$, then we must have $V_4 = V_3 = 0$. In this limit, the integral in equation (6.110) is Gaussian and we may write

$$\Gamma_1\left[\phi^A\right] = -i\hbar \ln \mathrm{Det}\left[S''\right]^{-1/2} + O\left(\hbar^2\right) \tag{6.117}$$

Of course, since the propagators themselves depend on the background fields, we do not mean that individual graphs are easy to compute. The loop expansion, however, provides us with a classification scheme to consider the different processes contributing to a given amplitude in order of increasing complexity.

6.4.3 The one-loop closed time path effective action for the $g\Phi^3$ theory

As an example, let us compute the one-loop approximation to the CTPEA in the familiar scalar field theory with cubic self-interaction.

We assume the simplest case of vacuum initial conditions specified in the distant past. The classical potential is given in equation (6.43), where m^2 is shifted to $m^2 - i\varepsilon$. The CTP action is given by equation (6.57) and the CTPEA by equation (6.109). To construct the new action S_r which appears in this equation, we write the old action in terms of a displaced field variable $\phi + \varphi$, and then discard constant and linear terms in φ. Therefore, splitting S_r into a free part and an interaction part as in equation (6.115),

$$\frac{1}{2}\frac{\delta^2 S}{\delta\phi^2}\varphi^2 = \int d^4x \left\{\frac{c_{ab}}{2}\left[-\partial\varphi^a\partial\varphi^b - m^2\varphi^a\varphi^b\right] - \frac{g}{2}\,c_{abc}\phi^a\varphi^b\varphi^c\right\} \tag{6.118}$$

$$S_Q\left[\varphi\right] = \int d^4x \left\{-\frac{g}{6}\,c_{abc}\varphi^a\varphi^b\varphi^c\right\} \tag{6.119}$$

where $c_{111} = -c_{222} = 1$, all other components being zero. In principle, $\Gamma_1\left[\phi^A\right]$ is the sum of all 1PI vacuum graphs for this new theory. The one-loop approximation consists of discarding S_Q, so that $\Gamma_1\left[\phi^A\right]$ reduces to

$$\Gamma_1\left[\phi^A\right] = -i\hbar \ln \int D\varphi^A \exp\left\{(i/\hbar)\int d^4x\,\varphi^a\left\{\frac{c_{ab}}{2}\left[\nabla^2 - m^2\right] - \frac{g}{2}\,c_{abc}\phi^c\right\}\varphi^b\right\}$$
$$\tag{6.120}$$

with the formal solution (6.117).

We shall have more to say about the full one-loop approximation later in the book, but for now let us simply use equation (6.120) to recover the quadratic part of the CTPEA.

Expanding equation (6.120) to quadratic order we get

$$
\Gamma_1\left[\phi^A\right] \sim \int d^4x \left(-\frac{g}{2}\right) c_{abc}\phi^c \left\langle \varphi^a \varphi^b \right\rangle_c
$$
$$
+\frac{i}{2\hbar}\left(\frac{g}{2}\right)^2 \int d^4x d^4x'\, c_{abc} c_{def}\phi^c\left(x\right)\phi^f\left(x'\right)\left\langle\left(\varphi^a\varphi^b\right)\left(x\right)\left(\varphi^d\varphi^e\right)\left(x'\right)\right\rangle_c
$$

$$(6.121)$$

where $\langle\,\rangle_c$ denotes the expectation value of path-ordered products of free field operators, keeping only the connected contributions. The first-order term will be canceled by the h term in the classical action, so we only need to worry about the second, which reads

$$
\Gamma_1\left[\phi^A\right] \sim \frac{i}{2\hbar}\left(\frac{g}{2}\right)^2 \int d^4x d^4x'\, \Big\{ \phi^1\left(x\right)\phi^1\left(x'\right)\left\langle T\left[\varphi^2\left(x\right)\varphi^2\left(x'\right)\right]\right\rangle_c
$$
$$
-\phi^2\left(x\right)\phi^1\left(x'\right)\left\langle \varphi^2\left(x\right)\varphi^2\left(x'\right)\right\rangle_c - \phi^1\left(x\right)\phi^2\left(x'\right)\left\langle \varphi^2\left(x'\right)\varphi^2\left(x\right)\right\rangle_c
$$
$$
+\phi^2\left(x\right)\phi^2\left(x'\right)\left\langle \tilde{T}\left[\varphi^2\left(x\right)\varphi^2\left(x'\right)\right]\right\rangle_c \Big\}
$$

$$(6.122)$$

Now write

$$
\phi^{1,2} = \phi_+ \pm \frac{1}{2}\phi_-
$$

$$(6.123)$$

to get

$$
\Gamma_1\left[\phi^A\right] \sim \frac{i}{2\hbar}\left(\frac{g}{2}\right)^2 \int d^4x d^4x'\, \Big\{ \phi_-\left(x\right)\phi_+\left(x'\right)\Big[\left\langle T\left[\varphi^2\left(x\right)\varphi^2\left(x'\right)\right]\right\rangle_c
$$
$$
+\left\langle \varphi^2\left(x\right)\varphi^2\left(x'\right)\right\rangle_c - \left\langle \varphi^2\left(x'\right)\varphi^2\left(x\right)\right\rangle_c - \left\langle \tilde{T}\left[\varphi^2\left(x\right)\varphi^2\left(x'\right)\right]\right\rangle_c\Big]
$$
$$
+\frac{1}{2}\phi_-\left(x\right)\phi_-\left(x'\right)\left\langle\left\{\varphi^2\left(x\right),\varphi^2\left(x'\right)\right\}\right\rangle_c \Big\}
$$

$$(6.124)$$

which, under the definitions (6.5) for temporal and anti-temporal order, yields

$$
\Gamma_1\left[\phi^A\right] \sim \frac{i}{4\hbar}\left(\frac{g}{2}\right)^2 \int d^4x d^4x'\, \Big\{ 4\phi_-\left(x\right)\phi_+\left(x'\right)\left\langle\left[\varphi^2\left(x\right),\varphi^2\left(x'\right)\right]\right\rangle_c \theta\left(x^0 - x'^0\right)
$$
$$
+\phi_-\left(x\right)\phi_-\left(x'\right)\left\langle\left\{\varphi^2\left(x\right),\varphi^2\left(x'\right)\right\}\right\rangle_c \Big\}
$$

$$(6.125)$$

Comparing with equations (6.97) and (6.99), we identify

$$
\mathbf{D}^{\text{full}}\left(x,x'\right) = \left[\nabla^2 - m^2\right]\delta\left(x,x'\right) - \Sigma_{\text{ret}}\left(x,x'\right)
$$

$$(6.126)$$

$$
i\Sigma_{\text{ret}}\left(x,x'\right) = \frac{1}{\hbar}\left(\frac{g}{2}\right)^2 \left\langle\left[\varphi^2\left(x\right),\varphi^2\left(x'\right)\right]\right\rangle_c \theta\left(x^0 - x'^0\right)
$$

$$(6.127)$$

$$
\mathbf{N}\left(x,x'\right) = \frac{1}{2\hbar}\left(\frac{g}{2}\right)^2 \left\langle\left\{\varphi^2\left(x\right),\varphi^2\left(x'\right)\right\}\right\rangle_c
$$

$$(6.128)$$

so the CTPEA takes the influence functional structure, as expected. Observe that $\mathbf{D}^{\text{full}}$ is real, and obviously causal. Expanding the expectation value using Wick's theorem, the dynamical equation (6.100) gives back equation (6.89).

Although we still found no use for the noise kernel (see Chapter 8), it is undeniably nonzero. To lowest order, we find

$$i\Sigma_{\text{ret}}\left(x,x'\right) = \frac{g^2}{2\hbar}\left[\left[\Delta_F\left(x,x'\right)\right]^2 - \left[\Delta^-\left(x,x'\right)\right]^2\right] \tag{6.129}$$

$$\mathbf{N}\left(x,x'\right) = \frac{g^2}{4\hbar}\left[\left(\Delta^+\left(x,x'\right)\right)^2 + \left(\Delta^-\left(x,x'\right)\right)^2\right] \tag{6.130}$$

They can be expressed in terms of the U, ν and μ kernels introduced in Chapter 5

$$-\Sigma_{\text{ret}} = \frac{g^2\hbar}{2}\left[\frac{U}{2} + \mu\right] \tag{6.131}$$

$$\mathbf{N} = \frac{g^2\hbar}{4}\nu \tag{6.132}$$

It is interesting to observe a relationship between the Fourier transforms of these kernels. Since Σ_{ret} is causal it satisfies the Kramers–Kronig relations (6.23),

$$\Sigma_{\text{ret}}\left(p\right) = \frac{1}{\pi}\int\frac{d\omega}{\omega - p^0 - i\varepsilon}\text{Im}\,\Sigma_{\text{ret}}\left(\omega,\mathbf{p}\right) + \text{ local terms} \tag{6.133}$$

The imaginary part comes from the Fourier transform of the μ kernel

$$\text{Im}\,\Sigma_{\text{ret}}\left(\omega,\mathbf{p}\right) = \text{sign}\,\left(\omega\right)\,\Pi\left(\omega^2 - \mathbf{p}^2\right)\,\theta\left(\omega^2 - \mathbf{p}^2 - 4m^2\right) \tag{6.134}$$

where

$$\Pi\left(\sigma^2\right) = \frac{g^2\hbar}{32\pi}\sqrt{1 - \frac{4m^2}{\sigma^2}} \tag{6.135}$$

Also

$$\frac{1}{\omega - p^0 - i\varepsilon} + \frac{1}{\omega + p^0 + i\varepsilon} = \frac{2\omega}{\omega^2 - \left(p^0 + i\varepsilon\right)^2} \tag{6.136}$$

Writing $\omega^2 - \mathbf{p}^2 = \sigma^2$ we have

$$\Sigma_{\text{ret}}\left(p\right) = \frac{1}{\pi}\int_{4m^2}^{\infty}\frac{d\sigma^2}{\left(p + i\varepsilon\right)^2 + \sigma^2}\Pi\left(\sigma^2\right) + \text{ local terms} \tag{6.137}$$

The Fourier transform of $\mathbf{N}$ comes from the ν kernel

$$\mathbf{N}\left(p\right) = \frac{g^2\hbar}{32\pi}\sqrt{1 - \frac{4m^2}{\left(-p^2\right)}}\theta\left(-p^2 - 4m^2\right) \tag{6.138}$$

Comparing with equation (6.135), we see that the noise kernel coincides up to a sign with the imaginary part (in frequency domain) of the dissipation kernel. We shall see in Chapter 8 that the imaginary part of the dissipation kernel describes the dissipation of the mean field by its interaction with quantum fluctuations. The relationship between the noise and dissipation kernels in this simple example is just a basic manifestation of the fluctuation–dissipation theorem at zero temperature.

6.4.4 The large N expansion

Computing Feynman graphs is easy. The harder question is how many Feynman graphs must be computed to achieve a prescribed accuracy.

The number N of replicas of essentially identical fields (like the N scalar fields in an $O(N)$ invariant theory, or the $N^2 - 1$ gauge fields in a $SU(N)$ invariant non-abelian gauge theory) suggests using $1/N$ as a natural small parameter, with a well-defined physical meaning. Unlike coupling constants, this is not subjected to renormalization or radiative corrections. By ordering the perturbative expansion in powers of this small parameter, several nonperturbative effects (in terms of coupling constants) may be systematically investigated.

The ability of the $1/N$ framework to address the nonperturbative aspects of quantum field dynamics has motivated a detailed study of the properties of these systems [CoJaPo74, Roo74]. In nonequilibrium situations, this formalism has been applied to the dynamics of symmetry breaking [HKMP96, CHKMPA94, CKMP95, BBHKP98, BVHS99a, LoMaRi03] and self-consistent semiclassical cosmological models (see Chapter 15).

In the case of the $O(N)$ invariant theory, in the presence of a nonzero background field (or an external gravitational or electromagnetic field interacting with the scalar field) we may distinguish the longitudinal quantum fluctuations in the direction of the background field, in field space, from the $N - 1$ transverse (Goldstone or pion) fluctuations perpendicular to it. To first order in $1/N$, the longitudinal fluctuations drop out of the formalism, so we effectively are treating the background field as classical. Likewise, quantum fluctuations of the external field are overpowered by the fluctuations of the N scalar fields. In this way, the $1/N$ framework provides a systematic and quantitative measure of the semiclassical approximation [HarHor81].

To leading order (LO), the theory reduces to $N - 1$ linear fields with a time-dependent mass, which depends on the background field and on the linear fields themselves through a gap equation local in time. This depiction of the dynamics agrees both with the Gaussian approximation for the density matrix [EbJaPi88, MazPaz89] and with the Hartree approximation [HKMP96].

For example, let us consider an $O(N)$ invariant scalar field theory, in the limit $N \to \infty$. The action is

$$S = \int d^4x \left\{ \frac{-1}{2} \partial_\mu \Psi^\alpha \partial^\mu \Psi^\alpha - \frac{1}{2} M^2 \Psi^\alpha \Psi^\alpha - \frac{\lambda}{8N} (\Psi^\alpha \Psi^\alpha)^2 \right\} \tag{6.139}$$

or by a rescaling $\Psi^\alpha = \sqrt{N} \Phi^\alpha$,

$$S = N \int d^4x \left\{ \frac{-1}{2} \partial_\mu \Phi^\alpha \partial^\mu \Phi^\alpha - \frac{1}{2} M^2 \Phi^\alpha \Phi^\alpha - \frac{\lambda}{8} (\Phi^\alpha \Phi^\alpha)^2 \right\} \tag{6.140}$$

whereby the classical equations are

$$\nabla^2 \phi^\alpha - \left[M^2 + \frac{\lambda}{2}\left(\phi^\beta \phi^\beta\right)\right]\phi^\alpha = 0 \tag{6.141}$$

To compute the 1PIEA we shift the field $\Phi \to \phi + \varphi$ and discard linear terms to get

$$S_r[\varphi] = N \int d^4x \left\{ \frac{-1}{2}\partial_\mu \varphi^\alpha \partial^\mu \varphi^\alpha - \frac{1}{2}M_{\alpha\beta}^2 \varphi^\alpha \varphi^\beta - \frac{\lambda}{2}\phi^\beta \varphi^\beta \varphi^\alpha \varphi^\alpha - \frac{\lambda}{8}\left(\varphi^\alpha \varphi^\alpha\right)^2 \right\} \tag{6.142}$$

where

$$M_{\alpha\beta}^2 = \left[M^2 + \frac{\lambda}{2}\phi^\gamma \phi^\gamma\right]\delta_{\alpha\beta} + \lambda \phi^\alpha \phi^\beta \tag{6.143}$$

We see that the fluctuation field in the direction of ϕ^α has a different propagator than the "pions," namely the fluctuations orthogonal to the mean field. However, since there are $N-1$ pions, they dominate the perturbative expansion, and we may think only of them (or even simpler, let us consider the loop expansion at $\phi = 0$).

In this theory, propagators carry a weight of N^{-1} and vertices a weight N. Therefore, an individual vacuum Feynman graph carries a weight N^C, where $C = 1 - L$ is the number of vertices minus the number of internal lines (L is the number of loops in the graph). However, when we sum over internal indices, it acquires an additional power of N for each independent choice of the $O(N)$ index α. For this reason, adding a tadpole to an internal line does not affect the overall power of N in the graph: although there are two more lines after the insertion, there is also one more vertex and one more $O(N)$ index to sum over. In particular, the "double-bubble" graph has an overall power of N (that is, the same scaling as the classical action itself) coming from two internal lines, one vertex, and two possible choices of the internal indices. By adding tadpoles to the double-bubble in all possible ways, we obtain an infinite family of graphs of weight N, the so-called "daisy" graphs.

If we wish to say something about the large N limit of the theory, we must be able to add up the daisy graphs. This will be achieved by a more powerful formalism, the so-called two-particle irreducible (2PI) or Cornwall–Jackiw–Tomboulis (CJT) effective action, to which we now turn.

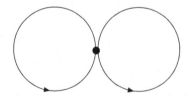

Figure 6.2 The "double-bubble" graph.

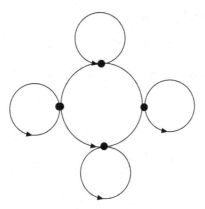

Figure 6.3 A "daisy" graph.

6.5 The two-particle irreducible (2PI) effective action

One clear advantage of working with the CTPEA rather than with the CTP generating functional is that the perturbative expansion is simpler: all connected graphs contribute to the latter, while only one-particle irreducible graphs contribute to the former. It is possible to simplify the perturbative expansion even more by writing Feynman graphs where internal lines represent the full propagators G^{ab}, rather than free propagators Δ^{ab} or some intermediate object. This means that graphs which just dress some internal line of some simpler graph must be disregarded, since all possible corrections are already taken into account in G^{ab}; the remaining graphs are those where no nontrivial subgraph can be isolated by cutting two internal lines, the so-called *two-particle irreducible* (2PI) graphs.

The basic idea is that, when computing the CTP generating functional, we want to constrain the deviations φ^A from the mean field ϕ^A so that not only their expectation value vanishes but also their fluctuations are known. We achieve this by adding suitable Lagrange multipliers: our already familiar source J^A, associated with the first constraint, and four new two-point sources $K^{AB} = K^{ab}(x, x')$ to enforce the second. The 2PI generating functional then reads, written as a path integral over full field Φ configurations

$$e^{(i/\hbar)W[J,K]} = \int D\Phi^A \exp\left\{(i/\hbar)\left[S\left[\Phi^A\right] + J_A\Phi^A + \frac{1}{2}K_{AB}\Phi^A\Phi^B\right]\right\} \quad (6.144)$$

The sources are connected to the mean fields and propagators through

$$\frac{\delta W}{\delta J_A} = \phi^A; \qquad \frac{\delta W}{\delta K_{AB}} = \frac{1}{2}\left[\phi^A\phi^B + G^{AB}\right] \quad (6.145)$$

The 2PI CTP EA is the full Legendre transform

$$\Gamma_2[\phi, G] = W[J, K] - J_A\phi^A - \frac{1}{2}K_{AB}\left[\phi^A\phi^B + G^{AB}\right]. \quad (6.146)$$

The equations of motion for mean fields and propagators are then

$$\frac{\delta \Gamma_2}{\delta \phi^A} = -J_A - K_{AB}\phi^B; \qquad \frac{\delta \Gamma_2}{\delta G^{AB}} = -\frac{1}{2}K_{AB} \qquad (6.147)$$

To implement the background field method, write

$$e^{i\Gamma_2/\hbar} = \int D\Phi^A \, e^{(i/\hbar)\left[S[\Phi^A]+J_A(\Phi^A-\phi^A)+\frac{1}{2}K_{AB}(\Phi^A\Phi^B-\phi^A\phi^B-G^{AB})\right]} \qquad (6.148)$$

The exponent becomes

$$S\left[\Phi^A\right] - \frac{\delta \Gamma_2}{\delta \phi^A}\left(\Phi^A - \phi^A\right) - \frac{\delta \Gamma_2}{\delta G^{AB}}\left[\left(\Phi^A - \phi^A\right)\left(\Phi^B - \phi^B\right) - G^{AB}\right] \qquad (6.149)$$

Write $\Phi = \phi + \varphi$ and expand the classical action as before,

$$S\left[\phi^A + \varphi^A\right] = S\left[\phi^A\right] + S_{,A}\varphi^A + \frac{1}{2}S_{,AB}\varphi^A\varphi^B + S_Q \qquad (6.150)$$

From our previous experience with the 1PI CTPEA, we know that the effective action will be equal to the classical action plus $O(\hbar)$ corrections, which will be given in terms of a Feynman path integral over the φ field. At the lowest order, this integral will be Gaussian, and will yield a term like $-(i\hbar/2)\ln\text{Det}\left\langle\varphi^A\varphi^B\right\rangle$. On the other hand, the formalism is set up so that $\left\langle\varphi^A\varphi^B\right\rangle \equiv G^{AB}$, thus we expect $\Gamma_2 = S[\phi] - (i\hbar/2)\ln\text{Det}G^{AB} + \ldots$. This effective action should generate the equations of motion for both the mean field and the propagators. To lowest order in $\hbar$, the Schwinger–Dyson equation for the propagators may be written as $(i/\hbar)S_{,AB} = -G^{-1}_{AB}$ (this is the statement that the Hessian of the effective action is the inverse of the propagators, specialized to lowest order). Since the variation of $-i\hbar\ln\text{Det}G^{AB}$ with respect to G^{AB} yields $-i\hbar G^{-1}_{AB}/2$, we get the right equation by adding a term whose variation is $S_{,AB}/2$. With these considerations in mind, we make the ansatz

$$\Gamma_2\left[\phi^A, G^{AB}\right] = S\left[\phi^A\right] + \frac{1}{2}S_{,AB}G^{AB} - \frac{1}{2}i\hbar\text{Tr}\ln G + \Gamma_Q - \frac{1}{2}i\hbar\delta_A^A \qquad (6.151)$$

(the final term does not affect the equations of motion and may be disregarded in practice) to get

$$e^{i\Gamma_Q/\hbar} = [\text{Det } G]^{-1/2} \int D\varphi^A$$

$$\times \exp\left\{\frac{-1}{2}G^{-1}_{AB}\varphi^A\varphi^B + (i/\hbar)\left[S_Q - \tilde{J}_A\varphi^A - \tilde{K}_{AB}\left(\varphi^A\varphi^B - G^{AB}\right)\right]\right\} \qquad (6.152)$$

where

$$\tilde{J}_A = \frac{1}{2}S_{,ABC}G^{BC} + \frac{\delta\Gamma_Q}{\delta\phi^A}; \qquad \tilde{K}_{AB} = \frac{\delta\Gamma_Q}{\delta G^{AB}} \qquad (6.153)$$

We see that the 2PI effective action is given, besides the terms already explicit in equation (6.151), by the sum of all *two-particle irreducible vacuum* graphs in a theory with action $(i/2)G^{-1}_{AB}\varphi^A\varphi^B + S_Q$. They are vacuum because there is

Figure 6.4 The setting sun graph.

Figure 6.5 Two tadpoles joined by one line.

Figure 6.6 The "horn" graph.

Figure 6.7 The fish graph.

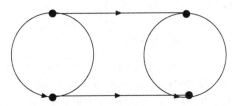

Figure 6.8 Two fishes joined by two lines.

no φ mean field, and 2PI because the nonlocal source $\tilde{K}_{AB}$ ensures that G^{AB} is the full propagator (we shall show this explicitly right away). The reduction from connected to 1PI to 2PI graphs entails a substantial increase in efficiency.

For example, for a $\lambda\Phi^4$ theory, S_Q contains both cubic and quartic vertices. The simplest Feynman graphs are the "double-bubble" and the "setting sun," both of which are of order $\hbar^2$. We may discard a graph consisting of two tadpoles joined by an internal line, since this is not 1PI. At the following order we may also discard graphs like the "horn," and two fishes joined by two internal lines, which are 1PI but not 2PI.

To conclude, let us verify that the sources enforce the proper constraints. As in our earlier discussion of the loop expansion in the 1PI CTPEA, this will follow from the invertibility of the relationship of sources to fields, namely, that the operator

$$\begin{pmatrix} \dfrac{\delta J_A}{\delta\phi^C} & \dfrac{\delta J_A}{\delta G^{CD}} \\[2ex] \dfrac{\delta K_{AB}}{\delta\phi^C} & \dfrac{\delta K_{AB}}{\delta G^{CD}} \end{pmatrix} \tag{6.154}$$

is nonsingular. In terms of derivatives of the 2PIEA, this becomes (minus)

$$\begin{pmatrix} \dfrac{\delta^2\Gamma_2}{\delta\phi^A\delta\phi^C} - 2\dfrac{\delta\Gamma_2}{\delta G^{AC}} - 2\dfrac{\delta^2\Gamma_2}{\delta G^{AB}\delta\phi^C}\phi^C & \dfrac{\delta^2\Gamma_2}{\delta\phi^A\delta G^{CD}} - 2\dfrac{\delta^2\Gamma_2}{\delta G^{AB}\delta G^{CD}}\phi^B \\[2ex] 2\dfrac{\delta^2\Gamma_2}{\delta G^{AB}\delta\phi^C} & 2\dfrac{\delta^2\Gamma_2}{\delta G^{AB}\delta G^{CD}} \end{pmatrix} \tag{6.155}$$

It is clear that this operator will be nonsingular if and only if the simpler matrix

$$\begin{pmatrix} \dfrac{\delta^2\Gamma_2}{\delta\phi^A\delta\phi^C} - 2\dfrac{\delta\Gamma_2}{\delta G^{AC}} & \dfrac{\delta^2\Gamma_2}{\delta\phi^A\delta G^{CD}} \\[2ex] 2\dfrac{\delta^2\Gamma_2}{\delta G^{AB}\delta\phi^C} & 2\dfrac{\delta^2\Gamma_2}{\delta G^{AB}\delta G^{CD}} \end{pmatrix} \tag{6.156}$$

also is. By taking variations of equation (6.152) with respect to ϕ^A and G^{AB}, and after some algebra, we obtain the set of equations

$$\begin{pmatrix} \dfrac{\delta^2\Gamma_2}{\delta\phi^A\delta\phi^C} - 2\dfrac{\delta\Gamma_2}{\delta G^{AC}} & \dfrac{\delta^2\Gamma_2}{\delta\phi^A\delta G^{CD}} \\[2ex] 2\dfrac{\delta^2\Gamma_2}{\delta G^{AB}\delta\phi^C} & 2\dfrac{\delta^2\Gamma_2}{\delta G^{AB}\delta G^{CD}} \end{pmatrix} \begin{pmatrix} \langle\varphi^C\rangle \\[2ex] \langle\varphi^C\varphi^D\rangle - G^{CD} \end{pmatrix} = 0 \tag{6.157}$$

and so the constraints are enforced. In practice, this means that we can forget about $\tilde{J}$ and $\tilde{K}$ when computing the nonlinear correction Γ_Q to the 2PIEA, provided that we omit all one- and two-particle reducible graphs, and use the full propagator G^{AB} in internal lines. The vertices, of course, are those contained in S_Q, and will generally depend on the mean fields.

As for the CTP method more generally, it is impossible to give a complete list of references for the 2PIEA. For some of the pioneering papers, see

[LutWar60, DomMar64a, DomMar64b, DahLas67, CoJaTo74]. This method has been generalized and applied to the establishment of a quantum kinetic field theory (Chapter 11). It has been applied to problems in gravitation and cosmology (Chapter 15), particles and fields (Chapter 14), Bose–Einstein condensates and condensed matter systems (Chapter 13) as well as to address the issues of thermalization and quantum phase transitions (Chapters 9 and 12). More generally, we may regard the 2PIEA as an implementation of the Φ-derivable approach to be discussed in Chapter 13.

6.5.1 The 2PI effective action in the $g\Phi^3$ theory

Let us test our understanding of this new object by applying it to the $g\Phi^3$ field theory. The 2PI CTPEA is given by equation (6.151), where, as before, the classical potential is given in equation (6.43), and m^2 is shifted to $m^2 - i\varepsilon$ on the first branch, $m^2 + i\varepsilon$ on the second. The second derivatives of the classical action may be read off from equation (6.118). We do not need an explicit knowledge of $\ln G$, other than the formal property

$$\frac{\delta \ln G}{\delta G^{AB}} = \left[G^{-1}\right]_{AB} \tag{6.158}$$

Γ_Q is the sum of all 2PI vacuum bubbles with vertices from equation (6.119) and full propagators G^{AB} in internal lines. Observe that in this model, Γ_Q is independent of the background fields, which is rather exceptional. The lowest order contribution to Γ_Q has two loops

$$\Gamma_Q \sim \frac{i}{2\hbar} \left(\frac{g^2}{6}\right) c_{abc} c_{def} \int d^4x d^4x' \, G^{ad}(x,x') \, G^{be}(x,x') \, G^{cf}(x,x') \tag{6.159}$$

The equations of motion are derived from the variations with respect to ϕ^A and G^{AB}. In the physical case where there are no external sources, we get

$$S_{,A} + \frac{1}{2} S_{,ABC} G^{BC} = 0 \tag{6.160}$$

$$\frac{1}{2} S_{,AB} - \frac{1}{2} i\hbar \left[G^{-1}\right]_{AB} + \frac{\delta \Gamma_Q}{\delta G^{AB}} = 0 \tag{6.161}$$

The equation for the propagators reduces to equation (6.82) after we identify

$$\frac{\delta \Gamma_Q}{\delta G^{AB}} = -\frac{\hbar}{2} \Sigma_{AB} \tag{6.162}$$

It is customary to rewrite it as

$$S_{,AB} G^{BC} - \hbar \Sigma_{AB} G^{BC} = i\hbar \delta_A^C \tag{6.163}$$

More explicitly (we write the equation for ϕ^1, after setting $\phi^1 = \phi^2 = \phi$)

$$\nabla^2 \phi(x) - m^2 \phi(x) + \frac{1}{2} g \left[\phi^2(x) + G^{11}(x,x)\right] = -h \tag{6.164}$$

$$\left[\nabla^2 - m^2\right] G^{ac}(x,y) + g c^{ag} c_{gdb} \phi^d(x) G^{bc}(x,y)$$
$$- \hbar c^{ag} \int d^4z \, \Sigma_{gb}(x,z) G^{bc}(z,y) = i\hbar c^{ac} \delta(x,y) \tag{6.165}$$

where

$$i\Sigma_{gb}(x,z) = \frac{g^2}{2\hbar^2} c_{gde} c_{fhb} G^{df}(x,z) G^{eh}(x,z) \qquad (6.166)$$

We may begin to see the power of the 2PIEA. Our linearized one-loop equation derived above from the 1PIEA is equivalent to neglecting the Σ kernels. In this case, the propagators decouple, and solving for the Feynman propagator in powers of ϕ, we recover the known results. Using the 2PIEA, although we are also doing a one-loop approximation to the Schwinger–Dyson equations, we have a much more complete description of the physics, including some of the nonlinear interactions between fluctuations (we shall return to this in Chapter 11).

The propagator equations are more transparent if we choose G_{ret} and G_1 as independent variables, rather than the four fundamental propagators. Recalling equation (6.9), we get

$$\left[\nabla^2 - m^2 + g\phi(x)\right] G_{\mathrm{ret}}(x,y) - \int d^4z\, \Sigma_{\mathrm{ret}}(x,z) G_{\mathrm{ret}}(z,y) = -\delta(x,y) \qquad (6.167)$$

where

$$\Sigma_{\mathrm{ret}}(x,z) = \hbar\left[\Sigma_{11}(x,z) + \Sigma_{12}(x,z)\right] \qquad (6.168)$$

and we have used that $G^{11} + G^{22} = G^{12} + G^{21}$. Observe that the kernel Σ_{ret} is causal. For the Hadamard propagator, we get

$$\left[\nabla^2 - m^2 + g\phi(x)\right] G_1(x,y)$$
$$- \int d^4z\, \Sigma_{\mathrm{ret}}(x,z) G_1(z,y) = -i\hbar \int d^4z\, \Sigma_1(x,z) G_{\mathrm{adv}}(z,y) \qquad (6.169)$$

$$\Sigma_1(x,z) = \hbar\left[\Sigma_{11}(x,z) + \Sigma_{22}(x,z)\right]$$
$$= -\hbar\left[\Sigma_{12}(x,z) + \Sigma_{21}(x,z)\right] \qquad (6.170)$$

We shall discuss further this equation in Chapter 8. For the time being, we remark that to lowest order in perturbation theory, $i\Sigma_1$ is just (twice) the noise kernel from the 1PIEA (cf. equation (6.130)).

6.5.2 *Large N expansion (suite)*

Let us return to the $O(N)$ invariant scalar field theory from the last section. After rescaling, the action is given by equation (6.140). We recall there was an infinite family of Feynman graphs all scaling as N in the large N limit, for which reason the 1PIEA was not easy to compute. As we shall see presently, the 2PI approach cures this problem. The reason is that all but one of the offending graphs are two-particle *reducible*, and therefore drop out of the effective action.

It follows that if we are satisfied with the leading order (LO) theory, we may simply write down the 2PIEA for the theory as given, and obtain a closed form expression. For simplicity, let us consider the *unbroken symmetry* case, $\phi^A = 0$. Therefore

$$\Gamma^{LO} = \frac{N}{2}\left\{\left[\nabla^2 - M^2\right]G^{A\alpha,A\alpha}\right\} - \frac{i\hbar}{2}\text{Tr}\,\ln G + \Gamma^{LO}_Q + \text{const.} \qquad (6.171)$$

$$\Gamma^{LO}_Q = -N\frac{\lambda_{abcd}}{8}\int d^4x\left\{G^{a\alpha,b\alpha}(x,x)\,G^{c\beta,d\beta}(x,x) + 2G^{a\alpha,b\beta}(x,x)\,G^{c\alpha,d\beta}(x,x)\right\} \qquad (6.172)$$

Of course, only the first term is truly LO. Discarding the terms which are not strictly LO we get

$$c_{ac}\left[\nabla^2 - M^2 - \frac{\lambda}{2}G^{\gamma\gamma}_F(x,x)\right]G^{c\alpha,b\beta}(x,x') = \frac{i\hbar}{N}\delta_{\alpha\beta}\delta^b_a\delta(x-x') \qquad (6.173)$$

These equations are all there is to leading order. Observe that the only difference with the equations for a free theory is the mass shift: the real mass of the theory is not M^2 but rather $M^2_{\text{phys}} = M^2 + (\lambda/2)\,G^{\gamma\gamma}_F(x,x)$. Since we may also solve the equation to obtain

$$G^{\alpha\beta}_F(x,x') = \frac{-i\hbar}{N}\delta_{\alpha\beta}\int\frac{d^4k}{(2\pi)^4}\frac{e^{ik(x-x')}}{k^2 + M^2_{\text{phys}} - i\varepsilon} \qquad (6.174)$$

this results in a nonlinear (gap) equation for the physical mass

$$M^2_{\text{phys}} = M^2 - \frac{i\hbar\lambda}{2}\int\frac{d^4k}{(2\pi)^4}\frac{1}{k^2 + M^2_{\text{phys}} - i\varepsilon} \qquad (6.175)$$

The name "gap equation" is adopted from condensed matter physics, to the fact that M here is the energy of an excitation with zero momentum.

To the next order, we find that strings of *fish* graphs are all of order N^0, since there are $l-1$ fishes in the graph, and each may carry an independent index. So to get a closed expression to NLO, we must use the Coleman–Jackiw–Politzer trick [CoJaPo74] of including an auxiliary field χ, by adding a term to the action, which now reads

$$S = N\int d^4x\left\{\frac{-1}{2}\partial_\mu\Phi^\alpha\partial^\mu\Phi^\alpha - \frac{1}{2}M^2\Phi^\alpha\Phi^\alpha - \frac{\lambda}{8}(\Phi^\alpha\Phi^\alpha)^2\right.$$
$$\left. + \frac{1}{2}\left(\frac{\chi}{\sqrt{\lambda}} - \frac{\sqrt{\lambda}}{2}(\Phi^\alpha\Phi^\alpha)\right)^2\right\} \qquad (6.176)$$

Figure 6.9 A string of three fishes.

Figure 6.10 The three-pointed star graph.

Expanding this, we get

$$S = N \int d^4x \left\{ \frac{-1}{2} \partial_\mu \Phi^\alpha \partial^\mu \Phi^\alpha - \frac{1}{2} M^2 \Phi^\alpha \Phi^\alpha + \frac{\chi^2}{2\lambda} - \frac{\chi}{2} \left(\Phi^\alpha \Phi^\alpha \right) \right\} \quad (6.177)$$

The new classical equations of motion are

$$\nabla^2 \phi^\alpha - \left[M^2 + \chi \right] \phi^\alpha = 0 \quad (6.178)$$

$$\chi = \frac{\lambda}{2} \phi^\beta \phi^\beta \quad (6.179)$$

which are seen to be identical to the old ones.

In this new action, strings of fish graphs beyond two loops are no longer 2PI. The next nontrivial graph is the *three-pointed star*, which scales as N^{-1}. Thus, once again, we obtain a closed form for NLO large N.

To obtain this explicit expression, we begin by shifting the field $\Phi \rightarrow \phi + \varphi$, $\chi \rightarrow \bar{\chi} + \delta\kappa$. As usual, we discard linear terms, so

$$\delta S = N \int d^4x \left\{ \frac{-1}{2} \partial_\mu \varphi^\alpha \partial^\mu \varphi^\alpha - \frac{1}{2} \left(M^2 + \bar{\chi} \right) \varphi^\alpha \varphi^\alpha \right.$$
$$\left. + \frac{\delta\kappa^2}{2\lambda} - \delta\kappa \left(\phi^\alpha \varphi^\alpha \right) - \frac{\delta\kappa}{2} \left(\varphi^\alpha \varphi^\alpha \right) \right\} \quad (6.180)$$

It is convenient to eliminate the quadratic cross-term, shifting $\delta\kappa = \delta\chi + \lambda \phi^\alpha \varphi^\alpha$. We get

$$\delta S = \frac{N}{2} \int d^4x \left\{ -(\partial\varphi)^2 - M^2_{\alpha\beta} \varphi^\alpha \varphi^\beta + \frac{\delta\chi^2}{\lambda} - \delta\chi \left(\varphi^\alpha \varphi^\alpha \right) - \lambda \phi^\alpha \varphi^\alpha \varphi^\beta \varphi^\beta \right\} \quad (6.181)$$

$M^2_{\alpha\beta} = \left(M^2 + \bar{\chi} \right) \delta_{\alpha\beta} + \lambda \phi_\alpha \phi_\beta$, whereby the 2PIEA,

$$\Gamma^{\text{NLO}} = S\left[\phi, \bar{\chi} \right] + \frac{N}{2} \left\{ \left[\nabla^2 \delta_{\alpha\beta} - M^2_{\alpha\beta} \right] G^{\alpha\beta} + \frac{H}{\lambda} \right\}$$
$$- \frac{i\hbar}{2} \left\{ \text{Tr} \ln H + \text{Tr} \ln G \right\} + \Gamma^{\text{NLO}}_Q + \text{const.} \quad (6.182)$$

$$\Gamma^{\text{NLO}}_Q = \frac{iN^2}{4\hbar} \int d^4x d^4x' \left\{ H(x,x') G^{\gamma\delta}(x,x')^2 + \lambda^2 \phi^\alpha(x) \phi^\beta(x') \Delta^{\alpha\beta}(x,x') \right\} \quad (6.183)$$

$$\Delta^{\alpha\beta}(x,x') = G^{\alpha\beta}(x,x') G^{\gamma\delta}(x,x')^2 + 2G^{\alpha\delta}(x,x') G^{\gamma\delta}(x,x') G^{\gamma\beta}(x,x') \quad (6.184)$$

Let us write the equations of motion for unbroken symmetry, leaving the CTP indices implicit

$$\left[\nabla^2 - M^2 - \chi\right]\delta_{\alpha\beta} - \frac{i\hbar}{N}G_{\alpha\beta}^{-1} + \frac{iN}{\hbar}H\left(x, x'\right)G^{\alpha\beta}\left(x, x'\right) = 0 \tag{6.185}$$

$$\chi - \frac{\lambda}{2}G^{\alpha\alpha}\left(x, x\right) = 0 \tag{6.186}$$

$$1 - \frac{i\lambda\hbar}{N}H^{-1} + \frac{i\lambda N}{2\hbar}G^{\gamma\delta}\left(x, x'\right)^2 = 0 \tag{6.187}$$

6.6 Handling divergences

As is well-known, the field theory of point particles is riddled with divergences. One needs to identify and remove them before one can begin to deal with physical applications. In this section we shall briefly summarize the most common types of divergences to be expected. By no means is this a complete treatment. As an example, we continue to use the $g\varphi^3$ theory to discuss its divergences.

6.6.1 Ultraviolet divergences

In field theories defined on flat spacetime and where the *in* and *out* vacuum agree to zeroth order in perturbation theory, to any finite order the 1PIEA is rendered free of ultraviolet divergences by renormalizing the parameters in the bare action in the same way one does for the *in–out* EA. This follows from the observation that a primitively divergent graph must have all its vertices on the same branch of the closed time path, and therefore, if we use free propagators in the internal lines, it is either equivalent to an *in–out* graph (all propagators are Feynman) or to its conjugate (all propagators are Dyson). By the assumed equivalence of the vacuum states, these are the same graphs appearing in the *in–out* EA.

If a graph does not have all vertices in the same branch, it cannot be primitively divergent. We say that a graph is primitively divergent when it diverges, and every subgraph is also divergent. Take a one-particle irreducible graph with vertices on both branches. Take one vertex, say, on the first branch, and consider the maximal set of vertices on the same branch which are connected to it. This set is not all the graph, because the graph has also second branch vertices. The maximal set is connected to the rest of the graph by at least two lines, because the graph is one-particle irreducible. These lines have mixed vertices at their ends, since otherwise they would be internal to the maximal set. When writing the corresponding amplitude, these lines will go on-shell, because both Δ^{21} and Δ^{12} are proportional to $\delta\left(p^2 + m^2\right)$. Now consider a loop including these two lines: it has to be finite, because there are two on-shell lines. Therefore the graph is not primitively divergent.

When working with the 2PIEA, one does not aim to make the effective action finite, but rather to show that the equations of motion admit finite solutions. To be ready for renormalization, we build the 2PIEA on the bare action

$$S_{\text{bare}}\left[\Phi\right] = \int d^d x \left[-\frac{1}{2}Z_\varphi \left(\partial\Phi\right)^2 - \frac{1}{2}m_b^2\Phi^2 + \frac{1}{6}Z_g g\Phi^3 + h_B\Phi\right] \qquad (6.188)$$

leading to the equations of motion

$$Z_\varphi\nabla^2\phi\left(x\right) - m_b^2\phi\left(x\right) + \frac{1}{2}Z_g g\left[\phi^2\left(x\right) + \frac{1}{2}G^{11}\left(x,x\right)\right] = -h_B \qquad (6.189)$$

$$\left[Z_\varphi\nabla^2 - m_b^2 + Z_g g\phi\left(x\right)\right]G_{\text{ret}}\left(x,y\right) - Z_g^2\int d^4z\,\Sigma_{\text{ret}}\left(x,z\right)G_{\text{ret}}\left(z,y\right) = -\delta\left(x,y\right) \qquad (6.190)$$

$$\left[Z_\varphi\nabla^2 - m_b^2 + Z_g g\phi\left(x\right)\right]G_1\left(x,y\right) - Z_g^2\int d^4z\,\Sigma_{\text{ret}}\left(x,z\right)G_1\left(z,y\right) = -\mathcal{K}\left(x,y\right) \qquad (6.191)$$

$$\mathcal{K}\left(x,y\right) = iZ_g^2\hbar\int d^4z\,\Sigma_1\left(x,z\right)G_{\text{adv}}\left(z,y\right) \qquad (6.192)$$

(The Σ's are defined by equations (6.166), (6.168) and (6.170).) In order to analyze the possible divergences in these equations, we adopt some kind of perturbative approach. Let us assume that ϕ is constant, and that in the Σ kernels we may approximate the propagators by free propagators, corresponding to a yet unknown mass M^2. The propagators will then be translation invariant, and we may Fourier transform all equations to get

$$m_b^2\phi\left(x\right) - \frac{1}{2}Z_g g\left[\phi^2 + \frac{1}{2}\int \frac{d^d p}{\left(2\pi\right)^d}G^{11}\left(p\right)\right] = -h_B \qquad (6.193)$$

$$\left[-Z_\varphi p^2 - m_b^2 + Z_g g\phi - Z_g^2\Sigma_{\text{ret}}\left(p\right)\right]G_{\text{ret}}\left(p\right) = -1 \qquad (6.194)$$

$$\left[-Z_\varphi p^2 - m_b^2 + Z_g g\phi - Z_g^2\Sigma_{\text{ret}}\left(p\right)\right]G_1\left(p\right) = -iZ_g^2\Sigma_1\left(p\right)G_{\text{adv}}\left(p\right) \qquad (6.195)$$

Since $\Sigma_1\left(p\right)$ is finite and $Z_g^2 = 1$ to lowest order, the third equation will be well defined if we can control the second.

At this point we need to relate the effective mass M^2 to the propagators. Two common choices are to define M^2 as the value of the inverse retarded propagator at $p = 0$, or else as the position of the pole of the retarded propagator as a function of $-p^2$. This second choice has a greater physical appeal, but it is harder to implement in practice.

Let us therefore define M^2 as the value of the inverse retarded propagator at $p = 0$. Recall that $\Sigma_{\text{ret}}\left(p\right)$ can be obtained from the results in Chapter 5, provided the Feynman prescription $p^2 \to p^2 - i\varepsilon = -p^{02} + \mathbf{p}^2 - i\varepsilon$ is replaced by the causal prescription $p^2 \to \left(p + i\varepsilon\right)^2 = -p^{02} + \mathbf{p}^2 - i\varepsilon \operatorname{sign} p^0$. It is convenient to parametrize m_b^2 in terms of the value M_0^2 of M^2 at $\phi = 0$ (which is always a solution of the equations of motion, by construction)

$$m_b^2 = M_0^2 + \frac{Z_g^2 g^2 \hbar}{16\pi^2} \left[\frac{1}{\varepsilon} + \text{constant} - \frac{1}{2} \ln \left(\frac{M_0^2}{4\pi\mu^2} \right) \right] \qquad (6.196)$$

The gap equation reads

$$M^2 - M_0^2 + Z_g g\phi + \frac{Z_g^2 g^2 \hbar}{32\pi^2} \ln \left(\frac{M^2}{M_0^2} \right) = 0 \qquad (6.197)$$

In this model, the gap equation is explicitly finite, so we may simply set $Z_g = 1$. Otherwise, we may use this further degree of freedom to control any remaining divergence.

The wavefunction renormalization Z_φ may be determined, for example, by requiring that

$$\left. \frac{\partial G_{\text{ret}}^{-1}}{\partial (-p^2)} \right|_{p^2 = 0} = -1 \qquad (6.198)$$

We get

$$1 = Z_\varphi + \frac{g^2 \hbar}{192\pi^2 M^2}$$

which is finite.

After these choices, we have exhausted our freedom to redefine the parameters in the classical action, so the mean field equation ought to be explicitly finite. The mean field equation reads (recall the tadpole from Chapter 5, equation (5.24))

$$h_B = \left\{ M_0^2 + \frac{g^2 \hbar}{16\pi^2} \left[\frac{1}{\varepsilon} + \text{constant} - \frac{1}{2} \ln \left(\frac{M_0^2}{4\pi\mu^2} \right) \right] \right\} \phi(x)$$

$$- \frac{1}{2} g \left\{ \phi^2 - \frac{\hbar M^2}{8\pi^2} \left[\frac{1}{\varepsilon} + \text{constant}' - \frac{1}{2} \ln \left(\frac{M^2}{4\pi\mu^2} \right) \right] \right\} \qquad (6.199)$$

Setting $\phi = 0$ we get

$$h_B = \frac{g\hbar M_0^2}{16\pi^2} \left[\frac{1}{\varepsilon} + \text{constant}' - \frac{1}{2} \ln \left(\frac{M_0^2}{4\pi\mu^2} \right) \right] \qquad (6.200)$$

so the coefficient of ε^{-1} is

$$\frac{g\hbar}{16\pi^2} \left[g\phi + M^2 - M_0^2 \right] \qquad (6.201)$$

which vanishes to lowest order by virtue of the gap equation.

For further discussion, we refer the reader to the literature [HeeKno02a, HeeKno02b, HeeKno02c, BIIaRe03].

6.6.2 Initial time singularities

As we have seen in the last subsection, the handling of ultraviolet singularities in the nonequilibrium formalism is not really different from the usual field theory methods. We shall now discuss a new class of singularities which are specific to nonequilibrium problems [Lin87, CooMot87, Baa98, Baa00a, Baa00b, HaMoMo99, BaBoVe01].

These singularities arise when one attempts to solve the mean field equations of motion with Cauchy data at some initial time (which we may choose as $t = 0$ without any loss of generality). In a perturbative scheme, it seems "natural," to lowest order, to use free propagators to compute the Feynman graphs in the effective action, and to assume an initial state uncorrelated with the initial conditions for the mean fields. But actually this is wrong: the switching on of the mean field (or equivalently, of the coupling constant) in an arbitrarily short time-scale always has an impact on the initial state of the quantum fluctuations. Neglect of this effect introduces an inconsistency in the theory, thereby the divergences.

Let us consider the mean field equations for the $g\varphi^3$ model, as derived above from the one-loop 1PIEA, equation (6.89). We are interested in finding the free evolution of the mean field, from given initial conditions at $t = 0$. We shall assume the local terms (including the ultraviolet singularities) in the quantum correction have been absorbed in the parameters of the equation. We also assume the mean field is spatially homogeneous, so we may write

$$\left[\nabla^2 - m^2\right]\phi(t) + \frac{g^2\hbar}{32\pi^2}\int\frac{d\omega}{2\pi}\int_0^t du\, e^{-i\omega(t-u)}\int_{4m^2}^\infty\frac{d\sigma^2\sqrt{1-\frac{4m^2}{\sigma^2}}}{4m^2 - (\omega+i\varepsilon)^2 + \sigma^2}\phi(u) = 0$$

(6.202)

Perform the integral over ω

$$\int\frac{d\omega}{2\pi}\frac{e^{-i\omega(t-u)}}{-(\omega+i\varepsilon)^2+\sigma^2} = \frac{\sin\left[\sigma(t-u)\right]}{\sigma}$$

(6.203)

$$\left[\nabla^2 - m^2\right]\phi(t) + \frac{g^2\hbar}{16\pi^2}\int_{4m^2}^\infty d\sigma\sqrt{1-\frac{4m^2}{\sigma^2}}\int_0^t du\,\sin\left[\sigma(t-u)\right]\phi(u) = 0$$

(6.204)

We improve the convergence of the σ integral with an integration by parts

$$0 = \left[\nabla^2 - m^2\right]\phi(t) + \frac{g^2\hbar}{16\pi^2}\int_{4m^2}^\infty\frac{d\sigma}{\sigma}\sqrt{1-\frac{4m^2}{\sigma^2}}\int_0^t du\left(\frac{d}{du}\cos\left[\sigma(t-u)\right]\right)\phi(u)$$

$$= \left[\nabla^2 - m^2\right]\phi(t) - \frac{g^2\hbar}{16\pi^2}\int_{4m^2}^\infty\frac{d\sigma}{\sigma}\sqrt{1-\frac{4m^2}{\sigma^2}}\int_0^t du\,\cos\left[\sigma(t-u)\right]\frac{d\phi}{du}$$

$$+ \delta m^2\phi(t) - \chi(t)\phi(0)$$

(6.205)

where

$$\delta m^2 = \frac{g^2 \hbar}{16\pi^2} \left[\int_{4m^2}^{\infty} \frac{d\sigma}{\sigma} \sqrt{1 - \frac{4m^2}{\sigma^2}} \right] \tag{6.206}$$

$$\chi(t) = \frac{g^2 \hbar}{16\pi^2} \left[\int_{4m^2}^{\infty} \frac{d\sigma}{\sigma} \sqrt{1 - \frac{4m^2}{\sigma^2}} \cos[\sigma t] \right] \tag{6.207}$$

The logarithmically divergent term δm^2 may be absorbed in m^2. At issue is the "source term" $\chi(t) \phi(0)$. At finite times, we may expect that the oscillatory behavior of the cosine will be enough to make the integral convergent. However, at $t = 0$ this improved convergence is lost, and $\chi(0)$ is ill defined. This is the initial time singularity.

In physical terms, it is as though we set $g = 0$ for $t < 0$, thereby allowing the quantum fluctuations to reach equilibrium as a free field (in this case, at zero temperature, but allowing for equilibrium at a finite temperature only makes the problem worse), and then suddenly we switch the interaction and the mean field on. This sudden transition will necessarily create particles, so it is inconsistent to assume that the state of the quantum fluctuations is the vacuum at any positive time, no matter how short.

The problem may be cured by adopting a more physical initial condition; we refer the reader to the literature for details [Lin87, CooMot87, Baa98, Baa00a, Baa00b, HaMoMo99, BaBoVe01].

6.6.3 Other divergences

Unfortunately, ultraviolet and initial time singularities are not the only problems to watch out for [CarKob98, CaKoPe98, Bed, Dad99, BoVeWa00, GeScSe01]. Among other common complications, we may mention infrared singularities, which appear when some quantum fluctuations are massless. Massless fields are rather common: they appear in problems related to unbroken gauge symmetries, at critical points in models of phase transitions, and as Goldstone bosons when a global symmetry is broken. For example, an $O(N)$ model in the broken symmetry phase has $N - 1$ massless fields in its spectrum. The spectrum of excitations above a homogeneous Bose–Einstein condensate also generally contains one massless mode, which arises from the breaking of global $U(1)$ invariance.

Although we shall connect with finite temperature field theory in a later chapter it is timely to mention that real time perturbation theory at finite temperature also has its peculiar kind of divergences. The free thermal propagators contain terms proportional to mass-shell delta functions $\delta(-p^2 - m^2)$, and so they produce singularities whenever two propagators are evaluated at collinear momenta in the same graph.

It is important to beware of singularities arising from a nonjudicious application of perturbation theory. Regardless of the formal order in g, $\hbar$ or N^{-1}, large corrections must be included to get consistent results. For example, weak damping of fluctuations due to higher order processes modifies the behavior of the propagators near the mass shell, and may cure some singularities. Judging the situation by the right physics is often the best way to handle the unfamiliar pathologies.

Part III

Gauge invariance, dissipation, entropy, noise and decoherence

7

Closed time path effective action for gauge theories

In this chapter we treat out-of-equilibrium behavior of gauge fields, particularly of the nonabelian kind. This is a broad topic, so we will only discuss some specific points.

Overall, we may distinguish two sets of features that make problems involving gauge fields different from those where only "matter" fields are present. On the one hand, there are "technical" differences associated with the fact that problems involving gauge fields usually abound with massless degrees of freedom. An important example is the so-called "hard thermal loop" problem, which is discussed in Chapter 10. We also consider "technical" difficulties associated with a particular symmetry breaking pattern or with the property of confinement, which clearly has a strong impact on the nonequilibrium phenomenology of QCD. Because of the rich variety of behavior, these problems are best treated on a case by case basis. In Chapter 14, for example, we give a brief account of nonequilibrium phenomena in relativistic heavy ion collisions.

On the other hand, there is an intrinsic difference between gauge and nongauge theories, coming from the fact that the "natural" description of the former in terms of spacetime fields is redundant. For example, the most efficient description of the Maxwell field is in terms of the potential 4-vector, but many different 4-vectors describe the same physical electromagnetic field. There is an intrinsic ambiguity in the equations of motion of the theory, which do not determine the evolution completely. At the same time there are restrictions on our freedom to choose Cauchy data for the physical fields; we say that the theory is "constrained."

In the quantum theory, the redundancy in the field variables is reflected in the fact that the "naive" Hilbert space of the theory is overlarge. The constraints of the classical theory become restrictions that the physical states must satisfy; these restrictions ensure that physical states respond to the physical part of the redundant field operators, but are impervious to the gauge part.

However, to get rid of ambiguities and constraints by reducing the theory to operators associated with measurable observables acting on physical states is, if at all possible, overwhelmingly inconvenient. These difficulties can be dealt with by formulating the theory in terms of the redundant, but natural, field variables. The subject of this chapter is to explore how nonequilibrium gauge theories are different from nongauge ones, because of this fundamental ambiguity.

The description of a gauge-invariant theory within quantum field theory techniques usually involves eliminating the gauge freedom by imposing "gauge fixing" conditions. These conditions are associated with new parameters, whose choice is arbitrary. To ensure the equivalence between the "gauge fixed" and the original theory, new fields (the so-called "ghost" fields) must be included, sometimes even with the wrong spin/statistics connection. It is expected that the predictions of the theory with respect to physical observables are independent of these manipulations: they do not change either if the fields are subject to a gauge transformation (gauge invariance) or if we change the chosen gauge fixing conditions (gauge independence). Nevertheless, oftentimes one is interested in computing objects that are not quite observable, such as a gluon correlation function. Then neither gauge invariance nor independence are guaranteed, and it becomes an important issue to decide which parts of the result really say something about the theory, and which are merely artifacts [Nie75, KobKun89, Bai92, KoKuRe91, GerReb03, ArrSmi02].

Our most important tool to investigate this issue is the observation that the constraints of the theory result in a number of restrictions on the structure of the Green functions such as vertices and propagators. These restrictions take the form of identities linking Green functions of different orders, the so-called Takahashi–Ward and Slavnov–Taylor identities. As it is often the case, there are two possible ways of looking at these identities. On the one hand, they are a check on the quality of a given approach to the problem: if important identities are violated (e.g. a field which ought to be massless is assigned a mass) then the approach is no good. On the other hand, these identities say things about the structure of the theory which may be used to motivate or to improve on a given approach (for example, by using an ansatz for the vertex functions which guarantees that the identities hold to a given order in perturbation theory).

The subject of gauge theory quantization is extremely rich and varied [HenTei92], and the addition of the nonequilibrium dimension only makes it even more so. Within the bounds of a single chapter, only a few of its avenues may be explored. Following the perspective developed in the early chapters, we shall adhere to the approach whereby nonequilibrium dynamics is followed through the evolution of the low-order Green functions. As in Chapter 6, we shall derive the dynamics of these Green functions self-consistently from a suitable closed time path action functional. As a matter of fact, if one is content to make a gauge choice from the start (e.g. to work within the "longitudinal" gauge), then the theory ceases to be "gauge" and the formalism from the earlier chapters may be applied straightforwardly [Gei96, Gei97, Gei99, Son97]. The problem is then whether any given result is valid generally, or limited to the given gauge choice. Our perspective in this chapter shall be the opposite, namely, leaving gauge choices as open and explicit as possible, and trying to learn about the deep structure of the theory from this very same freedom.

More concretely, our goal is to develop the 2PI approach to nonequilibrium gauge theories, as typical of approaches based on the evaluation of Green functions [Mot03, CaKuZa03, KraReb04]. We shall not discuss higher nPI effective actions, for which we refer the reader to the literature [Ber04a].

To set the stage for a discussion of the 2PIEA, we must begin by considering the essentials of the path integral quantization of gauge theories, and in particular how we set the initial conditions for gauge fields in a statistical state (such as a finite temperature one). For reasons of space and clarity we will restrict ourselves to Yang–Mills and to nonlinear abelian theories such as QED and SQED. We shall make no explicit attempt to discuss gravity, form fields or string theories [Wei00].

These self-imposed limitations in our aims here are correlated with some necessary technical choices. We shall discuss only the path integral Fadeev–Popov quantization of gauge theories. Although we shall use Becchi–Rouet–Stora–Tyutin (BRST) invariance at several stages, we shall not apply methods such as BRST or Batalin–Vilkovisky quantization, which really come on their own only in more demanding applications [Wei96]. We are deeply indebted to DeWitt's insights [DeW64, DeW79] and shall use his notation, but we shall not use the gauge-independent formulation of DeWitt and Vilkovisky [Vil84, DeW87] (on this subject, see the discussion in [Reb87]), nor more recent developments by DeWitt and collaborators [DeWMol98].

When gauge symmetries are unbroken, there are no preferred directions in gauge space, and all background fields will vanish identically. Therefore, the only degrees of freedom in the 2PI formalism shall be the propagators or two-point functions. Also, there will be no need to distinguish between the usual and the DeWitt–Abbott gauge invariant EA [DeW81, Abb81, Hart93, Alx99], nor to introduce gauge fixing conditions appropriate to the study of broken gauge theories [Wei96]. We shall only assume that the gauge fixing condition is linear on the quantum fields. On the other hand, we shall be completely general regarding group structure, matter content, (linear) gauge fixing condition and gauge fixing parameter.

As a word of caution, let us observe that symmetries that hold for the exact theory may be broken when the exact 2PI effective action is replaced by an approximated functional. In our case, this will manifest in violations of the Takahashi–Ward or Slavnov–Taylor identities. Usually this problem may be kept under control by working to a high enough order, by going over to a nPI approach with a large enough n, or simply by being careful about the approximations one uses. This problem is not actually exclusive of gauge theories; we will find it again when we attempt to make a consistent field theory of Bose–Einstein condensates, where the symmetry in question is the possibility of adding a phase to the condensate wavefunction. We shall discuss it in more detail in that simpler context, and refer the reader to the literature regarding gauge field theories [ReiSer06].

This chapter contains three sections besides this introduction. Section 7.1 summarizes the main results concerning the path integral quantization of gauge theories to be used in the following, including BRST invariance, the characterization of physical states and how to deal with nonvacuum initial conditions. We shall adopt the Kugo–Hata formalism, where ghost propagators acquire statistical corrections proper of a Bose field. Section 7.2 introduces the 2PIEA for gauge theories. Section 7.3 investigates the two main features of gauge theories which have no equivalent in their "normal" counterparts, namely, the issue of gauge dependence and the possibility of using gauge invariance arguments to investigate the structure of the theory. To develop our arguments, we shall introduce first the powerful tool of the Zinn-Justin identity, and then proceed to discuss these two problems in turn. The results we shall derive are well known in equilibrium settings; our goal is to express them in a way that holds even off-equilibrium.

We assume some familiarity with Grassmann calculus. For more details, we refer the reader to the monographs by Berezin [Ber66], DeWitt [DeW84] and Negele and Orland [NegOrl98].

7.1 Path integral quantization of gauge theories – an overview

7.1.1 Gauge theories

Due to the complexity of the subject, it becomes necessary to adopt a highly compressed notation. For starters, we shall do without explicit spacetime dependence. They may be thought of as so many "continuous" indices to be added to the string of discrete indices identifying each field within the theory.

A gauge theory contains "matter" fields ψ such that there are local (unitary) transformations g which are symmetries of the theory. The g's form a nonabelian (simple) group. Infinitesimal transformations may be written as $g = \exp[i\varepsilon]$, where the Hermitian matrix ε may be expanded as a linear combination of "generators" $\varepsilon = \varepsilon^A T_A$. The generators form a closed algebra under commutation

$$[T_A, T_B] = iC^C_{AB} T_C \tag{7.1}$$

The structure constants C^C_{AB} are antisymmetric on A, B and satisfy the Jacobi identity.

Gauge invariance of kinetic terms within the Lagrangian means that derivatives are written in terms of the gauge covariant derivative operator $D_\mu = \partial_\mu - iA_\mu$. The connection $A_\mu = A_{\mu A} T^A$ transforms upon an infinitesimal gauge transformation as

$$A_\mu \to A_\mu + D_\mu \varepsilon \tag{7.2}$$

where

$$D_\mu \varepsilon = \partial_\mu \varepsilon - i[A_\mu, \varepsilon] \tag{7.3}$$

Covariant derivatives do not commute, but their commutator contains no derivatives

$$[D_\mu, D_\nu] = -i F_{\mu\nu} \tag{7.4}$$

where F is the field tensor

$$F_{\mu\nu} = \partial_\mu A_\nu - \partial_\nu A_\mu - i [A_\mu, A_\nu] \tag{7.5}$$

Upon a gauge transformation

$$F_{\mu\nu} \to F_{\mu\nu} + i [\varepsilon, F_{\mu\nu}] \tag{7.6}$$

therefore the object

$$\mathcal{L} = \frac{-1}{4g^2} \mathrm{Tr}\, F^{\mu\nu} F_{\mu\nu} \tag{7.7}$$

is gauge invariant. This is the classical Lagrangian density for the gauge fields, g being the coupling constant. The total action is $S = S_0 + S_m$, where

$$S_0 = \int d^d x \left(\frac{-1}{4g^2} \right) \mathrm{Tr}\, F^{\mu\nu} F_{\mu\nu} \tag{7.8}$$

and S_m is the gauge-invariant action for the matter fields.

We may drop the distinction between gauge and matter fields, and consider a theory described by a string of fields ϕ^α invariant under infinitesimal transformations

$$\delta \phi^\alpha = T^\alpha_A [\phi] \, \varepsilon^A \tag{7.9}$$

The commutation rules are the statement that the commutator of two gauge transforms is also a gauge transform, namely

$$\frac{\delta T^\alpha_A [\phi]}{\delta \phi^\beta} T^\beta_B [\phi] - \frac{\delta T^\alpha_B [\phi]}{\delta \phi^\beta} T^\beta_A [\phi] = T^\alpha_C [\phi] \, C^C_{AB} \tag{7.10}$$

The classical equations of motion read

$$\frac{\delta S}{\delta \phi^\alpha} = 0 \tag{7.11}$$

and because of gauge invariance we must have the identity

$$\frac{\delta S}{\delta \phi^\alpha} T^\alpha_A [\phi] = 0 \tag{7.12}$$

7.1.2 Gauge symmetries and constraints

One important point regarding gauge theories is that a gauge theory is necessarily a constrained theory, and to a large extent vice versa [Dir50, Dir58b, BesKur90, Sun82].

To understand the reason why a gauge theory must have constraints, we observe that the dynamical information on the theory is carried by the canonical variables ϕ^α and their canonical momenta π^α. The information necessary to evolve these degrees of freedom in time are again the ϕ^α and their time derivatives $\dot\phi^\alpha$. Now the existence of gauge freedom means that knowledge of the canonical variables does not determine the evolution (the $\dot\phi^\alpha$ are determined only up to a gauge transformation). Therefore the relationship of the $\dot\phi^\alpha$ to the π^α is many-to-one. This relationship is usually given through a Lagrangian density $\mathcal{L}$ (for example, as in equation (7.7)). In the simplest case the Lagrangian is quadratic in the velocities, and

$$\pi_\alpha = \frac{\partial^2 \mathcal{L}}{\partial \dot\phi^\alpha \partial \dot\phi^\beta} \dot\phi^\beta \tag{7.13}$$

The ambiguity in the $\dot\phi^\alpha$ means that the operator at the left must have null directions

$$\frac{\partial^2 \mathcal{L}}{\partial \dot\phi^\alpha \partial \dot\phi^\beta} T_A^\beta [\phi] = 0 \tag{7.14}$$

We must therefore have a primary constraint

$$T_A^\beta [\phi] \, \pi_\beta = 0 \tag{7.15}$$

and since the primary constraint must hold over time we must also have the secondary constraint

$$\frac{d}{dt} \left[T_A^\beta [\phi] \, \pi_\beta \right] = 0 \tag{7.16}$$

Observe that each gauge freedom engenders two constraints.

Vice versa, assume a theory with fields ϕ and π and Hamiltonian H subject to a constraint $N = 0$. To enforce this constraint introduce a Lagrange multiplier λ and a new Hamiltonian $H + \lambda N$. The momentum Π conjugate to λ vanishes, which is our primary constraint. The secondary constraint is $N = 0$. The canonical equations of motion do not determine the evolution of (ϕ, λ) uniquely; the remaining freedom may be understood as resulting from gauge transformations generated by $\varepsilon N + \dot\varepsilon \Pi$, where ε is the gauge parameter.

7.1.3 The measure of integration

The main point in the path integral approach to the quantization of gauge theories is that the measure of integration is highly nontrivial, since it must count only physical histories of the field, each one being represented by many histories within the path integral. To motivate the measure of integration which does the trick let us look into the computation of the vacuum-to-vacuum amplitude.

In the quantum theory, we expect the vacuum-to-vacuum amplitude to be given by the *in–out* path integral

$$Z = \int D\phi \; e^{iS} \tag{7.17}$$

However this integral counts every history, and that means that each *physical* history is counted many times over. Not surprisingly, it is generally ill defined.

To cure this problem, let f^A be functionals in history space which are not gauge invariant. This means that if we begin from a history ϕ^α confined to the surface $f^A [\phi^\alpha] = 0$, then any infinitesimal gauge transform will take us out of that surface, unless the gauge transform is the trivial one $\varepsilon^A = 0$. In other words,

$$\frac{\delta f^A}{\delta \phi^\alpha} T^\alpha_B [\phi] \varepsilon^B = 0 \Rightarrow \varepsilon^A = 0 \tag{7.18}$$

which requires

$$\text{Det} \left[\frac{\delta f^A}{\delta \phi^\alpha} T^\alpha_B [\phi] \right] \neq 0 \tag{7.19}$$

Now let us call $\phi [\varepsilon]$ the result of applying a gauge transform parameterized by ε to the field configuration ϕ. Then we have the identity (which is an elaborate way of saying that a Dirac delta integrates to 1)

$$\int D\varepsilon \, \text{Det} \left[\frac{\delta f^A}{\delta \phi^\alpha} [\phi [\varepsilon]] T^\alpha_B [\phi [\varepsilon]] \right] \delta \left[f^A [\phi [\varepsilon]] - C^A \right] = 1 \tag{7.20}$$

where C^A may be anything, and by inserting this representation of the identity in the vacuum persistence amplitude we can write

$$Z = \int D\varepsilon \int D\phi \, \text{Det} \left[\frac{\delta f^A}{\delta \phi^\alpha} [\phi [\varepsilon]] T^\alpha_B [\phi [\varepsilon]] \right] \delta \left[f^A [\phi [\varepsilon]] - C^A \right] e^{iS[\phi]} \tag{7.21}$$

Of course, $S[\phi] = S[\phi[\varepsilon]]$, and

$$D\phi [\varepsilon] = D\phi \left\{ 1 + \varepsilon^A \text{Tr} \frac{\delta T^\alpha_A [\phi]}{\delta \phi^\beta} \right\} \tag{7.22}$$

so, provided

$$\text{Tr} \frac{\delta T^\alpha_A [\phi]}{\delta \phi^\beta} = 0 \tag{7.23}$$

we find, up to a constant

$$Z = \int D\phi \, \text{Det} \left[\frac{\delta f^A}{\delta \phi^\alpha} [\phi] T^\alpha_B [\phi] \right] \delta \left[f^A [\phi] - C^A \right] e^{iS[\phi]} \tag{7.24}$$

Since the C^A are arbitrary, any average over different choices will do too. For example, given a suitable metric we may take the Gaussian average

$$\int DC^A \, e^{-(i/2\xi)C^A C_A} \tag{7.25}$$

Integrating over C^A and after a Fourier transform we find

$$Z = \int D\phi Dh_A \, \text{Det} \left[\frac{\delta f^A}{\delta \phi^\alpha} [\phi] T^\alpha_B [\phi] \right] \exp \left\{ i \left[S[\phi] + h_A f^A [\phi] + \frac{\xi}{2} h^A h_A \right] \right\} \tag{7.26}$$

where h_A is the Nakanishi–Lautrup (N-L) field [Nak66] and ξ is the gauge fixing parameter.

We may write the determinant as a functional integral

$$Z = \int D\omega^B D\chi_A D\phi Dh_A \, \exp\left\{ i \left[S[\phi] + h_A f^A[\phi] + \frac{\xi}{2} h^A h_A + i\chi_A \Delta^A \right] \right\}$$

(7.27)

$$\Delta^A = \frac{\delta f^A}{\delta \phi^\alpha}[\phi] \, T_B^\alpha[\phi] \, \omega^B$$

(7.28)

The ω^B, χ_A are *independent* c-number Grassmann variables, namely the ghost and anti-ghost fields, respectively. Following Kugo and Ojima [KugOji79, HatKug80, Oji81], and unlike Weinberg [Wei96], we have included a factor of i in the ghost Lagrangian, which is consistent with taking the ghosts as formally "Hermitian" and demanding the action to be "real." We assign "ghost number" 1 to ω^B, and -1 to χ_A.

7.1.4 BRST invariance

Our goal is to investigate how to formulate a path integral when the initial state is not a vacuum. An important resource in this discussion is the observation that, after breaking the original gauge symmetry by adding gauge fixing conditions and ghosts, the resulting theory has a higher symmetry, the so-called BRST invariance.

We may regard the functional

$$S_{\text{eff}} = S[\phi] + h_A f^A[\phi] + \frac{\xi}{2} h^A h_A + i\chi_A \Delta^A$$

(7.29)

as the action of a new theory, built from the original by adding the N-L, ghost and anti-ghost fields. By construction, this action is not gauge invariant in the original sense. However, let us consider a gauge transform parameterized by $\theta \omega^B$, where θ is an anticommuting "constant," namely

$$\delta\phi^\alpha = \theta T_A^\alpha[\phi] \, \omega^A$$

(7.30)

Observe that, keeping the other fields invariant for the time being

$$\delta f^A[\phi] = \theta \Delta^A$$

(7.31)

$$\delta\Delta^A = \theta f_{,\alpha\beta}^A T_B^\alpha[\phi] \, T_C^\beta[\phi] \, \omega^B \omega^C + \theta f_{,\alpha}^A T_B^\alpha[\phi]_{,\beta} \, T_C^\beta[\phi] \, \omega^C \omega^B$$

(7.32)

Since the ghosts are Grassmann, the first term vanishes, and the second may be written in terms of the commutation relations (7.10), whereby this becomes

$$\delta\Delta^A = \frac{-1}{2} \theta f_{,\alpha}^A T_D^\alpha[\phi] \, C_{BC}^D \omega^B \omega^C$$

(7.33)

These results suggest extending the definition of the transformation to

$$\delta h_A = 0 \tag{7.34}$$

$$\delta \chi_A = i\theta h_A \tag{7.35}$$

$$\delta \omega^D = \frac{1}{2}\theta C^D_{BC}\omega^B \omega^C \tag{7.36}$$

Then S_{eff} is invariant under this "BRST" transformation. Let us define the operator Ω

$$\Omega[X] = \frac{d}{d\theta}\delta X \tag{7.37}$$

The operator Ω increases the "ghost number" by one. It is nilpotent ($\Omega^2 = 0$, see [Wei96]). Also, observe that

$$S_{\text{eff}} = S_0 + \Omega[F] \tag{7.38}$$

where

$$S_0 = S[\phi] \tag{7.39}$$

is BRST invariant, and F is the so-called "gauge fixing fermion"

$$F = -i\chi_A \left\{ f^A[\phi] + \frac{1}{2}\xi h^A \right\} \tag{7.40}$$

Recall that

$$\Omega[F] = -i\left(\Omega[\chi_A] \left\{ f^A[\phi] + \frac{1}{2}\xi h^A \right\} - \chi_A \Omega[f^A[\phi]] \right) \tag{7.41}$$

Also, observe that, *provided* $C^A_{AB} \equiv 0$ the functional volume element is also BRST invariant.

It follows from the above that any gauge fixing dependence (that is, dependence on the choice of the gauge fixing condition f^A, gauge fixing parameter ξ or the metric used to raise indices in the N-L field) may only come from a dependence upon changes in the functional F. Any such change induces a perturbation

$$\delta Z = i \int D\omega^B D\chi_A D\phi D h_A \, \Omega[\delta F] \exp\{iS_{\text{eff}}\} \tag{7.42}$$

Now, call X^r the different fields in the theory. Then

$$\Omega[\delta F] = (-1)^{g_r + 1} \delta F_{,r}\Omega[X^r] \tag{7.43}$$

where g_r is the corresponding ghost number. Integrating by parts (see [GoPaSa95]), and *provided the surface term vanishes*, we get

$$\delta Z = i \int D\omega^B D\chi_A D\phi D h_A \, \delta F \left\{ \frac{\delta}{\delta X^r} \exp\{iS_{\text{eff}}\} \Omega[X^r] \right\} \tag{7.44}$$

But the brackets vanish, because of BRST invariance of S_{eff} and because $\Omega\left[X^r\right]$ is divergence-free. Therefore the *physicality condition* is that the flux of any vector pointing in the direction of $\Omega\left[X^r\right]$ over the boundary of the space of field configurations must vanish.

This shows by the way that F could be any expression of ghost number -1, since S_{eff} must have ghost number zero.

7.1.5 Physical states

BRST invariance allows us to give a simple criterion for physical states. In this section, we shall consider the concrete case where $\phi^\alpha = A_\mu^A$, $f_{,\alpha}^A = \delta_B^A \partial_\mu$ and $T_B^\alpha = \delta_B^A \partial_\mu + C_{CB}^A A_\mu^C$. We can write S_{eff} explicitly:

$$
S_{\text{eff}} = \int d^4x \left\{ \frac{-1}{4g^2} F^{A\mu\nu} F_{A\mu\nu} - \partial_\mu h_A A^{\mu A} + \frac{\xi}{2} h^A h_A \right.
$$

$$
\left. - i\partial_\mu \chi_A \left[\delta_B^A \partial_\mu + i C_{CB}^A A_\mu^C \right] \omega^B \right\} \tag{7.45}
$$

If we take A_{Aa} $(a = 1, 2, 3)$, h_A, χ_A and ω^A as canonical variables, then we may identify the corresponding momenta [KugOji79, HatKug80, Oji81]

$$
p_\phi^{Aa} = \frac{1}{g^2} F^{Aa0}
$$

$$
p_h^A = -A^{A0}
$$

$$
p_\chi^A = -i \left[\delta_B^A \partial_0 + C_{CB}^A A_0^C \right] \omega^B
$$

$$
p_{\omega A} = i\partial_0 \chi_A \tag{7.46}
$$

and impose the ETCCRs

$$
\left[p_{Xr}, X^s \right]_{\mp} = -i\delta_s^r \tag{7.47}
$$

where we use anticommutators for ghost fields and momenta, and commutators for all other cases.

The BRST invariance of S_{eff} implies the conservation of the Noether current

$$
j^\mu = \Omega\left[X^r\right] \frac{\delta L_{\text{eff}}}{\delta \partial_\mu X^r} \tag{7.48}
$$

We define the BRST charge as

$$
\Omega = \int d^3x \, \Omega\left[X^r\right] p_{Xr} \tag{7.49}
$$

This is the generator of BRST transforms, since

$$
\delta X^r = \theta \Omega\left[X^r\right] = i\left[\theta\Omega, X^r\right] \tag{7.50}
$$

(Since θ is Grassmann, we use commutators throughout.) Then $\Omega^2 = 0$.

S_{eff} is also invariant upon the scale transformation

$$\omega^B \to e^\lambda \omega^B, \qquad \chi^B \to e^{-\lambda}\chi^B \tag{7.51}$$

The corresponding generator

$$Q = \int d^3x \left\{ \omega^B p_{\omega B} - \chi_A p^A_\chi \right\} \tag{7.52}$$

is the *ghost charge*. Ghost charge is bosonic, so $[Q, Q] = 0$. On the other hand, Ω has ghost charge 1, so

$$i[Q, \Omega] = \Omega \tag{7.53}$$

Both Q and $\theta\Omega$ commute with the effective Hamiltonian.

We say that a state $|\alpha\rangle$ is BRST closed if $\Omega |\alpha\rangle = 0$ and BRST exact if there is a $|\beta\rangle$ such that $|\alpha\rangle = \Omega |\beta\rangle$. Since $\Omega^2 = 0$, an exact state is necessarily closed but there may be closed states that are not exact. Observables are BRST invariant, and so they commute with $\theta\Omega$. Physical states are also BRST invariant, therefore annihilated by Ω. Physical states differing by a BRST transform are physically indistinguishable, in the sense that they lead to the same matrix elements for all observables. We therefore introduce an equivalence relation among states, $|\alpha\rangle \approx |\beta\rangle$ if $|\alpha\rangle - |\beta\rangle$ is BRST exact. A physical state is a representative of an equivalence class of states which are closed but not exact.

7.1.6 Initial conditions for nonvacuum states

We shall now use the above characterization of physical states to introduce a simple way (due to Hata and Kugo) of introducing initial conditions for nonvacuum states in the path integral.

We need one more result from BRST theory, namely, there is an operator R such that

(a) if $|\alpha\rangle$ is exact, $|\alpha\rangle = \Omega |\theta\rangle$, then $R|\alpha\rangle \approx |\theta\rangle$;
(b) if $|\alpha\rangle$ is not exact, then $R|\alpha\rangle \approx 0$.

Given such an operator, then the projector P' orthogonal to the space of physical states has the form $P' = \{\Omega, R\}$. Indeed, if $|\alpha\rangle$ is physical, then it is closed (so $R\Omega |\alpha\rangle = 0$) but not exact (so $\Omega R |\alpha\rangle = 0$). On the other hand, if $|\alpha\rangle$ is not physical, it is either exact or not closed. If $|\alpha\rangle$ is exact, then $R\Omega |\alpha\rangle = 0$ but $\Omega R |\alpha\rangle \approx |\alpha\rangle$. If $|\alpha\rangle$ is not closed, then $\Omega R |\alpha\rangle = 0$ but $R\Omega |\alpha\rangle \approx |\alpha\rangle$.

We may now deal with the construction of statistical operators in gauge theories. In principle, a physical statistical operator should shield nonzero probabilities only for physical states, and so it should satisfy $\rho = P\rho = \rho P$, where P projects over the space of physical states, $P = 1 - P'$. This is a much stronger requirement than BRST invariance $[\Omega, \rho] = 0$. So, given a BRST invariant density matrix ρ, we ought to define the physical expectation value of any (BRST invariant) observable C as

$$\langle C \rangle_{\text{phys}} = \text{Tr}\left[P \rho C \right] \tag{7.54}$$

However, Kugo and Hata [KugOji79, HatKug80, Oji81] (KH) have shown that the same expectation values may be obtained by using the statistical operator $e^{-\pi Q}\rho$. The key to the argument is that the commutation relation $[iQ, \Omega] = \Omega$ implies that, if $|N\rangle$ is an eigenstate of iQ with eigenvalue N, then $\Omega|N\rangle$ has eigenvalue $N + 1$. It follows that $\{e^{-\pi Q}, \Omega\} = 0$, since $e^{-\pi Q} = e^{i\pi(iQ)}$. We then find that, for any BRST invariant observable C

$$\langle C \rangle_{\text{phys}} = \text{Tr}\left[P \rho C \right] = \text{Tr}\left[P e^{-\pi Q} \rho C \right] = \text{Tr}\left[e^{-\pi Q}\rho C \right] - \text{Tr}\left[\{\Omega, R\} e^{-\pi Q}\rho C \right] \tag{7.55}$$

We must show that the second term vanishes, and this follows from $\{e^{-\pi Q}, \Omega\} = 0$ and $[\Omega, \rho C] = 0$.

This suggests we define the expectation value $\langle C \rangle$ of any observable as $\langle C \rangle = \text{Tr}\left[e^{-\pi Q}\rho C \right]$. Of course, this agrees with the physical expectation value only if C is BRST invariant. For example, the partition function computed from $e^{-\pi Q}\rho$ agrees with the partition function defined by tracing only over physical states, but the generating functionals obtained by adding sources coupled to non-BRST invariant operators will in general be different.

The advantages of the Kugo–Hata ansatz are clearly seen by considering the form of the KMS theorem appropriate to the ghost propagator. Let us define

$$G_{AB}^{ab}(x, x') = \left\langle P\left[\chi_A^a(x)\, \omega_B^b(x') \right] \right\rangle \tag{7.56}$$

where P is the usual (CTP)-ordering operator. Then

$$G_{AB}^{21}(x, x') = \langle \chi_A(x)\, \omega_B(x') \rangle \tag{7.57}$$

$$G_{AB}^{12}(x, x') = -\langle \omega_B(x')\, \chi_A(x) \rangle \tag{7.58}$$

(observe the sign change, owing to the anticommuting character of the ghost fields). The Jordan propagator is defined as $G = G^{21} - G^{12}$.

Had we omitted the KH $e^{-\pi Q}$ factor, we would reason, given $\rho = e^{-\beta H}$,

$$
\begin{aligned}
G_{AB}^{21}(x, x') &\approx \text{Tr}\left[e^{-\beta H}\chi_A(x)\, \omega_B(x') \right] \\
&= \text{Tr}\left[\chi_A(x + i\beta)\, e^{-\beta H}\omega_B(x') \right] \\
&= -G_{AB}^{12}(x + i\beta, x')
\end{aligned}
\tag{7.59}
$$

Therefore $G_{AB}^{21}(\omega) = -e^{\beta\omega} G_{AB}^{12}(\omega)$, leading to a Fermi–Dirac form of the thermal propagators. This reasoning is incorrect. The proper way is

$$G_{AB}^{21}(x, x') = \text{Tr}\left[e^{-\pi Q}\chi_A(x + i\beta)\, e^{-\beta H}\omega_B(x') \right] = G_{AB}^{12}(x + i\beta, x') \tag{7.60}$$

So $G_{AB}^{21}(\omega) = e^{\beta\omega} G_{AB}^{12}(\omega)$, which leads to the Bose–Einstein form.

The KH factor does not appear explicitly in the path integral representation; it only changes the boundary conditions on ghost fields from anti-periodic to periodic.

We conclude that in this formalism, unphysical degrees of freedom and ghosts get statistical corrections, both being of the Bose–Einstein form, in spite of the ghosts being fermions (for which reason ghost loops do get a minus sign). For an alternative formulation, see [LanReb92, LanReb93].

7.2 The 2PI formalism applied to gauge theories

7.2.1 The 2PI effective action

We can now move towards our real goal, namely, the application of the 2PI CTP formalism to gauge theories. We shall proceed with a fair amount of generality, only assuming that the gauge condition is linear, and that the gauge generators satisfy $T_A^\alpha[\phi] = T_{0A}^\alpha + T_{1A\beta}^\alpha \phi^\beta$. We shall develop the basic formulae in some detail, emphasizing the subtleties associated with having both normal and Grassmann degrees of freedom in the same theory.

The classical action is given by equation (7.29). To this we add sources coupled to the individual degrees of freedom and also to their products

$$X^r J_r + \frac{1}{2} X^r \mathbf{K}_{rs} X^s = j_\alpha x^\alpha + \theta^u \lambda_u + \frac{1}{2} \kappa_{\alpha\beta} x^\alpha x^\beta + \frac{1}{2} \sigma_{uv} \theta^u \theta^v + \theta^u \psi_{u\alpha} x^\alpha \tag{7.61}$$

where x^α represents the bosonic degrees of freedom (ϕ, h) and θ the Grassmann ones (ω, χ), and we introduce the definition $\mathbf{K}_{\alpha u} = -\mathbf{K}_{u\alpha}$. Observe that j, κ and σ are normal, while λ and ψ are Grassmann; σ is antisymmetric.

We therefore define the generating functional

$$e^{iW} = \int DX^r \exp\left\{ i \left[S_{\text{eff}} + X^r J_r + \frac{1}{2} X^r \mathbf{K}_{rs} X^s \right] \right\} \tag{7.62}$$

The information about the initial state is implicit in the integration measure and will reappear only as an initial condition on the equations of motion. We find

$$W \frac{\overleftarrow{\delta}}{\delta J_r} = \bar{X}^r \tag{7.63}$$

$$W \frac{\overleftarrow{\delta}}{\delta \mathbf{K}_{rs}} = \frac{\theta^{sr} \theta^s}{2} \left[\bar{X}^r \bar{X}^s + \mathbf{G}^{rs} \right] \tag{7.64}$$

where we introduce the bookkeeping device $\theta^r = (-1)^{q_r}$, where q_r is the ghost charge of the corresponding field, and $\theta^{rs} = (-1)^{q_r q_s}$.

We define the Legendre transform

$$\Gamma = W - \bar{X}^r J_r - \frac{1}{2} \bar{X}^r \mathbf{K}_{rs} \bar{X}^s - \frac{\theta^{sr} \theta^s}{2} \mathbf{G}^{rs} \mathbf{K}_{rs} \tag{7.65}$$

whereby

$$\frac{\delta}{\delta X^r}\Gamma = -J_r - \frac{1}{2}\mathbf{K}_{rs}\bar{X}^s - \frac{1}{2}\theta^r\bar{X}^s\mathbf{K}_{sr} \qquad (7.66)$$

Now observe that $\mathbf{K}_{sr} = \theta^r\theta^s\theta^{rs}\mathbf{K}_{rs}$. In the end

$$\frac{\delta}{\delta X^r}\Gamma = -J_r - \mathbf{K}_{rs}\bar{X}^s \qquad (7.67)$$

$$\frac{\delta}{\delta \mathbf{G}^{rs}}\Gamma = -\frac{\theta^{sr}\theta^s}{2}\mathbf{K}_{rs} \qquad (7.68)$$

In order to evaluate the 2PIEA, we make the ansatz

$$\Gamma = \bar{S}\left[\bar{X}^r\right] + \frac{1}{2}\theta^{sr}\theta^s\mathbf{G}^{rs}\mathbf{S}_{rs} - \frac{i}{2}\ln\mathrm{sdet}\left[\mathbf{G}^{rs}\right] + \Gamma_2 - \frac{i}{2}\theta^s\mathbf{G}^{rs}\mathbf{G}^{-1}_{Rsr} \qquad (7.69)$$

where

$$\mathbf{S}_{rs} = \left[\overrightarrow{\frac{\delta}{\delta\bar{X}^r}}\bar{S}\right]\overleftarrow{\frac{\delta}{\delta\bar{X}^s}} \qquad (7.70)$$

and $\bar{S}$ is the classical action (7.29), evaluated at the background fields. The generating functional Γ_2 is the sum of 2PI vacuum bubbles in a theory with free action $i\mathbf{G}^{-1}_{Lrs}$ and interacting terms coming from the cubic and quartic terms in the development of $\bar{S}$ around the mean fields. In spite of appearances, the new term $\theta^s\mathbf{G}^{rs}\mathbf{G}^{-1}_{Rsr}$ is a constant. It may therefore be discarded.

7.2.2 The 2PI Schwinger–Dyson equations

Let us now investigate the 2PI Schwinger–Dyson equations

$$\frac{\delta}{\delta X^r}\Gamma = 0$$

$$\frac{\delta}{\delta \mathbf{G}^{rs}}\Gamma = 0 \qquad (7.71)$$

From equation (7.69) we get

$$\frac{\delta}{\delta X^r}\bar{S}\left[\bar{X}^r\right] + \frac{1}{2}\theta^{pq}\theta^q\theta^{rp}\theta^{rq}\mathbf{G}^{pq}\frac{\delta}{\delta X^r}\mathbf{S}_{pq} + \frac{\delta}{\delta X^r}\Gamma_2 = 0$$

$$\theta^{sr}\theta^s\mathbf{S}_{rs} - i\theta^r\theta^{rs}\left(\mathbf{G}^{-1}_R\right)_{rs} + 2\frac{\delta}{\delta \mathbf{G}^{rs}}\Gamma_2 = 0 \qquad (7.72)$$

The second set of equations may be rewritten as

$$\theta^{sr}\theta^s\mathbf{S}_{rs} - i\theta^r\left(\mathbf{G}^{-1}_L\right)_{sr} + 2\frac{\delta}{\delta \mathbf{G}^{rs}}\Gamma_2 = 0 \qquad (7.73)$$

and finally as

$$\mathbf{S}_{rs} - i\left(\mathbf{G}^{-1}_L\right)_{rs} + 2\theta^{sr}\theta^s\frac{\delta}{\delta \mathbf{G}^{rs}}\Gamma_2 = 0 \qquad (7.74)$$

The classical action is given by equation (7.29). If we expand $X^r = \bar{X}^r + \delta X^r$, then the quadratic terms are

$$\bar{S}^{(2)} = S_0^{(2)}\left[\bar{\phi}, \delta\phi\right] + \delta h_A f_\alpha^A \delta\phi^\alpha + \frac{\xi}{2}\delta h^A \delta h_A + i\delta\chi_A f_\alpha^A T_B^\alpha\left[\bar{\phi}\right]\delta\omega^B$$
$$+ i\bar{\chi}_A f_\alpha^A T_{1B\beta}^\alpha \delta\phi^\beta \delta\omega^B + i\delta\chi_A f_\alpha^A T_{1B\beta}^\alpha \delta\phi^\beta \bar{\omega}^B \tag{7.75}$$

The cubic and quartic terms are

$$\bar{S}^{(3+)} = S_0^{(3+)}\left[\bar{\phi}, \delta\phi\right] + i\delta\chi_A f_\alpha^A T_{1B\beta}^\alpha \delta\phi^\beta \delta\omega^B \tag{7.76}$$

7.2.3 The reduced 2PI effective action

The introduction of the Nakanishi–Lautrup (N-L) field h has been useful to obtain a simple definition of the BRST transformation, but since it only appears quadratically in the action, there are no h field lines in Γ_2. To take advantage of this fact, it is convenient *not* to couple sources to the h field. In this way, Γ_2 is independent of the h field, and the respective variations are exact, namely

$$f_\alpha^A \bar{\phi}^\alpha + \xi \bar{h}^A = 0$$
$$\xi\delta_{AB} - i\left[\mathbf{G}_L^{-1}\right]_{hhAB} = 0$$
$$f_\alpha^A - i\left[\mathbf{G}_L^{-1}\right]_{h\phi\alpha}^A = 0$$
$$\left[\mathbf{G}_L^{-1}\right]_{h\omega B}^A = \left[\mathbf{G}_L^{-1}\right]_{h\chi B}^A = 0 \tag{7.77}$$

where a $L(R)$ superscript denotes a left (right) inverse. Moreover, from the N-L field being Gaussian

$$G_{hX}^{Ar} = \frac{-1}{\xi}f_\beta^A G_{\phi X}^{\beta r}, \qquad X = \phi, \chi, \omega \tag{7.78}$$

and

$$G_{hhA}^C = \frac{1}{\xi}\left[-f_{A\beta}G_{\phi h}^{\beta C} + i\delta_A^C\right] = \frac{1}{\xi}\left[\frac{1}{\xi}f_{A\beta}f_\gamma^C G_{\phi\phi}^{\beta\gamma} + i\delta_A^C\right] \tag{7.79}$$

We could use these formulae to actually eliminate the N-L field from the 2PIEA, thus obtaining a reduced effective action.

Let us explore the solutions to the equations of motion where all fields with nonzero ghost number vanish, i.e.

$$\bar{\omega} = \bar{\chi} = G_{\omega\omega} = G_{\chi\chi} = G_{\omega\phi} = G_{\omega h} = G_{\chi\phi} = G_{\chi h} = 0 \tag{7.80}$$

Since the effective action itself has zero ghost number, it cannot contain terms linear on any of the above, and therefore this condition is consistent with the equations of motion.

Given these conditions, we have, besides the equations determining the h propagators, the further equations

$$\left[\mathbf{G}^{-1}\right]_{\phi\phi\alpha\beta} G_{\phi\phi}^{\beta\gamma} + \left[\mathbf{G}^{-1}\right]_{\phi h\alpha B} G_{h\phi}^{B\gamma} = \delta_\alpha^\gamma$$

$$\left[\mathbf{G}^{-1}\right]_{\phi\phi\alpha\beta} G_{\phi h}^{\beta C} + \left[\mathbf{G}^{-1}\right]_{\phi h\alpha B} G_{hh}^{BC} = 0 \qquad (7.81)$$

The inverse propagators may be read off the variation of the 2PIEA, leading to the equation for the gluon propagator

$$S_{c,\alpha\beta} - \frac{1}{\xi} f_{B\alpha} f_\beta^B - i \left[G_{\phi\phi}^{-1}\right]_{\alpha\beta} + 2\frac{\delta\Gamma_2}{\delta G_{\phi\phi}^{\alpha\beta}} = 0 \qquad (7.82)$$

The other nontrivial equation is

$$-i f_\alpha^{A'} T_B^\alpha \left[\bar{\phi}\right] + i \left[\mathbf{G}_L^{-1}\right]_{\omega\chi B}^{A'} + 2\frac{\delta\Gamma_2}{\delta G_{\chi\omega A}^B} = 0 \qquad (7.83)$$

In deriving this equation we must consider $G_{\chi\omega A}^B$ and $G_{\omega\chi A}^B$ as independent quantities.

7.3 Gauge dependence and propagator structure

7.3.1 The Zinn-Justin equation

As we have noted in the introduction to this chapter, the most distinctive feature of gauge theories as opposed to "normal" ones is the existence of relationships among propagators of different orders, the so-called Takahashi–Ward or Slavnov–Taylor identities. The powerful BRST formulation allows us to derive them all from a single master identity, the so-called Zinn-Justin (Z-J) equation, which we shall now present.

The key observation is that under a BRST transform within the path integral which defines the generating functional (7.62), only the source terms are really transformed. Therefore

$$\langle \Omega\left[X^r\right]\rangle J_r + \frac{1}{2}\theta^{rs}\theta^s \langle \Omega\left[X^r X^s\right]\rangle \mathbf{K}_{rs} = 0 \qquad (7.84)$$

The sources are eliminated in terms of derivatives of the 2PIEA (cf. equation (7.68)), leading to

$$0 = \langle \Omega\left[X^r\right]\rangle \frac{\delta\Gamma}{\delta\bar{X}^r} + \left[\langle \Omega\left[X^r X^s\right]\rangle - 2\langle \Omega\left[X^r\right]\rangle \bar{X}^s\right]\left[\frac{\delta}{\delta\mathbf{G}^{rs}}\Gamma\right] \qquad (7.85)$$

Taking derivatives of this identity we obtain the desired relationships.

In the remainder of this section we shall give a simple example of how a concrete identity may be derived from equation (7.85).

For simplicity, we shall assume that all background fields vanish. Since the Z-J operator has ghost number 1, it makes no sense to assume that all quantities with nonzero ghost number vanish, as we have done in the previous section.

However, we may still "turn on" these quantities one by one, and thus obtain partial Z-J identities. For example, we get three identities relating quantities with zero ghost number by requiring that the coefficients of $\bar{\omega}$ and $G_{\omega\phi}$ vanish (we shall not investigate the first, as we are assuming no nonzero backgrounds, and we are working throughout with the reduced 2PIEA). This means that we may still set

$$\bar{\omega} = \bar{\chi} = G_{\omega\omega} = G_{\chi\chi} = G_{\chi\phi} = G_{\chi h} = 0 \tag{7.86}$$

and retain only terms linear in $G_{\omega\phi}$ and $G_{\omega h}$. In this approximation, terms with ghost number neither 0 or 1 must vanish identically, so

$$\langle \Omega \left[\omega^D \right] \rangle = \langle \Omega \left[\omega^D \omega^E \right] \rangle = \langle \Omega \left[h_A \omega^D \right] \rangle = \langle \Omega \left[\phi^\alpha \omega^D \right] \rangle = \langle \Omega \left[\chi_A \chi_B \right] \rangle = 0 \tag{7.87}$$

and

$$\frac{\delta\Gamma}{\delta\omega^D} = \frac{\delta\Gamma}{\delta G^{\alpha D}_{\phi\omega}} = \frac{\delta\Gamma}{\delta G^D_{h\omega A}} = \frac{\delta\Gamma}{\delta G^{DE}_{\omega\omega}} = \frac{\delta\Gamma}{\delta G_{\chi\chi AB}} = 0 \tag{7.88}$$

Also, since there are no preferred directions in gauge space, objects with a single gauge index must vanish out of symmetry, and therefore

$$\langle \Omega \left[\phi^\alpha \right] \rangle = \langle \Omega \left[h^A \right] \rangle = \langle \Omega \left[\chi_A \right] \rangle = \frac{\delta\Gamma}{\delta\bar{\phi}^\alpha} \left[0 \right] = 0 \tag{7.89}$$

Finally, observe that at zero external sources,

$$\frac{\delta\Gamma}{\delta h^A} = \frac{\delta\Gamma}{\delta G^\alpha_{\phi h B}} = \frac{\delta\Gamma}{\delta G_{h\chi AB}} = \frac{\delta\Gamma}{\delta G_{hhAB}} \equiv 0 \tag{7.90}$$

In other words, from the terms in equation (7.85) we keep the terms in $\phi\phi$, $\phi\chi$, and $\chi\omega$ only.

Equation (7.85) must vanish at the physical point, since each coefficient vanishes. What is remarkable is that it vanishes identically, even if $G^{\alpha A}_{\phi\omega} \neq 0$. Now $\delta\Gamma/\delta G^{\alpha\beta}_{\phi\phi}$ and $\delta\Gamma/\delta G^B_{\omega\chi A}$ have ghost number zero, and therefore contain no terms linear in $G^{\alpha A}_{\phi\omega}$. We conclude that, to linear order in $G^{\alpha A}_{\phi\omega}$, we may write

$$\langle \Omega \left[\phi^\alpha \chi_A \right] \rangle \frac{\delta\Gamma}{\delta G^{\alpha A}_{\phi\chi}} \approx 0 \tag{7.91}$$

Here $\approx$ means up to terms proportional to the equations of motion. Now

$$\frac{\delta\Gamma}{\delta G^{\alpha A}_{\phi\chi}} = \frac{-i}{2} \left[\mathbf{G}^{-1}_L \right]_{\phi\chi\alpha A} \tag{7.92}$$

Expanding the identity

$$\left[\mathbf{G}^{-1}_L \right]_{\phi X \alpha r} G^r_{X\omega B} = 0 \tag{7.93}$$

and using equations (7.77), (7.78), (7.79) and (7.81)

$$\left[\mathbf{G}_L^{-1}\right]_{\phi\chi\alpha A} = -\left[\left[\mathbf{G}_L^{-1}\right]_{\phi\phi\alpha\beta} + \frac{i}{\xi}f_{\alpha C}f_{\beta}^{C}\right]G_{\omega\phi}^{B\beta}\left[\mathbf{G}_R^{-1}\right]_{\chi\omega AB}$$

$$\approx -\left[G_{\phi\phi}^{-1}\right]_{\alpha\beta}G_{\omega\phi}^{C\beta}\left[\mathbf{G}_R^{-1}\right]_{\chi\omega AC} \tag{7.94}$$

Since $G_{\omega\phi}^{C\beta}$ can be anything and $\left[G_{\phi\phi}^{-1}\right]_{\alpha\beta}$ and $\left[\mathbf{G}_R^{-1}\right]_{\chi\omega AC}$ are regular, $\langle\Omega\left[\phi^{\alpha}\chi_A\right]\rangle$ must vanish:

$$\langle\Omega\left[\phi^{\alpha}\chi_A\right]\rangle = -\langle\chi_A\Omega\left[\phi^{\alpha}\right]\rangle - \frac{i}{\xi}f_{\beta}^{A}G_{\phi\phi}^{\beta\alpha} = 0 \tag{7.95}$$

The point is that this identity links the gluon and ghost propagators to a gluon–ghost–ghost vertex. To see this, observe that

$$\Omega\left[\phi^{\alpha}\chi_A\right] = T_B^{\alpha}\left[\phi\right]\omega^{B}\chi_A + i\phi^{\alpha}h_A \tag{7.96}$$

involves cubic terms so the missing expectation value may be written as

$$\langle\Omega\left[\phi^{\alpha}\chi_A\right]\rangle = G_{\omega\chi A}^{B}T_B^{\alpha}\left[0\right] - \frac{i}{\xi}G_{\phi\phi}^{\alpha\beta}f_{AB} + T_{B,\gamma}^{\alpha}\left[0\right]\langle\phi^{\gamma}\omega^{B}\chi_A\rangle \tag{7.97}$$

Below we shall use equation (7.95) to investigate the gauge dependence and structure of the propagators.

7.3.2 Gauge dependence of the propagators

There are two issues central to gauge theories with no analog in "normal" theories, namely, to what extent the results of the theory depend on all the machinery associated with the gauge fixing procedure, and second, how the Zinn-Justin identity may be exploited to glean certain facts about the theory over and beyond actual computation. We shall begin by discussing the first issue, taking as case in point how the propagators depend on the gauge fixing conditions.

To investigate the gauge dependence of the 2PIEA, recall equations (7.38), (7.39) and (7.40). Consider a change δF in the gauge fermion F (cf. equation (7.40))

$$\delta F = -i\chi_A\left\{\delta f^A\left[\phi\right] + \frac{1}{2}\delta\xi h^A\right\} \tag{7.98}$$

Holding the background fields constant, we get

$$\delta\Gamma|_{\bar{X}^r,\mathbf{G}^{rs}} = \delta W|_{J_r,\mathbf{K}_{rs}} \tag{7.99}$$

The variation of the generating functional is computed as in equations (7.42), (7.43) and (7.44). However, now the "action" is not BRST invariant, because it includes the source terms, and we get a nontrivial result

$$\delta\Gamma|_{\bar{X}^r,\mathbf{G}^{rs}} = i\left\{\langle\delta F\Omega\left[X^r\right]\rangle J_r + \frac{1}{2}\theta^{rs}\theta^{s}\langle\delta F\Omega\left[X^rX^s\right]\rangle\mathbf{K}_{rs}\right\} \tag{7.100}$$

Again we use equation (7.68) to get

$$\delta\Gamma|_{\bar{X}^r,\mathbf{G}^{rs}} = (-i)\left\{ \langle\delta F\Omega\,[X^r]\rangle \frac{\delta\Gamma}{\delta\bar{X}^r} \right.$$
$$\left. + \left[\langle\delta F\Omega\,[X^r X^s]\rangle - 2\langle\delta F\Omega\,[X^r]\rangle\,\bar{X}^s\right]\left[\frac{\delta}{\delta\mathbf{G}^{rs}}\Gamma\right] \right\} \quad (7.101)$$

As before, we shall assume that all background fields vanish and that at such a point $\Gamma_{,r}$ vanishes identically, so the above expression simplifies to

$$\delta\Gamma|_{\bar{X}^r,\mathbf{G}^{rs}} = -Y^{rs}\frac{\delta}{\delta\mathbf{G}^{rs}}\Gamma; \qquad Y^{rs} = i\langle\delta F\Omega\,[X^r X^s]\rangle \quad (7.102)$$

At the physical point, the Schwinger–Dyson equations now read

$$\frac{\delta}{\delta\mathbf{G}^{tu}}\Gamma - Y^{rs}\frac{\delta^2}{\delta\mathbf{G}^{tu}\delta\mathbf{G}^{rs}}\Gamma = 0 \quad (7.103)$$

Of course, the solution is now $\mathbf{G}^{tu} + \delta\mathbf{G}^{tu}$, so

$$\left(\frac{\delta^2}{\delta\mathbf{G}^{tu}\delta\mathbf{G}^{rs}}\Gamma\right)[\delta\mathbf{G}^{rs} - Y^{rs}] = 0 \quad (7.104)$$

Since the Hessian is supposed to be invertible, we must have $\delta\mathbf{G}^{rs} = Y^{rs}$.

Let us also assume that all propagators with nonzero ghost number vanish. Then

$$\delta G_{\phi\phi}^{\alpha\beta} = i\langle\delta F\Omega\,[\phi^\alpha\phi^\beta]\rangle$$
$$= \left\langle \chi_A\left\{\delta f^A\,[\phi] + \frac{1}{2}\delta\xi h^A\right\}\left(T_C^\alpha\,[\phi]\,\omega^C\phi^\beta + (\alpha\leftrightarrow\beta)\right)\right\rangle \quad (7.105)$$

Assume δf^A is also linear and use the Gaussianity of h^A to get

$$\delta G_{\phi\phi}^{\alpha\beta} = \left[\delta f_\gamma^A - \frac{\delta\xi}{2\xi}f_\gamma^A\right]\langle\chi_A\phi^\gamma\left(T_C^\alpha\,[\phi]\,\omega^C\phi^\beta + (\alpha\leftrightarrow\beta)\right)\rangle \quad (7.106)$$

To lowest order, we find

$$\langle\chi_A\phi^\gamma\left(T_C^\alpha\,[\phi]\,\omega^C\phi^\beta + (\alpha\leftrightarrow\beta)\right)\rangle \sim G_{\phi\phi}^{\beta\gamma}\langle\chi_A\Omega\,[\phi^\alpha]\rangle + (\alpha\leftrightarrow\beta) \quad (7.107)$$

Now recall equation (7.95)

$$\delta G_{\phi\phi}^{\alpha\beta} = \frac{(-i)}{\xi}\left[\delta f_\gamma^A - \frac{\delta\xi}{2\xi}f_\gamma^A\right]f_\delta^A G_{\phi\phi}^{\delta\alpha}G_{\phi\phi}^{\beta\gamma} + (\alpha\leftrightarrow\beta) \quad (7.108)$$

or else

$$\delta G_{\phi\phi\alpha\beta}^{-1} = \frac{i}{\xi}f_\alpha^A\left[\delta f_\beta^A - \frac{\delta\xi}{2\xi}f_\beta^A\right] + (\alpha\leftrightarrow\beta) \quad (7.109)$$

This is the result we wanted to show. We will use it below to analyze the structure of the propagators.

7.3.3 Transverse and longitudinal gluon propagators

We now turn to the second issue outlined above, namely, how we can turn the gauge dependence identities around to investigate the structure of the theory. For simplicity, we shall consider only a pure (nonabelian) Yang–Mills theory to two-loop accuracy.

To this order, variation of the 2PIEA yields the equation for the ghost propagator

$$\left[\mathbf{G}_L^{-1}\right]_{\omega\chi B}^A = f_\alpha^A \left[T_B^\alpha\left[\bar{\phi}\right] - f_{\alpha'}^C T_{1B\beta}^{\alpha'} T_{1B'\beta'}^\alpha G_{\phi\phi}^{\beta\beta'} G_{\omega\chi C}^{B'}\right] \tag{7.110}$$

Multiplying on the right by $G_{\omega\chi C}^B$ we get

$$f_\alpha^A L_B^\alpha = \delta_B^A \tag{7.111}$$

where

$$L_C^\lambda = \left[T_B^\lambda\left[\bar{\phi}\right] - f_{\alpha'}^A T_{1B\gamma}^{\alpha'} T_{1B'\beta'}^\lambda G_{\phi\phi}^{\gamma\beta'} G_{\omega\chi A}^{B'}\right] G_{\omega\chi C}^B \tag{7.112}$$

This suggests defining

$$P_{L\beta}^\alpha = L_C^\alpha f_\beta^C \tag{7.113}$$

which is a projection operator

$$P_{L\beta}^\alpha P_{L\gamma}^\beta = P_{L\gamma}^\alpha \tag{7.114}$$

Now let us return to equation (7.95), which is a consequence of the Z-J identity (7.85). An explicit calculation to two-loop accuracy yields

$$L_C^\alpha = -\langle\chi_C\Omega\left[\phi^\alpha\right]\rangle = \frac{i}{\xi} f_{C\beta} G_{\phi\phi}^{\beta\alpha} \tag{7.115}$$

Multiply again by L_δ^C to get

$$P_{L\delta\beta} G_{\phi\phi}^{\beta\alpha} = -i\xi L_\delta^C L_C^\alpha \tag{7.116}$$

Therefore we have a decomposition of the gluon propagators into "transverse" and "longitudinal" parts

$$G_{\phi\phi}^{\lambda\beta} = G_{T\phi\phi}^{\lambda\beta} - i\xi L_C^\lambda L^{C\beta}, \qquad P_{L\lambda}^\gamma G_{T\phi\phi}^{\lambda\beta} = 0 \tag{7.117}$$

The corresponding decomposition for the inverse propagators is

$$\left[G_{\phi\phi}^{-1}\right]_{\alpha\lambda} = \left[G_{T\phi\phi}^{-1}\right]_{\alpha\lambda} + \frac{i}{\xi} f_\lambda^A f_{A\alpha} \tag{7.118}$$

Comparing with equation (7.109) we see that the transverse part $\left[G_{T\phi\phi}^{-1}\right]_{\gamma\lambda}$ is gauge-fixing independent to two-loop order. This is the desired result, laying out the gauge dependence of the propagators in its most explicit form. Of course, the projector $P_{L\beta}^\alpha$ is just the generalization of the usual $k^\mu k^\nu / k^2$ to a nonequilibrium setting.

8

Dissipation and noise in mean field dynamics

In Chapter 6 we presented the main computational schemes to derive the dynamical laws for the mean field, including the back-reaction from quantum fluctuations. These equations may be derived from the variation of the CTPEA. The result of this approach is a semiclassical theory of a c-number condensate interacting with a quantized fluctuation field.

This approach developed at this level of sophistication is limited as it offers no description of the fluctuations themselves. In most applications the magnitude of the fluctuations can be comparable and at times dominates the effects of the mean field in the semiclassical description. One possible way to incorporate fluctuations is to use the 2PI formalism, where the propagators describing the fluctuations are considered as dynamical variables evolving along with the mean fields.

In this chapter we shall explore a different strategy, which is to allow for a stochastic component in the mean field. This component arises from both the uncertainty of the initial configuration of the mean field, and from the fluctuations in the back-reaction from the quantized excitations. Both sources of randomness combine so that stochastic averages in the noisy theory reproduce suitable quantum averages in the underlying quantum field theory.

Formally, this approach lifts the seemingly overladen CTPEA. So far in this generally complex object, only the real part is enlisted in the derivation of the relevant equations of motion of the mean field. By regarding the CTPEA as a kind of influence functional, we shall see that the imaginary part contains the information about the stochastic sources.

The material in this chapter also clarifies the relationship between the CTP and the influence functional approaches. This issue has been addressed by Su *et al.* [SCYC88] and the authors [CalHu94]. In Chapter 5, we derived a Langevin equation for the long-wavelength modes of a quantum field, viewed as an open system interacting with the environment made up of short-wavelength modes. The system–environment divide we shall assume in this chapter is more elusive, since it depends on the c vs. q-number nature rather than on the value of a "hard" observable such as wavelength. In the end, as we shall discuss in detail, the physics is very much the same in one or the other approach. The stochastic mean field approach we shall discuss in this chapter has the redeeming feature that it does not force us to choose an a priori separation between modes which go into the system and which are relegated to the environment. In this sense, it is more pliable to the demands of a particular application: for example, if higher modes are generated through nonlinear effects, we run no risk of them crossing into the environment. This versatility will allow the stochastic mean field approach to retain full information about certain quantum correlations, as opposed to only their long-wavelength components.

An equivalent approach is to write down a Fokker–Planck equation describing the probability density function for the stochastic mean field. We will discuss only the Langevin equation approach; the translation to other formalisms is straightforward with the tools presented in the early chapters of this book.

Also, to facilitate comparison with the illustrated groundwork laid down in Chapter 6, we shall continue with the example of a $g\Phi^3$ relativistic quantum theory at zero temperature. The addition of statistical fluctuations over and above the quantum ones, as well as applications to more realistic theories, will be discussed in the forthcoming chapters.

The themes we shall develop in this chapter are:

(a) The complex terms in the retarded propagator in frequency domain $G_{\text{ret}}(\omega)$ imply dissipation.

(b) Underlying the dissipation of the mean field is the effect of particle creation arising from the amplification of quantum fluctuations by the time-dependent mean field. Dissipation results from the back-reaction of particle creation on the mean field. We shall see this to order g^2 by a direct derivation of the number of created particles.

(c) There are fluctuations in the number of created particles, which brings forth fluctuations in the back-reaction effect. These fluctuations may be incorporated into the dynamics of the mean field or condensate by introducing a stochastic source in the right-hand side of the equation of motion. We shall show that the noise autocorrelation is given precisely by the noise kernel in the 1PI CTPEA. The stochastic c-number field $\phi(t)$ does not represent the expectation value of the Heisenberg field anymore; we shall refer to it as the stochastic condensate. In the linearized theory, the stochastic average of the condensate gives back the quantum average which is the mean field.

(d) The resulting stochastic theory is a nontrivial extension of mean field dynamics, in the sense that, at least for linear theories, the stochastic formulation reproduces some quantum correlation functions of the full theory. This result, which is similar to one already proven for quantum open systems, shows that the identification of the CTPEA as an influence functional – and therefore of the condensate as an effectively open system – is not merely a formal device.

(e) It is clear from their perturbative expressions that the noise and dissipation kernels are closely related to each other. We may now show that, if we allow the condensate to equilibrate under the effect of the noise, then the relationship between the noise and dissipation kernels becomes the fluctuation–dissipation theorem. Alternatively, one may use the fluctuation–dissipation relation to find the noise kernel given the dissipation kernel, and vice versa.

(f) While one can envisage many situations where a quantum field may be split into a system field and an environment, it is not obvious that it is justified to treat the former as classical. We will show that particle creation is also central to this issue, by deriving an expression for the decoherence functional between two system histories in terms of the Bogoliubov coefficients describing particle creation in the environment. In short, system and environment get entangled through particle creation, and decoherence occurs when it is efficient.

(g) From the linear theory results it may seem that these effects are restricted to high frequencies $\omega > 2m$. We shall see that this limitation is lifted by nonlinear effects. In particular, we shall show that a coherent condensate oscillation will create particles even if the frequency is below threshold, through the process of parametric amplification. The difference is that parametric amplification is an essentially nonperturbative phenomenon, and it is exponentially suppressed as we move away from the threshold. So dissipation and fluctuation are generic properties of condensate dynamics.

Of course, a simple oscillation will not in general be a solution of the free equations of motion, precisely because it will dissipate through particle creation. The problem of evolution under back-reaction from quantum fluctuations is rather complex. It involves not only finding long-time solutions to the equations, but also the harder problem of making sure that the equations contain the relevant physics in the different time ranges. For example, fluctuation–fluctuation interactions, which are totally ignored in the one-loop or leading order $1/N$ approximations, are crucial on scales of the order of the thermalization time. We shall discuss these issues in later chapters.

Since dissipation and noise are central elements in nonequilibrium evolutions, a complete set of references for this chapter would be coextensive with the literature on nonequilibrium field theory itself. Our discussion will loosely follow [CalHu89, CalHu94, CalHu95, CalHu97]. See also [Hu89, HuSin95, CamVer96]. The latter two papers, when read as a sequel to [CalHu87] give a clear example of how dissipation and noise can be identified from the CTPEA with the help of the influence action, and Langevin equations (in that context, the Einstein–Langevin equations) can be derived for the stochastic mean field (semiclassical) dynamics. Stochastic equations for classical systems arising from the decoherence functional formalism have been discussed by Gell-Mann and Hartle [HarGel93]. The formal analysis of the Einstein–Langevin equations developed by Hu and Matacz [HuMat95], Lombardo and Mazzitelli [LomMaz97], Martin and Verdaguer [MarVer99a, MarVer99b, MarVer99c, MarVer00] and Roura and Verdaguer [RouVer99, Rou02] (see reviews [HuVer02, HuVer03, HuVer04]) could be adapted (or rather, simplified) to provide a foundation for the stochastic equations of scalar field theory below. The computation of full quantum correlations from the stochastic formulation is elaborated by Calzetta *et al.* [CaRoVe03].

A partial list of references for further reading on this subject is [Law89, Law92, Law99, LawKer00, BerRam01, RamNav00, BeGlRa98, HosSak84, MorSas84, Mor86, Mor90, Paz90a, Paz90b, Bet01, GleRam94, GreMul97, ABBCFJ99, LeeBoy93, Mos02]. See also those mentioned in the chapters on applications to atom–optical physics (13), nuclear–particle physics (14) and gravitation–cosmology (15).

8.1 Preliminaries

We return to the $g\varphi^3$ theory to illustrate the ideas highlighted above. The classical action with a cubic potential as in Chapter 6, equation (6.43) is

$$S\left[\Phi\right] = \int d^4x \left\{ -\frac{1}{2} \left(\partial\Phi\right)^2 - V\left[\Phi\left(x\right)\right] \right\} \tag{8.1}$$

We shall begin by considering the regression of the mean field towards its equilibrium value. To this end it is enough to consider the linearized equations of motion. The quadratic effective action is given in Chapter 6; see Sections 6.3.1 and 6.4.3 there. We have already seen that, after ultraviolet singularities have been disposed of, and assuming the initial conditions are laid out in the distant past to avoid initial time singularities, the free linearized evolution of the mean field is described by an equation of the form

$$\left[-\frac{d}{dt^2}^2 - m^2 \right] \phi\left(t\right) + \int_{}^{t} ds - \Sigma_{\text{ret}}\left(t-s\right)\phi\left(s\right) = 0 \tag{8.2}$$

where we are assuming a spatially homogeneous mean field, and

$$\Sigma_{\text{ret}}\left(t\right) = \int d^3x\, \Sigma_{\text{ret}}\left(t, \mathbf{x}\right) = \int \frac{d\omega}{2\pi} e^{-i\omega t} \Sigma_{\text{ret}}\left(\omega, \mathbf{p} = 0\right) \tag{8.3}$$

From now on, we shall omit writing the $\mathbf{p}$ argument when it is zero.

The fundamental solution of the equation of motion is the (space averaged) retarded propagator

$$G_{\text{ret}}\left(t\right) = \int \frac{d\omega}{2\pi} e^{-i\omega t} G_{\text{ret}}\left(\omega\right) \tag{8.4}$$

$$G_{\text{ret}}\left(\omega\right) = (-1) \left[\left(\omega + i\varepsilon\right)^2 - m^2 + -\Sigma_{\text{ret}}\left(\omega\right) \right]^{-1} \tag{8.5}$$

The physical mass M^2 is defined by the requirement that the retarded propagator has a simple pole at $\omega = \pm M$,

$$M^2 - m^2 + \frac{1}{\pi} \int_{4m^2}^{\infty} \frac{d\sigma^2}{\sigma^2 - M^2} \Pi\left(\sigma^2\right) = 0 \tag{8.6}$$

The function $\Pi\left(\sigma^2\right)$ was introduced in Chapter 6, equation (6.135). We shall assume M^2 is positive. The retarded propagator has a branch cut for $\omega^2 > 4m^2$. If M^2 exists, it must be less than $4m^2$; otherwise the retarded propagator has no first sheet poles.

8.2 Dissipation in the mean field dynamics

Let us begin by showing that the existence of an imaginary component in $G_{\text{ret}}(\omega)$ implies that the dynamics of mean fields is dissipative.

The simplest way to show this is by looking at the response of the mean field to an impulse, that is, adding a source $-\delta(t)$ to the right-hand side of equation (8.2). The solution is

$$\phi(t) = G_{\text{ret}}(t) = \int \frac{d\omega}{2\pi} e^{-i\omega t} G_{\text{ret}}(\omega) \tag{8.7}$$

As we know, the integrand has poles at $\omega = \pm M$ and branch cuts for $|\omega| > 2m$. Separating these contributions, we get

$$\phi(t) = \frac{1}{ZM} \sin Mt + \frac{1}{\pi} \int_{4m^2}^{\infty} d\sigma^2 \frac{\sin \sigma t}{\sigma} \Pi(\sigma^2) |G_{\text{ret}}(\sigma)|^2 \tag{8.8}$$

where Z comes from the residue at the pole. Since the integrand in the second term is regular, this term goes to zero as $t \to \infty$.

A less rigorous argument is based on a Breit–Wigner approximation for $G_{\text{ret}}(\omega)$. We simply approximate $\text{Re}G_{\text{ret}}^{-1}(\omega) \sim \omega^2 - M^2$; for the imaginary part, we write

$$\text{Im}G_{\text{ret}}^{-1}(\omega) = \Pi(\omega^2) \, \text{sign}(\omega) \sim 2\gamma\omega \tag{8.9}$$

$$\gamma \sim \frac{g^2\hbar}{128\pi m} \tag{8.10}$$

Therefore

$$\phi(t) = \frac{1}{M} \sin Mt \, e^{-\gamma t} \tag{8.11}$$

This approximation, which amounts to writing $\Pi(\omega^2) \sim$ constant, $\text{sign}(\omega) \sim \omega/2m$, cannot be valid at very short times $t^{-1} \gg m$, nor at very late times $t^{-1} \leq m$, but it does show that there is an approximately exponential decay in between. The decay turns to a power law at later times.

As a final argument, let us regard the nonlocal term in the equation of motion as a friction force acting on the mean field. Suppose we act on the mean field with an external source $j(t)$ such that it follows a given trajectory $\phi(t)$ vanishing both in the distant past and future. Therefore the total energy exchanged with the mean field vanishes. The instantaneous power is of course (minus) the product of force times velocity. The total power extracted from the mean field is

$$0 = \int_{-\infty}^{\infty} dt \left\{ m^2\phi(t) - \int^t ds - \Sigma_{\text{ret}}(t-s)\phi(s) - j(t) \right\} \frac{d\phi}{dt} \tag{8.12}$$

so we must have

$$Q = -\int_{-\infty}^{\infty} dt \int^t ds - \Sigma_{\text{ret}}(t-s)\phi(s) \frac{d\phi}{dt} \tag{8.13}$$

where Q is the work extracted from the source. In terms of Fourier transforms

$$Q = \int \frac{d\omega}{2\pi} - \Sigma_{\text{ret}}(\omega)(-i\omega)\phi(\omega)\phi(-\omega) \tag{8.14}$$

It is clear that $\phi(\omega)\phi(-\omega) = |\phi(\omega)|^2$ is an even function of ω, so only the odd part of $\Sigma_{\text{ret}}(\omega)$ may contribute to Q. Since the real part of Σ_{ret} is even, we are left with

$$Q = \frac{1}{\pi}\int_{2m}^{\infty} \omega d\omega\, \Pi\left(\omega^2\right)|\phi(\omega)|^2 \qquad (8.15)$$

which is clearly positive. We may think of this as work which is transferred from the external source to the mean field and then transformed into "heat," since it is not returned to the source nor stored in the mean field. We shall show in the next section that this work was transferred to the quantum fluctuations above the condensate.

8.3 Dissipation and particle creation

We have seen in the last section that along its evolution the mean field dissipates an amount of heat Q given by equation (8.15). We shall now show that this energy is actually spent in creating particles in the quantum field above the condensate.

Let us consider the Heisenberg equation of motion as given in Section 4.1.2 of Chapter 4. Split the quantum field Φ into a (c-number) mean field ϕ and a quantum fluctuation field φ

$$\Phi = \phi + \varphi \qquad (8.16)$$

where

$$\langle\varphi\rangle = 0 \qquad (8.17)$$

The expectation value of the Heisenberg equation yields

$$\partial^2\phi - m^2\phi + \frac{1}{2}g\phi^2 + \frac{1}{2}g\left[\langle\varphi^2\rangle_\phi - \langle\varphi^2\rangle_{\phi=0}\right] = 0 \qquad (8.18)$$

where $\langle\varphi^2\rangle_\phi$ denotes the expectation value computed in the presence of the mean field. A linear expansion of (8.18) around $\phi = 0$ takes us back to (8.2). Subtracting (8.18) from the Heisenberg equation we find the equation for the fluctuations

$$\partial^2\varphi - m^2\varphi + g\phi\varphi + \frac{1}{2}g\left[\varphi^2 - \langle\varphi^2\rangle_\phi\right] = 0 \qquad (8.19)$$

The one-loop approximation amounts to leaving out the last term

$$\partial^2\varphi - m^2\varphi + g\phi\varphi = 0 \qquad (8.20)$$

We see that, in this model, the one-loop approximation reduces to the Hartree approximation. If the mean field is spatially independent, we may expand the fluctuation field in modes as in Chapter 4

$$\varphi(t,\mathbf{x}) = \int \frac{d^3\mathbf{k}}{(2\pi)^{3/2}}\, e^{i\mathbf{k}\mathbf{x}}\varphi_{\mathbf{k}}(t) \qquad (8.21)$$

Each mode is a harmonic oscillator with a time-dependent natural frequency

$$\frac{d^2\varphi_{\mathbf{k}}}{dt^2} + \omega_k^2\varphi_{\mathbf{k}} - g\phi(t)\varphi_{\mathbf{k}} = 0; \qquad \omega_k^2 = \mathbf{k}^2 + m^2 \qquad (8.22)$$

Given two complex independent solutions f_k, f_k^* of equation (8.22), we may write

$$\varphi_{\mathbf{k}}(t) = f_k(t) a_{\mathbf{k}} + f_k^*(t) a_{-\mathbf{k}}^\dagger \tag{8.23}$$

where $a_{\mathbf{k}}$ is the usual destruction operator. Let us solve for the modes in powers of g. To zeroth order in g the Minkowski modes $f_k(t)$ introduced in Chapter 4 are the single global positive frequency solution. To first order in g we have a choice: either the *in* positive frequency solution

$$f_k^{in}(t) = f_k(t) + g \int_{-\infty}^t ds \, \frac{\sin \omega_k (t-s)}{\omega_k} \phi(s) f_k(s) \tag{8.24}$$

or the *out* positive frequency wave

$$f_k^{out}(t) = f_k(t) + g \int_t^\infty ds \, \frac{\sin \omega_k (s-t)}{\omega_k} \phi(s) f_k(s) \tag{8.25}$$

If the mean field is well behaved, then at very late times we have $f_k^{out}(t) \sim f_k(t)$ whereas $f_k^{in}(t)$ is obtained through a Bogoliubov transformation

$$f_k^{in}(t) = \alpha_k f_k^{out}(t) + \beta_k \left[f_k^{out}(t) \right]^* \tag{8.26}$$

Conversely, the destruction operators in the distant past and future are related through

$$a_{\mathbf{k}}^{out} = \alpha_k a_{\mathbf{k}}^{in} + \beta_k^* a_{-\mathbf{k}}^{in\dagger} \tag{8.27}$$

As we saw in Chapter 4, if the initial state is the *in* vacuum, at late times we find a nonzero population density of created particles $|\beta_k|^2$. From the explicit expression, we find

$$\beta_k = \frac{(-ig)}{2\omega_k} \int_{-\infty}^\infty ds \, \phi(s) e^{-2i\omega_k s} = \frac{(-ig)}{2\omega_k} \left[\phi(2\omega_k) \right]^* \tag{8.28}$$

Since each particle carries an energy $\hbar \omega_k$, the total energy density in the fluctuations is

$$\rho = \int \frac{d^3\mathbf{k}}{(2\pi)^3} \frac{g^2 \hbar}{4\omega_k} |\phi(2\omega_k)|^2$$
$$= \frac{g^2 \hbar}{4\pi} \int_{2m}^\infty \omega d\omega \, \nu(\omega) |\phi(\omega)|^2$$
$$= Q \tag{8.29}$$

where

$$\nu(\omega) = \frac{1}{8\pi} \sqrt{1 - \frac{4m^2}{\omega^2}} \theta(\omega^2 - 4m^2) \tag{8.30}$$

was already introduced in Chapter 5. We see that the energy extracted from the source is being transferred to the fluctuations. For completeness, we observe that the kernel ν in (8.30) is related to the kernel Π in (8.6) through

$$\Pi(\omega^2) = \frac{g^2 \hbar}{4} \nu(\omega) \tag{8.31}$$

8.4 Particle creation and noise

We have seen in the last section that the mean field loses energy which is spent in exciting the quantum fluctuations of the vacuum into particles. The back-reaction from this process is experienced by the mean field as dissipation. We now observe that particle creation from the vacuum has an intrinsic stochastic character: there are always fluctuations in the number of created particles. These fluctuations affect the mean field through its back-reaction. The dynamics of the mean field thus acquires a stochastic element. Of course, at this point it ceases to be the "mean" field: it is a c-number field which represents the evolution of the condensate component of the full Heisenberg field.

To obtain a measure of the fluctuations in particle creation, let us consider the correlations between particles created in different modes

$$\langle N_{\mathbf{p}} N_{\mathbf{q}} \rangle = \langle 0in \,| a_{\mathbf{p}}^{out\dagger} a_{\mathbf{p}}^{out} a_{\mathbf{q}}^{out\dagger} a_{\mathbf{q}}^{out} |\, 0in \rangle$$

$$= V^2 |\beta_p|^2 |\beta_q|^2 + V |\beta_p|^2 |\alpha_p|^2 \left[\delta\,(\mathbf{p} - \mathbf{q}) + \delta\,(\mathbf{p} + \mathbf{q}) \right] \qquad (8.32)$$

It follows that the fluctuations in the energy density are

$$\langle \delta\rho^2 \rangle = \langle \rho^2 \rangle - \langle \rho \rangle^2$$

$$= \frac{2\hbar^2}{V} \int \frac{d^3\mathbf{p}}{(2\pi)^3}\, \omega_p^2 |\beta_p|^2 |\alpha_p|^2 \qquad (8.33)$$

To lowest order we may approximate $|\alpha_p|^2 = 1$. Using the explicit expression for the Bogoliubov coefficients, we get,

$$\langle \delta\rho^2 \rangle = \frac{g^2\hbar^2}{2V} \int \frac{d^3\mathbf{k}}{(2\pi)^3}\, |\phi\,(2\omega_k)|^2$$

$$= \frac{g^2\hbar^2}{4\pi V} \int_{2m}^{\infty} \omega^2 d\omega\, \nu\,(\omega)\, |\phi\,(\omega)|^2$$

$$= \delta Q^2 \qquad (8.34)$$

We may account for these fluctuations by adding a stochastic term $\zeta\,(t, \mathbf{x})$ to the right-hand side of the mean field equations of motion. For a homogeneous condensate, they reduce to

$$\left[-\frac{d}{dt^2}^2 - m^2 \right] \phi\,(t) + \int^t ds - \Sigma_{\mathrm{ret}}\,(t - s)\, \phi\,(s) = -\frac{g}{2} \Xi\,(t) \qquad (8.35)$$

where

$$\phi\,(t) = \frac{1}{V} \int d^3\mathbf{x}\, \phi\,(t, \mathbf{x}) \qquad (8.36)$$

$$\Xi\,(t) = \frac{1}{V} \int d^3\mathbf{x}\, \zeta\,(t, \mathbf{x}) \qquad (8.37)$$

We assume ζ is a Gaussian noise with zero average $\langle \zeta\,(t, \mathbf{x}) \rangle_s = 0$, where, hereafter, the subscript s will denote stochastic averages over the noise distribution function, and (possibly colored) autocorrelation $\langle \zeta (t, \mathbf{x}) \zeta (s, \mathbf{y}) \rangle_s = \nu_s (t - s, \mathbf{x} - \mathbf{y})$. For a prescribed trajectory $\phi (t)$, the work done by the random source is

$$Q_s = \frac{g}{2} \int dt \, \Xi (t) \frac{d}{dt} \phi (t) \qquad (8.38)$$

Assuming independence of $\phi (t)$ and $\Xi (t)$, $\langle Q_s \rangle_s = 0$ and

$$\langle Q_s^2 \rangle_s = \frac{g^2}{4V} \int dt ds \, \frac{d}{dt} \phi (t) \frac{d}{ds} \phi (s) \int d^3 \mathbf{x} \, \nu_s (t - s, \mathbf{x}) \qquad (8.39)$$

Introducing the Fourier transform

$$\nu_s (t, \mathbf{x}) = \int \frac{d^4 k}{(2\pi)^4} e^{ikx} \nu_s (k) \qquad (8.40)$$

$$\langle Q_s^2 \rangle_s = \frac{g^2}{4V} \int_{-\infty}^{\infty} \frac{d\omega}{2\pi} \omega^2 |\phi (\omega)|^2 \nu_s (\omega) = \frac{g^2}{4V} \int_0^{\infty} \frac{d\omega}{\pi} \omega^2 |\phi (\omega)|^2 \nu_s (\omega) \quad (8.41)$$

where as usual we write $\nu_s (\omega) = \nu_s (\omega, \mathbf{p} = 0)$ and we have used the obvious symmetry condition that ν_s is even. If we request that $\langle Q_s^2 \rangle_s$ accounts for the fluctuations δQ^2, equation (8.34), then $\nu_s = \hbar^2 \nu$.

Since we are discussing a Lorentz invariant theory, this result determines $\nu (k)$ everywhere. We of course recognize the noise kernel introduced in Chapters 5 and 6. In other words, we could arrive at the same Langevin type equation for the mean field simply by arguing that the CTPEA may be regarded as a Feynman–Vernon influence functional for an open system (the condensate) interacting with an environment (the quantum fluctuations) and adopting the usual interpretation that the imaginary part of the influence action (IA) describes noise.

This point of view is validated by the fact that the stochastic formulation allows us to compute certain quantum expectation values in the original theory. Before developing this point further, let us show briefly yet another way to arrive at the same stochastic equation. If we consider the full Heisenberg equation and subtract the equation for the fluctuations we see that the Langevin equation (8.35) amounts to the replacement

$$\left[\varphi^2 - \langle \varphi^2 \rangle_\phi \right] \leftrightarrow \zeta \qquad (8.42)$$

Of course we cannot simply interpose an identity, because we have on the left a Heisenberg quantum operator, and on the right a c-number stochastic field. To give meaning to the connection between the two, we adopt the Landau prescription that the symmetric quantum expectation value of the left-hand side equals (twice) the stochastic expectation value of the right-hand side, or

$$\nu_s (t - s, \mathbf{x} - \mathbf{y}) = \frac{1}{2} \left[\langle \{ \varphi^2 (t, \mathbf{x}), \varphi^2 (s, \mathbf{y}) \} \rangle_\phi - 2 \langle \varphi^2 (t, \mathbf{x}) \rangle_\phi \langle \varphi^2 (s, \mathbf{y}) \rangle_\phi \right]$$

$$(8.43)$$

An explicit evaluation at $\phi = 0$ gives again the noise kernel from the CTPEA, as we have seen in Chapter 6. We see how this approach leading to a Langevin equation is an improvement over the usual mean field theory, which simply disregards $\varphi^2 - \left\langle \varphi^2 \right\rangle_\phi$ entirely.

8.5 Full quantum correlations from the Langevin approach

As a simple application of the Langevin approach, we shall show how it may be used to compute the Hadamard propagator for the underlying field theory. This is the field theory counterpart of a method applicable more generally to quantum open systems, and therefore reinforces the view of the CTPEA as the IA for the mean field.

Let us begin by connecting the propagators of the theory to the CTPEA. In the condensed notation from Chapter 6, the full propagators $G^{AB} = \left\langle \varphi^A \varphi^B \right\rangle$ in the equilibrium state are given by

$$G^{AB} = -i\hbar \left. \frac{\delta^2 W}{\delta J_A \delta J_B} \right|_{J=0} \tag{8.44}$$

where W is the CTP generating functional. As usual we identify $G^{11} = G_{\mathrm{F}}$, $G^{12} = G^-$, $G^{21} = G^+$ and $G^{22} = G_{\mathrm{D}}$. On the other hand, $W^{,AB} = \delta\phi^A/\delta J_B = -\left(\Gamma_{,AB}\right)^{-1}$, so we obtain an equation relating the propagators to the second variation of the CTPEA

$$\Gamma_{,AB} G^{BC} = i\hbar \delta_A^C \tag{8.45}$$

Observe that if the field theory is defined only for $t > t_0$, rather than on the whole Minkowski space, then the intermediate integral is equally restricted:

$$\phi_B \psi^B \equiv \int d^3\mathbf{x} \int_{t_0}^\infty dt\, \phi_b\left(t, \mathbf{x}\right) \psi^b\left(t, \mathbf{x}\right) \tag{8.46}$$

We have seen in Chapter 6 that the quadratic part of the CTPEA must have the structure of equation (6.97), where the kernels $\mathbf{D}^{\mathrm{full}}$ and $\mathbf{N}$ are real, and $\mathbf{D}^{\mathrm{full}}$ is causal. We may further split $\mathbf{D}^{\mathrm{full}}$ into its symmetric and antisymmetric parts, $\mathbf{D}^{\mathrm{full}} = \mathbf{D}_s^{\mathrm{full}} + \boldsymbol{\Gamma}$, respectively. The Hessian $\Gamma_{,AB}$ becomes

$$\Gamma_{,AB} = \begin{pmatrix} \mathbf{D}_s^{\mathrm{full}} + i\mathbf{N} & \boldsymbol{\Gamma} - i\mathbf{N} \\ -\boldsymbol{\Gamma} - i\mathbf{N} & -\mathbf{D}_s^{\mathrm{full}} + i\mathbf{N} \end{pmatrix} \tag{8.47}$$

Since the equilibrium state is translation invariant, the propagators (as well as the $\mathbf{D}_s^{\mathrm{full}}$, $\boldsymbol{\Gamma}$ and $\mathbf{N}$ kernels) are functions of the difference variable $x - x'$ alone, and equations (8.45) are algebraic equations for their Fourier transforms. Setting $a = 1$ in equations (8.45) and using the matrix form (8.47), we obtain

$$\left(\mathbf{D}_s^{\mathrm{full}} + i\mathbf{N}\right) G^{11} + \left(\boldsymbol{\Gamma} - i\mathbf{N}\right) G^{21} = i\hbar \tag{8.48}$$

$$\left(\mathbf{D}_s^{\mathrm{full}} + i\mathbf{N}\right) G^{12} + \left(\boldsymbol{\Gamma} - i\mathbf{N}\right) G^{22} = 0 \tag{8.49}$$

Subtracting these two equations, and writing the fundamental propagators in terms of G_{ret}, G_{adv} and G_1, we get $\mathbf{D}^{\mathrm{full}} G_{\mathrm{ret}} = -1$. This is just the statement that the retarded propagator is the fundamental solution to the linearized equations of motion for the mean field, $\mathbf{D}^{\text{full}}\phi = -J$.

Let us go back to equation (8.48) to get

$$\mathbf{D}^{\text{full}}G_1 + 2\hbar \mathbf{N}G_{\text{adv}} = 0 \tag{8.50}$$

Since the equation $\mathbf{D}^{\text{full}}\phi = 0$ admits plane waves of momentum p as homogeneous solutions, provided $(-p^2) = M^2$, the solution to this equation reads

$$G_1 = C\delta\left(-p^2 - M^2\right) + 2\hbar G_{\text{ret}}\mathbf{N}G_{\text{adv}} \tag{8.51}$$

We are using the fact that G_1 must be Lorentz invariant, so C must be a simple constant.

In the Langevin approach we postulate an equation for the stochastic condensate (absorbing coupling constants into the stochastic source, i.e. $\xi = g\zeta/2$)

$$\mathbf{D}^{\text{full}}\phi = -\xi \tag{8.52}$$

where

$$\langle \xi\left(x\right)\xi\left(y\right)\rangle_s = \hbar \mathbf{N}\left(x, y\right) \tag{8.53}$$

Suppose we set the initial conditions for this equation at some time t_0. Then

$$\phi\left(x\right) = \phi_{\text{hom}}\left(x\right) + \int_{y^0 > t_0} d^4y\, G_{\text{ret}}\left(x, y\right)\xi\left(y\right) \tag{8.54}$$

where $\phi_{\text{hom}}\left(x\right)$ is determined by the Cauchy data at t_0

$$\phi_{\text{hom}}\left(t, \mathbf{x}\right) = \int d^3y\, \left\{ G_{\text{ret}}\left(t - t_0, \mathbf{x} - \mathbf{y}\right)\frac{d}{dt_0}\phi\left(t_0, \mathbf{y}\right)\right.$$
$$\left. + \frac{d}{dt}G_{\text{ret}}\left(t - t_0, \mathbf{x} - \mathbf{y}\right)\phi\left(t_0, \mathbf{y}\right)\right\} \tag{8.55}$$

The stochastic average, assuming independence between the initial conditions and the noise sources, becomes

$$\langle \phi\left(x\right)\phi\left(y\right)\rangle_s = \langle \phi_{\text{hom}}\left(x\right)\phi_{\text{hom}}\left(y\right)\rangle_s$$
$$+ \hbar \int_{z^0, z'^0 > t_0,} d^4z\, d^4z'\, G_{\text{ret}}\left(x, z\right)\mathbf{N}\left(z, z'\right)G_{\text{adv}}\left(z', y\right) \tag{8.56}$$

Twice this is a solution of equation (8.50), and therefore if the Cauchy data for $2\langle \phi\left(x\right)\phi\left(y\right)\rangle_s$ and $G_1\left(x, y\right)$ are chosen to be the same, they will remain equal everywhere.

This shows that the stochastic approach may reproduce the Hadamard propagator of the underlying quantum theory. Observe that both random initial conditions and noise sources are required.

8.6 The fluctuation–dissipation theorem

Before we show how the above analysis may be generalized to the nonlinear regime, it is interesting to pause for the following observation. We have just shown that the Hadamard propagator for quantum fluctuations may be obtained as a stochastic average over a random c-number field. This field may be decomposed into a homogeneous solution of the linearized mean field equations of motion plus an extra term, induced by the effect of a particular Gaussian noise.

We have seen at the beginning of this chapter that solutions of the mean field equations are partially dissipated away as they evolve. But the Hadamard propagator is time-translation invariant. So the noise sources must be injecting the precise amount of fluctuations necessary to compensate for the dissipation of the free part. The quantitative statement of this observation is the (zero temperature) fluctuation dissipation relation [CalWel51, LaLiPi80a, Ma76, BooYip91]. This is a simple application of a deeper, generic relationship between noise and dissipation in the CTPEA, whose origin is that both arise from particle creation in the fluctuation field. Here we are using the term "particle creation" also to denote such phenomena as the transfer of atoms from a condensate to higher modes, as in the Bose–Nova experiment [Don01].

To quantify this statement, let us return to the expression for the heat dissipated during the whole evolution of the field

$$Q = \int d^4 x \left\{ \left[-\nabla^2 + m^2 \right] \phi(x) - \int_{y^0 < x^0} d^4 y - \Sigma_{\text{ret}}(x, y) \phi(y) - \xi(x) \right\} \frac{d\phi}{dt}(x)$$

(8.57)

Since the spectrum of fluctuations is stationary, we must have $\langle Q \rangle_s = 0$. The first terms have been analyzed in the beginning of this chapter, with the only difference that now we do not assume a homogeneous condensate. Using the results of the last section to replace field averages by the Hadamard propagator, we get

$$\int d^4 x \left\langle \xi(x) \frac{d\phi}{dt}(x) \right\rangle_s = VT \int \frac{d^4 k}{(2\pi)^4} \, k^0 \Pi\left(-k^2\right) G_1(k) \theta\left(k^0\right)$$

(8.58)

where VT is the 4-volume of spacetime. Since we are assuming a Gaussian noise and a linearized equation of motion,

$$\int d^4 x \left\langle \xi(x) \frac{d\phi}{dt}(x) \right\rangle_s = \int d^4 x d^4 y \, \langle \xi(x) \xi(y) \rangle_s \, \frac{d}{dt} \frac{\delta\phi(x)}{\delta j(y)}$$

$$= -iVT\hbar \int \frac{d^4 k}{(2\pi)^4} \, k^0 \mathbf{N}(k) G_{\text{ret}}(k)$$

(8.59)

Since $\mathbf{N}(k)$ is even, this becomes

$$-i\frac{VT\hbar}{2} \int \frac{d^4 k}{(2\pi)^4} \, \left|k^0\right| \mathbf{N}(k) \left[G_{\text{ret}}(k) - G_{\text{ret}}(-k) \right] \theta\left(k^0\right)$$

(8.60)

But $G_{\text{ret}}(k) - G_{\text{ret}}(-k) = iG(k)$, where G is the Jordan propagator. Therefore, defining

$$G_1(k) = \rho(k) G(k) \operatorname{sign}(k^0) \qquad (8.61)$$

then

$$\hbar N(k) = 2\Pi(-k^2) \rho(k) \qquad (8.62)$$

This is the fluctuation–dissipation theorem at zero temperature (cf. Einstein's relation from Chapter 2). By the way, for free fields $\rho = 1$, as we saw in Chapter 6.

8.7 Particle creation and decoherence

At this point it is interesting to go back to the beginning and question whether it is consistent to treat the system field ϕ as classical. One possible answer is to consider two different histories for the ϕ field, leaving the environment field φ unspecified, and to compute their decoherence functional $\mathcal{D}$ (introduced in Chapter 3). If $|\mathcal{D}| \ll 1$, the classical approximation is warranted.

The basic observation is that to compute the decoherence functional we must perform a CTP path integral over all histories of the field φ^a, adding in each branch an external source to enforce the constraint that $\langle \varphi^a \rangle = 0$. The result is that the path integral defining $\mathcal{D}$ is identical to the one defining the CTPEA, and we find the relationship

$$\mathcal{D}\left[\phi^1, \phi^2\right] = \exp\left\{\frac{i\Gamma\left[\phi^1, \phi^2\right]}{\hbar}\right\} \qquad (8.63)$$

It is clear that the classical part of the CTPEA does not contribute to decoherence. Let us consider the one-loop term Γ_1 (cf. Chapter 6). In canonical terms, Γ_1 measures the overlap between the state which evolves from the *in* vacuum under the influence of the external field $\phi^1(x)$ and the state which evolves under $\phi^2(x)$, as measured in the far future. For simplicity, let us assume that the background fields are homogeneous, in which case we may decompose the fluctuation fields in plane waves, to find

$$\Gamma_1\left[\phi^A\right] = \sum_{k^0 > 0} \Gamma_{1k}\left[\phi^A\right]$$

$$\Gamma_{1k}\left[\phi^A\right] = -i\hbar \ln \int D\varphi_k^a D\varphi_{-k}^a$$

$$\times \exp\left\{-\frac{i}{\hbar} \int dt\, \varphi_{-k}^a \left[c_{ab}\left(\frac{d^2}{dt^2} + \omega_k^2\right) + g\, c_{abc}\phi^c\right] \varphi_k^b\right\} \qquad (8.64)$$

Interposing a complete set of *out* modes, we may write

$$\Gamma_{1k}\left[\phi^A\right] = -i\hbar \ln \sum_n \left\langle 0in \,\middle|\, n_k, n_{-k}, out \right\rangle_{\phi^2} \left\langle n_k, n_{-k}, out \,\middle|\, 0in \right\rangle_{\phi^1} \qquad (8.65)$$

where we are using the fact that particles may be created in pairs only; the subscript ϕ indicates the external field under which the quantum field evolves. Since the quantum field on each branch is a free Klein–Gordon field with a time-dependent frequency, the *in* and *out* destruction operators are related through a Bogoliubov transform. The relevant brackets are given in Chapter 4, and after a simple summation, we arrive at

$$\Gamma_{1k}\left[\phi^A\right] = i\hbar \ln \left[\alpha_k^2 \alpha_k^{1*} - \beta_k^2 \beta_k^{1*}\right] \tag{8.66}$$

where α_k^i, β_k^i denote the Bogoliubov coefficients for the corresponding branch.

One can check that this expression complies with the basic expectations regarding the CTPEA. It is clear that Γ_{1k} vanishes if $\phi^1 = \phi^2$. If the two fields are exchanged, the real part changes sign, while the imaginary part is unchanged.

To clarify the meaning of equation (8.66) let us observe that it is invariant if we subject both pairs of Bogoliubov coefficients to the same Bogoliubov transformation. In other words, the effective action is independent of the choice of *out* particle model in equation (8.65). Therefore we may assume without loss of generality that $\beta_k^2 = 0$. This implies $|\alpha_k^2| = 1$, and so, in this representation,

$$|\mathcal{D}| = \frac{1}{|\alpha_k^1|} = \frac{1}{\sqrt{1 + |\beta_k^1|^2}} \tag{8.67}$$

As expected, particle creation is necessary to suppress coherence. Of course, the physical mechanism behind this result is the entanglement of the system and environment fields through the particle creation process.

The relation between particle creation and decoherence was given in [CalMaz90]. The expression of the CTP effective action or the influence functional in terms of the Bogoliubov coefficients was given in [CalHu94, HKMP96, RaStHu98].

8.8 The nonlinear regime

So far we have demonstrated the presence of noise and dissipation for far off-shell modes ($-p^2 > 4M^2$). We shall now see that particle creation, and therefore dissipation and noise, is restricted by a lower threshold only in the linearized theory. In the nonlinear regime, the possibilities are much richer: Schwinger proved the existence of particle creation from static electric fields, shown as an example in Chapter 4. So dissipation and noise in particle creation are the rule rather than the exception.

Let us attempt a nonperturbative evaluation of the one-loop effective action as given in Chapter 6, equation (6.120). Observe that

$$\phi^1 \left(\varphi^1\right)^2 - \phi^2 \left(\varphi^2\right)^2 = \phi_+ \left(\left(\varphi^1\right)^2 - \left(\varphi^2\right)^2\right) + \frac{\phi_-}{2} \left(\left(\varphi^1\right)^2 + \left(\varphi^2\right)^2\right) \tag{8.68}$$

This suggests expanding

$$\Gamma_1 = \left(-\frac{g}{2}\right) \int d^4x \langle \varphi^2 \rangle_{\phi_+}(x)\phi_- + \Delta S\left[\phi_+, \phi_-\right] \tag{8.69}$$

where $\Delta S[\phi_+, \phi_-] \sim O(\phi_-^2)$. To compute $\langle \varphi^2 \rangle_{\phi_+}(x)$ we consider the fluctuation field φ as a free quantum field with the equation of motion

$$[\partial^2 - m^2 - g\phi_+]\varphi = 0 \qquad (8.70)$$

In other words, φ is a quantum field propagating on the dynamic background ϕ_+, a situation we have already analyzed in Chapter 4.

To see the effect of ΔS on the equation of motion for ϕ, we perform a functional Fourier transform

$$\exp\{i\Delta S[\phi_+, \phi_-]/\hbar\} = \int D\xi \, e^{i\hbar^{-1}\int \xi\phi_-} P[\xi, \phi_+] \qquad (8.71)$$

Calling $\langle \ldots \rangle_s = \int D\xi \ldots P[\xi, \phi_+]$, we find

$$\langle \xi(x) \rangle_s = \left. \frac{\delta \Delta S}{\delta \phi_-} \right|_{\phi_- = 0} = 0 \qquad (8.72)$$

$$\langle \xi(x)\xi(x') \rangle_s \equiv \hbar \mathbf{N}(x, x')$$
$$= \left(\frac{g^2}{8}\right) \left[\langle \{\varphi^2(x), \varphi^2(x')\} \rangle_{\phi_+} - 2\langle \varphi^2 \rangle_{\phi_+}(x)\langle \varphi^2 \rangle_{\phi_+}(x') \right] \qquad (8.73)$$

This is to be contrasted with the result in the perturbative treatment.

The functional $P[\xi, \phi_+]$ must be real (as follows from $\Delta S[\phi_+, -\phi_-] = -\Delta S[\phi_+, \phi_-]^*$) and it is nonnegative to the one-loop approximation. We may think of it as a functional Wigner transform of the effective action [Hab92], and thereby as a probability density "for all practical purposes." Observe that P will not be Gaussian in general.

In the limit $\phi_- \to 0$, $\phi_+ \to \phi$, we obtain the equation of motion for the mean field

$$(-\partial^2 + m^2)\phi(x) + \frac{1}{2}g\phi^2 + \left(\frac{g}{2}\right)\left[\langle \varphi^2 \rangle_\phi - \langle \varphi^2 \rangle_0\right](x) = \xi(x) \qquad (8.74)$$

A linear expansion of (8.74) around $\phi = 0$ and the assumption of a homogeneous condensate give back (8.35). Our goal is to show that the noise $\xi(x)$ is not restricted to modes above threshold. To this end, we shall assume a simple mean field configuration, homogeneous in space and harmonic in time, i.e.

$$\phi(t) = \phi_0 \cos \gamma t \qquad (8.75)$$

where $\gamma \leq 2M$, so we are below threshold. We shall see that in spite of this the noise is nonzero. Moreover, the noise itself is not restricted to the high-frequency domain, but it has low-frequency components as well.

To compute the nonperturbative noise kernel, we expand the quantum field φ in normal modes. The amplitude functions of the normal modes are complex, with

$$\varphi_{-\mathbf{k}} = \varphi_{\mathbf{k}}^\dagger \qquad (8.76)$$

They obey a mode equation where the time-dependent natural frequency of the kth mode is

$$\omega_k^2 = \mathbf{k}^2 + m^2 - g\phi(t) \qquad (8.77)$$

Here we shall disregard the possibility of ω becoming imaginary through a large negative light field, i.e. we assume $g\phi_0 \leq m^2$. We assume the fluctuation field is in the vacuum state at some initial time $t = 0$ (we assume the coupling constant g is switched on adiabatically, so initial time singularities do not arise). Since φ is a free field, Wick's theorem holds, and our problem is to relate the field at arbitrary times to the initial creation and destruction operators.

The general relationship we seek is

$$\varphi_k(t) = f_k(t)a_k(0) + f_k^*(t)a_{-k}^\dagger(0) \tag{8.78}$$

where f_k is the positive frequency mode associated with the *in* particle model. For the given mean field evolution (8.75) the mode equation is in the form of Mathieu's equation, with a periodically driven field in the narrow resonance regime. The results of Chapter 4 will apply here if we identify $\omega_{k0}^2 = \mathbf{k}^2 + m^2 - g\phi_0$, $\omega_1^2 = g\phi_0$. The mode function f_k may be written as a linear combination of WKB solutions.

$$f_k(t) = \alpha_k(t) f_k^+(t) + \beta_k(t) f_k^-(t) \tag{8.79}$$

Let us consider the case where we are in the neighborhood of the ℓth resonance band, namely $\omega_{k0} = \gamma(\ell + \delta_k)$ (remember in Chapter 4 we set $\gamma = 1$, so now we must re-insert γ in all the equations). Then

$$\alpha_k(t) = \left[\alpha_{k0}^{(+)} e^{\mu_k \gamma t} + \alpha_{k0}^{(-)} e^{-\mu_k \gamma t}\right] e^{i\sigma_k \gamma t}$$

$$\beta_k(t) = \left[\beta_{k0}^{(+)} e^{\mu_k \gamma t} + \beta_{k0}^{(-)} e^{-\mu_k \gamma t}\right] e^{-i\sigma_k \gamma t} \tag{8.80}$$

where

$$\sigma_k = \frac{\omega_1^2}{2\gamma\omega_{0k}} + \delta_k$$

$$\kappa_k \sim \frac{1}{(2\ell-1)!} \frac{\gamma}{4\omega_{k0}} \left(\frac{\omega_1^2}{2\gamma\omega_{0k}}\right)^{2\ell}$$

$$\mu_k = \sqrt{\kappa_k^2 - \sigma_k^2}$$

$$\beta_{k0}^{(\pm)} = \frac{1}{\kappa_k}[\sigma_k \mp i\mu_k]\,\alpha_{k0}^{(\pm)} \tag{8.81}$$

We are particularly interested in the case where μ_k is real. Let us write

$$\frac{\sigma_k + i\mu_k}{\kappa_k} = e^{i\vartheta_k} \tag{8.82}$$

(outside of the resonant region, ϑ_k becomes imaginary). Imposing the boundary conditions $\alpha_k(0) = 1$, $\beta_k(0) = 0$, we find

$$\alpha_k(t) = \frac{\sinh[\mu_k\gamma t + i\vartheta_k]}{i\sin\vartheta_k} e^{i\sigma_k\gamma t}$$

$$\beta_k(t) = \frac{\sinh[\mu_k\gamma t]}{i\sin\vartheta_k} e^{-i\sigma_k\gamma t} \tag{8.83}$$

Finally, let us define

$$f_k^+(t)e^{i\sigma_k\gamma t} = \hbar^{1/2}\frac{e^{-i\ell\gamma t}}{\sqrt{2\ell\gamma}}g_k(t) \tag{8.84}$$

where $g_k(t) \sim 1$ with great accuracy. We can now write the mode functions as

$$f_k(t) = \hbar^{1/2}\left\{\frac{\sinh[\mu_k\gamma t + i\vartheta_k]}{i\sin\vartheta_k}\frac{e^{-i\ell\gamma t}}{\sqrt{2\ell\gamma}}g_k(t) + \frac{\sinh[\mu_k\gamma t]}{i\sin\vartheta_k}\frac{e^{i\ell\gamma t}}{\sqrt{2\ell\gamma}}g_k^*(t)\right\} \tag{8.85}$$

Three features stand out, namely (1) the generation of the negative frequency components, which is the physical basis for vacuum particle creation; (2) the exponential amplification due to ongoing particle creation; and (3) the phase-locking of a whole range of wavelengths at the resonance frequency $\ell\gamma$. As we shall now see, phase locking allows the generation of a low-frequency, inhomogeneous stochastic field. This is the main physical indication of the new features of dissipation and fluctuation below threshold we want to highlight.

In order to find the noise kernel, let us decompose the Heisenberg operator φ^2 into a c-number, a diagonal (D) and a nondiagonal (ND) (in the particle number basis) part

$$\varphi^2 = \langle\varphi^2\rangle_\phi + \varphi_D^2 + \varphi_{ND}^2 \tag{8.86}$$

where

$$\langle\varphi^2\rangle_\phi = \int\frac{d^3\mathbf{k}}{(2\pi)^3}\,|f_k(t)|^2 \tag{8.87}$$

and the (D) and (ND) components are

$$\varphi_D^2 = \int\frac{d^3\mathbf{k}}{(2\pi)^3}\frac{d^3\mathbf{k'}}{(2\pi)^3}\,e^{i(\mathbf{k}+\mathbf{k'})\mathbf{x}}\left\{f_k(t)f_{k'}^*(t)a_{-\mathbf{k'}}^\dagger a_\mathbf{k} + f_k^*(t)f_{k'}(t)a_{-\mathbf{k}}^\dagger a_{\mathbf{k'}}\right\} \tag{8.88}$$

$$\varphi_{ND}^2 = \int\frac{d^3\mathbf{k}}{(2\pi)^3}\frac{d^3\mathbf{k'}}{(2\pi)^3}\,e^{i(\mathbf{k}+\mathbf{k'})\mathbf{x}}\left\{f_k(t)f_{k'}(t)a_\mathbf{k}a_{\mathbf{k'}} + f_k^*(t)f_{k'}^*(t)a_{-\mathbf{k}}^\dagger a_{-\mathbf{k'}}^\dagger\right\} \tag{8.89}$$

Observe that, assuming vacuum initial conditions,

$$\langle\varphi_D^2\rangle_\phi = \langle\varphi_{ND}^2\rangle_\phi = \langle\varphi_D^2\varphi_{ND}^2\rangle_\phi = \langle\varphi_{ND}^2\varphi_D^2\rangle_\phi = \langle\varphi_D^2\varphi_D^2\rangle_\phi \equiv 0 \tag{8.90}$$

Therefore

$$\hbar\mathbf{N}(x,x') = \left(\frac{g^2}{8}\right)\langle\{\varphi_{ND}^2(x),\varphi_{ND}^2(x')\}\rangle_{\phi+}$$

$$= \left(\frac{g^2}{2}\right)\int\frac{d^3\mathbf{k}}{(2\pi)^3}\frac{d^3\mathbf{k'}}{(2\pi)^3}\,e^{i(\mathbf{k}+\mathbf{k'})(\mathbf{x}-\mathbf{x'})}\mathrm{Re}\left\{f_k(t)f_{k'}(t)f_k^*(t')f_{k'}^*(t')\right\} \tag{8.91}$$

If no particle creation occurred, the noise kernel would contain frequencies above threshold only. However, in the presence of frequency-locking and a negative frequency part of the mode functions f, the noise kernel also contains a steady component

$$\hbar N_S(x, x') = \left(\frac{g^2 \hbar^2}{8 \ell^2 \gamma^2}\right) \int' \frac{d^3 \mathbf{k}}{(2\pi)^3} \frac{d^3 \mathbf{k}'}{(2\pi)^3} \, e^{i(\mathbf{k}+\mathbf{k}')(\mathbf{x}-\mathbf{x}')} \mathbf{F}_{kk'}(t, t') \qquad (8.92)$$

where the integral is restricted to those modes where μ_k is real, and

$$\mathbf{F}_{kk'}(t, t') = \text{Re}\left\{F_{kk'}(t) F^*_{kk'}(t')\right\} \qquad (8.93)$$

$$F_{kk'}(t) = \frac{1}{\sin \vartheta_k \sin \vartheta_{k'}} \left[\sinh\left[\mu_k \gamma t + i\vartheta_k\right] \sinh\left[\mu_{k'} \gamma t\right] g_k(t) g^*_{k'}(t) + (k \leftrightarrow k')\right]$$

$$(8.94)$$

It is important to notice that $\mathbf{F}$ is slowly varying not only with respect to the frequency $\ell\gamma$, but also with respect to the background frequency γ itself. Of course we do not observe the noise kernel directly, but only through its effect on the mean field. However, since the steady part of the stochastic source is slowly varying in space and time, to a first approximation it induces a stochastic mean field ϕ_S which is simply proportional to it:

$$\phi_S \sim \left(\frac{1}{m^2}\right) \xi_S; \qquad \langle \phi_S \phi_S \rangle \sim \left(\frac{1}{m^2}\right)^2 \hbar N_S \qquad (8.95)$$

One can deduce the noise and its auto-correlation in this way.

Since κ_k decays exponentially with l, it is clear that only the lowest possible resonance band makes a meaningful contribution. So we may assume that $\mathbf{k}^2 \ll m^2$ and approximate $\omega_{k0} = \omega_{00} + (\mathbf{k}^2/2\omega_{00})$, where $\omega_{00}^2 = m^2 - g\phi_0$. Therefore

$$\delta_k = \delta_0 + \frac{\mathbf{k}^2}{2\gamma\omega_{00}}$$

$$\sigma_k \sim \sigma_0 + \frac{\mathbf{k}^2}{2\gamma\omega_{00}}$$

$$\kappa_k \sim \kappa_0 e^{-\ell \mathbf{k}^2/\gamma\omega_{00}} \qquad (8.96)$$

The limit of the resonance band is reached at some wavenumber k_0, and we may approximate

$$\mu_k = \mu_0 \left(1 - \frac{\mathbf{k}^2}{k_0^2}\right) \qquad (8.97)$$

If $\ell \kappa_0^2 \ll 1$, $k_0^2 \sim 2\gamma\omega_{00}\kappa_0 \ll 2\gamma^2$. At short times, we may approximate

$$\frac{\sinh\left[\mu_k \gamma t\right]}{\sin \vartheta_k} \sim \frac{\mu_k \gamma t}{\sin \vartheta_k} = \kappa_k \gamma t \qquad (8.98)$$

$$F_{kk'}(t) = \gamma t \left[\kappa_k + \kappa_{k'}\right] \qquad (8.99)$$

Initially the stochastic source grows linearly in time and is coherent over distances of order k_0^{-1}. At late times

$$F_{kk'}(t) = \frac{e^{(\mu_k + \mu_{k'})\gamma t}}{2\sin\vartheta_k \sin\vartheta_{k'}}\left(e^{i\vartheta_k} + e^{i\vartheta_{k'}}\right) \sim F_{00}e^{2\mu_0\gamma t}\exp\left[-\frac{\mu_0\gamma t}{k_0^2}\left(\mathbf{k}^2 + \mathbf{k}'^2\right)\right]$$

$$(8.100)$$

so not only the strength of the stochastic source grows exponentially, with a time constant defined by the Floquet exponent, but also the size of the coherent domains grows as a power of time (in this simple model, $t^{1/2}$).

8.9 Final remarks

In this chapter, we have analyzed dissipation and fluctuations in the mean field by viewing it as an effectively open system, interacting with the environment provided by the quantum fluctuations of the same fundamental field. We shall conclude by mentioning some concrete problems where this way of thinking is fruitful in understanding their behaviors.

Physically, a quantum field develops a nontrivial expectation value through the process of condensation. By including fluctuations in its dynamics, we see the distinction between a condensate field and a mean field. The condensate is now regarded as a classical subsystem, interacting with a quantum environment and acquiring a stochastic component as a consequence.

Since in practice only long-wavelength–low-frequency modes condensate, one may attempt to draw a sharp distinction between condensate and fluctuations by defining an a priori separation between a long-wavelength condensate band, and a short-wavelength fluctuation band. Then the former may be described as a quantum open system. Eventually, if it actually condensates, the quantum fluctuations in the condensate band may be neglected. This kind of approach to condensate dynamics has been proposed by Gardiner and Anglin [GaAnFu01], Gardiner and Davis [GarDav03], and by Stoof [Sto99] in the context of Bose–Einstein condensates (BEC) (Chapter 13). Another example is in nuclear-particle physics. The effect of high-frequency modes in the quark–gluon plasma (QGP) on the (soft) gluon dynamics can be described by a Langevin equation, the so-called "Boedeker equation" of a similar construct (Chapter 10). Both the BEC and the QGP problems can be described by the coarse-grained effective action (or its equivalent) discussed in Chapter 5, with better built-in self-consistency.

In a truly dynamical setting, any a priori separation between a condensate and a fluctuations band may turn out to be artificial. Nonlinear effects in the condensate will tend to create short-wavelength features even out of smooth initial conditions, as shown dramatically in the phenomena of condensate collapse. Therefore it is better to avoid such a rigid distinction, but stress instead the difference between a c-number (albeit stochastic) component and the environment provided by the remnant q-number fluctuations. This is the approach taken in this chapter. An example we mentioned at the beginning of this chapter is stochastic semiclassical gravity [HuVer03, HuVer04] the arena where many of these ideas were developed and advanced. There, the Einstein–Langevin equation arises from incorporating the fluctuations of the quantum field as a noise term in the semiclassical Einstein equation. By implication this views Einstein's theory as a mean-field theory, a novel conception which can shed some new light on a radically different approach towards quantum gravity, via kinetic theory and stochastic dynamics. For further exposition of these ideas, see [Hu99, Hu02, Hu05].

Although the c-number part is not quite an open system – since no a priori criteria for separation between system and environment have been established – in practice it amounts to very much the same thing. Formally this is reflected by the close analogy between the CTPEA and an influence functional. We therefore say that, by adopting a description based on the fluctuating condensate, we turn the original problem into an effectively open system.

More generally, the fact that the Langevin approach allows us to reproduce the correct quantum Hadamard propagator makes this kind of approach useful in any situation where the amplitudes of fluctuations, rather than their coherence properties, are the main concern. In this light, the decoherence of the mean field is both a subject of theoretical and practical interest – theoretical for reasons stated above, practical because many physical phenomena originate from such processes. Examples are cosmological structure formation and quantum phase transitions. These topics will be treated in later chapters.

9

Entropy generation and decoherence of quantum fields

In Chapter 4 we studied particle creation in an external field, building from the basic concepts and techniques of quantum field theory in a dynamical background field or spacetime to the point where we can recognize that particle creation is in general a non-Markovian process. We derived a quantum Vlasov equation for the rate of particle creation in a changing electric field, and discussed cosmological particle creation from a changing background spacetime. In these processes we pointed out an intrinsic relation between the number and phase of a system in a particular quantum state. We presented a squeezed-state description of particle creation and discussed the conditions under which particle number may increase and others when it may decrease. These discussions bring out some basic issues in the statistical mechanics of quantum fields. In this chapter we will discuss two of these, entropy generation from particle creation and decoherence of quantum fields in the transition from quantum to classical. We will show that dissipation and fluctuations (or noise) in quantum field systems are the primary causes responsible in each of these processes.

In this chapter we shall adopt natural units $\hbar = c = k_B = 1$.

9.1 Entropy generation from particle creation

In discussing the problem of entropy generation from cosmological particle creation [Park69, Zel70, Hu82] we are confronted by the following apparent paradox: on the one hand textbook formulae suggest that entropy (S) is proportional to the number (N) of particles produced (e.g. $S \propto N$ for photons). On the other hand, from quantum field theory, particle pairs created in the vacuum will remain in a pure state and one should not expect any entropy generation. Inquiry into this paradox led to serious subsequent investigations into the statistical properties of particles and fields [Hu84, HuKan87, HuPav86, Kan88a, Kan88b]. These early inquiries in the 1980s of the theoretical meaning of entropy of quantum fields were conducive to gaining a better understanding of the statistical mechanical properties of quantum fields and useful for practical calculations such as for a relativistic plasma of particles and fields in heavy ion experiments, or in finding the entropy content of primordial gravitons in the early universe.

9.1.1 Choice of representations and initial conditions

Many different schemes were proposed in the 1990s for entropy generation from particle production. Brandenberger, Mukhanov and Prokopec [BrMuPr92, BrMuPr93] suggested a coarse graining of the field by integrating out the rotation angles in the probability functional, while Gasperini and Giovannini [GasGio93, GasGioVen93] considered a squeezed vacuum in terms of new variables which give the maximum and minimum fluctuations, and suggested a coarse graining by neglecting information about the subfluctuant variable (defined in Section 4.2). Keski-Vakkuri [Kes94] studied entropy generation from particle creation with many particle mixed initial states. Matacz [Mat94] considered a squeezed vacuum of a harmonic oscillator system with time-dependent frequency, and, motivated by the special role of coherent states, modeled the effect of the environment by decohering the squeezed vacuum in the coherent state representation. Kruczenski, Oxman and Zaldarriaga [KrOxZa94] used a procedure of setting the off-diagonal elements in the density matrix to zero before calculating the entropy. Despite the variety of coarse-graining measures used, in the large squeezing limit (late times) these approaches all give an entropy of $S = 2r$ per mode, where r is the squeezing parameter. This result which gives the number of particles created at late times agrees with that obtained earlier by Hu and Pavon [HuPav86].

Noteworthy in this group of work is that the *representation of the state* of the quantum field and the *coarse graining in the field* are stipulated, not derived. What is implicitly assumed or glossed over in these approaches is the important process of decoherence – the suppression of the off-diagonal components of a reduced density matrix in a certain basis. It is a necessary condition for realizing the quantum-to-classical transition, see [Zur81, Zur82, Zur91, JooZeh85, CalLeg85, UnrZur89, HuPaZh92, Zur93]. The deeper issues are to show explicitly how the entropy of particle creation depends on the choice of specific initial state and/or particular ways of coarse graining, and to understand how natural or how plausible these choices of the initial state representation or the coarse-graining measure are in different realistic physical conditions [Hu94a].[1] To answer these questions, one needs to work with a more basic theoretical framework incorporating statistical mechanics and quantum fields. We shall treat the decoherence and entropy/uncertainty issues with the quantum open system concept [Davies76, LinWes90, Wei93] and the influence functional formalism introduced in Chapters 3 and 5. Our discussion of the different ways of defining the entropy of quantum fields is adapted from [CaHuRa00], while our open systems

---

[1] This includes conditions when, for example, the quantum field is at a finite temperature or is out of equilibrium, interacting with other fields, or that its vacuum state is dictated by some natural choice, for example, in the earlier quantum cosmology regime such as the Hartle–Hawking boundary condition leading to the Bunch–Davies vacuum in de Sitter spacetime.

treatment of entropy generation follows that of Koks *et al.* [Kok96, KoMaHu97]. Notable later work on related subjects includes that of Kiefer *et al.* [KiPoSt00] and Campo and Parentani [CamPar04].

9.1.2 Coarse graining the environment in an open system

In the quantum Brownian motion paradigm the role of the Brownian particle can be played by a detector, a designated mode of a quantum field, such as the homogeneous inflaton field, or the scale factor of the background spacetime (as in minisuperspace quantum cosmology), while the bath could be a set of coupled oscillators, a quantum field, or just the high-frequency sector of the field, as in stochastic inflation. The statistical properties of the system are depicted by the reduced density matrix (rdm) formed by integrating out the details of the bath. One can use the rdm or the associated Wigner function to calculate the statistical average of physical observables of the system, such as the uncertainty or the entropy functions. The von Neumann entropy of an open system is given by

$$S_{\mathrm{CG}} = -\mathrm{Tr}[\rho_R(t) \ln \rho_R(t)], \tag{9.1}$$

The entropy function constructed from the reduced density matrix (or the reduced Wigner function) of a particular state measures the information loss of the system in that state to the environment (or, in the phraseology of [ZuHaPa03], the "stability" characterized by the loss of predictive power relative to the classical description). One can study the entropy increase for a specific state, or compare the entropy at each time for a variety of states characterized by the squeeze parameter. Interaction with the environment changes the system's dynamics from unitary to dissipative, the energy loss being measured by the viscosity function, which governs the relaxation of the system into equilibrium with the environment. The entropy function for such open systems can also be used [AndHal93, Hal93, AnaHal95, HuZha93b, HuZha95, ZuHaPa03] as a measure of how close different quantum states can lead to a classical dynamics. For example, the coherent state being the state of minimal uncertainty has the smallest entropy function [ZuHaPa03] and a squeezed state in general has a greater uncertainty function [HuZha93b, HuZha95]. One can thus use the uncertainty to measure how classical or "nonclassical" a quantum state is.

With regard to the issue of entropy of quantum fields raised at the beginning, we can ask, what is the difference of this more rigorous definition based on open-system dynamics and those obtained with more *ad hoc* prescriptions?

9.1.3 Differences in various definitions of entropy

Consider, for example a representative list of papers on the entropy of quantum fields, such as [Hu84, HuPav86, HuKan87, Kan88a, Kan88b, BrMuPr92,

BrMuPr93, GasGio93, GasGioVen93]. We see that in some cases the entropy refers to that of the field, and is obtained by coarse graining some information of the field itself, such as making a random phase approximation, adopting the number basis, or integrating over the rotation angles. The entropy of [HuZha93b, HuZha95, AndHal93, Hal93, AnaHal95, ZuHaPa03], on the other hand, refers to that of the open system and is obtained by coarse graining the environment. Why is it that for certain generic models in some common limit (late time, high squeezing), both groups of work obtain the same result? Under what conditions would they differ? Understanding this relation could provide a more solid theoretical foundation for the intuitively argued definitions of field entropy.

At the formal level, supposing we have some system which has been decomposed into two subsystems, it is well known (e.g. [Pag93]) that between the entropies S_1, S_2 of the two subsystems, and that of the total system, S_{12}, a triangle inequality holds:

$$|S_1 - S_2| \leq S_{12} \leq S_1 + S_2 \qquad (9.2)$$

In particular, if the total system is closed and so in a pure state, then it has zero entropy, so that the two subsystems necessarily have equal entropies.[2] Hence, asking for the entropy change of a system is equivalent to asking for the entropy change of the environment it couples to, if the overall closed system is in a pure state. Now consider the case of the system as a detector (or a single mode of a field) and the environment as the field. The information lost in coarse graining the field which was used to define the field entropy in the above examples is precisely the information lost as registered in the particle detector, which shows up in the calculation of entropy from the reduced density matrix. The bilinear coupling between the system and the bath as used in the simple quantum Brownian motion models also ensures that the information registered in both sectors is directly commutable. This explains the commonalities. However, not all coarse graining and coupling will lead to the same results, as we shall explicitly demonstrate in some examples below.

Another important feature of the entropy function obtained in this more rigorous open-systems definition which is not obvious in other *ad hoc* approaches is that it depends nonlocally on the entire history of the squeezing parameter. This can be seen from the fact that the rate of particle creation varies in time and its effect is history dependent [HarHu79, CalHu87]. We have seen this behavior in Chapter 4. Existing methods of calculating the entropy generation give results which only depend on the squeezing parameter at the time when a particular

---

[2] This could be the reason why the derivation of black hole entropy (e.g. [Bek94]) can be obtained equivalently by computing the entropy of the radiation (e.g. [FroNov93]) emitted by the black hole, or by counting the internal states (if one knows how) of the black hole (e.g. [ZurTho85, Bek83, BekMuk95, StrVaf96]). Physically one can view what happens to the particle as a probe into the state of the field.

coarse graining (or dropping the off-diagonal components of the density matrix) is implemented.

In Chapter 3 we introduced open quantum systems in terms of influence functionals, following the treatment of [HuPaZh92, HuMat94]. In Chapter 4 we introduced squeezed quantum system, using a general oscillator Hamiltonian as an example, following the treatment of [HuKaMa94, HuMat94]. Here we apply these methods to calculate the entropy and uncertainty functions and then specialize to an oscillator system, recovering en route the results of [HuZha93b, HuZha95, AndHal93, Hal93, AnaHal95, HuZha93b, HuZha95] for the uncertainty function at finite temperature, and of [ZuHaPa03] on the entropy of coherent states. These results are also useful for the consideration of entropy of particles created in the early universe (see, e.g. [KoMaHu97] for a minimally coupled scalar field mimicking a graviton field in a de Sitter universe).

9.2 Entropy of quantum fields

Our discussion in this chapter started with the posing of a deceptively simple question: Is there entropy generation in particle creation? Attempting to answer this question uncovers a host of basic issues in the statistical mechanics of quantum fields. Here we briefly describe the entropy functions obtained from two different types of considerations and operations: The first type is for particle creation in free quantum fields. The main point is the choice of representations and the specification of the initial state. The second type is for particle creation in interacting quantum fields.

To begin with, we note that for a closed system with a unitarily evolving quantum field its dynamics is governed by the quantum Liouville equation, and the von Neumann entropy constructed from the density matrix of the closed system,

$$S_{\text{VN}} = -\text{Tr}[\rho(t) \ln \rho(t)], \qquad (9.3)$$

is exactly conserved. One can introduce approximations or assumptions to render a closed system open or effectively open (see Chapter 1). We distinguish two situations: If there is a justifiable separation of macroscopic and microscopic time-scales, one can adopt the theoretical framework of quantum kinetic field theory. If one assumes an initial factorization condition for the density matrix (as in the "molecular chaos" assumption), one obtains a relativistic Boltzmann equation. The Boltzmann entropy S_{B} defined in terms of the phase space distribution $f(k, X)$ for quasiparticles can in this case be shown to satisfy a relativistic H-theorem [GrLeWe80, CalHu88]. We want to generalize this to a correlation entropy for interacting quantum fields.

However, in the case where there does *not* exist such a separation of time-scales, how does one define the entropy of a quantum field? For nonperturbative truncations of the dynamics of interacting quantum fields, this is a nontrivial question [HuKan87]. Intuitively, one expects that any coarse graining which leads to an effectively open system with irreversible dynamics will also lead to the growth of entropy. These operations can be systematically carried out by way of the projection operator techniques. A projection operator P projects out the *irrelevant* degrees of freedom from the total system described by the density operator ρ, yielding the reduced density matrix ρ_R

$$\rho_R(t) = P\rho(t) \tag{9.4}$$

There exists a well-developed formalism for deriving the equation of motion of the *relevant* degrees of freedom, and in terms of it, the behavior of the coarse-grained entropy (9.1), which will in general not be conserved [Nak58, Zwa60, Zwa61, Mor65, WilPic74, Gra82, Kam85, GoKaZi04, GorKar04, Bal75]. The projection operator formalism can be used to express the slaving of higher correlation functions in the correlation hierarchy. From it one can define an entropy in effectively open systems (see, e.g. [Ana97a, Ana97b]). (So far it has only been implemented within the framework of perturbation theory.) Another powerful method adept to field theory is the Feynman–Vernon influence functional formalism developed in Chapter 5. We shall use it to illustrate how to define the entropy functions for quantum open systems [KoMaHu97].

9.2.1 Entropy special to choice of representation and initial conditions

We begin with the simpler yet more subtle case of a free quantum field. Take for example particle creation in a time-varying background field or in an expanding universe studied in Chapter 4. Entropy is generated in the particle production process from the parametric amplification of vacuum fluctuations. The focal point is a wave equation with a time-dependent natural frequency for the amplitude function of a normal mode. (The same condition arises for an interacting field, such as the $\lambda\Phi^4$ theory in the Hartree–Fock approximation or the $O(N)$ field theory at leading order in the large-N expansion.) Since the underlying dynamics is clearly unitary and time-reversal invariant in this case, a suitable coarse graining leading to entropy growth is not trivially evident. Hu and Pavon [HuPav86] made the observation that a coarse graining is implicitly incorporated when one chooses to depict particle numbers in the n-particle Fock (or "N") representation or to depict the phase coherence in the phase (or "P") representation. This idea has been further developed and clarified [Kan88a, Kan88b, KoMaHu97, KlMoEi98]. The source of entropy generation for free fields is very different from that of interacting fields (e.g. the growth of correlational entropy, described below) in that particle creation from parametric amplification depends sensitively on the choice of representation for the state space of the parametric oscillators, and the specificity of the initial conditions.

9.2.2 Entropy from projecting out irrelevant variables

In contrast to entropy growth resulting from parametric *particle creation* from
the vacuum, entropy growth due to *particle interactions* in quantum field theory
has a very different physical origin. A coarse graining scheme was proposed by Hu
and Kandrup [HuKan87] for these processes. Expressing an interacting quantum
field in terms of a collection of coupled parametric oscillators, their proposal is
to define a reduced density matrix by projecting the full density operator onto
each oscillator's single-oscillator Hilbert space in turn,

$$\varrho(\mathbf{k}) \equiv \mathrm{Tr}_{\mathbf{k}' \neq \mathbf{k}} \rho \tag{9.5}$$

and defining the reduced density operator as the tensor product Π of the projected single-oscillator density operators $\varrho(\mathbf{k})$,

$$\rho_R \equiv \Pi_{\mathbf{k}} \varrho(\mathbf{k}) \tag{9.6}$$

The coarse-grained (Hu–Kandrup) entropy by projection is then just given by
equation (9.1), from which we obtain

$$S_{\mathrm{HK}} = -\sum_{\mathbf{k}} \mathrm{Tr}[\varrho(\mathbf{k}) \ln \varrho(\mathbf{k})] \tag{9.7}$$

It is interesting to observe that for a spatially translation-invariant density matrix
for a quantum field theory which is Gaussian in the position basis, this entropy
is just the von Neumann entropy of the full density matrix, because the spatially translation-invariant Gaussian density matrix separates into a product over
density submatrices for each $\mathbf{k}$ oscillator. This projection (Hu–Kandrup) coarse
graining, like the correlation-hierarchy (Calzetta–Hu) coarse-graining scheme
described below, does not choose or depend on a particular representation for the
single-oscillator Hilbert space. It is sensitive to the establishment of correlations
through the explicit couplings.

9.2.3 Entropy from slaving of higher correlations

We presented in Chapter 6 a general procedure for obtaining coupled equations for the correlation functions at any order l in the correlation hierarchy,
which involves a truncation of the *master effective action* at a finite order in
the loop expansion [NorCor75, CalHu95a, CalHu00, Ber04a]. By working with
an l loop-order truncation of the master effective action, one obtains a closed,
time-reversal invariant set of coupled equations for the first $l + 1$ correlation
functions, $\hat{\phi}, G, C_3, \dots, C_{l+1}$. In general, the equation of motion for the highest
order correlation function will be linear, and thus can be formally solved using
Green's function methods. The existence of a unique solution depends on sup-
planting this with some causal boundary conditions. When the resulting solution
for the highest correlation function is back-substituted into the evolution equations for the other lower-order correlation functions, the resulting dynamics is

not time-reversal invariant, but generically dissipative, as measured by the correlation entropy. Thus, as was described before, with the slaving of the higher-order correlation functions we have rendered a closed system (the truncated equations for correlation functions) into an *effectively open system*. This coarse-graining scheme and the associated correlation entropy defined for an interacting quantum field has the benefit that it can be implemented in a nonperturbative manner. In addition to dissipation, one expects that an effectively open system will manifest noise/fluctuations [NorCor75, CalHu95a, CalHu00, Ber04a] arising from the slaving of the four-point function to the two-point function in the symmetry-unbroken $\lambda\Phi^4$ field theory. Thus a framework exists for exploring the irreversibility and fluctuations within the context of a unitary quantum field theory, using the truncation and slaving of the correlation hierarchy. The effectively open system framework is useful for precisely those situations, where a separation of macroscopic and microscopic time-scales (which would permit an effective kinetic theory description) does *not* exist, such as is encountered in the thermalization issue.

9.3 Entropy from the (apparent) damping of the mean field

We shall discuss these two situations in more detail with examples in the following two sections. In the first case we consider entropy generation in a closed system of a free quantum field, following the treatment of [HKMP96, KlMoEi98]. In the next section we consider entropy generation in an open system interacting with an environment.

Consider the dynamics of a closed system comprising of a mean field and the fluctuation fields. The time evolution of a closed system is Hamiltonian. The general time-dependent Gaussian density matrix of the system may be parameterized by the canonical variables, as we have seen in Chapter 4. Yet, the evolution in some circumstances can manifest apparent irreversible energy flow from the coherent mean fields to the fluctuating quantum modes and give the appearance of quantum decoherence of the mean field.

So what causes the appearance of damping in the dynamics of the mean field of such systems? To highlight the essential physics we note that this process is analogous to the Landau damping of collective modes in a collisionless electromagnetic plasma described by the Vlasov equation. One can understand this damping and decoherence as the result of *dephasing* of the rapidly varying fluctuations and particle production in the time-varying mean field, as shown in Chapter 4. There, when we show the derivation of the quantum Vlasov equation for the semiclassical scalar QED following [KlMoEi98], we encounter a typical situation in nonequilibrium statistical mechanics, namely, if there is a clear separation of time-scales amongst various processes going on in a system, we can seek an effective description of a particular subsystem by coarse graining or "projecting-out" the other subsystems. In the example at hand if we are only interested in the behavior of the slowly varying particle number $\mathcal{N}_{\mathbf{k}}$, the fast changing correlations $\mathcal{C}_{\mathbf{k}}$ can be projected out.[3] The effect of the environment on the open system is calculated through its back-action on the subsystem of interest. Here we focus on the statistical mechanics of particle creation, highlighting the non-Markovian nature of these processes, and seek a physical definition of entropy for such quantum field systems.

9.3.1 Time-scales

The essential physical ingredient in passing from the quantum evolution of the particle-field system to the kinetic description by the quantum Vlasov equation is the *dephasing* phenomenon, i.e. the near exact cancellation of the rapidly varying phases of the quantum mode functions contributing to the mean electric current of the created pairs. This cancellation depends in turn upon a clean separation of the following time-scales (refer to Chapter 4 for notation): [KlMoEi98]

(1) τ_{qu}, the inverse of the *natural frequency* of a normal mode, rapidly oscillating. It is the shortest time-scale reflecting the microscopic quantum theory.
(2) τ_{cl}, the inverse of $\dot{\mathcal{N}}_{\mathbf{k}}$, measures the slowly varying *mean number* of particles in the adiabatic number basis.
(3) τ_{pl}, of the collective *plasma oscillations* of the electric current and mean electric field produced by those particles.

In the limit $\tau_{\mathrm{qu}} \ll \tau_{\mathrm{cl}}$ quantum coherence (reflected in the phase or correlations) between the created pairs can be neglected because of efficient dephasing and a (semi)classical local kinetic approximation to the underlying quantum theory becomes possible. In the limit $\tau_{\mathrm{cl}} \ll \tau_{\mathrm{pl}}$ the electric field may be treated as approximately *constant* over the interval of particle creation. Thus when both inequalities apply we can replace the true nonlocal source term which describes particle creation in field theory by one that depends only on the instantaneous value of the quasi-stationary electric field, at least over very long intervals of time.

---

[3] Note that projecting out or coarse graining does not mean elimination or truncation. The information of the "irrelevant" variables in the other subsystems (constituting the environment) is retained fully in the integro-differential equation for the subsystem of special interest to us (the "relevant" variable), where the nonlocal kernels retain all the information about the subsystem and the environment. One can attempt to solve it, but because of the memory functions, it requires complicated and elaborate integration procedures. Depending on what specific physical information is targeted, one can devise coarse-graining measures to describe the effect of the environment on the system thus leading to a simplification of this integro-differential equation. One extreme yet familiar example is a heat bath where the environment is so grossly coarse grained that only temperature enters into the overall effect on the system (thus making it possible to use the canonical ensemble in equilibrium statistical mechanics, and linear response theory in near-equilibrium conditions).

Making use of these local approximations, one can find [KlMoEi98] an exact analytic expression for the spontaneous pair creation rate $\frac{d}{dt}\mathcal{N}_{\mathbf{k}}(t)$ for a *constant* electric field in real time, in agreement with the Schwinger result [Sch51] in both its exponential and nonexponential factors. Then by making use of an asymptotic expansion of the exact analytic result for constant fields, uniformly valid everywhere on the real time axis, one obtains a useful *local* approximation to the spontaneous pair creation rate for the slowly varying electric fields. A numerical comparison [KlMoEi98] between the quantum and local kinetic approaches to the dynamical back-reaction problem shows remarkably good agreement, even in quite strong electric fields, $eE \simeq m^2 c^3/\hbar$, over a large range of times.

9.3.2 Density matrix

After the elimination of the rapid variables $\mathcal{C}_{\mathbf{k}}$ defined in (4.113) in favor of the slow variables $\mathcal{N}_{\mathbf{k}}$ one can construct the density matrix in the adiabatic number basis easily [KlMoEi98]. In a pure state (setting $\zeta = 1$ in equation (4.48)) the only nonvanishing matrix elements of ρ are in uncharged pair states with equal numbers of positive and negative charges, $\ell_{\mathbf{k}} = n_{\mathbf{k}}^{(+)} = n_{\mathbf{k}}^{(-)}$, with $\ell_{\mathbf{k}}$ the number of pairs in the mode $\mathbf{k}$, *viz.*

$$\langle 2\ell_{\mathbf{k}}'|\rho|2\ell_{\mathbf{k}}\rangle_{\text{pure}} = e^{i(\ell_{\mathbf{k}}'-\ell_{\mathbf{k}})\vartheta_{\mathbf{k}}(t)} \operatorname{sech}^2 r_{\mathbf{k}}(t) \; (\tanh r_{\mathbf{k}}(t))^{\ell_{\mathbf{k}}'+\ell_{\mathbf{k}}} \qquad (9.8)$$

where the magnitude of the Bogoliubov transformation, $r_{\mathbf{k}}(t)$, is defined in equation (4.27) and its phase, $\vartheta_{\mathbf{k}}(t)$, is determined by

$$\alpha_{\mathbf{k}}\beta_{\mathbf{k}}^* e^{-2i\Theta_{\mathbf{k}}} = -\sinh r_{\mathbf{k}} \cosh r_{\mathbf{k}} \; e^{i\vartheta_{\mathbf{k}}} \qquad (9.9)$$

Hence the off-diagonal matrix elements $\ell' \neq \ell$ of ρ are rapidly varying on the time-scale τ_{qu} of the quantum mode functions, while the diagonal matrix elements $\ell' = \ell$ depend only on the adiabatic invariant average particle number via

$$\langle 2\ell_{\mathbf{k}}|\rho|2\ell_{\mathbf{k}}\rangle_{\text{pure}} \equiv \rho_{2\ell_{\mathbf{k}}} = \operatorname{sech}^2 r_{\mathbf{k}} \tanh^{2\ell_{\mathbf{k}}} r_{\mathbf{k}} = \frac{|\beta_{\mathbf{k}}|^{2\ell_{\mathbf{k}}}}{(1+|\beta_{\mathbf{k}}|^2)^{\ell_{\mathbf{k}}+1}} = \left.\frac{\mathcal{N}_{\mathbf{k}}^{\ell_{\mathbf{k}}}}{(1+\mathcal{N}_{\mathbf{k}})^{\ell_{\mathbf{k}}+1}}\right|$$

$$(9.10)$$

and are therefore much more slowly varying functions of time. The average number of positively charged particles (or negatively charged antiparticles) in this basis is given by

$$\sum_{\ell_{\mathbf{k}}=0}^{\infty} \ell_{\mathbf{k}}\rho_{2\ell_{\mathbf{k}}} = \mathcal{N}_{\mathbf{k}} \qquad (9.11)$$

Thus the diagonal and off-diagonal elements of the density matrix in the adiabatic particle number basis stand in precisely the same relationship to each other and contain the same information as the particle number $\mathcal{N}_{\mathbf{k}}$ and pair correlation $\mathcal{C}_{\mathbf{k}}$ respectively.

9.3.3 Entropy generation

In the density matrix (9.10) the diagonal elements $\rho_{2\ell_{\mathbf{k}}}$ may be interpreted (for a pure state) as the independent probabilities of creating $\ell_{\mathbf{k}}$ pairs of charged particles with canonical momentum $\mathbf{k}$ from the vacuum. This corresponds to disregarding the intricate quantum phase correlations between the created pairs in the unitary Hamiltonian evolution. When physics is expressed in the adiabatic particle number basis (the Fock or N representation) the phase information is ignored. The quantum density matrix in this representation produces an entropy function which reflects that associated with particle creation but says nothing about the evolution of the quantum phase or correlation. This illustrates the crucial role played by the choice of representations in the definition of entropy associated with particle creation [HuPav86].

Results obtained from neglecting quantum phase are known to be quite accurate for long intervals of time in the back-reaction of the current on the electric field producing the pairs, because when the current is summed over all the $\mathbf{k}$ modes, the phase information in the pair correlations cancels very efficiently. Thus for practical purposes one can approximate the full Gaussian density matrix over large time intervals by its diagonal elements only, in this basis.

Let us examine the reduced von Neumann entropy constructed from the diagonal density matrix (9.10)

$$S_{\mathcal{N}}(t) = -\sum_{\mathbf{k}} \sum_{\ell_{\mathbf{k}}=0}^{\infty} \rho_{2\ell_{\mathbf{k}}} \ln \rho_{2\ell_{\mathbf{k}}} \tag{9.12}$$

Upon substituting (9.10) into this, the sums over $\ell_{\mathbf{k}}$ are geometric series which are easily performed. The von Neumann entropy of this reduced density matrix

$$S_{\mathcal{N}}(t) = \sum_{\mathbf{k}} \{(1+\mathcal{N}_{\mathbf{k}}) \ln(1+\mathcal{N}_{\mathbf{k}}) - \mathcal{N}_{\mathbf{k}} \ln \mathcal{N}_{\mathbf{k}}\} \tag{9.13}$$

is precisely equal to the Boltzmann entropy of the single particle distribution function $\mathcal{N}_{\mathbf{k}}(t)$. Hence

$$\frac{d}{dt} S_{\mathcal{N}} = \sum_{\mathbf{k}} \ln\left(\frac{1+\mathcal{N}_{\mathbf{k}}}{\mathcal{N}_{\mathbf{k}}}\right) \frac{d}{dt} \mathcal{N}_{\mathbf{k}} \tag{9.14}$$

increases if the mean particle number increases. This is always the case *on average* for bosons if one starts with vacuum initial conditions, since $|\beta_{\mathbf{k}}|^2$ is necessarily nonnegative and can only increase if it is zero initially [Kan88a, Kan88b]. Locally, or once particles are present in the initial state, particle number or the entropy (9.14) does not necessarily increase monotonically in time.

Hence the notion of entropy associated with particle creation, and the lore that it increases in time, is only valid for spontaneous production of bosons from an initial vacuum state. This function associated with fermions, and that associated with stimulated production of both boson and fermions, can decrease in time. This we have remarked in Chapter 4.

9.3.4 Decoherence functional

Decoherence is also addressable within the same framework. Consider the case, where $\omega(t)$ is a function of one external degree of freedom, the mean field $A(t)$. If only the evolution of A is of interest, then the fluctuating modes described by $f(t)$ may be treated as the "environment." To solve for the evolution of the reduced density matrix of A, we compute the influence functional of two trajectories $A_1(t)$ and $A_2(t)$ corresponding to two different evolution operators $U_1(t)$ and $U_2(t)$ defined by

$$F_{12}(t) \equiv \exp(i\Gamma_{12}(t)) \equiv \mathrm{Tr}\left(U_1(t)\rho(0)U_2^\dagger(t)\right) \tag{9.15}$$

Explicit evaluation may be carried out using (4.1.53). Restricting again to pure states with vanishing $\bar{q}$ mean fields we find

$$\Gamma_{12}\bigg|_{\substack{\varsigma=1 \\ \bar{q}=\bar{p}=0}} = \frac{-i}{2}\ln\left\{\frac{i\hbar}{|f_1 f_2|}\left(\frac{f_1 f_2^*}{f_1 \dot{f}_2^* - \dot{f}_1 f_2^*}\right)\right\} \tag{9.16}$$

in terms of the two sets of mode functions $f_1(t)$ and $f_2(t)$ which satisfy (4.54) and (4.17). This Γ_{12} is precisely the closed time path (CTP) effective action functional which generates the connected real time n-point vertices in the quantum theory [CHKMPA94]. For a pure initial state, the absolute value of F_{12} measures the overlap of the two different evolutions at some time t, beginning with the same initial $|\psi(0)\rangle$. In mean field theory, instead of evaluating Γ_{12} for two arbitrary trajectories, the evaluation is over trajectories determined by the *self-consistent* evolution of the closed system, beginning with two different initial mean fields. The intimate relation between the CTP effective action functional and the decoherence functional was pointed out by [CalHu93, CalHu95a, CalHu94, HuMat94, Ana97a, Ana97b].

9.4 Entropy of squeezed quantum open systems

In Chapter 3 we studied quantum open systems with the harmonic oscillator Brownian motion model (QBM). In Chapter 4 we studied squeezed quantum systems as exemplified by particle creation in a dynamical background (with a Lagrangian (4.233)) and squeezed quantum open system exemplified by a parametric oscillator QBM (with Lagrangian (3.133)). Now we inquire about the entropy of squeezed quantum open systems. We seek a definition of the entropy S and the uncertainty function of a squeezed system interacting with a thermal bath, and study how they change in time by following the evolution of the reduced density matrix in the influence functional formalism. As examples, we calculate the entropy of two exactly solvable squeezed systems: an inverted harmonic oscillator and a scalar field mode evolving in an inflationary universe. For the inverted oscillator with weak coupling to the bath, at both high and low temperatures, $S \to r$, where r is the squeeze parameter defined in equation (4.217). For a massless minimally coupled scalar field in the de Sitter universe, $S \to (1-c)r$ at high temperatures where $c = \gamma_0/H$, γ_0 is the coupling to the bath and H the Hubble constant. These two cases confirm previous results based on more *ad hoc* prescriptions for calculating entropy. But for such a scalar field at low temperatures, the de Sitter entropy $S \to (1/2-c)r$ is noticeably different. This result, obtained from a more rigorous treatment, shows that factors usually ignored by the conventional approaches, i.e. the nature of the environment and the coupling strength between the system and the environment, are important. Our treatment here is based on the results obtained in Chapter 5, Section 5.4, derived from the work of Hu, Koks and Matacz [KoMaHu97, HuMat94].

9.4.1 Entropy from the evolutionary operator for reduced density matrix

Consider again the quantum Brownian model discussed in Chapter 3. Our system is modeled by a harmonic oscillator (with coordinate x) with time-dependent mass (M), cross-term $(\mathcal{E})$ and natural frequency (Ω) coupled bilinearly with an environment modeled by many oscillators (with coordinates q_n) of similar nature $(m_n, \varepsilon_n, \omega_n)$. The total Lagrangian is given by equation (3.133).

Assume the systems are initially in the vacuum state, so that their density matrix is Gaussian. Starting with an initial Gaussian reduced density matrix in the form

$$\rho_r(x_i \, x_i' \, t_i) \propto e^{-\lambda x_i^2 + \lambda_\times x_i x_i' - \lambda^* x_i'^2} \tag{9.17}$$

it is evolved by action of the evolutionary operator $\mathcal{J}_r$ for the reduced density matrix of the parametric quantum Brownian oscillator defined in (3.49) into

$$\rho_r(x, x', t) = N e^{-Au^2 - 2iBXu - 4CX^2} \tag{9.18}$$

where $x, x' = X \pm (u/2)$ and the A, B, C functions enter into the evolutionary operators $\mathcal{J}_r$ given by (3.135). They are in turn dependent on the a_{ij}, b_k coefficients given by (4.294), which are solutions to the differential equations for the coefficients of the generalized master equation (3.150) [HuPaZh92, HuPaZh93a]. Here,

$$N = 2\sqrt{C/\pi} \tag{9.19}$$

$$A = a_{22} + \frac{1}{D}\left\{[(2\lambda_r + \lambda_\times)/4 + a_{11}]\,b_3^2 + (2\lambda_i + b_4)\,a_{12}\,b_3 - (2\lambda_r - \lambda_\times)a_{12}^2\right\} \tag{9.20}$$

$$B = -b_1/2 + \frac{1}{D}[(\lambda_i + b_4/2)\,b_2\,b_3 - (2\lambda_r - \lambda_\times)a_{12}\,b_2] \tag{9.21}$$

$$C = \frac{1}{4D}(2\lambda_r - \lambda_\times)\,b_2^2 \tag{9.22}$$

$$D = 4|\lambda|^2 - \lambda_\times^2 + 4\,(2\lambda_r - \lambda_\times)a_{11} + 4\,\lambda_i\,b_4 + b_4^2 \tag{9.23}$$

where λ_r, λ_i are the real and imaginary parts of λ. These expressions form the basis of calculations for squeezed quantum open systems. The reduced density matrix can be obtained by using the expressions above, which depend on $\chi = \alpha + \beta$, the sum of the Bogoliubov coefficients for the effective oscillator. For more details refer to Chapter 4, Section 4.7 [HuMat94, KoMaHu97].

The entropy of a field mode has been calculated by Joos and Zeh [JooZeh85, BKLS86] and others. It can be derived from the reduced density matrix at time t by using the von Neumann entropy (9.3), and is given by

$$S = \frac{-1}{w}[w \ln w + (1 - w) \ln(1 - w)] \simeq 1 - \ln w \quad \text{if } w \to 0 \tag{9.24}$$

where

$$w \equiv \frac{2\sqrt{C/A}}{1 + \sqrt{C/A}} \tag{9.25}$$

A simpler quantity to use is the linear entropy:

$$S_{\text{lin}} \equiv -\text{Tr}\,\rho^2 = -\sqrt{C/A} \tag{9.26}$$

and $S = 0 \to \infty$ is equivalent to $S_{\text{lin}} = -1 \to 0$, both strictly increasing. Then if $S_{\text{lin}} \to 0$ we have

$$S \to -\ln|S_{\text{lin}}| + 1 - \ln 2, \quad \text{i.e. } S_{\text{lin}} \to -\frac{1}{2}(e^{1-S}) \tag{9.27}$$

As an example, suppose we have a system in an initially pure Gaussian state ($\lambda_\times = 0$), so that noise and dissipation are absent: $\gamma_0 = 0$, defined in (3.142). In this case, we have

$$a_{11} = a_{12} = a_{22} = 0 \tag{9.28}$$

so that $C/A = 1$ and hence $S = 0$, as expected.

9.4.2 Measures of fluctuations and coherence

At this point it is useful to supplement our presentations of squeezed quantum open systems in Chapters 3–5 by a discussion of the relation between fluctuations, coherence and entropy. In some cases the description for the dynamics of a squeezed (closed) quantum system can be simplified by expressing the density matrix in terms of the so called super- and subfluctuant variables $u_{\text{SF}}, v_{\text{SF}}$ obtained as real linear combinations of the canonical variables q, p:

$$u_{\text{SF}} = -\kappa \sin \phi\, q + \cos \phi\, p \tag{9.29}$$

$$v_{\text{SF}} = \cos \phi\, q + \frac{\sin \phi}{\kappa}\, p \tag{9.30}$$

This rotation eliminates the cross-terms in the Wigner function. We fix the linear combinations such that one variable (u, the superfluctuant) grows exponentially while the other decays exponentially.

Writing the density matrix in the u_{SF} basis, e.g. $\rho(u_{SF}, u'_{SF})$, one can then compute the fluctuations in u_{SF} and v_{SF} as (see Section IIIC of [KoMaHu97] for details)

$$\Delta u_{SF}^2 = \langle u_{SF}^2 \rangle - \langle u_{SF} \rangle^2 = \frac{\varsigma \varpi}{2C} \quad \Delta v_{SF}^2 = \frac{\varsigma}{2C} \tag{9.31}$$

where $\varphi, \varsigma, \varpi$ are defined as

$$\varphi \equiv \frac{\kappa}{2} \cot \phi, \quad \varsigma \equiv \frac{\sin^2 \phi}{\kappa^2} \left[4AC + (B - \varphi)^2 \right] \tag{9.32}$$

$$\varpi \equiv \frac{4AC + (4\varphi\varsigma + B - \varphi)^2}{4\varsigma^2} \tag{9.33}$$

As a measure of coherence we note from (9.18) that a large A coefficient means that the density matrix is strongly peaked along its diagonal, i.e. there is very little coherence in the system. A measure of coherence was defined in [Mat94] as a squared coherence length L^2, equal to the coefficient of $-u^2$ divided by 8, so that a large L^2 means a high degree of coherence in the system. With this definition of L^2, we have

$$L_u^2 = \frac{\varsigma \varpi}{2A}, \quad L_v^2 = \frac{\varsigma}{2A} \tag{9.34}$$

We can thus relate the coherence lengths and fluctuations to the entropy of the system by

$$\frac{L_u^2}{\Delta u_{SF}^2} = \frac{L_v^2}{\Delta v_{SF}^2} = S_{\text{lin}}^2 = \frac{C}{A} \tag{9.35}$$

(Note that linear entropy is negative by definition in order for it to increase with S. Then as S_{lin} increases, S_{lin}^2 will decrease.) Also the uncertainty relation for u_{SF}, v_{SF} becomes

$$\Delta u_{SF}^2 \Delta v_{SF}^2 = \frac{1}{S_{\text{lin}}^2} \left[\frac{1}{4} + \frac{(4\varphi\varsigma + B - \varphi)^2}{16AC} \right] \tag{9.36}$$

For the free field the last term in the square brackets is zero while $S_{\text{lin}} = -1$ (since $S = 0$), so that $\Delta u_{SF} \Delta v_{SF} = 1/2$.

9.4.3 Entropy and uncertainty functions of an inverted oscillator

We can now demonstrate how the previous results are used. An oscillator with time-independent frequency Ω coupled to a thermal ohmic bath of like oscillators has local dissipation (i.e. $\mathbf{D} \propto \delta'(t - t')$), and at $T \to \infty$ the noise becomes white ($\mathbf{N} \propto \delta(t - t')$). The entropy in this simple case is easily compared with known results in equilibrium statistical mechanics: the entropy at high temperature is

$$S \to 1 + \ln \frac{T}{\Omega} \tag{9.37}$$

We can also use this entropy expression to investigate the claim by Zurek, Habib and Paz [ZuHaPa03] (in the small γ_0 limit by using a Wigner function approach) that for large times the state of least entropy for the oscillator (with a time-independent natural frequency) is the coherent state, at least for white noise and local dissipation. Since the coherent state is the "most classical-like" quantum state, this was invoked as an indication of quantum to classical transition. Equivalently one can use the uncertainty function as a measure. This was shown by Hu and Zhang [HuZha93b, HuZha95], and Anderson, Anastopoulos and Halliwell [AndHal93, Hal93, AnaHal95].

The static inverted harmonic oscillator (IHO) is perhaps the simplest squeezed system. It has been used as a model to study quenching in a quantum phase transition (see the next section). It also models the zero mode of the inflaton field in new inflation (see Chapter 15). Its Lagrangian is:

$$L(t) = \frac{1}{2}[\dot{x}^2 + \Omega^2 x^2] \tag{9.38}$$

We touched on this case in Chapter 4, Section 4.7 as an example of a squeezed quantum system. Suppose this system is coupled to the usual environment of harmonic oscillators in a thermal state, with coupling constant $c(s) = 1$. Then the equivalent oscillator we consider has unit mass, no cross-term and frequency

$$\Omega^2_{eff} = -\Omega^2 - \gamma_0^2 \equiv -\kappa^2 \tag{9.39}$$

so that from (4.239) the sum of its Bogoliubov coefficients is (taking $t_i = 0$, recall $z \equiv \kappa t$, $\sigma \equiv \kappa s$)

$$\chi(t) = \cosh z - i \sinh z \tag{9.40}$$

Hence we have

$$\alpha = \cosh z, \quad \beta = -i \sinh z \tag{9.41}$$

so that at late times as $z \to \infty$, $r \to z$. To determine the entropy generated we need to calculate the various quantities in the propagator coefficients. For white noise these coefficients have analytic solutions, but for zero temperature we need to calculate them numerically.

The b_i's are independent of the temperature, and are found to be (where here and elsewhere a carat will denote division by κ)

$$b_{\{\substack{1 \\ 4}\}} = \kappa(\pm \coth z - \hat{\gamma}_0), \quad b_{\{\substack{2 \\ 3}\}} = \frac{\pm \kappa e^{\pm \hat{\gamma}_0 z}}{\sinh z} \tag{9.42}$$

High temperature

White noise is given by $\mathbf{N}(s, s') = 4\gamma_0 T \, \delta(s - s')$, or $\mathbf{N}(\sigma, \sigma') = 4\hat{\gamma}_0 \kappa^2 T \delta(\sigma - \sigma')$. Using these, Kok, Matacz and Hu derived the expressions for the a_{ij} coefficients. Note that $\hat{\gamma}_0 = \gamma_0/\kappa < 1$; however if we assume small dissipation ($\hat{\gamma}_0 \ll 1$) we can write down large time limits of these quantities:

$$a_{11} \to \frac{T\hat{\gamma}_0}{1 - \hat{\gamma}_0}, \quad a_{12} \to \frac{2Te^{-(1-\hat{\gamma}_0)z}}{1 + \hat{\gamma}_0}, \quad a_{22} \to \frac{T\hat{\gamma}_0}{1 + \hat{\gamma}_0}$$

$$b_{\{\substack{1 \\ 4}\}} \to \kappa(\pm 1 - \hat{\gamma}_0), \quad b_{\{\substack{2 \\ 3}\}} \to \pm 2\kappa \, e^{-(1\mp\hat{\gamma}_0)z} \tag{9.43}$$

We can then calculate large time limits of the density matrix coefficients from (9.19):

$$A \to a_{22}, \quad B \to -b_1/2, \quad C \to \frac{b_2^2}{16a_{11}} \tag{9.44}$$

These coefficients are independent of the initial conditions, which might be expected since the dissipation is acting to damp out any late time dependence on these initial conditions. We have

$$S_{\text{lin}} = -\sqrt{\frac{C}{A}} \to \frac{-\kappa^2 e^{-z}}{2\gamma_0 T} \tag{9.45}$$

From (9.27) and the fact that $r \to z$ as $z \to \infty$ we obtain

$$S \to r + 1 + \ln \frac{T\gamma_0}{\kappa^2} \tag{9.46}$$

Zero temperature

At $T = 0$, the action of the environment is due to quantum effects only. Analytic expressions for the a_{ij}, b_k coefficients in this case can be found in [KoMaHu97].

At $T = 0$, for weak dissipation, $\hat{\gamma}_0 \ll 1$ we have at late times,

$$A \to a_{22}, \quad B \to -b_1/2, \quad C \to \frac{b_2^2}{16a_{11}} \tag{9.47}$$

Again the coefficients are independent of the initial conditions. Since b_2 is unchanged from the high-temperature case and a_{11}, a_{22} tend toward constants, we see that

$$S_{\text{lin}} \to \frac{-\kappa e^{-z}}{2\sqrt{a_{11}a_{22}}} \tag{9.48}$$

and so again from (9.27) and since at late times, $r \to z$,

$$S \to r + 1 + \ln \frac{\sqrt{a_{11}a_{22}}}{\kappa} \tag{9.49}$$

In conclusion, approaching the problem of entropy and uncertainty from the open system viewpoint enables one to see explicitly their dependence on the coarse graining of the environment and the system–environment couplings. It also exposes the relation between quantum and classical descriptions – it is through decoherence that the quantum field becomes classical [CalHu94, AngZur96]. This is the subject of the next section.

9.4.4 Entropy from graviton production in de Sitter spacetime

We now turn to an example in cosmology, that of an inflationary universe (see Chapter 15). We want to calculate the entropy of a massless scalar field minimally coupled to gravity in a de Sitter spacetime by examining the evolution of the density matrix. As we shall see, it is a generally solvable squeezed system.

Consider a massless minimally coupled scalar field in de Sitter space,

$$L_{\text{new}}(\eta) = \sum \frac{1}{2}\left[\chi'_{\mathbf{k}}\chi_{-\mathbf{k}} + \frac{2}{\eta}\chi_{\mathbf{k}}\chi'_{-\mathbf{k}} - \chi_{\mathbf{k}}\chi_{-\mathbf{k}}\left(k^2 - \frac{1}{\eta^2}\right)\right] \tag{9.50}$$

We also use a spectral density [Wei93] of the form

$$I(\omega,\eta,\eta') = \frac{2\gamma_0}{\pi H}\frac{\omega}{\sqrt{\eta\eta'}} \tag{9.51}$$

so that $c(\eta) = 1/\sqrt{-H\eta}$. This corresponds to an ohmic bath with a time-dependent coupling to the system. Since γ_0/H is dimensionless we rewrite it as c, not to be confused with $c(\eta)$. Incorporating the bath gives the equivalent oscillator with $M = 1$, $\mathcal{E} = 1/\eta$ and frequency

$$\Omega_{\text{eff}}^2 = k^2 - \frac{1+c^2}{\eta^2} \tag{9.52}$$

With $z = k\eta$, $\sigma = ks$ we can write the dynamical equation for the quantity χ introduced in Chapter 4, Section 4.7 as

$$\chi''(z) + \left(1 - \frac{2+c^2}{z^2}\right)\chi = 0$$

$$\chi(z_i) = 1, \quad \chi'(z_i) = -i - 1/z_i \tag{9.53}$$

where $z < 0$. The solution of this equation can be constructed using Bessel functions whose index is a function of c; however since we are interested in small c we take the solution to be approximately that of the same equation but with c set to zero. This simplifies things greatly:

$$\chi(z) = \left(1 + \frac{i}{2z_i}\right)f(z) + \frac{i}{2z_i}f^*(z) \tag{9.54}$$

where

$$f(z) \equiv \left(1 - \frac{i}{z}\right)e^{i(z_i - z)} \tag{9.55}$$

We can further simplify χ by using a very early initial time, setting $z_i \to -\infty$. We also disregard the phase in the resulting expression for χ, since this is not expected to make any difference to physical quantities. In this case we obtain a new function which we rename χ:

$$\chi(z) \to \left(1 - \frac{i}{z}\right)e^{-iz} \tag{9.56}$$

The Bogoliubov coefficients can now be found from (4.241):

$$\alpha = \left(1 - \frac{i}{2z}\right)e^{-iz}, \quad \beta = \frac{-i}{2z}e^{-iz} \tag{9.57}$$

and so at late times

$$r \to -\ln|z| \tag{9.58}$$

This result was also obtained in [Mat94] using a different method.

In [KoMaHu97] the expressions for a_{ij}, b_k were derived to leading order in z, and from them the authors show that the coefficients A, B, C tend to the same form as for the static oscillator.

High temperature

We begin by writing

$$\mathbf{N} = 4cc^2(s)T\delta(s - s') \qquad (9.59)$$

$$= \frac{-4ck^2T}{\sigma}\,\delta(\sigma - \sigma') \qquad (9.60)$$

From the expressions given in [KoMaHu97] for a_{ij}, b_k at high temperature one can obtain their behavior as $z \to 0$. Since in this case the coefficients A, B, C tend to the same form as for the static oscillator, thus

$$S_{\text{lin}} \to \frac{-|b_2|}{4\sqrt{a_{11}a_{22}}} = O|z|^{1-c}. \qquad (9.61)$$

Using (9.27) and (9.58) we obtain

$$S \to (1 - c)r + \text{constant} \qquad (9.62)$$

Finite temperature

In this case

$$A \to a_{22} - \frac{a_{12}^2}{4a_{11}}, \quad B \to -b_1/2, \quad C \to \frac{b_2^2}{16a_{11}} \qquad (9.63)$$

and so

$$S_{\text{lin}} \to O|z|^{1/2-c} \qquad (9.64)$$

Then with (9.27) and (9.58) we have

$$S \to (1/2 - c)r + \text{constant} \qquad (9.65)$$

9.4.5 Discussion

In the last two sections we calculated the entropy of two physical and exactly solvable squeezed systems: an inverted harmonic oscillator and a scalar field mode evolving in a de Sitter inflationary universe. To compare these results with that obtained from the more *ad hoc* approaches, we must bear in mind that for a field mode that could be split into two independent sine and cosine (standing wave) components, the result will be twice that obtained here, namely, $S = 2r$ (rather than r in here)

For the inverted oscillator, in both temperature regimes with weak coupling, we obtained $S \to r + \text{constant}$. In the de Sitter case, the high-temperature result is $S \to (1 - c)r + \text{constant}$. In these three examples the results obtained for the entropy from the more *ad hoc* approaches comply with the first principles results presented here. However at lower temperatures the de Sitter entropy is $S \to (1/2 - c)r + \text{constant}$. This last result requires us to look more closely at A and C which together give the entropy.

From (9.26) and (9.27), and neglecting the added constants which are always implied, we find that in the high squeezing limit the entropy behaves as $S \to \frac{1}{2} \ln A - \frac{1}{2} \ln C$. When the system–environment coupling is weak, all of the above cases give $-1/2 \ln C \to r$, which is the expected result. The dominant contribution to C always comes from b_2 in the high squeezing limit. This parameter is determined by the squeezing of the system and is essentially independent of the nature of the environment and its coupling to the system. We can therefore conclude that the $\ln C$ contribution to the entropy represents entropy intrinsic to the squeezed system itself. This should be true quite generally for squeezed systems. However these results fail to take into account the contributions to the entropy from the $\ln A$ term. This contribution is determined by the a_{ij} factors which strongly depend on the nature of the environment and its coupling to the system. There is no a priori reason to expect this contribution to be small, as illustrated by the finite temperature de Sitter example where $1/2 \ln A \to -r/2$. This highlights the danger in trusting the more *ad hoc* approaches. The crucial point is that the entropy of a system depends not only on the system itself but also on the nature of the environment and how it is coupled to the system.

9.5 Decoherence in a quantum phase transition

Quantum phase transitions [Sac99] refer to phase transitions mediated by quantum fluctuations or parameters of a quantum nature, as opposed to classical fields or parameters (such as temperature or magnetic fields) in classical phase transitions. It is an area of active current research in condensed matter physics. Interestingly enough, this subject has also been the focus of theoretical cosmology – the inflationary universe proposal highlights the vital role played by phase transitions in determining the state and dynamics of the early universe. The essential quantum nature in these phase transitions comes about because the vacuum expectation value of the quantum inflaton field is what drove the universe to a period of rapid expansion and its quantum fluctuations acted as seeds to structure formation in the later universe. Topological defects [VilShe00] appearing in the field configurations, such as magnetic monopoles, cosmic strings and domain walls, may often be of quantum field origins. Unfortunately the existing theories for phase transition, structure and defect formation have largely been built on classical field models. Such existing classical theories may not be naively adaptable for the description of these quantum phenomena without careful scrutiny. Overall, we know that any reliable investigation of these processes should entail both the quantum field and the nonequilibrium (dynamical) aspects. A number of basic issues common to them need be addressed from both the conceptual and the technical levels. Foremost is the question of how the quantum field comes to behave like a classical field, and how the quantum fluctuations become classical stochastic sources. These are the issues of decoherence and noise of quantum fields respectively. We will discuss in this section the issue of decoherence and quantum-to-classical transitions in the context of a second-order phase transition for an interacting quantum field, and revisit both issues in the context of structure formation from quantum fluctuations in the early universe in Chapter 15.

The key question we wish to seek an answer to is the emergence of a classical order parameter field after a second-order phase transition described in quantum field theory language [Cal89]. The system field can be the long-wavelength modes of a quantum field and the environment field can be its own short-wavelength modes, or a different set of quantum fields. We have given a thorough treatment of these two cases in Chapter 5, with a derivation of the influence action, the master equations, and an analysis of the dissipation and noise kernels. Here we show how those results can be of use for tackling this problem. The goal here is to compute the decoherence times for the system-field modes and place them in relation to the other time-scales in the model. If it is shorter than all the other relevant physical time-scales then it may provide some justification for viewing the system quantum field as a classical order parameter field, thus providing a justification for the common practices in existing theories of classical phase transitions. If not, then one has to work out the theory of quantum phase transitions from first principles to highlight the differences from their classical counterparts.

Criteria for decoherence

Correlations peaking around the classical trajectory in the phase space, as indicated by the Wigner function showing such behavior (for a long time being perceived as the closest analog to a classical distribution function), were once believed to be a sufficient criterion of classicality [Hal89], but was shown to be inadequate by Habib and Laflamme [HabLaf90]. As we mentioned in Chapter 5, the Wigner function contains just as much information as the density matrix, and thus one needs to demonstrate by some mechanism the diminishing of the phase information in the quantum system to begin to possess some classical attributes. Since a quantum system almost always interacts with its environment, according to the environment-induced decoherence viewpoint, one can use the diminishing or vanishing of the off-diagonal elements of the reduced density matrix in a suitable basis (such as the "pointer basis" of Zurek) as an indication of, or criterion for, decoherence and the transition to classicality. Likewise one needs to do this for the Wigner function.

Models for quench transition

We focus on quenching which is a second-order quantum phase transition. For a quantum field ϕ with infinite degrees of freedom undergoing a continuous transition, the field ordering after the transition begins is due to the growth in amplitude of its unstable long-wavelength modes. A quench transition can in general be characterized by the quench transition time-scale t_q. Physically this is the time by which the order parameter field has sampled the degenerate ground states. One can take the field to be classical by the time it is ordered as such. This has implications [RiLoMa02] for the formation of the defects that are a necessary by-product of transitions.

A simple model for quench transition is an inverted harmonic oscillator (IHO) which we studied in some detail in an earlier section in this chapter. This is also the simplest model which depicts the evolution of the inflaton field and the growth of inhomogeneities in the early universe. To see why, recall that the normal modes of a massless free scalar field propagating in a Friedmann–Lemaitre–Robertson–Walker universe satisfy the equation

$$\phi_k'' + \left(k^2 - \frac{a''}{a} \right) \phi_k = 0. \tag{9.66}$$

For sufficiently long wavelengths $(k^2 \ll a''/a)$, this equation describes an unstable oscillator.

Guth and Pi [GutPi85] used the IHO model to study the evolution of the inflaton field. They assumed that at the onset of inflation the universe was in a Gaussian quantum state centered on the maximum of the potential. It is easy to show from the solution of the (functional) Schrödinger's equation that the initial wave packet spreads quickly in time but maintains its Gaussian shape (due to the linearity of the model). The initial Gaussian state becomes highly squeezed and indistinguishable from a classical stochastic process. Since the wavefunction is Gaussian, the Wigner function is positive for all times and peaks on the classical trajectories in phase space as the wavefunction spreads. In these situations the Wigner function can be interpreted as a classical probability distribution for coordinates and momenta, showing sharp classical correlations at long times. But the harmonic oscillator is a special case where this condition holds. As remarked above, this criterion based on correlations in phase space is not sufficient to prove the transition from quantum to classical. One needs to also show how the phase information in the quantum system disappears, such as by invoking an environment-induced decoherence process.

Open systems

Guth and Pi did not expound the decoherence and quantum to classical transition issues in depth, but simply invoked the uncertainty principle as an indication of such a transition. Uncertainty principle at a finite temperature was studied by Hu and Zhang [HuZha93b, HuZha95] (see also Anderson, Anastopoulos and Halliwell [AndHal93, Hal93, AnaHal95, HuZha93b, HuZha95]) using a harmonic oscillator bath at finite temperature as the environment. They showed explicitly how a quantum oscillator system evolves from a quantum- to a thermal-dominated state which marks such a transition. Independently Zurek,

Habib and Paz [ZuHaPa03] showed that a quantum system interacting with a high-temperature ohmic bath will most likely evolve to a coherent state, which is known as the quantum state with the most classical features. This was invoked as a criterion for classicality. In an earlier section we have shown how these two criteria, i.e. uncertainty at finite temperature and a quantum state evolving to a coherent state, are actually two sides of the same coin in the environment-induced decoherence perspective.

Interacting fields

The feature of a Gaussian wavefunction maintaining its Gaussian nature in evolution are special to linear systems and the linear instabilities described above are valid only for free fields. For example in an inverted anharmonic oscillator, it has been shown [LoMaMo00] by numerically evolving the Schrödinger equation that an initially Gaussian wavefunction becomes non-Gaussian, the Wigner function develops negative parts, and its interpretation as a classical probability breaks down.[4] A similar argument for quantum mechanics, but for *open systems*, was also presented in [LoMaMo00]. Coupling an inverted oscillator with an anharmonic potential to a high-temperature environment, the authors showed that it becomes classical very quickly, even before the wavefunction probes the nonlinearities of the potential. Being an early time event, the quantum-to-classical transition can be studied perturbatively. Lombardo, Mazzitelli and Rivers (LMR) [LoMaRi01] have extended this to field theory by considering a system field comprising the long wavelengths of the order parameter interacting with a large number of environmental fields, including its own short wavelengths. Assuming weak coupling and high critical temperature, they showed that decoherence is a short time event, shorter than the quench transition time $t_{\rm sp}$. As a result, perturbative calculations are justified. Subsequent dynamics can be described by a stochastic Langevin equation, the details of which are only known for early times.[5]

---

[4] In this connection we mention numerical computations of quantum mechanical models and of different approximations to interacting field theory (see Chapter 12). In such calculations, classical correlations do appear in some field theory models [CHKMPA94, BoVeHo99]. However, since such decoherence (in a time-averaged sense) takes place at long times after the transition has been achieved initially, when the mean field approximation has broken down, this may be an artifact of the Gaussian-like approximations [LoMaMo00].

[5] A remark on the relation with thermal field theory is in place here. As pointed out by [LoMaRi01] there are similarities and differences between this quantum open system approach and the well-established classical behavior of thermal scalar field theory [AarBer02, AarSmit98] at high temperature. It is known that at high temperatures, the behavior of long-wavelength modes is determined by classical statistical field theory. The effective classical theory is obtained after integrating out the hard modes with $k \geq gT$. The "classical behavior" in this soft thermal mode analysis is defined through the coincidence of the quantum and the statistical correlation functions. Thermal equilibrium is assumed to hold at all times and the cut-off that divides system and environment depends on the temperature, which is externally fixed. In phase transitions, the quantum-to-classical transition is defined by the effective diagonalization of the reduced density matrix, which is not assumed to be thermal and the separation between long and short wavelengths is determined by their stability, which depends on the parameters of the potential.

Using a model with biquadratic coupling between the system and the environment, LMR first [LoMaRi01] considered the case of an instantaneous quench, then a slow quench [RiLoMa02]. The consideration of slow quenches is very important since the Kibble–Zurek mechanism [Kib80, Kib88, Zur85, Zur96] predicts the relation between the subsequent domain structure and the quench time (by indirectly counting defects [LagZur97, LagZur98, YatZur98]). The authors of [LoMaRi03] worked out the theories for other couplings but show that the biquadratic coupling is the most relevant for the quantum-to-classical transition. Also, since all relevant time-scales depend only logarithmically on the parameters of the theory, they also showed the necessity of keeping track of $O(1)$ prefactors carefully. In the next section we illustrate a quench quantum transition following their treatment.

9.6 Spinodal decomposition of an interacting quantum field

The model we discuss contains a real system field ϕ, which undergoes a transition, coupled biquadratically to other scalar fields χ_a ($a = 1, 2, \ldots, N$), which constitute the external part of the environment (the internal environment is provided by the short-wavelength modes of the field ϕ itself). The influence functional and the master equation obtained from integrating out the environmental fields have been derived in Chapter 5. We focus on the diffusion coefficients central to the process of decoherence and evaluate upper bounds on the decoherence time t_D for slow quenches. The general conclusion is that the decoherence time is typically shorter than the quench transition time.

The model

We consider for the system a self-interacting scalar ϕ-field which describes the order parameter, whose $\mathcal{Z}_2$ symmetry is broken by a double-well potential, and an environment comprising N free scalar fields χ_a with classical action

$$S[\phi, \chi] = S_{\text{syst}}[\phi] + S_{\text{env}}[\chi] + S_{\text{int}}[\chi_a, \phi] \tag{9.67}$$

where (with $\mu^2, m^2 > 0$)

$$S_{\text{syst}}[\phi] = \int d^4x \left\{ -\frac{1}{2}\partial_\mu\phi\partial^\mu\phi + \frac{1}{2}\mu^2\phi^2 - \frac{\lambda}{4!}\phi^4 \right\}$$

$$S_{\text{env}}[\chi_a] = \sum_{a=1}^{N} \int d^4x \left\{ -\frac{1}{2}\partial_\mu\chi_a\partial^\mu\chi_a - \frac{1}{2}m_a^2\chi_a^2 \right\}$$

The most important interactions will turn out to be of the biquadratic form

$$S_{\text{int}}[\chi_a, \phi] = S_{\text{qu}}[\phi, \chi] = -\sum_{a=1}^{N} \frac{g_a}{8} \int d^4x \, \phi^2(x)\chi_a^2(x) \tag{9.68}$$

Physical conditions

To keep our calculations tractable, we need a significant part of the environment to have a strong impact upon the system field, but not vice versa, from which we can bound t_D. The simplest way to implement this is to take a large number $N \gg 1$ of scalar χ_a fields with comparable masses $m_a \simeq \mu$ weakly coupled to the ϕ, with λ, $g_a \ll 1$. Thus, at any step, there are N weakly coupled environmental fields influencing the system field, but only one weakly self-coupled system field to back-react upon the explicit environment.

For one-loop consistency at second order in our calculation of the diffusion coefficient (that enforces classicality) it is sufficient, at order of magnitude level, to take identical $g_a = g/\sqrt{N}$. Further, at the same order of magnitude level, we take $g \simeq \lambda$.[6]

For small g the model has a continuous transition at a temperature T_c. The environmental fields χ_a reduce T_c and, in order that $T_c^2/\mu^2 = 24/(\lambda + \sum g_a) \gg 1$, we must take $\lambda + \sum g_a \ll 1$, whereby $1 \gg 1/\sqrt{N} \gg g$. Further, with this choice the dominant hard loop contribution of the ϕ-field to the χ_a thermal masses (see Chapter 10) is

$$\delta m_T^2 = O(gT_c^2/\sqrt{N}) = O(\mu^2/N) \ll \mu^2 \tag{9.69}$$

Similarly, the two-loop (setting sun) diagram which is the first to contribute to the discontinuity of the χ-field propagator is of magnitude

$$g^2 T_c^2/N = O(g\mu^2/N^{3/2}) \ll \delta m_T^2 \tag{9.70}$$

in turn. That is, the effect of the thermal bath on the propagation of the environmental χ-fields is ignorable. In particular, the infinite N limit does not exist. Dependence on N is implicit through T_c as well as through the couplings, for initial temperatures $T_0 = O(T_c)$. $\eta = \sqrt{6\mu^2/\lambda}$ determines the position of the minima of the potential and the final value of the order parameter. As has been shown in [LoMaRi01] this choice of coupling and environments gives the hierarchy of scales necessary for establishing a reliable approximation scheme.

We shall assume that the initial states of the system and environment are both thermal, at a high temperature $T_0 > T_c$. We then imagine a change in the global environment (e.g. expansion in the early universe) that can be characterized by a change in temperature from T_0 to $T_f < T_c$. That is, we do not attribute the transition to the effects of the environment fields. As initial conditions of the

---

[6] This is very different from the more usual large-N $O(N+1)$-invariant theory with one ϕ-field and N χ_a fields, dominated by the $O(1/N)$ $(\chi^2)^2$ interactions, that has been the standard way to proceed for a *closed* system. With our choice there are no direct χ^4 interactions, and the indirect ones, mediated by ϕ loops, are depressed by a factor $g/\sqrt{N}$. In this way the effect of the external environment qualitatively matches the effect of the internal environment provided by the short-wavelength modes of the ϕ-field, but in a more calculable way.

open system we take a factorized density matrix at temperature T_0 of the form $\hat{\rho}[T_0] = \hat{\rho}_\phi[T_0] \times \hat{\rho}_\chi[T_0]$.[7]

Provided the change in temperature is not too slow the exponential instabilities of the ϕ-field grow so fast that the field has populated the degenerate vacua well before the temperature has dropped significantly below T_c. Since the temperature T_c has no particular significance for the environment fields, for these early times we can keep the temperature of the environment fixed at $T_\chi \approx T_c$. For simplicity the χ_a masses are fixed at the common value $m \simeq \mu$.

9.6.1 The quench transition time

To describe the physics of the quenching transition we show the estimation of the quench transition time t_{sp} defined from $\langle \phi^2 \rangle_{t=t_{sp}} \sim \eta^2$. We assume that the quench begins at $t = 0$ and ends at time $t = 2\tau_q$, with $\tau_q \gg t_r \sim \mu^{-1}$. At the qualitative level at which we are working it is sufficient to take $m_\phi^2(T_0) = \mu^2$ exactly. Most simply, we consider a quench linear in time, with temperature $T(t)$, for which the mass function is of the following form [BowMom98][8]

$$m^2(t) = m_\phi^2(T(t)) = \begin{cases} \mu^2 & \text{for } t \leq 0 \\ \mu^2 - \dfrac{t\mu^2}{\tau_q} & \text{for } 0 < t \leq 2\tau_q \\ -\mu^2 & \text{for } t \geq 2\tau_q \end{cases} \qquad (9.72)$$

The field behaves as a free field in an inverted parabolic potential for an interval of approximately t_{sp} [KarRiv97], where

$$\langle \phi^2 \rangle_{t_{sp}} \sim \eta^2 \qquad (9.73)$$

The equation of motion for the mode $u_k(t)$, with wavenumber k is, in the quench period,

$$\left[\frac{d^2}{ds^2} + k^2 + \mu^2 - \frac{\mu^2 s}{\tau_q} \right] u_k(s) = 0 \qquad (9.74)$$

---

[7] Given our thermal initial conditions it is not the case that the full density matrix has ϕ and χ fields uncorrelated initially, since it is the interactions between them that lead to the restoration of symmetry at high temperatures. Rather, incorporating the hard thermal loop "tadpole" diagrams of the χ (and ϕ) fields in the ϕ mass term leads to the effective action for ϕ quasi-particles,

$$S_{syst}^{eff}[\phi] = \int d^4x \left\{ -\tfrac{1}{2}\partial_\mu\phi\partial^\mu\phi - \tfrac{1}{2}m_\phi^2(T_0)\phi^2 - \tfrac{\lambda}{4!}\phi^4 \right\} \qquad (9.71)$$

where $m_\phi^2(T) \propto (T/T_c - 1)$ for $T \approx T_c$. As a result, we can take an initial factorized density matrix at temperature T_0 of the form $\hat{\rho}[T_0] = \hat{\rho}_\phi[T_0] \times \hat{\rho}_\chi[T_0]$, where $\hat{\rho}_\phi[T_0]$ is determined by the quadratic part of $S_{syst}^{eff}[\phi]$ and $\hat{\rho}_\chi[T_0]$ by $S_{env}[\chi_a]$. That is, the many χ_a fields have a large effect on ϕ, but the ϕ-field has negligible effect on the χ_a.

[8] Note that the τ_q of [LoMaRi03] is the inverse quench rate $T_c^{-1}dT/dt|_{T=T_c}$, and so differs from that of [BowMom98] by a factor of 2.

subject to the boundary condition $u_k(t) = e^{-i\omega t}$ for $t \leq 0$, where $\omega^2 = \mu^2 + k^2$. Instead of the simple exponentials of the instantaneous quench, $u_k(t)$ has solution [BowMom98]

$$u_k(t) = a_k \mathrm{Ai}\left(\frac{\Delta_k(t)}{\bar{t}}\right) + b_k \mathrm{Bi}\left(\frac{\Delta_k(t)}{\bar{t}}\right) \tag{9.75}$$

with $\mathrm{Ai}(s)$, $\mathrm{Bi}(s)$ the Airy functions; $\Delta_k(t) = t - \omega^2\bar{t}^3$ and $\bar{t} = (\tau_q/\mu^2)^{1/3}$. Note that $\Delta_0(t) = t - \tau_q$, the time since the onset of the transition. In the causal analysis of Kibble [Kib80] $\bar{t}$ ($\mu^{-1} \ll \bar{t} \ll \tau_q$) is the time at which the adiabatic field correlation length collapses at the speed of light, the earliest time in which domains could have formed. The analysis of [LoMaRi03] suggests that this earliest time is not $\bar{t}$, but $t_{\rm sp}$.

It is straightforward to establish a relationship between $\bar{t}$ and $t_{\rm sp} > \bar{t}$. The constants of integration in (9.75) are

$$a_k = \pi[\mathrm{Bi}'(-\omega^2\bar{t}^2) + i\omega\bar{t}\,\mathrm{Bi}(-\omega^2\bar{t}^2)] \tag{9.76}$$
$$b_k = -\pi[\mathrm{Ai}'(-\omega^2\bar{t}^2) + i\omega\bar{t}\,\mathrm{Ai}(-\omega^2\bar{t}^2)]$$

It follows that, when $\Delta_k(t)/\bar{t}$ is large, then

$$|u_k(t)|^2 \approx \omega\bar{t}\left(\frac{\bar{t}}{\Delta_k(t)}\right)^{1/2} \exp\left[\frac{4}{3}\left(\frac{\Delta_k(t)}{\bar{t}}\right)^{3/2}\right]$$
$$\approx \mu\bar{t}\left(\frac{\bar{t}}{\Delta_0(t)}\right)^{1/2} \exp\left[\frac{4}{3}\left(\frac{\Delta_0(t)}{\bar{t}}\right)^{3/2}\right] e^{-k^2/\bar{k}^2} \tag{9.77}$$

where $\bar{k}^2 = \bar{t}^{-3/2}(\Delta_0(t))^{-1/2}/2$.

For large initial temperature $T_0 = O(T_c)$, the power spectrum for field fluctuations peaked around $\bar{k}$, and

$$\langle\phi^2\rangle_t \approx \frac{T_0}{2\pi^2\mu^2} \int k^2\,dk\,|u_k(t)|^2 \approx \frac{CT_0}{\mu\bar{t}^2}\left(\frac{\Delta_0(t)}{\bar{t}}\right)^{-5/4} \exp\left[\frac{4}{3}\left(\frac{\Delta_0(t)}{\bar{t}}\right)^{3/2}\right] \tag{9.78}$$

The prefactor C is included to show that terms, nominally $O(1)$, can in fact be large or small (in this case $C = (64\sqrt{2}\pi^{3/2})^{-1} = O(10^{-3})$). Note that, although the unstable modes have a limited range of k-values, increasing in time, this is effectively no restriction when $\Delta_0(t)/\bar{t}$ is significantly larger than unity.

Finally, we obtain

$$\frac{\eta^2}{C'} \simeq \frac{T_c}{\mu\bar{t}^2} \exp\left[\frac{4}{3}\left(\frac{\Delta_0(t_{\rm sp})}{\bar{t}}\right)^{3/2}\right] \tag{9.79}$$

where $C' = C[\ln(\mu\bar{t}^2\eta^2/CT_c)^{-5/6}]$. Since the effect on $t_{\rm sp}$ only arises at the level of "ln ln" terms, $C' \approx C$ is a good estimation in all that follows. (Since this choice underestimates $t_{\rm sp}$ it only strengthens the claim that $t_{\rm sp} > t_{\rm D}$.)

9.6.2 Decoherence time

We now turn to the question of whether decoherence proceeds faster than spinodal decomposition. Rather than attempt a full estimate of the decoherence time t_D (see [LoMaRi03]), we shall run a simple test.

As we have already remarked, at early times the system field may be described as an inverted harmonic oscillator. The evolution is then well approximated by an ensemble of classical trajectories, but there remains the question of whether two different classical histories are consistent in the Gell-Mann–Hartle sense [RivLom05, LoRiVi07].

A classical history displays spatial structure as well as time evolution. We are helped by the observation that the ordering of the field is due to the growth of the long-wavelength unstable modes. Unstable long-wavelength modes start growing exponentially as soon as the quench is performed, whereas short-wavelength modes will oscillate. As a result, the field correlation function rapidly develops a peak (Bragg peak) at wavenumber $k = \bar{k} \ll \mu$. Specifically [KarRiv97], initially as $\bar{k}^2 = \mathcal{O}(\mu/\sqrt{t\tau_q})$, where τ_q^{-1} is the quench rate. Assuming that a classical description can be justified *post hoc*, a domain structure forms quickly with a characteristic domain size $O(\bar{k})$, determined from the position of this peak. (As an example, see the numerical results of [LagZur97, LagZur98, YatZur98], where this classical behavior has been assumed through the use of stochastic equations – see later.) With this in mind, we adopt an approximation in which the system-field contains only one Fourier mode with $\mathbf{k} = \mathbf{k}_0 = O(\bar{k})$, characteristic of the domain size. For simplicity, we shall further assume $\mathbf{k} = \mathbf{k}_0 = 0$ (we refer the reader to [LoMaRi03] for a more complete analysis).

We may simplify the issue further by considering only trajectories which begin from $\phi = 0$ at $t = \tau_q$. Two such trajectories are distinguished by the value of $\dot{\phi}$ at the initial time. We shall ask what is the minimum speed difference at the initial time that ensures consistency by $t = t_{sp}$. If this minimum difference is much smaller than the natural spread $\sim (T_c/V)^{1/2}$, then the conclusion that decoherence is faster than spinodal decomposition is upheld [RivLom05].

We will calculate the decoherence functional to lowest nontrivial order (two vertices) for large N. Again, we assume weak coupling $\lambda \simeq g \ll 1$, where we have defined g by the order of magnitude relations $g_a \simeq g/\sqrt{N}$. As such we may expand the logarithm of the decoherence functional up to second nontrivial order in coupling strengths.

As "trial" classical solutions, we take

$$\phi(\mathbf{x}, s) = \dot{\phi}\, u(s), \quad \phi'(\mathbf{x}, s) = \dot{\phi}'\, u(s) \tag{9.80}$$

where $u(s)$ is the solution of the mode equation with boundary conditions $u(\tau_q) = 0$, $\dot{u}(\tau_q) = 1$. Since we are neglecting the self-interaction term, our conclusions are only trustworthy for $t \leq t_{sp}$.

The solution of the equations of motion for the mode functions is given by

$$u(t) = \frac{\pi \bar{t}}{3^{2/3}\Gamma(2/3)} \left[\sqrt{3}\mathrm{Ai}\left(\frac{t - \tau_q}{\bar{t}}\right) + \mathrm{Bi}\left(\frac{t - \tau_q}{\bar{t}}\right) \right] \tag{9.81}$$

where $\mathrm{Ai}(s)$ and $\mathrm{Bi}(s)$ are the Airy functions, and we have used $\Delta_0(t) = t - \tau_q$.

The procedure outlined above is quite general and applies to a range of couplings [LoMaRi03]. We now specialize to biquadratic coupling. The modulus of the decoherence functional is given by

$$|\mathcal{D}[\phi^1, \phi^2]|^2 \sim \exp\left\{ -\frac{g^2}{32} \int d^4x \int d^4y\, \phi_-^{(2)}(x) N_q(x,y) \phi_-^{(2)}(y) \right\} \tag{9.82}$$

where $N_q(x - y) = \mathrm{Re}G_F^2(x,y)$ is the noise (diffusion) kernel. G_F is the relevant Feynman propagator of the χ-field at temperature T_0. We have defined $\phi_-^{(2)} = (\phi^1)^2 - (\phi^2)^2$. For our chosen classical histories, and at times $t \sim t_{\mathrm{sp}}$ it becomes

$$|\mathcal{D}[\phi^1, \phi^2]|^2 \sim \exp\left\{ -\frac{g^2}{32} \left[(\dot\phi^1)^2 - (\dot\phi^2)^2 \right]^2 DV \right\} \tag{9.83}$$

where V is the volume of space and

$$D = 2 \int_0^{t_{\mathrm{sp}}} dt \int_0^t ds\, u^2(t) F(t - s) u^2(s) \tag{9.84}$$

$$F(t) = \frac{27}{512} \mathrm{Re}G_F^2(0; t) \tag{9.85}$$

where $G_F^2(k, t)$ is the Fourier transform of the square of the Feynman χ propagator. It is only in $u(s)$ that the slow quench is apparent.

In the high-temperature limit $(T \gg \mu)$, LMR obtain the explicit expression for the kernels

$$\mathrm{Re}G_F^2(0; t) = \frac{T_c^2}{64\pi^2} \int_0^\infty dp\, \frac{p^2}{(p^2 + \mu^2)^2} \cos\left(2\sqrt{p^2 + \mu^2}\, t\right) \tag{9.86}$$

where μ is the thermal χ-field mass at temperature $T \sim T_c$. In this scheme, it is approximately the cold χ mass. Because the χ-field propagator is unaffected by the ϕ-field interactions one can obtain the detail of the expression in (9.86).

We see that, for times $\mu t \geq 1$, the behavior of D is dominated by the exponential growth of $u(s)$, and the integral in equation (9.84) by the interval $s \approx t$. We will assume large $\Delta_0(t)$ (and $\Delta_0(s)$), which means $\Delta_0(t), \Delta_0(s) \gg \bar{t}$. This condition is satisfied provided s is larger than and not too close to $\omega_0^2 \tau_q / \mu^2$, and allows us to use the asymptotic expansions of the Airy functions and their derivatives for the evaluation of $u(s)$. This will be justified *post hoc*. In particular,

$$u(t) = \left(\frac{\sqrt{\pi \bar{t}}}{3^{2/3}\Gamma[2/3]} \right) \left(\frac{\bar{t}}{t} \right)^{1/4} \exp\left[\frac{2}{3}\left(\frac{(t - \tau_q)}{\bar{t}} \right)^{\frac{3}{2}} \right] \tag{9.87}$$

Keeping only the parametric dependence, we obtain

$$|\mathcal{D}[\phi^1, \phi^2]|^2 \sim \exp\left\{-g^2 V \bar{t}^5 \left[(\dot{\phi}^1)^2 - (\dot{\phi}^2)^2\right]^2 \frac{T_c^2 \bar{t}^2}{\mu t_{sp}} \exp\left[\frac{8}{3}\left(\frac{t_{sp} - \tau_q}{\bar{t}}\right)^{3/2}\right]\right\}$$

(9.88)

Now we can use the relations

$$\left[(\dot{\phi}^1)^2 - (\dot{\phi}^2)^2\right]^2 \sim \frac{T_c}{V}(\dot{\phi}^1 - \dot{\phi}^2)^2$$

(9.89)

$$\exp\left[\frac{4}{3}\left(\frac{\Delta_0(t_{sp})}{\bar{t}}\right)^{3/2}\right] \sim \frac{\mu \bar{t}^2 \eta^2}{T_c},$$

(9.90)

to get

$$|\mathcal{D}[\phi^1, \phi^2]|^2 \sim \exp\left\{-g^2 T_c \frac{\bar{t}^{11}}{t_{sp}} \mu \eta^4 (\dot{\phi}^1 - \dot{\phi}^2)^2\right\}$$

(9.91)

Unless the self-coupling is exceedingly small or the space volume too big (in which case it is not appropriate to disregard the spatial structure of the relevant classical evolutions), strong enough decoherence follows from the observation that $\tau_q/t_{sp} \gg 1$.

When these bounds are satisfied the minimum wavelength for which the modes decohere by time t_{sp} can be shown [RiLoMa02] to be shorter than that which characterizes domain size at that time. Although one can talk loosely, but sensibly, about a classical domain structure at time t_{sp} one cannot yet talk about classical defects on their boundaries, as the naive picture might suggest. Defects (in this case, walls) are described by shorter wavelength modes ($k \le \mu$). Nonetheless, the classical domain structure is sufficient to determine their density [RiLoMa02].

The emphasis has been on the many weak environments because of the control that this gives us on establishing a robust upper bound on t_D. However, LMR noted that their total contribution at one loop was qualitatively that of the short-wavelength modes of the ϕ field alone without assuming the action of the environmental fields. So it seems that rapid decoherence is a general feature.

In Chapter 5 we have also shown how for a general class of system–environment interactions (such as the $\phi^2 \chi^m$ types studied), the effect of the environment is largely equivalent to the presence of a stochastic source term in the dynamics of the classical system field, with the correlation functions obeyed by the noise $\xi_m(x)$ corresponding to the specific type of couplings. In particular, for the linear interaction with the environment (to the exclusion of self-interaction) LMR recovered the *additive* noise that has been the basis for stochastic equations in relativistic field theory that confirm the scaling behavior of Kibble's and Zurek's analysis. For times later than t_{sp}, neither perturbation theory nor more general non-Gaussian methods are valid. Also LMR found that the role initially attributed by Kibble (and subsequently by others, e.g. [BraMag99]) to the Ginzburg regime is just not present.

9.7 Decoherence of the inflaton field

As another example of the application of the coarse-grained effective action and the influence functional formalism, we consider the decoherence of the inflaton field in the early universe. The key ingredient in this consideration is the noise associated with quantum fluctuations. We have seen how it is defined from the influence action for an interacting field in Chapter 5. Some background material on inflationary cosmology can be gleaned from the first part of Chapter 15, which the uninitiated readers may wish to consult before reading this section.

As noted earlier, an inverted harmonic oscillator model was used by Guth and Pi [GutPi85] to describe the dynamics of the inflaton field. Though useful for intuitive reasoning, it is over-simplistic in addressing the quantum-to-classical transition issue. This model has also been used by many authors to describe the appearance of classical inhomogeneities from quantum fluctuations in the inflationary era [PolSta96, LePoSt97]. Due to the linearity of the model and the Gaussian form of the wavefunction [Hab04] the quantum–classical correspondence is straightforward. In more general circumstances, the Wigner function can be negative and the simple identification with the one-particle classical distribution function no longer holds. One needs to consider decoherence of the (reduced) Wigner function by an environment [Hab90, HabLaf90], just as we have done for the reduced density matrix in similar considerations.

Turning our attention briefly to cosmology, the proposal to view the long-wavelength sector quantum field as classical, such as demanded by stochastic inflation (in fact, commonly assumed in most theories of structure formation), can only be justified by showing that some decoherence mechanism applies to the inflaton field. Interaction of a quantum system with an environment may bring about decoherence, as we have seen in model problems (such as the QBM) discussed in Chapters 3 and 5. The effectiveness of an environment to bring about quantum-to-classical transition depends on many factors, such as the type of coupling (bilinear, nonlinear), the nature of the bath (spectral density, temperature) and how the interaction determines the pointer basis. Quantitatively, decoherence is usually described by the diagonalization of the reduced density matrix, but this is only meaningful (since a symmetric matrix can always be diagonalized) by specifying or, better yet, showing the likely existence of a pointer basis, which is a physical rather than a mathematical issue. There is by now a huge literature on decoherence (see, e.g. the reviews [GKJKSZ96, Paz00, Zur03]), both in terms of conceptual discussions and model calculations. Here we will limit our discussion only to some attributes of decoherence, and in the context of quantum processes in the early universe.

What in a realistic situation could play the role of the environment field? One can consider either one interacting field partitioned into two sectors, the low-frequency sector as the system and the high-frequency sector as the environment, as in the stochastic inflation scheme for the inflaton field; or two separate self-interacting scalar fields coupled biquadratically, each assuming a full spectrum of modes. Both cases have been treated in Chapter 5 using the CTP CGEA in flat space. The environment field can also be referring to other fields present besides the inflaton field. Only the quantum fluctuations of such fields need be present to generate the noise which seeds the galaxies. Even if one assumes nothing, there is always the gravitational field itself which the inflaton field is coupled to, and the vacuum gravitational fluctuations can also seed the structures in our universe [MuFeBr92, CalHu95, CalGon97, Mat97a, Mat97b]. (Note that in such cases the coupling is of a derivative form rather than a polynomial form. Noise arising from a derivative type of coupling has been studied before in the context of minisuperspace quantum cosmology [SinHu91].)

We add a cautionary note that the simple criterion of classicality derived from the study of linear systems (e.g. free fields) fails when interactions are taken into account. Indeed, as shown in simple quantum mechanical models (e.g. the anharmonic inverted oscillator [LoMaMo00]), an initially Gaussian wavefunction becomes non-Gaussian when evolved under the Schrödinger equation. The Wigner function will develop negative parts, and its interpretation as a classical probability breaks down.

Assuming weak self-coupling constant (a nearly flat inflaton potential) Lombardo and Nacir [LomNac05] have shown that decoherence is an event shorter than the time t_{end}, which is a typical time-scale for the duration of inflation.

9.7.1 Noise from interacting quantum fields

From the influence functional for an interacting field in a de Sitter universe given in Chapter 5 for the Minkowski spacetime, or the conformally related theory in de Sitter spacetime, we learned how to identify the noise (in both cases our treatment follows [Zha90, HuPaZh93b, LomMaz96, CaHuMa01]). Now we use it to consider decoherence and structure formation in stochastic inflation.

For illustrative purposes, in discussing the issue of decoherence, we shall derive the master equation from this influence functional only for a special case. This equation and its associated Langevin or Fokker–Planck equation will enable us later to calculate the fluctuation spectrum as a problem in classical stochastic dynamics.

Consider a real, gauge singlet, massive, $\lambda\Phi^4$ self-interacting scalar field in a de Sitter spacetime. In the inflationary regime of interest, the scale factor $a(t)$ expands exponentially in cosmic time t

$$a(t) = a_0 \exp Ht \qquad (9.92)$$

We split the classical action of the inflaton field $\Phi(x)$ as

$$S[\Phi] = S_0[\Phi] + S_I[\Phi] \qquad (9.93)$$

where $S_0[\Phi]$ is that part of the classical action describing a free, massless, conformally coupled scalar field, and

$$S_I[\Phi] = \int d^n x \sqrt{-g(x)} \left\{ -\frac{1}{2} \left[m^2 \Phi^2 + (\xi_n - \xi_c) R(x) \Phi^2 \right] - \frac{1}{4!} \lambda \Phi^4 \right\} \quad (9.94)$$

contains the remaining (interactive) terms with contributions from nonzero mass m, self-interaction λ, and ξ_c, the coupling between the field and the spacetime curvature scalar R. Here, $\xi_c = 1/6$ for conformal coupling and $\xi_c = 0$ for minimal coupling in four dimensions, $\xi_n = (n-2)/4(n-1)$ is a constant equal to $1/6$ in four dimensions, and $\sqrt{-g(x)} = a^{n-1}(t) = a^n(\eta)$.

In the stochastic inflation scheme, one makes a system–bath field splitting

$$\Phi(\mathbf{x}, t) = \phi(\mathbf{x}, t) + \psi(\mathbf{x}, t) \quad (9.95)$$

such that the system field is defined by

$$\phi(\mathbf{x}, t) = \int\limits_{|\mathbf{k}| < \Lambda} \frac{d^3 \mathbf{k}}{(2\pi)^3} \, \Phi(\mathbf{k}, t) \, e^{i\mathbf{k} \cdot \mathbf{x}} \quad (9.96)$$

and the bath field is defined by

$$\psi(\mathbf{x}, t) = \int\limits_{|\mathbf{k}| > \Lambda} \frac{d^3 \mathbf{k}}{(2\pi)^3} \, \Phi(\mathbf{k}, t) \, e^{i\mathbf{k} \cdot \mathbf{x}} \quad (9.97)$$

where Λ is the cut-off wavenumber determined by the horizon size. The system field $\phi(x)$ contains the long-wavelength modes, which undergo a slow roll-over phase transition in the inflation period, while the bath field ψ contains the short-wavelength modes, which are the quantum fluctuations.

With this splitting, we find the following effective action from expanding the influence action for $\chi = \phi a$, $\chi' = \phi' a$ to one-loop order in $\hbar$ and second order in S_I. We consider only the biquadratic coupling here, which corresponds to the limit where the system field is homogeneous.

The computation of the effective action follows the lines of Chapter 5 with conformal time here replacing cosmic time there. The dissipation is of a nonlinear nonlocal type, and there is a multiplicative (nonlinearly coupled) colored noise. The fluctuation–dissipation theorem for this field model in de Sitter space has the same form as that in Minkowski space.

9.7.2 Decoherence in two interacting fields model

The functional quantum master equation for this field-theoretical model with general nonlinear nonlocal dissipation and nonlinearly coupled colored noise has a complicated form in cosmic time t. However, in conformal time η it has the same form as in Minkowski spacetime, derived in Chapter 5, following the work of [Zha90, HuPaZh93a, HuPaZh93b, Paz94]. We will consider a simpler case here, where one can get an explicit form of the functional quantum master equation, i.e. by making a local truncation in the effective action. Setting

$$V(x - x') = v_0(\eta)\delta^4(x - x') \tag{9.98}$$

$$\mu(x - x') = \frac{\partial}{\partial(\eta - \eta')}\left\{\gamma_0(\eta)\delta^4(x - x')\right\} \tag{9.99}$$

$$\nu(x - x') = v_0(\eta)\delta^4(x - x') \tag{9.100}$$

and using the same procedure as outlined in Chapter 5, we can derive the functional quantum master equation in the local truncation approximation [Zha90]:

$$i\frac{\partial}{\partial\eta}\,\rho_r[\chi^1, \chi^2, \eta] = \hat{H}_\rho[\chi^1, \chi^2, \eta]\,\rho_r[\chi^1, \chi^2, \eta] \tag{9.101}$$

where

$$
\begin{aligned}
H_\rho[\chi^1, \chi^2, \eta] = \int d^3\mathbf{x}\,\Big\{ &\hat{h}_r(\chi^1) - \hat{h}_r(\chi^2) + 3\lambda^2\gamma_0(\eta)\Big[(\chi^1(\mathbf{x}))^4 - (\chi^2(\mathbf{x}))^4\Big] \\
&+ 2\lambda^2\gamma_0(\eta)\Big[(\chi^1(\mathbf{x}))^2 - (\chi^2(\mathbf{x}))^2\Big]\Big[\chi^1(\mathbf{x})\frac{\delta}{\delta\chi^1(\mathbf{x})} - \chi^2(\mathbf{x})\frac{\delta}{\delta\chi^2(\mathbf{x})}\Big] \\
&- (i/2)\lambda^2 v_0(\eta)\Big[(\chi^1(\mathbf{x}))^2 - (\chi^2(\mathbf{x}))^2\Big]\Big\}
\end{aligned}
\tag{9.102}
$$

and

$$
\begin{aligned}
\hat{h}_r(\phi) = &-\frac{1}{2}\frac{\delta^2}{\delta\chi^2(\mathbf{x})} + \frac{1}{2}\big[\nabla\chi(\mathbf{x})\big]^2 + \frac{1}{2}a^2(\eta)\Big[m_r^2 + \frac{1 + \xi_r}{6}R(\eta)\Big]\chi^2(\mathbf{x}) \\
&+ \frac{1}{4!}\lambda_r\chi^4(\mathbf{x}) + \delta m^2(\eta)a^2(\eta)\chi^2(\mathbf{x}) - \frac{1}{2}\lambda^2 v_0(\eta)\chi^4(\mathbf{x})
\end{aligned}
\tag{9.103}
$$

This functional quantum master equation and its associated Langevin equation or Fokker–Planck–Wigner equation can be used to analyze the dynamics of the system field (long-wavelength modes in the stochastic inflation scheme) for studying the decoherence and structure formation processes in the early universe [HuPaZh93b]. Instead of solving these equations in detail, we can get some qualitative information on how the system decoheres by analyzing the behavior of the diffusion term in the master equation.

Diffusive effects are generated by the last term in the effective action, the variation of which produces the following contribution on the right-hand side of the master equation for $\rho[\chi^1, \chi^2]$:

$$\dot{\rho}[\chi^1, \chi^2, \eta] \propto -\big[(\chi^1)^2 - (\chi^2)^2\big] * \nu(\eta) * \big[(\chi^1)^2 - (\chi^2)^2\big] \times \rho[\chi^1, \chi^2, \eta] \tag{9.104}$$

Here the symbol $*$ denotes the convolution product and χ represents a configuration of the scalar field in a surface of constant conformal time. The diffusion "coefficient" ν is therefore a nonlocal kernel that can be written in terms of its spatial Fourier transform as

$$\nu(\mathbf{x}, \mathbf{y}, \eta) = \int \frac{d^3\mathbf{k}}{(2\pi)^3} \nu_\mathbf{k}(\eta)\, e^{i\mathbf{k}\cdot(\mathbf{x}-\mathbf{y})} \tag{9.105}$$

To justify treating the long-wavelength modes classically, a minimal check is to see if the diffusive effects are stronger for long-wavelength modes than they are for short ones. To do so, note that [HuPaZh93b] the coefficient in (9.104) can be written in terms of the product of the Fourier transform (9.105) and that of the field ϕ^2:

$$[(\phi^1)^2 - (\phi^2)^2] * \nu(t) * [(\phi^1)^2 - (\phi^2)^2] = \int d\mathbf{k}\, [(\phi^1)^2 - (\phi^2)^2]_\mathbf{k} D_\mathbf{k} [(\phi^1)^2 - (\phi^2)^2]_\mathbf{k} \tag{9.106}$$

We want to examine the dependence on $k = |\mathbf{k}|$ of the function $D_\mathbf{k}$ entering in (9.106). This function can be written in terms of the physical wave vector $\mathbf{p} = \mathbf{k}/a$ as

$$D_\mathbf{k}(\eta) = \frac{a^3}{4\pi}\lambda^2 \left[1 - \frac{H}{p} f\left(\frac{p}{H}\right) + g\left(\frac{p}{H}\right)\right] \tag{9.107}$$

where

$$f(x) = \frac{1}{2\pi} \int_0^{2x} dx\, [-\sin x\, \mathrm{Ci}(x) + \cos x\, \mathrm{Si}(x)] \tag{9.108}$$

$$g(x) = \frac{1}{2\pi} \int_0^{2x} dx\, [\cos x\, \mathrm{Ci}(x) + \sin x\, \mathrm{Si}(x)] \tag{9.109}$$

and $\mathrm{Si}(x)$, $\mathrm{Ci}(x)$ are the usual integral trigonometric functions. A plot of $D_\mathbf{k}(\eta)$ for a fixed value of the conformal time as a function of p/H, i.e. the ratio between the horizon size and the physical wavelength can be found in [HuPaZh93b]. The function has a strong peak in the infrared region of the spectrum suggesting that diffusion effects (decoherence is one of them) are indeed more pronounced for long-wavelength modes and weaker for wavelengths shorter than the horizon size.

We learned from earlier discussions that noise of quantum origin arising from nonlinear fields is under general circumstances both multiplicative and colored (see, e.g. [HuPaZh93a]). Noise could generate fluctuations which could give rise to non-Gaussian galaxy distributions (NGD).[9]

As for the present scheme, since the value of λ is restricted to be very small ($< 10^{-12}$) in the standard inflationary models (so that the magnitude of the

---

[9] There are, of course, simpler ways to generate NGD. A changing Hubble rate $H = \dot{a}/a$ as in a "slow-roll" transition, or an exponential potential $V(\phi)$ [LucMat85] will do. However, such mechanisms only generate NGD at very long wavelengths, much longer than the horizon size to be relevant to the observable spectrum. See, e.g. Proceedings of ICTP meeting, July 2006 [SelCre06].

density contrast is compatible with the observed value $\delta\rho/\rho \approx 10^{-4}$ when the fluctuation mode enters the horizon), the constituency of the colored portion of the noise is accordingly small. The effect of nonlinear coupling on the generation of inhomogeneities is an active research topic at the accumulation of increasingly detailed observational data. Details of galaxy formation analysis from the stochastic equations of motion derived here with different types of colored noise and realistic physical parameters will come from solutions to these stochastic equations for galaxy formation considerations. We will have more discussions on the effect of quantum noise on structure formation in Chapter 15.

9.7.3 Partitioning one interacting field: noise from high frequency modes

In an earlier section we have discussed the appearance of classical features in a quantum phase transition. There the separation between long and short wavelengths is determined by their stability, which depends on the parameters of the potential. For our present consideration of quantum-to-classical transition in inflationary cosmology, this separation is conveniently set by the existence of the Hubble radius. Modes crossing the horizon during their evolution are usually treated as classical. The rationale for it can only come from a detailed study of decoherence, such as identifying the conditions whereby the behavior of a quantum fluctuation field can be adequately described by a classical stochastic field. We now discuss this issue.

The influence functional and the density matrix

For this case, we consider a massless quantum scalar field minimally coupled to a de Sitter spacetime. We choose the initial time η_i to be when $a(\eta_i) = 1(\eta_i = -H^{-1})$. Perform a system–environment field splitting [LomMaz96]

$$\chi = \chi_< + \chi_> \qquad (9.110)$$

where the system field $\chi_<$ contains the modes with wavelength longer than the partition scale $\ell_c \equiv 2\pi/\Lambda$, while the environment field $\chi_>$ contains modes with wavelength shorter than ℓ. As we set $a(\eta_i) = 1$, a physical length $\ell_{\text{phys}} = a(\eta)\ell$ coincides with the corresponding comoving length ℓ_i at the initial time. Therefore, the splitting between the system and the environment defines a system sector containing all the modes with physical wavelengths longer than the partition scale ℓ_c at the initial time η_i.

The influence functional for a similar problem has been computed in Chapter 5, Section 5.1, except that here a is a function of time. If there is a natural separation of the real and imaginary terms in this functional (as illustrated in the QBM model discussed in Chapter 3) one can then identify a noise and dissipation kernel related by a categorical fluctuation–dissipation relation. Assuming that the initial state $\hat{\rho}_>[\eta_i]$ is the Bunch–Davies vacuum state, the real and imaginary parts of the influence action are given by

$$\text{Re}S_{IF} = -\lambda \int d^4x_1 \left\{ \chi_-^{(4)}(x_1) - 6\chi_-^{(2)}(x_1)G_F^\Lambda(x_1, x_1) \right\}$$

$$+ \lambda^2 \int d^4x_1 \int d^4x_2 \, \theta(\eta_1 - \eta_2) \left\{ 32\chi_+^{(3)}(x_1)\text{Im}G_F^\Lambda(x_1, x_2)\chi_-^{(3)}(x_2) \right.$$

$$\left. -144\chi_+^{(2)}(x_1)\text{Im}[G_F^\Lambda(x_1, x_2)]^2\chi_-^{(2)}(x_2) \right\}, \qquad (9.111)$$

$$\text{Im}S_{IF} = \lambda^2 \int d^4x_1 \int d^4x_2 \left\{ 8\chi_-^{(3)}(x_1)\text{Re}G_F^\Lambda(x_1, x_2)\chi_-^{(3)}(x_2) \right.$$

$$\left. + 36\chi_-^{(2)}(x_1)\text{Re}[G_F^\Lambda(x_1, x_2)]^2\chi_-^{(2)}(x_2) \right\}, \qquad (9.112)$$

$\theta(x)$ is the Heaviside step function, and the integrations in time run from η_i to η. $G_F^\Lambda(x_1, x_2) \equiv \langle T\chi_>^1(x_1)\chi_>^1(x_2)\rangle_0$ is the relevant short-wavelength closed time path correlator (it is proportional to the Feynman propagator of the environment field, where the integration over momenta is restricted by the presence of the partition momentum Λ), and we have defined

$$\chi_-^{(n)} = (\chi_<^1)^n - (\chi_<^2)^n, \quad \chi_+^{(n)} = \frac{1}{2}[(\chi_<^1)^n + (\chi_<^2)^n] \qquad (9.113)$$

with $n = 1, 2, 3$.

Master equation and diffusion coefficients

As we learned by the QBM model (Chapter 3) and the field theory example (Chapter 5) once one obtains the evolutionary operator $\mathcal{J}_r$ for the reduced density matrix one can derive the master equation for the reduced density matrix. These expressions for a quantum scalar field in the de Sitter universe were obtained by Zhang [Zha90].

To get a qualitative idea of decoherence, as noted earlier, one could just focus on the behavior of the diffusion "coefficients" (actually nonlocal functions) related to the noise kernel obtained from the imaginary part of the influence action. Making the further simplification that the system field contains only one mode k_0, Lombardo and Nacir showed that the terms in the master equation relevant to decoherence are [LomNac05]

$$i\partial_\eta \rho_r[\phi_{<f}^1|\phi_{<f}^2; \eta] = \langle \phi_{<f}^1|[\hat{H}_{\text{ren}}, \hat{\rho}_r]|\phi_{<f}^2\rangle$$

$$- i\left[\Gamma_3 D_3(\mathbf{k_0}, \eta, \Lambda) + \Gamma_2 D_2(\mathbf{k_0}, \eta, \Lambda)\right]\rho_r[\phi_{<f}^1|\phi_{<f}^2; \eta] + \cdots$$

where $\Gamma_2 = \frac{\lambda^2 V}{4}[(\phi_{<f}^1)^2 - (\phi_{<f}^2)^2]^2$ and $\Gamma_3 \equiv \frac{\lambda^2 V}{H^2}[(\phi_{<f}^1)^3 - (\phi_{<f}^2)^3]^2$. (The subscripts 2, 3 refer to the order of the system field $\phi_{<f} = \chi_{<f}/a(\eta_f)$.) The ellipsis denotes additional terms coming from the time derivative that do not contribute to the diffusive effects.

This equation contains time-dependent diffusion coefficients D_j. Up to one loop, only D_2 and D_3 survive. Coefficient D_2 is related to the interaction term

$\phi_<^2\phi_>^2$ while D_3 is related to $\phi_<^3\phi_>$. These coefficients can be (formally) written as

$$D_2(\mathbf{k_0},\eta,\Lambda) = 36 \int_{\eta_i}^{\eta} d\eta'\, a^2(\eta)a^2(\eta')F_{\mathrm{cl}}^2(\eta,\eta',k_0) \tag{9.114}$$

$$\times \left\{ \mathrm{Re}[G_F^>(\eta,\eta',2\mathbf{k_0})]^2 + 2\,\mathrm{Re}[G_F^>(\eta,\eta',0)]^2 \right\},$$

and

$$D_3(\mathbf{k_0},\eta,\Lambda) = -\frac{H^2}{2}\int_{\eta_i}^{\eta} d\eta'\, a^3(\eta)a^3(\eta')F_{\mathrm{cl}}^3(\eta,\eta',k_0) \tag{9.115}$$

$$\times \mathrm{Re}G_F^>(\eta,\eta',3\mathbf{k_0})\,\theta(3k_0-\Lambda)$$

with the function F_{cl} defined by

$$F_{\mathrm{cl}}(\eta,\eta_i,k_0) = \frac{\sin[k_0(\eta-\eta_i)]}{k_0\eta} + \frac{\eta_i\cos[k_0(\eta-\eta_i)]}{\eta} \tag{9.116}$$

Note that only the effect of normal diffusion terms are included in our considerations here. It is known from QBM studies [Zha90, HuPaZh92, HuPaZh93a, PaHaZu93, Paz94, HalYu96] that anomalous diffusion terms can also be relevant at zero temperature.

Using these expressions for the two diffusion functions and placing the physical parameters relevant to successful inflationary models, Lombardo and Nacir [LomNac05] calculated the decoherence times t_{d_2} and t_{d_3} associated with D_2 and D_3. They conclude that if one sets $\Lambda \le H$, the decoherence time-scale for the system field is shorter than the minimal duration of inflation for all the wavevectors in the system sector. This is by far the most detailed and thorough study of the decoherence of the inflaton.

Part IV

Thermal, kinetic and hydrodynamic regimes

10

Thermal field and linear response theory

Thermal field theory, or finite temperature quantum field theory, deals with quantum systems in equilibrium. This is likely a familiar subject to the readers. Given the huge literature on this subject (see e.g. [LaLiPi80b, LeB91, LeB96, Kap89, Par88, AbGoDz75, Mil69, LanWer87, Ber74, DolJac74, Wei74, KuToHa91]), we will not attempt to give a full treatment of it *per se,* but rather emphasize how the CTP approach actually unifies the study of equilibrium and nonequilibrium systems. We will discuss thermal perturbation theory from a CTP perspective, including a discussion of screening and damping in quantum fields at finite temperature. In Chapter 11 we shall consider quantum kinetic field theory, including a derivation of its centerpiece, the Kadanoff–Baym equations, and in Chapter 12 the issue of thermalization, namely the processes which bring about equilibrium to systems out-of-equilibrium.

10.1 The thermal generating functional

A thermal state is a mixed state described by the density matrix

$$\rho = e^{\beta F} e^{-\beta H} \tag{10.1}$$

where H is the Hamiltonian and F the *free energy,* defined in terms of the partition function

$$e^{-\beta F} \equiv Z = \text{Tr } e^{-\beta H} \tag{10.2}$$

We shall assume the trace exists. Hence we are treating systems in equilibrium in a canonical ensemble; the generalization to grand canonical ensemble is straightforward. Observe that the equilibrium state is stationary but not Lorentz invariant, since the form (10.1) of the density matrix holds only in a preferred frame, the so-called *rest frame* of the field.

To investigate the matrix elements of the thermal density matrix, it is natural to proceed by analogy with the evolution operator $U = e^{-itH/\hbar}$ introduced in Chapter 4. In general, we may consider an evolution operator in complex time $U(z) = e^{-izH/\hbar}$. Given two field eigenvectors the matrix element is

$$\left\langle \psi \left| e^{-izH/\hbar} \right| \varphi \right\rangle = \sum_n e^{-izE_n/\hbar} \langle \psi | n \rangle \langle n | \varphi \rangle \tag{10.3}$$

Since energies are positive, the sum will converge when Im $z \leq 0$ (we are mostly concerned with z's of the form $-i\beta$, of course). Also we have the semigroup property

$$U(z + z') = U(z) U(z') \tag{10.4}$$

provided all terms exist, namely, that Im $z, z' \leq 0$. In general, given a complex path $\gamma(u)$ connecting 0 to z, if the imaginary part of $\gamma(u)$ is nonincreasing throughout we may decompose $U(z)$ as a product of infinitesimal evolution operators, and obtain the path integral representation

$$\left\langle \psi \left| e^{-izH/\hbar} \right| \varphi \right\rangle = \int_{\varphi(0)=\varphi, \varphi(z)=\psi} D\varphi \, e^{(i/\hbar)S[\varphi]} \tag{10.5}$$

where the integration is carried over field configurations defined on the complex path. The fundamental property of the representation (10.5) is path independence: the path integral is independent of the choice of γ, as long as the difference between the endpoints is z, and the imaginary part is nonincreasing [Mil69, McL72a, McL72b].

In order to obtain the partition function, we set $\psi = \varphi$ and integrate over all choices. Thus Z is given by a path integral over *periodic* (for Bose–Einstein statistics) or antiperiodic (for Fermi–Dirac statistics) [Ber66, NegOrl98]) field configurations defined on a complex time path going from an arbitrary point z to $z - i\beta\hbar$ with nonincreasing imaginary parts.

In practice, different choices of time path lead to different formulations. If the main object is the computation of the partition function itself, then possibly the simplest choice is the so-called *Matsubara contour*, which goes straight down from 0 to $-i\beta\hbar$ [LaLiPi80b, KuToHa91]. If the goal is to compute real time correlations, then it is convenient to include (patches of) the real time axis into the contour. For example, if we choose a time path going along the real axis from $-T$ to T', then down to $T' - i\beta\hbar/2$, back on a reverse time line to $-T - i\beta\hbar/2$, and finally down again to the endpoint $-T - i\beta\hbar$, we get a functional representation of Umezawa's *thermo field dynamics* [UmMaTa82, NieSem84a, NieSem84b]. Although these different formulations are of course equivalent from the physical point of view, one or the other could be more adept to particular perturbative calculations.

From the point of view of making contact with the CTP approach to nonequilibrium dynamics, the natural choice of contour is from $-T$ to T', then immediately back to $-T$ and straight down to $-T - i\beta\hbar$ [CSHY85, LanWer87]. Comparing with the CTP generating functional, we see that this procedure is equivalent to replacing the density matrix by its path integral representation. The thermal CTP generating functional is given by

$$e^{(i/\hbar)W_\beta[J]} = e^{\beta F} \int D\varphi^A \exp\left\{ (i/\hbar) \left[S\left[\varphi^A \right] + J_A \varphi^A \right] \right\} \tag{10.6}$$

with the branch index $a = 1, 2, 3$, with the new value 3 corresponding to the downward imaginary branch, and we adopt the convention $J^3 = 0$. Observe that the thermal generating functional shares with the vacuum one the property that the choice of quantum state has been encoded into the time path. The formal manipulations leading from the generating functional to the effective action, as well as the perturbative set-up for the evaluation of this latter object, are the same as in the zero-temperature theory.

10.2 Linear response theory

A profound property of the thermal state is that the dynamic response of a system in thermal equilibrium to small external perturbations can be described rigorously in terms of *equilibrium* expectation values, by means of the so-called *linear response theory* (LRT) [KuToHa91]. We shall describe very briefly the basics of LRT, and then show how it can be trivially derived from the CTP generating functional just introduced.

The basic set-up for LRT is a system which at time $t = 0$ is in equilibrium (namely, its density matrix is given by equation (10.1)) and is subsequently perturbed by an addition of a time-dependent term $-\sigma(t) P(t)$ to the Hamiltonian, where σ is a c-number external source, and P some Heisenberg operator acting on the system. We wish to follow the nonequilibrium, driven evolution of the field, through the time-dependent expectation value of some other observable $Q(t)$.

To this end, it is most efficient to adopt an interaction picture approach. The density matrix at time t then is given by

$$\rho(t) = T\left[e^{(i/\hbar)\int_0^t dt' \sigma P}\right] \rho(0) T\left[e^{-(i/\hbar)\int_0^t dt' \sigma P}\right] \tag{10.7}$$

The desired expectation value is $\langle Q\rangle_J(t) = \text{Tr}\, Q(t) \rho(t)$. For small sources, we may linearize

$$\langle Q\rangle(t) = \langle Q\rangle_0(t) + \int_0^t dt'\, \mathcal{R}(t, t') \sigma(t') \tag{10.8}$$

where $\langle Q\rangle_0(t)$ is the expectation value for $Q(t)$ in equilibrium, and $\mathcal{R}$ is the *response function*

$$\mathcal{R}(t, t') = \left(\frac{i}{\hbar}\right) \langle [Q(t), P(t')]\rangle_0\, \theta(t - t') \tag{10.9}$$

The general relationship between the retarded and Jordan propagators introduced before is but a particular case of this general identity. Furthermore, certain transport coefficients may be written in terms of time integrals of response functions, by means of the Kubo formulae, to be discussed in Chapter 12. Then equation (10.9) may be used to link those transport coefficients to equilibrium expectation values of Heisenberg operators.

To obtain equations (10.8) and (10.9) in a CTP framework, we first introduce a generating functional for $\langle Q \rangle (t)$

$$e^{(i/\hbar)W_Q[J]} = e^{\beta F} \int D\varphi^A \exp\left\{(i/\hbar)\left[S_\sigma\left[\varphi^A\right] + J_A Q^A\right]\right\} \tag{10.10}$$

so that the desired expectation value is

$$\langle Q \rangle (t) = \left.\frac{\delta W_Q[J]}{\delta J^1(t)}\right|_{J=0} \tag{10.11}$$

(observe that we *doubled the degrees of freedom*). In this equation, the action S_σ contains the σ-dependent term $\int dt\, \sigma(t)\left(P^1 - P^2\right)(t)$; it is unnecessary to add CTP indices to σ, since this is a physical source, and in any case we would obtain $\sigma^1 = \sigma^2 = \sigma$. Since σ turns on for $t > 0$ only, the free energy is independent of it. We may therefore expand

$$W_Q[J] = W_Q[J]\big|_{\sigma=0} + \frac{\int D\varphi^A\, e^{\left\{(i/\hbar)\left[S_0\left[\varphi^A\right] + J_A Q^A\right]\right\}} \int dt'\, \left(P^1 - P^2\right)\sigma(t')}{\int D\varphi^A\, e^{\left\{(i/\hbar)\left[S_0\left[\varphi^A\right] + J_A Q^A\right]\right\}}} \tag{10.12}$$

The path integral in the numerator of the second term vanishes at $J = 0$. Performing the J derivative in equation (10.11), we see that equations (10.8) and (10.9) are a simple consequence of the time ordering properties of the path integral.

10.3 The Kubo–Martin–Schwinger theorem

Since we have succeeded in incorporating the information about the state in the time path, we may adopt *verbatim* the perturbative approaches already discussed in Chapter 6. In particular, any expectation value may be developed as a sum of Feynman graphs, carrying in internal legs the *thermal propagators*

$$G_\beta^{11}(x, x') \equiv G_{\beta F}(x, x') = e^{\beta F}\mathrm{Tr}\left\{e^{-\beta H}T\left[\Phi(x)\,\Phi(x')\right]\right\} \tag{10.13}$$

etc. Since the thermal time path has three branches, it may seem that now we need nine different propagators to carry out perturbation theory. However, the path independence of the path integral may be invoked to push the third branch arbitrarily into the distant past, whereby it decouples from the other branches, and the usual set-up, based on four propagators, is sufficient. Nevertheless, the third branch is essential in enforcing the fundamental property of the thermal propagators, namely, that they can be analytically continued to complex time, and when so continued, they obey certain periodicity conditions, embodied in the so-called *Kubo–Martin–Schwinger* (KMS) *theorem* [Kub57, KuYoNa57, MarSch59]. The KMS theorem plays such a central role in thermal perturbation theory that it is often adopted as the *definition* of the thermal propagators.

To state the KMS theorem, let us consider the thermal positive frequency propagator

$$G_\beta^{21}(x, x') \equiv G_\beta^+(x, x') = e^{\beta F} \text{Tr} \left\{ e^{-\beta H} \Phi(x) \Phi(x') \right\} \tag{10.14}$$

and insert energy–momentum eigenstates

$$G_\beta^+(x, x') = e^{\beta F} \sum_{n,m} e^{i(\mathbf{P}_m - \mathbf{P}_n)(\mathbf{x} - \mathbf{x'})/\hbar} e^{-\beta E_n} e^{-i(E_m - E_n)(t - t')/\hbar} |\langle n | \Phi(0) | m \rangle|^2 \tag{10.15}$$

Given suitable conditions on the matrix elements of the field operators, we may regard the sum in equation (10.15) as converging when the integrand is exponentially suppressed, namely for $0 \geq \text{Im}\,(t - t') \geq -\beta$. In this strip, equation (10.15) defines a complex variable function, which is the definition of the thermal propagator for complex time. G_β^+ may be analytically continued beyond the strip, of course, but this constitutes its fundamental domain of definition.

If we apply the same reasoning to the "negative frequency" propagator we obtain

$$G_\beta^-(x, x') = e^{\beta F} \sum_{n,m} e^{i(\mathbf{P}_m - \mathbf{P}_n)(\mathbf{x} - \mathbf{x'})/\hbar} e^{-\beta E_m} e^{-i(E_m - E_n)(t - t')/\hbar} |\langle n | \Phi(0) | m \rangle|^2 \tag{10.16}$$

Comparing equations (10.15) and (10.16), it is immediate that

$$G_\beta^+((t, \mathbf{x}), x') = G_\beta^-((t + i\hbar\beta, \mathbf{x}), x') \tag{10.17}$$

This is the KMS theorem for Bose–Einstein fields. The KMS theorem for Fermi–Dirac fields will be discussed in a later section.

A shorter argument may serve as a mnemotechnic device, and also underscores the generality of this result. Let us define the complex time Heisenberg operators

$$\Phi(z) = e^{izH/\hbar} \Phi(0) e^{-izH/\hbar} \tag{10.18}$$

and observe the identity

$$e^{-\beta H} \Phi(t) = \Phi(t + i\hbar\beta) e^{-\beta H} \tag{10.19}$$

Then a simple cyclic permutation under the trace yields

$$G_\beta^+(t, t') = e^{\beta F} \text{Tr} \left\{ e^{-\beta H} \Phi(t) \Phi(t') \right\}$$
$$= e^{\beta F} \text{Tr} \left\{ e^{-\beta H} \Phi(t') \Phi(t + i\hbar\beta) \right\} = G_\beta^-(t + i\hbar\beta, t') \tag{10.20}$$

QED

Given the importance of the KMS theorem for this subject, we shall show a third proof of it, now based on the properties of the path integral representation. The propagator $G_\beta^+(t, t') = G_\beta^{21}(t, t')$ may be computed by inserting the product $\varphi^2(t)\varphi^1(t')$ inside the path integral (10.6); namely, the field at t is put on the second branch, and at t' on the first branch. Given the path independence, we may choose a time contour beginning at t' and ending at $t' - i\beta\hbar$. Because field configurations are periodic, we obtain $\varphi^2(t)\varphi^1(t') = \varphi^3(t' - i\beta\hbar)\varphi^2(t)$. Since the path integral automatically sets field operators on the third branch to the left of field operators in the second branch, when we integrate we obtain the negative frequency propagator $G_\beta^-(t, t' - i\beta\hbar)$. Thermal propagators being translation invariant, this is again the KMS theorem.

The KMS theorem implies a new relationship among the Fourier transforms of the thermal propagators, namely

$$G_\beta^+(\omega) = e^{\beta\hbar\omega}G_\beta^-(\omega) \tag{10.21}$$

Together with the universal relationship $G_\beta^+ - G_\beta^- = G_\beta$, the latter being the thermal Jordan propagator, we obtain

$$G_\beta^+(\omega) = \frac{e^{\beta\hbar\omega}}{e^{\beta\hbar\omega} - 1}G_\beta(\omega); \quad G_\beta^-(\omega) = \frac{1}{e^{\beta\hbar\omega} - 1}G_\beta(\omega) \tag{10.22}$$

The two formulae can be combined into

$$G_\beta^\pm(\omega) = 2\pi\hbar\left[\theta(\pm\omega) + f_0(\omega)\right]\mathcal{D}(\omega) \tag{10.23}$$

where

$$2\pi\hbar\mathcal{D}(\omega) = \text{sign}(\omega)G_\beta(\omega) = G_\beta(|\omega|) \tag{10.24}$$

and f_0 is the Bose–Einstein distribution

$$f_0(\omega) = \frac{1}{e^{\beta\hbar|\omega|} - 1} \tag{10.25}$$

These formulae generalize the vacuum Lehmann representation. As in the vacuum case, they allow us to find all the propagators, once the Jordan propagator is known. For example, the Hadamard propagator becomes

$$G_{\beta1}(\omega) = 2\pi\hbar\left[1 + 2f_0(\omega)\right]\mathcal{D}(\omega) \tag{10.26}$$

For a free field the commutator of two field operators is a c-number, and therefore its expectation value is independent of the state. In this case the zero-temperature Jordan propagator remains valid at all temperatures.

The linear response equation (10.9) allows us to connect the Jordan and retarded propagators

$$G_{\beta\text{ret}}(x, x') = \left(\frac{i}{\hbar}\right)G_\beta(x, x')\theta(t - t') \tag{10.27}$$

which is the same relationship as at zero temperature. Fourier transforming, we obtain as in Chapter 3

$$G_\beta(p) = 2\hbar\text{Im}G_{\beta\text{ret}}(p) \tag{10.28}$$

and therefore equation (10.26) may be regarded as a statement of the fluctuation–dissipation theorem.

10.4 Thermal self-energy: Screening

As remarked earlier, the CTP perturbation theory at finite temperature is formally identical to its zero-temperature counterpart, only now propagators must be consistent with the KMS theorem. In particular, the 2PIEA is formally the same, but it is now evaluated at a different set of propagators. We may even give a direct proof of this statement, by noting that thermal field theory is the usual field theory on a three-branched path. We may push the third branch to the remote past, and at the same time switch interactions off adiabatically, so that the action in the third branch corresponds to a free theory. Then the path integral over Euclidean configurations will be Gaussian, and the net effect will be to add a two-point source $K_{\beta AB}$ (concentrated at the initial time), as we did when we formally constructed the 2PIEA. The construction will then go as in Chapter 6, only that the final equation for the thermal propagators becomes

$$\frac{\delta \Gamma_2}{\delta G_\beta^{AB}} = -\frac{1}{2} K_{\beta AB} \tag{10.29}$$

Because the right-hand side turns on only at the initial time, in practice we may ignore it; its only role is to enforce the KMS initial conditions. We may therefore write down the thermal 2PI Schwinger–Dyson equation: it is just the same as the vacuum equation we discussed in Chapter 6.

Because Lorentz invariance is lost for thermal fields, we cannot define the mass of the field from the location of the pole of the propagators. Luckily, this problem only appears at the three-loop level in the 2PIEA, so we may still make progress by analyzing the one-loop gap equation.

The analysis is in fact the same as in Chapter 6, only now the tadpole is computed with a thermal Feynman propagator or, what is the same at the coincidence limit, a thermal Hadamard propagator. Since the free Jordan propagator is independent of temperature, this means that there is a new term, corresponding to the integration over the thermal correction demanded by the KMS theorem. So the gap equation becomes

$$M^2 = m_b^2 + m_V^2 + \frac{\lambda_b \hbar}{2} M_T^2 \tag{10.30}$$

where

$$M_T^2 = \int \frac{d^4 p}{(2\pi)^3} \delta(\Omega_0) f_0(p) = \frac{1}{\pi^2} \int_M^\infty d\omega \frac{\sqrt{\omega^2 - M^2}}{e^{\beta \omega} - 1} \tag{10.31}$$

and $\Omega_0 = p^2 + M^2$. The second term m_V^2 is a vacuum tadpole.

Although strictly speaking the gap equation must be renormalized before it makes sense, we may learn some of its implications from the following simple arguments. First, observe that there exists a critical temperature T_c (maybe imaginary) such as to make M^2 vanish. If we use dimensional regularization, the massless tadpole vanishes as well, and the massless thermal contribution gives the usual value of $M_{T_c}^2 \sim (1/6)\left(k_B T_c / \hbar\right)^2$, so we get $m_b^2 = (-\lambda_b / 12\hbar)\left(k_B T_c\right)^2$. If we adopt T_c rather than m_b as the fundamental object, then $M^2 \sim O\left(\lambda T^2\right) \ll T^2$, which justifies neglecting the tadpole term and computing the thermal mass as for a massless theory. We then get the simple gap equation

$$M^2 = \frac{\lambda k_B^2}{12\hbar}\left[T^2 - T_c^2\right] \tag{10.32}$$

The first and obvious consequence of the thermal gap equation is that a field theory which is massless at some temperature ($T = T_c = 0$, say) will not be massless at other temperatures. If we associate a massless field (such as the photon) with a long-range interaction, then at finite temperature the same interaction will be short range (of course, Maxwell's theory is a gauge theory, and we must be careful [LeB96]). This phenomenon is called *screening*, and because it fixes the screening length M^{-1}, M^2 is sometimes called the *Debye mass* M_D^2.

If T_c is real, the gap equation admits negative solutions at low temperature (meaning that the symmetric point $\phi = 0$ corresponds to an unstable configuration of the field, and we should not be doing perturbation theory around it) and regular solutions above T_c. This is the phenomenon of *symmetry restoration*, and T_c is the critical temperature which marks the destabilization of the symmetric point [DolJac74, Wei74].

In theories with multiple fields the thermal mass matrix may not be positive definite. In this case, there may be inverse symmetry restoration (namely, a symmetry is broken at high temperature) or else symmetry nonrestoration (a symmetry is never restored) [Wei74, PinRam06].

The generation of the Debye mass and the corresponding screening length means that the behavior of thermal propagators is totally different from their zero temperature counterparts, at least for *soft momenta* $p \leq M_D$. In this range, to assume that one can develop a meaningful perturbation theory without a careful consideration of screening is simply wrong. At the very least, one should use the physical mass throughout. Since a mass shift in the inverse propagator is equivalent to resumming an infinite number of graphs in the perturbative expansion of the propagator itself, and the shift may be seen as coming mostly from the high momentum sector where all masses can be neglected, the techniques necessary to implement a consistent perturbation theory are generally known as *hard thermal loops resummation*.

10.5 Landau damping

In addition to screening, at finite temperature a collective excitation will be damped by scattering off quanta in the heat bath. Therefore, there are decay channels unavailable at zero temperature. The most important of these decay processes is the so-called Landau damping, originally discussed by Landau in the context of collisionless plasma theory [LifPit81].

In this section, we shall discuss Landau damping through a concrete example, namely, the damping of a Maxwell field interacting with a Dirac quantum field at finite temperature. To stress the physics involved, we shall begin with a brief review of Landau damping in its original plasma physics context.

10.5.1 Landau damping in a relativistic collisionless plasma

Before we consider the issue of damping in the equation for a Maxwell background field coupled to quantum fluctuations in a Dirac spinor field, let us discuss the same issue in the simpler context of a classical relativistic collisionless plasma.

Under the collisionless approximation, particles evolve independently under the electromagnetic field, which in turn is sourced by the average charge density and current. The dynamics of each particle is determined by the Hamiltonian (we set the speed of light $c = 1$)

$$p^0 = \left[(\mathbf{p} - e\mathbf{A})^2 + m^2 \right]^{1/2} + eA^0 \tag{10.33}$$

Therefore we have the velocity

$$\mathbf{v} = \nabla_{\mathbf{p}} p^0 = \frac{(\mathbf{p} - e\mathbf{A})}{(p^0 - eA^0)} \tag{10.34}$$

and the Lorentz force (in a somewhat unusual notation)

$$[p_i]^{\cdot} = -\partial_i p^0 = e \left[v^j A_{j,i} - \partial_i A^0 \right] \tag{10.35}$$

The charge and current densities are given by

$$j^0 = e \int \frac{d^3 \mathbf{p}}{(2\pi)^3} f(\mathbf{x}, \mathbf{p}, t) \tag{10.36}$$

$$\mathbf{j} = e \int \frac{d^3 \mathbf{p}}{(2\pi)^3} \mathbf{v}\, f(\mathbf{x}, \mathbf{p}, t) \tag{10.37}$$

Charge is conserved as a consequence of Hamilton's equations, provided f satisfies the Vlasov equation

$$\frac{\partial f}{\partial t} + \mathbf{v} \cdot \nabla_{\mathbf{x}} f + \dot{\mathbf{p}} \cdot \nabla_{\mathbf{p}} f = 0 \tag{10.38}$$

Explicitly

$$(p - eA)^{\mu} \frac{\partial f}{\partial x^{\mu}} + e \left[(p - eA)^{\nu} A_{\nu,j} \right] \frac{\partial f}{\partial p_j} = 0 \tag{10.39}$$

We are interested in linearized Maxwell fields, so we may expand $f = f^0 + f^1 + \ldots$ in powers of A. Assume $f^0 = f^0(\mathbf{p})$. The first-order terms read

$$p^{\mu} \frac{\partial f^1}{\partial x^{\mu}} + e \left[p^{\nu} A_{\nu,j} \right] \frac{\partial f^0}{\partial p_j} = 0 \tag{10.40}$$

If the Maxwell background corresponds to a plane wave

$$A_\mu = A_{k\mu} e^{ikx} \tag{10.41}$$

we may seek a solution

$$f^1 = f_k^1 e^{ikx} \tag{10.42}$$

where, adopting Landau's causal boundary conditions,

$$f_k^1 = e \left[\frac{p^\nu A_{k\nu}}{-p^\lambda (k + i\varepsilon)_\lambda} \right] k_j \frac{\partial f^0}{\partial p_j} \tag{10.43}$$

We may now write the charge density (10.36). Using our perturbative solution and discarding the zeroth-order term this becomes

$$j^0 = j_k^0 e^{ikx} \tag{10.44}$$

$$j_k^0 = e^2 \int \frac{d^3 \mathbf{p}}{(2\pi)^3} \left[\frac{p^\nu A_{k\nu}}{-p^\lambda (k + i\varepsilon)_\lambda} \right] k^i \frac{\partial f^0}{\partial p^i} \tag{10.45}$$

Observe that the induced current is gauge invariant. Essentially, what we have done is to compute the conductivity of the plasma. The important point is that equation (10.45) develops an imaginary part when there are particles whose momenta satisfy $p^\lambda k_\lambda = 0$. The imaginary part reads

$$\text{Im}\left[j_k^0 \right] = \frac{-\pi e^2}{(k^0)^2} \left[k^0 A_k^j - k^j A_{k0} \right] k^i \int \frac{d^3 \mathbf{p}}{(2\pi\hbar)^3} \delta\left[p^0 - \frac{\mathbf{k}.\mathbf{p}}{k^0} \right] p_j \frac{\partial f^0}{\partial p^i} \tag{10.46}$$

If we use this as a source for the Maxwell equations, the imaginary part in the charge density will engender the so-called Landau damping of the background plane wave. As can be seen from equation (10.46), Landau damping occurs when there are charged particles moving at the phase speed of the wave. These particles see the wave as a time-independent field, and may extract energy from it. Actually, the expression for $\text{Im}\left[j_k^0 \right]$ does not have a definite sign; however, damping obtains generally for isotropic distributions [LifPit81].

10.5.2 A nonequilibrium problem with fermions: The case of QED

As a simple example of fermionic nonequilibrium field theory, we wish to consider the linearized equations of motion for a Maxwell background field coupled to a Dirac spinor (representing the electron field). The action is given by

$$S = S_\text{M} + S_\text{D} + S_\text{int} \tag{10.47}$$

where S_M is the free Maxwell action

$$S_\text{M} = \left(\frac{-1}{4} \right) \int d^4 x \, F^{\mu\nu} F_{\mu\nu} \tag{10.48}$$

$$F_{\mu\nu} = \partial_\mu A_\nu - \partial_\nu A_\mu \tag{10.49}$$

is the Maxwell field tensor, and A_μ the photon field. S_D is the free Dirac action

$$S_D = \int d^4x \, \overline{\psi} \, (i\gamma^\mu \partial_\mu - m) \, \psi \tag{10.50}$$

where ψ is a four-component Dirac spinor, the γ^μ are the Dirac matrices obeying

$$\{\gamma^\mu, \gamma^\nu\} = -2g^{\mu\nu} \tag{10.51}$$

and $\overline{\psi}$ is the Dirac adjoint spinor

$$\overline{\psi} = \psi^\dagger \gamma^0 \tag{10.52}$$

The interaction term is

$$S_{\text{int}} = \int d^4x \, eA_\mu \overline{\psi} \gamma^\mu \psi \tag{10.53}$$

We wish to compute the lowest order $(O\,(e^2))$ linearized equation of motion for the Maxwell background field. To this order, photon quantum fluctuations are decoupled, so we need not consider them further. The CTPEA for the Maxwell field reads

$$\Gamma\left[A_\mu^a\right] = S_M\left[A_\mu^a\right] + \Gamma_1\left[A_\mu^a\right] \tag{10.54}$$

$$S_M\left[A_\mu^a\right] = S_M\left[A_\mu^1\right] - S_M\left[A_\mu^2\right] \tag{10.55}$$

$$\Gamma_1\left[A_\mu^a\right] = -i\hbar \ln \int D\psi^a \, D\overline{\psi}^b \, e^{(i/\hbar)\{S_D + S_{\text{int}}\}} \tag{10.56}$$

where the path integral is over Grassmann fields [Ber66, NegOrl98] defined on the closed time path, and the actions in the integrand also are CTP actions.

To quadratic order in the external field,

$$\Gamma_1\left[A_\mu^a\right] = \frac{ie^2}{2\hbar} \, c_{acd} c_{bef} \int d^4x d^4x' \, A_\mu^a\,(x) \, A_\nu^b\,(x')$$
$$\left\langle \left[\overline{\psi}^c\,(x)\,\gamma^\mu \psi^d\,(x)\right] \left[\overline{\psi}^e\,(x')\,\gamma^\nu \psi^f\,(x')\right] \right\rangle_0 \tag{10.57}$$

where

$$\langle O \rangle_0 = \int D\psi^a \, D\overline{\psi}^b \, e^{iS_D} \, O \tag{10.58}$$

and we have used the fact that

$$\left\langle \left[\overline{\psi}\,(x)\,\gamma^\mu \psi\,(x)\right] \right\rangle_0 = 0 \tag{10.59}$$

(see below). Since the integration measure is Gaussian, Wick's theorem holds and

$$\left\langle \left[\overline{\psi}^c\,(x)\,\gamma^\mu \psi^d\,(x)\right] \left[\overline{\psi}^e\,(x')\,\gamma^\nu \psi^f\,(x')\right] \right\rangle_0 = -\gamma^\mu G^{de}\,(x,x')\,\gamma^\nu G^{fc}\,(x',x) \tag{10.60}$$

where

$$G^{de}(x, x') = \left\langle \psi^d(x) \overline{\psi}^e(x') \right\rangle_0 \tag{10.61}$$

and we have used the fact that in this problem

$$\left\langle \psi^d(x) \psi^e(x') \right\rangle_0 = 0 \tag{10.62}$$

which follows directly from the Gaussian integration.

The G^{de} are the CTP free Dirac propagators. From the ordering properties of the CTP path integral

$$G^{21}(x, x') = \left\langle \hat{\psi}(x) \hat{\overline{\psi}}(x') \right\rangle \tag{10.63}$$

$$G^{12}(x, x') = -\left\langle \hat{\overline{\psi}}(x') \hat{\psi}(x) \right\rangle \tag{10.64}$$

where $\hat{\psi}$ is the Heisenberg picture field operator, and the brackets denote vacuum expectation value. We also have

$$G^{11}(x, x') = G^{21}(x, x')\,\theta(t - t') + G^{12}(x, x')\,\theta(t' - t) \tag{10.65}$$

$$G^{22}(x, x') = G^{21}(x, x')\,\theta(t' - t) + G^{12}(x, x')\,\theta(t - t') \tag{10.66}$$

The Heisenberg equation for the free field operator is the Dirac equation, whereby

$$(i\gamma^\mu \partial_\mu - m)\,G^{21}(x, x') = 0 \tag{10.67}$$

and similarly for $G^{12}(x, x')$, while

$$(i\gamma^\mu \partial_\mu - m)\,G^{11}(x, x') = i\hbar\delta(x - x') \tag{10.68}$$

$$(i\gamma^\mu \partial_\mu - m)\,G^{22}(x, x') = -i\hbar\delta(x - x') \tag{10.69}$$

The solution to these equations with the proper CTP boundary conditions is

$$G^{ab}(x, x') = (i\gamma^\mu \partial_\mu + m)\,\Delta^{ab}(x, x') \tag{10.70}$$

where Δ^{ab} are the Klein–Gordon CTP propagators. With the representation (10.70) it is immediate to obtain equation (10.59).

The new term in the CTPEA induces a new term in the wave equation for the photon field

$$\frac{e^2}{\hbar} \int d^4x'\, \Pi^{\mu\nu}(x, x')\, A_\nu(x') \tag{10.71}$$

$$\Pi^{\mu\nu}(x, x') = (-i)\gamma^\mu \left\{ G^{11}(x, x')\gamma^\nu G^{11}(x', x) - G^{12}(x, x')\gamma^\nu G^{21}(x', x) \right\} \tag{10.72}$$

As in the case of the scalar field, the first term is what we would have found from the "in-out" EA, and the second term enforces reality and causality.

Using the representation (10.70) the calculation of this term is a standard exercise in quantum field theory [Ram80, PesSch95] and we will not repeat it. The photon retarded propagator acquires an imaginary part for off-shell momenta $-p^2 \geq 4m^2$. This represents damping of a photon wave from pair creation. We have seen in Chapter 4 that damping also goes on below threshold, but it is exponentially suppressed.

10.5.3 KMS and thermal Fermi propagators

The proof of the KMS theorem works as well for Fermi fields, but now we must take into account the proper relation between the Fermi propagators and their expression as averages of Heisenberg fields. If we introduce the thermal positive and negative frequency propagators $G_\beta^+ \equiv G_\beta^{21}$ and $G_\beta^- \equiv G_\beta^{12}$ the KMS condition becomes

$$G_\beta^+ (\omega) = -e^{\beta \hbar \omega} G_\beta^- (\omega) \tag{10.73}$$

We may introduce a Fermi Jordan propagator

$$G_\beta = G^+ - G^- = \langle \{\psi, \overline{\psi}\} \rangle_\beta \tag{10.74}$$

(for a free field, G_β is independent of the temperature) and a density of states

$$\mathcal{D}_F (\omega) = \frac{1}{2\pi\hbar} \operatorname{sign} (\omega) G_\beta (\omega) \tag{10.75}$$

Then the KMS condition becomes

$$G_\beta^{\pm} (\omega) = 2\pi\hbar \{\theta (\pm\omega) - f_{\mathrm{FD}} (\omega)\} \mathcal{D}_F (\omega) \tag{10.76}$$

where f_{FD} is the Fermi–Dirac distribution

$$f_{\mathrm{FD}} (\omega) = \left[e^{\beta\hbar|\omega|} + 1\right]^{-1} \tag{10.77}$$

All other propagators may be built from these two.

10.5.4 Induced charge density from a finite temperature Dirac quantum field

We now return to the quantum field problem. A nontrivial Maxwell background induces a current (cf. equation (10.71))

$$j^\mu (x) = \frac{\delta\Gamma_1}{\delta A_\mu (x)} = \frac{e^2}{\hbar} \int d^4 x' \, \Pi^{\mu\nu} (x, x') A_\nu (x') \tag{10.78}$$

where $\Pi^{\mu\nu} (x, x')$ is defined in equation (10.72). As in our simple plasma example, we shall look into the induced charge density only. If the background is a single plane wave as in equation (10.41), then

$$j_k^0 = \frac{e^2}{\hbar} \Pi_k^{0\nu} A_{k\nu} \tag{10.79}$$

$$\Pi_k^{0\nu} = (-i) \gamma^0 \int \frac{d^4 p}{(2\pi)^4} \left\{ G^{11} (p) \gamma^\nu G^{11} (p - k) - G^{12} (p) \gamma^\nu G^{21} (p - k) \right\} \tag{10.80}$$

To continue, let us decompose all propagators into a zero-temperature and a statistical component

$$G^{ab}(p) = G_0^{ab}(p) - G_{\text{stat}}(p) \tag{10.81}$$

where $G_0^{ab}(p)$ has the form of the zero-temperature propagators, but maybe with temperature-dependent coefficients, and

$$G_{\text{stat}}(p) = 2\pi\hbar \, f_{\text{FD}}(p^0) \left[-\gamma^\mu p_\mu + m\right] \delta\left(-p^2 - m^2\right) \tag{10.82}$$

is the same for all basic propagators. Clearly

$$\Pi_k^{0\nu} = \Pi_{k0}^{0\nu} + \Pi_{k\text{stat}}^{0\nu} \tag{10.83}$$

where the first term is the zero-temperature contribution. As we know, this term describes, among other things, damping from pair creation out of the vacuum. Our interest here is the other term

$$\Pi_{k\text{stat}}^{0\nu} = \hbar\gamma^0 \int \frac{d^4 p}{(2\pi)^4} J^\nu \tag{10.84}$$

$$J^\nu = G_{\text{ret}}\left(p + \frac{k}{2}\right)\gamma^\nu G_{\text{stat}}\left(p - \frac{k}{2}\right) + G_{\text{stat}}\left(p + \frac{k}{2}\right)\gamma^\nu G_{\text{adv}}\left(p - \frac{k}{2}\right) \tag{10.85}$$

The exact expression for $\Pi_{k\text{stat}}^{0\nu}$ is involved and shall not be discussed further. It becomes simpler at high temperature, where we may argue that the leading contribution to the integral comes from momenta $p \approx T \gg k, m$. The leading contribution in this limit is

$$4\hbar^2 \int \frac{d^4 p}{(2\pi)^3} \frac{p^0 p^\nu}{p(k + i\varepsilon)} J \tag{10.86}$$

$$J = f_{\text{FD}}\left(p - \frac{k}{2}\right)\delta\left(\left(p - \frac{k}{2}\right)^2 + m^2\right) - f_{\text{FD}}\left(p + \frac{k}{2}\right)\delta\left(\left(p + \frac{k}{2}\right)^2 + m^2\right) \tag{10.87}$$

If $k \ll p$, we may expand inside the brackets. The leading contribution to the imaginary part comes from a term

$$4\hbar^2 \int \frac{d^4 p}{(2\pi)^3} \frac{2p^0 \delta\left(-p^2 - m^2\right)\theta\left(p^0\right)}{[-p(k + i\varepsilon)]} p^\nu k^0 \frac{\partial f_{\text{FD}}}{\partial p^0} \tag{10.88}$$

The momenta which contribute to the imaginary part satisfy $p \cdot k = 0$, and so

$$k^0 \frac{\partial f_{\text{FD}}}{\partial p^0} = \frac{k^0 p^0}{p^0} \frac{\partial f_{\text{FD}}}{\partial p^0} = \frac{\mathbf{k} \cdot \mathbf{p}}{p^0} \frac{\partial f_{\text{FD}}}{\partial p^0} = \mathbf{k}\nabla_{\mathbf{p}} f_{\text{FD}} \tag{10.89}$$

We see a quantum field theory version of the Vlasov equation, only now we have a factor of 4 reflecting the presence of electrons and positrons with two spin states each, and $\hbar f_{\text{FD}}$ instead of the classical one-particle distribution function.

10.6 Hard thermal loops

In this section we show how the above machinery can be applied to one set of important problems – the dynamics of long-wavelength or "soft" modes of a nonlinear quantum field, as affected by the short-wavelength or "hard" modes towering over them. Historically, the techniques of this subject were developed in an attempt to understand the physics of soft modes in non-abelian gauge theories, for application to the quark–gluon plasma in relativistic heavy ion collisions (RHICs), and to topology change in electroweak theory. This specific context imposed a number of constraints, such as the need to deal with gauge invariance, derivative couplings, ghost fields, etc. We wish to isolate the basic physical ideas relating to nonequilibrium and statistical behavior from the technical devices specific to a given application, and therefore we shall not follow the historical path, but rather present a fictitious toy model which retains sufficient fundamental physics.

The important lesson to be gleaned from this is that formal questions, such as which is the best perturbative scheme or to which order should it be pursued, cannot be separated from physical questions. As illustrated in this example, we get different answers depending on whether we wish to discuss soft or ultrasoft modes. In either case, simply counting powers of coupling constants gives the wrong result. It is necessary to analyze the contents of the theory to make sure that what looks small is indeed small, and to realize that different scales pertain to different physics.

In this section we shall use some tools of quantum kinetic field theory which will be discussed in detail in Chapter 11. The reader unfamiliar with these techniques may return to this section after getting acquainted with them.

10.6.1 The model

The essential elements we need to keep from the physics of non-abelian gauge fields are the presence of massless fields and a derivative cubic coupling. For massless fields radiative corrections are infrared sensitive, and will require special care to evaluate them. There is a term in the action containing three gauge fields and one derivative. In momentum space, the derivative becomes one momentum component. At finite temperature T, typical momenta are of the order of T, and so the effective coupling strength increases with temperature.

Therefore we postulate as our toy model a massless scalar field theory with cubic interaction

$$S = \int d^4x \left\{ -\frac{1}{2}\partial_\mu \Phi \partial^\mu \Phi - \frac{gT}{6}\Phi^3 + h\Phi \right\} \tag{10.90}$$

with the constitutive relation $h = gT^3/12$. We assume $g \ll 1$. The linear term is necessary to cancel a tadpole term later on; non-abelian gauge theories are protected against such terms by gauge invariance. Of course, this model is not by itself a viable theory, since it has no stable ground state; we shall use it only as an ersatz for a fully consistent, but necessarily more involved, non-abelian gauge theory.

The basic problem, as already stated, is to find the dynamics of the "soft" modes as modified by the virtual "hard" particles, the so-called "hard thermal loops" (HTLs). Here a "soft" mode corresponds to wavenumbers $k \sim gT$, while a "hard" mode has a much larger wavenumber $k \sim T$. Taking the open systems approach as introduced in Chapter 5, we partition the scalar field $\Phi = \phi + \varphi$, where ϕ is the soft field, and φ is the hard field. The soft and hard field equations are, respectively

$$\partial^2 \phi - \frac{gT}{2}\phi^2 - \frac{gT}{2}\left.\varphi^2\right|_{\text{soft}} + h = 0 \tag{10.91}$$

$$\partial^2 \varphi - gT\phi\varphi - \frac{gT}{2}\left.\varphi^2\right|_{\text{hard}} = 0 \tag{10.92}$$

In a naive perturbation expansion, we would argue that since the hard modes appear in equation (10.91) in a $O(g)$ term, we only need to solve equation (10.92) up to $O(1)$ accuracy. We would neglect the second and third terms in equation (10.92), proceeding to treat φ as a massless Klein–Gordon field. The treatment simplifies even further if we actually think of the soft modes as a classical field (which may be justified on the grounds of the large occupation numbers prevalent in the soft sector of the theory). Then we may replace $\left.\varphi^2\right|_{\text{soft}}$ in equation (10.91) by the thermal expectation value appropriate for a massless field $\langle\varphi^2\rangle \sim T^2/6$. In this approximation the $\langle\varphi^2\rangle$ term is canceled by the h term, by design, and we find that to leading order in g, the hard modes have no effect on the soft modes.

10.6.2 Hard thermal loops

Braaten and Pisarski [BraPis90a, BraPis90b, BraPis92] and Frenkel and Taylor [FreTay90] were the first to point out that this argument is not only naive, but actually wrong. The reason is resonance. Assume for simplicity that the soft modes undergo a homogeneous oscillation $\phi = \phi_0 \cos\omega t$. Assume also a perturbative expansion $\varphi = \varphi_0 + \varphi_1 + \ldots$ for the hard modes, where $\varphi_n \propto g^n$, and neglect interactions between hard modes. Then $\partial^2 \varphi_0 = 0$, and we may expand (in this section, we use natural units)

$$\varphi_0 = \int \frac{d^3\mathbf{k}}{(2\pi)^3} \frac{e^{i\mathbf{kx}}}{\sqrt{2k}}\varphi_{0\mathbf{k}}; \qquad \varphi_{0\mathbf{k}} = a_{0\mathbf{k}}e^{-ikt} + a_{0-\mathbf{k}}^\dagger e^{ikt} \tag{10.93}$$

with

$$\langle a_{0\mathbf{k}}^\dagger a_{0\mathbf{p}}\rangle = n_k \delta(\mathbf{k} - \mathbf{p}) \tag{10.94}$$

where n_k is the Bose–Einstein distribution

$$n_k = \left[e^{k/T} - 1\right]^{-1} \tag{10.95}$$

For not-so-hard modes, $n_k \sim T/k$.

Now consider the first-order correction. Assuming a Fourier decomposition

$$\varphi_1 = \int \frac{d^3\mathbf{k}}{(2\pi)^3} \frac{e^{i\mathbf{k}\mathbf{x}}}{\sqrt{2k}} \varphi_{1\mathbf{k}} \tag{10.96}$$

we have

$$\frac{\partial^2}{\partial t^2} \varphi_{1\mathbf{k}} + k^2 \varphi_{1\mathbf{k}} = -\left[gT\phi_0 \cos \omega t\right] \varphi_{0\mathbf{k}} \tag{10.97}$$

Neglecting the homogeneous solution, we get, in the limit of soft ω,

$$\varphi_{1\mathbf{k}} = \frac{gT\phi_0}{2k^2\omega} \sin \omega t \frac{\partial}{\partial t} \varphi_{0\mathbf{k}} \tag{10.98}$$

Glossing over the details of actually computing the expectation values, we see that a priori we expect a correction to the soft equation of motion of order of magnitude $gT \langle \varphi_0 \varphi_1 \rangle \sim g^2 (T/\omega) T^2 \phi_0$. For $\omega \sim gT$, this is larger than the classical term $\omega^2 \phi_0$ by a factor of g^{-1}, and it cannot possibly be neglected.

In diagrammatic terms, we may represent $\langle \varphi^2 \rangle$ as a tadpole graph (cf. Fig. 6.1 in Chapter 6). By considering a soft field insertion, it turns into a fish graph (Fig. 6.7 in Chapter 6). Explicitly

$$\langle \varphi^2 \rangle = \Delta(x,x) + gT \int d^4y \, \Delta^2(x,y) \phi(y) + \cdots \tag{10.99}$$

where Δ represents a massless scalar propagator. We postpone the question on exactly which propagator is involved, and work for now with the Feynman propagators (as would be the case in the "in-out" formulation). Fourier transforming,

$$\langle \varphi^2 \rangle = \int \frac{d^4p}{(2\pi)^4} \Delta(p) + gT \int \frac{d^4k}{(2\pi)^4} \int \frac{d^4p}{(2\pi)^4} \Delta(p) \Delta(p-k) \phi(k) \tag{10.100}$$

We are interested in the contribution from the second term. Consider the contribution from the thermal part in $\Delta(p)$. Then p is on-shell. Since $p - k$ cannot be on-shell, $\Delta(p - k) \sim (p - k)^{-2} \sim 1/2p^0\omega$. The presence of an inverse power of ω in the integral invalidates the naive perturbation theory when ω is parametrically small. If in particular $\omega \sim gT$, the "correction" is a priori as large as the leading term.

Moreover, the problem appears with *every* soft insertion. Adding a soft insertion to a pre-existing graph adds a power of $gT\phi$ but also a power of $(p^0\omega)^{-1}$. If we are considering large field amplitudes $\phi \sim T$, hard momenta $p \sim T$, and soft frequencies $\omega \sim gT$, then $gT\phi \sim p^0\omega \sim gT^2$. The overall amplitude of the graph with the insertion is not smaller than without it, and we must sum over all soft field insertions at once. The result is called a HTL resummed perturbation theory.

10.6.3 Hard thermal loops from the 2PI CTP effective action

To derive the HTL resummed theory, we shall use the 2PI CTP formalism. Since graphs with more than one external field insertion cannot be two-particle irreducible, they do not appear explicitly in the 2PI effective action (that is, the graph with a single insertion represents them all). The CTP technique warrants causality of the resulting equations of motion.

To simplify things, we shall make the semiclassical approximation for the soft field (that is, we shall not include soft field propagators) and set the hard background field to zero (that is, the hard field will be represented only by its propagators). The 2PI CTP effective action Γ_2 is thus a functional of a CTP soft background field ϕ^A (as before, the index A comprises a branch index $a = 1, 2$ and a spacetime location x) and hard propagators G^{AB}, and takes the form

$$\Gamma_2 = S\left[\phi^A\right] + \frac{1}{2}S_{,AB}\left[\phi^A\right]G^{AB} - \frac{i}{2}\mathrm{Tr}\,\ln G + \Gamma_Q \tag{10.101}$$

where $S\left[\phi^A\right] = S\left[\phi^1\right] - S\left[\phi^2\right]$, and

$$S_{,AB}\left[\phi^A\right] = \left\{c_{ab}\partial_x^2 - c_{abc}gT\phi^c\left(x\right)\right\}\delta\left(x - y\right) \tag{10.102}$$

(the tensors $c_{abc...}$ take the value 1 when all their indices are 1, -1 when all the indices are 2, and vanish otherwise). Γ_Q is the sum of all 2PI vacuum bubbles with a cubic vertex and G propagators. It represents the hard field's self-interactions, and we shall disregard it for the time being.

Variation of Γ_2 yields the 2PI Schwinger–Dyson equation for G

$$\left\{c_{ab}\partial_x^2 - c_{abc}gT\phi^c\left(x\right)\right\}G^{bd}\left(x,y\right) = i\delta_a^d\delta\left(x - y\right) \tag{10.103}$$

(where $\phi^1\left(x\right) = \phi^2\left(x\right) = \phi\left(x\right)$) and the field equation for the soft modes

$$\partial^2\phi\left(x\right) - \frac{gT}{2}\phi^2\left(x\right) - \frac{gT}{2}G^{11}\left(x,x\right) + h = 0 \tag{10.104}$$

In principle, one is to solve equation (10.103) for G and plug into equation (10.104) for ϕ. In practice, solving equation (10.103) is nontrivial, because of the spacetime dependence in $\phi\left(x\right)$. One possibility is to take advantage of the slow variation of the soft field to write the hard propagator in terms of a Wigner function. Since there are no hard self-interactions, the result is a Vlasov equation for the hard-field Wigner function.

10.6.4 The Vlasov equation for hard modes

Observe that the soft field only couples to the hard Hadamard propagator, which, neglecting hard self-interactions, obeys the simple equation

$$\left(\partial_x^2 - gT\phi\left(x\right)\right)G_1\left(x,y\right) = 0 \tag{10.105}$$

To avoid formal problems particular to the cubic interaction, we shall assume $\phi\left(x\right) > 0$ throughout.

Following the usual quantum kinetic theory approach, to be discussed further in Chapter 11, we decompose

$$G_1(x,y) = \int \frac{d^4p}{(2\pi)^4} e^{ipu} G_1(X,p) \tag{10.106}$$

$$u = x - y; \qquad X = \frac{1}{2}(x+y) \tag{10.107}$$

We also expand

$$\phi(x) \sim \phi(X) + \frac{u}{2}\partial\phi(X) + \cdots \tag{10.108}$$

Keeping the first derivatives only (formally, $\partial\phi \sim gT\phi$) we obtain the mass shell condition

$$\left[p^2 + gT\phi(X)\right] G_1(X,p) = 0 \tag{10.109}$$

therefore

$$G_1(X,p) = 2\pi\delta\left(p^2 + gT\phi(X)\right)\left[1 + 2f(X,p)\right] \tag{10.110}$$

where the distribution function f obeys the Vlasov equation

$$\left(p\frac{\partial}{\partial X} - \frac{1}{2}gT\frac{\partial\phi}{\partial X}\frac{\partial}{\partial p}\right)f = 0 \tag{10.111}$$

We solve this equation perturbatively

$$f = f_0 + f_1 + \cdots \tag{10.112}$$

off the thermal distribution

$$f_0 = n_{p^0} \tag{10.113}$$

Then

$$p\frac{\partial}{\partial X}f_1 = \frac{1}{2}g\frac{\partial\phi}{\partial X^0}n_{p^0}\left(1 + n_{p^0}\right) \tag{10.114}$$

which admits the particular solution

$$f_1 = \frac{(-i)}{2}gn_{p^0}\left(1 + n_{p^0}\right)\int d^4Y \int \frac{d^4Q}{(2\pi)^4} \frac{e^{iQ(X-Y)}}{p(Q+i\varepsilon)}\frac{\partial\phi}{\partial Y^0} \tag{10.115}$$

To find the soft equation of motion we must compute the coincidence limit $G_1(x,x)$

$$G_1(x,x) = G_1^0(x,x) + G_1^1(x,x) + \cdots \tag{10.116}$$

$$G_1^0(x,x) = \int \frac{d^4p}{(2\pi)^3}\delta\left(p^2 + gT\phi(x)\right)\left[1 + 2n_{p^0}\right] \sim \frac{T^2}{6} + O(g) \tag{10.117}$$

$$G_1^1(x,x) = (-i)g\int d^4Y \int \frac{d^4Q}{(2\pi)^4}e^{iQ(X-Y)}\frac{\partial\phi}{\partial Y^0}$$

$$\times \int \frac{d^4p}{(2\pi)^3}\delta\left(p^2 + gT\phi(x)\right)\frac{n_{p^0}\left(1 + n_{p^0}\right)}{p(Q+i\varepsilon)} \tag{10.118}$$

Assume for simplicity that ϕ is spatially homogeneous, then the integral over the space components $\mathbf{Y}$ is immediate, and we obtain

$$G_1^0(x,x) = ig \int dY \int \frac{dQ \, e^{-iQ(X^0-Y)}}{(2\pi)(Q+i\varepsilon)} \frac{\partial\phi}{\partial Y} \int \frac{d^4p}{(2\pi)^3}$$

$$\times \, \delta\left(p^2 + gT\phi(x)\right) \frac{n_{p^0}(1+n_{p^0})}{p^0} \tag{10.119}$$

Now

$$\int \frac{d^4p}{(2\pi)^3} \, \delta\left(p^2 + gT\phi(x)\right) \frac{n_{p^0}(1+n_{p^0})}{p^0} \sim \alpha\sqrt{\frac{T^3}{g\phi}} \tag{10.120}$$

where α is some numerical constant. In the other integral, we integrate by parts

$$\int dY \int \frac{dQ}{(2\pi)(Q+i\varepsilon)} \, e^{-iQ(X^0-Y)} \frac{\partial\phi}{\partial Y} \sim -i\phi(X) \tag{10.121}$$

Finally, we retrieve the equation for the soft modes

$$\partial^2\phi(x) - \frac{gT}{2}\phi^2(x) - \frac{\alpha gT^2}{4}\sqrt{gT\phi} = 0 \tag{10.122}$$

The correction is actually larger than the classical potential term when $(\phi/T) < g^{1/3}$; in any case, the corrections behave like $g^{3/2}$ rather than the expected g^2. The infrared sensitivity of the theory invalidates naive perturbation theory.

10.6.5 Ultrasoft modes and Boltzmann equation

We now return to address a possible concern about the consistency of neglecting interactions among hard modes. A moment's reflection shows that for soft modes $\omega \sim gT$ Feynman graphs containing hard cubic vertices are indeed of higher order. For example, if we compare the graph in Fig. 10.1 to the fish graph, we see that both lead to one power of ω^{-1}, but Fig. 10.1 has four powers of gT against 2 in the fish. Even the graph in Fig. 10.2 is safe, because although it scales as ω^{-2}, it also has four powers of gT in the denominator.

Figure 10.2 becomes unsafe, however, if we push the theory to deal with *ultrasoft* modes $\omega \sim g^2T$. To include it into the model, we must reconsider the role of Γ_Q. In particular, we obtain Fig. 10.2 if we approximate Γ_Q by the setting-sun

Figure 10.1

Figure 10.2

graph (Fig. 6.4 in Chapter 6). The three loops graph in Fig. 6.10 of Chapter 6 will lead to graphs of higher order even in the ultrasoft regime.

Adding a nontrivial Γ_Q has the important effect that the equation for the hard propagators becomes nonlinear. We may still make use of Wigner function techniques, but now the transport equation acquires a collision term – for the cubic self-interaction, the corresponding kinetic equation has been worked out by Danielewicz in the early days of NEqQFT [Dan84a, Dan84b], and it is not quite Boltzmann's. Of course, if we are only interested in small oscillations, we may linearize the collision term around the equilibrium solution for zero soft background. Even after linearization, the presence of the collision term affects the physics in important ways.

To get an idea of the changes brought by the collision term, we may adopt the simple "collision time approximation" [Lib98], and write the full kinetic equation as

$$\left(p\frac{\partial}{\partial X} - \frac{1}{2}gT\frac{\partial\phi}{\partial X}\frac{\partial}{\partial p}\right)f = -\frac{T}{\tau}(f - f_0) \tag{10.123}$$

where τ is the relaxation time. From naive power counting and dimensional analysis, we see $\tau \sim (g^2 T)^{-1}$. The unperturbed solution is still a Bose–Einstein distribution, but now the first correction is

$$\left(p\frac{\partial}{\partial X} + \frac{T}{\tau}\right)f_1 = \frac{1}{2}g\frac{\partial\phi}{\partial X^0}n_{p^0}\left(1 + n_{p^0}\right) \tag{10.124}$$

and we may approximate

$$f_1 = \frac{g\tau}{2T}\frac{\partial\phi}{\partial X^0}n_{p^0}\left(1 + n_{p^0}\right) \tag{10.125}$$

This introduces a dissipative term in the equation for ultrasoft modes

$$\partial^2\phi(x) - \frac{gT}{2}\phi^2(x) - 2\gamma\frac{\partial\phi}{\partial X^0} = 0 \tag{10.126}$$

$$\gamma = \frac{g^2\tau}{8}\int\frac{d^4p}{(2\pi)^3}\,\delta\left(p^2 + gT\phi(x)\right)n_{p^0}\left(1 + n_{p^0}\right) \tag{10.127}$$

An explicit calculation yields

$$\gamma \sim \alpha'g^2T^2\tau\,\ln\left[\frac{T}{g\phi}\right] \tag{10.128}$$

where α' is (another) numerical constant.

Once again, the contribution from HTLs is of order $g^2 T^2 \phi$, much larger than the classical term $\partial^2 \phi / \partial X^{02} \sim g^4 T^2 \phi$.

10.6.6 Langevin dynamics of ultrasoft modes

We have seen in the previous section that the dynamics of ultrasoft modes is dissipative. From the discussion in Chapter 8, we know that it will be noisy as well. We shall use this as a model problem, by deriving the fluctuations in three different ways.

Method 1: fluctuation–dissipation theorem for ultrasoft modes

The simplest approach is to apply to the ultrasoft equation of motion (10.126) the fluctuation-dissipation theorem as discussed in Chapter 8. We modify equation (10.126) to read

$$\partial^2 \phi \left(x \right) - \frac{gT}{2} \phi^2 \left(x \right) - 2\gamma \frac{\partial \phi}{\partial X^0} = \xi \left(x \right) \qquad (10.129)$$

$$\langle \xi \left(x \right) \xi \left(x' \right) \rangle = \sigma^2 \delta \left(x - x' \right) \qquad (10.130)$$

The fluctuation–dissipation theorem yields

$$\sigma^2 = 4\gamma T \qquad (10.131)$$

Method 2: fluctuations from the CTPEA

Our second approach to the derivation of noise in the dynamics of ultrasoft modes will be based on the derivation of equation (10.129) from the CTPEA, a problem we already confronted in Chapter 8. The equation of motion, as derived from the CTPEA, reads

$$\partial^2 \phi \left(x \right) - \frac{gT}{2} \phi^2 \left(x \right) - \frac{gT}{2} \left[\langle \varphi^2 \rangle_\phi - \langle \varphi^2 \rangle_0 \right] \left(x \right) = \frac{gT}{2} \zeta \left(x \right) \qquad (10.132)$$

We have shown in Chapter 8 that

$$\langle \zeta \left(x \right) \zeta \left(x' \right) \rangle = \frac{1}{2} \left[\langle \{ \varphi^2 \left(x \right), \varphi^2 \left(x' \right) \} \rangle_\phi - 2 \langle \varphi^2 \rangle_\phi \left(x \right) \langle \varphi^2 \rangle_\phi \left(x' \right) \right] \qquad (10.133)$$

To compare these expressions to the explicit derivation above, we recall the linear response theory result (discussed earlier in this chapter)

$$\langle \varphi^2 \rangle_{\phi + \delta\phi} \left(x \right) = \langle \varphi^2 \rangle_\phi \left(x \right) - \frac{igT}{2} \int d^4 x' \ \langle [\varphi^2 \left(x \right), \varphi^2 \left(x' \right)] \rangle_\phi \ \theta \left(x^0 - x'^0 \right) \delta\phi(x') \qquad (10.134)$$

Comparison with equation (10.129) yields

$$\langle [\varphi^2 \left(x \right), \varphi^2 \left(x' \right)] \rangle_\phi = \frac{16i\gamma}{g^2 T^2} \frac{\partial}{\partial x^0} \delta \left(x - x' \right) \qquad (10.135)$$

On the other hand, commutator and anticommutator are related through the KMS theorem (also discussed earlier in this chapter). Introducing the Fourier decomposition

$$\langle [\varphi^2(x), \varphi^2(x')] \rangle_\phi = \int \frac{d^4p}{(2\pi)^4} e^{ip(x-x')} R(p) \tag{10.136}$$

$$\langle \{\varphi^2(x), \varphi^2(x')\} \rangle_\phi = \int \frac{d^4p}{(2\pi)^4} e^{ip(x-x')} R_1(p) \tag{10.137}$$

then

$$R(p) = \frac{16\gamma}{g^2 T^2} p^0 \tag{10.138}$$

and the fluctuation–dissipation theorem yields

$$R_1(p) = \frac{16\gamma}{g^2 T^2} (1 + 2n_{p^0}) |p^0| \tag{10.139}$$

In the high-temperature limit, this leads to the classical result

$$\langle \zeta(x)\zeta(x') \rangle = \frac{16\gamma}{g^2 T} \delta(x - x') \tag{10.140}$$

which agrees with the results from Method 1 after identifying $\xi = gT\zeta/2$.

Method 3: fluctuations in the Boltzmann equation

Yet another method to derive the fluctuations in the ultrasoft modes is to keep to the derivation of the ultrasoft equation of motion in the previous section, but now using for the hard modes the full Boltzmann equation which, as discussed in Chapter 2, must contain stochastic terms over and above the usual collision term, thus becoming a Boltzmann–Langevin equation. This means we replace equation (10.123) by

$$\left(p \frac{\partial}{\partial X} - \frac{1}{2} gT \frac{\partial \phi}{\partial X} \frac{\partial}{\partial p} \right) f = -\frac{T}{\tau} (f - f_0) + J(X, \mathbf{p}) \tag{10.141}$$

where it is understood that p^0 is given as a function of the spatial components $\mathbf{p}$ through the mass shell condition. The noise self-correlation has been derived in Chapter 2. Under the collision time approximation (10.123) for the collision integral, we get

$$\langle J(X, \mathbf{p}) J(Y, \mathbf{q}) \rangle = 2(2\pi)^3 \delta(X - Y) \delta(\mathbf{p} - \mathbf{q}) \frac{T}{\tau} p^0 n_{p^0} (1 + n_{p^0}) \tag{10.142}$$

For the ultrasoft components of the distribution function, we obtain

$$f_1 = f_{1\text{det}} + \frac{\tau}{T} J \tag{10.143}$$

where $f_{1\text{det}}$ is the deterministic solution given in equation (10.125). The equation of motion for ultrasoft modes (10.126) is transformed into equation (10.129), where now we have a model for the noise

$$\xi(x) = \frac{gT}{2} \int \frac{d^4p}{(2\pi)^3}\, \delta\left(p^2 + gT\phi(x)\right) \left[\frac{\tau}{T} J(x, \mathbf{p})\right] \tag{10.144}$$

leading to the self-correlation

$$\langle \xi(x)\, \xi(x')\rangle = \frac{g^2 T \tau}{2} \delta(x - x') \int \frac{d^4p}{(2\pi)^3}\, \delta\left(p^2 + gT\phi(x)\right) n_{p^0}\left(1 + n_{p^0}\right)$$

$$\tag{10.145}$$

Comparing this to equation (10.127), we recover equation (10.130)

10.6.7 A note on the literature

As already mentioned, Braaten, Pisarsky, Frenkel and Taylor were the first to point out the need to restructure perturbation theory to account for the physics at different scales. Their work was motivated by the need to derive a reliable estimate of the decay constants of various fields in a hot non-abelian plasma. In this context, there are a number of issues associated with the gauge nature of the fundamental theory (such as whether the right decay constants are automatically gauge invariant) which have no analog in our toy model. We have only attempted to give a flavor of the physical ideas behind the formalism.

The subsequent literature on hard thermal loops is voluminous. Le Bellac's book [LeB96] has a nice chapter on this subject. Our presentation here is mostly a retelling of work by Bödeker [Bod98, Bod99] and by Arnold, Moore, Son and Yaffe [Son97, ArSoYa99a, ArSoYa99b, ArMoYa00] (of course, any flaw incurred in our attempt to "simplify" their discussion is our own). Important contributions from Blaizot and Iancu are summarized in the *Physics Reports* review article by these authors [BlaIan02]. The Boltzmann–Langevin equation for Φ^4 field was investigated by the authors in [CalHu00] while for non-abelian plasmas by Litim and Manuel. We recommend their review article as a good entry point to the literature [LitMan02].

We shall return to some of these issues in later chapters.

11

Quantum kinetic field theory

11.1 The Kadanoff–Baym equations

Quantum kinetic field theory is the theme of this chapter. In this section we get right to the heart of it by showing a derivation of the celebrated Kadanoff–Baym (KB) equations [KadBay62]. The basic idea is that close to equilibrium, propagators are nearly translation invariant. It is possible to define a partial Fourier transform with respect to the relative position of the arguments. The Kadanoff–Baym equations then determine how the partial Fourier transform depends on the average (or "center of mass") of the arguments in the original propagators.

Besides the presentation in Kadanoff and Baym's textbook, there are several derivations of these equations in the literature [Dub67, Dan84a, MroDan90, MroHei94, ZhuHei98]. We shall follow [CalHu88, CaHuRa00]. See also [Hen95, IvKnVo00, KnIvVo01, Nie02, Koi02]. References [BoVeWa00, WBVS00] follow a different path towards quantum kinetic theory, based on the so-called dynamical renormalization group.

11.1.1 The model

To better appreciate the main points in this derivation, we shall consider a simple model, namely, the KB equations for the theory of a single real self-interacting $\lambda\Phi^4$ scalar field, in the absence of background fields. Actually, the key ideas are not sensitive to the particular models, but for concreteness it will be helpful to have a model in mind. The classical action is given by equation (6.106).

A translation-invariant propagator G^{ab} depends on its arguments x and x' only through the so-called "relative" variable $u = x - x'$. The Fourier transform with respect to u yields the momentum representation

$$G^{ab}(x, x') = \hbar \int \frac{d^d k}{(2\pi)^d} \, e^{iku} \, G^{ab}(k) \tag{11.1}$$

We have discussed in Chapter 6 the basic properties of these Fourier transforms.

We say that a G^{ab} is almost translation invariant if, when partially Fourier transformed with respect to u, the Fourier transform is weakly dependent on the "center of mass" variable $X = (x + x')/2$, i.e.

$$G^{ab}(x, x') = \int \frac{d^d k}{(2\pi)^d} \, e^{iku} G^{ab}(X, k) \tag{11.2}$$

The precise definition of what "weakly dependent" means depends up to a certain point on the problem at hand. For example, in a hard thermal loop scheme such as discussed in Chapter 10, we may find a situation where $\partial_x G^{ab}(x, x') \approx T G^{ab}$, while $\partial_X G^{ab}(X, k) \approx g^n T G^{ab}$ with $n \geq 1$. In such a case, the propagators are almost translation invariant in the weak coupling limit.

On the other hand, beware that in gauge theories the same object may be almost translation invariant in some gauges and not in others (with a corresponding problem in relativistic theories with respect to changes of coordinates). We also mention that in the presence of external gauge background fields, or in curved spacetimes, some care must be taken to define precisely the Fourier transform in (11.2) [Hei83, Win85, CaHaHu88, Fon94]. We shall discuss these issues later in this chapter.

Irrespective of whether the assumption of almost translation-invariance holds, expressions involving $G^{ab}(X, k)$ may be classified according to their "adiabatic order," namely, the number of X derivatives appearing in the expression. We call this the "adiabatic expansion." When almost translation-invariance is satisfied, we may further reject all terms above a given adiabatic order. We call such a truncation of an adiabatic expansion an "adiabatic approximation." In other words, the adiabatic order is used as a tag to bunch together certain terms in the equations of motion in accordance to their derivative orders and the adiabatic approximation determines how many of those terms are kept.

Our aim is to analyze the adiabatic expansion of the 2PI Schwinger–Dyson equations for the propagators. These are deduced from the 2PI CTPEA (cf. Chapter 6)

$$\Gamma = \frac{1}{2Z_B} \int d^d x d^d y \, c_{ab} D(x, y) G^{ab}(x, y) - \frac{i\hbar}{2} \text{Tr} \, \ln G + \Gamma_Q \qquad (11.3)$$

where $c_{11} = -c_{22} = 1$, $c_{12} = c_{21} = 0$,

$$D(x, y) = \left[\partial_x^2 - m_b^2 \right] \delta(x - y) \qquad (11.4)$$

Γ_Q is the sum of all 2PI vacuum bubbles. Taking variations of the 2PI CTPEA we find the equations of motion

$$\frac{1}{Z_B} c_{ab} D(x, y) - i\hbar \left[G^{-1} \right]_{ab}(x, y) - \hbar \Sigma_{ab}(x, y) = 0 \qquad (11.5)$$

$$\hbar \Sigma_{ab}(x, y) = -2 \frac{\delta \Gamma_Q}{\delta G^{ab}(x, y)} \qquad (11.6)$$

These are the exact equations we must solve. We assume there are known relations expressing Σ_{ab} in terms of the propagators. These can be found, for example, by adopting one of the perturbative schemes discussed in Chapter 6.

Observe that to determine $G(X, k)$ as the inverse Fourier transform of $G(x, x')$ we must know the whole evolution of the correlation, both to the past and future of the event X. It is possible to present an alternative formulation where only equal time correlations are Fourier transformed, thus more in keeping with the spirit of causality [ZhaHei96a, ZhaHei96b, ABZH96, ZhuHei98].

In general, it will be necessary to add nonlocal sources to the classical action to account for nontrivial correlations at the initial time. We consider these sources are included into the $\Sigma_{ab}(x, y)$.

Let us continue with the analysis of (11.5). Our first task is to find an efficient parameterization for the propagators. It is clear that the four basic propagators (Feynman, Dyson, positive and negative frequency) are not independent. As we shall see, there are essentially two (phase space) functions which contain the relevant information from which all propagators may be reconstructed. One of these functions plays the role of a (position-dependent) density of states, and the other one of the nonequilibrium one-particle distribution function. To be able to write all propagators in terms of these two functions we must consider first the so-called Keldysh representation of the propagators, in which the four basic propagators are written in terms of the Hadamard, retarded and advanced propagators as

$$G^{ab} = \begin{pmatrix} G^{11} & G^{12} \\ G^{21} & G^{22} \end{pmatrix} = \frac{1}{2} \begin{pmatrix} 1 & 1 \\ -1 & 1 \end{pmatrix} \begin{pmatrix} 0 & -i\hbar G_{\text{adv}} \\ -i\hbar G_{\text{ret}} & G_1 \end{pmatrix} \begin{pmatrix} 1 & -1 \\ 1 & 1 \end{pmatrix} \quad (11.7)$$

with inverse

$$\begin{pmatrix} 0 & -i\hbar G_{\text{adv}} \\ -i\hbar G_{\text{ret}} & G_1 \end{pmatrix} = \frac{1}{2} \begin{pmatrix} 1 & -1 \\ 1 & 1 \end{pmatrix} \begin{pmatrix} G^{11} & G^{12} \\ G^{21} & G^{22} \end{pmatrix} \begin{pmatrix} 1 & 1 \\ -1 & 1 \end{pmatrix} \quad (11.8)$$

11.1.2 Density of states and distribution function

Let us introduce the density of states $\mathcal{D}(X, k)$ out of the Fourier transform of the Jordan propagator $G = G^{21} - G^{12}$

$$G(X, k) \equiv 2\pi\hbar\mathcal{D}(X, k) \ \text{sign}\left(k^0\right) \quad (11.9)$$

Observe that by symmetry we must have $G(X, (0, \mathbf{k})) = 0$. We shall assume G is continuous there, implying $\mathcal{D}(X, (0, \mathbf{k})) = 0$.

The Jordan and retarded propagators are related through

$$\begin{aligned} G(X, k) &= 2\hbar \ \text{Im} G_{\text{ret}}(X, k) \\ &= -2\hbar \left|G_{\text{ret}}(X, k)\right|^2 \text{Im}\left[G_{\text{ret}}(X, k)\right]^{-1} \end{aligned} \quad (11.10)$$

This suggests defining a new kernel $\gamma(X, k)$ such that

$$\mathcal{D}(X, k) = \left|G_{\text{ret}}(X, k)\right|^2 \gamma(X, k) \quad (11.11)$$

$$\gamma(X, k) = \frac{1}{\pi} \text{Im}\left[-G_{\text{ret}}(X, k)\right]^{-1} \text{sign}\left(k^0\right) \quad (11.12)$$

Observe that per earlier assumptions, $\gamma(X, (0, \mathbf{k})) = 0$.

We now define the (dimensionless) distribution function $f(X, k)$ through the partial Fourier transform of the Hadamard propagator

$$G_1(X, k) \equiv 2\pi\hbar\mathcal{D}(X, k)\ F_1(X, k) \tag{11.13}$$

$$F_1(X, k) = 1 + 2f(X, k) \tag{11.14}$$

It follows that

$$G^{21(12)}(X, k) = 2\pi\hbar\ F^{21(12)}(X, k)\mathcal{D}(X, k) \tag{11.15}$$

where

$$F^{21(12)}(X, k) = \theta\left(\pm k^0\right) + f(X, k) \tag{11.16}$$

In equilibrium, f is the Bose–Einstein distribution function (KMS theorem). It can be assumed that (11.13) serves as the definition of the function f, valid to all orders in perturbation theory. Observe that, since the relevant Fourier transforms are distributions (e.g. in free theories), this definition may only be applied if both Fourier transforms have the same singularity structure, which amounts to a restriction on allowed quantum states. In what follows, we shall assume these restrictions are met.

While this definition of the one-particle distribution function will prove to be very convenient in practice, and it is guaranteed to give the right result in equilibrium, it is not tied to any fundamental definition of what a particle is. It is also possible to take an alternative route, where one introduces a physically motivated particle destruction operator, builds the corresponding particle number operator, and finally derives an equation of motion for the latter (cf. the discussion of the quantum Vlasov equation in Chapter 4) [GreLeu98].

11.1.3 The dissipation and noise kernels

As we have seen, the information content of the almost translation-invariant propagators can be encoded in just two functions $\mathcal{D}(X, k)$ and $f(X, k)$. To proceed, we must perform a similar compression of the self-energies Σ_{ab}. We do this by writing both propagators and self-energies in terms of the dissipation $\mathbf{D}$ and the noise kernel $\mathbf{N}$, which appear in the Hessian of the one-particle irreducible (1PI) effective action. These two kernels are largely independent of each other, and have a distinct physical meaning, with $\mathbf{D}$ carrying the dynamical information and $\mathbf{N}$ the statistical information. This division of labor is most clearly seen in a free theory. Together $\mathbf{D}$ and $\mathbf{N}$ are a more compact description of the theory than the propagators themselves.

We have two different ways of relating $\mathbf{D}$ and $\mathbf{N}$ to the propagators. On one hand, they are constructed from Feynman diagrams which carry propagators in their internal legs. Which diagrams must be considered depends on which approximation is being used. On the other hand, the Schwinger–Dyson (SD) equations allow us to express the propagators in terms of **D** and **N**. For a true solution, these two paths must be equivalent. This consistency requirement yields the most efficient representation of the dynamics.

The dissipation and noise kernels **D** and **N** appear in the linearized CTP 1PI EA Γ_{1PI} (cf. Chapter 6)

$$\Gamma_{1PI} = \frac{1}{2} \int d^d x d^d y \left\{ \left[\varphi^1 - \varphi^2 \right] (x) \left[\frac{1}{Z_B} D(x,y) + \mathbf{D}(x,y) \right] \left[\varphi^1 + \varphi^2 \right] (y) \right.$$
$$\left. + i \left[\varphi^1 - \varphi^2 \right] (x) \mathbf{N}(x,y) \left[\varphi^1 - \varphi^2 \right] (y) \right\} \tag{11.17}$$

D is causal and **N** is even, and both are real. The causality of **D** allows for a more efficient parameterization. Introduce the kernels

$$\mathbf{D}_{even}(x,y) = \frac{1}{2} \left[\mathbf{D}(x,y) + \mathbf{D}(y,x) \right]; \qquad \mathbf{\Gamma}(x,y) = \frac{1}{2} \left[\mathbf{D}(x,y) - \mathbf{D}(y,x) \right] \tag{11.18}$$

then

$$\mathbf{D} = 2\mathbf{\Gamma}\,\theta\left(x^0 - y^0\right) \tag{11.19}$$
$$\mathbf{D}_{even} = \mathbf{\Gamma}\,\mathrm{sign}\left(x^0 - y^0\right) \tag{11.20}$$

D and **N** are related to the G^{ab} through the identity

$$\frac{\mathcal{D}^2 \Gamma_{1PI}}{\mathcal{D}\varphi^a \mathcal{D}\varphi^b} = i\hbar \left[G^{-1} \right]_{ab} \tag{11.21}$$

The inverse propagators may be read off the Schwinger–Dyson equations, and we get

$$-\hbar \Sigma_{11} = \mathbf{D}_{even} + i\mathbf{N} \tag{11.22}$$
$$-\hbar \Sigma_{12} = \mathbf{\Gamma} - i\mathbf{N} \tag{11.23}$$
$$-\hbar \Sigma_{21} = -\mathbf{\Gamma} - i\mathbf{N} \tag{11.24}$$
$$-\hbar \Sigma_{22} = -\mathbf{D}_{even} + i\mathbf{N} \tag{11.25}$$

Since $\mathbf{D}(x,y)$ is real, we know the real part of $\mathbf{D}(X,k)$ is even and the imaginary part is odd: $\mathbf{D}(X,k) = \mathbf{D}(X,-k)^*$, so

$$\mathbf{D}_{even}(X,k) = \mathrm{Re}\,\mathbf{D}(X,k) \tag{11.26}$$
$$\mathbf{\Gamma}(X,k) = i\,\mathrm{Im}\,\mathbf{D}(X,k) \tag{11.27}$$

and (11.20) transforms into the Kramers–Kronig relations for the causal kernel **D**.

11.1.4 The retarded and advanced propagators

We have seen how to relate the dissipation and noise kernels $\mathbf{D}$ and $\mathbf{N}$ to the inverse propagators. To relate them to the propagators themselves, let us first investigate the SD equations in the Keldysh representation.

Write

$$
\hbar\left[G^{-1}\right]_{ab} = \frac{1}{2}\begin{pmatrix} 1 & 1 \\ -1 & 1 \end{pmatrix}\begin{pmatrix} \hbar^{-1}G_{\mathrm{ret}}^{-1}G_1 G_{\mathrm{adv}}^{-1} & iG_{\mathrm{ret}}^{-1} \\ iG_{\mathrm{adv}}^{-1} & 0 \end{pmatrix}\begin{pmatrix} 1 & -1 \\ 1 & 1 \end{pmatrix}
\tag{11.28}
$$

The equations of motion now read

$$
\begin{pmatrix} \hbar^{-1}G_{\mathrm{ret}}^{-1}G_1 G_{\mathrm{adv}}^{-1} & iG_{\mathrm{ret}}^{-1} \\ iG_{\mathrm{adv}}^{-1} & 0 \end{pmatrix} = (-i)\begin{pmatrix} 2i\mathbf{N} & \frac{D}{Z_B} + \mathbf{D} \\ \frac{D}{Z_{z\mathrm{B}}} + \mathbf{D}_{\mathrm{adv}} & 0 \end{pmatrix}
\tag{11.29}
$$

where

$$
\mathbf{D}_{\mathrm{adv}}\left(x,y\right) = \mathbf{D}_{\mathrm{even}} - \boldsymbol{\Gamma} = \mathbf{D}\left(y,x\right)
\tag{11.30}
$$

These equations show that G_1 and G_{ret} may be considered functionals of $\mathbf{D}$ and $\mathbf{N}$. The formulae above upon partial Fourier transform become

$$
\mathbf{D}\left(X,k\right) = \left\{-G_{\mathrm{ret}}^{-1}\left(X,k\right) + \frac{1}{Z_B}\left(k^2 + m_b^2\right)\right\}
\tag{11.31}
$$

To relate $G_{\mathrm{ret}}^{-1}\left(X,k\right)$ to $\left[G_{\mathrm{ret}}\left(X,k\right)\right]^{-1}$ we recall the formula for the partial Fourier transform of a convolution

$$
\left[f*g\right]\left(X,k\right) = f\left(X,k\right)g\left(X,k\right) - \frac{i}{2}\left\{f,g\right\} \frac{-1}{8}\left\{\frac{\partial^2 f}{\partial X^\mu \partial X^\nu}\frac{\partial^2 g}{\partial k_\mu \partial k_\nu}\right.
$$
$$
\left. + \frac{\partial^2 g}{\partial X^\mu \partial X^\nu}\frac{\partial^2 f}{\partial k_\mu \partial k_\nu} - 2\frac{\partial^2 f}{\partial X^\mu \partial k_\nu}\frac{\partial^2 g}{\partial k_\mu \partial X^\nu}\right\} + \dots
\tag{11.32}
$$

where we use the Poisson bracket (cf. Chapter 2)

$$
\left\{f,g\right\} = \frac{\partial f}{\partial k}\frac{\partial g}{\partial X} - \frac{\partial f}{\partial X}\frac{\partial g}{\partial k}
\tag{11.33}
$$

We obtain the adiabatic expansion of $G_{\mathrm{ret}}^{-1}\left(X,k\right)$ by applying equation (11.32) to the obvious statement that the convolution of G_{ret}^{-1} and G_{ret} is the identity operator. To simplify the resulting expression, we assume that in the second-order terms we may approximate $G_{\mathrm{ret}}^{-1}\left(X,k\right)$ by its quasi-particle approximation form $\left[(k+i\epsilon)^2 + M^2\right]$.

More generally, the so-called quasi-particle approximation consists in replacing the actual propagators for those of a free field (see Chapter 5) with a yet-to-be-determined mass M^2. The physical basis of this approximation is that one expects the most interesting dynamics may be described in terms of localized excitations which, in between collisions, propagate as free particles with a well-defined mass. This leads to propagators concentrated on a sharp mass shell, which can be well approximated by free propagators. The quasi-particle approximation is expected to hold when the mean free path for quasi-particles is long compared with the Debye length M^{-1}. We warn the reader beforehand that important processes, such as thermalization, are not well described within this approximation (see the next chapter).

Computing the required derivatives and rearranging, we obtain

$$
1 = \left\{ \frac{1}{Z_B} \left(k^2 + m_b^2 \right) - \mathbf{D}\left(X, k \right) \right\} G_{\text{ret}} \left(X, k \right)
$$
$$
- \frac{1}{4} \nabla_X^2 G_{\text{ret}} \left(X, k \right) - \frac{1}{8} \frac{\partial^2 M^2}{\partial X^\mu \partial X^\nu} \frac{\partial^2 G_{\text{ret}} \left(X, k \right)}{\partial k_\mu \partial k_\nu} \tag{11.34}
$$

Over and above the need to renormalize (11.34), observe that this is a full-fledged evolution equation for the Fourier transform of the retarded propagator. To make the approach more definite, we may request that the quasi-particle approximation for G_{ret} actually becomes exact as $k \to 0$. We shall discuss mass renormalization in more detail below.

Ideally one would seek simultaneous solutions for (11.34) and the transport equation to be derived below, but these are hard (and may be impossible) to find [Mro97]. In such a case, one simply regards (11.34) as a way to generate the adiabatic expansion of G_{ret}.

In the approximation where only terms linear in the gradients of the Fourier transforms of the propagators are retained, it is possible to write down a non-perturbative (in the coupling constant) expression for the retarded and Jordan propagators. The advantage of this approach is that it goes beyond the quasi-particle approximation. In particular, it is sufficient for the discussion of the transition to hydrodynamics and the computation of transport functions.

It is convenient to introduce a real kernel

$$
R\left(X, k \right) = \frac{1}{Z_B} \left(k^2 + m_b^2 \right) - \mathbf{D}_{\text{even}} \left(X, k \right) \tag{11.35}
$$

The required expression is

$$
\left[G_{\text{ret}} \left(X, k \right) \right]^{-1} = R - \boldsymbol{\Gamma} \tag{11.36}
$$

From (11.27)

$$
\boldsymbol{\Gamma}\left(X, k \right) = (-i) \ \text{Im} \ \left[G_{\text{ret}} \left(X, k \right) \right]^{-1} = i\pi \ \gamma \left(X, k \right) \text{sign} \left(k^0 \right) \tag{11.37}
$$

and finally, from (11.11),

$$
\mathcal{D}\left(X, k \right) = \frac{\gamma \left(X, k \right)}{R^2 - \boldsymbol{\Gamma}^2} \tag{11.38}
$$

Recall that since we assume we know how to express R and $\boldsymbol{\Gamma}$ (and therefore also γ) in terms of propagators, this is really a consistency condition linking the density of states and the distribution function. Also recall that in deriving it we have neglected terms of second adiabatic order and higher.

11.1.5 The off-shell kinetic equation

To obtain the dynamics of the distribution function f, we make use of the remaining equation involving the noise kernel

$$\mathbf{N}(x, y) = \frac{1}{2\hbar} \int d^d z_1 d^d z_2 \, G_{\text{ret}}^{-1}(x, z_1) \, G_1(z_1, z_2) \, G_{\text{adv}}^{-1}(z_2, y) \qquad (11.39)$$

Iterating the formula for the partial Fourier transform of a convolution (11.32), and dropping second-order terms and higher, we get

$$2\hbar \mathbf{N} = \left[G_{\text{ret}}^{-1} G_{\text{adv}}^{-1} - \frac{i}{2} \{ G_{\text{ret}}^{-1}, G_{\text{adv}}^{-1} \} \right] G_1$$

$$- \frac{i}{2} \left[G_{\text{adv}}^{-1} \{ G_{\text{ret}}^{-1}, G_1 \} - G_{\text{ret}}^{-1} \{ G_{\text{adv}}^{-1}, G_1 \} \right] \qquad (11.40)$$

Observe that by performing the adiabatic expansion on this form of the SD equations we avoid the appearance of $\mathbf{N}$ within Poisson brackets. This choice is related to the so-called Botermans and Malfliet approach [BotMal90, IvKnVo00, KnIvVo01, IvKnVo03]. Next, write

$$G_1 = 2\pi \hbar \gamma G_{\text{ret}} G_{\text{adv}} F_1 \qquad (11.41)$$

$$\mathbf{N} = \left[\pi\gamma \left(1 - \frac{i}{2} G_{\text{ret}} G_{\text{adv}} \{ G_{\text{ret}}^{-1}, G_{\text{adv}}^{-1} \} \right) \right.$$

$$\left. - \frac{i\pi}{2} \left(G_{\text{ret}} \{ G_{\text{ret}}^{-1}, \gamma \} - G_{\text{adv}} \{ G_{\text{adv}}^{-1}, \gamma \} \right) \right] F_1$$

$$- \frac{i\pi\gamma}{2} \left[G_{\text{ret}} \{ G_{\text{ret}}^{-1}, F_1 \} - G_{\text{adv}} \{ G_{\text{adv}}^{-1}, F_1 \} \right] \qquad (11.42)$$

Recall that

$$\{ \mathbf{\Gamma}, \gamma \} = i\pi \, \gamma(X, p) \{ \text{sign}(p^0), \gamma \} = i\pi \, \delta(p^0) \frac{\partial}{\partial X^0} \gamma^2(X, p) = 0 \qquad (11.43)$$

because $\gamma(X, (0, \mathbf{p})) = 0$. Therefore

$$\{ G_{\text{ret}}^{-1}, G_{\text{adv}}^{-1} \} = \{ R - \mathbf{\Gamma}, R + \mathbf{\Gamma} \} = 2 \{ R, \mathbf{\Gamma} \} \qquad (11.44)$$

$$\{ G_{\text{ret}}^{-1}, \gamma \} = \{ G_{\text{adv}}^{-1}, \gamma \} = \{ R, \gamma \} \qquad (11.45)$$

$$\mathbf{N} = \left[\pi\gamma \left(1 - 2i G_{\text{ret}} G_{\text{adv}} \{ R, \mathbf{\Gamma} \} \right) \right] F_1$$

$$- \frac{i\pi\gamma}{2} \left[G_{\text{ret}} \{ G_{\text{ret}}^{-1}, F_1 \} - G_{\text{adv}} \{ G_{\text{adv}}^{-1}, F_1 \} \right] \qquad (11.46)$$

Introduce the *collision integral*

$$I_{\text{col}} \equiv [\mathbf{N} - \pi\gamma F_1] \, \text{sign}(k^0)$$

$$= \frac{-i\hbar}{2} \left[(\Sigma_{12} + \Sigma_{21}) (F^{21} - F^{12}) + (\Sigma_{12} - \Sigma_{21}) (F^{21} + F^{12}) \right]$$

$$= -i\hbar \left[\Sigma_{12} F^{21} - \Sigma_{21} F^{12} \right] \qquad (11.47)$$

Then we obtain the kinetic equation

$$A\left\{R, F_1\right\} + B\left\{\boldsymbol{\Gamma}, F_1\right\} + CF_1 = I_{\mathrm{col}}\,\mathrm{sign}\left(k^0\right) \tag{11.48}$$

where

$$A = -\frac{\boldsymbol{\Gamma}^2}{R^2 - \boldsymbol{\Gamma}^2} \tag{11.49}$$

$$B = \frac{R\boldsymbol{\Gamma}}{R^2 - \boldsymbol{\Gamma}^2} \tag{11.50}$$

$$C = -2\frac{\boldsymbol{\Gamma}}{R^2 - \boldsymbol{\Gamma}^2}\left\{R, \boldsymbol{\Gamma}\right\} \tag{11.51}$$

Equation (11.48) is the key result of this chapter.

11.1.6 Weakly coupled theories and the Boltzmann equation

For weakly coupled theories, a series of approximations allow us to reduce (11.48) to the more familiar Boltzmann kinetic equation (cf. Chapter 2).

We observe that in terms of the coupling constant λ we have, for a generic momentum p, $R \sim O\left(1\right)$ while $\boldsymbol{\Gamma} \sim O\left(\lambda^2\right)$. The C term in (11.48) combines both space derivatives (assumed small) and radiative corrections. It is therefore expected to be smaller than the other terms in the equation, and thus neglected (approximations of this kind are further discussed in Section 11.1.9). An alternative, which we shall not follow, is to consider these terms as parts of the collision integral, in which case we could regard them as a first-order approximation to a more general, non-Markovian kinetic equation [KBKS97, Ike04].

A second observation is that in general $\boldsymbol{\Gamma}$, which involves the coupling constants, will be much smaller than R for a generic choice of p. When the coupling constants go to zero $\boldsymbol{\Gamma} \to 0$, but the retarded propagator has a well-defined asymptotic value, and (11.36) becomes

$$G_{\mathrm{ret}} \sim \mathrm{PV}\frac{1}{R} + i\pi\,\mathrm{sign}(k^0)\delta(R) \tag{11.52}$$

From equations (11.9) and (11.10), the density of states

$$\mathcal{D} = \delta(R) \tag{11.53}$$

In this limit the propagators are insensitive to the behavior of the distribution function "off shell" (i.e. when $R \neq 0$), because the distribution function is always multiplied by the density of states, and this is very small there. Therefore, only "on shell" modes (i.e. those for which $R = 0$) really contribute to the field correlation functions. If our only concern is to follow the evolution of the distribution function on shell, we are allowed to replace the A and B coefficients in (11.48) by their "on shell" values, namely $A = 1$ and $B = 0$. We thus obtain the Kadanoff–Baym equations

$$\left\{R, F_1\right\} = -i\hbar\,\mathrm{sign}\left(k^0\right)\left[\Sigma_{12}F^{21} - \Sigma_{21}F^{12}\right] \tag{11.54}$$

Observe that in this argument we first took the weakly coupled limit, and then went on-shell. Also we assumed that somehow the adiabatic expansion and the expansion in powers of the coupling constant were linked; otherwise the C coefficient would be found to be comparable to A and B. A way to put this on a systematic basis is the hard thermal loop expansion discussed in Chapter 10.

After these approximations (in keeping with the weak coupling assumption, we are entitled to keep only $O(\lambda)$ terms in R as well), the nontrivial content of (11.54) is given by the form of the collision integral, namely, which Feynman graphs contribute to the self-energies.

The Kadanoff–Baym equations are formally valid to all orders in the coupling constant. It is convenient to consider the loop expansion of the self-energies to reduce this equation to a more familiar form. However, even now we recognize the structure of the collision term as the difference between a gain and a loss term for particles moving in or out of a phase space cell around the point (X, p) per unit time. Taking $p^0 > 0$ for simplicity, we see that $\Sigma_{12} F^{21}$ is the gain term, with $F^{21} = 1 + f$ accounting for stimulated emission of particles into the cell, while the other term is the loss term, which is proportional to the number of particles $F^{12} = f$ already there.

Let us consider the expansion of the self-energies in terms of Feynman graphs of increasing loop order, as a means of obtaining a definite expression for the collision term in the kinetic equation. Since we have the relationship $\Sigma_{21}(p) = \Sigma_{12}(-p)$ it is enough to analyze only the expansion of Σ_{12}. Physically this means considering only the gain processes, which produce a particle within a given phase space cell. The collision term is then obtained by subtracting the loss processes, which remove a particle therein.

The first term in the expansion is the setting-sun graph. To this order,

$$\Sigma_{12}(x, y) = \frac{i}{6} \lambda^2 \hbar \mathcal{G}(x, y) \tag{11.55}$$

$$\mathcal{G} = \frac{1}{\hbar^3} \left[G^{12}(x, y) \right]^3 \tag{11.56}$$

In momentum space, dealing with the propagators as if they were translation invariant, and using the definition of F^{12}, we get

$$\mathcal{G}(p) = (2\pi)^4 \int \frac{d^4 r \mathcal{D}(r)}{(2\pi)^3} \frac{d^4 s \mathcal{D}(s)}{(2\pi)^3} \frac{d^4 t \mathcal{D}(t)}{(2\pi)^3} \delta(p - r - s - t) F^{12}(r) F^{12}(s) F^{12}(t) \tag{11.57}$$

We also replace $\mathcal{D}$ by its quasi-particle form $\mathcal{D}_0 = \delta(p^2 + M^2)$. We must then find sets of four on-shell momenta adding up to zero. If $p^0 > 0$, this means that two of the r, s, t momenta must be future oriented, and the third past oriented. Using the symmetries of this expression, we obtain

$$\mathcal{G}(p) = 3(2\pi)^4 \int Dr\, Ds\, Dt\, \delta(p + r - s - t) \left[1 + f(r) \right] f(s) f(t) \tag{11.58}$$

where

$$Dp = \frac{d^4 p \mathcal{D}(p)\theta\left(p^0\right)}{(2\pi)^3} \tag{11.59}$$

It is fairly obvious that the resulting kinetic equation is just Boltzmann's.

An important if simple consequence of this fact is that the usual arguments showing that the only stationary solutions of the Boltzmann equation are thermal distributions carry over to the Kadanoff–Baym equations. In other words, the only translation-invariant propagators which solve the Kadanoff–Baym equations, or, for that matter, the 2PI Schwinger–Dyson equations, to this order in perturbation theory are thermal propagators. This fact is relevant to the discussion of thermalization in quantum field theory [JuCaGr04].

The basic formalism we presented here can be extended in several ways, such as including higher terms in the derivative expansion [Mro97, Jak02], higher correlations [WanHei02] or non-Markovian effects [MorRop99, SeKrBo00]. Another important generalization consists of explicitly incorporating the effects of quantum fluctuations in higher composite operators by including a stochastic source besides the collision integral [ReiToe94, AARS96, CalHu00]. We may regard this so-called Boltzmann–Langevin equation (see Chapter 2) as the quantum kinetic analog of the Langevin approach we discussed in Chapter 8.

The classical limit

It is interesting to consider the classical limit of the Boltzmann equation. Naively, we have, in powers of $\hbar$, that $\{R, F_1\} \sim O(1)$ and $I_{\text{col}} \sim O(\hbar^2)$. However, in the classical limit we must have that the Jordan propagator $G \to 0$ but the Hadamard propagator G_1 remains finite. To allow for a nonzero limit we must have

$$f = \hbar^{-1} f_{\text{cl}} \tag{11.60}$$

and counting powers of $\hbar$ we get

$$\{R, f_{\text{cl}}\} = I_{\text{col}}^{(3)}[f_{\text{cl}}] \tag{11.61}$$

where $I_{\text{col}}^{(3)}[f_{\text{cl}}]$ contains all terms in the collision integral that are cubic in f_{cl}.

The conclusion is that the correlation functions for a weakly interacting classical *field*, in the nearly translation-invariant limit, may be captured by a kinetic equation describing two by two scattering of on-shell excitations [MueSon04].

The classical Boltzmann equation describing interacting *particles* has a collision term quadratic in the distribution function. To obtain equation (11.61) instead, we must include the Bose enhancement factors, although of course this is a classical theory [Ein17].

11.1.7 The Vlasov equation

To lowest order in λ, our theory reduces to the Vlasov equation, namely, a transport theory for collisionless particles interacting with a self-consistent field.

This theory is obtained by neglecting the $O\left(\lambda^2\right)$ terms in our equations. The unperturbed equations are

$$R = \Omega_0 = p^2 + M^2 \tag{11.62}$$

and

$$\mathcal{D}(p) = \delta\left(p^2 + M^2\right) + O\left(\lambda^2\right) \tag{11.63}$$

The kinetic equation reduces to

$$0 = \mathcal{D}\left[p\frac{\partial}{\partial X} - \frac{1}{2}\partial_X M^2 \partial_p\right] f \tag{11.64}$$

which is indeed in the form of a Vlasov equation. The mass is defined through the self-consistent gap equation

$$M^2 = m_b^2 + m_V^2 + \frac{\lambda_b \hbar}{2} M_T^2 \tag{11.65}$$

where

$$M_T^2 = \int \frac{d^4 p}{(2\pi)^3} \delta\left(\Omega_0\right) f\left(X, p\right) \tag{11.66}$$

$$m_V^2 = \frac{\lambda_b \hbar}{4} \int \frac{d^4 p}{(2\pi)^3} \delta\left(\Omega_0\right) \tag{11.67}$$

This second quantity is actually divergent, so to evaluate it we need to regularize it first. We shall use dimensional regularization, writing (cf. Chapter 5)

$$m_V^2 = -\frac{\lambda_b \hbar M^2}{16\pi^2}\left[z - \frac{1}{2}\ln\left(\frac{M^2}{4\pi\mu^2}\right)\right] \tag{11.68}$$

$$z \equiv \frac{\Gamma\left[1 + \frac{\varepsilon}{2}\right]}{\varepsilon\left[1 - \frac{\varepsilon}{2}\right]} = \frac{1}{\varepsilon} + \frac{1}{2}\left(1 - \gamma\right) + \dots \tag{11.69}$$

$(\gamma = 0.5772\dots)$.

We go back to the gap equation and write it as

$$M^2\left\{1 + \frac{\lambda_b \hbar}{16\pi^2}\left[z - \frac{1}{2}\ln\left(\frac{M^2}{4\pi\mu^2}\right)\right]\right\} = m_b^2 + \frac{\lambda_b \hbar}{2} M_T^2 \tag{11.70}$$

which implies

$$\frac{1}{2}\frac{dM_T^2}{dM^2} = \frac{1}{\lambda_b \hbar} + \frac{1}{16\pi^2}\left[z - \frac{1}{2} - \frac{1}{2}\ln\left(\frac{M^2}{4\pi\mu^2}\right)\right] \tag{11.71}$$

Since the left-hand side is finite, the expression

$$\frac{1}{\lambda_b \hbar} + \frac{1}{16\pi^2}\left[z - \frac{1}{2}\right] \equiv \frac{1}{\lambda \hbar} \tag{11.72}$$

must also be finite, and the differential gap equation becomes

$$\frac{1}{2}\frac{dM_T^2}{dM^2} = \frac{1}{\lambda\hbar} - \frac{1}{32\pi^2}\ln\left(\frac{M^2}{4\pi\mu^2}\right) \tag{11.73}$$

Mass renormalization entails defining an initial condition for this differential equation, such as $M_T^2(M^2 = 0) = T^2/6\hbar^2$.

What is the small parameter?

Since reducing the theory to just the Vlasov equation means dropping terms of order λ^2, it might appear that one ought to replace the physical mass by the solution of the gap equation to the same order, namely

$$M^2 = M_0^2 + \frac{\lambda\hbar}{32\pi^2}M_0^2\ln\left(\frac{M_0^2}{4\pi\mu^2}\right) \tag{11.74}$$

where

$$M_0^2 = \frac{\lambda\hbar}{2}\left(M_T^2 - \frac{T_c^2}{6}\right) \tag{11.75}$$

(we gloss over the fact that M_T^2 itself depends on M^2; at high enough temperature M_T^2 stabilizes at a value of $T^2/6$, as in the massless theory). However, a moment's thought shows that, at least in the high-temperature limit, this is not a good idea. For high enough temperature, the second term in our expansion is of the order of the first term, meaning the breakdown of naive perturbation theory.

However, we can also proceed differently. In the regime where the derivation of the gap equation is valid, we can also write it as

$$M^2 = \frac{M_0^2}{1 - \frac{\lambda\hbar}{32\pi^2}\ln\left(\frac{M_0^2}{4\pi\mu^2}\right)} \tag{11.76}$$

(of course, this expression also blows up when the denominator vanishes, but that is a pathology of the $\lambda\phi^4$ theory, which is not asymptotically free). If we replace back equation (11.74) into the gap equation, we see that there is an error term of order

$$2\left[\frac{\lambda\hbar}{32\pi^2}\ln\left(\frac{M_0^2}{4\pi\mu^2}\right)\right]^2 M_0^2 \tag{11.77}$$

If we repeat the same with the expression (11.76), we see that the error has been reduced to

$$\left[\frac{\lambda\hbar}{32\pi^2}\right]^2\ln\left(\frac{M_0^2}{4\pi\mu^2}\right)M_0^2 \tag{11.78}$$

that is, an improvement by a factor of $\left(\ln\left[M_0^2/4\pi\mu^2\right]\right)^{-1}$. An analysis of the perturbative expansion shows that the terms from higher order Feynman graphs are also of this order (see [CaJaPA86] and references therein).

In other words, by adopting expression (11.76) we obtain a nonperturbative (in powers of the coupling constant) approximation to the physical mass, which is equivalent to summing all terms of the form $\left(\lambda\hbar\ln\left[M_0^2/4\pi\mu^2\right]\right)^p$ in the perturbative expansion (the so-called *leading logs*) while leaving aside terms of the form $(\lambda\hbar)^p\left(\ln\left[M_0^2/4\pi\mu^2\right]\right)^q$ with $q < p$. In this sense, the true small parameter in our expansion is not the coupling constant, but rather $\left(\ln\left[M_0^2/4\pi\mu^2\right]\right)^{-1}$ [ArSoYa99a, ArSoYa99b].

11.1.8 Time reversal invariance

Time reversal invariance means that for *any* solution $G^{ab}(x,y)$ of the equations of motion the time-reversed expression $G_{\text{rev}}^{ab}(x,y)$ is also a solution. The form of G_{rev}^{ab} is determined by the time reversal operation appropriate to the underlying field theory. In our case, time reversal transforms an expectation value $\left\langle\Phi\left(\mathbf{x},x^0\right)\Phi\left(\mathbf{y},y^0\right)\right\rangle$ into $\left\langle\Phi\left(\mathbf{y},-y^0\right)\Phi\left(\mathbf{x},-x^0\right)\right\rangle$ (see Streater and Wightman [StrWig80] and T.D. Lee [Lee81]). If $x=\left(\mathbf{x},x^0\right)$, write $\bar{x}=\left(\mathbf{x},-x^0\right)$ (observe that $\bar{x}^\mu = x_\mu = \eta_{\mu\nu}x^\nu$, where $\eta_{\mu\nu}$ is the Minkowski metric); therefore

$$G_{\text{rev}}^{21}(x,y) = G^{21}(\bar{y},\bar{x}) \tag{11.79}$$

$$G_{\text{rev}}^{12}(x,y) = G^{12}(\bar{y},\bar{x}) \tag{11.80}$$

For the Feynman propagator, we have

$$\begin{aligned}G_{\text{rev}}^{11}(x,y) &= \theta\left(x^0-y^0\right)G_{\text{rev}}^{21}(x,y) + \theta\left(y^0-x^0\right)G_{\text{rev}}^{12}(x,y)\\ &= \theta\left(-y^0-\left(-x^0\right)\right)G^{21}(\bar{y},\bar{x}) + \theta\left(-x^0-\left(-y^0\right)\right)G^{12}(\bar{y},\bar{x})\\ &= G^{11}(\bar{y},\bar{x})\end{aligned} \tag{11.81}$$

Similarly,

$$G_{\text{rev}}^{22}(x,y) = G^{22}(\bar{y},\bar{x}) \tag{11.82}$$

These formulae are summarized by

$$G_{\text{rev}}^{ab}(x,y) = G^{ab}(\bar{y},\bar{x}) \tag{11.83}$$

It is convenient to introduce the notation: for any kernel $A(x,y)$, we define the kernel $\bar{A}(x,y) = A(\bar{y},\bar{x})$. Therefore the time reversal operation means changing G^{ab} into $\bar{G}^{ab}$. Observe that a spherically symmetric translation invariant solution is automatically a fixed point under time reversal.

In terms of the partial Fourier transform we get for any kernel A that

$$\bar{A}(X,p) = A(\bar{X},-\bar{p}) \tag{11.84}$$

and as a consequence the first-order terms on the left-hand side of the kinetic equations (11.48) change sign (there is always one derivative that does), while the right-hand side $I_{\text{col}}\text{sign}\left(k^0\right)$ does not. So equations (11.48) are *not* time reversal invariant, unless $I_{\text{col}}^{(0)}$ vanishes (which implies a thermal solution). On the other hand, a local thermal solution cannot be a solution to first order, because then $I_{\text{col}}^{(0)}$ vanishes, but just four degrees of freedom $\beta_\mu(X)$ are not enough to kill the left-hand side terms identically in p.

Since (11.48) is the result of a systematic adiabatic expansion of the original 2PI SD equations, and the expansion itself would not break time reversal invariance, we conclude that the equations derived from the 2PIEA, unlike the Heisenberg equations themselves, break time reversal symmetry. This is to be expected, since these equations result from the slaving of higher correlations to the two-point functions [IvKnVo99, CalHu00].

However, we must not conclude that the observed thermalization in solutions of the evolution equations derived from the 2PIEA [Ber02, Ber04b, JuCaGr04, ArSmTr05] is an artifact of the approach. Thermalization is also observed in classical field theories, where the wave equation is directly solved [BoDeVe04]. We shall discuss this important issue in Chapter 12.

11.1.9 The limits of the kinetic approach

The derivation of the kinetic equations in this chapter is important for both practical and fundamental reasons. The fact that it can be done, as we have seen, already shows that the equations derived from the 2PIEA are not time reversal invariant. We will see in the next chapter that the Kadanoff–Baym equations play an important role in the derivation of the transport coefficients for a quantum field, and that kinetic equations may be used to describe an important stage in the thermalization process.

However, whether the kinetic equations, and more generally the adiabatic approximation, are quantitatively accurate, is a difficult issue and should not be taken lightly. We present an example, taken from [Mro97], which clearly displays the dangers at hand.

Consider a simple free Klein–Gordon field. The Heisenberg equations may be solved exactly, and the field decomposed in creation and destruction operators:

$$\Phi = \int \frac{d^3\mathbf{k}}{(2\pi)^{3/2}} \frac{e^{i\mathbf{k}x}}{\sqrt{2\omega_k}} \left\{ a_{\mathbf{k}} e^{-i\omega_k t} + a_{\mathbf{k}}^\dagger e^{i\omega_k t} \right\} \tag{11.85}$$

where $\omega_k^2 = k^2 + M^2$. Thus we may write any correlation function in terms of the expectation values of products of $a_{\mathbf{k}}$ and $a_{\mathbf{k}}^\dagger$. Let us assume for simplicity a spatially homogeneous and isotropic state, so that

$$\langle a_{\mathbf{p}}^\dagger a_{\mathbf{q}} \rangle = f_p \delta(\mathbf{p} - \mathbf{q}) \tag{11.86}$$

$$\langle a_{\mathbf{p}} a_{\mathbf{q}} \rangle = \langle a_{\mathbf{p}}^\dagger a_{\mathbf{q}}^\dagger \rangle^* = g_p \delta(\mathbf{p} + \mathbf{q}) \tag{11.87}$$

The Jordan propagator is of course state independent and translation invariant. The Hadamard propagator is

$$G_1\left(x, x'\right) = \int \frac{d^3\mathbf{k}}{(2\pi)^3} \frac{e^{i\mathbf{k}(\mathbf{x}-\mathbf{x}')}}{\omega_k} \left\{ [1 + 2f_k]\cos\omega_k\left(t - t'\right) + g_k e^{-2i\omega_k T} + g_k^* e^{2i\omega_k T} \right\}$$

(11.88)

where $T = (t + t')/2$. Observe that no nontrivial choice of the g_k makes this almost translation invariant. The propagator is either *exactly* translation invariant (if all $g_k = 0$) or else strongly T dependent; in particular, no modification of the distribution function f in a neighborhood of the mass-shell may account for the g_k terms.

We find the kinetic theory formalism is of no help in this problem, except in the case where it is unnecessary, since the state is time independent.

11.2 Quantum kinetic field theory on nontrivial backgrounds

11.2.1 The scalar Wigner function in scalar quantum electrodynamics (SQED)

The application of quantum kinetic field theory methods to fields defined on nontrivial backgrounds (both abelian and non-abelian gauge fields and gravitational backgrounds) presents special features which are not found in the general formulation presented above. We shall now discuss some of these characteristics.

For simplicity we shall concentrate on the basic issues of how to define a Wigner transform on a nontrivial background, the nature of the object so introduced and the "transport" part of the kinetic equation. Once these difficulties are overcome, the construction of the "collision" term of the kinetic equation follows the general guidelines presented above. For the remainder of this Chapter we set $\hbar = 1$.

The first difficulty encountered in applying the formalism of quantum kinetic field theory to a scalar field on an electromagnetic background is also the most obvious. Quantum kinetic theory assumes the two-point functions of the theory are nearly translation invariant. But this is not a gauge-invariant statement. For example, the Green function for a charged scalar field

$$G_1\left(x, x'\right) = \left\langle \left\{ \hat{\phi}\left(x\right), \hat{\phi}^\dagger(x') \right\} \right\rangle$$

(11.89)

becomes

$$G_1\left(x, x'\right) \rightarrow e^{i\left\{\varepsilon(x) - \varepsilon(x')\right\}} G_1\left(x, x'\right)$$

(11.90)

under a gauge transformation. It is clear that a nearly translation-invariant kernel in one gauge may seem to be arbitrarily far from translation invariance in some other gauge.

As we shall show, it is possible nevertheless to associate a "Wigner function" with the propagators under a well-defined gauge transformation law. For an abelian theory, the Wigner function is actually gauge invariant; in the non-abelian case, it transforms as an element of the adjoint representation.

The price to be paid is to relinquish the identification of the Wigner function as the partial Fourier transform of a propagator. That relationship will hold only if both the propagator and the Wigner function are expressed in the Fock–Schwinger gauge to be introduced momentarily [Foc37, Sch70, Jac02]. Observe that this is not a very original procedure: it is the same logic by which one specifies the temperature of a fluid by identifying the frame in which it should be measured, namely the rest frame.

Concretely, let X be a spacetime event at which we want to define the Wigner function $F(X, p)$ for the charged scalar field. We want to specify a gauge in which the gauge aspects of the background are suppressed as much as possible. As a start, we demand $A_\mu^{(X)}(X) = 0$, where $A_\mu^{(X)}$ is the background abelian gauge field in the special gauge around X. We cannot remove all the derivatives of the abelian field by gauge transformations (unless the field is trivial to begin with), but we can and will set to zero the symmetric combination

$$A_{(\mu,\nu)}^{(X)} = \frac{1}{2}\left[A_{\mu,\nu}^{(X)} + A_{\nu,\mu}^{(X)}\right] \tag{11.91}$$

In general, we define

$$A_{(\mu,\nu_1...\nu_n)} = \frac{1}{n+1}\left[A_{\mu,\nu_1...\nu_n} + \sum_{i=1}^{n} A_{\nu_i,\mu\overline{\nu_1}...\nu_n}\right] \tag{11.92}$$

where the overbar means that ν_i is omitted. Then the Fock–Schwinger gauge is defined by the conditions

$$A_{(\mu,\nu_1...\nu_n)}^{(X)}(X) = 0 \tag{11.93}$$

Without loss of generality we may take $X = 0$. The above equation (11.93) reduces to

$$u^\mu A_\mu^{(0)}(tu) = 0 \tag{11.94}$$

where t is just a parameter, unrelated to time. We may now define the Wigner function as

$$F(0, p) = \int d^4u\, e^{-ipu} G_1^{(0)}(u/2, -u/2) \tag{11.95}$$

where

$$G_1^{(0)}(x, x') = \langle\{\hat{\phi}^{(0)}(x), \hat{\phi}^{\dagger(0)}(x')\}\rangle \tag{11.96}$$

is the Hadamard propagator in the Fock–Schwinger gauge. On the other hand, suppose the background field in the gauge we happen to be working in (which we shall refer to as "the gauge," for short) is A_μ. There must exist a gauge parameter $\epsilon^{(0)}(x)$ such that

$$A_\mu^{(0)}(u) = A_\mu(u) + \frac{\partial}{\partial u^\mu}\varepsilon^{(0)} \tag{11.97}$$

The Fock–Schwinger gauge condition (11.94) becomes an equation for $\epsilon^{(0)}$

$$\frac{d}{dt}\varepsilon^{(0)}(tu) = -u^\mu A_\mu(tu) \tag{11.98}$$

with solution

$$\varepsilon^{(0)}(u) = \varepsilon^{(0)}(0) - \int_0^1 dt\, u^\mu A_\mu(tu) \tag{11.99}$$

By the same token

$$G_1^{(0)}(u/2, -u/2) = G_1(u/2, -u/2)\exp\left\{\left(-\frac{i}{2}\right)\int_{-1}^1 dt\, u^\mu A_\mu\left(\frac{tu}{2}\right)\right\} \tag{11.100}$$

Performing a simultaneous gauge transformation of G_1 and A_μ in (11.100), we see that $G_1^{(0)}$, and therefore also F, are gauge invariant. Also, observe that the constant of integration in the gauge parameter (11.99) drops out. In the non-abelian case, the constant of integration matters, and the Wigner function will be merely gauge covariant, rather than invariant.

The next step is to invert (11.95), that is, to express the Hadamard propagator in terms of the Wigner function. Let x and x' be the points at which we want to evaluate the Hadamard propagator. Let $X(x, x')$ be the midpoint and $u(x, x')$ the relative variable

$$X^\mu = \frac{1}{2}(x^\mu + x'^\mu); \quad u^\mu = x^\mu - x'^\mu \tag{11.101}$$

Then

$$G_1(x, x') = \exp\left\{\left(\frac{i}{2}\right)\int_{-1}^1 dt\, u^\mu A_\mu\left(X + \frac{tu}{2}\right)\right\}\int\frac{d^4p}{(2\pi)^4}e^{ipu}F(X, p) \tag{11.102}$$

The transport equation

We shall use (11.102) to obtain the transport equation for the Wigner function F. Observe that since we already know F is a gauge-invariant object, the transport equation we are looking for must be gauge invariant. This observation will be useful in our search.

If we disregard scalar field self-interactions, the field operators obey the Heisenberg equations

$$\left[D^\mu D_\mu - m^2\right]\hat{\phi} = 0 \tag{11.103}$$

where D is the covariant derivative

$$D_\mu = \partial_\mu - iA_\mu \tag{11.104}$$

For a non-self-interacting theory the Hadamard propagator obeys the same equation. From (11.102) we find

$$D_\mu G_1 = e^{\left(\frac{i}{2}\right)\int_{-1}^1 dt\, u^\mu A_\mu\left(X + \frac{tu}{2}\right)}\mathbf{D}_\mu \tag{11.105}$$

where

$$\mathbf{D}_\mu = \int \frac{d^4p}{(2\pi)^4} e^{ipu} \left\{ ip_\mu + \frac{1}{2}\frac{\partial}{\partial X^\mu} + i\mathcal{P}_\mu \right\} F(X,p) \qquad (11.106)$$

$$\mathcal{P}_\mu = \frac{1}{2}\int_{-1}^1 dt \left[A_\mu\left(X + \frac{tu}{2}\right) + \frac{1+t}{2}u^\lambda A_{\lambda,\mu}\left(X + \frac{tu}{2}\right)\right] - A_\mu\left(X + \frac{u}{2}\right) \qquad (11.107)$$

We now make a crucial observation. The object $\mathbf{D}_\mu$ defined in (11.106) is gauge invariant, so we may evaluate it in any gauge, and in particular in the Fock–Schwinger gauge around X. Similarly, now that all derivatives have been made explicit, there is no harm done if we set $X = 0$. We add the assumption that the background field tensor is slowly varying, so we may approximate the background field (in the special gauge) by its Taylor expansion. Up to two derivatives, we get

$$A_\nu^{(0)}\left(\frac{u}{2}\right) = \frac{u^\lambda}{4}F_{\lambda\nu}(0) + \frac{u^\lambda u^\rho}{24}\left[F_{\lambda\nu,\rho} + F_{\rho\nu,\lambda}\right](0) + \dots \qquad (11.108)$$

$$\mathbf{D}_\mu = \int \frac{d^4p}{(2\pi)^4} e^{ipu} \left\{ ip_\mu + \frac{1}{2}\left[F_{\lambda\mu}\frac{\partial}{\partial p_\lambda} + \frac{\partial}{\partial X^\mu}\right] \right\} F(X,p) \qquad (11.109)$$

Since this expression is gauge invariant, it holds in any gauge.

We now observe that $D_\mu G_1$ has the same structure as G_1 itself, namely, with the factor

$$e^{\left(\frac{i}{2}\right)\int_{-1}^1 dt\, u^\mu A_\mu\left(X + \frac{tu}{2}\right)} \qquad (11.110)$$

multiplying a Fourier integral over momentum space of a gauge-invariant quantity. Therefore, we immediately find

$$D^\mu D_\mu G_1 = e^{\left(\frac{i}{2}\right)\int_{-1}^1 dt\, u^\mu A_\mu\left(X + \frac{tu}{2}\right)}\mathbf{D}^2 \qquad (11.111)$$

for some operator $\mathbf{D}^2$. This operator, like $\mathbf{D}_\mu$, has both real and imaginary parts. The former, together with the mass term in the Klein–Gordon equation, give rise to the mass-shell constraint of the theory, while the latter yields the transport equation

$$p^\mu\left[\frac{\partial}{\partial X^\mu} + F_{\lambda\mu}\frac{\partial}{\partial p_\lambda}\right] F(X,p) = 0 \qquad (11.112)$$

The transport equation describes the evolution of a swarm of particles acted upon by the Lorentz force. A similar calculation yields the conserved current

$$j^\mu(x) = \left(\frac{-i}{2}\right)\left\{\left[D_x^\mu - D_{x'}^\mu\right]G_1(x,x')\right\}_{x=x'} \qquad (11.113)$$

Using our previous results for the covariant derivatives we get

$$j^\mu(X) = \int \frac{d^4p}{(2\pi)^4} p^\mu F(X,p) \qquad (11.114)$$

whose conservation follows from the transport equation.

We see that both the transport equation and the conserved current agree (to this order) with the corresponding expressions in classical kinetic theory, while the mass-shell condition begins to show traces of nonlocality.

11.2.2 Scalar Wigner functions on non-abelian backgrounds

We now consider the case in which the scalar field forms a multiplet minimally coupled to a non-abelian gauge field. Observe that now the Hadamard propagator carries group indices, transforming as $\hat{\phi}$ at x and as $\hat{\phi}^+$ at x'.

The first issue we must confront is whether the Fock–Schwinger gauge condition (11.94) can be realized [Cro80]. We now have

$$A_\mu^{(0)}(u) = g\left[A_\mu(u) - ig^{-1}\frac{\partial g}{\partial u^\mu}\right]g^{-1} \tag{11.115}$$

where $g = \exp\{i\epsilon\}$ is a group element and $\epsilon = \epsilon^A T_A$ belongs to the group algebra. If we impose the condition (11.94) we get

$$\frac{d}{dt}g(tu) = -ig\, u^\mu A_\mu(tu) \tag{11.116}$$

whose solution is

$$g(u) = g(0)\tilde{T}\left[e^{-i\int_0^1 dt\, u^\mu A_\mu(tu)}\right] \tag{11.117}$$

where the operator $\tilde{T}$ anti-orders with respect to the parameter t. Recall that $A_\mu = A_\mu^A T_A$ are matrices, so they may not commute at different values of t.

This shows that the Fock–Schwinger gauge exists. However, the constant of integration is no longer irrelevant. When we express the Hadamard propagator in terms of the propagator in the Fock–Schwinger gauge we find

$$G_1 = T\left[e^{\left(\frac{i}{2}\right)\int_0^1 dt\, u^\mu A_\mu\left(X+\frac{tu}{2}\right)}\right]g^{-1}(0)\,G_1^{(0,g)}g(0)\,\tilde{T}\left[e^{\left(\frac{i}{2}\right)\int_0^1 dt\, u^\mu A_\mu\left(X-\frac{tu}{2}\right)}\right] \tag{11.118}$$

To obtain the right gauge properties for G_1 we must assume $g^{-1}(0)\,G_1^{(0,g)}y(0) = G_1^{(0)}$ Is independent of g. This means that $G_1^{(0)}$ transforms as an element of the adjoint representation at X.

We adopt the same definition (11.95) for the Wigner function as in the abelian case; now F is an element of the adjoint representation. The inverse relationship reads

$$G_1 = T\left[e^{\left(\frac{i}{2}\right)\int_0^1 dt\, u^\mu A_\mu\left(X+\frac{tu}{2}\right)}\right]\int\frac{d^4p}{(2\pi)^4}\,e^{ipu}\,F(X,p)\,\tilde{T}\left[e^{\left(\frac{i}{2}\right)\int_0^1 dt\, u^\mu A_\mu\left(X-\frac{tu}{2}\right)}\right] \tag{11.119}$$

To compute the transport equation, observe that

$$\frac{\partial}{\partial X^\lambda}T\left[e^{\left(\frac{i}{2}\right)\int_0^1 dt\, u^\mu A_\mu\left(X+\frac{tu}{2}\right)}\right]$$

$$= T\left[e^{\left(\frac{i}{2}\right)\int_0^1 dt\, u^\mu A_\mu\left(X+\frac{tu}{2}\right)}\right]\left(\frac{i}{2}\right)\int_0^1 dt\, u^\mu \hat{A}_{\mu,\lambda}\left(X+\frac{tu}{2}\right) \tag{11.120}$$

where

$$\hat{A}_{\mu,\lambda}\left(X+\frac{tu}{2}\right) = \tilde{T}\left[e^{\left(\frac{-i}{2}\right)\int_0^t dr\, u^\mu A_\mu\left(X+\frac{ru}{2}\right)}\right]$$

$$\times A_{\mu,\lambda}\left(X+\frac{tu}{2}\right)T\left[e^{\left(\frac{i}{2}\right)\int_0^t dr\, u^\mu A_\mu\left(X+\frac{ru}{2}\right)}\right] \quad (11.121)$$

also

$$A_\lambda\left(X+\frac{u}{2}\right)T\left[e^{\left(\frac{i}{2}\right)\int_0^1 dt\, u^\mu A_\mu\left(X+\frac{tu}{2}\right)}\right] = T\left[e^{\left(\frac{i}{2}\right)\int_0^1 dt\, u^\mu A_\mu\left(X+\frac{tu}{2}\right)}\right]\hat{A}_\lambda\left(X+\frac{u}{2}\right)$$

$$(11.122)$$

and so

$$D_\mu G_1 = T\left[e^{\left(\frac{i}{2}\right)\int_0^1 dt\, u^\mu A_\mu\left(X+\frac{tu}{2}\right)}\right]D_\mu\tilde{T}\left[e^{\left(\frac{i}{2}\right)\int_0^1 dt\, u^\mu A_\mu\left(X-\frac{tu}{2}\right)}\right] \quad (11.123)$$

where

$$\mathcal{D}_\mu = \int \frac{d^4p}{(2\pi)^4} e^{ipu}\left\{ip^\mu F + \frac{1}{2}\frac{\partial}{\partial X^\mu}F(X,p) - i\hat{A}_\mu\left(X+\frac{u}{2}\right)F(X,p)\right.$$

$$+ \left(\frac{i}{2}\right)\int_0^1 dt\left[\hat{A}_\mu\left(X+\frac{tu}{2}\right) + \frac{u^\rho}{2}(1+t)\hat{A}_{\rho,\mu}\left(X+\frac{tu}{2}\right)\right]F(X,p)$$

$$+ \left.\left(\frac{i}{2}\right)\int_0^1 dt\left[\hat{A}_\mu\left(X-\frac{tu}{2}\right) + \frac{u^\rho}{2}(1-t)\hat{A}_{\rho,\mu}\left(X-\frac{tu}{2}\right)\right]F(X,p)\right\}$$

$$(11.124)$$

Since $\mathcal{D}_\mu$ has definite gauge transformation properties (it belongs to the adjoint representation) it is enough to evaluate it in the Fock–Schwinger gauge, where $\hat{A} = A$. Moreover, we replace the background fields by their Taylor expansion around $X = 0$, which, since $A^{(0)}(0) = 0$, is formally identical to the expansion (11.108), taking into account that now the field tensor is a matrix. In the same way that in the abelian case the covariant derivative of the propagator decomposes into real and imaginary parts, here the covariant derivative is the sum of Hermitian and anti-Hermitian terms

$$\mathcal{D}_\mu^{(0)} = \int \frac{d^4p}{(2\pi)^4} e^{ipu}\left\{ip^\mu F + \frac{1}{2}\frac{\partial}{\partial X^\mu}F + \frac{1}{4}\left\{F_{\lambda\mu}, \frac{\partial F}{\partial p_\lambda}\right\} + \frac{1}{8}\left[F_{\lambda\mu}, \frac{\partial F}{\partial p_\lambda}\right]\right\}$$

$$(11.125)$$

This expression is valid in the Fock–Schwinger gauge. To obtain the corresponding expression in an arbitrary gauge we must replace $\partial F/\partial X^\mu$ by the covariant derivative for an element of the adjoint representation

$$D_\mu F = \frac{\partial F}{\partial X^\mu} - i[A_\mu, F] \quad (11.126)$$

The double covariant derivative $D_\mu D^\mu$ in the wave equation may be analyzed in the same terms. Its anti-Hermitian part gives rise to the transport equation

$$0 = p^\mu \left(D_\mu F + \frac{1}{2} \left\{ F_{\lambda\mu}, \frac{\partial F}{\partial p_\lambda} \right\} \right)$$
$$- \frac{i}{8} \left(\left[F_{\lambda\mu}, \frac{\partial^2 F}{\partial p_\lambda \partial X^\mu} \right] + \frac{1}{2} \left[F^\mu_{\lambda,\mu}, \frac{\partial F}{\partial p_\lambda} \right] + \frac{1}{4} \left[\{ F_{\lambda\mu}, F^\mu_\rho \}, \frac{\partial^2 F}{\partial p_\lambda \partial p_\rho} \right] \right)$$

(11.127)

The conserved current is $-i$ times the momentum integral of the anti-Hermitian part of the covariant derivative of the propagator. Discarding total derivatives it is given formally by the same expression (11.114) as in the abelian case.

Classical limit and the Wong equations

The issue of the classical limit in the kinetic theory of particles on a non-abelian background is subtler than in the abelian case, because at first sight the objects involved are of a quite different nature. In the quantum case, as we have seen, the distribution function is a Hermitian matrix $F(X,p)$ belonging to the adjoint representation of the group; in the classical case, particles carry a non-abelian charge q^A which may rotate within the group manifold, and the distribution function $f(X,p,q^A)$ is then an ordinary function with extra arguments.

One simple way of connecting these two objects is by demanding that the sequence of moments of both distributions are the same. The moments are defined as

$$M^Q_{A_1...A_n} = \mathrm{Tr}\, \{ T_{A_1} \cdots T_{A_n} F \}$$

(11.128)

in the quantum case, and as

$$m^c_{A_1...A_n} = \int dq\, q_{A_1} \cdots q_{A_n}\, f$$

(11.129)

where dq is the invariant measure on the group manifold. Observe that because of the group algebra only a few quantum moments are truly independent. We find no such restriction in the classical case, which underlines the difference between both approaches.

We shall carry the comparison in the "near-equilibrium" case where F is close to a diagonal matrix in color space

$$F = f^0(X,p)\, 1 + f^A(X,p)\, T_A$$

(11.130)

Let us assume the trace relations

$$\mathrm{Tr}\, T_A = 0; \quad \mathrm{Tr}\, T_A T_B = \frac{1}{2} \delta_{AB}$$

(11.131)

The first few moments are then

$$M^Q_0 = N f^0(X,p); \quad M^Q_A = \frac{1}{2} f_A(X,p)$$

(11.132)

where N is the dimension of the representation. If we have the corresponding moments

$$\int dq = N; \quad \int dq\, q^A = 0; \quad \int dq\, q^A q^B = \frac{1}{2}\delta^{AB} \tag{11.133}$$

then we are led to suggest

$$f = f^0\left(X, p\right) + f^A\left(X, p\right) q_A \tag{11.134}$$

Our recipe meant the replacement of the identity matrix 1 by the number 1, and the group generators T_A by q_A. Assume further the multiplication table

$$T_A T_B = \frac{1}{2N}1\delta_{AB} + \frac{1}{2}\left(K_{AB}^C + iC_{AB}^C\right)T_C \tag{11.135}$$

The C_{AB}^C are the structure constants; the K_{AB}^C vanish for $SU\left(2\right)$, but not for $SU\left(3\right)$. We insert the quantum distribution function (11.130) into the transport equation (11.127), with the added assumption that the f^A, being small, can be neglected in terms involving the field tensor. Applying our recipe of replacing generators by classical charges we obtain the classical transport equation

$$p^\mu \frac{\partial f}{\partial X^\mu} + q_C C_{AB}^C \left(p^\mu A_\mu^A\right) \frac{\partial f}{\partial q_B} + p^\mu \left(q_A F_{\lambda\mu}^A\right) \frac{\partial f}{\partial p_\lambda} = 0 \tag{11.136}$$

If we wish to interpret this as a conservation equation for the number of particles in a phase-space volume, then we must conclude that these particles move along worldlines whose tangent is proportional to p^μ, and whose momenta and charge evolve according to

$$p^\mu \frac{\partial p_\lambda}{\partial X^\mu} = \left(q_A F_{\lambda\mu}^A\right) p^\mu \tag{11.137}$$

$$p^\mu \frac{\partial q_B}{\partial X^\mu} = \left(p^\mu A_\mu^A\right) C_{AB}^C q_C \tag{11.138}$$

These are the so-called Wong equations [Won70, LitMan02], which form the basis for a classical theory of non-abelian plasmas.

In most problems of interest the back-reaction of the particles described by the distribution function on the background fields is not negligible and one must seek a self-consistent dynamical framework. One possibility is to couple the transport equation for the particle distribution function to the Yang–Mills equations for the soft part of the background fields. On general grounds [LitMan02] one expects that such an approach is reliable when the plasma parameter ϵ is small. The plasma parameter is the inverse to the number of particles within a sphere whose radius is the screening length (see Chapter 10). In a gluon plasma, for example (see below for the application of quantum kinetic theory to the gauge fields themselves), the density scales as T^3 and the screening length as $(gT)^{-1}$, so $\epsilon \approx g^3$. In this case, this scheme works for theories with weak coupling.

11.2.3 Quantum kinetic theory in curved spacetimes

Quantum kinetic field theory in curved spacetimes has both similarities with and important differences from the transport theory in non-abelian field backgrounds. To begin with, there is one more layer of structure, because besides the Riemann tensor (to be defined momentarily), which is the natural analog to the field tensor, and the Christoffel symbols, which are the analogs to the field 4-vector, there is the metric tensor itself, which has no analog in non-abelian gauge theory. In particular, we shall carry the derivation of the transport equation up to two derivatives of the metric, which means only one derivative of the connection and no derivative of the Riemann tensor.

Let us begin by summarizing the useful definitions and conventions. The metric tensor appears in the expression of Pythagoras' theorem appropriate to the spacetime in question: the geodesic distance between two events whose coordinates differ by infinitesimal amounts dx^μ is $ds^2 = g_{\mu\nu}dx^\mu dx^\nu$ (we adopt the MTW conventions [MiThWh72] throughout this book). The connection appears in the covariant derivative for a contravariant vector field A^μ

$$\nabla_\nu A^\mu = A^\mu_{;\nu} = A^\mu_{,\nu} + \Gamma^\mu_{\nu\lambda}A^\lambda \qquad (11.139)$$

We shall adopt the so-called Levi-Civita connection, whose components are the Christoffel symbols

$$\Gamma^\mu_{\nu\lambda} = \frac{1}{2}g^{\mu\rho}\{g_{\nu\rho,\lambda} + g_{\lambda\rho,\nu} - g_{\nu\lambda,\rho}\} \qquad (11.140)$$

The Riemann tensor is the commutator of two covariant derivatives

$$[\nabla_\nu, \nabla_\rho] A^\mu = R^\mu_{\lambda\nu\rho}A^\lambda \qquad (11.141)$$

It is related to the connection through

$$R^\mu_{\lambda\nu\rho} = \partial_\nu\Gamma^\mu_{\lambda\rho} - \partial_\rho\Gamma^\mu_{\lambda\nu} + \Gamma^\mu_{\nu\sigma}\Gamma^\sigma_{\lambda\rho} - \Gamma^\mu_{\rho\sigma}\Gamma^\sigma_{\lambda\nu} \qquad (11.142)$$

As in our earlier discussions, start from an event P on the spacetime manifold at which we wish to define the Wigner function. We will build a special coordinate system in a neighborhood of P: the so-called Riemann normal coordinates (RNC) centered at P [Pet69] comes in handy. In this system, the coordinates of P are $X^\mu = 0$. We also perform a linear change of variables such that the metric tensor at P becomes $g_{\mu\nu} = \eta_{\mu\nu}$. We now consider a second point P' and assume there is a unique geodesic joining P and P' (we say P' belongs to a normal neighborhood of P). Moreover we parameterize this geodesic as $P'(t)$, such that $P'(0) = P$ and $P'(1) = P'$. We define the RNC of P' as the components u^μ of the tangent vector to this geodesic at $t = 0$. Observe that $\eta_{\mu\nu}u^\mu u^\nu$ gives the geodesic distance $\sigma(P', P)$ between P and P'.

In RNC the line tu^μ is by definition a geodesic. Substituting it into the geodesic equation we obtain the identity

$$u^\nu u^\rho \Gamma^\mu_{\nu\rho}(tu) = 0 \qquad (11.143)$$

This allows us to express the connection in terms of the Riemann tensor

$$\Gamma^{\mu}_{\nu\rho}(u) = \frac{1}{3}\left[R^{\mu}_{\nu\lambda\rho} + R^{\mu}_{\rho\lambda\nu}\right]u^{\lambda} + \dots \tag{11.144}$$

and similarly for the metric

$$g_{\mu\lambda} = \eta_{\mu\nu} - \frac{1}{3}R_{\mu\nu\lambda\rho}u^{\nu}u^{\rho} + \dots \tag{11.145}$$

Now we define the Wigner function $F(P,p)$ by demanding that in RNC the Hadamard propagator evaluated at opposite points may be represented as

$$G_1\left(\frac{u}{2}, -\frac{u}{2}\right) = \frac{\Delta_{\mathrm{VM}}^{1/2}\left(\frac{u}{2}, -\frac{u}{2}\right)}{\sqrt{-g(P)}} K(P,u) \tag{11.146}$$

where

$$K(P,u) = \int \frac{d^4p}{(2\pi)^4} e^{ipu} F(P,p) \tag{11.147}$$

and Δ_{VM} is the Van Vleck–Morette determinant [Vle28, Mor51]

$$\Delta_{\mathrm{VM}}(x,x') = \frac{1}{16\sqrt{-g(x)}\sqrt{-g(x')}}\det\left[\frac{\partial^2\sigma(x,x')}{\partial x \partial x'}\right] \tag{11.148}$$

Δ_{VM} is a biscalar, that is, a scalar both at x and x'. It is included so that the lowest order adiabatic expansion of the propagator agrees with its WKB approximation.

The factor $\sqrt{-g(P)} = 1$, but we have made it explicit for the following reason. The Hadamard propagator is a biscalar. If we make a coordinate transformation from coordinates x to coordinates x', then u transforms as a contravariant vector at P. To make the product pu a scalar, p must transform as a covariant vector at P, and in this case, $d^4p/\sqrt{-g(P)}$ is the invariant measure. So we get the right transformation properties, provided $F(P,p_{\mu})$ transforms into

$$F'(P,p'_{\mu}) = F\left(P, \frac{\partial x'^{\lambda}}{\partial x^{\mu}}p'_{\lambda}\right) \tag{11.149}$$

The representation (11.146) may be generalized to the case when the propagator is evaluated at two arbitrary points. Consider three points x, y and z in a normal neighborhood of P, and let $x(s)$ be the geodesic going from $x(0) = z$ to $x(1) = x$. Then in an adiabatic expansion we have $x(s) = xs + z(1-s) + \xi(s)$. Plug this into the geodesic equation to get

$$\frac{d^2\xi^{\mu}}{ds^2} = \frac{-1}{3}\left[R^{\mu}_{\nu\lambda\rho} + R^{\mu}_{\rho\lambda\nu}\right](xs + z(1-s))^{\lambda}(x-z)^{\nu}(x-z)^{\rho} + \dots$$

$$= \frac{-z^{\lambda}}{3}\left[R^{\mu}_{\nu\lambda\rho} + R^{\mu}_{\rho\lambda\nu}\right](x-z)^{\nu}(x-z)^{\rho} \tag{11.150}$$

$$x^{\mu}(s) = x^{\mu}s + z^{\mu}(1-s) + \frac{z^{\lambda}}{6}\left[R^{\mu}_{\nu\lambda\rho} + R^{\mu}_{\rho\lambda\nu}\right](x-z)^{\nu}(x-z)^{\rho}s(1-s) \tag{11.151}$$

The tangent at $s = 0$ is

$$t^\mu (x, z) = (x - z)^\mu + \frac{z^\lambda}{6} \left[R^\mu_{\nu \lambda \rho} + R^\mu_{\rho \lambda \nu} \right] (x - z)^\nu (x - z)^\rho \qquad (11.152)$$

To obtain the RNC of x around z we should make a linear coordinate transform so that the metric tensor at z assumes its Minkowski value. However, this last step is nonessential for obtaining the representation of the propagator, because it is compensated by a change of variables in the momentum integral and the $\sqrt{-g(P)}$ factor.

If the point $\bar{z}$ is the geodesic midpoint between x and y, then $t^\mu (x, \bar{z}) = -t^\mu (y, \bar{z})$. We get

$$\bar{z}^\mu = \frac{(x + y)^\mu}{2} + \frac{(x + y)^\lambda}{48} \left[R^\mu_{\nu \lambda \rho} + R^\mu_{\rho \lambda \nu} \right] (x - y)^\nu (x - y)^\rho + \dots \qquad (11.153)$$

$$t^\mu (x, \bar{z}) = \frac{(x - y)^\mu}{2} + \dots \qquad (11.154)$$

and the representation of the propagator is

$$G_1 (x, y) = \frac{\Delta^{1/2}_{VM} (x, y)}{\sqrt{-g(\bar{z})}} K (\bar{z}, 2t^\mu (x, \bar{z})) \qquad (11.155)$$

where K was defined in (11.147). To this adiabatic order we may approximate $\bar{z} = (x + y)/2$ within the K function.

We can now evaluate

$$\nabla^\mu \partial_\mu G_1 (x, x') = g^{\mu \nu} (x) \partial_\mu \partial_\nu G_1 (x, y) - g^{\mu \nu} (x) \Gamma^\lambda_{\mu \nu} (x) \partial_\lambda G_1 (x, y)$$

$$= \eta^{\mu \nu} \partial_\mu \partial_\nu G_1 (x, y) + \frac{\Delta^{1/2}_{VM} (x, y)}{\sqrt{-g(\bar{z})}}$$

$$\times \left[\frac{1}{3} R^{\mu \ \nu}_{\lambda \ \rho} x^\lambda x^\rho \partial_\mu \partial_\nu K - \frac{2}{3} R^\lambda_\sigma x^\sigma \partial_\lambda K \right] \qquad (11.156)$$

Observe that

$$\partial_\nu G_1 (x, y) = \frac{\Delta^{1/2}_{VM} (x, y)}{\sqrt{-g(\bar{z})}} \left[\partial_\nu K + \left(\frac{1}{2} \partial_\nu \ln [\Delta_{VM} (x, y)] - \frac{1}{2} \partial_\nu \ln [-g(\bar{z})] \right) K \right] \qquad (11.157)$$

$$\eta^{\mu \nu} \partial_\mu \partial_\nu G_1 (x, y) = \eta^{\mu \nu} \frac{\Delta^{1/2}_{VM} (x, y)}{\sqrt{-g(\bar{z})}} \mathcal{J}_{\mu \nu} \qquad (11.158)$$

$$\mathcal{J}_{\mu \nu} = \partial_\mu \partial_\nu K + (\partial_\nu \ln [\Delta_{VM} (x, y)] - \partial_\nu \ln [-g(\bar{z})]) \partial_\mu K$$

$$+ \frac{1}{2} (\partial_\mu \partial_\nu \ln [\Delta_{VM} (x, y)] - \partial_\mu \partial_\nu \ln [-g(\bar{z})]) K \qquad (11.159)$$

Now that all derivatives have been made explicit, there is no loss of generality if we specialize to the case $x = -y = u/2$. We have

$$\partial_\nu \ln\left[\Delta_{\text{VM}}\left(\frac{u}{2}, \frac{-u}{2}\right)\right] = \frac{1}{3} R_{\nu\sigma} u^\sigma \tag{11.160}$$

$$\partial_\nu \ln\left[-g\left(\bar{z}\right)\right] = 0 \tag{11.161}$$

$$\eta^{\mu\nu} \partial_\mu \partial_\nu \ln\left[\Delta_{\text{VM}}\left(x, y\right)\right] = \frac{1}{3} R \tag{11.162a}$$

$$\eta^{\mu\nu} \partial_\mu \partial_\nu \ln\left[-g\left(\bar{z}\right)\right] = \frac{-1}{6} R \tag{11.162b}$$

$$\nabla^\mu \partial_\mu G_1 \left(\frac{u}{2}, \frac{-u}{2}\right) = \frac{\Delta_{\text{VM}}^{1/2}\left(\frac{u}{2}, -\frac{u}{2}\right)}{\sqrt{-g\left(P\right)}} \left\{ \eta^{\mu\nu} \partial_\mu \partial_\nu K + \frac{1}{4} R K + \frac{1}{12} R^{\mu\,\nu}_{\,\lambda\,\rho} u^\lambda u^\rho \partial_\mu \partial_\nu K \right\}$$

$$= \frac{\Delta_{\text{VM}}^{1/2}\left(\frac{u}{2}, -\frac{u}{2}\right)}{\sqrt{-g\left(P\right)}} \int \frac{d^4 p}{(2\pi)^4} e^{ipu} \left\{ -p^2 + ip^\mu \frac{\partial}{\partial X^\mu} \right.$$

$$+ \frac{\eta^{\mu\nu}}{4} \frac{\partial^2}{\partial X^\mu \partial X^\nu} \frac{1}{6} R + \frac{1}{12} R^{\mu\,\nu}_{\,\lambda\,\rho} p_\mu p_\nu \frac{\partial}{\partial p_\lambda} \frac{\partial}{\partial p_\rho}$$

$$\left. - \frac{1}{6} R^\nu_{\,\lambda} p_\nu \frac{\partial}{\partial p_\lambda} \right\} F\left(X, p\right)_{X=0} \tag{11.163}$$

Therefore the mass-shell constraint and the transport equation, evaluated at the origin of a RNC system, read

$$\left[-p^2 - m^2 - \left(\xi - \frac{1}{6}\right) R + \frac{1}{12} R^{\mu\,\nu}_{\,\lambda\,\rho} p_\mu p_\nu \frac{\partial}{\partial p_\lambda} \frac{\partial}{\partial p_\rho} \right.$$

$$\left. - \frac{1}{6} R^\nu_{\,\lambda} p_\nu \frac{\partial}{\partial p_\lambda} + \frac{\eta^{\mu\nu}}{4} \frac{\partial^2}{\partial X^\mu \partial X^\nu} \right] F = 0 \tag{11.164}$$

$$p^\mu \frac{\partial}{\partial X^\mu} F = 0 \tag{11.165}$$

To obtain the corresponding expressions in an arbitrary coordinate system, we must replace the ordinary derivatives by the covariant derivatives

$$\nabla_\mu F = \left[\frac{\partial}{\partial X^\mu} + \Gamma^\lambda_{\mu\rho} p_\lambda \frac{\partial}{\partial p_\rho} \right] F \tag{11.166}$$

$$\nabla_\nu \nabla_\mu F = \partial_\nu \nabla_\mu F - \Gamma^\lambda_{\nu\mu} \nabla_\lambda F + \Gamma^\lambda_{\nu\rho} p_\lambda \frac{\partial}{\partial p_\rho} \nabla_\mu F \tag{11.167}$$

At the origin of the RNC

$$\nabla_\nu \nabla_\mu F\left(X, p\right)_{X=0} = \partial_\nu \partial_\mu F + \frac{1}{3} \left[R^\lambda_{\,\mu\nu\rho} + R^\lambda_{\,\rho\nu\mu} \right] p_\lambda \frac{\partial}{\partial p_\rho} \tag{11.168}$$

so the covariant mass-shell constraint is

$$\left[-p^2 - m^2 - \left(\xi - \frac{1}{6}\right) R + \frac{1}{12} \left[R^{\mu\,\nu}_{\,\lambda\,\rho} p_\mu p_\nu \frac{\partial}{\partial p_\lambda} \frac{\partial}{\partial p_\rho} - R^\nu_{\,\lambda} p_\nu \frac{\partial}{\partial p_\lambda} \right] \right.$$

$$\left. + \frac{g^{\mu\nu}}{4} \nabla_\nu \nabla_\mu \right] F = 0 \tag{11.169}$$

and the covariant transport equation is

$$p^\mu \frac{\partial F}{\partial X^\mu} + p^\mu \Gamma^\lambda_{\mu\rho} \, p_\lambda \frac{\partial F}{\partial p_\rho} = 0 \qquad (11.170)$$

If we think of this as a classical Liouville equation, it describes particles moving along the geodesics of the background spacetime. The geodesics are parameterized by $s = \tau/m$, where m is the mass of the particles and τ is their proper time, and the 4-velocity is $u^\mu = p^\mu/m$.

Higher spin fields

We now discuss the generalization of the quantum kinetic theory for scalar fields in curved spacetimes to fields of higher spin. For concreteness, we shall discuss the case of a Dirac spinor, but the central ideas apply to fields of any spin.

To begin, the notion of a local Lorentz transformation is introduced to define spinor fields in curved spacetimes. To do this, we need a moving frame, or, in four-dimensional spacetimes, a vierbein. A vierbein is a set of four vector fields e^μ_a such that at every point

$$g_{\mu\nu} e^\mu_a e^\nu_b = \eta_{\mu\nu} \quad \text{and} \quad \eta^{ab} e^\mu_a e^\nu_b = g^{\mu\nu} \qquad (11.171)$$

The components of the vierbein transform as contravariant vectors under general coordinate transformations. The vierbein changes under a local Lorentz transformation as

$$e^\mu_a \to \xi^\mu_a = \Lambda^b_a e^\mu_b \qquad (11.172)$$

A Dirac spinor ψ is a set of four (world) scalar fields which transform as a spinor under the local Lorentz transformation Λ^b_a. In general the quantity obtained by taking the ordinary derivatives of a spinor field is not a spinor. We define instead the covariant derivative

$$\nabla_\mu \psi = \partial_\mu \psi - \Gamma_\mu \psi \qquad (11.173)$$

where

$$\Gamma_\mu = \frac{1}{2} \Sigma^{ab} e_{a\nu} e^\nu_{b;\mu} \qquad (11.174)$$

Σ^{ab} is the Lorentz generator appropriate to the representation to which ψ belongs. $\nabla_\mu \psi$ is a spinor of the same order as ψ.

The propagator $S(x, x')$ transforms as the product $\psi(x) \bar\psi(x')$. We want to express it in terms of a Wigner function F defined at the geodesic midpoint $\bar z$ between x and x', which transforms as $\psi(\bar z) \bar\psi(\bar z)$. To do this, we introduce the so-called parallel transport matrices $A(x, \bar z)$, which transform as $\psi(x) \bar\psi(\bar z)$, and write

$$S(x, x') = A(x, \bar z) S^{(\bar z)}(x, x') \overline{A(x', \bar z)} \qquad (11.175)$$

The matrices $A\left(x, \bar{z}\right)$ are parallel transported along the geodesic from $\bar{z}$ to x. In RNC around $\bar{z}$ this means

$$x^{\mu} A\left(x, 0\right)_{;\mu} = 0 \qquad (11.176)$$

with the boundary condition $A\left(0, 0\right) = 1$. This equation allows us to write the parallel transport matrices in terms of the spin connection

$$A\left(x, 0\right) = 1 + x^{\mu} \Gamma_{\mu}\left(\bar{z}\right) + \ldots \qquad (11.177)$$

The object $S^{(\bar{z})}\left(x, x'\right)$ may be treated with the methods we have used for scalar fields; indeed, it is a world biscalar, though a local bispinor at $\bar{z}$. We refer the reader to [CaHaHu88] for further details.

Higher spin fields in non-abelian theories require the combination of methods presented in all sections of this chapter. The gauge fields themselves pose a particular problem, since their transformation law is not homogeneous. In this case the simplest strategy is the so-called background field method [DeW81, Abb81, Hart93, Alx99, PesSch95]. The gauge field A^{μ} is split into a c-number background V^{μ} and a quantum fluctuation W^{μ}. Under a gauge transformation, V^{μ} transforms as a gauge field, and W^{μ} as a field on the adjoint representation. A gauge-fixing term

$$\frac{1}{2\alpha} \left(D_V^{\mu} W_{\mu}\right)^2 \qquad (11.178)$$

is added to the action, where α is the gauge-fixing parameter and $D_V^{\mu} W_{\nu}$ is the gauge covariant derivative with connection V^{μ}, namely

$$D_{V\mu} W^{\nu} = \partial_{\mu} W^{\nu} - i\left[V_{\mu}, W^{\nu}\right] \qquad (11.179)$$

The action (where we must also add the corresponding ghost terms) is invariant under joint gauge transformations of V^{μ} and W^{μ}, but non-invariant under gauge transformations of W^{μ} alone. This is enough to make the W propagator well defined.

The quantum field W has a homogeneous transformation law, and may be handled as any other higher spin field. In curved spacetime, of course, we would not be concerned with the world-vector W^{μ} but with the four world-scalars $W^a = e^a_{\mu} W^{\mu}$, which transform as a vector under local Lorentz transformations.

11.2.4 A note on the literature

For original literature on Wigner functions in gauge backgrounds we recommend Heinz [Hei83] and Winter [Win84], and for Wigner functions in curved space-times, Winter [Win85] and Calzetta, Habib and Hu [CaHaHu88]. These methods were elaborated by many authors; other relevant references are [ElGyVa86, Mro89, Fon94, Gei96, Gei97, Son97, BlaIan99, BlaIan02, LitMan02]. Our exposition on gauge backgrounds is greatly influenced by [WRSG02, WRSG03], and on curved spacetimes by [CaHaHu88].

We recommend [BirDav82] as an entry point to the literature on higher spin fields in curved spacetimes, and [ChWiDi77] for further information on geometry and analysis on group manifolds.

12

Hydrodynamics and thermalization

Since the systems described by quantum fields are by definition extended, it is natural to think that in some limit they may reasonably well be approximated as fluids. This means that the state of the system is parameterized by a few locally well-defined fields such as temperature or energy density, obeying a set of hyperbolic equations of motion. A concrete example is the extensive use of fluid models to describe high-energy collisions [BelLan56, CarDuo73, CoFrSc74, Bjo83, CarZac83].

Our earlier derivation of quantum kinetic theory suggests a way to put this insight on a formal basis. Within its range of validity, the Boltzmann equation will drive the one-particle distribution function towards local thermal equilibrium. On scales much larger than the local thermalization scale, we expect to see hydrodynamical behavior [BeCoPa02]. This is, after all, the usual way of deriving hydrodynamics from kinetic theory [Hua87]. Beware, notwithstanding, that even at the level of classical kinetic theory there are still open questions regarding the cross-over from the kinetic to the hydro regime [KarGor02, KarGor03].

If we understand hydrodynamics as stated in the first paragraph of this chapter, then a system defined in terms of a quantum field may not have a hydrodynamic limit. This has been shown in [Elz02] for the case of a free Fermi field. However, since the hydrodynamic description seems justifiable when applied to the physics of quark–gluon plasmas (see the discussion in Chapter 14) and early universe cosmology [Hu82, Hu83, CalGra02], we shall accept as a working hypothesis that for "interesting" systems whose fundamental description involves quantum fields there is a local thermal equilibrium limit where the system may be described as a fluid. The specifics of quantum fields are manifested through the gap equation and constitutive relations, whose derivation will be our main goal in this chapter.

Let us begin, however, with a brief review of basic thermodynamics, and then its relativistic generalization. The subject of hydrodynamics is one where the generally covariant formulation is actually simpler than the flat spacetime one, and much simpler than the nonrelativistic version. Therefore, it is worth investing some initial effort to familiarize ourselves with the generally covariant approach from scratch. Our presentation follows the review articles by Israel [Isr72, Isr88]; see also [Cal98].

12.1 Classical relativistic hydrodynamics

12.1.1 A primer on thermodynamics

The basic tenets of thermodynamics we need to keep in mind are the following: we have a (simple) system described by some intensive parameters (temperature T, chemical potential μ, pressure p, etc.) whose meaning we take for granted (e.g. we already know everything about the zeroth law) and extensive parameters, such as energy U, entropy S, volume V, and particle number N (we use particle number for concreteness, but any – or several – conserved charge(s) would serve just as well). In equilibrium, all these quantities are position independent. Their first deviations from equilibrium are related by the first law

$$TdS = dU + pdV - \mu dN \tag{12.1}$$

Extensive quantities are homogeneous functions of each other, so we must have

$$TS = U + pV - \mu N \tag{12.2}$$

From the differential of this second identity we obtain the Gibbs–Duhem relation

$$dp = sdT + nd\mu \tag{12.3}$$

where $s = S/V$ and $n = N/V$ are the entropy and particle number densities, respectively. This means

$$\left.\frac{\partial p}{\partial T}\right|_\mu = s = \frac{\rho + p - \mu n}{T}; \qquad \left.\frac{\partial p}{\partial \mu}\right|_T = n \tag{12.4}$$

where $\rho = U/V$ is the energy density.

Actually, it is convenient to adopt as independent variables T and $\alpha = \mu/T$, whereby

$$\left.\frac{\partial p}{\partial T}\right|_\alpha = \frac{\rho + p}{T}; \qquad \left.\frac{\partial p}{\partial \alpha}\right|_T = Tn \tag{12.5}$$

We may also write

$$S = \Phi + \left(\frac{1}{T}\right) U - \alpha N \tag{12.6}$$

where the *thermodynamic potential* $\Phi = pV/T$ is the logarithm of the grand-canonical partition function. Finally we have the second law

$$TdS \geq dQ \tag{12.7}$$

This concludes our mini-tutorial on nonrelativistic thermodynamics.

12.1.2 Covariant hydrostatics

We now generalize the above framework of thermodynamics to a relativistic fluid evolving in a spacetime with an arbitrary metric $g_{\mu\nu}$. After overcoming some

initial threshold, the reader will be rewarded in the long run by the economy
and exactitude of this formulation. All derivatives shall be covariant derivatives
with respect to the Levi-Civita connection, so that $g_{\mu\nu;\rho} \equiv 0$. We use MTW
conventions [MiThWh72].

The divergence of a (contravariant) vector X^μ is defined by

$$X^\mu_{;\mu} = \frac{1}{\sqrt{-g}} \partial_\mu \left(\sqrt{-g} X^\mu \right) \tag{12.8}$$

(where ∂ denotes an ordinary derivative), and the flux of a vector through a
hypersurface Σ is

$$\int d^3\mathbf{x} \sqrt{^{(3)}g} n_\mu X^\mu \tag{12.9}$$

where $^{(3)}g_{ab}$ is the induced metric on the surface and n_μ is the outer normal.
If the surface element is space like (that is, the normal is a time-like vector)
we adopt the convention that $n_0 < 0$ (so that $n^0 > 0$, recall that $g^{00} < 0$ in any
frame). For example, in ordinary Minkowski space we say a $t = $ constant surface
is space like, and its normal $n^\mu = (\partial/\partial t)^\mu = (1,0,0,0)$. Then we obtain Gauss'
theorem

$$\int_V d^4x \sqrt{-g} \, X^\mu_{;\mu} = \int_{\partial V} d^3\mathbf{x} \sqrt{^{(3)}g} \, (\varepsilon n_\mu) \, X^\mu \tag{12.10}$$

where $\varepsilon = 1$ if the normal is space like, and -1 if time-like.

To simplify matters we will describe the construction of a covariant theory in
terms of a set of rules:

(a) *Intensive quantities* (T, p, μ) are associated with scalars, which represent the
value of the quantity at a given event, as measured by an observer at rest
with respect to the fluid.

(b) *Extensive quantities* (S, V, N) are associated with vector currents S^μ, u^μ, N^μ,
such that for any given space like surface element $d\Sigma_\mu = n_\mu d\Sigma$, then the
amount of quantity X within the volume $d\Sigma$ as measured by an observer
with velocity n^μ is given by $-X^\mu d\Sigma_\mu$. Therefore $x_n = -n_\mu X^\mu$ is the density
of the quantity X measured by such an observer. If the quantity X is conserved, then $X^\mu_{;\mu} = 0$. The quantity u^μ associated with volume is the fluid
4-velocity, and obeys the additional constraint $u^2 = -1$. We call density *tout
court* the density measured by an observer comoving with the fluid, namely
$x = -u_\mu X^\mu$.

(c) Energy and momentum are combined into a single extensive quantity
described by an energy–momentum tensor $T^{\mu\nu}$ which is symmetric. The
energy current, properly speaking, is $U^\mu = -T^{\mu\nu} u_\nu$, and the energy density
$\rho = T^{\mu\nu} u_\mu u_\nu$.

We wish to describe a fluid in a state of *equilibrium*. However, we do not assume
that the metric is stationary; at the very least, we must allow for the possibility

that the metric appears time dependent in our chosen coordinates. An example is a static de Sitter universe whose metric appears as an expanding spatially flat universe if written locally in Robertson–Walker metric form. Because of the (possibly) changing metric, we cannot expect that the relevant quantities are position independent. We shall only assume that the fluid is isotropic in the rest frame. This means that in the equilibrium state all vector currents are collinear with the velocity, the mixed components of $T^{\mu\nu}$ vanish in the rest frame, and the spatial components (namely, the momentum flux) are isotropic. In other words, in equilibrium we may decompose

$$N^\mu = nu^\mu, \qquad T^{\mu\nu} = \rho u^\mu u^\nu + p\Delta^{\mu\nu}, \qquad \Delta^{\mu\nu} = g^{\mu\nu} + u^\mu u^\nu \qquad (\Delta^{\mu\nu}u_\nu = 0) \tag{12.11}$$

where p is the equilibrium or hydrostatic pressure of the fluid. The entropy current S^μ is given by $TS^\mu = -T^{\mu\nu}u_\nu + pu^\mu - \mu N^\mu$, which we rewrite as

$$S^\mu = \Phi^\mu - \beta_\nu T^{\mu\nu} - \alpha N^\mu \tag{12.12}$$

Here, we have introduced the thermodynamic potential current $\Phi^\mu = p\beta^\mu$, and the inverse temperature vector $\beta^\mu = T^{-1}u^\mu$. Observe that $T^{-2} = -\beta^\mu\beta_\mu$. Contracting with u_μ we get $Ts = p + \rho - \mu n$, so locally we recover the usual thermodynamics. The first law becomes

$$TdS^\mu = -d(T^{\mu\nu}u_\nu) + pdu^\mu - \mu dN^\mu = -u_\nu dT^{\mu\nu} - \mu dN^\mu \tag{12.13}$$

Contracting the Gibbs–Duhem relation $(dT)\,S^\mu = (dp)\,u^\mu - (d\mu)\,N^\mu$ with the 4-velocity, we recover the derivatives (12.5). If we regard the thermodynamic potential as a function of β^μ and α, then

$$\frac{\partial \Phi^\mu}{\partial \beta^\nu} = p\delta^\mu_\nu + (p+\rho)\,u^\mu u_\nu = T^\mu_\nu; \qquad \frac{\partial \Phi^\mu}{\partial \alpha} = nu^\mu = N^\mu \tag{12.14}$$

so

$$S^\mu_{;\mu} = -\beta_\nu T^{\mu\nu}_{;\mu} - \alpha N^\mu_{;\mu} \tag{12.15}$$

This means that entropy production vanishes in equilibrium, provided the conservation laws of energy–momentum and particle number hold. Now, linear deviations from equilibrium are constrained by the first law (12.13). If we consider a state which deviates linearly from equilibrium, but where the conservation laws still hold, then in this state the entropy production must be

$$d\left(S^\mu_{;\mu}\right) = (dS^\mu)_{;\mu} = -\beta_{\nu;\mu}dT^{\mu\nu} - \alpha_{,\mu}dN^\mu \tag{12.16}$$

On the other hand, entropy production must be stationary at equilibrium, so the linear variation must vanish, whatever the deviations $dT^{\mu\nu}$ and dN^μ might be. In equilibrium the inverse temperature vector must be a *Killing field* $(\beta_{(\nu;\mu)} = 0)$ and α must be constant. Being a Killing field means that a coordinate transformation of the type $x^\mu \to x^\mu + \varepsilon\beta^\mu$ is a symmetry of the underlying spacetime.

Observe that not every spacetime or field theory may support an equilibrium state.

If we consider the variation of the entropy content within some spatial region Σ as a function of time, the second law demands that the increase in entropy should be higher than the entropy flow through the boundary $\delta\Sigma$. Thus the covariant statement of the second law is that entropy production must be positive, i.e. $S^\mu_{;\mu} \geq 0$.

12.1.3 Ideal and real fluids

In order to generalize the above framework to hydrodynamics (rather than hydrostatics) let us first introduce the concept of an *ideal fluid*, namely a fluid where the decomposition (12.11) is always available, not just under equilibrium conditions. Everything we said about equilibrium states is valid for an ideal fluid even away from equilibrium; this applies in particular to the vanishing of entropy production.

The equations of motion for the perfect fluid are the conservation laws for energy–momentum and particle number. Suppose we know α to be constant (for example, $\alpha = 0$). Then energy–momentum conservation implies the identities (recall that $u_\mu u^\mu_{,\nu} = 0$)

$$\rho_{,t} - (\rho + p)\, u^\nu_{;\nu} = 0; \qquad -(\rho + p)\, u^\mu_{,t} + \Delta^{\mu\nu} p_{,\nu} = 0 \qquad (12.17)$$

where $X_{,t} \equiv -u^\mu X_{,\mu}$. The second equation is the Navier–Stokes equation for a fluid without viscosity. Since ρ and p become space dependent only through their temperature dependence, we may write $\rho_{,t} = \rho_{,T} T_{,t}$, and similarly for p. Using the identity (12.5), equations (12.17) simplify to

$$\frac{1}{T} T_{,t} - c_s^2 u^\nu_{;\nu} = 0; \qquad -u^\mu_{,t} + \frac{1}{T}\Delta^{\mu\nu} T_{,\nu} = 0 \qquad (12.18)$$

which can be reduced in a standard way to the wave equation with

$$c_s = \sqrt{p_{,T}/\rho_{,T}} \qquad (12.19)$$

denoting the speed of sound.

We are interested in weakly nonideal fluids, namely, fluids which are not ideal, but whose properties remain close to a reference ideal fluid. The first obstacle we encounter is an ambiguity in the concept of the velocity of the fluid.

In effect, if the decomposition (12.11) fails, then the motion of mass does not agree with the motion of the conserved charges. In other words, heat transfer implies energy transfer, and therefore mass transfer, even if there is no charge flow. Therefore we must define what we mean by velocity. In practice, two different conventions have proved useful, namely the Eckart and the Landau–Lifshitz prescriptions [Wei72, LanLif59].

In the Eckart prescription, velocity and particle number densities are defined from the particle number current through the equations $N^\mu = nu^\mu$, $u^2 = -1$, and the energy density is read off the energy–momentum tensor $\rho = T^{\mu\nu}u_\mu u_\nu$. In the Landau–Lifshitz prescription, the velocity is defined as the (only) normalized time-like eigenvector of the energy–momentum tensor, and the energy density is (minus) the corresponding eigenvalue: $T^{\mu\nu} = \rho u^\mu u^\nu + T_T^{\mu\nu}$, with $T_T^{\mu\nu}u_\nu = 0$. The number density is read off the number current, $n = -u_\mu N^\mu$. In either case, the reference ideal fluid is chosen as having the same velocity, energy and particle number densities as the actual flow. The equation of state prescribes the pressure, $p_0 = p_0(\rho, n)$ of the reference fluid, and we may parameterize $T_T^{\mu\nu} = (p_0 + \pi)\Delta^{\mu\nu} + T_{TT}^{\mu\nu}$, where $T_{TT}^{\mu\nu}u_\nu = T_{TT\mu}^\mu = 0$. Here π measures the deviation of the isotropic part of the stress tensor from its local equilibrium value. By definition, it is the product of the *bulk viscosity* of the real fluid times the local rate of expansion.

We must emphasize that, although so far we may think of the Eckart and Landau–Lifshitz prescriptions as simply different conventions, later on, when we impose constitutive relations, say, linking viscous stresses to velocity gradients, these different prescriptions will lead to different physical models of the fluid. For reasons which will become clear in due time, we adopt the Landau–Lifshitz prescription.

Let us write the energy–momentum and particle number currents for a real fluid as

$$T^{\mu\nu} = T_0^{\mu\nu} + \tau^{\mu\nu}; \qquad N^\mu = N_0^\mu + j^\mu \tag{12.20}$$

where $T_0^{\mu\nu} = \rho u^\mu u^\nu + p_0\Delta^{\mu\nu}$ and $N_0^\mu = nu^\mu$ are the energy–momentum and particle number current of a reference ideal fluid. We emphasize that, while ρ and n have a direct operational meaning, p_0 is a theoretical construct. Concretely, p_0 results from using ρ and n as inputs in the equation of state for the ideal fluid. Recall that $\tau^{\mu\nu}u_\nu = j^\mu u_\mu = 0$.

In order to complete the specification of the model for the real fluid, we must describe also the entropy current. To do this, let us go back to our discussion of hydrostatics. There we saw that, if we consider an arbitrary state departing by amounts $dT^{\mu\nu}$ and dN^μ from an equilibrium state, then, *to first order in departures from equilibrium*, the entropy production is given by $(dS^\mu)_{;\mu} = -\beta_{\nu;\mu}dT^{\mu\nu} - \alpha_{,\mu}dN^\mu$.

We shall apply this formula to two nonequilibrium states of two different fluids (namely, the real and reference ideal ones) rather than to nonequilibrium and equilibrium states of the same fluid, while disregarding the higher order corrections. Suffice it to be forewarned that these assumptions are not rigorous (we shall return to this point later) and will get us in trouble.

We therefore adopt as our model for a real fluid the entropy production formula

$$S_{;\mu}^\mu = -\beta_{\nu;\mu}\tau^{\mu\nu} - \alpha_{,\mu}j^\mu \tag{12.21}$$

which may be easily integrated to a formula for the entropy current, namely

$$S^\mu = \Phi_0^\mu - \beta_\nu T^{\mu\nu} - \alpha N^\mu \tag{12.22}$$

where $\Phi_0^\mu = p_0\beta^\mu$. This formula may look like a simple extension of the corresponding one for the equilibrium entropy current, but it is not, since it depends on keeping the ideal form for the thermodynamic potential but replacing the other two terms by their real counterparts.

The second law, in the form of positivity of entropy production, allows us to put further restrictions on the form of $\tau^{\mu\nu}$ and j^μ. For example, if we demand that $-\alpha_{,\mu}j^\mu$ be nonnegative while j^μ is transverse, then we are led to

$$j^\mu = -\kappa\Delta^{\mu\nu}\alpha_{,\nu} \tag{12.23}$$

and the second law implies $\kappa \geq 0$. This coefficient is related to heat conductivity.

In the other term, it is convenient to decompose $\beta_{\nu;\mu}$ in its components along the direction transverse to the velocity, symmetrize, and, in the transverse part, extract the trace part to obtain

$$\beta^{(\mu;\nu)} = \{P_L + P_{LT} + P_T + P_{TT}\}^{\mu\nu\rho\sigma}\beta_{\rho;\sigma} \tag{12.24}$$

where

$$P_L^{\mu\nu\rho\sigma} = u^\mu u^\nu u^\rho u^\sigma \tag{12.25}$$

$$P_{LT}^{\mu\nu\rho\sigma} = \frac{-1}{2}\left[u^\mu u^\rho\Delta^{\nu\sigma} + u^\nu u^\sigma\Delta^{\mu\rho} + u^\nu u^\rho\Delta^{\mu\sigma} + u^\mu u^\sigma\Delta^{\nu\rho}\right] \tag{12.26}$$

$$P_T^{\mu\nu\rho\sigma} = \frac{1}{3}\Delta^{\mu\nu}\Delta^{\rho\sigma} \tag{12.27}$$

$$P_{TT}^{\mu\nu\rho\sigma} = \frac{1}{2}\left[\Delta^{\mu\rho}\Delta^{\nu\sigma} + \Delta^{\mu\sigma}\Delta^{\nu\rho} - \frac{2}{3}\Delta^{\mu\nu}\Delta^{\rho\sigma}\right] \tag{12.28}$$

Observe that the P's are symmetric, mutually orthogonal projectors. Since moreover $P_L T = P_{LT} T = 0$, and $P_T T = \pi\Delta^{\rho\sigma}$, we obtain

$$-\beta_{\nu;\mu}T^{\mu\nu} = -\pi\Delta^{\rho\sigma}\beta_{\rho;\sigma} - T_{TT}^{\mu\nu}\left[P_{TT}\beta_{\rho;\sigma}\right]_{\mu\nu} \tag{12.29}$$

which leads us to

$$T_{TT}^{\mu\nu} = -2\eta T\left[P_{TT}\beta_{\rho;\sigma}\right]^{\mu\nu}; \qquad \pi = -\zeta T\Delta^{\rho\sigma}\beta_{\rho;\sigma} \tag{12.30}$$

where the coefficients η and ζ are, respectively, the *shear* (or first) and *bulk* (or second) viscosities, and they must be nonnegative.

The reason why we have introduced factors of temperature explicitly in the above formulae is that they cancel the corresponding powers of T^{-1} in

$$\beta_{\rho;\sigma} = \frac{1}{T}\left\{-\frac{T_{,\sigma}}{T}u_\rho + u_{\rho;\sigma}\right\} \tag{12.31}$$

The first term vanishes under the projectors, and the second is purely transverse ($u^\rho u_{\rho;\sigma} = 0$), so

$$T_{TT}^{\mu\nu} = -\eta H^{\mu\nu}; \qquad \pi = -\zeta u^\sigma_{;\sigma} \tag{12.32}$$

$$H^{\mu\nu} = 2\left[P_{TT}u_{\rho;\sigma}\right]^{\mu\nu} \tag{12.33}$$

which are the most often quoted forms. It is easy to see that with these constitutive relations, energy–momentum conservation leads to the Navier–Stokes equation in covariant form.

This outcome is very good from the phenomenological point of view, but troublesome from a theoretical point of view, as it can be proven that the covariant Navier–Stokes theory admits no stable solutions [HisLin83, HisLin85]. Eventually we shall learn to live with this contradiction, but let us elaborate on it a little further, to the point where at least we understand what we have settled for.

12.1.4 Stability and the Landau–Lifshitz theory

Rather than attempting a direct study of stability in the Navier–Stokes equations, we shall show that the theory for a real fluid just constructed is a particular case of a class of theories which satisfy the essential condition of causality, whereas the Landau–Lifshitz theory does not. These are the *divergence type theories* of Geroch and Lindblom [GerLin90]. To simplify our discussion, we shall assume $\alpha \equiv 0$ throughout, since this is the relevant case to compare against the quantum theory of a real scalar field.

In a divergence type theory, the degrees of freedom X^A of the theory are used to construct currents T_A^μ, which are assumed to be *ultralocal* functions of the X^A (that is, they depend on the X^A at each point, but not on their derivatives). The equations of motion take the form of conservation laws for the currents $T_{A;\mu}^\mu = I_A$. For simplicity, let us assume linear production terms, $I_A = -V_{AB}X^B$ with a nonnegative matrix V_{AB} (this will be the relevant case below).

Suppose we consider a linear departure δX from some solution to the equations of motion, say $X = 0$. To make it even simpler, suppose that the Cauchy data are homogeneous in space, so that δX depends only on time. Then the equations of motion take the form $M_{AB}\delta\dot{X}^B = -V_{AB}\delta X^B$, where $M_{AB} = \partial T_A^\mu/\partial X^B$. If the solution we are starting from is stable, then δX must regress, and since the matrix V is nonnegative, for this to be true the matrix M must be positive definite. In a covariant theory, moreover, this must be true for any choice of time variable, and so we conclude that the matrix $-t_\mu \partial T_A^\mu/\partial X^B$ must be positive definite for every future-oriented, time-like vector t^μ, or, equivalently, the vector

$$\frac{\partial T_A^\mu}{\partial X^B}\delta X^A \delta X^B \tag{12.34}$$

must be time-like and future oriented for any choice of the δX displacements. This is our simple stability criterion.

The Landau–Lifshitz prescription does not directly fit in this scheme, because under the obvious choice $X^A \to \beta^\mu$, $T_A^\mu \to T^{\mu\nu}$, the energy–momentum tensor is not ultralocal. We must first extend the number of degrees of freedom, so that we can write the theory as first order throughout.

Concretely we introduce a new traceless, symmetric tensor $\zeta^{\mu\nu}$ ($\zeta_\mu^\mu = 0$) and write (we shall omit indices, for simplicity) the viscous stress tensor as $\tau = C\zeta$.

In time, we want this to be equivalent to the constitutive relations (12.30), which we write as $\tau = -B\nabla\beta$. To this end, we introduce a new current $A^{\mu\nu\rho} = (T^2/2)[\Delta^{\mu\nu}\beta^\rho + \Delta^{\mu\rho}\beta^\nu]$. The divergence of this current may be expressed in terms of first derivatives of the velocity, $A^{\mu\nu\rho}_{,\rho} = D\nabla\beta$. Then, by imposing the conservation law $\nabla A = -(DB^{-1}C)\zeta$, we see that indeed the new form $\tau = C\zeta$ of the viscous stress is equivalent to the old one.

So now we have a theory of fields β^μ and $\zeta^{\mu\nu}$, and ultralocal currents $T^{\mu\nu}$ and $A^{\mu\nu\rho}$, and are ready to apply the stability criterion. But then we realize that, while $T^{\mu\nu}$ depends on all fields, $A^{\mu\nu\rho}$ depends on β^μ *only*. The vector in equation (12.34) cannot possibly be time-like for *every* displacement, since a whole diagonal block is missing from $\partial T^\mu_A/\partial X^B$. So the Landau–Lifshitz theory fails the stability criterion.

The failure of the Landau–Lifshitz approach to depict real fluids may be attributed to two unwarranted assumptions, namely, that the real fluid could be described with the same set of variables and with the same entropy current as its perfect counterpart. As a matter of fact, all that equilibrium thermodynamics suggests is that, whatever extra variables are brought in to describe the nonequilibrium state, they must vanish in equilibrium, and the entropy current must match its equilibrium value up to first order in the deviations from equilibrium. In other words, when we write the entropy production as $S^\mu_{;\mu} = -\beta_{\nu;\mu}T^{\mu\nu} - \alpha_{,\mu}j^\mu$ we are neglecting second-order deviations from equilibrium. But under the Landau–Lifshitz constitutive relations, the two terms we are retaining are second order themselves. The inconsistency of keeping only *some* second-order terms is the root cause of our problems.

The fact remains that the Navier–Stokes theory is highly successful phenomenologically. The answer to this riddle seems to be that the would-be runaway perturbations of the Landau–Lifshitz theory are in reality high-frequency oscillations around the Navier–Stokes solutions. These oscillations cancel out if evolution is averaged over macroscopic time-scales, and therefore they do not appear in actual observations [NaOrRe94, KrNaOrRe97]. With this understanding, we shall carry on with the Landau–Lifshitz theory [Ger01].

12.2 Quantum fields in the hydrodynamic limit

12.2.1 Quantum hydrodynamic models

Since thermodynamics alone cannot provide further information on the transport functions, to proceed, we must place the above discussion in the context of a more fundamental description of the field, namely, the quantum kinetic field theory based on the Kadanoff–Baym equations. Let us begin with analyzing the equilibrium states.

Since we shall only discuss the theory of a real scalar field, we may also set $\alpha = 0$ from scratch. This reflects the fact that a real scalar field is its own antiparticle. Thus our problem is to connect the hydrostatic equilibrium states, described by an ideal energy–momentum tensor $T_0^{\mu\nu} = \rho u^\mu u^\nu + p_0 \Delta^{\mu\nu}$, with the equilibrium kinetic theory states, described by the Bose–Einstein distribution $f_0 = [\exp|\beta p| - 1]^{-1}$.

To begin with, let us identify the energy–momentum tensor with the *expectation value* of the corresponding Heisenberg operator. This is derived from the CTPEA Γ. The arguments of Γ are field configurations on a closed time path and in general we will have different metrics $g_{\mu\nu}^{(1)}$ and $g_{\mu\nu}^{(2)}$ in the forward and backward branches, respectively. The energy–momentum tensor is defined by the formula (valid for a general state)

$$T^{\mu\nu} = \frac{2}{\sqrt{-g}} \frac{\delta\Gamma}{\delta g_{\mu\nu}^{(1)}} \tag{12.35}$$

where only the derivative with respect to the metric in the first time branch is taken. After the derivative is taken we identify $g_{\mu\nu}^{(1)}$ and $g_{\mu\nu}^{(2)}$ with the physical metric $g_{\mu\nu}$. The effective action itself is given by

$$\Gamma[G] = -\frac{i\hbar}{2}\mathrm{Tr}\ln G + \frac{1}{2}S_{,AB}G^{AB} + \Gamma_2[G], \tag{12.36}$$

where the functional Γ_2 is $-i\hbar$ times the sum of all two-particle-irreducible diagrams with lines given by G and vertices given by the quartic interaction. The first term $\mathrm{Tr}\ln G$ does not depend on the metric. Written in full, the second term reads

$$\frac{1}{2}\int d^4x \left\{ \sqrt{-g^{(1)}}\left(\partial_x^2 - m_b^2\right)G^{11}\left(x, x'\right)\Big|_{x'=x} - (1 \to 2) \right\} \tag{12.37}$$

As usual

$$\frac{\delta\sqrt{-g}}{\delta g_{\mu\nu}} = \frac{1}{2}\sqrt{-g}g^{\mu\nu}; \qquad \frac{\delta g^{\mu\nu}}{\delta g_{\rho\sigma}} = -g^{\mu\rho}g^{\nu\sigma} \tag{12.38}$$

and so the contribution from this term to $T^{\mu\nu}$ is

$$\left[\partial^\mu\partial^\nu + \frac{1}{2}\eta^{\mu\nu}\left(\partial_x^2 - m_b^2\right)\right]G^{11}\left(x, x'\right)\Big|_{x'=x} \tag{12.39}$$

In the third term, the metric appears through the $\sqrt{-g}$ factors multiplying the coupling constants. The contribution to $T^{\mu\nu}$ takes the form

$$-\frac{\lambda_b}{8}\eta^{\mu\nu}\left[G^{11}\left(x, x\right)\right]^2 - \tilde{\Lambda}_b\eta^{\mu\nu} \tag{12.40}$$

where $\tilde{\Lambda}_b$ contains all the higher order contributions. To the accuracy desired, $\tilde{\Lambda}_b$ is position independent, and we shall not analyze it further. Adding the two nontrivial contributions we get

$$T^{\mu\nu} = -\left[\partial^\mu\partial^\nu - \frac{1}{2}\eta^{\mu\nu}\partial_x^2\right]G^{11}\left(x, x'\right)\Big|_{x'=x}$$
$$-\frac{\eta^{\mu\nu}}{2}\left[m_b^2 + \frac{\lambda_b}{4}G^{11}\left(x, x\right)\right]G^{11}\left(x, x\right) - \tilde{\Lambda}_b\eta^{\mu\nu} \tag{12.41}$$

We assume the quasi-particle approximation for G^{11}

$$G^{11} = \frac{(-i\hbar)}{p^2 + M^2 - i\varepsilon} + 2\pi\hbar\delta\left(p^2 + M^2\right) f\left(X, p\right) \qquad (12.42)$$

f is the solution to a kinetic equation of the form

$$p^\mu \frac{\partial f}{\partial X^\mu} - \frac{1}{2} M^2_{,\mu} \frac{\partial f}{\partial p_\mu} = I_{\text{col}}\left(X, p\right) \qquad (12.43)$$

We only assume energy–momentum conservation

$$\int Dp\, p^\mu I_{\text{col}}\left(X, p\right) = 0; \qquad Dp = \frac{d^4 p}{(2\pi)^3} \theta\left(p^0\right) \delta\left(\Omega_0\right) \qquad (12.44)$$

where $\Omega_0 = p^2 + M^2$. Observe that equation (12.42) implies that f must be real and even in p. In turn, the effective mass M^2 is the solution of the gap equation (11.65) given in Chapter 11.

To write the energy–momentum tensor in terms of the distribution function, observe that $\partial_x \to ip + \frac{1}{2}\partial_X$. We must neglect second derivative terms, and observe that terms involving $p\partial_X$ eventually vanish because $G^{11}\left(X, p\right)$ is even in p. So

$$T^{\mu\nu}\left(X\right) = \int \frac{d^4 p}{(2\pi)^4} \left[p^\mu p^\nu - \frac{1}{2}\eta^{\mu\nu} p^2\right] G^{11}\left(X, p\right)$$

$$- \frac{1}{2}\eta^{\mu\nu} \left[m^2_b + \frac{\lambda_b}{4}G^{11}\right] G^{11} - \tilde{\Lambda}_b \eta^{\mu\nu} \qquad (12.45)$$

Let us isolate

$$T^{\mu\nu}_V = -i\hbar \int \frac{d^d p}{(2\pi)^d} \frac{\left[p^\mu p^\nu - \frac{1}{2}\eta^{\mu\nu} p^2\right]}{p^2 + M^2 - i\varepsilon} \equiv -\eta^{\mu\nu}\Lambda_1 \qquad (12.46)$$

where

$$\Lambda_1 = \left(\frac{(d-2)}{2d}\right) \eta^{\mu\nu} \mu^\varepsilon \hbar \int \frac{d^d p}{(2\pi)^d} \frac{(-i)\, p^2}{p^2 + M^2 - i\varepsilon} = \frac{M^4 \hbar}{32\pi^2} \left[z - \frac{1}{4} - \frac{1}{2}\ln\left(\frac{M^2}{4\pi\mu^2}\right)\right] \qquad (12.47)$$

and z was defined in (11.69) of Chapter 11. Also write

$$\hbar \int \frac{d^4 p}{(2\pi)^4} p^\mu p^\nu 2\pi\delta\left(\Omega_0\right) f\left(X, p\right) \equiv T^{\mu\nu}_T \qquad (12.48)$$

and observe that

$$\int \frac{d^4 p}{(2\pi)^4} \left(-\frac{1}{2}p^2\right) 2\pi\delta\left(\Omega_0\right) f\left(X, p\right) = \frac{1}{2}M^2 M^2_T \qquad (12.49)$$

so we get $T^{\mu\nu} = T^{\mu\nu}_T - \Lambda_b \eta^{\mu\nu}$, with

$$\Lambda_b = \Lambda_1 + \frac{1}{2}\left[m^2_b + \frac{\lambda_b}{4}G^{11}\right] G^{11} + \tilde{\Lambda}_b - \frac{\hbar}{2}M^2 M^2_T \qquad (12.50)$$

If we regard $G^{11}(x,x)$ and M_T^2 as functions of M^2 defined by the gap equation, then Λ_b is a function of M^2 too, meaning that there is no explicit state dependence other than through M^2.

Our first concern is to eliminate the formal divergences from these expressions, following the procedure outlined in Chapter 11. With respect to the renormalization of the cosmological constant Λ_b term, we observe that any M^2 independent term may be absorbed in the gravitational action (even if it is formally infinite). So we only need to show that $d\Lambda_b/dM^2$ is finite. Now, the gap equation yields $dG^{11}/dM^2 = 2/\lambda_b$, and

$$\frac{d\Lambda_b}{dM^2} = \frac{-\hbar}{2} M_T^2 \tag{12.51}$$

Consistency requires that we actually neglect the $O(\lambda^2)$ terms in $\tilde\Lambda_b$, or at least that we consider them as a true (temperature independent) constant. Equation (12.51) then implies that energy–momentum conservation follows from the transport equation. Henceforth we shall assume that any constant contribution has been subtracted, and drop the b subscript.

To summarize, what we have done in this section is to introduce a class of theories which, although they receive some support as the long-wavelength limit of an underlying quantum field theory, may be – indeed, should be – studied as *bona fide* models of physical systems. These theories describe fluids with energy–momentum tensor $T^{\mu\nu} = T_T^{\mu\nu} - \Lambda g^{\mu\nu}$, where the first term is defined from a one-particle distribution function f in equation (12.48), and Λ is the solution to equation (12.51), with M_T^2 also defined in terms of f. The construction of the model is completed by stipulating the collision term in the Boltzmann equation for f and the gap equation, given by

$$M^2 - \varphi\left(M^2, \mu^2\right) = \frac{\hbar\lambda}{2} M_T^2 \tag{12.52}$$

The two functions I_{col} and φ allow us to incorporate some higher order effects into the same general scheme.

Charge conservation may be introduced as in classical hydrodynamics through the corresponding currents, provided the collision term has the required symmetry. The entropy current was defined in Chapter 2, equation (2.98). From this, entropy production is given by

$$S_{;\mu}^\mu = 2 \int Dp \left[\ln\frac{(1+f)}{f}\right] I_{\mathrm{col}} \tag{12.53}$$

where Dp was introduced in (12.44). The positivity of this integral is an expression of the H-theorem.

12.2.2 Thermal equilibrium states

Our next task is to investigate the equation of state for an equilibrium state described by a Bose–Einstein distribution function f_0. The energy–momentum tensor takes the perfect fluid form. The thermal component $T_T^{\mu\nu}$ admits a similar decomposition

$$T_{0T}^{\mu\nu} = 2\hbar \int Dp \, p^\mu p^\nu f_0 \left(X, p\right) = \rho_T u^\mu u^\nu + p_T \Delta^{\mu\nu} \tag{12.54}$$

where

$$\rho_T = 2\hbar \int Dp \, \left(up\right)^2 f_0 \left(X, p\right) \tag{12.55}$$

Since ρ_T and M_T^2 are scalars, we may compute them in the rest frame

$$\rho_T = \frac{\hbar}{2\pi^2} \int_M^\infty d\omega \, \omega^2 f_0 \left(\omega\right) \sqrt{\omega^2 - M^2} \tag{12.56}$$

$$M_T^2 = \frac{1}{2\pi^2} \int_M^\infty d\omega \, f_0 \left(\omega\right) \sqrt{\omega^2 - M^2} \tag{12.57}$$

For the thermal pressure, we find $3p_T - \rho_T = -\hbar M^2 M_T^2$, so

$$p_T = \frac{1}{3} \left(\rho_T - \hbar M^2 M_T^2\right) = \frac{\hbar}{6\pi^2} \int_M^\infty d\omega \, \left[\omega^2 - M^2\right]^{3/2} f_0 \tag{12.58}$$

The total energy density and pressure are then $\rho = \rho_T + \Lambda$ and $p = p_T - \Lambda$.

The equilibrium entropy current takes the form

$$S_0^\mu = p\beta^\mu - T_0^{\mu\nu}\beta_\nu = (\rho + p)\,\beta^\mu = (\rho_T + p_T)\,\beta^\mu \tag{12.59}$$

On the other hand, equation (2.98) yields $S_0^\mu = \Phi_{0T}^\mu - T_{0T}^{\mu\nu}\beta_\nu$, where

$$\Phi_{0T}^\mu = -2 \int Dp \, p^\mu \ln \left[1 - e^{-|\hbar\beta_\mu p^\mu|}\right] \tag{12.60}$$

This form of the thermodynamic potential recalls another equivalent expression for the thermal pressure

$$\frac{p_T}{T} = \frac{-1}{2\pi^2} \int_M^\infty d\omega \, \omega \sqrt{\omega^2 - M^2} \ln \left[1 - e^{-\hbar\beta\omega}\right] \tag{12.61}$$

Together (12.51) and (12.61) imply the thermodynamic relationship $dp/dT = (p + \rho)/T$. (Here and hereafter, we shall use d/dT to denote a total temperature derivative, which accounts for the explicit temperature dependence through f_0 as well as the implicit dependence through M^2. We shall use $\partial/\partial T$ when we mean only the former.) Indeed, equation (12.61) implies

$$T\frac{dp_T}{dT} = p_T + \rho_T - \frac{\hbar M_T^2}{2} T \frac{dM^2}{dT} \tag{12.62}$$

but $p_T + \rho_T = \rho + p$, and

$$T\frac{dp}{dT} = T\frac{dp_T}{dT} - T\frac{d\Lambda}{dT} = T\frac{dp_T}{dT} + \frac{\hbar M_T^2}{2} T \frac{dM^2}{dT} \tag{12.63}$$

Observe that for $T^2 \gg M^2$ we recover the Stefan–Boltzmann law and $p = \rho/3$, as expected (in this regime, the cosmological constant $\sim M^2 T^2$ is negligible compared to $\rho \sim T^4$). This concludes our study of the equilibrium states.

12.2.3 Local equilibria

We now extend this analysis to local equilibrium states. The idea is to generate a solution to the transport equation as a formal expansion "in derivatives of" β^μ, replace this solution in the definition of $T^{\mu\nu}$, and to compare the result to the Landau–Lifshitz energy–momentum tensor for a real fluid. The first point to realize is that it is not possible to assume arbitrary values for the derivatives of the temperature 4-vector at a given point; they must satisfy constraints derived from the symmetries of the transport equation. These constraints may be used to eliminate the time derivatives of the inverse temperature 4-vector from the equations.

Let us recall the transport equation (12.43). Write $f = f_0 + f_1$, where f_1 is "first order," and observe that, since the collision integral involves no derivatives, $I_{\text{col}}[f_0] = 0$. Therefore, to first order, we may write $I_{\text{col}}[f] = \tilde{K}[f_1]$, where operator $\tilde{K}$ is linear. To analyze the left-hand side, let us assume $p^0 > 0$, so that

$$f_0 = \frac{1}{e^{-\hbar \beta_\mu p^\mu} - 1} \tag{12.64}$$

Then

$$\hbar f_0 (1 + f_0) \left\{ p^\mu p^\nu \beta_{\mu;\nu} - \frac{1}{2} \beta^\mu M^2_{,\mu} \right\} = \tilde{K}[f_1] \tag{12.65}$$

Our goal is to solve for f_1. However, we must realize there are integrability conditions derivable from (12.44), so a solution exists only when

$$\hbar \int Dp \, f_0 (1 + f_0) \left\{ p^\kappa p^\mu p^\nu \beta_{\mu;\nu} - \frac{1}{2} p^\kappa \beta^\mu M^2_{,\mu} \right\} = 0 \tag{12.66}$$

The idea is to use the integrability conditions to eliminate time derivatives from the linearized transport equations, thereby obtaining an equation relating f_1 to spatial derivatives of the inverse temperature tensor only.

Since the integrability conditions are clearly covariant, we may write them down in any frame, in particular, the rest frame. In general, we have $\beta^\mu = (1/T\sqrt{1 - v^2})(1, \mathbf{v})$. In the rest frame, $\mathbf{v} = 0$, the above equations result in

$$\langle \omega^3 \rangle \frac{\dot{T}}{T^2} + \frac{1}{3} \langle \omega (\omega^2 - M^2) \rangle \frac{\nabla \mathbf{v}}{T} - \frac{1}{2} \langle \omega \rangle T M^2_{,T} \frac{\dot{T}}{T^2} = 0 \quad (\kappa = 0) \tag{12.67}$$

$$\frac{1}{3T} \langle \omega (\omega^2 - M^2) \rangle \left[(\mathbf{v})^\cdot + \frac{\nabla T}{T} \right] = 0 \quad (\kappa = 1, 2, 3) \tag{12.68}$$

In these expressions, we have introduced the notation

$$\langle X \rangle = \int Dp \, f_0 (1 + f_0) X$$

$$= \frac{1}{2\pi^2} \int_M^\infty d\omega \, \sqrt{\omega^2 - M^2} f_0 (1 + f_0) X \tag{12.69}$$

To simplify the integrability conditions, recall that

$$\frac{d\rho}{dT} = \frac{d\rho_T}{dT} + \frac{d\Lambda}{dT} = \frac{\partial\rho_T}{\partial T} + M_{,T}^2 \left[\frac{\partial\rho_T}{\partial M^2} - \frac{\hbar}{2}M_T^2\right] \tag{12.70}$$

$$\frac{\partial\rho_T}{\partial T} = \hbar^2 \frac{\langle\omega^3\rangle}{T^2} \tag{12.71}$$

$$\begin{aligned}
\frac{\partial\rho_T}{\partial M^2} &= \left(\frac{-\hbar}{2}\right)\frac{1}{2\pi^2}\int_M^\infty d\omega \frac{\omega^2}{\sqrt{\omega^2 - M^2}}f_0 \\
&= \left(\frac{-\hbar}{2}\right)\frac{1}{2\pi^2}\int_M^\infty d\omega \left(\frac{d}{d\omega}\sqrt{\omega^2 - M^2}\right)\omega f_0 = \left(\frac{\hbar}{2}\right)\left[M_T^2 - \frac{\hbar\langle\omega\rangle}{T}\right]
\end{aligned} \tag{12.72}$$

so

$$\langle\omega^3\rangle - \frac{1}{2}\langle\omega\rangle TM_{,T}^2 = \frac{T^2}{\hbar^2}\frac{d\rho}{dT} \tag{12.73}$$

On the other hand,

$$\frac{dp_0}{dT} = \frac{dp_{0T}}{dT} - \frac{d\Lambda}{dT} = \frac{\partial p_{0T}}{\partial T} + M_{,T}^2 \left[\frac{\partial p_{0T}}{\partial M^2} + \frac{\hbar}{2}M_T^2\right] \tag{12.74}$$

$$\frac{\partial p_{0T}}{\partial M^2} = -\frac{\hbar}{2}M_T^2 \tag{12.75}$$

$$\frac{\partial p_{0T}}{\partial T} = \frac{\hbar^2}{3T^2}\langle\omega(\omega^2 - M^2)\rangle \tag{12.76}$$

Also, recall that

$$\frac{dp_0}{dT} = \frac{p_0 + \rho}{T} \tag{12.77}$$

so finally

$$\langle\omega^3\rangle - M^2\langle\omega\rangle = \frac{3T}{\hbar^2}(p_0 + \rho) \tag{12.78}$$

The integrability conditions are simply the conservation equations for the ideal energy–momentum tensor built out of f_0. These equations determine the dynamics of local equilibrium states.

We may regard (12.73) and (12.78) as a system of equations for the two unknowns $\langle\omega\rangle$ and $\langle\omega^3\rangle$, which yields

$$\begin{aligned}
\langle\omega^3\rangle &= \frac{T^2}{\hbar^2}\frac{d\rho}{dT}\frac{\left[M^2 - \frac{3}{2}TM_{,T}^2 c_s^2\right]}{\left[M^2 - \frac{1}{2}TM_{,T}^2\right]} \\
\langle\omega\rangle &= \frac{T^2}{\hbar^2}\frac{d\rho}{dT}\frac{\left[1 - 3c_s^2\right]}{\left[M^2 - \frac{1}{2}TM_{,T}^2\right]}
\end{aligned} \tag{12.79}$$

where c_s^2 is the speed of sound (12.19).

12.3 Transport functions in the hydrodynamic limit

While in equilibrium the energy–momentum tensor for the quantum fields takes the ideal fluid form, for mere local equilibrium this will not be so. In general, we may seek a solution of the transport equation as a formal series "in derivatives of the hydrodynamic variables." The first order in this series is given by the solution to the linearized equation (12.65). When the corrected distribution function is employed to compute the energy–momentum tensor, we get non-ideal terms which are, by construction, linear in gradients. By matching these terms to the Landau–Lifshitz template, we may read off the transport functions, thereby "deriving" the constitutive relations for the quantum real fluid. This is, of course, the traditional way of deriving the transport functions from kinetic theory [ChaCow39, GrLeWe80, Lib98, Hei94]; what is new is the unconventional form of the collision integral. Our treatment here follows [CaHuRa00].

It is amusing to observe that, while in deriving the Kadanoff–Baym equations we had to justify at every step the neglect of higher gradient terms (and were admittedly not always quite convincing), the transport terms are lifted from terms in the energy–momentum tensor which are linear in gradients *by definition*. So many approximations which may be controversial at the quantitative level, are fully legitimate in the context of the derivation of the constitutive relations. We shall not discuss the further issue of whether a first-order theory is a good description of the quantum field in the hydrodynamic limit.

Let us begin by eliminating time derivatives from the left-hand side of the linearized transport equation (12.65). In the rest frame

$$\beta_{\mu,\nu} = \frac{1}{T}\begin{pmatrix} \frac{\dot{T}}{T} & \frac{\nabla T}{T} \\ \dot{\mathbf{v}} & v_{i,j} \end{pmatrix} \tag{12.80}$$

Using the integrability conditions we get

$$\beta_{\mu,\nu} = \frac{1}{T}\begin{pmatrix} -c_s^2\nabla\mathbf{v} & \frac{\nabla T}{T} \\ -\frac{\nabla T}{T} & v_{i,j} \end{pmatrix} \tag{12.81}$$

Obviously only the symmetric part contributes to the linearized transport equation. Also

$$\frac{1}{2}M_{,\mu}^2\beta^\mu = \frac{1}{2}M_{,T}^2\frac{\dot{T}}{T} = \frac{-c_s^2}{2}M_{,T}^2\nabla\mathbf{v} \tag{12.82}$$

Splitting $v_{(i,j)}$ into the diagonal and the traceless parts, and reverting to the covariant form, we get the left-hand side of (12.65) as

$$\hbar f_0\left(1+f_0\right)\left[\frac{1}{T}p_\mu p_\nu H^{\mu\nu} - \frac{1}{T}\left\{(p.u)^2\left[c_s^2 - \frac{1}{3}\right] + \frac{M^2}{3} - \frac{c_s^2}{2}TM_{,T}^2\right\}u_{,\lambda}^\lambda\right] \tag{12.83}$$

where $H^{\mu\nu}$ was defined in (12.33).

12.3.1 The collision term

On the right-hand side of the transport equation the collision integral has the structure of the balance between a gain and a loss term [FiGaJe06]. Let us consider a collision process whereby n reactant particles are transformed into m product ones. We get a gain when one of the product particles has the moment p_1, say, where we wish to evaluate I_{col}. Let the other product particles have momenta $p_2, \ldots p_m$, and the reactants have momenta $q_1, \ldots q_n$. Then the gain term is

$$\hbar \sigma_{n,m}^2 [\mathbf{q}, \mathbf{p}] \, \delta \left(\sum_{j=1}^{m} p_j - \sum_{i=1}^{n} q_i \right) \prod_{j=1}^{m} (1 + f(p_j)) \prod_{i=1}^{n} f(q_i) \tag{12.84}$$

where we have made explicit use of the energy–momentum conservation and placed properly the Bose enhancement factor. The corresponding loss term is

$$\hbar \sigma_{n,m}^2 [\mathbf{q}, \mathbf{p}] \, \delta \left(\sum_{j=1}^{m} p_j - \sum_{i=1}^{n} q_i \right) \prod_{i=1}^{n} (1 + f(q_i)) \prod_{j=1}^{m} f(p_j) \tag{12.85}$$

We make the micro-reversibility assumption that the cross-section σ^2 is the same for both processes. The collision integral is

$$I_{\text{col}} [p_1] = \hbar \sum_{n,m} \int \prod_{i=1}^{n} Dq_i \prod_{j=2}^{m} Dp_j \, \sigma_{n,m}^2 [\mathbf{q}, \mathbf{p}] \, \delta \left(\sum_{j=1}^{m} p_j - \sum_{i=1}^{n} q_i \right)$$
$$\times \left\{ \prod_{j=1}^{m} (1 + f(p_j)) \prod_{i=1}^{n} f(q_i) - \prod_{i=1}^{n} (1 + f(q_i)) \prod_{j=1}^{m} f(p_j) \right\} \tag{12.86}$$

In equilibrium, each (n, m) term vanishes independently. We assume the cross-sections are invariant under permutations of the reactants and products, separately. For reasons made clear below, we are interested in collision integrals which *do not* conserve particle number, meaning

$$\int Dp_1 \, I_{\text{col}} [p_1] \neq 0 \tag{12.87}$$

Explicitly, this says

$$\sum_{n>m} \int \prod_{i=1}^{n} Dq_i \prod_{j=1}^{m} Dp_j \, \left[\sigma_{n,m}^2 [\mathbf{q}, \mathbf{p}] - \sigma_{m,n}^2 [\mathbf{p}, \mathbf{q}] \right] \delta \left(\sum_{j=1}^{m} p_j - \sum_{i=1}^{n} q_i \right)$$
$$\left\{ \prod_{j=1}^{m} (1 + f(p_j)) \prod_{i=1}^{n} f(q_i) - \prod_{i=1}^{n} (1 + f(q_i)) \prod_{j=1}^{m} f(p_j) \right\} \neq 0 \tag{12.88}$$

so in general we request $\sigma_{n,m}^2 [\mathbf{q}, \mathbf{p}] \neq \sigma_{m,n}^2 [\mathbf{p}, \mathbf{q}]$ if $n \neq m$. In an explicit perturbative calculation, we find that, to order λ^2, only $\sigma_{2,2}^2$ is not zero, yielding the usual Boltzmann collision integral. To order λ^4, both $\sigma_{2,4}^2$ and $\sigma_{4,2}^2$ are activated, and the inequality may be explicitly verified (in fact, $\sigma_{2,4}^2 \sim 2\sigma_{4,2}^2$).

12.3.2 The linearized transport equation

Writing $f = f_0 + f_1$,

$$f_1 = f_0 \left(1 + f_0\right) \chi; \qquad \tilde{K}\left[f_1\right] = \hbar f_{0p} \left(1 + f_{0p}\right) K\left[\chi\right] \tag{12.89}$$

$$D_\beta p = Dp \, f_{0p} \left(1 + f_{0p}\right) \tag{12.90}$$

we obtain

$$K\left[\chi\right] = -\sum_{n,m} \int \prod_{i=1}^n D_\beta q_i \prod_{j=2}^m D_\beta p_j \, \sigma_{n,m}^2 \left[\mathbf{q}, \mathbf{p}\right] \delta \left(\sum_{j=1}^m p_j - \sum_{i=1}^n q_i\right)$$

$$\times \frac{\left\{\chi\left(p_1\right) + (m-1)\,\chi\left(p_2\right) - n\chi\left(q_1\right)\right\}}{\prod_{i=1}^n \left(1 + f_0\left(q_i\right)\right) \prod_{j=1}^m f_0\left(p_j\right)} \tag{12.91}$$

Thus far we have reduced our problem to that of solving the linear integral equation

$$K\left[\chi\right] = \frac{1}{T} p_\mu p_\nu H^{\mu\nu} - \frac{1}{T} \left\{ (p \cdot u)^2 \left[c_s^2 - \frac{1}{3}\right] + \frac{M^2}{3} - \frac{c_s^2}{2} T M_{,T}^2 \right\} u_{,\lambda}^\lambda \equiv R \tag{12.92}$$

Let us write $K = K_{\mathrm{B}} + K_1$, where the former is the lowest order (Boltzmann's) collision operator. K_{B} is a Hermitian operator in the space of functions defined on the positive energy mass shell with inner product

$$\langle \varsigma \mid \chi \rangle = \int D_\beta p \, \varsigma^* \left(p\right) \chi\left(p\right); \qquad \langle \chi \rangle \equiv \langle 1 \mid \chi \rangle \tag{12.93}$$

and $\langle \chi \rangle$ agrees with the expectation value introduced earlier (12.69).

K_1 will not be symmetric, in general. There is a basis of eigenvectors $|\chi_n\rangle$ of K_{B}, with eigenvalues a_n. Four eigenvectors correspond to the functions p^μ, with eigenvalue zero. Because of momentum conservation, these are also eigenvectors of the full collision operator. K_{B} admits a fifth null eigenvector, namely a constant: this follows from particle number conservation in Boltzmann's theory. Let us call $|\chi_0\rangle$ this (normalized) eigenvector (in conventional notation, $|\chi_0\rangle = \langle 1 \rangle^{-1/2}$). We observe that the inhomogeneous term R in the linearized transport equation (12.92) is orthogonal to the null eigenvectors p^μ (*not* to $|\chi_0\rangle$). We shall ignore the former, that is, we shall restrict our considerations to the orthogonal space to the p^μ's.

Writing the unknown χ in Dirac's notation as $|\chi\rangle = \sum |\chi_n\rangle \langle \chi_n \mid \chi\rangle$, we get

$$\langle \chi_0 \mid K_1 \mid \chi_0 \rangle \langle \chi_0 \mid \chi \rangle + \sum_{n \geq 1} \langle \chi_0 \mid K_1 \mid \chi_n \rangle \langle \chi_n \mid \chi \rangle = \langle \chi_0 \mid R \rangle \tag{12.94}$$

$$\sum_m \left(a_n \delta_{nm} + \langle \chi_n | K_1 | \chi_m \rangle\right) \langle \chi_m \mid \chi \rangle = \langle \chi_n \mid R \rangle \tag{12.95}$$

if $n \geq 1$. From the second equation, we see that $\langle \chi_m \mid \chi \rangle \sim O(\lambda^{-2})$ for $m \neq 0$; instead, the first equation suggests that $\langle \chi_0 \mid \chi \rangle$ is much larger $(O(\lambda^{-4}))$. We are therefore led to the approximation

$$\langle \chi_0 \mid \chi \rangle = \frac{\langle \chi_0 \mid R \rangle}{\langle \chi_0 \mid K_1 \mid \chi_0 \rangle}; \qquad \langle \chi_n \mid \chi \rangle = \frac{1}{a_n} \langle \chi_n \mid R \rangle \qquad (12.96)$$

As a matter of fact, R may be split into a term R_s proportional to the shear tensor $H^{\mu\nu}$ and a term R_b proportional to $u^\lambda_{,\lambda}$, and therefore so will the solution. Actually, $\langle \chi_0 \mid R_s \rangle = 0$, so solving the "shear" problem involves only the Boltzmann collision operator. The eigenvalues of this operator are of order T/τ_{rel}, where τ_{rel} is the mean free time, and so the shear linear correction to the distribution function is $\chi_s \sim \left(-\tau_{\text{rel}}/T^2 \right) p_\mu p_\nu H^{\mu\nu}$.

The mean free time may be identified by writing the Boltzmann equation in the collision time approximation, where $f \sim f_{\text{eq}} + \delta f$ and $\delta \dot{f} \sim -\delta f / \tau_{\text{rel}}$. On power counting and dimensional arguments, we find $\tau_{\text{rel}} \sim 1/\lambda^2 T$.

On the other hand, $\langle \chi_0 \mid R_b \rangle$ is not zero. It follows that the component of the "bulk" solution in the direction of $|\chi_0 \rangle$ is much larger than in any other direction, and we may approximate

$$|\chi_b\rangle = \frac{|\chi_0\rangle \langle \chi_0 \mid R_b \rangle}{\langle \chi_0 \mid K_1 \mid \chi_0 \rangle} = \frac{\langle R_b \rangle}{\langle K_1 [1] \rangle} = \text{constant} \qquad (12.97)$$

Expanding in the rest frame

$$\chi_b \equiv c_0 = \frac{-1}{T \langle K_1 [1] \rangle} \left\{ \langle \omega^2 \rangle \left[c_s^2 - \frac{1}{3} \right] + \langle 1 \rangle \left[\frac{M^2}{3} - \frac{c_s^2}{2} T M_{,T}^2 \right] \right\} u^\lambda_{,\lambda} \qquad (12.98)$$

12.3.3 The temperature shift and the bulk stress

As we have seen, the correction to the distribution function has two components. The one associated with the $H^{\mu\nu}$ tensor contributes to the shear stress, but it does not induce a change in the energy density. Therefore it is compatible with the Landau–Lifshitz matching conditions. The constant shift of χ by c_0, on the other hand, affects in principle both the energy density and the thermal mass M_T. So, to enforce the Landau–Lifshitz conditions, it must be partially compensated by a temperature shift. Concretely, if we call T the temperature of the fiducial equilibrium state, such that $\rho(T)$ is equal to the energy density in the nonequilibrium state, then the temperature appearing in the local equilibrium distribution function f_0 must be $T_0 = T + \delta T$. The effect of this temperature shift is the same as adding another term proportional to ω in the first-order correction χ.

The distribution function and temperature shifts in turn produce a shift δM^2 in the physical mass, which likewise does not affect the transport equation. However, both δT and δM^2 enter in the consideration of the bulk stress. Observe that there is no shift in the four velocity u^μ.

The three displacements c_0 (12.98), δT and δM^2 are related by the constraints that the gap equation must hold, and the total energy density in the nonequilibrium state must be the same as in the local equilibrium state. Writing the gap equation as in (12.52), the linearized equation then reads

$$\left[1 - \varphi' - \frac{\hbar\lambda}{2}\frac{\partial M_T^2}{\partial M^2}\right]\delta M^2 = \frac{\hbar\lambda}{2}\left[\frac{\partial M_T^2}{\partial T}\delta T + c_0\langle 1\rangle\right] \qquad (12.99)$$

In fact, $\partial M_T^2/\partial T = \hbar\langle\omega\rangle/T^2$, so $\delta M^2 = M_{,T}^2\delta T + M_{,c}^2 c_0$, where $M_{,c}^2 = T^2 M_{,T}^2\langle 1\rangle/\hbar\langle\omega\rangle$. Since the gap equation is enforced, we can look at the (cosmological) constant Λ as a function of M^2, so $\delta\Lambda_f = -\hbar M_T^2\delta M^2/2$. Then

$$\delta\rho = \frac{d\rho}{dT}\delta T + \left[\frac{\partial\rho_T}{\partial M^2} - \frac{\hbar}{2}M_T^2\right]M_{,c}^2 c_0 + \hbar\langle\omega^2\rangle c_0 \qquad (12.100)$$

Actually

$$\frac{\partial\rho_T}{\partial M^2} = \frac{\hbar}{2}M_T^2 - \hbar^2\frac{\langle\omega\rangle}{2T} \qquad (12.101)$$

so

$$\delta\rho = \frac{d\rho}{dT}\delta T + \left[\langle\omega^2\rangle - \frac{\langle 1\rangle}{2}TM_{,T}^2\right]\hbar c_0 \qquad (12.102)$$

And, since the total energy remains the same,

$$\frac{d\rho}{dT}\delta T = -\hbar c_0\left[\langle\omega^2\rangle - \frac{\langle 1\rangle}{2}TM_{,T}^2\right] \qquad (12.103)$$

Let us apply the same reasoning to the bulk stress, which results from both the deviation of the pressure from $p(T)$ and the direct contribution from the new terms in the distribution function

$$\tau = c_s^2\frac{d\rho}{dT}\delta T + \left[\frac{\partial p_T}{\partial M^2} + \frac{\hbar}{2}M_T^2\right]M_{,c}^2 c_0 + \frac{1}{3}\left[\langle\omega^2\rangle - M^2\langle 1\rangle\right]\hbar c_0 \qquad (12.104)$$

Now $\partial p_T/\partial M^2 = -\hbar M_T^2/2$, so

$$\tau = -\hbar c_0\left\{\left[c_s^2 - \frac{1}{3}\right]\langle\omega^2\rangle + \left[\frac{M^2}{3} - \frac{c_s^2}{2}TM_{,T}^2\right]\langle 1\rangle\right\} \qquad (12.105)$$

Using (12.98) and the expressions for $\langle\omega\rangle$ and $\langle\omega^3\rangle$ from the last section, we get

$$\tau = -\frac{\hbar^5 u_{,\lambda}^\lambda\left[M^2 - \frac{1}{2}TM_{,T}^2\right]^2}{9T^5\left(\frac{d\rho}{dT}\right)^2}\frac{\{\langle\omega^3\rangle\langle 1\rangle - \langle\omega^2\rangle\langle\omega\rangle\}^2}{|\langle K[1]\rangle|} \qquad (12.106)$$

where we have used the fact that an explicit calculation shows that $\langle K[1]\rangle < 0$ to lowest nontrivial order.

12.3.4 Shear stress and bulk viscosity

The shear stress can be read off directly from the new terms in $T_T^{\mu\nu}$. In the rest frame, we get $\chi_s = \left(-\tau_{\text{rel}}/T^2\right)p_\mu p_\nu H^{\mu\nu}$

$$\tau^{ij} = \frac{-\hbar\tau_{\text{rel}}}{T^2}H^{kl}\langle p^i p^j p_k p_l\rangle \sim \frac{-\hbar\tau_{\text{rel}}}{T^2}H^{ij}\langle p^4\rangle \qquad (12.107)$$

from where we can read out the shear viscosity η. To estimate η, it is enough to keep only the leading (binary scattering) contributions, so $\eta \sim \lambda^{-2}$. On dimensional grounds, we recover the usual result, $\eta \sim T^3/\lambda^2$.

As expected, things are not so simple with the bulk viscosity. We can read it off from equation (12.106). However, in evaluating it we must consider that $\langle 1 \rangle$ is logarithmically divergent in the massless limit, so we must correct the sheer dimensional estimate to $\langle 1 \rangle \sim T^2 \ln(M/T)$. As for the size of $|\langle K[1] \rangle|$, observe that the integral is dominated by the Rayleigh–Jeans tail, where $f_0 \sim T/\omega \gg 1$. Thus $|\langle K[1] \rangle| \sim \hbar^3 \lambda^4 T^6 F(M^2)$. Since the overall units are $[\text{mass}]^4$, it must be that $|\langle K[1] \rangle| \sim T^6/M^2$. For the remaining elements we may use the conventional estimates $\langle \omega^3 \rangle \sim T^5$, $\rho \sim \hbar T^4$, and thus obtain

$$\zeta \sim \frac{M^2}{\lambda^4 T^3} \left[M^2 - \frac{1}{2} T M_{,T}^2 \right]^2 \ln^2(M/T) \tag{12.108}$$

which is the folk result. In the limit in which the bare mass vanishes, or equivalently in the $T \to \infty$ limit, we may write on dimensional grounds

$$M^2 - \frac{1}{2} T M_{,T}^2 \equiv \frac{1}{2} \mu M_{,\mu}^2 \sim \lambda M^2 \tag{12.109}$$

and since $M^2 \sim \lambda T^2$ itself, equation (12.108) reduces to $\zeta \sim \lambda T^3 \ln^2(\lambda)$.

12.3.5 Transport functions for non-abelian plasmas

Although the calculation of transport coefficients in field theories follows the general strategy we have exemplified with a self-interacting scalar field, it is important to keep in mind the particularities of specific theories when aiming for a derivation of those coefficients good enough for a sensible comparison against experimental data.

In this sense, the most important scenario where an estimate of transport coefficients is of crucial relevance is the physics of relativistic heavy ion collisions (RHICs) [Ris98, BaRoWi06a, BaRoWi06b] (which we shall discuss in greater detail in Chapter 14), and correspondingly great effort has been devoted to the derivation of transport functions for hot non-abelian plasmas.

While we shall be content to refer the reader to the comprehensive set of papers by Arnold, Moore and Yaffe on this subject [ArMoYa00, ArMoYa03a, ArMoYa03b, ArDoMo06], we also wish to point out some aspects where the derivation of transport functions for non-abelian plasma differs from the equivalent study in scalar field theory.

First, there is the issue of momentum-dependent interactions and small denominator effects. Because of these, the actual weight of a given diagram may be very different from naive power-counting estimates. We have already encountered this phenomenon in Chapter 10, in our discussion of hard thermal loop resummation for a toy model scalar field.

As in the scalar field case, particle number-changing interactions play a central role in the derivation of the bulk viscosity coefficient [ArDoMo06]. However, the relevant processes are different. In particular, for a hot gluon field the most important contributions to bulk viscosity come from "2 to 1" processes, namely gluon splitting and joining.

In a conformally invariant theory, such as classical pure Yang–Mills theory, the bulk viscosity vanishes. We can see this in two related ways [ArDoMo06]. First, bulk viscosity is related to the departure of the trace part of the stress tensor from its equilibrium value upon isotropic expansion. In a conformally invariant theory, such expansion does not drive the system out of equilibrium – for the Maxwell case, see [Pla59] – so there is no departure. Second, in a conformally invariant theory the energy–momentum tensor must be traceless. This leaves no room for deviations of the trace part of the stress tensor from the value prescribed by the equilibrium equation of state $p = \rho/3$.

Therefore, the bulk viscosity in Yang–Mills theory is linked to the trace anomaly of the energy–momentum tensor [Fuj80]. In non-abelian gauge theories the trace anomaly is proportional to the β function which describes the running with scale of the gauge coupling [CoDuJo77]. Arnold, Dogan and Moore [ArDoMo06] observe that, since in principle the β function can have either sign (in a theory with matter fields included), while the bulk viscosity must be positive because of the second law, the bulk viscosity must be related to the *square* of the β function.

We must also mention the Landau–Pomeranchuk–Migdal (LPM) effect [LanPom53a, LanPom53b, Mig56, BaiKAt03]. This effect concerns the suppression of the emission probability for low-frequency photons, and correspondingly the suppression of exchange interactions in the low-frequency sector. It also affects gluon emission at both low and high frequency.

It is crucial that the kinetic equation one takes as take-off point be consistent with the LPM effect. Consistency can be achieved by an explicit calculation of the relevant cross-sections in the collision integral [ArMoYa00, ArMoYa03a, ArMoYa03b]. For an estimate of transport coefficients one does not often require a detailed knowledge of the cross-sections, but rather of certain integrals of them, for which there exist sum rules [AuGeZa02]. In such a case, it is enough to incorporate the LPM effect through the relevant sum rules [BBGM06].

An exciting new development is the possibility of an absolute lower bound for the ratio of shear viscosity to entropy density [KoSoSt05]. A low value for this ratio is usually an indication of a strongly coupled theory (compare with (12.107)). If this "viscosity bound conjecture" is confirmed, it would open up new avenues for the investigation of transport coefficients in a variety of strongly coupled systems, ranging from RHICs to cold atomic gases [Coh07].

More generally, the method of AdS/CFT correspondence [Mal99] is a new tool which is playing an increasing role in the study of strongly coupled gauge theories. A crucial step is the generalization of the correspondence for the computation of Schwinger–Keldysh (as opposed to Euclidean) propagators [HerSon03]. Similar tools have been used to study the hydrodynamic limit of M theory [Her02, Her03]. Since this field is growing exponentially at the time of writing we cannot even aim to provide a comprehensive list of references. However, see [PoSoSt02a, PoSoSt02b, KoSoSt05, HelJan07] for some key developments.

This concludes our study of the hydrodynamic limit from the kinetic field theory. There are several interesting directions to extend these results, such as including higher order effects [CaDeKo01]. We shall discuss some of these developments in later chapters in the context of applications to concrete problems.

To gain a broader perspective it is instructive to show the derivation of the transport coefficients from a different approach, namely, that of linear response theory, which we now turn to.

12.4 Transport functions from linear response theory

Linear response theory aims to provide exact representations for the transport functions as equilibrium expectation values of current correlations. The actual evaluation of these expressions may be technically rather subtle. The reader should consult the literature for details. However, the fact that one has, in principle, a rigorous definition of the transport functions opens up the possibility of implementing nonperturbative techniques, such as extracting the relevant correlations from numerical simulations [AarBer01, AarMar02]. Moreover, the fact that the linear response theory program may be carried through is a beautiful illustration of the deep connection between equilibrium and near-equilibrium dynamics such as embodied in the fluctuation–dissipation theorem, as well as in the stochastic approach to NEqQFT discussed in Chapter 8.

In the literature, there are several equivalent derivations of the linear response expressions for the transport functions. With some over-simplification, they can be traced back to the work of Mori [Mor58, HorSch87], Zubarev [Zub74, HoSata84] and Kadanoff and Martin [KadMar63]. The work of Jeon [Jeo93, Jeo95] and Jeon and Yaffe [JeoYaf96] is also of substantive value. For later developments, see [WaHeZh96, CaDeKo00, WanHei99, WRSG03, Koi07].

Following the presentation of Kadanoff and Martin [KadMar63], we shall first demonstrate this approach with the simpler case of the spin diffusion coefficient for an Ising-like model of a ferromagnetic material, and then derive the linear response theory expressions for the viscosity coefficients η and ζ.

12.4.1 The spin diffusion coefficient

We consider a model of some ferromagnetic material where the spin density is described by a continuous scalar quantum field $\mathbf{m}(t, \mathbf{x})$. (Here we use bold face to denote quantum fields – not for a vector field – with light face for classical fields.) The model is nonrelativistic, and for brevity we consider the symmetric phase only. Since the total magnetization is conserved, the Heisenberg equation of motion for the spin density takes the form of a continuity equation

$$\frac{\partial}{\partial t} \mathbf{m}\left(t, \mathbf{x}\right) + \nabla \mathbf{J}\left(t, \mathbf{x}\right) = 0 \qquad (12.110)$$

As we have seen in Chapter 8, it is possible to introduce a classical stochastic field $m\left(t, \mathbf{x}\right)$ such that

$$\langle m\left(t, \mathbf{x}\right) m\left(0, \mathbf{y}\right) \rangle_s = \frac{1}{2} \langle \{ \mathbf{m}\left(t, \mathbf{x}\right), \mathbf{m}\left(0, \mathbf{y}\right) \} \rangle \qquad (12.111)$$

where we have a stochastic average on the left-hand side, and a quantum average on the right. m satisfies a Langevin equation

$$\frac{\partial}{\partial t} m\left(t, \mathbf{x}\right) + \nabla \mathbf{J}\left(t, \mathbf{x}\right) = H_s\left(t, \mathbf{x}\right) \qquad (12.112)$$

To linear order in m, for slowly varying fields, with consideration of Galilei invariance, we must have

$$\mathbf{J}\left(t, \mathbf{x}\right) = -D \nabla m\left(t, \mathbf{x}\right) + \ldots \qquad (12.113)$$

where D is the spin diffusion coefficient we want to determine.

Since in this approximation the dynamics is linear and space translation invariant, it is convenient to introduce Fourier transforms

$$m\left(t, \mathbf{x}\right) = \int \frac{d^3 \mathbf{k}}{(2\pi)^3} e^{i\mathbf{k}\mathbf{x}} m_{\mathbf{k}}\left(t\right) \qquad (12.114)$$

If the value of the amplitude at $t = 0$ is $m_{\mathbf{k}}\left(0\right)$, then for $t > 0$

$$m_{\mathbf{k}}\left(t\right) = m_{\mathbf{k}}\left(0\right) e^{-Dk^2 t} + m_{\mathbf{k}}^S\left(t\right) \qquad (12.115)$$

where $m_{\mathbf{k}}^S\left(t\right)$ depends on the noise between 0 and t. If the noise and $m_{\mathbf{k}}\left(0\right)$ are uncorrelated, then

$$\langle m_{\mathbf{k}}\left(t\right) m_{\mathbf{k}'}\left(0\right) \rangle_s = e^{-Dk^2 t} \langle m_{\mathbf{k}}\left(0\right) m_{\mathbf{k}'}\left(0\right) \rangle_s \qquad (t > 0) \qquad (12.116)$$

From Onsager's principle of microscopic reversibility [LaLiPi80a], we know that the correlation is even in t, so this equation determines its value for $t < 0$ as well, namely,

$$\langle m_{\mathbf{k}}\left(t\right) m_{\mathbf{k}'}\left(0\right) \rangle_s = e^{-Dk^2 |t|} \langle m_{\mathbf{k}}\left(0\right) m_{\mathbf{k}'}\left(0\right) \rangle_s \qquad (12.117)$$

We now compute the inverse Fourier transform

$$m_{\mathbf{k}}\left(\omega\right) = \int dt \, e^{i\omega t} m_{\mathbf{k}}\left(t\right) \qquad (12.118)$$

$$\frac{2Dk^2}{\omega^2 + (Dk^2)^2} \langle m_{\mathbf{k}}\left(0\right) m_{\mathbf{k}'}\left(0\right) \rangle_s = \frac{1}{2} \int dt \, e^{i\omega t} \langle \{ \mathbf{m}_{\mathbf{k}}\left(t\right), \mathbf{m}_{\mathbf{k}'}\left(0\right) \} \rangle \qquad (12.119)$$

To obtain a prediction for D from this formula, we take the limits $k \to 0$ and $\omega \to 0$ *in this order* [KuToHa91] to get

$$D \langle m_{\mathbf{k}}(0) \, m_{\mathbf{k'}}(0) \rangle_s = \frac{\omega^2}{4k^2} \int dt \, e^{i\omega t} \langle \{ \mathbf{m_k}(t), \mathbf{m_{k'}}(0) \} \rangle \qquad (12.120)$$

Since the equilibrium correlation on the right-hand side is time-translation invariant, we may write

$$\omega^2 \int dt \, e^{i\omega t} \langle \{ \mathbf{m_k}(t), \mathbf{m_{k'}}(0) \} \rangle = \int dt \, e^{i\omega t} \left\langle \left\{ \frac{\partial}{\partial t} \mathbf{m_k}(t), \frac{\partial}{\partial t} \mathbf{m_{k'}}(0) \right\} \right\rangle$$

$$= - \int dt \, e^{i\omega t} k_i k'_j \left\langle \left\{ \mathbf{J}^i_{\mathbf{k}}(t), \mathbf{J}^j_{\mathbf{k'}}(0) \right\} \right\rangle \qquad (12.121)$$

It only remains to compute $\langle m_{\mathbf{k}}(0) \, m_{\mathbf{k'}}(0) \rangle_s$. As $k \to 0$,

$$m_{\mathbf{k}}(t) \to M(t) = \int d^3\mathbf{x} \, m(t, \mathbf{x}) \qquad (12.122)$$

where M is the total magnetization of the sample. Recall that if we turn on an external magnetic field H, then the Hamiltonian $\mathbf{H}$ acquires a new term $-HM$. Therefore, at constant temperature

$$M = - \left. \frac{\partial F}{\partial H} \right|_T \qquad (12.123)$$

where F is the free energy

$$e^{-\beta F} = \mathrm{Tr} \, e^{-\beta \mathbf{H}} \qquad (12.124)$$

Taking two derivatives we get

$$\langle M^2 \rangle = -k_{\mathrm{B}} T \left. \frac{\partial^2 F}{\partial H^2} \right|_T = k_{\mathrm{B}} T V \chi \qquad (12.125)$$

where χ is the susceptibility

$$\chi = \frac{1}{V} \left. \frac{\partial M}{\partial H} \right|_T \qquad (12.126)$$

(where V is the volume of the sample). We get

$$D\chi = \frac{-1}{4V k_{\mathrm{B}} T} \int dt \, e^{i\omega t} \frac{k_i k'_j}{k^2} \left\langle \left\{ \mathbf{J}^i_{\mathbf{k}}(t), \mathbf{J}^j_{\mathbf{k'}}(0) \right\} \right\rangle \qquad (k, k', \omega \to 0) \quad (12.127)$$

By using the symmetries of the correlator, we may simplify this expression to obtain

$$D\chi = \frac{1}{4k_{\mathrm{B}} T} \int d^3\mathbf{x} \int dt \, e^{i(\omega t - k x)} \frac{k_i k_j}{k^2} \left\langle \left\{ \mathbf{J}^i(t, \mathbf{x}), \mathbf{J}^j(0,0) \right\} \right\rangle \qquad (k, \omega \to 0)$$

$$(12.128)$$

We may also use the KMS theorem to express the anticommutator in terms of a commutator.

We shall now use this calculation as a model for the derivation of the viscosity coefficients.

12.4.2 The bulk and shear viscosity coefficients

We return to the calculation of the viscosity coefficients in scalar quantum field theory. We wish to write them in terms of equilibrium correlations of Heisenberg operators. Observe that in the Landau–Lifshitz prescription there is no heat flux, and for real scalar field theory there is no particle number conservation law. So we have no heat conductivity or particle number diffusion constants. The transport functions to be determined are the shear and bulk viscosities η and ζ.

In this subsection we shall not use different types for q or c number quantities. The basic dynamical law, both in the fundamental quantum field theory and in the stochastic field theory formulation, is the conservation of energy–momentum

$$T^{\mu\nu}_{;\nu} = 0 \qquad (12.129)$$

Decomposing the energy–momentum tensor as in equation (12.20), we get

$$\dot{\rho} + (\rho + p)\, u^{\lambda}_{;\lambda} - \tau^{\mu\nu} u_{\mu;\nu} = 0 \qquad (12.130)$$

$$(\rho + p)\, \dot{u}^{\mu} + \Delta^{\mu\nu} \left(p_{,\nu} + \tau^{\lambda}_{\nu;\lambda} \right) = 0 \qquad (12.131)$$

In the local rest frame of the fluid, when terms of second order in deviations from equilibrium are neglected, they reduce to

$$\frac{\partial \rho}{\partial t} + (\rho + p)\, u_{i,i} = 0 \qquad (12.132)$$

$$(\rho + p)\, \frac{\partial u_i}{\partial t} + p_{,i} + \tau_{ij,j} = 0 \qquad (12.133)$$

They have the form of continuity equations with currents $J^i_\rho = (\rho + p)\, u^i$ and $\left(J_{u^i} \right)^j = p\delta^j_i + \tau^j_i$.

Let us now consider the stochastic description. As in the spin diffusion case, the noise terms will not affect the final result, so we will not consider them. We may now parameterize

$$\tau_{ij} = -\eta \left(u_{i,j} + u_{j,i} \right) - \left(\zeta - \frac{2}{3}\eta \right) \delta_{ij} u_{s,s} \qquad (12.134)$$

The second conservation equation becomes

$$(\rho + p)\, \frac{\partial u_i}{\partial t} + p_{,i} - \eta u_{i,jj} - \left(\zeta + \frac{1}{3}\eta \right) (u_{j,j})_{,i} = 0 \qquad (12.135)$$

We may decompose the velocity field $u = u^L + u^T$, where $\nabla \times u^L = \nabla u^T = 0$. The transverse part decouples from the energy fluctuations, and obeys the simple heat equation

$$(\rho + p)\, \frac{\partial u^T_i}{\partial t} - \eta u^T_{i,jj} = 0 \qquad (12.136)$$

This is the same as in the spin diffusion case, with D there replaced by $D^T_u =$

$\eta / (\rho + p)$ here. We therefore write

$$\frac{\eta}{(\rho + p)} \left\langle u_{ik}^T (0) \, u_{jk'}^T (0) \right\rangle_s = \frac{\omega^2}{4k^2} \int dt \; e^{i\omega t} \left\langle \{u_{ik}^T (t) , u_{jk'}^T (0)\} \right\rangle \qquad (k, k', \omega \to 0)$$

(12.137)

For the longitudinal part, observe that $u_{i,jj}^L = (u_{j,j}^L)_{,i}$. So we may write

$$\frac{\partial \rho}{\partial t} + (\rho + p) \, u_{i,i}^L = 0 \qquad (12.138)$$

$$(\rho + p) \frac{\partial u_i}{\partial t} + p_{,i} - \left(\zeta + \frac{4}{3}\eta\right) u_{i,jj}^L = 0 \qquad (12.139)$$

Introduce the velocity potential $u^L = -\nabla \phi$, the sound speed $p_{,i} = c_s^2 \rho_{,i}$, and Fourier transform

$$\frac{\partial \rho_{\mathbf{k}}}{\partial t} + (\rho + p) \, k^2 \phi_{\mathbf{k}} = 0 \qquad (12.140)$$

$$(\rho + p) \frac{\partial \phi_{\mathbf{k}}}{\partial t} - c_s^2 \rho_{\mathbf{k}} + \left(\zeta + \frac{4}{3}\eta\right) k^2 \phi_{\mathbf{k}} = 0 \qquad (12.141)$$

These are the equations of a damped harmonic oscillator

$$\frac{\partial^2 \phi_{\mathbf{k}}}{\partial t^2} + k^2 c_s^2 \phi_{\mathbf{k}} + 2\Gamma k^2 \frac{\partial \phi_{\mathbf{k}}}{\partial t} = 0 \qquad (12.142)$$

where

$$\Gamma = \frac{\left(\zeta + \frac{4}{3}\eta\right)}{2\left(\rho + p\right)} \qquad (12.143)$$

The secular equation

$$\omega^2 - k^2 c_s^2 - 2i\Gamma k^2 \omega = 0 \qquad (12.144)$$

has solutions

$$\omega_{\pm} = i\Gamma k^2 \pm \sqrt{k^2 c_s^2 - \Gamma^2 k^4} \qquad (12.145)$$

If $\Gamma^2 k^2 \ll c_s^2$, we may expand

$$\omega_{\pm} = \pm k c_s + i\Gamma k^2 + O\left(k^3\right) \qquad (12.146)$$

so the general solution is

$$\phi_{\mathbf{k}}(t) = e^{-\Gamma k^2 t} \left[\phi_{\mathbf{k}}(0) \cos\left(k c_s t\right) + A_k \sin\left(k c_s t\right)\right] \qquad (t > 0) \qquad (12.147)$$

At $t = 0$, we find

$$\frac{\partial \phi_{\mathbf{k}}}{\partial t} = k c_s A_k - \Gamma k^2 \phi_{\mathbf{k}}(0) \qquad (12.148)$$

so

$$(\rho + p) \, k c_s A_k - c_s^2 \rho_{\mathbf{k}}(0) + \left(\zeta + \frac{4}{3}\eta\right) \frac{k^2}{2} \phi_{\mathbf{k}}(0) = 0 \qquad (12.149)$$

$$A_k = \frac{c_s^2 \rho_{\mathbf{k}}(0)}{(\rho + p) \, k c_s} - \frac{\Gamma k}{c_s} \phi_{\mathbf{k}}(0) \qquad (12.150)$$

Assuming that the equal-time potential and energy fluctuations are uncorrelated (see below), we get

$$\langle \phi_{\mathbf{k}}(t)\,\phi_{\mathbf{k}'}(0)\rangle = e^{-\Gamma k^2 |t|}\left[\cos\left(kc_s t\right) - \frac{\Gamma k}{c_s}\sin\left(kc_s\,|t|\right)\right]\langle \phi_{\mathbf{k}}(0)\,\phi_{\mathbf{k}'}(0)\rangle \quad (12.151)$$

Upon Fourier transforming

$$\frac{\left(\zeta + \frac{4}{3}\eta\right)}{(\rho + p)}\left\langle u_{i\mathbf{k}}^L(0)\,u_{j\mathbf{k}'}^L(0)\right\rangle_s = \frac{\omega^2}{4k^2}\int dt\, e^{i\omega t}\left\langle\left\{u_{i\mathbf{k}}^L(t),u_{j\mathbf{k}'}^L(0)\right\}\right\rangle \quad (12.152)$$

$(k, k', \omega \to 0)$. From symmetry considerations, we expect

$$\left\langle\left\{u_{i\mathbf{k}}^L(t),u_{j\mathbf{k}'}^T(0)\right\}\right\rangle = 0 \quad (12.153)$$

and so we may combine the longitudinal and transverse correlations into a single expression

$$\eta\left\langle u_{i\mathbf{k}}^T(0)\,u_{j\mathbf{k}'}^T(0)\right\rangle_s + \left(\zeta + \frac{4}{3}\eta\right)\left\langle u_{i\mathbf{k}}^L(0)\,u_{j\mathbf{k}'}^L(0)\right\rangle_s$$

$$= (\rho + p)\frac{\omega^2}{4k^2}\int dt\, e^{i\omega t}\left\langle\left\{u_{i\mathbf{k}}(t),u_{j\mathbf{k}'}(0)\right\}\right\rangle \quad (12.154)$$

$(k, k', \omega \to 0)$. We now have to compute the equal-time averages on the left-hand side. Let us begin by computing the velocity–velocity correlation. Recall that if the center of mass of the system is moving with velocity $\mathbf{V}$, then in the statistical operator we must add a new term $-\mathbf{V}\mathbf{P}$ to the Hamiltonian $\mathbf{H}$, where $\mathbf{P}$ is the total momentum. Therefore

$$\langle \mathbf{P}_i\rangle = -\frac{\partial F}{\partial \mathbf{V}^i} \quad (12.155)$$

and

$$\langle \mathbf{P}_i\mathbf{P}_j\rangle = k_B T\frac{\partial\langle\mathbf{P}_i\rangle}{\partial\mathbf{V}^j} \quad (12.156)$$

To transform this into velocity correlations, we simply observe that from the equilibrium energy–momentum tensor

$$\langle\mathbf{P}_i\rangle = V\left(\rho + p\right)\mathbf{V}_i \quad (12.157)$$

so

$$\langle\mathbf{V}_i\mathbf{V}_j\rangle = \frac{k_B T}{V\left(\rho + p\right)}\delta_{ij} \quad (12.158)$$

The longitudinal part of the velocity may be obtained from the total velocity by projection

$$u_{i\mathbf{k}}^L(0) = \frac{k_i k^j}{k^2}u_{j\mathbf{k}}(0) \quad (12.159)$$

We do likewise for $u_{j\mathbf{k}'}^L$, observe that the correlation must be proportional to $\delta\left(\mathbf{k} + \mathbf{k}'\right)$ by translation invariance, and that in the limit $k \to 0$, $u_{i\mathbf{k}} \to V\mathbf{V}_i$, we get

$$\left\langle u^L_{i\mathbf{k}}(0) \, u^L_{j\mathbf{k}'}(0) \right\rangle_{\mathrm{s}} = \frac{k_i k_j}{k^2} \frac{V k_{\mathrm{B}} T}{(\rho + p)} \qquad (k, k' \to 0) \qquad (12.160)$$

which also implies

$$\left\langle u^T_{i\mathbf{k}}(0) \, u^T_{j\mathbf{k}'}(0) \right\rangle_{\mathrm{s}} = \left[\delta_{ij} - \frac{k_i k_j}{k^2} \right] \frac{V k_{\mathrm{B}} T}{(\rho + p)} \qquad (k, k' \to 0) \qquad (12.161)$$

Thereby we find

$$\eta \left[\delta_{ij} + \frac{1}{3} \frac{k_i k_j}{k^2} \right] + \zeta \frac{k_i k_j}{k^2} = \frac{(\rho + p)^2}{V k_{\mathrm{B}} T} \frac{\omega^2}{4k^2} \int dt \, e^{i\omega t} \left\langle \{ u_{i\mathbf{k}}(t), u_{j\mathbf{k}'}(0) \} \right\rangle \quad (12.162)$$

$(k, k', \omega \to 0)$. As in the previous case of the spin diffusion coefficient, this may be reduced to an expression involving correlations of the energy–momentum tensor alone. First, use the expressions for T^{0i} and the conservation laws to write this as

$$\eta \left[\delta^{ij} + \frac{1}{3} \frac{k^i k^j}{k^2} \right] + \zeta \frac{k^i k^j}{k^2}$$

$$= \frac{1}{V k_{\mathrm{B}} T} \frac{k_m k_n}{4k^2} \int dt \, e^{i\omega t} \left\langle \left\{ T^{im}_{\mathbf{k}}(t), T^{jn}_{\mathbf{k}'}(0) \right\} \right\rangle \qquad (k, k', \omega \to 0) \quad (12.163)$$

Next, separate $T^{im}_{\mathbf{k}}$ into scalar and traceless components

$$T^{im}_{\mathbf{k}} = \mathcal{P}_{\mathbf{k}} \delta^{im} + \tau^{im}_{\mathbf{k}}; \qquad \tau^i_{\mathbf{k}i} = 0 \qquad (12.164)$$

In the limit $\mathbf{k} \to 0$, the tensor structure of the correlations can be expressed in terms of the isotropic tensor δ^{ij} alone. By symmetry, we must have

$$\left\langle \{ \mathcal{P}_0, \tau^{im}_0 \} \right\rangle = 0 \qquad (12.165)$$

$$\frac{1}{V} \left\langle \{ \tau^{im}_0(t), \tau^{im}_0(0) \} \right\rangle = A(t) \, \delta^{im} \delta^{jn} + B(t) \left(\delta^{ij} \delta^{mn} + \delta^{in} \delta^{mj} \right) \quad (12.166)$$

This last expression must be traceless with respect to (im), so $3A + 2B = 0$, and

$$\frac{1}{V} \left\langle \{ \tau^{im}_0(t), \tau^{im}_0(0) \} \right\rangle = A(t) \left[\delta^{ij} \delta^{mn} + \delta^{in} \delta^{mj} - \frac{2}{3} \delta^{im} \delta^{jn} \right] \qquad (12.167)$$

Contracting (ij) and (mn) we get

$$\sigma^2(t) \equiv \frac{1}{V} \left\langle \{ \tau^{im}_0(t), \tau^{im}_0(0) \} \right\rangle = 10 A(t) \qquad (12.168)$$

Substituting this back in equation (12.163) we get

$$\eta = \frac{1}{40 k_{\mathrm{B}} T} \int dt \, e^{i\omega t} \sigma^2(t) \qquad (\omega \to 0) \qquad (12.169)$$

$$\zeta = \frac{1}{4 k_{\mathrm{B}} T} \int dt \, e^{i\omega t} \left\langle \{ \mathcal{P}_0(t), \mathcal{P}_0(0) \} \right\rangle \qquad (\omega \to 0) \qquad (12.170)$$

which are the familiar expressions [Jeo95].

12.5 Thermalization

Perhaps the single most important demand on a theory of nonequilibrium quantum fields is that it should describe the means by which equilibrium is reached and sustained by those systems. The process of thermalization plays an important role in all the applications of the theory, such as the behavior of order-parameter fluctuations after a quench (Chapter 9), the dynamics of Bose–Einstein condensates and their associated noncondensed atomic clouds (Chapter 13), the early stages of relativistic heavy ion collisions (Chapter 14) and the physics of reheating after inflation (Chapter 15). In this chapter, we will deploy the knowledge gained so far in the physics of nonequilibrium fields to describe some general features of the thermalization process; then we will discuss some of these applications indicated above.

In spite of this ubiquity, the thermalization process is very hard to access experimentally. Usually all one can actually observe are relics superposed on the equilibrated thermal background, such as topological defects after a nonequilibrium phase transition or the ratios between different particle species after hadronization of the quark–gluon liquid. For this reason, a good deal of our understanding of the thermalization process comes from large-scale numerical simulations. We shall not discuss these simulations *per se*, but will point out below the key entry points to the literature.

Before we proceed, a word is in order about what thermalization is. Quantum field theory is unitary; quantum field theoretic evolution in a closed system cannot create entropy, and so a quantum field starting from a pure state, say, cannot thermalize in the strict thermodynamic sense (unless, e.g. it is coupled to a heat bath, see next section). By thermalization we mean that a restricted set of observables (correlation functions, hydrodynamics variables such as energy density and pressure, equation of state, field configurations over regions of space small compared to the total available volume) evolve in time towards stable, near-stationary values which are robust against changes in the initial conditions and may be approximated by thermal distributions with suitable intensive parameters (temperature, chemical potentials, etc.) [BoDeVe04].

If we talk about thermalization in the context of quantum field theories, the problem becomes slightly academic because nobody has ever solved the full unitary evolution (unless in trivial cases, which do not thermalize). One solves instead the equations of motion for the correlation functions derived, e.g. by some n-PI effective action functional with a finite n, which are not time-reversal invariant [IvKnVo99]. However, for classical field theories one can, in principle, actually solve the field equations. Then thermalization in the strict sense is impossible. For example, in a thermal state one should be able to observe arbitrarily high values of the total energy in the field (though large values will be very unlikely); in a numerically correct calculation, one should never see energy values outside the range defined by the initial conditions. Nevertheless, in the thermodynamic limit the behavior of local observables becomes indistinguishable from equilibrium. This means that thermalization is obtained "for all practical purposes" (FAPP) in the sense defined above. In this chapter, we shall adhere to this use of the term thermalization.

12.5.1 A toy model of thermalization

Although our goal is to describe thermalization (FAPP) in an isolated quantum field, it is instructive to consider first the case in which the field is thermalized (*strictu sensu*) by bringing it in contact with a heat bath. This problem was analyzed by Schwinger [Sch61]. The reservoir may be described by one or several quantum fields, and the action will be expanded by adding the action describing these fields, plus the new term describing the system–bath interaction. Probably the most mysterious empirical fact about thermodynamics is that the long-term equilibrium state, if achieved, is totally independent of the details of the bath dynamics and interaction. Therefore we shall leave open the details of the bath, and simply write an interaction term of the form $g\varphi^A\Psi_A$, where Ψ is some (generally composite) bath operator and g is a coupling constant. We shall assume the usual set-up where system and bath are brought into contact at some initial time $t = 0$. We also assume the initial condition is spatially homogeneous, which, neglecting the system's self-interactions, allows us to decompose it into independent spatial modes. We consider the thermalization of each mode, and in so doing reduce the original theory to a $1 + 0$ field theory.

As we have seen in Chapter 11, the equations for the (system) Jordan and Hadamard propagators are determined by the dissipation and noise kernels in the 1PI CTPEA for the system field. To lowest order in the system–bath coupling constant g, they are

$$\mathbf{D}\left(t - t'\right) = ig^2\hbar^{-1}\theta\left(t - t'\right)\left\langle\left[\Psi\left(t\right), \Psi\left(t'\right)\right]\right\rangle$$

$$\mathbf{N}\left(t - t'\right) = \frac{1}{2}g^2\hbar^{-1}\left\langle\left\{\Psi\left(t\right), \Psi\left(t'\right)\right\}\right\rangle \tag{12.171}$$

Here, a common or implicit assumption is that the bath is always kept in equilibrium at some temperature T, and that any back-reaction from the system is negligible. So, writing

$$\left\langle\left[\Psi\left(t\right), \Psi\left(t'\right)\right]\right\rangle = \hbar\int\frac{d\omega}{2\pi}\, e^{-i\omega\left(t-t'\right)}\text{sign}\left(\omega\right)\mathbf{R}\left(\omega\right) \tag{12.172}$$

where $\mathbf{R}\left(\omega\right)$ must be even and positive, and following the analysis in Chapter 11, the imaginary part of the retarded propagator becomes

$$\text{Im}\, G_{\text{ret}}^{-1} = \left(\frac{-g^2}{2}\right)\text{sign}\left(\omega\right)\mathbf{R}\left(\omega\right) \tag{12.173}$$

The full equation is determined by causality

$$\left[\omega^2 - m_b^2 - g^2 \int_0^\infty \frac{d\sigma^2}{2\pi} \frac{\mathbf{R}(\sigma)}{(\omega + i\varepsilon)^2 - \sigma^2}\right] G_{\text{ret}} = -\mathbf{1} \qquad (12.174)$$

where m_b is the bare mass of the system field.

If $\omega^2 = \xi + i\eta$, the inverse propagator develops an imaginary part

$$\eta \left[1 + g^2 \int_0^\infty \frac{d\sigma^2}{2\pi} \frac{\mathbf{R}(\sigma)}{(\xi - \sigma^2)^2 + \eta^2}\right] \qquad (12.175)$$

The expression in brackets is positive, so any (first sheet) zero must have $\eta = 0$. But on the real axis, the inverse propagator has a cut, with a discontinuity $g^2 \mathbf{R}(\xi)$ in the imaginary part. So, unless $\mathbf{R}(\sigma)$ vanishes below some threshold, the inverse propagator cannot be zero. We shall assume this is the case, which means that all excitations of the system are unstable against decay into the bath.

Besides damping, the bath also provides screening. The Debye mass is defined as the closest thing to a zero of the inverse propagator, namely a zero of the real part of the inverse propagator. Therefore

$$M_D^2 - m_b^2 - g^2 PV \left[\int_0^\infty \frac{d\sigma^2}{2\pi} \frac{\mathbf{R}(\sigma)}{M_D^2 - \sigma^2}\right] = 0 \qquad (12.176)$$

We shall assume the physically reasonable condition that $M_D^2 \geq 0$. A sufficient condition for this is that the left-hand side of the gap equation changes sign as we go from $M_D = 0$ to ∞. If $\mathbf{R}(\sigma)$ is well behaved (which may require that we perform a subtraction beforehand) the left-hand side is dominated by the first term M_D^2 when this is large, and so it is positive. The sufficient condition for screening (as opposed to anti-screening) boils down to

$$m_b^2 \geq g^2 PV \left[\int_0^\infty \frac{d\sigma^2}{2\pi} \frac{\mathbf{R}(\sigma)}{\sigma^2}\right] \qquad (12.177)$$

We may now write the equation for the retarded propagator as

$$\left\{(\omega^2 - M_D^2)\left[1 + g^2 PV \int_0^\infty \frac{d\sigma^2}{2\pi} \frac{\mathbf{R}(\sigma)}{(\omega^2 - \sigma^2)(M_D^2 - \sigma^2)}\right]\right.$$
$$\left. + \frac{ig^2}{2} \frac{\omega}{|\omega|} \mathbf{R}(\omega)\right\} G_{\text{ret}} = -\mathbf{1} \qquad (12.178)$$

The nice thing about this expression is that it makes it easy to identify the mean life of a field excitation. Indeed, the inverse propagator has a near Ornstein-Zernike structure

$$\left[\omega^2 - M_D^2 + 2i\gamma\omega\right] G_{\text{ret}} = -\mathbf{1} \qquad (12.179)$$

where $\gamma \sim g^2 \mathbf{R}(M_D)/4M_D$ is the damping constant, and so we may conclude that the decay of an excitation will be nearly exponential (at very long times it may turn to power law, depending on the behavior of $\mathbf{R}(\omega)$ as $\omega \to 0$). Incidentally, this equation also fixes the field density of states

$$\mathcal{D}(\omega) = \pi^{-1} |\text{Im}G_{ret}| = \left(\frac{g^2}{2\pi}\right) |G_{ret}|^2 \mathbf{R}(\omega) \qquad (12.180)$$

In particular, the Ornstein-Zernike approximation for the retarded propagator implies a corresponding approximation for the density of states

$$\mathcal{D}(\omega) \sim \frac{1}{\pi} \frac{2\gamma |\omega|}{(\omega^2 - M_D^2)^2 + 4\gamma^2 \omega^2} \qquad (12.181)$$

So far we have analyzed how the interaction with the heat bath affects the system dynamics, but we have not addressed thermalization *per se*. To do this, it is not efficient to look at the retarded propagator, because this propagator is very robust against thermal corrections. We look instead at the Hadamard propagator, which obeys the equation

$$G_{ret}^{-1} G_1 = 2\hbar \mathbf{N} G_{adv} \qquad (12.182)$$

We have arrived at the crucial point. The inhomogeneous equation (12.182) admits a particular solution $G_1 = 2\hbar \mathbf{N} |G_{ret}|^2$ and also homogeneous solutions which carry the information about the initial conditions. But the homogeneous solutions decay, so after a time long compared to the mean life γ^{-1}, only the particular solution remains. Now, the bath propagators are subject to the KMS theorem (it being insensitive to whether the field Ψ is fundamental or composite)

$$\mathbf{N} = \frac{g^2}{2} [1 + 2f_0(\omega)] \mathbf{R}(\omega) \qquad (12.183)$$

Therefore the asymptotic Hadamard propagator obeys

$$G_1 = 2\pi\hbar [1 + 2f_0(\omega)] \mathcal{D}(\omega) \qquad (12.184)$$

This is just the KMS theorem for the *field* (as opposed to the bath) Hadamard propagator.

In conclusion, the essential elements of the thermalization process are that there must be a heat bath, capable of transmitting the KMS condition to the system, and at the same time a damping mechanism so that the field initial conditions may be forgotten in time. Of course, damping does not cease when equilibrium is finally reached, but at late times it is exactly compensated by the inhomogeneous term in the equation for the Hadamard propagator. Thus we arrive at yet another perspective on the KMS theorem, now as a detailed balance condition which enforces the stability of the thermal state.

It is remarkable that when we come to view (12.184) as a relationship between the field Hadamard propagator and density of states, any direct reference to the bath has disappeared. The bath is necessary to validate the KMS theorem on the system, but once this task is accomplished, it can go free. In fact, any bath can perform this function (although the relaxation times will be different) as long as it is a *good* bath, meaning that it is able to sustain a constant temperature in the face of back-reaction, and that it provides efficient dissipation in all scales. Of course, this is precisely the condition for thermodynamics to prevail.

12.5.2 Thermalization of isolated fields

Let us now turn our attention to isolated fields, and examine whether in any sense they fulfill the two conditions above. The answer is yes in both cases. The KMS theorem is built in the Kadanoff–Baym equations, because field configurations consistent with the KMS theorem have slower dynamics, and eventually outlive those that do not. As for dissipation, beware that by restricting ourselves to thermalization in the FAPP sense, we are *de facto* turning the problem into an effectively open system. The interaction between the relevant system and irrelevant sectors (the environment, with its large capacity) brings dissipation and decoherence (or its classical analog, dephasing) to the system, by which the memory of initial conditions is lost. These mechanisms work either for quantum or classical field theories, although we expect them to be more efficient in the quantum case. For example, quantum particle creation may emanate from the vacuum, while classical parametric amplification can only work from a pre-existing seed; thus pumping energy from a heat bath or a classical background into an unpopulated region of the spectrum is easier in quantum theories (for Fermi systems, of course, we have to take Pauli blocking into consideration).

By now, there is a mounting body of (numerical) evidence in support of these statements. Numerical work has focused mostly on scalar field theories with quartic self-interactions, with either one single field or else N fields with $O(N)$ symmetry in the large N limit. Numerical investigations of the equations of motion as derived from the Kadanoff–Baym equations were pioneered by Danielewicz [Dan84a, Dan84b].

As we mentioned in the Introduction, it is not our aim to discuss numerical approaches in detail. However, it is important to know what has been achieved. To this end, it is useful to classify the mounting literature on the subject into the four basic categories of quantum mechanical, classical, semiclassical and quantum field models.

Quantum mechanical models

The complexity of the field theoretic equations led to the search for simpler systems where at least the basic approximations could be tested. One possible simplification is to consider a field theory in $0 + 1$ dimensions, namely quantum mechanics. For example, Cooper *et al.* [CDHR98] showed that while the evolution of the coefficients in the quantum mechanical Hartree and leading order (LO) large N approximations may be described as a chaotic Hamiltonian system, in truth chaos is an artifact of the approximation. Another work using quantum mechanical systems as a testing ground is [BetWet98]. [MACDH00] matches Hartree, LO and next-to-LO (NLO) large N against numerical solutions of the Schrödinger equation. [MiDaCo01] analyzes the so-called bare-vertex and dynamic Debye screening approximations. [Hab04] shows that the Gaussian approximation in a closed system leads to the same dynamics for the Wigner and the distribution functions, irrespective of whether the system is quantum or classical.

Classical field models

Another direction in which the theory may be simplified is by taking the classical limit. Thermalization in classical ϕ^4 theory was investigated in [AaBoWe00b, Aar01, BoDeVe04]. Oftentimes a classical field theory arises from a mean field approximation to a quantum problem. In particular, the nonequilibrium dynamics of Bose–Einstein condensates has been thoroughly investigated as described by the Gross–Pitaevskii equation [GaFrTo01, SanShl02, UedSai03, BajaMa04, Adh04].

Both classical field theories and the time-dependent Ginzburg–Landau equation have been investigated as models of defect formation after an instantaneous quench [AntBet97, DzLaZu99, Ste00]. Adding a $U(1)$ gauge field leads to the Gorkov equations for a type II superconductor [YatZur98, IbaCal99, StBeZu02]. To simulate a quench at a finite rate, it is possible to introduce interaction with a heat bath by adding ohmic dissipation and white noise [Kib80, Kib88, Zur85, Zur96, Riv01, RiKaKa00, LagZur97, LagZur98, YatZur98, AnBeZu99, BeHaLy99, HabLyt00, BeAnZu00, FASA05, AGRS06]. There have also been analyses of classical theories in expanding universes, motivated by the problem of reheating; see [KhlTka96, KoLiSt97, FelTka00, FeKoLi01, FelKof01, FGGKLT01, MicTka04, PFKP06].

Classical field has been extensively used as a test bench for different approximations, for example, the use of a scalar field in 1+1 to compare the Hartree, LO and NLO large N approximations in [AaBoWe00a].

Semiclassical field models

One step up in the ladder of increasing complexity we find semiclassical models, often arising from Hartree or leading order $1/N$ approximations to the full quantum field models [BVHLS95].

This category also includes external field problems beyond the test field approximation (cf. Chapter 4). Among these, the most studied have been electromagnetic and gravitational backgrounds. [KESCM92] compares the semiclassical evolution to a quantum Vlasov equation incorporating Schwinger's pair creation from the electric field. [CEKMS93] generalizes the above by including the effect of an expanding background geometry. [CHKMPA94] investigates a symmetric scalar $O(N)$ theory and QED with N fermion fields. See also [KlMoEi98] and [AarSmi99], which deals with the abelian Higgs model with fermions in 1+1 dimensions.

Fully numerical solution of semiclassical cosmological models presents enormous difficulties, not only because of the intrinsic complexity of general relativity but also because most schemes lead to wildly unstable dynamical equations [ParSim93]. Some questions have been investigated, though, most notably the back-reaction effect of trace anomalies of quantum fields and particle creation leading to avoidance of cosmological singularity and anisotropy damping [FiHaHu79, HarHu79, HarHu80, Har80, Har81].

Another source of semiclassical problems has been the development of spinodal decomposition [CHKM97]. [BBHKP98] considers initial conditions relevant to a relativistic heavy ion collision. See also [BVHS99a, SCHR99]. In particular, the possibility of actually observing disoriented chiral condensates in relativistic heavy ion collisions has focused much attention on the specifics of this problem [CKMP95, LaDaCo96, CoKlMo96, BeRaSt01].

The problem of reheating after inflation combines aspects of both semiclassical theory on curved spacetime backgrounds and spinodal instability. [BoVeHo94] formulates the Hartree and one-loop approximations in an expanding background within the test field approximation. [BVHS96] discusses the effect of anharmonicity on the background field dynamics and the structure of resonances. [RamHu97a, RamHu97b] incorporate fully the back-reaction of quantum fluctuations on the dynamics of the inflaton field as well as the dynamics of the expanding background spacetime. See also [ZiBrSc01].

The problem of condensate collapse within the Hartree–Fock–Bogoliubov approximation has been studied in [WuHoSa05].

Full quantum field models

At the top of the complexity ladder we find the full quantum field models. Of course, the field theoretic Heisenberg equations being unassailable, some kind of perturbative scheme is necessary. Much of our present understanding of nonequilibrium quantum fields comes from the analysis of $O(N)$ scalar fields to NLO in the large N approximation [Ber02, AarBer02, BerSer03a] in $1 + 1$ dimensions. This work is reviewed in [Ber04b, BerSer03b, BerSer04, BerBol06].

The $\lambda\Phi^4$ theory to two loops and beyond leads to a similar phenomenology. It has been investigated in one [AarBer01], two [JuCaGr04] and three space dimensions [ArSmTr05]. Going beyond scalar fields, [BeBoSe03] studies the abelian Higgs model in 3+1 dimensions at two loops. Other approximation schemes have been explored [BaaHei03a, BaaHei03b]. A radical new approach to numerical nonequilibrium field theory has been proposed in [BerSta05, BBSS06]. Nonequilibrium Bose–Einstein condensates have been analyzed from a 2PI perspective both in a large N expansion and to second order in the interaction strength [RHCRC04, GBSS05].

The fields of lattice QCD and hydrodynamical and kinetic models of relativistic collisions targeting the hadronization process are beyond the scope of this book. See [Shu88, Cse94, Wan97, Ris98, BaRoWi06a, BaRoWi06b, TeLaSh01,

HKHRV01, KolRap03, HirTsu02, HirNar04, HeiKol02b]. There is also some numerical work on processes which may speed up thermalization in the early stages of the collision [ArMoYa05, ArnMoo05, Moo05, RomVen06].

By way of summary

This brisk enumeration should convince the reader that by now a wide variety of cases has been studied, with a matching diversity of means. The important point is that a coherent picture emerges, since the phenomenology observed in the different cases is consistent. In the remainder of this chapter, we shall tell the prototype thermalization story, by combining the insights gained from these numerical experiments, supporting it whenever possible by analytical arguments. In the remaining chapters of the book we shall contrast this theory with the findings and demands of concrete applications.

12.5.3 The stages of thermalization

Summarizing the results of both numerical and analytical work, we may say that typically the thermalization process in an isolated quantum field goes through three distinctive stages [Son96]:

(a) early stage;
(b) intermediate stage;
(c) late stage.

The early stage: Preheating and prethermalization

Description of the earliest stage of thermalization varies a lot from one model to another. It is generally characterized by an explosive pumping of energy into the field, usually because of instabilities. For example, in a quench from a stable to an unstable phase (cf. Chapters 4 and 9), the infrared modes become unstable and begin to grow explosively. A similar phenomenon marks the growth of fluctuations around a collapsing condensate, or the growth of large-scale magnetic fields from an anisotropic distribution of hard gluons after a relativistic heavy ion collision (RHIC). Without involving an actual instability, parametric amplification by a dynamical background is also an efficient way to transfer energy to the field; this occurs in the so-called preheating stage in reheating after inflation.

It is possible to reach an analytic understanding of the early phase if a set of modes may be identified as a linear field on an evolving background. In this case, the early stage may be analyzed within a one-loop or Hartree type approximation. One generic phenomenon is that of decoherence (or dephasing) brought about through quantum diffusion [HKMP96]. As a result, quantities involving contributions from many modes (like the energy density or the pressure) quickly lose memory about the initial conditions, and the equation of state stabilizes to its near-thermal form. This is the phenomenon of pre-thermalization [BeBoWe04].

One basic difficulty in formulating a model of this early stage is accounting for the system–environment interaction which is what drives the system out of equilibrium to begin with. This is usually done by assuming an *ad hoc* time dependence in the field parameters, and/or adding dissipation and noise to the equations of motion. However this procedure is hard to justify on a first principle basis.

There is however an environment which is easy to include into the equations and can bring about the desired effect – a dynamical background spacetime. In an expanding universe only conformally invariant fields may hold on to thermal equilibrium. Once conformal invariance is broken, field modes are relentlessly red-shifted by the expansion, bringing about an effective (in both senses of the word) cooling. We have analyzed this problem in Chapter 4.

The initial stage concludes at the point the infrared peak becomes nonlinear. After a period of parametric amplification, the end result is a nonthermal spectrum with a narrow band of highly populated modes. In the next stage, this far-from-equilibrium spectrum evolves into a Planck distribution through the process of turbulent thermalization.

Intermediate stage: Turbulent thermalization and kinetic equilibration

The second stage of the thermalization process is characterized by nonlinear interactions among quantum modes, bringing about an effective thermalization in the energy spectrum, as measured from the Fourier transform of the two-point functions.

As we have seen, the early stage may be described as a theory of linear fields evolving on a classical background. However, as the quantum field amplitude grows, there is a point where a linear model ceases to make sense. We emphasize that the breakdown of linear models beyond a certain point goes over and above formal problems, such as the existence of secular terms [Ber04b, BerSer03b, BerSer04]. If that were the case, it would be enough to resum those terms, for example, by using dynamical renormalization group techniques [BVHS99b, KunTsu06]. The point is that the model of linear fields on a background misses an essential part of the physics, for which there is no formal remedy.

To estimate the point at which the linear approximation is no longer valid, we could for example consider the case of a Bose–Einstein condensate. The Heisenberg operator $\psi(x)$ which destroys a noncondensate atom at point x may be regarded as nonrelativistic scalar field theory with a quartic self-interaction. There is a scattering length $a = UM/4\pi\hbar^2$ pertaining to the strength U of the self-interaction and the mass of the atom M. The cross-section for scattering between noncondensate atoms is $\sigma \sim a^2$. If we split the density into its condensate and noncondensate (or anomalous) parts $n = \Phi^2 + \tilde{n}$, then the mean free path for a noncondensate atom is $\lambda \sim 1/\tilde{n}\sigma$. On the other hand, let L be a characteristic length of the problem. It could be the distance over which the condensate varies, or the size of a causal horizon as measured from the start of the nonequilibrium evolution. Once $\lambda \leq L$, self-interactions can no longer be ignored, which yields

$$\tilde{n} \geq \frac{\left(4\pi\hbar^2\right)^2}{U^2 M^2 L} \tag{12.185}$$

This may come about because $\tilde{n}$ gets large or L gets large, for example, if L grows linearly in time. We observe that for a self-interacting field, there is a nonzero anomalous density even at zero temperature.

$\lambda \leq L$ is also the condition for the validity of the quantum kinetic theory approach, since we expect λ to set the scale for the decay of correlations with respect to the relative variable, and L to describe the dependence of the correlations on the center-of-mass variable. This affords an enormous simplification of the problem.

Observe, however, that in the same way that it is wrong to apply a linear model in the intermediate stage, it would be wrong to apply a quantum kinetic theory scheme in the early stage [BerBol06, BeBoWe05]. The simplest kinetic theories assume, besides $\lambda \leq L$, that all initial non-Gaussian correlations have decayed, and that effectively the initial time may be chosen as in the asymptotic distant past. These restrictions are removed in more complex approaches, but they also set limits to the applicability of quantum kinetic theory at early times.

In a typical problem, the early stage concludes with most of the energy in the field concentrated in a narrow set of modes, and the intermediate stage sees the spread of energy over the full spectrum. There is an initial stage where a cascade is formed between the initial scale k_0 and a moving front $k_{\max}(t)$. Within these limits, there is a constant energy current towards higher wavenumbers. This phenomenon closely resembles Kolmogorov's 1941 scenario for fully developed turbulence, and hence the name of turbulent thermalization [Fri95, McC94, LanLif59, Hin75, Bat59, ZaLvFa92].

Turbulent thermalization ends with a self-similar particle number spectrum (as defined by the Fourier transform of the Hadamard propagator, see Chapter 11) $f(k) \sim k^{-\alpha}$, with $\alpha > 1$. The following stage, or kinetic equilibration, sees the evolution of the spectrum towards the Rayleigh–Jeans form $f(k) \sim (k_B T) k^{-1}$. We are assuming of course a high-temperature, weakly coupled scalar field, with high occupation numbers.

The basic features of turbulent thermalization may be understood in terms of a simple quantum kinetic theory model [ZaLvFa92, FelKof01, MicTka04, MuShWo07]. Assume a kinetic equation

$$\omega_k \frac{\partial f}{\partial t}(t, k) = I_{\text{col}} \tag{12.186}$$

For the collision kernel, write

$$I_{\text{col}} = \int \prod_{i=1}^{m} \frac{d^d p_i}{2\omega_i} \sigma_{m,j}^2 \left\{ \prod_{l=0}^{j} (1 + f(p_l)) \prod_{l'=j+1}^{m} f(p_{l'}) \right.$$

$$\left. - \prod_{l=0}^{j} f(p_l) \prod_{l'=j+1}^{m} (1 + f(p_{l'})) \right\} \delta^{(d+1)} \left[\sum_{l=0}^{j} p_l - \sum_{l'=j+1}^{m} p_{l'} \right] \quad (12.187)$$

where d is the number of spatial dimensions and $p_0 = k$. For example, for elastic $2 \to 2$ scattering we have $m = 3$, $j = 1$. For large occupation numbers, this simplifies to

$$I_{\text{col}} = \int \prod_{i=1}^{m} \frac{d^d p_i}{2\omega_i} \sigma_{m,j}^2 \left[\prod_{l=0}^{m} f(p_l) \right] \left[\sum_{l=0}^{j} f^{-1}(p_l) - \sum_{l'=j+1}^{m} f^{-1}(p_{l'}) \right]$$

$$\times \delta \left[\sum_{l=0}^{j} p_l - \sum_{l'=j+1}^{m} p_{l'} \right] \quad (12.188)$$

Observe that $\omega_k f(t, k)$ is also the energy density in wavenumber space, and so the energy current $\mathbf{J}_E$ obeys

$$\nabla_k \mathbf{J}_E = -I_{\text{col}} \quad (12.189)$$

Let us assume an isotropic situation $\mathbf{J}_E = J_E \hat{k}$. The total energy flux through a shell of radius k is $K_E = rk^{d-1} J_E$, where r is a constant pure number. Therefore, if I_{col} scales as $k^{-\beta}$, then J_E scales as $k^{1-\beta}$ and $K_E \sim k^{d-\beta}$. It follows that turbulent thermalization requires $\beta = d$.

On the other hand, if $f(k) \sim k^{-\alpha}$ then from equation (12.188) we get (assuming $\sigma_{m,j}^2$ does not scale)

$$\beta = (d+1) + m(\alpha + 1 - d) \quad (12.190)$$

and finally

$$\alpha = d - 1 - \frac{1}{m} \quad (12.191)$$

The numerical result for $d = 3$ is $\alpha = 3/2$, which corresponds to $m = 2$ [MicTka04]. This is obtained for a $g\phi^3$ theory or for a $\lambda\phi^4$ theory in the presence of a background field. Observe that for $d = 1$ we get $\alpha < 0$ whatever the value of m. Therefore there are no turbulent UV cascades in $1 + 1$, also consistent with numerical results. The observed cascade has the Rayleigh–Jeans spectrum $\alpha = 1$ corresponding to $K_E = 0$ (cf. equation (12.188)).

The evolution of the wave front $k_{\text{max}}(t)$ depends upon further details such as whether the total energy (or else the total particle number) contained in the cascade may be considered constant.

As the spectrum spreads it also loses amplitude, and at some point the typical occupation numbers are no longer large. At this point turbulent thermalization ceases. The subsequent relaxation to Rayleigh–Jeans equilibrium may be described by nonequilibrium renormalization group methods. The time-scale for kinetic equilibration may be estimated from a simple Boltzmann equation approach as $\tau_{\text{rel}} \sim 1/\lambda^2 T$.

Late stage: chemical equilibration

The latest stage concerns chemical, rather than kinetic, equilibration. This means that the energy and particle number spectra already have (local) equilibrium forms, and now the issue is the equilibration among different species. For a real scalar field theory, the two species involved (particles and antiparticles) are identical, and chemical equilibration means the vanishing of the chemical potential.

As in the earlier stages, the basic problem is to find the right tool for the right job. For example, the simplest Boltzmann equation with $2 \to 2$ scattering may be successfully used to describe kinetic equilibration. Nevertheless, it fails to describe chemical equilibration, because it has a spurious particle number conservation law built in. To describe chemical equilibration we must go beyond this lowest order kinetic equation, either by considering a more general density of states (as opposed to a sharp mass-shell) or/and by considering higher order terms in the loop or $1/N$ expansions [CaHuRa00, Wei05b, FiGaJe06]. The relevant terms have been analyzed earlier in this chapter.

Concrete applications may demand other departures from the simple Boltzmann approach. For example, in dealing with a quark–gluon plasma, the relevant kinetic equation is not Boltzmann's, but rather a Landau-type kinetic equation incorporating the effects of grazing collisions [ChaCow39, Lib98, LifPit81, Mue00a, Mue00b, BjoVen01]. Also we must take into account the color degree of freedom, for example, by analyzing the Wong equations [LitMan02].

Another possibility is to go over directly to a hydrodynamic description. Since the underlying quantum field theory is obviously causal, one expects that the correct hydrodynamic theory would not be a first-order theory (in the classification of Hiscock and Lindblom) but rather a Israel–Stewart or a divergence type theory [CalThi00, CalThi03].

As a concrete example, let us analyze the regression towards zero of the chemical potential in a self-interacting scalar field theory. We assume we are close enough to equilibrium that the chemical potential may be regarded as a linear perturbation, and use the kinetic equation with higher order terms already discussed in this chapter. Assuming the chemical potential is a function of time only, for simplicity, we may write the equation in the spatially translation-invariant case, namely

$$\left[\omega \frac{\partial}{\partial t} - \frac{1}{2} \left(\frac{dM^2}{dt} \right) \frac{\partial}{\partial \omega} \right] f = I_{\text{col}} \left[f, M^2 \right] \qquad (12.192)$$

The mass is given by the gap equation

$$M^2 - \varphi\left(M^2, \mu^2\right) = \frac{\lambda}{2} M_T^2 \tag{12.193}$$

where

$$M_T^2 = \int \frac{d^4p}{(2\pi)^3} \delta\left(p^2 + M^2\right) f\left(X, p\right) \tag{12.194}$$

Now write

$$f = f_0 + f_0\left(1 + f_0\right)\chi \tag{12.195}$$

$$M^2 = M_0^2 + \delta M^2 \tag{12.196}$$

$$I_{\text{col}}\left[f, M^2\right] = f_0\left(1 + f_0\right) K\left[\chi\right] \tag{12.197}$$

$$\omega\frac{\partial\chi}{\partial t} + \frac{\beta}{2}\frac{d\delta M^2}{dt} = K\left[\chi\right] \tag{12.198}$$

$$\left[1 - \varphi' - \frac{\lambda}{2}\frac{\partial M_T^2}{\partial M^2}\right]\delta M^2 = \lambda\left\langle\chi\right\rangle \tag{12.199}$$

The perturbation χ may be expanded in eigenfunctions of the linearized Boltzmann collision operator

$$\chi = \beta\left[\mu + \frac{\delta T}{T}\omega + \bar{\chi}\right] \tag{12.200}$$

$$\bar{\chi} = \sum_{n=1} c_n \chi_n \tag{12.201}$$

so

$$\delta M^2 = 2\omega_0\left[\mu + \frac{\langle\omega\rangle}{\langle 1\rangle}\frac{\delta T}{T}\right] \tag{12.202}$$

$$\left[\omega + \omega_0\right]\frac{d\mu}{dt} + \frac{1}{T}\left[\omega^2 + \frac{\langle\omega\rangle}{\langle 1\rangle}\omega_0\right]\frac{d\delta T}{dt} + \omega\frac{\partial\bar{\chi}}{\partial t} = \mu K\left[1\right] + K\left[\bar{\chi}\right] \tag{12.203}$$

$$\omega_0 = \frac{\lambda\beta\langle 1\rangle}{2\left[1 - \varphi' - \frac{\lambda}{2}\frac{\partial M_T^2}{\partial M^2}\right]} \tag{12.204}$$

Since we know that $\mu = $ constant is a solution if we keep only the lowest order Boltzmann collision term, we expect μ to decay on time-scales of the order of λ^{-4} at least. $\bar{\chi}$ will have a slow part, that will track μ, and a fast part, that will decay on time-scales of the order of ω^{-1}. Clearly only the slow part is relevant to our discussion, and so we may neglect the $\partial\bar{\chi}/\partial t$ term.

Observe that now we have an equation of the form

$$K\left[\bar{\chi}\right] = A\omega^2 + B\omega + C \tag{12.205}$$

If we look at the right-hand side as a function of ω, then the dominant term is the first. The solution is $\bar{\chi} \sim -\left(\tau_{\text{rel}}/T\right) A\omega^2 + \nu\omega + \nu'$, where the linear and constant terms enforce the constraints $\langle\bar{\chi}\rangle = \langle\omega\bar{\chi}\rangle = 0$. In other words, the effect of the $K[\bar{\chi}]$ term in equation (12.203) is to compensate the $d\delta T/dt$ one. Canceling those two, we get the regression equation for μ as the average of equation (12.203) (recall that $\langle K[1]\rangle < 0$)

$$[\langle\omega\rangle + \omega_0\langle 1\rangle]\frac{d\mu}{dt} = -\mu|\langle K[1]\rangle| \tag{12.206}$$

From the estimates in this chapter, we conclude that the characteristic time-scale for chemical equilibration is $\tau_{\text{chem}} \sim M^2/\lambda^4 T^3$, parametrically larger than τ_{rel}.

12.5.4 Coda

In this section we have painted a broad outline of the thermalization process, going as far as possible without invoking the specifics of modeling or features of concrete applications. We have seen that a general picture indeed emerges, and that it reveals the communion between quantum field theory and other parts of physics, represented in this chapter by fluid and wave turbulence theory. As all portraits of its kind, it emphasizes more the generalities than the specifics, and so case by case considerations are still useful. In the remainder of this book we shall do just that, at least for the most conspicuous and developed applications of nonequilibrium quantum field theory.

Part V

Applications to selected current research

13
Nonequilibrium Bose–Einstein condensates

Bose–Einstein condensation was predicted in 1925 [Ein24, Ein25] but, except for its indirect manifestation in the superfluidity of He^4 [Lon38, Kap38, AllMis38], it remained a purely theoretical construct until 1995, when condensation in alkali gases was achieved in the laboratory [CorWie02, Ket02]. Since then, a great deal of the theoretical work in the previous 70 years has been put to the experimental test, while new avenues have been opening up, such as the superfluid–insulator or Mott transition [FWGF89, GMEHB02, CaHuRe06], the BEC-BCS cross-over [Leg06, Reg04] and the Tonks gas regime [Gir60, Par04]. Because of the great experimental control over the relevant parameters and the deep understanding of the fundamental physics, BECs have become a field of choice to perform experiments of interest not just in atomic and molecular physics, but also in quantum optics, condensed matter physics, quantum critical dynamics, even field theory, gravitation, cosmology and black hole physics [CoEnWi99, PetSmi02, Sou02, BonSen04, ParZha93, ParZha95, UnrSch07, BaLiVi05]. Moreover, cold Bose gases on optical lattices have been proposed as a possible implementation of a quantum information processing (QIP) device [JakZol04], boosting new interest in these systems.

As introductions and reviews on this fast evolving subject abound, it is perhaps more fitting for us to focus on certain aspects of the nonequilibrium field theory of BEC, specifically [And04] the application of quantum field theoretic methods described in this book to the description of nonequilibrium evolution of condensed gases in magneto-optical traps [ChCoPh95].[1]

Of course NEqQFT is not the only possible description. Reflecting on the characteristics of this field as a current attractor of different subdisciplines listed above, the literature presents an almost bewildering array of possibilities. However, there are a few basic criteria that any successful description must meet: it must be faithful to the presence of gapless excitations above the condensate [HohMar65, Gri96, ShiGri98], and must respect the basic conservation laws of particle number and energy–momentum [Kra60, BayKad61, Bay62]. These requirements are sealed at the roots of a quantum field theoretical formulation

---

[1] This means we would have to sacrifice the description of important topics like the physics of cold atoms in optical lattices, which has a rapidly expanding literature [ChCoPh95], and vortices in BECs and their associated phenomena [ElKrVo06], purely from space limitation considerations.

and, as such, provide a benchmark and standard against which other approximations may be compared. It also provides a systematic way to develop a perturbative expansion to arbitrary order [PRSC02, SPRS02, BFGR01, BaFrRa02, Boy02].

Realistically, once one gets to the point of actually writing down a nonrelativistic field-theoretic action to describe the second-quantized atomic gas, the functional approach developed earlier in this book in the context of relativistic scalar field theories works well in every detail we have considered so far. This is one of the strengths of this approach. For this reason we will concentrate on the first stage, namely, how to get from the physical model of the trapped gas to a nonrelativistic field theory. In the process, we shall attempt to give a model-independent characterization of the two requirements mentioned above, and to discuss how they enter into the functional method.

Current experimental work on BECs presents a variety of nonequilibrium problems, including the dynamics of condensation itself and the response of the BEC to changes in its environment (temperature and trapping fields) and particle interactions [KaSuSh96, KaSuSh97]. Probably the most extreme demonstration of far-from-equilibrium behavior is the so-called Bose–Nova experiment [Don01, Cla03a, Cla03b, CoThWi06, SaiUed03, SanShl02, BajaMa04, Adh04, GaFrTo01, SaRoHo03, WuHoSa05, Yur02, CalHu03, WDBDBH07], where a sudden sign change in the interatomic interaction triggers the implosion of the condensate. The possible use of cold gases in optical lattices in QIP poses, among others, two specific challenges for a nonequilibrium theory: the detailed description of the initialization of the device [Rey04, Bre05, Pup04, ReBlCl03], and an accurate estimation of static and dynamic decoherence times [SaOHTh97, Oos02, PuWiPr06, Rei05].

The plan for this chapter is as follows: starting from the second-quantized version of the weakly interacting Bose gas Hamiltonian in a closed time path (CTP) framework, we shall present the basic (symmetry-breaking) formulation in a model-independent way. We shall give a precise formulation to the requirements of a gapless spectrum (the so-called Hugenholtz–Pines theorem) and particle number conservation in the mean. In the process, we shall introduce the class of Φ-derivable theories as a broad framework for viable models of BEC dynamics.

Then we shall introduce the 1PI and 2PI effective action descriptions of the BEC, as described in Chapter 6 [LutWar60, DomMar64a, DomMar64b, CoJaTo74, LunRam02]. This means that we opt to follow the evolution of the condensate through the unfolding of correlation functions, as opposed, for example, to obtaining a time-dependent wavefunction for the many-body system [KohBur02, PrBuSt98, GKGB04]. We shall show that, *in principle,* the 2PIEA leads to a Φ-derivable theory which is both gapless and conserving. However, the appearance of many models derived from truncations of the 2PIEA in different degrees may tell a different story. We shall show how the familiar theories arise from such truncations and examine them in detail. They are the

Gross–Pitaevskii (GP), Bogoliubov, one-loop, Hartree–Fock–Bogoliubov (HFB), Popov and two-loop approximations. We shall show that the two-loop approximation yields a minimal theory which is both gapless (to the required order in perturbation theory) and conserving. We shall not discuss other approximation schemes, like the $1/N$ approximation because they can be analyzed in terms similar to those introduced in Chapter 6 [TemGas06, GBSS07].

Our next goal will be to discuss two specific predictions of the two-loop theory, namely, that the evolution of condensate fluctuations is dissipative and stochastic. This is in accord with the fluctuation–dissipation theorem discussed earlier in the book. In particular, to discuss fluctuations we shall adopt a coarse-grained effective action scheme where high-energy "noncondensate" modes act as an environment for the low-energy "condensate band" modes, where condensation takes place. We shall concentrate on the derivation of the noise terms coupled to the Gross–Pitaevskii equation, yielding a stochastic GP equation. Of course, in so doing the initial conditions for the condensate can also become stochastic, which is an important consideration in actual applications.

In the regime where modes above the condensate are highly populated – not macroscopically, of course – relaxation is efficient enough that a kinetic theory description becomes possible, leading eventually to a two-fluid hydrodynamic. Since we have discussed quantum kinetic theory in detail earlier in the book, we shall focus here only on those features which are characteristic of the BEC environment.

Finally, we shall close the chapter with a brief description of the so-called *particle number conserving* formalism. The symmetry breaking approach described so far has the drawback that strictly speaking it cannot be applied to a system with a finite number of particles. The particle number conserving formalism overcomes this difficulty. In particular, we shall discuss a functional implementation of this formalism, which makes it as flexible as the better known symmetry breaking approach.

13.1 The closed time path integral approach to BECs

In this section we put together the basic formulae for the coherent state representation [NegOrl98] of the causal or CTP path integral method (introduced in Chapters 3, 5 and 6) to compute the expectation values of physical observables. Let ρ_i be the density matrix describing the initial state of the system at $t = t_i$. Then expectation values with respect to ρ_i may be obtained from the CTP generating functional (cf. Chapter 6)

$$e^{iW} = \mathrm{Tr} \left\{ U_2^{-1} \left(t_f, t_i \right) U_1 \left(t_f, t_i \right) \rho \left(t_i \right) \right\} \tag{13.1}$$

where

$$U_{1,2} \left(t_f, t_i \right) = T \left[e^{-i \int_{t_i}^{t_f} dt \, H^{1,2}(t)} \right] \tag{13.2}$$

We shall use the well-known coherent state representation [NegOrl98] in the construction of a path integral representation of the generating functional in the next subsection. The CTP boundary conditions will be introduced in the following subsection.

13.1.1 The coherent state representation

For simplicity, we consider a single one-particle state. There is a basis made of occupation number eigenstates $|n\rangle$

$$N \left|n\right\rangle = n \left|n\right\rangle \tag{13.3}$$

where N is the number operator (in particular, $n = 0$ is the vacuum state $|0\rangle$). These states are orthonormal and complete

$$\langle m \left|n\right\rangle = \delta_{mn} \tag{13.4}$$

$$\sum \left|n\right\rangle \langle n| = 1 \tag{13.5}$$

The destruction and creation operators relate states of different occupation numbers

$$\hat{a} \left|n\right\rangle = \sqrt{n} \left|n-1\right\rangle ; \qquad \hat{a}^\dagger \left|n\right\rangle = \sqrt{n+1} \left|n+1\right\rangle \tag{13.6}$$

Therefore

$$\hat{a}^\dagger \hat{a} = N; \qquad [\hat{a}, \hat{a}^\dagger] = 1 \tag{13.7}$$

A coherent state $|a\rangle$ is an eigenstate of the destruction operator

$$\hat{a} \left|a\right\rangle = a \left|a\right\rangle \tag{13.8}$$

It follows that

$$\langle n \left|a\right\rangle = \frac{1}{\sqrt{n}} \langle n-1| \hat{a} \left|a\right\rangle = \frac{a}{\sqrt{n}} \langle n-1 \left|a\right\rangle \tag{13.9}$$

Adopting the normalization

$$\langle 0 \left|a\right\rangle = 1 \tag{13.10}$$

one gets

$$\langle n \left|a\right\rangle = \frac{a^n}{\sqrt{n!}} \tag{13.11}$$

Or else,

$$\left|a\right\rangle = \sum \frac{a^n}{\sqrt{n!}} \left|n\right\rangle = \sum \frac{a^n \hat{a}^{\dagger n}}{n!} \left|0\right\rangle = \exp\left\{a\hat{a}^\dagger\right\} \left|0\right\rangle \tag{13.12}$$

Observe that

$$\hat{a}^\dagger \left|a\right\rangle = \frac{\partial}{\partial a} \left|a\right\rangle \tag{13.13}$$

Let $|b\rangle$ be a second coherent state; then

$$b\langle a\,|b\rangle = \langle a|\,\hat{a}\,|b\rangle = \frac{\partial}{\partial a^*}\,\langle a\,|b\rangle \tag{13.14}$$

and

$$\langle a\,|b\rangle = \exp\{a^*b\} \tag{13.15}$$

The constant is determined by recognizing that the vacuum is the coherent state with $a = 0$. From this point on, we shall omit the hats on operators whenever there is no risk of confusion.

While not orthogonal, the coherent states are complete, in the following sense

$$\int \frac{da^*da}{2\pi i}\exp\{-a^*a\}\,|a\rangle\,\langle a| = 1 \tag{13.16}$$

We may use the completeness relationship to write down the trace of an operator A

$$\mathrm{Tr}A = \sum \langle n|\,A\,|n\rangle = \int \frac{da^*da}{2\pi i}\exp\{-a^*a\}\,\langle a|\,A\,|a\rangle \tag{13.17}$$

Now consider the transition amplitude between the state $|a_i\rangle$ at time $t_i = 0$ and the state $|\bar{a}_f\rangle$ at time t_f. We have (setting $\hbar = 1$)

$$|\bar{a}_f\rangle = e^{iHt_f}\,|\bar{a}\rangle \tag{13.18}$$

and

$$\langle \bar{a}_f\,|a_i\rangle = \langle \bar{a}|\,e^{-iHt_f}\,|a_i\rangle \tag{13.19}$$

Note that $|\bar{a}_f\rangle$ is *not* a solution of the Schrödinger equation, but an eigenstate of the Heisenberg operator $a\,(t_f)$ with proper value $\bar{a}$. Since $a\,(t_f) = e^{iHt_f}ae^{-iHt_f}$, we have $a\,(t_f)\,|\bar{a}_f\rangle = \bar{a}\,|\bar{a}_f\rangle$.

Let N be some large number and $\varepsilon = t_f/N$. Write $a_i = a_0$, $\bar{a} = a_N$. Then, inserting $N - 1$ identity operators, we have

$$\langle \bar{a}_f\,|a_i\rangle = \int \left\{\prod_{n=1}^{N-1} \frac{da_n^*da_n}{2\pi i}\exp\{-a_n^*a_n\}\,\langle a_{n+1}|\,e^{-iH\varepsilon}\,|a_n\rangle\right\}\langle a_1|\,e^{-iH\varepsilon}\,|a_0\rangle \tag{13.20}$$

which may be written as (assuming the Hamiltonian $H = H(a^\dagger, a)$ is in a normal form)

$$\langle a_N\,|a_0\rangle = \int [Da]_{N-1}\,\exp\{iS_N\,[a^*, a]\}\,e^{a_N^*a_N} \tag{13.21}$$

where

$$[Da]_{N-1} = \prod_{n=1}^{N-1} \frac{da_n^*da_n}{2\pi i} \tag{13.22}$$

$$S_N\,[a^*, a] = \sum_{n=1}^{N}\{ia_n^*\,(a_n - a_{n-1}) - \varepsilon H\,(a_n^*, a_{n-1})\} \tag{13.23}$$

Going to the continuum limit, where $a_n - a_{n-1} \sim \varepsilon \partial a/\partial t$, we get

$$\langle \bar{a}_f \, | a \rangle = \int [Da] \, \exp \{ iS [a^*, a] \} \, e^{a^* a(t_f)} \tag{13.24}$$

$$S [a^*, a] = \int dt \left\{ ia^* \frac{\partial a}{\partial t} - H (a^*, a) \right\} \tag{13.25}$$

The integration is over paths which interpolate between $a(0) = a$ and $a^*(t_f) = \bar{a}^*$.

13.1.2 The closed time path boundary conditions

We now have all the necessary elements to evaluate the CTP generating functional (13.1). The idea is that the initial density matrix ρ is propagated forwards in time with some Hamiltonian H^1 and then backwards with a Hamiltonian H^2. Insert three identity operators in (13.1) to obtain

$$e^{iW} = \int \frac{da_N^* da_N}{2\pi i} \frac{da_0^{1*} da_0^1}{2\pi i} \frac{da_0^{2*} da_0^2}{2\pi i} \exp \left\{ - \left(a_N^* a_N + a_0^{1*} a_0^1 + a_0^{2*} a_0^2 \right) \right\}$$
$$\times \langle a_N | U_2 (t_f, t_i) | a_0^2 \rangle^* \langle a_N | U_1 (t_f, t_i) | a_0^1 \rangle \langle a_0^1 | \rho (t_i) | a_0^2 \rangle \tag{13.26}$$

Now use the corresponding path integral representations

$$e^{iW} = \int \frac{da_N^* da_N}{2\pi i} \frac{da_0^{1*} da_0^1}{2\pi i} \frac{da_0^{2*} da_0^2}{2\pi i} \exp \left\{ a_N^* a_N - a_0^{1*} a_0^1 - a_0^{2*} a_0^2 \right\} \langle a_0^1 | \rho (t_i) | a_0^2 \rangle$$
$$\times \int [Da^2]_{N-1}^* \, \exp \left\{ -iS_N^2 \left[a^{2*}, a^2 \right]^* \right\} \int [Da^1]_{N-1} \, \exp \left\{ iS_N^1 \left[a^{1*}, a^1 \right] \right\} \tag{13.27}$$

The configuration on the forward branch has $a^1(0) = a_0^1$ and $a^{1*}(t_f) = a_N^*$. On the backward branch, we have $a^{2*}(0) = a_0^{2*}$ and $a^2(t_f) = a_N$. Once W is known, causal expectation values may be computed by differentiation. Equation (13.27) is the main result of this section.

13.2 The symmetry-breaking approach to BECs

For a field-theoretic description of BECs we begin with a second-quantized field operator $\Psi(\mathbf{x}, t)$ which removes an atom at the location $\mathbf{x}$ at times t. It obeys the canonical commutation relations

$$[\Psi(\mathbf{x}, t), \Psi(\mathbf{y}, t)] = 0 \tag{13.28}$$

$$[\Psi(\mathbf{x}, t), \Psi^\dagger(\mathbf{y}, t)] = \delta(x - y) \tag{13.29}$$

The dynamics of this field is given by the Heisenberg equations of motion

$$-i\hbar \frac{\partial}{\partial t} \Psi = [\mathbf{H}, \Psi] \tag{13.30}$$

where $\mathbf{H}$ is the Hamiltonian. The theory is invariant under a global phase change of the field operator

$$\Psi \to e^{i\theta}\Psi, \qquad \Psi^\dagger \to e^{-i\theta}\Psi^\dagger \tag{13.31}$$

The constant of motion associated with this invariance through Noether's theorem is the total particle number.

To motivate the symmetry-breaking approach to BECs we observe that there is a special one-particle state, with wavefunction ϕ_0, which, upon condensation, acquires a macroscopic occupation number N_0, comparable to the total number of particles N. We call this state the "condensate." We regard ϕ_0 as the first element of a complete basis of one-particle states, and expand $\Psi = a_0\phi_0 + \cdots$. The operator a_0 is the destruction operator for the condensate.

Let $|N, N_0\rangle$ be the state of the gas with N particles, N_0 of which are in the condensate. Then

$$a_0 |N, N_0\rangle = \sqrt{N_0} |N - 1, N_0 - 1\rangle \tag{13.32}$$

If N and N_0 are both very large, then the state does not change much. We see that the condensed state is very close to a coherent state for a_0. Taking the actual state for a coherent state is an excellent approximation when both N and N_0 are macroscopic (but an approximation nonetheless). We shall return to this point below, in Section 13.3.

Under the approximation

$$a_0 |N, N_0\rangle \approx \sqrt{N_0} |N, N_0\rangle \tag{13.33}$$

the expectation value of the field operator is no longer zero

$$\langle \Psi \rangle \equiv \Phi \approx \sqrt{N_0}\phi_0 \tag{13.34}$$

Because the field operator develops an expectation value, the symmetry (13.31) is spontaneously broken. (Beware that the actual relationship between the expectation value and the wavefunction of the condensate is more complex than a simple proportionality, see Section 13.3.)

In the symmetry-breaking approach to BEC dynamics, one relegates this motivation to the background and views the condensation as a resultant of the spontaneous breakdown of symmetry (13.31). Upon symmetry breaking Ψ develops a nonzero expectation value Φ (c-number). We introduce a background field decomposition for Ψ

$$\Psi = \Phi + \psi \tag{13.35}$$

where ψ (q-number) is the field operator corresponding to quantum fluctuations with zero mean $\langle \psi \rangle$. Various approaches differ on how to handle the dynamics of these two constituents.

To progress further, we need a specific model for the atom–atom interactions. In principle, we should specify the atom–atom interaction potential. However, in many applications it is enough to know the cross-section σ for low-energy spherically symmetric scattering of two identical bosons. We introduce the scattering length a through $\sigma \equiv 8\pi a^2$, where the factor 8π involves both integration over all scattering angles and Bose enhancement factors. We shall adopt as a model atom–atom interaction a contact potential $U\delta(\mathbf{x})$. To reproduce the right scattering length we need $U = 4\pi\hbar^2 a/M$, where M is the mass of the atoms. A positive value of a means a repulsive interaction; we adopt the convention that an attractive interaction is described by a negative value of a.

We observe that from the expectation value Φ and the scattering length a it is possible to build a new characteristic length, the healing length ξ, as $\xi^{-2} \equiv a\Phi^2$. Physically, suppose we introduce a condition such as a boundary into the condensate forcing $\Phi = 0$ there. Then ξ is the distance from the boundary where Φ grows back to its asymptotic value. The healing length also plays an important role in the spectrum of fluctuations above the condensate, as we shall show below.

Assuming a contact atom–atom potential we get then the Hamiltonian

$$\mathbf{H} = \int d^d\mathbf{x} \left\{ \Psi^\dagger H \Psi + \frac{U}{2}\Psi^{\dagger 2}\Psi^2 \right\} \tag{13.36}$$

The single-particle Hamiltonian H is given by

$$H\Psi = -\frac{\hbar^2}{2M}\nabla^2\Psi + V_{\text{trap}}(\mathbf{x})\Psi \tag{13.37}$$

where $V_{\text{trap}}(\mathbf{x})$ denotes a confining trap potential. Then the Heisenberg equation of motion

$$i\hbar\frac{\partial}{\partial t}\Psi = H\Psi + U\Psi^\dagger\Psi^2 \tag{13.38}$$

is also the classical equation of motion derived from the action

$$S = \int d^{d+1}x\, i\hbar\Psi^*\frac{\partial}{\partial t}\Psi - \int dt\, \mathbf{H} \tag{13.39}$$

For later use, it is convenient to introduce a single field doublet $\Psi^A = \left(\Psi, \Psi^\dagger\right)$. Recall the Pauli matrices

$$\sigma_1 = \begin{pmatrix} 0 & 1 \\ 1 & 0 \end{pmatrix} \tag{13.40}$$

$$\sigma_2 = \begin{pmatrix} 0 & -i \\ i & 0 \end{pmatrix} \tag{13.41}$$

$$\sigma_3 = \begin{pmatrix} 1 & 0 \\ 0 & -1 \end{pmatrix} \tag{13.42}$$

We also include spatial and temporal position in the A indices (repeated indices are added over discrete indices and integrated over spacetime). The classical action reads (we also set $\hbar = 1$)

$$S = \frac{1}{2}\sigma_{2AB}\Psi^A\frac{\partial}{\partial t}\Psi^B - \frac{1}{2}\sigma_{1AB}\Psi^A H\Psi^B - \frac{U_{ABCD}}{24}\Psi^A\Psi^B\Psi^C\Psi^D \tag{13.43}$$

where

$$\sigma_{iAB} \to \sigma_{iAB} \delta \left(\mathbf{x}_A - \mathbf{x}_B \right) \delta \left(t_A - t_B \right) \tag{13.44}$$

$$U_{ABCD} \to U \left[\sigma_{1AB} \sigma_{1CD} + \sigma_{1AC} \sigma_{1DB} + \sigma_{1AD} \sigma_{1BC} \right]$$
$$\delta \left(\mathbf{x}_A - \mathbf{x}_B \right) \delta \left(\mathbf{x}_A - \mathbf{x}_C \right) \delta \left(\mathbf{x}_A - \mathbf{x}_D \right)$$
$$\delta \left(t_A - t_B \right) \delta \left(t_A - t_C \right) \delta \left(t_A - t_D \right) \tag{13.45}$$

is totally symmetric. $U_{ABCD} = 2U$ if $(ABCD)$ is a permutation of (2211), and zero otherwise. The Heisenberg equations become

$$\sigma_{2AB} \frac{\partial}{\partial t} \Psi^B - \sigma_{1AB} H \Psi^B - \frac{U_{ABCD}}{6} \Psi^B \Psi^C \Psi^D = 0 \tag{13.46}$$

From the expectation value of the Heisenberg equations we find the mean field equation

$$\sigma_{2AB} \frac{\partial}{\partial t} \Phi^B - \sigma_{1AB} H \Phi^B - \eta_A = 0 \tag{13.47}$$

where we parameterize

$$\eta_A = \frac{U_{ABCD}}{6} \left\langle \Psi^C \Psi^D \Psi^B \right\rangle \tag{13.48}$$

We adopt the convention that whenever different operators evaluated at the same time appear within an expectation value, they must be normal ordered. Therefore, in expanded notation

$$\eta_2 = \eta_1^\dagger = U \left\langle \Psi^\dagger \Psi^2 \right\rangle \tag{13.49}$$

In Section 13.2.9 we will relate η to the chemical potential.

The fluctuations around the mean field will be described through the correlation functions

$$\left\langle T \left[\Psi^A \left(t, \mathbf{x} \right) \Psi^B \left(t', \mathbf{y} \right) \right] \right\rangle \equiv \Phi^A \left(t, \mathbf{x} \right) \Phi^B \left(t', \mathbf{y} \right) + G^{AB} \left((t, \mathbf{x}), (t', \mathbf{y}) \right) \tag{13.50}$$

$$G^{AB} = \left\langle T \left[\psi^A \left(t, \mathbf{x} \right) \psi^B \left(t', \mathbf{y} \right) \right] \right\rangle \tag{13.51}$$

These include the so-called normal and anomalous densities

$$\tilde{n} \left(t, \mathbf{x} \right) = \left\langle \psi^\dagger \psi \right\rangle \left(t, \mathbf{x} \right) = G^{21} \left((t, \mathbf{x}), (t, \mathbf{x}) \right) \tag{13.52}$$

$$\tilde{m} \left(t, \mathbf{x} \right) = \left\langle \psi^2 \right\rangle \left(t, \mathbf{x} \right) = G^{11} \left((t, \mathbf{x}), (t, \mathbf{x}) \right) \tag{13.53}$$

The fluctuation field ψ inherits the ETCR

$$\left[\psi^A \left(\mathbf{x}, t \right), \psi^B \left(\mathbf{y}, t \right) \right] = i \sigma_2^{AB} \delta \left(\mathbf{x} - \mathbf{y} \right) \tag{13.54}$$

From the usual formulae

$$\left\langle T \left[\psi^A \left(t, \mathbf{x} \right) \psi^B \left(t', \mathbf{y} \right) \right] \right\rangle = \theta \left(t - t' \right) \left\langle \psi^A \left(t, \mathbf{x} \right) \psi^B \left(t', \mathbf{y} \right) \right\rangle$$
$$+ \theta \left(t' - t \right) \left\langle \psi^B \left(t', \mathbf{y} \right) \psi^A \left(t, \mathbf{x} \right) \right\rangle \tag{13.55}$$

and the equation of motion for the fluctuations (which is obtained by subtracting the mean field from the Heisenberg equations)

$$\sigma_{2AB}\frac{\partial}{\partial t}\psi^B - \sigma_{1AB}H\psi^B - \frac{U_{ABCD}}{6}\Psi^B\Psi^C\Psi^D + \eta_A = 0 \qquad (13.56)$$

the equations of motion for the propagators read

$$0 = \sigma_{2AB}\frac{\partial}{\partial t}G^{\mathrm{BE}} - \sigma_{1AB}HG^{\mathrm{BE}} - \frac{U_{ABCD}}{6}\left\langle T\left(\Psi^C\Psi^D\Psi^B\psi^E\right)\right\rangle - i\delta_A^E \qquad (13.57)$$

which we parameterize as

$$0 = \sigma_{2AB}\frac{\partial}{\partial t}G^{\mathrm{BE}} - \sigma_{1AB}HG^{\mathrm{BE}} - \Sigma_{AB}G^{\mathrm{BE}} - i\delta_A^E \qquad (13.58)$$

$$\Sigma_{AB}G^{\mathrm{BE}} = \frac{U_{ABCD}}{6}\left\langle T\left(\Psi^C\Psi^D\Psi^B\psi^E\right)\right\rangle \qquad (13.59)$$

Let us define the "free" propagators D^{BE} as the solutions to

$$0 = \sigma_{2AB}\frac{\partial}{\partial t}D^{\mathrm{BE}} - \sigma_{1AB}HD^{\mathrm{BE}} - i\delta_A^E \qquad (13.60)$$

Observe that

$$D_{AB}^{-1} = -i\left.\frac{\delta^2 S}{\delta\Phi^A\delta\Phi^B}\right|_{\Phi=0} \qquad (13.61)$$

or, more explicitly,

$$D_{AB}^{-1} = (-i)\begin{pmatrix} 0 & D^{-1*} \\ D^{-1} & 0 \end{pmatrix} \qquad (13.62)$$

$$D^{-1} = i\hbar\partial_t + \frac{\hbar^2}{2M}\nabla^2 - V(\mathbf{x}) \qquad (13.63)$$

We may then write this equation as

$$G_{AB}^{-1} = D_{AB}^{-1} + i\Sigma_{AB} \qquad (13.64)$$

Of course we cannot compute η_A and Σ_{AB} in closed form. Different theories arise from different ansatz for these unknowns as functionals of the mean fields and propagators, thus closing the system.

13.2.1 *A relationship between η_A and Σ_{AB}*

Observe that all of the above remains valid if we consider fields defined on a closed time path. If we need to differentiate among the branches of the time path, we shall make them explicit in the time argument. We shall write t for a generic point on the time path, or else t^a, where $a = 1$ denotes a point on the first (forward) branch, and $a = 2$ a point on the second (backward) one.

One way of generating a vacuum expectation value for the field is coupling it to an external source. The mean field is obtained from the derivatives of a generating functional

$$\Phi^A = \frac{\delta W[J]}{\delta J_A} \tag{13.65}$$

$$e^{iW} = \left\langle e^{i\int J_A \Psi^A} \right\rangle \tag{13.66}$$

In the presence of the sources, the Heisenberg equations now read

$$\sigma_{2AB}\frac{\partial}{\partial t}\Psi^B - \sigma_{1AB}H\Psi^B - \frac{U_{ABCD}}{6}\Psi^B\Psi^C\Psi^D = -J_A \tag{13.67}$$

so taking the expectation value we obtain

$$\sigma_{2AB}\frac{\partial}{\partial t}\Phi^B - \sigma_{1AB}H\Phi^B - \eta_A = -J_A \tag{13.68}$$

Since we have not committed ourselves as to the nature of η_A, this statement is totally general.

We now have the linear response theory result

$$\frac{\delta\Phi^A}{\delta J_E} = iG^{AE} \tag{13.69}$$

whereby

$$i\delta_A^E = \sigma_{2AB}\frac{\partial}{\partial t}G^{BE} - \sigma_{1AB}HG^{BE} - \frac{d\eta_A}{d\Phi^B}G^{BE} \tag{13.70}$$

We use d for the variational derivative of η_A in the last term to emphasize that we mean the *full* derivative. We shall return to this point below. Comparing with (13.58) we see that in the *exact* theory there is a connection

$$\Sigma_{AB} = \frac{d\eta_A}{d\Phi^B} \tag{13.71}$$

Any approximation which does not respect this will get into trouble at some point.

13.2.2 *Gaplessness and phase invariance*

It is a property of the Heisenberg equations that if $\Psi^A = \left(\Psi, \Psi^\dagger\right)$ is a solution, then

$$\exp\left(i\sigma_{3AB}\theta\right)\Psi^B = \left(e^{i\theta}\Psi, e^{-i\theta}\Psi^\dagger\right) \tag{13.72}$$

where θ is a constant, is also a solution. In the exact theory, this property is inherited by the mean field equations, and so the small fluctuations equations must always admit a solution $\delta\Phi^A = \sigma_{3AB}\Phi^B$. This means that the fundamental solutions $-iG^{AB}$ must have a pole.

In equilibrium, time-translation invariance means that Φ^A must have the form

$$\Phi^A = e^{-i\sigma_{3AB}\mu t}\Phi_0^B = \left(e^{-i\mu t}\Phi_0^1, e^{i\mu t}\Phi_0^2\right) \tag{13.73}$$

where Φ_0^B is constant and may be chosen as real, $\Phi_0^1 = \Phi_0^2$. Now recall the mean field equations and write

$$\eta_A = e^{i\sigma_{3AB}\mu t}\eta_{A0} = \left(e^{i\mu t}\eta_{10}, e^{-i\mu t}\eta_{20}\right) \tag{13.74}$$

Then

$$\eta_{10} = \eta_{20} = (\mu - H)\,\Phi_0^1 \tag{13.75}$$

For a homogeneous trap $V(\mathbf{x}) = 0$, Φ_0^1 is a constant and $H\Phi_0^1 = 0$.

The linearized equations are

$$\left[\sigma_{2AB}\frac{\partial}{\partial t} - \sigma_{1AB}H - \Sigma_{AB}\right]\delta\Phi^B = 0 \tag{13.76}$$

The requirement that these equations must admit a solution where $\delta\Phi_A$ is a constant times a simple harmonic factor means that the operator in brackets has a zero, but this is the same as saying that the two-point functions G^{AB} have a pole. Therefore, provided the relationship above between the self-energy Σ_{AB} and the "force" η_A holds, the theory must be gapless.

Actually, substituting $\delta\Phi_A = \sigma_{3AC}e^{-i\sigma_{3CB}\mu t}\Phi_0^B = \left(e^{-i\mu t}\Phi_0, -e^{i\mu t}\Phi_0\right)$ and

$$\Sigma_{AB} = \begin{pmatrix} \Sigma_{11}^0 e^{i\mu(t+t')} & \Sigma_{12}^0 e^{i\mu(t-t')} \\ \Sigma_{21}^0 e^{-i\mu(t-t')} & \Sigma_{22}^0 e^{-i\mu(t+t')} \end{pmatrix} \tag{13.77}$$

implies

$$(\mu - H)\,\Phi_0(\mathbf{x}) - \int dt' d^3\mathbf{y}\,\left[\Sigma_{21}^0 - \Sigma_{22}^0\right]((t,\mathbf{x}),(t',\mathbf{y}))\,\Phi_0(\mathbf{y}) = 0 \tag{13.78}$$

If $V(\mathbf{x}) = 0$, the constant Φ_0 cancels out and we obtain a connection between μ and the Σ_{AB}. This is the Hugenholtz–Pines theorem [HugPin59, Gold61]

$$\mu = \int dt' d^3\mathbf{y}\,\left[\Sigma_{21}^0 - \Sigma_{22}^0\right]((t,\mathbf{x}),(t',\mathbf{y})) \tag{13.79}$$

13.2.3 Conserving and Φ-derivable theories

A theory is called conserving if particle number is conserved in the mean. The theory is called Φ-derivable if there is a functional Φ of Φ^A and G^{AB} such that

$$\eta_A = \frac{\delta\Phi}{\delta\Phi^A} \tag{13.80}$$

$$\Sigma_{AB} = 2\frac{\delta\Phi}{\delta G^{AB}} \tag{13.81}$$

We shall now show that a Φ-derivable theory is necessarily conserving provided the Φ functional is invariant under *time-dependent* phase changes

$$\Phi^A \to \exp\left[i\sigma_{3AB}\theta(t_A)\right]\Phi^B \tag{13.82}$$

$$G^{AB} \to \exp\left[i\sigma_{3AC}\theta(t_A)\right]\exp\left[i\sigma_{3BD}\theta(t_B)\right]G^{CD} \tag{13.83}$$

This is a more demanding requirement than the global phase invariance of the classical action. To see this, introduce a tensor

$$c_{ABC} = \frac{1}{2}\delta\left(t_A - t_B\right)\sigma_{1BC} \tag{13.84}$$

The particle number operator is

$$N_A = c_{ABC}\Psi^B\Psi^C \tag{13.85}$$

Global particle number conservation means that

$$\frac{\partial}{\partial t}\langle N_A\rangle = 0 \tag{13.86}$$

But

$$\frac{\partial}{\partial t}\langle N_A\rangle = c_{ABC}\left[2\Phi^C\frac{\partial\Phi^B}{\partial t_B} + \frac{\partial}{\partial t_B}G^{BC} + \frac{\partial}{\partial t_C}G^{BC}\right] \tag{13.87}$$

By symmetry in the $SU(2)$ indices and a spatial integration by parts the mean field equations imply

$$2c_{ABC}\Phi^C\frac{\partial\Phi^B}{\partial t_B} = i\delta\left(t_A - t_B\right)\sigma_{3BC}\Phi^C\eta_B \tag{13.88}$$

Analogously

$$0 = \frac{\partial}{\partial t_B}G^{BC} + i\sigma_3^{BD}H_DG^{DC} - \sigma_2^{BD}\Sigma_{DE}G^{EC} - i\sigma_2^{BC} \tag{13.89}$$

$$0 = \frac{\partial}{\partial t_C}G^{BC} + i\sigma_3^{DC}H_DG^{BD} + \sigma_2^{EC}\Sigma_{DE}G^{BD} + i\sigma_2^{BC} \tag{13.90}$$

Therefore, a conserving theory must obey

$$0 = i\delta\left(t_A - t_C\right)\left\{\sigma_{3BC}\Phi^C\eta_B + \left[\sigma_3^{DC}G^{EC} + \sigma_3^{EC}G^{CD}\right]\frac{\Sigma_{DE}}{2}\right\} \tag{13.91}$$

which for a Φ-derivable theory is just the invariance statement above.

By extending the symmetry properties of the Φ functional it is possible to enforce energy and momentum conservation as well. The discussion is similar to the general proof of energy–momentum conservation in the mean in relativistic theories, and we shall not repeat it here.

We emphasize that in the exact theory, particle number is strongly conserved, not only in the mean. Strong particle number conservation implies an infinite chain of identities which several correlation functions must obey; in a Φ-derivable theory, they cannot all be satisfied.

Gapless, conserving and Φ-derivable theories

To summarize, Φ-derivable theories are always conserving if the Φ functional is invariant under time-dependent simultaneous phase changes of the mean fields and propagators. They are also gapless if

$$\Sigma_{AB} = 2\frac{\delta\mathbf{\Phi}}{\delta G^{AB}} = \frac{d\eta_A}{d\Phi^B} = \frac{\delta^2\mathbf{\Phi}}{\delta\Phi^A\delta\Phi^B} + \frac{1}{2}\frac{\delta\Sigma_{CD}}{\delta\Phi^A}\frac{dG^{CD}}{d\Phi^B} \qquad (13.92)$$

In the last term, the propagators are regarded as functionals of the mean fields through their equation of motion, namely

$$G_{AB}^{-1} = D_{AB}^{-1} + i\Sigma_{AB} \qquad (13.93)$$

implies

$$\left[G_{EG}^{-1}G_{HF}^{-1} + i\frac{\delta\Sigma_{EF}}{\delta G^{GH}}\right]\frac{dG^{GH}}{d\Phi^B} = (-i)\frac{\delta\Sigma_{EF}}{\delta\Phi^B}, \qquad (13.94)$$

so the condition for a gapless theory becomes

$$\Sigma_{AB} = 2\frac{\delta\mathbf{\Phi}}{\delta G^{AB}} = \frac{\delta^2\mathbf{\Phi}}{\delta\Phi^A\delta\Phi^B} - \frac{i}{2}\frac{\delta\Sigma_{GH}}{\delta\Phi^A}\left[G_{EG}^{-1}G_{HF}^{-1} + i\frac{\delta\Sigma_{EF}}{\delta G^{GH}}\right]^{-1}\frac{\delta\Sigma_{EF}}{\delta\Phi^B} \qquad (13.95)$$

Observe that in general the term in brackets is nonlocal, so either

$$\frac{\delta\Sigma_{GH}}{\delta\Phi^A} = 0 \qquad (13.96)$$

or else the self-energy Σ_{AB} must contain a nonlocal part. This observation will be crucial below.

13.2.4 The full 2PI effective action as a $\mathbf{\Phi}$-derivable approach

In this subsection, we shall discuss the 2PIEA as a $\mathbf{\Phi}$-derivable approach, assuming one knows the full effective action.

As shown in Chapter 6, the 2PIEA is given by [RHCRC04]

$$\Gamma_2\left[\Phi^A, G^{AB}\right] = S\left[\Phi^A\right] + \frac{1}{2}S_{,AB}G^{AB} - \frac{1}{2}i\hbar\text{Tr}\,\ln G + \Gamma_Q \qquad (13.97)$$

where Γ_Q is the sum of all 2PI vacuum bubbles for a theory with propagators G^{AB} and vertices

$$\frac{U_{ABCD}}{24}\psi^A\psi^B\psi^C\psi^D \qquad \text{and} \qquad \frac{U_{ABCD}}{6}\Phi^A\psi^B\psi^C\psi^D \qquad (13.98)$$

The equations of motion are

$$S_{,A} + \frac{1}{2}S_{,ABC}G^{BC} + \frac{\delta\Gamma_Q}{\delta\Phi^A} = 0 \qquad (13.99)$$

$$-iS_{,AB} - 2i\frac{\delta\Gamma_Q}{\delta G^{AB}} = \left[G^{-1}\right]_{AB} \qquad (13.100)$$

Therefore

$$\eta_A = \frac{U_{ABCD}}{6}\Phi^B\Phi^C\Phi^D + \frac{U_{ABCD}}{2}\Phi^B G^{CD} - \frac{\delta\Gamma_Q}{\delta\Phi^A} \qquad (13.101)$$

$$\Sigma_{AB} = \frac{U_{ABCD}}{2}\Phi^C\Phi^D - 2\frac{\delta\Gamma_Q}{\delta G^{AB}} \qquad (13.102)$$

which follow from the functional

$$\Phi = \frac{U_{ABCD}}{24} \Phi^A \Phi^B \Phi^C \Phi^D + \frac{U_{ABCD}}{4} \Phi^A \Phi^B G^{CD} - \Gamma_Q \qquad (13.103)$$

Conservation follows from the fact that Γ_Q is made out of graphs where the same number of 1 and 2 fields enter at each vertex.

Since we are assuming Γ_Q contains all graphs, the theory must be gapless. Nevertheless, it is interesting to seek a direct proof. We must verify the identity (13.95). To do this, let us put back the external sources in the equations of motion

$$iD_{AB}^{-1} \Phi^B - \eta_A = -J_A - K_{AB} \Phi^B \qquad (13.104)$$

$$iD_{AB}^{-1} - i\left[G^{-1}\right]_{AB} - \Sigma_{AB} = -K_{AB} \qquad (13.105)$$

Taking variations we get

$$iD_{AB}^{-1} \frac{\delta \Phi^B}{\delta J_C} - \frac{\delta \eta_A}{\delta \Phi^B} \frac{\delta \Phi^B}{\delta J_C} - \frac{\delta \eta_A}{\delta G^{BD}} \frac{\delta G^{BD}}{\delta J_C} = -\delta_A^C \qquad (13.106)$$

$$i\left[\left[G^{-1}\right]_{AD} \left[G^{-1}\right]_{EB} + i\frac{\delta \Sigma_{AB}}{\delta G^{DE}}\right] \frac{\delta G^{DE}}{\delta J_C} - \frac{\delta \Sigma_{AB}}{\delta \Phi^D} \frac{\delta \Phi^D}{\delta J_C} = 0 \qquad (13.107)$$

In any Φ-derivable approach,

$$\frac{\delta \eta_A}{\delta G^{BD}} = \frac{1}{2} \frac{\delta \Sigma_{BD}}{\delta \Phi^A} \qquad (13.108)$$

and we still have the LRT result (13.69), from which we get

$$\frac{\delta G^{DE}}{\delta J_C} = \left[\left[G^{-1}\right]_{AD} \left[G^{-1}\right]_{EB} + i\frac{\delta \Sigma_{AB}}{\delta G^{DE}}\right]^{-1} \frac{\delta \Sigma_{AB}}{\delta \Phi^F} G^{FC} \qquad (13.109)$$

so the first equation (13.106) becomes

$$\left[G^{-1}\right]_{AB} = D_{AB}^{-1} + i\frac{\delta \eta_A}{\delta \Phi^B}$$

$$+ \frac{1}{2} \frac{\delta \Sigma_{FD}}{\delta \Phi^A} \left[\left[G^{-1}\right]_{GD} \left[G^{-1}\right]_{EF} + i\frac{\delta \Sigma_{GE}}{\delta G^{DF}}\right]^{-1} \frac{\delta \Sigma_{GE}}{\delta \Phi^B} \qquad (13.110)$$

QED

This shows that, *in principle*, the 2PIEA yields a theory which is *both* gapless and conserving. In reality, though, one does not known the full effective action, and truncations may spoil either of these features, or both.

13.2.5 *Varieties of theories from truncations of the 2PI effective action*

Let us expand on this last statement by looking at some common approaches to nonequilibrium BECs as truncations of the 2PIEA. For simplicity, in this section we assume $V(\mathbf{x}) = 0$.

(1) The simplest, and surprisingly useful, approach is the **Gross–Pitaevskii** **(GP)** one: just write the classical equations of motion for Φ, and forget about G. However, this approach is incomplete, because it says nothing about fluctuations.

(2) The next simplest approach is **Bogoliubov**'s, which is based on the identifications

$$\eta_A^{\text{Bog}} = \frac{U_{ABCD}}{6} \Phi^B \Phi^C \Phi^D \tag{13.111}$$

$$\Sigma_{AB}^{\text{Bog}} = \frac{U_{ABCD}}{2} \Phi^C \Phi^D \tag{13.112}$$

or, in expanded notation

$$\eta_2^{0\text{Bog}} = U\Phi_0^3 \tag{13.113}$$

$$\Sigma_{21}^{0\text{Bog}} = 2\Sigma_{22}^{0\text{Bog}} = 2U\Phi_0^2 \delta\left(t - t'\right) \delta\left(\mathbf{x} - \mathbf{y}\right) \tag{13.114}$$

Here, the mean fields obey the Gross–Pitaevskii equation

$$i\frac{\partial}{\partial t}\Phi = H\Phi + U\Phi^\dagger \Phi^2 \tag{13.115}$$

and the fluctuations the linearized equation

$$i\frac{\partial}{\partial t}\psi = H\psi + U\psi^\dagger \Phi^2 + 2U\Phi^\dagger \Phi\psi \tag{13.116}$$

Write $\Phi = e^{-i\mu t}\Phi_0$, $\psi = e^{-i\mu t}\psi_{\text{phys}}$ to get

$$\mu = U\Phi_0^2 \tag{13.117}$$

$$i\frac{\partial}{\partial t}\psi_{\text{phys}} = H\psi_{\text{phys}} + U\Phi_0^2 \left[\psi_{\text{phys}}^\dagger + \psi_{\text{phys}}\right] \tag{13.118}$$

Bogoliubov's approach is not Φ-derivable; however, it is gapless, because the equation for the propagators is defined to be identical to the first variation of the equation for the mean fields, and this is phase invariant. Equivalently, we see that the Bogoliubov approach is consistent with the Hugenholtz–Pines theorem.

The Bogoliubov approach is not conserving. This may be seen from the analysis above, but it is probably simplest to give a direct proof. Since the equation for the mean field is just the classical equation, its contribution to particle number is conserved, so the only question is about the number of particles in the fluctuation field. From the equations above we find

$$\frac{d\langle N\rangle}{dt} = (-i)\,U\Phi_0^2 \int d^d\mathbf{x}\, \left[\left\langle\psi_{\text{phys}}^{\dagger 2}\right\rangle - \left\langle\psi_{\text{phys}}^2\right\rangle\right] \tag{13.119}$$

which does not vanish identically.

The simplest Φ-derivable extension of the Bogoliubov approach is the one-loop theory, where $\Gamma_Q = 0$

$$\Phi^{1\,\text{loop}} = \frac{U_{ABCD}}{24} \Phi^A \Phi^B \Phi^C \Phi^D + \frac{U_{ABCD}}{4} \Phi^A \Phi^B G^{CD} \tag{13.120}$$

From the above analysis, one loop is obviously conserving, but it is not gapless. This can be seen from the fact that Σ_{AB} is purely local, while to satisfy the gapless condition it must also include nonlocal terms.

Alternatively, we can check that the one-loop approximation violates the Hugenholtz–Pines theorem. The one-loop self-energies are the same as in the Bogoliubov approach, but the forces are different

$$\eta_2^{01\text{loop}} = \left[U\Phi_0^2 + U\left(2\tilde{n} + \tilde{m}\right) \right] \Phi_0 \tag{13.121}$$

leading to

$$\mu = U\Phi_0^2 + U\left(2\tilde{n} + \tilde{m}\right) \tag{13.122}$$

(3) The so-called **Hartree–Fock–Bogoliubov** approximation (HFB) is another local Φ-derivable approach, where Γ_Q is reduced to the double-bubble diagram

$$\Phi^{\text{HFB}} = \Phi^{1\,\text{loop}} + \frac{U_{ABCD}}{8} G^{AB} G^{CD} \tag{13.123}$$

HFB is conserving but not gapless, for the same reasons as the one-loop approach. The HFB forces are the same as in the one-loop approach, while the self-energies are

$$\Sigma_{22}^{0\text{HFB}} = U\left(\Phi_0^2 + \tilde{m}\right) \delta\left(t - t'\right) \delta\left(\mathbf{x} - \mathbf{y}\right) \tag{13.124}$$

$$\Sigma_{21}^{0\text{HFB}} = 2U\left(\Phi_0^2 + \tilde{n}\right) \delta\left(t - t'\right) \delta\left(\mathbf{x} - \mathbf{y}\right) \tag{13.125}$$

Observe that the Hugenholtz–Pines theorem is violated because of the $\tilde{m}$ term. This suggests a simple way to modify HFB so that it becomes gapless, though no longer conserving. In the HFB approach, the equations for the mean fields are

$$i\frac{\partial}{\partial t} \Phi = H\Phi + U\Phi^{\dagger}\Phi^2 + 2U \langle \psi^{\dagger}\psi \rangle \Phi + U \langle \psi^2 \rangle \Phi^{\dagger} \tag{13.126}$$

and the fluctuations obey the linearized equation

$$i\frac{\partial}{\partial t} \psi = H\psi + U\psi^{\dagger}\left(\Phi^2 + \langle \psi^2 \rangle\right) + 2U\left(\Phi^{\dagger}\Phi + \langle \psi^{\dagger}\psi \rangle\right)\psi \tag{13.127}$$

(4) In the so-called **Popov** approximation, one neglects the "anomalous" density in both equations (13.126), (13.127). Writing $\Phi = e^{-i\mu t}\Phi_0$, $\psi = e^{-i\mu t}\psi_{\text{phys}}$, we get

$$\mu = U\left[\Phi_0^2 + 2\langle \psi^{\dagger}\psi \rangle\right] \tag{13.128}$$

$$i\frac{\partial}{\partial t}\psi_{\text{phys}} = H\psi_{\text{phys}} + U\Phi_0^2\left[\psi_{\text{phys}}^{\dagger} + \psi_{\text{phys}}\right] \tag{13.129}$$

which is easily verified to give a gapless spectrum (see the next subsection). The first equation is the Hugenholtz–Pines theorem reduced to this approximation.

The spectrum under the Popov approximation

We now investigate more closely the spectrum which results from the Popov approximation. To this end, let us reinstate $\hbar$ into the equation, and assume a homogeneous condensate in a three-dimensional normalizing box of volume V. Φ_0 is a constant, and ψ_{phys} may be expanded

$$\psi_{\text{phys}} = \sum_{\mathbf{k}} \frac{e^{ikx}}{\sqrt{V}} \psi_{\mathbf{k}}(t) \tag{13.130}$$

$$i\hbar \frac{\partial}{\partial t} \psi_{\mathbf{k}} = \frac{\hbar^2 k^2}{2M} \psi_{\mathbf{k}} + U\Phi_0^2 \left[\psi_{-\mathbf{k}}^\dagger + \psi_{\mathbf{k}} \right] \tag{13.131}$$

We seek a solution

$$\psi_{\mathbf{k}} = \alpha_k A_{\mathbf{k}} e^{-i\omega_k t} - \beta_k A_{-\mathbf{k}}^\dagger e^{i\omega_k t} \tag{13.132}$$

where α_k and β_k are real and spherically symmetric, and $\alpha_k^2 - \beta_k^2 = 1$. Collecting positive and negative frequency terms we get

$$\left(\frac{\hbar^2 k^2}{2M} + U\Phi_0^2 - \hbar\omega_k \right) \alpha_k - U\Phi_0^2 \beta_k = 0 \tag{13.133}$$

$$U\Phi_0^2 \alpha_k - \left(\frac{\hbar^2 k^2}{2M} + U\Phi_0^2 + \hbar\omega_k \right) \beta_k = 0 \tag{13.134}$$

leading to the dispersion relation

$$\hbar^2 \omega_{\mathbf{k}}^2 = \left(\frac{\hbar^2 k^2}{2M} + U\Phi_0^2 \right)^2 - U^2 \Phi_0^4 \tag{13.135}$$

and to a gapless spectrum, as expected. Let us write

$$\frac{\hbar^2 k^2}{2M} + U\Phi_0^2 = \hbar\omega_k \cosh 2\varphi \tag{13.136}$$

$$U\Phi_0^2 = \hbar\omega_k \sinh 2\varphi \tag{13.137}$$

Then

$$\alpha_k = \cosh\varphi, \qquad \beta_k = \sinh\varphi \tag{13.138}$$

It follows that

$$\tanh\varphi = \frac{U\Phi_0^2}{\left(\frac{\hbar^2 k^2}{2M} + U\Phi_0^2 + \hbar\omega_k \right)} \tag{13.139}$$

and so

$$\beta_k = \frac{U\Phi_0^2}{\left[\left(\frac{\hbar^2 k^2}{2M} + U\Phi_0^2 + \hbar\omega_k \right)^2 - U^2 \Phi_0^4 \right]^{1/2}} \tag{13.140}$$

Let us introduce the scattering length a through $U \sim \hbar^2 a/M$ and the healing length ξ through $a\Phi_0^2 = \xi^{-2}$. Then the dispersion relation reads

$$\omega_k = c_{\mathrm{s}} k \sqrt{1 + \frac{1}{4}\left(\xi k\right)^2} \qquad (13.141)$$

where the speed of sound is

$$c_{\mathrm{s}} = \frac{\hbar}{M\xi} \qquad (13.142)$$

We see that there are roughly two set of modes, *hard* modes with $k > \xi^{-1}$ which remain mostly undisturbed by the condensate, and soft modes $k < \xi^{-1}$ which lose their particle-like character and become phonon-like.

The above analysis does not apply to the homogeneous mode, which by construction has zero frequency. Therefore the amplitude of the zero mode will grow linearly in time, and eventually it will invalidate perturbation theory. The point is that within the symmetry-breaking approach this is unavoidable. Of course the zero mode is physically different from other modes, being closer to a collective variable [Raj87] than to a true physical degree of freedom. Therefore it is justified to treat it in a different way than other modes [MCBE98, SiCaWi06]. But when we do so we move beyond the symmetry-breaking approach. A possible strategy is the particle conserving formulation, to be discussed later in this chapter.

The equation obtained from the Popov approximation may be used to clarify one important point, namely, what the physical small parameter in the loop expansion is. As a measure of the size of the higher corrections, let us compare the density of noncondensate particles $\tilde{n} = \left\langle \psi_{\mathrm{phys}}^\dagger \psi_{\mathrm{phys}} \right\rangle$ against the condensate density Φ_0^2. In the continuous approximation

$$\tilde{n} = \int d^3\mathbf{k}\, \beta_{\mathbf{k}}^2 = 4\pi \int_0^\infty dk\, k^2 \beta_{\mathbf{k}}^2 \qquad (13.143)$$

The integral converges at both limits. For small k, $\omega_{\mathbf{k}} \sim k \left[U\Phi_0^2/M\right]^{1/2}$ and

$$\beta_{\mathbf{k}}^2 \sim \left(MU\Phi_0^2\right)^{1/2}/\hbar k \qquad (13.144)$$

Replacing $U \sim \hbar^2 a/M$ we get $k^2 \beta_{\mathbf{k}}^2 \sim a^{-2} (ka) \left(a^{3/2}\Phi_0\right)$. In the opposite limit of large k, we have $\hbar\omega_{\mathbf{k}} \sim \hbar^2 k^2/2M$ and $k^2 \beta_{\mathbf{k}}^2 \sim a^{-2} (ka)^{-2} \left(a^{3/2}\Phi_0\right)^4$. The largest contribution comes from the cross-over region where $k \sim a^{-1}\left(a^{3/2}\Phi_0\right)$.

The resulting estimate yields $\tilde{n} \sim \Phi_0^2 \left(a^{3/2}\Phi_0\right)$. We see that the physical small parameter in the expansion is $\sqrt{N_a}$, where $N_a = a^3\Phi_0^2 \sim a^3 N/V$ is the number of particles within a scattering length of a given particle. The loop expansion is therefore a dilute gas approximation.

13.2.6 Higher gapless approximations

We see from the previous discussion that, while the 2PIEA yields a theory which is truly both gapless and conserving, in practice truncations of the effective action lead to approaches where one or the other feature must be sacrificed. To prevent this, we must stick to approximations to the 2PIEA which satisfy the gapless condition

$$
-2\frac{\delta \Gamma_Q}{\delta G^{AB}} = \frac{U_{ABCD}}{2}G^{CD} - \frac{\delta \Gamma_Q}{\delta \Phi^A \delta \Phi^B}
$$

$$
-\frac{i}{2}\left[U_{GHAD}\Phi^D - 2\frac{\delta \Gamma_Q}{\delta G^{GH}\delta \Phi^A}\right]\left[G_{EG}^{-1}G_{HF}^{-1} - 2i\frac{\delta^2 \Gamma_Q}{\delta G^{EF}\delta G^{GH}}\right]^{-1}
$$

$$
\times \left[U_{EFBJ}\Phi^J - 2\frac{\delta \Gamma_Q}{\delta G^{EF}\delta \Phi^B}\right] \tag{13.145}
$$

This nonlinear equation in functional derivatives of Γ_Q is too complex to admit a closed-form solution, but it can be solved iteratively: we start by replacing some value of $\Gamma_Q^{(n)}$ on the right-hand side, and find $\Gamma_Q^{(n+1)}$ by one integration with respect to G. We thereby generate a family of theories which are gapless within a prescribed accuracy.

Choosing as starting point the Bogoliubov approximation $\Gamma_Q^{(1)} = 0$, we obtain the first nontrivial approximation

$$
\Gamma_Q^{(2)} = -\frac{U_{ABCD}}{8}G^{AB}G^{CD} + \frac{i}{12}U_{GHAD}\Phi^D G^{AB}G^{GE}G^{HF}U_{EFBJ}\Phi^J \tag{13.146}
$$

which is the full two-loop approximation to the 2PIEA, including the double-bubble and setting sun graphs. This approximation was first explored by Beliaev [Bel58a, Bel58b].

We note that other approximation schemes have been explored in the literature, most notably the $1/N_f$ expansion in a theory with N_f "flavors" or equivalent Bose fields. If going over to a nonlocal approximation is considered too involved, another possibility is to depart from the 2PIEA approach, adding *ad hoc* terms, for example, to restore gaplessness in an otherwise conserving theory.

As we have discussed in detail in earlier chapters, the two-loop approximation leads to self-energies which are in general complex, signaling damping of the condensate fluctuations. At zero temperature, the leading damping mechanism is the decay of a condensate fluctuation into two noncondensate excitations. This so-called Beliaev damping [Bel58a, Bel58b] has been discussed in detail in Chapter 8, in the simpler context of a $g\phi^3$ scalar field theory. At finite temperature a new mechanism appears, the so-called Landau damping where a condensate fluctuation is absorbed by a noncondensate excitation, which transmutes into a higher energy excitation. The imaginary parts of the thermal self-energy have been discussed in Chapter 10, to which we refer the reader for details.

Finally, we observe that for cold gases in an optical lattice, gaplessness may actually become a problem, if one is interested in describing the Mott regime. Let us return to $\Gamma_Q^{(2)} = \Gamma_Q^{\text{HFB}} + \delta\Gamma_Q$. In a more natural notation,

$$\delta\Gamma_Q = \frac{iU^2}{2} \left\langle \left\{ \int dt d^3\mathbf{x} \left[\Phi^* \psi^\dagger \psi^2 + \Phi \psi^{\dagger 2} \psi \right] \right\}^2 \right\rangle \tag{13.147}$$

where the expectation value is computed under a Gaussian approximation and only 2PI terms are kept. Recall that the time integration runs over the closed time path, and that products of fields are path ordered, or normal ordered if the path ordering prescription is ambiguous. Expanding

$$\delta\Gamma_Q = \frac{iU^2}{2} \int dt d^3\mathbf{x} \int dt' d^3\mathbf{y}$$

$$\times \left\{ \Phi^* (t, \mathbf{x}) \, \Phi^* (t', \mathbf{y}) \left[2 \langle \psi^\dagger \psi^{\dagger\prime} \rangle \langle \psi\psi' \rangle^2 + 4 \langle \psi^\dagger \psi' \rangle \langle \psi\psi^{\dagger\prime} \rangle \langle \psi\psi' \rangle \right] \right.$$

$$+ \Phi^* (t, \mathbf{x}) \, \Phi (t', \mathbf{y}) \left[4 \langle \psi^\dagger \psi^{\dagger\prime} \rangle \langle \psi\psi' \rangle \langle \psi\psi^{\dagger\prime} \rangle + 2 \langle \psi^\dagger \psi' \rangle \langle \psi\psi^{\dagger\prime} \rangle^2 \right]$$

$$+ \Phi (t, \mathbf{x}) \, \Phi^* (t', \mathbf{y}) \left[2 \langle \psi\psi^{\dagger\prime} \rangle \langle \psi^\dagger \psi' \rangle^2 + 4 \langle \psi\psi' \rangle \langle \psi^\dagger \psi' \rangle \langle \psi^\dagger \psi^{\dagger\prime} \rangle \right]$$

$$\left. + \Phi (t, \mathbf{x}) \, \Phi (t', \mathbf{y}) \left[2 \langle \psi^\dagger \psi^{\dagger\prime} \rangle^2 \langle \psi\psi' \rangle + 4 \langle \psi^\dagger \psi' \rangle \langle \psi\psi^{\dagger\prime} \rangle \langle \psi^\dagger \psi^{\dagger\prime} \rangle \right] \right\} \tag{13.148}$$

Although the model is built to be gapless to $O\left(U^2\right)$, it is interesting to give a direct check. We consider only the zero temperature case. Observe that

$$\eta_2 = \eta_2^{\text{HFB}} + \delta\eta_2 \tag{13.149}$$

$$\delta\eta_2 = iU^2 \int dt' d^3\mathbf{y}$$

$$\times \left\{ \Phi^* (t', \mathbf{y}) \left[2 \langle \psi^\dagger \psi^{\dagger\prime} \rangle \langle \psi\psi' \rangle^2 + 4 \langle \psi^\dagger \psi' \rangle \langle \psi\psi^{\dagger\prime} \rangle \langle \psi\psi' \rangle \right] \right.$$

$$\left. + \Phi (t', \mathbf{y}) \left[4 \langle \psi^\dagger \psi^{\dagger\prime} \rangle \langle \psi\psi' \rangle \langle \psi\psi^{\dagger\prime} \rangle + 2 \langle \psi^\dagger \psi' \rangle \langle \psi\psi^{\dagger\prime} \rangle^2 \right] \right\} \tag{13.150}$$

In equilibrium, and after extracting the phases, this leads to a chemical potential

$$\mu = \mu^{\text{HFB}} + \delta\mu \tag{13.151}$$

with

$$\delta\mu = iU^2 \int dt' d^3\mathbf{y} \left[2 \langle \psi^\dagger \psi^{\dagger\prime} \rangle \langle \psi\psi' \rangle^2 + 4 \langle \psi^\dagger \psi' \rangle \langle \psi\psi^{\dagger\prime} \rangle \langle \psi\psi' \rangle \right.$$

$$\left. + 4 \langle \psi^\dagger \psi^{\dagger\prime} \rangle \langle \psi\psi' \rangle \langle \psi\psi^{\dagger\prime} \rangle + 2 \langle \psi^\dagger \psi' \rangle \langle \psi\psi^{\dagger\prime} \rangle^2 \right] \tag{13.152}$$

To $O(U^2)$ we may compute the expectation values as pertaining to a free field. At zero temperature $\delta\mu$ vanishes.

Similarly we compute the self-energies

$$\Sigma_{AB} = \Sigma_{AB}^{\text{HFB}} + \delta\Sigma_{AB} \tag{13.153}$$

$$\delta\Sigma_{21}^0 = -2iU^2\Phi_0^2 \left\langle \psi\psi^{\dagger\prime} \right\rangle^0 \left[2\left\langle \psi\psi^\prime \right\rangle^0 + \left\langle \psi\psi^{\dagger\prime} \right\rangle^0 + 2\left\langle \psi^\dagger\psi^\prime \right\rangle + 2\left\langle \psi^\dagger\psi^{\dagger\prime} \right\rangle^0 \right] \tag{13.154}$$

$$\delta\Sigma_{22}^0 = -2iU^2\Phi_0^2 \left\langle \psi\psi^\prime \right\rangle^0 \left[\left\langle \psi\psi^\prime \right\rangle^0 + 2\left\langle \psi\psi^{\dagger\prime} \right\rangle^0 + 2\left\langle \psi^\dagger\psi^\prime \right\rangle^0 + 2\left\langle \psi^\dagger\psi^{\dagger\prime} \right\rangle^0 \right] \tag{13.155}$$

In equilibrium, $\delta\Sigma_{22}^0 = 0$ and

$$\delta\Sigma_{21}^0 = -2iU^2\Phi_0^2 \left(\left\langle \psi\psi^{\dagger\prime} \right\rangle^0 \right)^2 \tag{13.156}$$

The gaplessness condition reads

$$2U\tilde{m} = \int dt^\prime d^3\mathbf{y} \; \delta\Sigma_{21}^0 \tag{13.157}$$

to lowest order in U.

To compute the left-hand side we expand the destruction operators as

$$\psi^0 = \sum \frac{e^{i(\mathbf{kx}-\omega_k t)}}{\sqrt{V}} a_\mathbf{k} \tag{13.158}$$

where

$$\omega_k = \frac{\hbar k^2}{2M} \tag{13.159}$$

Therefore, after separating the contributions from both branches of the closed time path

$$\int dt^\prime d^3\mathbf{y} \; \delta\Sigma_{21}^0 = -\frac{2U^2\Phi_0^2}{V} \sum_{\mathbf{p},\mathbf{q}} \frac{\delta_{\mathbf{p}+\mathbf{q}}}{\omega_p + \omega_q} = -\frac{U^2\Phi_0^2}{V} \sum_\mathbf{p} \frac{1}{\omega_p} \tag{13.160}$$

On the other hand, at zero temperature

$$\tilde{m} = \frac{-1}{V} \sum_\mathbf{p} \alpha_p \beta_p \tag{13.161}$$

To lowest order we have $\alpha_p = 1$

$$\beta_p = \frac{U\Phi_0^2}{2\hbar\omega_p} \tag{13.162}$$

QED

13.2.7 Damping

The fact that under the above approximation there are nonlocal terms in the equations of motion for both the mean field and propagators suggest that they already include damping effects. Indeed, this has been proved by Beliaev [Bel58a, Bel58b].

Let us consider the evolution of a mean field fluctuation $e^{-i\mu t}\delta\Phi$. The linearized equation of motion is

$$i\frac{\partial}{\partial t}\delta\Phi = (H - \mu)\,\delta\Phi + U\Phi_0^2\left[\delta\Phi^\dagger + 2\delta\Phi\right] + 2U\left\langle\psi^\dagger\psi\right\rangle_0\delta\Phi + U\left\langle\psi^2\right\rangle_0\delta\Phi^\dagger$$
$$+ U\Phi_0\left[2\delta\left\langle\psi^\dagger\psi\right\rangle + \delta\left\langle\psi^2\right\rangle\right] + \delta\eta_2\left[\delta\Phi\right] \tag{13.163}$$

We see that there are two types of nonlocal terms, the terms coming from the modification of the fluctuating field propagators, and terms from the second variation of the effective action. The former will be shown to be proportional to $U\Phi_0^2$ and therefore will dominate at low temperatures, where almost all particles are condensed. Conversely, we expect the direct variation terms to dominate immediately below the critical temperature. We consider only the former case.

Since the perturbed propagators appear already in $O(U)$ terms, we only need to compute them to $O(U)$ accuracy. At this level, it is enough to consider the Heisenberg equation

$$i\frac{\partial}{\partial t}\psi_{\text{phys}} = (H - \mu)\,\psi_{\text{phys}} + U\psi_{\text{phys}}^\dagger\left(\Phi_0^2 + 2\Phi_0\delta\Phi\right)$$
$$+ 2U\left(\Phi_0^2 + \Phi_0\left(\delta\Phi^\dagger + \delta\Phi\right) + \left\langle\psi^\dagger\psi\right\rangle_0\right)\psi_{\text{phys}} \tag{13.164}$$

To the desired order

$$\mu = U\left[\Phi_0^2 + 2\left\langle\psi^\dagger\psi\right\rangle\right] \tag{13.165}$$

and

$$i\frac{\partial}{\partial t}\psi_{\text{phys}} = H\psi_{\text{phys}} + U\psi_{\text{phys}}^\dagger\left(\Phi_0^2 + 2\Phi_0\delta\Phi\right)$$
$$+ U\left(\Phi_0^2 + 2\Phi_0\left(\delta\Phi^\dagger + \delta\Phi\right)\right)\psi_{\text{phys}} \tag{13.166}$$

Let us write

$$\psi_{\text{phys}} = \psi_{\text{phys}}^{\text{eq}} + \delta\psi \tag{13.167}$$

where

$$\psi_{\text{phys}}^{\text{eq}} = \sum\frac{e^{ikx}}{\sqrt{V}}\left[A_\mathbf{k}e^{-i\omega_k t} - \frac{U\Phi_0^2}{2\hbar\omega_p}A_{-\mathbf{k}}^\dagger e^{i\omega_k t}\right] \tag{13.168}$$

and expand

$$\delta\Phi = \int\frac{d\omega}{2\pi}\sum_k\frac{e^{i(\mathbf{k}\mathbf{x}-\omega t)}}{\sqrt{V}}f_\mathbf{k}\left(\omega\right) \tag{13.169}$$

Keeping only up to $O(U)$ terms

$$i\frac{\partial}{\partial t}\delta\psi - H\delta\psi = 2\Phi_0 U\left[\psi_{\text{phys}}^{\text{eq}\dagger}\delta\Phi + \left(\delta\Phi^\dagger + \delta\Phi\right)\psi_{\text{phys}}^{\text{eq}}\right] \tag{13.170}$$

and so

$$\delta\psi = 2\Phi_0 U \int \frac{d\omega}{2\pi} \sum_{\mathbf{p},\mathbf{q}} \frac{e^{i(\mathbf{p}+\mathbf{q})\mathbf{x}}}{V}$$

$$\times \left\{ \frac{f_{\mathbf{p}}(\omega) A^\dagger_{-\mathbf{q}} e^{-i(\omega-\omega_q)t}}{\omega - \omega_q - \omega_{|\mathbf{p}+\mathbf{q}|} + i\varepsilon} + \frac{\left[f_{\mathbf{p}}(\omega) + f^*_{-\mathbf{p}}(-\omega)\right] A_{\mathbf{q}} e^{-i(\omega+\omega_q)t}}{\omega + \omega_q - \omega_{|\mathbf{p}+\mathbf{q}|} + i\varepsilon} \right\}$$

$$(13.171)$$

We may now compute

$$\delta \left\langle \psi^\dagger \psi \right\rangle = \left\langle \psi^{\text{eq}\dagger}_{\text{phys}} \delta\psi \right\rangle + \left\langle \delta\psi^\dagger \psi^{\text{eq}}_{\text{phys}} \right\rangle = O\left(U^2\right) \qquad (13.172)$$

$$\delta \left\langle \psi^2 \right\rangle = \left\langle \psi^{\text{eq}}_{\text{phys}} \delta\psi \right\rangle = 2\Phi_0 U \int \frac{d\omega}{2\pi} \sum_{\mathbf{p},\mathbf{q}} \frac{e^{i(\mathbf{p}\mathbf{x}-\omega t)}}{V^{3/2}} \frac{f_{\mathbf{p}}(\omega)}{\omega - \omega_q - \omega_{|\mathbf{p}+\mathbf{q}|} + i\varepsilon}$$

$$(13.173)$$

The equation for the fluctuation is then

$$i\frac{\partial}{\partial t}\delta\Phi = H\delta\Phi + U\Phi_0^2\left[\delta\Phi^\dagger + \delta\Phi\right] + U\left\langle\psi^2\right\rangle_0 \delta\Phi^\dagger + U\Phi_0\delta\left\langle\psi^2\right\rangle \qquad (13.174)$$

or, after Fourier transformation,

$$\left[\omega - \omega_p - U\Phi_0^2 - \frac{2U^2\Phi_0^2}{V}\sum_{\mathbf{q}} \frac{1}{\left(\omega - \omega_q - \omega_{|\mathbf{p}+\mathbf{q}|} + i\varepsilon\right)}\right] f_{\mathbf{p}}(\omega)$$

$$-U\left[\Phi_0^2 + \tilde{m}\right] f^*_{-\mathbf{p}}(-\omega) = 0 \qquad (13.175)$$

Changing $\mathbf{p} \to -\mathbf{p}$, $\omega \to -\omega$ and conjugating we find the second equation

$$U\left[\Phi_0^2 + \tilde{m}\right] f_{\mathbf{p}}(\omega)$$

$$+ \left[\omega + \omega_p + U\Phi_0^2 - \frac{2U^2\Phi_0^2}{V}\sum_{\mathbf{q}} \frac{1}{\left(\omega + \omega_q + \omega_{|\mathbf{p}+\mathbf{q}|} + i\varepsilon\right)}\right] f^*_{-\mathbf{p}}(-\omega) = 0$$

$$(13.176)$$

Up to $O(U^2)$ the secular equation is

$$0 = \omega^2 - \left(\omega_p + U\Phi_0^2\right)^2 - \frac{2U^2\Phi_0^2}{\hbar^2 V}\left(\omega - \omega_p\right)\sum_{\mathbf{q}} \frac{1}{\left(\omega + \omega_q + \omega_{|\mathbf{p}+\mathbf{q}|} + i\varepsilon\right)}$$

$$- \frac{2U^2\Phi_0^2}{\hbar^2 V}\left(\omega + \omega_p\right)\sum_{\mathbf{q}} \frac{1}{\left(\omega - \omega_q - \omega_{|\mathbf{p}+\mathbf{q}|} + i\varepsilon\right)} + U^2\Phi_0^4 \qquad (13.177)$$

We expect that the solution will be close to ω_p, but if there is a $\mathbf{q}$ such that $\omega_p \sim \omega_q + \omega_{\mathbf{p}+\mathbf{q}}$ then the $O(U^2)$ terms become large and perturbation theory breaks down. What is going on is that the free evolution of condensate fluctuations cannot be described as oscillations with a small number of fundamental frequencies. This is clearly seen in the continuum limit, where we may replace

$$\frac{1}{V}\sum_{\mathbf{q}} \rightarrow \int \frac{d^3\mathbf{q}}{(2\pi)^3} \tag{13.178}$$

The resulting integrals have an imaginary part and the frequencies for the free evolution of condensate fluctuations become complex, $\omega \sim \omega_p - i\Gamma$

$$\Gamma \sim \frac{U^2\Phi_0^2 M}{8\pi\hbar^3}p \sim \left(\frac{a}{\xi}\right)c_{\mathrm{s}}p \tag{13.179}$$

The underlying mechanism is that the energy of a condensate fluctuation carrying momentum $\mathbf{p}$ is spent in exciting two particles out of the condensate, one of momentum $-\mathbf{q}$ and another of momentum $\mathbf{p}+\mathbf{q}$. Of course, this mechanism requires the presence of a condensate. The term $\delta\eta_2[\delta\Phi]$ contains additional channels describing the direct decay of the condensate fluctuation into three particles.

Also, we have assumed that the mode p was hard enough that it fell into the "particle-like" part of the spectrum. In practice, damping is very sensitive to the shape of the dispersion relation and to the number of spatial dimensions [TsuGri03, TsuGri05, Rob05, RHCC05]. A more detailed calculation shows, for example, that the mechanism we have described does not work in one dimension, because it is not possible to satisfy energy conservation. In such a case damping becomes a higher order effect.

13.2.8 The stochastic Gross–Pitaevskii equation

If the evolution of condensate fluctuations is damped, then from fluctuation–dissipation relation considerations we must expect it will also be stochastic. This is indeed the case. The resulting "stochastic Gross–Pitaevskii equation" has been investigated by Stoof [Sto99], Duine and Stoof [DuiSto01] and specially by Gardiner and collaborators [GaAnFu01, GarDav03, Jaigar04, BrBlGa05]. Our treatment is essentially a translation of the discussion by Gardiner, Anglin and Fudge [GaAnFu01] into the language of this book [CaHuVe07]. It is interesting to compare our treatment of this problem with [DaDzOn02, KKHOSK06] and [DomRit02].

The simplest way to identify the stochastic elements in the evolution of the condensate is to adopt a coarse-grained effective action scheme (cf. Chapter 5) where the single-particle modes are divided into a "condensate band" (system) of low-lying modes, where most of the condensation takes place, and a "noncondensate" band (environment) of higher modes which act as an environment for the system. In the open system treatment (see Chapters 5 and 8) the quantum fluctuations of the higher band can be represented as classical stochastic fluctuations in the lower band through the nonlinear coupling between the two bands.

A second source of stochasticity is in the random initial conditions appropriate to the condensate [Ste98, ScHuGa06, NobaGa05, NobaGa06].

Since the basic formalism and its physical content have been discussed in detail in the quoted chapters, we shall only review here the simplest scenario. We consider a bosonic gas confined to a box of volume V with periodic boundary conditions, and assume the condensate band to contain just the homogeneous mode, namely

$$\Psi\left(\mathbf{x},t\right) = \phi_c\left(t\right) + \chi\left(\mathbf{x},t\right) \tag{13.180}$$

where ϕ_c is the condensate band field operator. Note the subscript "c" here denotes condensate, not classical, thus this is not quite the background field–quantum field split we have considered so far because ϕ_c, unlike the mean field Φ, is a q-number, and the noncondensate band operator χ, unlike the fluctuation field ψ, has no zero mode.

We compute the influence functional (equivalent to the coarse-grained closed time path effective action) for the ϕ field to order U^2, to which order the field χ is just a nonrelativistic free bosonic field. Let $\phi_c{}^1$ and $\phi_c{}^2$ be the fields in the first and second branch, respectively, and write $(\phi^n)_- = \left(\phi_c{}^1\right)^n - \left(\phi_c{}^2\right)^n$, $(\phi^n)_+ = \left(\left(\phi_c{}^1\right)^n + \left(\phi_c{}^2\right)^n\right)/2$. Then

$$\begin{aligned}
S_{\mathrm{IF}}\left[\phi_c{}^1, \phi_c{}^2\right] = {} & S\left[\phi_c{}^1\right] - S\left[\phi_c{}^2\right] \\
& + \frac{iU^2}{2}\int dt\, dt'\Big\{ \left(\phi^{\dagger 2}\right)_- (t)\left(\phi^2\right)_+ (t')\, \nu\left(t - t'\right)\theta\left(t - t'\right) \\
& - \left(\phi^2\right)_- (t)\left(\phi^{\dagger 2}\right)_+ (t')\, \nu\left(t' - t\right)\theta\left(t - t'\right) \\
& + \frac{1}{2}\left(\phi^{\dagger 2}\right)_- (t)\left(\phi^2\right)_- (t')\, \nu\left(t - t'\right)\Big\}
\end{aligned} \tag{13.181}$$

where

$$\nu\left(t - t'\right) = \sum_{\mathbf{p}} c^{-2i\omega_p\left(t - t'\right)}, \qquad \omega_p = \frac{\hbar p^2}{2M} \tag{13.182}$$

The last line in the influence functional may be traded for two stochastic sources

$$\begin{aligned}
\exp\Big\{ & \frac{-U^2}{4\hbar}\int dt\, dt'\,\left(\phi^{\dagger 2}\right)_- (t)\left(\phi^2\right)_- (t')\, \nu\left(t - t'\right)\Big\} \\
& = \int D\xi D\xi^*\, P\left[\xi, \xi^*\right]\exp\Big\{\frac{iU}{2\hbar}\int dt\,\left[\xi\left(t\right)\left(\phi^2\right)_- (t) + \xi^*\left(t\right)\left(\phi^{\dagger 2}\right)_- (t)\right]\Big\}
\end{aligned} \tag{13.183}$$

where P is a Gaussian measure defined by the correlations

$$\begin{aligned}
\langle\xi\left(t\right)\rangle &= \langle\xi\left(t\right)\xi\left(t'\right)\rangle = 0 \\
\langle\xi^*\left(t\right)\xi\left(t'\right)\rangle &= \hbar\nu\left(t - t'\right)
\end{aligned} \tag{13.184}$$

Variation of the influence functional yields the stochastic GPE for the condensate field

$$i\hbar\phi_{c,t} - U\phi_c^\dagger\phi_c^2 + \frac{iU^2}{V}\phi_c^\dagger \int^t dt' \, \nu\,(t-t')\,\phi_c^2\,(t') = -\frac{U}{V}\xi^*\,(t)\,\phi_c^\dagger\,(t) \quad (13.185)$$

As a check, let us seek the equilibrium solution (neglecting the stochastic term). In equilibrium,

$$\phi_c = \sqrt{\frac{N}{V}} \, e^{-i\mu t/\hbar} \quad (13.186)$$

so the only unknown is the chemical potential

$$\mu = \frac{UN}{V} - \frac{U^2N}{2V^2}\sum_p \frac{1}{\omega_p - \mu - i\varepsilon} \quad (13.187)$$

which is equivalent to the one-loop result.

We see that in general the condensate will undergo non-Markovian dynamics driven by multiplicative colored noise. The generalization of (13.185) for a trap of arbitrary shape is given in [CaHuVe07].

13.2.9 The hydrodynamic and quantum kinetic approach to BECs

So far we have described in some detail the equilibrium and linear response regimes of the condensate, but a nonequilibrium approach has not shown its worth unless it can tackle also the out-of-equilibrium evolution. Of course, the truly far from equilibrium case is as hard to handle as with all other quantum fields we have discussed in this book; see e.g. Chapter 12. However, there is one case where one should be able to make progress, namely, when both the condensate and noncondensate densities are high enough to enforce efficient local thermalization. Then a quantum kinetic theory approach along the lines of Chapter 11 ought to be viable.

The quantum kinetic theory approach to BECs was introduced by Kane and Kadanoff [KanKad65] and elaborated in two series of papers by Gardiner, Zoller and collaborators and Holland, Wachter, Walser and collaborators [GarZol97, JaGaZo97, GarZol98, JGGZ98, GarZol00a, WWCH99, WaCoHo00, WWCH01, WWCH02a, BhWaHo02, WWCH02b]. The derivation of quantum kinetic theory from the 2PIEA is discussed in [BaiSto04, RHCC05]. We follow the latter reference.

There are two basic differences between the quantum kinetic theory applied to BECs and to a generic scalar field theory as discussed in Chapter 11. First, there are two fundamental quantum fields (ψ and $\psi^\dagger$) and therefore the number of propagators is higher. This poses only formal difficulties and we will not discuss it in detail (similar problems arise in the application of the quantum kinetic theory approach to gauge theories, see Chapter 11).

Second, the quantum kinetic theory approach assumes that all mean fields are slowly varying on the scale of the wavelength of the relevant quantum modes, so that an adiabatic expansion is feasible. In the case of BECs, this assumption can be made for the condensate density, but the condensate phase may show strong position dependence.

A solution to this problem is suggested by the long known fact that the evolution of the condensate as described by the GPE is equivalent to the evolution of an irrotational fluid. The idea is that the kinetic description will be valid when the hydrodynamic variables (rather than the condensate wavefunction itself) are slowly varying functions of position.

Let us begin by briefly reviewing the hydrodynamic formulation. Unlike the relativistic theories described in Chapter 12, the condensate is represented as a nonrelativistic (super) fluid. Since the superfluid carries no entropy, the energy density ϵ, pressure p, number density ρ, chemical potential μ, superfluid velocity $\mathbf{v}$ and momentum density π are linked through the relationship

$$\varepsilon + p - \rho\mu - \mathbf{v} \cdot \pi = 0 \tag{13.188}$$

This implies the Gibbs–Duhem relation

$$dp - \rho d\mu - \pi \cdot d\mathbf{v} = 0 \tag{13.189}$$

If we assume the usual relationship $\pi = M\rho\mathbf{v}$, this suggests

$$\mu = \mu_0 - \frac{1}{2}M\mathbf{v}^2 \tag{13.190}$$

where the relationship between μ_0 and p is the usual one for a fluid at rest

$$dp = \rho d\mu_0 \tag{13.191}$$

We have to make contact between this fluid description and the usual one in terms of a condensate wavefunction. Let us write the mean field as [Mad27, Hal81, Cas04]

$$\Phi = e^{i\Theta(\mathbf{x},t)}\sqrt{\rho(\mathbf{x},t)} \tag{13.192}$$

(observe the position dependence of the phase), whereby we have a microscopic interpretation of the density, and the propagators as

$$G^{AB}((\mathbf{x},t),(\mathbf{y},t')) = e^{i\sigma_{3AC}\Theta(\mathbf{x},t)}e^{i\sigma_{3BD}\Theta(\mathbf{y},t')}\bar{G}^{CD}((\mathbf{x},t),(\mathbf{y},t')) \tag{13.193}$$

Observe that since Γ_Q is built out of Feynman graphs based on local interactions it has no explicit dependence on the phases $\Theta(\mathbf{x},t)$. Therefore the force η_2 will transform as

$$\eta_2[\Phi] = e^{i\Theta(\mathbf{x},t)}\bar{\eta}[\rho,\bar{G}^{AB}] \tag{13.194}$$

Now the mean field equation is given by

$$e^{-i\Theta(\mathbf{x},t)}\left[i\hbar\frac{\partial}{\partial t} + \frac{\hbar^2}{2M}\nabla^2\right]\left\{e^{i\Theta(\mathbf{x},t)}\sqrt{\rho(\mathbf{x},t)}\right\} - V(\mathbf{x})\sqrt{\rho} - \bar{\eta} = 0 \tag{13.195}$$

Its imaginary part reads

$$\frac{\partial \rho}{\partial t} + \frac{\hbar}{M} \nabla \left[\rho \nabla \Theta \right] = 2\sqrt{\rho} \text{Im} \, \bar{\eta} \qquad (13.196)$$

This allows us to identify

$$\mathbf{v} = \frac{\hbar}{M} \nabla \Theta \qquad (13.197)$$

as the superfluid velocity, which is therefore (locally) irrotational by definition. There may be global rotation, if the volume occupied by the condensate is not simply connected.

The real part of the mean field equation reads

$$-\hbar \frac{\partial \Theta}{\partial t} = \frac{M}{2} \mathbf{v}^2 + V(x) + \frac{\text{Re} \, \bar{\eta}}{\sqrt{\rho}} - \frac{\hbar^2}{2M \sqrt{\rho}} \nabla^2 \sqrt{\rho} \qquad (13.198)$$

This leads to the evolution equation for the superfluid velocity

$$\frac{\partial v^i}{\partial t} + \left(v^j \nabla_j \right) v^i = \frac{-1}{M} \nabla^i \left[V(x) + \frac{\text{Re} \, \bar{\eta}}{\sqrt{\rho}} - \frac{\hbar^2}{2M \sqrt{\rho}} \nabla^2 \sqrt{\rho} \right] \qquad (13.199)$$

where we have used the assumption that the superfluid velocity is irrotational. For the momentum density we get

$$\frac{\partial M \rho v^i}{\partial t} + \nabla_j \left[M \rho v^j v^i \right] + \rho \nabla^i \frac{\text{Re} \, \bar{\eta}}{\sqrt{\rho}} = -\rho \nabla^i \left[V(x) - \frac{\hbar^2}{2M \sqrt{\rho}} \nabla^2 \sqrt{\rho} \right]$$
$$+ 2M v^i \sqrt{\rho} \text{Im} \, \bar{\eta} \qquad (13.200)$$

The usual hydrodynamic equation would read

$$\frac{\partial M \rho v^i}{\partial t} + \nabla_j T^{ij} = F^i \qquad (13.201)$$

where T^{ij} is the nonrelativistic momentum flux tensor

$$T^{ij} = M \rho v^j v^i + p \delta^{ij} \qquad (13.202)$$

Comparing the hydrodynamic and the microscopic forms of the equation for the superfluid velocity we may identify the pressure. Assume $\text{Re} \, \bar{\eta}$ is a function of ρ. Then

$$\frac{dp}{d\rho} = \rho \frac{d}{d\rho} \left[\frac{\text{Re} \, \bar{\eta}}{\sqrt{\rho}} \right] \qquad (13.203)$$

It is interesting to observe that also

$$\frac{dp}{d\rho} = M c_{\text{s}}^2 \qquad (13.204)$$

defines the speed of sound in the condensate. Going back to the Gibbs–Duhem relation we find

$$\mu_0 = \frac{\text{Re} \, \bar{\eta}}{\sqrt{\rho}} \qquad (13.205)$$

and so the equation for the time dependence of the phase is

$$-\hbar\frac{\partial\Theta}{\partial t} = \frac{1}{2}M\mathbf{v}^2 + V(x) + \mu_0 - \frac{\hbar^2}{2M\sqrt{\rho}}\nabla^2\sqrt{\rho} \qquad (13.206)$$

To close this system we need the equations for the propagators. From the decompositions

$$G_{AB}^{-1}((\mathbf{x},t),(\mathbf{y},t')) = e^{-i\sigma_3{}_{AC}\Theta(\mathbf{x},t)}e^{-i\sigma_3{}_{BD}\Theta(\mathbf{y},t')}\bar{G}_{CD}^{-1}((\mathbf{x},t),(\mathbf{y},t')) \qquad (13.207)$$

$$\Sigma_{AB}((\mathbf{x},t),(\mathbf{y},t')) = e^{-i\sigma_3{}_{AC}\Theta(\mathbf{x},t)}e^{-i\sigma_3{}_{BD}\Theta(\mathbf{y},t')}\bar{\Sigma}_{CD}((\mathbf{x},t),(\mathbf{y},t')) \qquad (13.208)$$

we get

$$\bar{G}_{AB}^{-1} = \bar{D}_{AB}^{-1} + i\bar{\Sigma}_{AB} \qquad (13.209)$$

where

$$\bar{D}_{AB}^{-1} = e^{i\sigma_3{}_{AC}\Theta(\mathbf{x},t)}D_{CD}^{-1}e^{i\sigma_3{}_{BD}\Theta(\mathbf{y},t')} \qquad (13.210)$$

Concretely,

$$\bar{D}_{AB}^{-1} = (-i)\begin{pmatrix} 0 & \bar{D}^{-1*} \\ \bar{D}^{-1} & 0 \end{pmatrix} \qquad (13.211)$$

$$\bar{D}^{-1} = e^{-i\Theta}\left[i\hbar\partial_t + \frac{\hbar^2}{2M}\nabla^2 - V(x)\right]e^{i\Theta}$$

$$= i\hbar\left(\partial_t + \mathbf{v}.\nabla + \frac{(\nabla.\mathbf{v})}{2}\right) + \frac{\hbar^2}{2M}\nabla^2 + \mu_0 - \frac{\hbar^2\left(\nabla^2\sqrt{\rho}\right)}{2M\sqrt{\rho}} \qquad (13.212)$$

From this point on, the derivation of the quantum kinetic equation for the non-condensate particles follows the lines of Chapter 11. For a discussion of nontrivial hydrodynamic behavior in BECs see [HACCES06].

13.3 The particle number conserving formalism

The symmetry-breaking approach described above has the disturbing feature that, strictly speaking, symmetry breaking only occurs in the thermodynamic limit. We therefore have a formalism that assumes the number of particles is essentially infinite. Most actual experiments deal with situations where particle number is bounded. Under this circumstance a condensate as described above simply cannot happen.

In this section we shall describe an alternative formulation which is designed to deal with gases with fixed particle numbers. We shall call this formulation the particle number conserving formalism, PNC for short. See [GirArn59, GirArn98, CasDum97, CasDum98, MorCas03, Gar97, GJDCZ00, Mor04, Mor99, Mor00, Idz05a, Dzi05b, GarMor07]. Let us begin by discussing how is it possible to speak of a BEC in a situation where there is no symmetry breaking.

13.3.1 Problems with the symmetry-breaking approach

Recall in the symmetry-breaking (SB) approach to BEC, condensation is signaled by a spontaneous breakdown of phase invariance (13.31), whereby Ψ develops a nonzero expectation value Φ. We can therefore employ a background field decomposition [NegOrl98, PetSmi02] around Φ (c-number): $\Psi = \Phi + \psi$ where ψ (q-number) is the field operator describing quantum fluctuations (see equation (13.35)).

A common feature of these approaches is that the total particle number

$$\mathbf{N} = \int d^d\mathbf{x}\, \Psi^\dagger \Psi \tag{13.213}$$

is not fixed. For example, let us assume that the condensate is confined within a homogeneous box of volume V, condensation occurring in the lowest (translation-invariant) mode. Let $a_\mathbf{k}$ be the operator that destroys an atom in the $\mathbf{k}$ mode. Then we may approximate (see the more careful discussion below)

$$\psi(\mathbf{x}, t) = \sum_{\mathbf{k} \neq 0} \frac{e^{i\mathbf{k}\mathbf{x}}}{\sqrt{V}} a_\mathbf{k} \tag{13.214}$$

Even if we treat ψ as a linear perturbation on the condensate, the Hamiltonian is not diagonal on the $a_\mathbf{k}$. To diagonalize it, we must introduce phonon destruction operators $A_\mathbf{k}$ and perform a Bogoliubov transformation

$$a_\mathbf{k} = \alpha_\mathbf{k} A_\mathbf{k} + \beta_\mathbf{k} A^\dagger_{-\mathbf{k}} \tag{13.215}$$

At zero temperature, the state is the phonon vacuum, $A_\mathbf{k}|0\rangle = 0$ for all $\mathbf{k} \neq 0$. We find

$$\langle \mathbf{N} \rangle = \int d^d\mathbf{x}\, \langle \Psi^\dagger \Psi \rangle = V\left[|\Phi|^2 + \tilde{n} \right] \tag{13.216}$$

where

$$\tilde{n} = \langle \psi^\dagger \psi \rangle = \frac{1}{V} \sum_{\mathbf{k} \neq 0} \langle a^\dagger_\mathbf{k} a_\mathbf{k} \rangle = \frac{1}{V} \sum_{\mathbf{k} \neq 0} |\beta_\mathbf{k}|^2 \tag{13.217}$$

but

$$\langle \mathbf{N}^2 \rangle = V^2 \left[\left(|\Phi|^2 \right)^2 + |\Phi|^2 \left(4\tilde{n} + \frac{1}{V} \right) + \Phi^{*2}\tilde{m} + \Phi^2 \tilde{m}^* + \dots \right] \tag{13.218}$$

where

$$\tilde{m} = \langle \psi^2 \rangle = \frac{1}{V} \sum_{\mathbf{k} \neq 0} \langle a_{-\mathbf{k}} a_\mathbf{k} \rangle = \frac{1}{V} \sum_{\mathbf{k} \neq 0} \alpha_\mathbf{k} \beta_\mathbf{k} \tag{13.219}$$

The Bogoliubov coefficients $\alpha_\mathbf{k}$ and $\beta_\mathbf{k}$ cannot be equal, because the canonical (Bose) commutation relations imply $|\alpha_\mathbf{k}|^2 - |\beta_\mathbf{k}|^2 = 1$, and so also $\tilde{m} \neq \tilde{n}$. We conclude that necessarily $\langle \mathbf{N}^2 \rangle \neq \langle \mathbf{N} \rangle^2$ in the symmetry-breaking approach, signaling the presence of particle number fluctuations.

13.3.2 *The one-body density matrix and long-range coherence*

We consider as above a second-quantized Bose field Ψ. The state of the many-body system is an eigenstate of total particle number operator (13.213). There is no particle exchange with the environment.

In this case of a finite system, there is no symmetry breaking. The symmetry-broken state is essentially a coherent state and thus a coherent superposition of states with arbitrarily large total particle number. Nevertheless, there are situations where there is long-range coherence across the system, thus capturing the essential feature of the condensed states. Sometimes these situations are referred to as quasi-condensates, but we shall not make this distinction, just referring to them as the symmetry-broken siblings of BECs.

To characterize the BEC state, let us introduce the one-body density matrix [PenOns56]

$$\sigma\left(\mathbf{x}, \mathbf{y}, t\right) = \left\langle \Psi^\dagger\left(\mathbf{x}, t\right) \Psi\left(\mathbf{y}, t\right) \right\rangle \tag{13.220}$$

Long-range coherence appears when σ fails to decay as x and y are taken apart. Observe that σ is Hermitian and nonnegative, in the sense that for any function f

$$\int d^d\mathbf{x} d^d\mathbf{y}\ f^*\left(\mathbf{x}\right)\sigma\left(\mathbf{x}, \mathbf{y}, t\right) f\left(\mathbf{y}\right) \geq 0 \tag{13.221}$$

Therefore it admits a basis of eigenfunctions

$$\int d^d\mathbf{x}\ \sigma\left(\mathbf{x}, \mathbf{y}, t\right)\phi_\alpha\left(\mathbf{y}, t\right) = n_\alpha \phi_\alpha\left(\mathbf{x}, t\right) \tag{13.222}$$

where the eigenvalues n_α are real and nonnegative. We assume the ϕ_α are normalized

$$\left(\phi_\alpha, \phi_\beta\right) = \delta_{\alpha\beta} \tag{13.223}$$

$$\left(f, g\right) = \int d^d\mathbf{x}\ f^*g \tag{13.224}$$

and complete

$$\sum_\alpha \phi_\alpha^*\left(\mathbf{x}, t\right)\phi_\alpha\left(\mathbf{y}, t\right) = \delta\left(\mathbf{x} - \mathbf{y}\right) \tag{13.225}$$

The field operator may be expanded in this basis

$$\Psi\left(\mathbf{x}, t\right) = \sum_\alpha a_\alpha\left(t\right)\phi_\alpha\left(\mathbf{x}, t\right) \tag{13.226}$$

The Bose commutation relations imply

$$\left[a_\alpha\left(t\right), a_\beta^\dagger\left(t\right)\right] = \delta_{\alpha\beta} \tag{13.227}$$

The $a_\alpha\left(t\right)$ are operators which, at time t, destroy a particle in the one-particle state α whose wavefunction is $\phi_\alpha\left(\mathbf{x}, t\right)$. From the definition of σ we find

$$\left\langle a_\alpha^\dagger\left(t\right) a_\beta\left(t\right)\right\rangle = n_\alpha\left(t\right)\delta_{\alpha\beta} \tag{13.228}$$

Therefore the eigenvalues $n_\alpha(t)$ are the mean number of particles in the one-body state α at time t. We also have the strong identity

$$N = \sum_\alpha a_\alpha^\dagger(t)\, a_\alpha(t) \tag{13.229}$$

Condensation occurs when one of the n_α, say $\alpha = 0$, becomes comparable with N itself. Then we have, for large separations

$$\sigma(\mathbf{x}, \mathbf{y}, t) \sim n_0 \phi_0^*(\mathbf{x}, t)\, \phi_0(\mathbf{y}, t) \tag{13.230}$$

which displays long-range coherence, as expected. Here $\phi_0(\mathbf{x}, t)$ is the condensate wavefunction. We stress that this is the fundamental definition; $\phi_0(\mathbf{x}, t)$ is not necessarily proportional to the mean field Φ introduced in the symmetry-breaking approach.

13.3.3 The particle number conserving approach

We shall now discuss the dynamics of the condensate wavefunction $\phi_0(\mathbf{x}, t)$ and the condensate occupation number N_0 (we switch to a capital N to emphasize its macroscopic character). We envisage a situation in which N is finite but large, and will seek equations of motion as an expansion in inverse powers of N. In preparation for this, it is convenient to scale the interaction term, writing $U = u/N$.

As we have seen above, in the symmetry-breaking approach the condensate state (for an interacting gas) is seen as a coherent superposition of particle pairs, each pair having zero total momentum. The basic insight of the PNC approach is that each particle above the condensate corresponds to a hole in the condensate, so we may speak of *particle–hole* (PH) pairs. Of course, introducing a PH into the system does not change the total number of particles.

Following Arnowitt and Girardeau, let us introduce the operator

$$\beta = \frac{1}{\sqrt{\hat{N}_0 + 1}} a_0 = a_0 \frac{1}{\sqrt{\hat{N}_0}} \tag{13.231}$$

where

$$\hat{N}_0 = N - \sum_{\alpha \neq 0} a_\alpha^\dagger a_\alpha \tag{13.232}$$

is the condensate number Heisenberg operator. Observe that for a number eigenstate $\beta\,|N_0\rangle = |N_0 - 1\rangle$ unless $N_0 = 0$, in which case $\beta\,|0\rangle = 0$. Therefore β preserves the norm for all states orthogonal to the state with no particles in the zeroth mode (which is much stronger than not having a condensate). If there is a condensate, any physically meaningful state will satisfy this requirement, and β may be considered a unitary operator, with inverse

$$\beta^\dagger = \frac{1}{\sqrt{\hat{N}_0}} a_0^\dagger = a_0^\dagger \frac{1}{\sqrt{\hat{N}_0 + 1}} \tag{13.233}$$

We now introduce the destruction operator of a PH with the particle in mode α

$$\lambda_\alpha = \beta^\dagger a_\alpha \tag{13.234}$$

If we consider the β's as unitary, then the λ's satisfy bosonic canonical commutation relations. This relationship may be inverted:

$$a_\alpha = \beta \lambda_\alpha \tag{13.235}$$

The number of particles in a given mode is equal to the number of PH

$$a_\alpha^\dagger a_\alpha = \lambda_\alpha^\dagger \lambda_\alpha \tag{13.236}$$

We write the field operator restricted to the subspace with a well-defined total number of particles N as $\Psi = \sqrt{N}\beta\phi$

$$\phi = \phi_0(\mathbf{x},t) + \frac{1}{\sqrt{N}}\lambda(\mathbf{x},t) - \frac{1}{2N}f[\delta n(t)]\phi_0(\mathbf{x},t) \tag{13.237}$$

where

$$\lambda(\mathbf{x},t) = \sum_{\alpha \neq 0} \lambda_\alpha(t)\phi_\alpha(\mathbf{x},t) \tag{13.238}$$

$$\delta n(t) = \int d^3\mathbf{x}\, \lambda^\dagger \lambda \tag{13.239}$$

$$f(x) = 2N\left[1 - \sqrt{1 - \frac{x}{N}}\right] \sim x + O\left(N^{-1}\right) \tag{13.240}$$

Within our approximations β commutes with ϕ. Finally we have the relationship

$$0 = \left\langle a_0^\dagger(t) a_\alpha(t) \right\rangle = \left\langle a_0^\dagger(t)\beta\lambda_\alpha(t) \right\rangle = \sqrt{N}\left\langle \left[\sqrt{1 - \frac{1}{N}\sum_{\gamma \neq 0}\lambda_\gamma^\dagger\lambda_\gamma}\right]\lambda_\alpha(t) \right\rangle \tag{13.241}$$

which implies

$$\langle \lambda \rangle = \frac{1}{2N}\langle f[\delta n(t)]\lambda \rangle \tag{13.242}$$

The idea is to seek a solution of the Heisenberg equations of motion for Ψ where β and the λ_α's have developments in inverse powers of N. Define a "q-number" chemical potential $\hat{\mu}$ from

$$\beta^\dagger \frac{d\beta}{dt} = \frac{-i\hat{\mu}}{\hbar} \tag{13.243}$$

We have

$$i\hbar\frac{\partial}{\partial t}\phi = (H - \hat{\mu})\phi + u\phi^\dagger\phi^2 \tag{13.244}$$

We then find

$$0 = -i\hbar\phi_{0,t} + (H - \hat{\mu})\phi_0 + u\phi_0^3$$
$$+ \frac{1}{\sqrt{N}}\left[-i\hbar\lambda_{,t} + (H - \hat{\mu})\lambda + u\phi_0^2\left(2\lambda + \lambda^\dagger\right)\right] + O\left(N^{-1}\right) \quad (13.245)$$

Taking the expectation value we find

$$0 = -i\hbar\phi_{0,t} + (H - \langle\hat{\mu}\rangle)\phi_0 + u\phi_0^3 - \frac{1}{\sqrt{N}}\langle\hat{\mu}\lambda\rangle + O\left(N^{-1}\right) \quad (13.246)$$

Recall that ϕ_0 is real (if the condensate is nondegenerate) and $\hat{\mu}$ is Hermitian. So we may decompose this equation into

$$0 = (H - \langle\hat{\mu}\rangle)\phi_0 + u\phi_0^3 - \frac{1}{2\sqrt{N}}\langle\hat{\mu}\lambda + \lambda^\dagger\hat{\mu}\rangle + O\left(N^{-1}\right) \quad (13.247)$$

and

$$0 = -i\hbar\phi_{0,t} - \frac{1}{2\sqrt{N}}\langle\hat{\mu}\lambda - \lambda^\dagger\hat{\mu}\rangle + O\left(N^{-1}\right) \quad (13.248)$$

This is consistent with the normalization condition

$$\int \phi_0\phi_{0,t} = 0 \quad (13.249)$$

Subtracting the expectation value from the Heisenberg equation, we get

$$0 = (\langle\hat{\mu}\rangle - \hat{\mu})\phi_0 + \frac{1}{\sqrt{N}}\left[-i\hbar\lambda_{,t} + (H - \hat{\mu})\lambda + u\phi_0^2\left(2\lambda + \lambda^\dagger\right)\right] + \frac{1}{\sqrt{N}}\langle\hat{\mu}\lambda\rangle$$
$$+ O\left(N^{-1}\right) \quad (13.250)$$

The orthogonality of ϕ_0 and λ implies

$$\int (\phi_0\lambda_{,t} + \phi_{0,t}\lambda) = 0 \quad (13.251)$$

and from (13.250), (13.248) and (13.247) we get

$$0 = \langle\hat{\mu}\rangle - \hat{\mu} + \frac{u}{\sqrt{N}}\left(J_3 + J_3^\dagger\right) + O\left(N^{-1}\right) \quad (13.252)$$

where

$$J_n = \int \phi_0^n\lambda \quad (13.253)$$

Observe that this implies

$$\langle\hat{\mu}\lambda\rangle = O\left(N^{-1/2}\right) \quad (13.254)$$

The equation for λ simplifies into

$$0 = -i\hbar\lambda_{,t} + (H - \hat{\mu})\,\lambda + u\phi_0^2\lambda + Q\left[u\phi_0^2\left(\lambda + \lambda^\dagger\right)\right] + O\left(N^{-1/2}\right) \quad (13.255)$$

where

$$Q\,[X] = X - \phi_0 \int \phi_0 X \quad (13.256)$$

The homogeneous case

To get a feeling of the working of the PMC approach, let us apply it to the simplest case of a BEC in a homogeneous box of volume V, with periodic boundary conditions.

In equilibrium, by symmetry, the condensate wavefunction must be homogeneous, and by normalization we must have $\phi_0 = V^{-1/2}$. This equation holds to all orders in $1/N$. Therefore

$$\langle\hat{\mu}\rangle = \frac{u}{V} + O\left(N^{-1}\right) \quad (13.257)$$

This gives $\langle\hat{\mu}\rangle = UN/V + \ldots$. By contrast, in the Bogoliubov approximation the chemical potential is $\mu^{Bog} = UN_0/V$ and in the Popov approximation $\mu^{Pop} = (U/V)\,(2N - N_0)$. We also have

$$\hat{\mu} = \langle\hat{\mu}\rangle + O\left(N^{-1/2}\right) \quad (13.258)$$

and so the lowest order equation for the inhomogeneous mode is

$$0 = -i\hbar\lambda_{,t} + H\lambda + \frac{UN}{V}\left(\lambda + \lambda^\dagger\right) \quad (13.259)$$

These are the Popov equations with N in place of N_0, and so we know the spectrum will be gapless. Moreover, in this case there is no zero mode divergence.

After solving these equations it is simple to compute the higher corrections to $\hat{\mu}$.

13.3.4 Particle number conserving functional approach

One problem with the PNC approach as presented so far is that it is not cast within a functional approach, and therefore lacks the flexibility which has been key to most of the applications of NEqQFT in this book. To be able to give a functional PNC approach, we must revise the measure of integration in the path integral expression for the generating functional we have considered so far. The idea is to define a new generating functional which will agree with the old one in the computation of expectation values for particle number conserving operators, but will lead to different results otherwise. In particular, the expectation value of the field operator in the new approach will be identically zero, as it must be in a system with a finite number of particles.

The quantum theory of the BEC may be regarded as the quantization of the nonrelativistic classical field theory defined by the action functional (13.39), where the canonical variables are $\Psi(\mathbf{x}, t)$ and its conjugate momentum $i\hbar\Psi^*$. This theory conserves particle number (13.213), and we are interested in the case in which particle number takes on a definite value N. We may reinforce this point by adding a constraint on the theory. This is achieved by introducing a Lagrange multiplier $\mu_q(t)$, whereby the action becomes

$$S = \int d^{d+1}x \left[i\hbar\Psi^* \frac{\partial}{\partial t}\Psi + \hbar\mu_q(t)\left(\Psi^*\Psi - \frac{N}{V}\right)\right] - \int dt\, \mathbf{H} \qquad (13.260)$$

The original action (13.39) is invariant under a global transformation (13.31) but the new action (13.260) is invariant under the local (in time) transformations (a familiar theory with local U(1) gauge symmetry is electromagnetism)

$$\Psi \to e^{i\theta(t)}\Psi, \qquad \Psi^\dagger \to e^{-i\theta(t)}\Psi^\dagger, \qquad \mu_q \to \mu_q + \frac{d\theta}{dt} \qquad (13.261)$$

provided θ vanishes both at the initial and final times (when θ is infinitesimal, these are just the canonical transformations generated by the constraint) [Dir50, Dir58b]. Therefore it must be quantized using the methods developed for gauge theories, such as the Fadeev–Popov method [PesSch95].

The need for a further refinement of the functional measure comes from the fact that now the path integral is redundant, since we may transform the fields as in (13.261). The Fadeev–Popov approach fixes the redundancy by factoring out the gauge group. Choose some function $f_\theta = f\left[\mu_{q\theta}, \Psi_\theta, \Psi_\theta^\dagger\right]$, such that $df_\theta/d\theta \neq 0$. Then

$$1 = \int \frac{df_\theta}{d\theta} d\theta\, \delta\left(f_\theta - c\right) \qquad (13.262)$$

Inserting this into the vacuum persistence amplitude and averaging over c with a weight $e^{ic^2/2\sigma}$ we get

$$Z_0 = \Theta \cdot \int D\Psi D\mu_q\, e^{iS_{\mu_q,\sigma}/\hbar} \text{Det}\left[\frac{\delta f_\theta}{\delta \theta}\right]_{\theta=0} \qquad (13.263)$$

where

$$\Theta = \int D\theta \qquad (13.264)$$

is the volume of the gauge group we wish to factor out;

$$S_{\mu,\sigma} = S + \frac{\hbar}{2\sigma}\int dt\, f_0^2 \qquad (13.265)$$

where S is defined in (13.260). The determinant is expressed as a path integral over Grassmann fields ζ, η (see Chapter 7)

$$\text{Det}\left[\frac{\delta f_\theta}{\delta \theta}\right]_{\theta=0} = \int D\zeta D\eta\, e^{-\frac{1}{\hbar}\int dt\, \zeta\frac{\delta f_\theta}{\delta \theta}\eta} \qquad (13.266)$$

To finalize the set-up, we need to choose the gauge fixing function f_0. Possibly the simplest choice is the "covariant" gauge

$$f_0 = \frac{d\mu_q}{dt} \tag{13.267}$$

which makes the ghost fields decouple. This gauge is employed in [CaHuRe06] to explore the critical regime in the Mott transition. Other choices are also available, and in fact the freedom to choose the gauge fixing condition is one of the main strengths of the approach [DeuDru02, DrDeKh04].

14

Nonequilibrium issues in RHICs and DCCs

14.1 Relativistic heavy ion collisions (RHICs)

14.1.1 In the beginning

The goal of this chapter is to provide a short summary of the main points where nonequilibrium field theory may contribute to our understanding of relativistic heavy ion collisions. We skip over details of strong interaction processes, but focus on those aspects which are directly related to the nonequilibrium features of the (collective) dynamics.

The relevant experiments are the Super Proton Synchroton (SPS) (CERN) and the Relativistic Heavy-Ion Collider (RHIC) (Brookhaven), with the Large Hadron Collider (LHC) coming on line soon. SPS accelerates lead ions ($Z = 82$, $A = 207$) to energies of 17 GeV per nucleon in the center-of-mass frame; RHIC collides gold ($Z = 79$, $A = 197$) at energies of 130 to 200 GeV per nucleon. The RHIC experiments are described in detail in the so-called "white papers," which are possibly the most reliable source on the subject [BRAHMS05, PHOBOS05, STAR05, PHENIX05]. Other basic references are [Cse94, Won94, Gyu01, Shu88].

We shall work in natural units, the characteristic scale for strong interactions being 1 fm $= 10^{-15}$ m $= (200\,\text{MeV})^{-1}$. The strength of the interaction is measured by the structure constant $\alpha_S = g^2/4\pi$, where g is the coupling constant. (We assume that the symmetry group is $SU(3)$ with eight gluons.) In the perturbative regime $E \gg \Lambda_{\text{QCD}} \sim 200\,\text{MeV}$, the structure constant runs with scale as

$$\alpha_S\left(E\right) = \frac{12\pi}{(33 - 2n_f)\ln\left[\frac{E^2}{\Lambda_{\text{QCD}}^2}\right]} \tag{14.1}$$

where n_f is the number of flavors (6) and Λ_{QCD} is the QCD energy scale. This means that for scales of the order of the proton mass $m_p \sim 1\,\text{GeV}$, $\alpha_S \sim 0.5$. Because of the logarithmic fall off, it will not get much smaller in the relevant range of energies.

The most abundant product from the heavy ion collisions are the lightest mesons, the pions $\pi^\pm$ and π^0 with masses $m_\pi \sim 140\,\text{MeV}$. Pions are pseudo-scalars, so they do not have different polarization states. The proton, on the other hand, comes in two different spin states – this will be important in what follows.

One of the goals of the RHIC program was to probe into possible new phases of nuclear matter at higher energies such as a conjectured deconfined phase. In such a high-energy phase, matter would most likely be a plasma of gluons and (massless) quarks (quark–gluon plasma, QGP). Remember that for relativistic particles each bosonic degree of freedom contributes

$$\epsilon_B = \frac{\pi^2}{30}T^4 \tag{14.2}$$

to the energy density in equilibrium, while (neglecting chemical potentials) each massless fermionic degree of freedom contributes $\epsilon_F = (7/8)\,\epsilon_B$. We have eight different gluons with two polarization degrees of freedom each, and four effectively massless quarks (u, d and their antiparticles) coming in three colors and two spin states each. Therefore the energy density in the deconfined phase is

$$\epsilon_{\mathrm{plasma}} = \frac{37}{30}\pi^2 T^4 \tag{14.3}$$

and the pressure is $p_{\mathrm{plasma}} = \epsilon_{\mathrm{plasma}}/3$.

In the low-temperature phase, only the pions are effectively massless. These pions live on a quark condensate which enforces confinement. Therefore the energy density is $\epsilon_{\mathrm{hadron}} = \epsilon_{\mathrm{pions}} + \epsilon_{\mathrm{condensate}}$, where

$$\epsilon_{\mathrm{pions}} = \frac{3}{30}\pi^2 T^4 \tag{14.4}$$

and $\epsilon_{\mathrm{condensate}} \equiv -B$, where $B \sim \Lambda_{\mathrm{QCD}}^4$ is known as the *bag constant*. The pressure of the confined phase is $p_{\mathrm{hadron}} = \epsilon_{\mathrm{pion}}/3 + B$.

At the coexistence point, both phases have the same pressure, and so the critical temperature obeys

$$\frac{34}{90}\pi^2 T_c^4 = B = \Lambda_{\mathrm{QCD}}^4 \tag{14.5}$$

namely $T_c = \left(90/34\pi^2\right)^{1/4} \Lambda_{\mathrm{QCD}} \sim 0.72\Lambda_{\mathrm{QCD}} \sim 150\,\mathrm{MeV}$. This means that to enter the deconfined phase, we need a mininal energy density of $\epsilon_{\mathrm{crit}} = (3 \cdot 37/34)$ $\Lambda_{\mathrm{QCD}}^4 = 650\,\mathrm{MeV/fm}^3$.

Of course, this is the transition point at zero chemical potential only; in general, we have a coexistence curve in the (μ, T) plane, so that the critical temperature may be lowered by increasing the baryon number density.

Nevertheless, evidence seems to suggest that the QGP has not been created at RHIC [BRAHMS05, PHOBOS05, STAR05, PHENIX05]. The high-energy collisions have created what seems to be a new state of dense nuclear matter, whose description in terms of purely hadronic degrees of freedom seems inadequate. This suggests the presence of unscreened color charges over distances larger than the size of a nucleon. However, the system seems to be strongly interacting throughout, with properties closer to a liquid than to a plasma.

14.1.2 The Bjorken scenario

Virtually all the field-theoretic analyses of RHICs assume a spacetime picture of collision provided by the Bjorken model [Bjo83]. The colliding nuclei are seen as slabs of quark and gluon matter. In the center-of-mass frame, both slabs approach each other at near light speed. Upon collision, the two slabs of matter will mostly go through each other, leaving behind a wake of hot plasma. We may then distinguish three different regions: the two fragmentation regions corresponding to the receding slabs, and the central region corresponding to the plasma in between. We are interested in phenomena in the central region.

At the time of crossing a number of hard scattering processes will occur, whose products will directly reach the detectors. These hard processes are unrelated to the nonequilibrium dynamics of the plasma; and may presumably be predicted on perturbative QCD grounds. In what follows, we will assume this hard component has been isolated despite great difficulty to achieve this in reality.

The hot plasma will expand and cool, and eventually fragment into ordinary particles in flight intercepted by the detectors. We wish to predict the number of particles of each species to be detected, as a function of the angle θ between the direction of flight and the direction z of the beam. It is remarkable that with this simple picture we can state a first observable prediction already.

Indeed, because of Lorentz contraction, we may think of the approaching slabs as infinitely thin in the direction of motion z, and in a first approach to the problem, as infinite and homogeneous in the transverse directions x and y. This picture is invariant under boost in the z direction, and so is the final distribution of particles. So if we parameterize the momentum of an out-going particle as $p^0 = E$, $p^3 = p$ and $(p^1, p^2) = p_\perp$, then the distribution of particles may depend only upon the transverse momentum and $E^2 - p^2 = m^2 + p_\perp^2$. In particular, it must be independent of θ, since $\cos\theta \sim p/E$ is not invariant. It is conventional to plot the yield of the collision in terms of the *rapidity* y, defined by $p/E \equiv \tanh y$, or rather the pseudo-rapidity $\eta = -\ln\tan[\theta/2]$, $\tanh\eta = p/|\mathbf{p}|$. Rapidity and pseudo-rapidity agree at momenta which are large compared to the mass of the particle. Then the prediction in this picture is that there is a *plateau* in the (pseudo) rapidity distribution, at least for small rapidity ($|\eta| \to \infty$ corresponds to the fragmentation rather than the central region).[1]

We may elaborate on the Bjorken picture further. Let us assume that the plasma is formed on the plane $z = 0$ at the time $t = 0$ of the collision, and then expands along the z direction. A given plasma element will cool according to its own proper time τ. Now, as in the twin paradox, the proper time lapse will be less for those elements which move faster, which are also those which reach farther if we compare the relative positions at a given fixed time, as measured, say, in the center-of-mass frame. Thus the plasma will be hotter in the outer layers than in the center. This situation resembles the dessert known as *baked Alaska*, made by briefly putting an ice-cream ball in the oven, thereby the outer crust heats up while the center remains frozen.

---

[1] This prediction is not clearly borne out by the RHIC data [PHOBOS05]. Therefore, it remains a possibility that analyses based on the Bjorken model are not so relevant to current experiments compared to future higher energy collisions.

Eventually, at some given constant τ surface, the plasma will be cold enough (and/or dilute enough) to break up into hadrons. Assuming that the product hadrons are thermally distributed, massless and at zero chemical potential, the Bose–Einstein distribution predicts that the energy per particle is $\epsilon/n = 2.7\,T$. Since temperature is constant on the break-up surface, this means that in all collisions particles should have the same average energy. Indeed, it is observed that the energy per particle is about 0.8 GeV, regardless of the center-of-mass energy and impact parameter.

Another important observation is that for transverse momenta higher than 2 GeV, the number of emitted protons is actually higher than pions. This can be explained as a consequence of hydrodynamic behavior [HeiKol02a], or else, at very large momenta, through a recombination mechanism [MulNag06]. If no chemical potentials were involved, then equality would obtain (at $p_T = 2$ GeV) for a temperature of about 340 MeV. In reality, pions do not have chemical potential, but protons do, associated with baryon number conservation. Adding a chemical potential $\mu \sim 40$ MeV for the protons reduces the crossing temperature to 280 MeV [HeiKol02a].

To obtain a more quantitative description of the process, we may describe the plasma as a relativistic ideal fluid [BelLan56, CarDuo73, CarZac83]. The assumption of a homogeneous plasma in the transverse direction is too simplistic, and a full four-dimensional solution must be sought, which requires numerical methods [KoSoHe00, MolGyu00, Hir01, MMNH02, HeiWon02, TeLaSh01, HirTsu02, KolRap03, HeiKol02b, HKHRV01]. To close the hydrodynamic system of equations we must provide the equation of state. The central feature of this is the "softening" near the critical point, meaning that the speed of sound $c_{\mathrm{s}}^2 = \partial p/\partial \epsilon \to 0$ as we approach the transition point. The softening of the equation of state affects the evolution of the fireball, which then becomes a signal of whether the transition point has been reached or not.

Since perfect fluids conserve entropy, the total entropy within the fireball remains constant, and T scales as $V^{-1/3}$. So, if the expansion is one-dimensional, and we consider the volume enclosed between two fixed rapidities, then $T \sim \tau^{-1/3}$, where τ is the proper time. In particular, the energy density scales as $\tau^{-4/3}$ rather than τ^{-1}, as in our earlier estimate. This leads to a slight increase in the estimate of the initial plasma temperature.

For treatment of RHICs beyond perfect fluids, see [GyRiZh96a, GyRiZh97, Ris98, Tea03].

14.1.3 Break-up

We now consider more closely the phenomenon of break-up [CooFry74, SiAkHa02]. Assume this occurs on a three-dimensional surface Σ defined by some equation $\Sigma\,(x^\mu) = 0$. If x_0 is a solution, then the normal vector at x_0 is $n_\mu = (-\alpha)\,\Sigma_{,\mu}$, $\alpha = (-\Sigma_{,\mu}\Sigma^{,\mu})^{-1/2}$. We shall assume that n_μ is time-like. For a more realistic scenario where the surface has both time-like and space-like regions, see [Bug03]. The invariant measure on Σ is given by $d^3\sigma = d^4x\,\delta\,(\Sigma)\,\alpha^{-1}$.

Let us assume that both before and after break-up, we can describe matter as a perfect relativistic fluid. Let $K_a = \partial/\partial x^a$ be the four Killing vectors of Minkowski space. Then Gauss' theorem shows that the quantities $n_\mu K_{a\nu}T^{\mu\nu}$ and $n_\mu N^\mu$ are continuous across the break-up surface (we shall consider only one conserved current, corresponding to, say, the baryon number). These conditions plus the equation of state of the hadronic phase define the energy density, pressure, baryon number density (or equivalently, the temperature and chemical potential) and the 4-velocity of the hadrons at break-up. The detailed spectrum is found by assuming that the hadrons are thermally distributed.

In principle we could distinguish between matter having a thermal distribution of momenta (kinetic equilibrium) and in chemical equilibrium. Correspondingly, there is a kinetic freeze-out, and a chemical freeze-out, which are not necessarily simultaneous. This permits some extra freedom in matching models to data.

The total number of emitted particles is

$$\int d^3\mathbf{x}\,K_{0\mu}N^\mu_{\text{had}} \tag{14.6}$$

where the integral is over some $t = $ constant surface well to the future of the collision. Because of Gauss' theorem, we may replace the integral by an integral over the break-up surface (we may have to complete this surface to get a Cauchy surface, but the particle density flux will vanish on these additions anyway). But then we may use the matching conditions to express this integral in terms of the particle current before break-up. We obtain the total number of emitted particles as

$$\int d^4x\,\delta\,(\Sigma)\,\Sigma_{,\mu}N^\mu_{\text{hydro}} \tag{14.7}$$

In practice, we may wish to smear a little the position of the break-up surface, thus writing the total number of emitted particles as

$$\int d^4x\,\left[\frac{e^{-\Sigma^2/2(\Delta\Sigma)^2}}{\sqrt{2\pi}\,(\Delta\Sigma)}\right]\Sigma_{,\mu}N^\mu_{\text{hydro}} \tag{14.8}$$

The total number of particles of species i with momentum p^μ is

$$g_i\int\frac{d^4x}{C\,(x)}\,\left[\frac{e^{-\Sigma^2/2(\Delta\Sigma)^2}}{\sqrt{2\pi}\,(\Delta\Sigma)}\right]\Sigma_{,\mu}N^\mu_{\text{hydro}}\delta\,(p_i^2 - m_i^2)\,\frac{U^{\text{had}}_\lambda p_i^\lambda}{[\exp\,(-\beta_\nu p_i^\nu - \mu b_i) - \varepsilon_i]} \tag{14.9}$$

where

$$C\,(x) = \sum_i g_i\int\frac{d^4p_i}{(2\pi)^4}\,\delta\,(p_i^2 - m_i^2)\,\frac{U^{\text{had}}_\mu p_i^\mu}{[\exp\,(-\beta_\nu p_i^\nu - \mu b_i) - \varepsilon_i]} \tag{14.10}$$

The two basic observables are the total number of particles with transverse (with respect to the beam axis) momentum $p_\perp$, which is usually given in terms of the transverse mass $m_\perp^2 = m^2 + p_\perp^2$, and the elliptic flow coefficient v_2, which results from fitting the particle spectrum in the transverse plane to a second harmonic $(1 + 2v_2(p_\perp)\cos 2\phi)$, where ϕ is the angle measured from the reaction plane. This is equivalent to considering an elliptic fireball, in which case v_2 measures the eccentricity of the ellipse. The first harmonic is called directed flow, and would represent a shifted spherical fireball in the transverse plane [VolZha96].

In our simplified discussion we have not considered the possibility that some particles produced at break-up may actually decay before reaching the detectors, so that the one-to-one correspondence we have assumed is not strictly valid. Also, because of long-range interactions, the propagation of charged particles from break-up to detection is not quite free. Both phenomena must be considered for a meaningful contrast between theory and experiment. Finally, observe that the form of the distribution function we have used is not a solution of the transport equation if there are gradients of the hydrodynamical variables. If these gradients are important, one may consider using an improved distribution function [Sin99].

The agreement of predictions from hydrodynamical simulations with experimental data is good, provided the simulation is started very early (earlier than 1 fm/c after the collision). If one believes that the validity of hydrodynamics preassumes (local) equilibration, this very short time is somewhat of a puzzle. However, as we shall see presently, not all is well with hydro simulations. This is the main area where NEqQFT may be relevant to understanding RHICs.

14.1.4 Measuring the fireball

We shall describe a method of data analysis from heavy ion collisions which, in principle, yields direct information on the geometry of the fireball at break-up. It pertains to studying the simultaneous detection of pairs of identical particles, rather than individual ones [GyKaWi79, Hei96, WieHei99].

Let us make the simplifying assumption that the only particles produced at break-up are pions, and that these may be treated as a free Klein–Gordon field. The Heisenberg pion field operator obeys a wave equation

$$\partial^2 \Phi(x) - m^2 \Phi(x) = -J(x) \tag{14.11}$$

where the external c-number source $J(x)$ represents the particle creating current at break-up. Under the action of this source, the pion vacuum state $|0\rangle$ evolves (in the interaction picture) into

$$|J\rangle = T\left\{ \exp\left[i \int d^4x\, J(x)\, \Phi_0(x) \right] \right\} |0\rangle \tag{14.12}$$

where $\Phi_0(x)$ is a free Klein–Gordon field. $\Phi_0(x)$ may be expanded into positive and negative frequency parts

$$\Phi_0(x) = \int \frac{d^3\mathbf{p}}{(2\pi)^3} \frac{e^{i\mathbf{p}\mathbf{x}}}{\sqrt{2\omega_p}} \left\{ e^{-i\omega_p t} a_\mathbf{p} + e^{i\omega_p t} a_{-\mathbf{p}}^\dagger \right\} \tag{14.13}$$

where $\omega_p^2 = \mathbf{p}^2 + m^2$. The state $|J\rangle$ is a coherent state

$$a_{\mathbf{p}} |J\rangle = \frac{iJ_{(\mathbf{p},\omega_p)}}{\sqrt{2\omega_p}} |J\rangle \qquad (14.14)$$

where

$$J_p = \int d^4x \, e^{-ipx} J(x) \qquad (14.15)$$

The number of particles with momentum p in the final state is then

$$N_p = \frac{|J_p|^2}{2p^0} \qquad (14.16)$$

Let us introduce the emission function

$$S(x,p) = \int d^4y \, e^{-ipy} J^* \left(x - \frac{y}{2}\right) J \left(x + \frac{y}{2}\right) \qquad (14.17)$$

whence

$$|J_p|^2 = \int d^4x \, S(x,p) \qquad (14.18)$$

Comparing (14.9) and (14.16), one may be strongly tempted to write

$$S(x,p) = \frac{g}{C(x)} \left[\frac{e^{-\Sigma^2/2(\Delta\Sigma)^2}}{\sqrt{2\pi}\,(\Delta\Sigma)}\right] \Sigma_{,\mu} N_{\text{hydro}}^{\mu} \delta \left(p^0 - \omega_p\right) \frac{U_\lambda^{\text{had}} p^\lambda}{[\exp\left(-\beta_\nu p^\nu - \mu\right) - 1]} \qquad (14.19)$$

The number of pairs of particles, one with spatial momentum $\mathbf{p}$ and another with spatial momentum $\mathbf{q}$, is

$$N_{\mathbf{pq}} = \langle J| a_{\mathbf{p}}^\dagger a_{\mathbf{q}}^\dagger a_{\mathbf{q}} a_{\mathbf{p}} |J\rangle \qquad (14.20)$$

For a coherent source, such as discussed so far, $N_{\mathbf{pq}} = N_{\mathbf{p}}N_{\mathbf{q}}$, which is not terribly interesting.

However, let us consider the case in which the source is an incoherent superposition of elementary sources

$$J(x) = \sum_i e^{i\theta_i} J_i(x), \qquad J_i(x) = e^{ip_i(x-x_i)} J_0(x-x_i) \qquad (14.21)$$

meaning that the identical elementary sources J_0 are translated, boosted and phased in different ways, with the x_i, p_i, θ_i all random mutually independent variables. In this case, the emission function reads

$$S(x,p) = \sum_{i,j} \int d^4y \, e^{-ipy} e^{i(\theta_i - \theta_j)} J_j^* \left(x - \frac{y}{2}\right) J_i \left(x + \frac{y}{2}\right) \qquad (14.22)$$

Averaging over the unknown phase of each source, we get

$$S(x,p) = \sum_i S_i(x,p) \qquad (14.23)$$

Let us consider again the average number of pairs

$$N_{pq} = \frac{1}{4\omega_p\omega_q} \left\langle J_p^* J_q^* J_q J_p \right\rangle$$

$$= \frac{1}{4\omega_p\omega_q} \sum_{ijkl} e^{i(\theta_i+\theta_j-\theta_k-\theta_l)} \left\langle J_{k,p}^* J_{l,q}^* J_{i,q} J_{j,p} \right\rangle \qquad (14.24)$$

The average over phases vanishes unless $i = l$, $j = k$ or $i = k$, $j = l$ (we neglect the possibility of $i = j = k = l$ simultaneously). Therefore

$$N_{pq} = N_p N_q + \frac{1}{4\omega_p\omega_q} \left| \sum_i \left\langle J_{i,q}^* J_{i,p} \right\rangle \right|^2 \qquad (14.25)$$

The second term shows the existence of correlations among the created particles. This is the so-called pion bunching, or HBT (for Hanbury-Brown/Twiss) correlations. In the real world, the sources are neither totally coherent nor totally incoherent; we may account for this by adding a fudge factor to the second term in (14.25) (for a more sophisticated treatment, see [AkLeSi01]). A similar factor may arise from the superposition of particle emission from a collision *core* and a *halo* of long-lived resonances [NiCsKi98].

Introducing

$$P = \frac{p+q}{2}, \qquad \xi = p - q \qquad (14.26)$$

then

$$\sum_i \left\langle J_{i,q}^* J_{i,p} \right\rangle = \int d^4x \, e^{-i\xi x} S(x, P) \qquad (14.27)$$

and we see that it is possible to express the HBT correlations in terms of the emission function, for which we already have the ansatz (14.19). In practice, this is too involved to attempt a direct comparison with data. Rather, the usual procedure is, for a given P, to evaluate the moments of the emission function

$$\int d^4x \, S(x, P) = 2\omega_P N_P \qquad (14.28)$$

$$\bar{x}^\mu = \frac{1}{2\omega_P N_P} \int d^4x \, x^\mu S(x, P) \qquad (14.29)$$

$$R^{\mu\nu} = \frac{1}{2\omega_P N_P} \int d^4x \, x^\mu x^\nu S(x, P) - \bar{x}^\mu \bar{x}^\nu \qquad (14.30)$$

Let us assume the source is axisymmetric and $\mathbf{P}$ points in the x-direction (z being the beam direction). In the center-of-mass frame we have $R^{0i} = R^{ij} = 0$ for $i \neq j$. The values of the momenta suggest the approximation

$$S(x, P) = \frac{2\omega_P N_P}{(2\pi)^2 (\det[R^{\mu\nu}])^{1/2}} \exp\left\{ -\frac{1}{2} \left[\frac{(t-\bar{t})^2}{R^{00}} + \frac{x^2}{R^{11}} + \frac{y^2}{R^{22}} + \frac{z^2}{R^{33}} \right] \right\}$$

$$(14.31)$$

It is important to realize that the $R^{\mu\nu}$ are not the moments of the source as a whole, since the emission function is weighted by P dependent factors. We may think of the emission functions as the probability of a particle with momentum P being emitted at point x. The $R^{\mu\nu}$ then measure the size of the region where emission is most likely. This expression for the emission function is simple enough that we may compute the HBT correlations.

One last point: If the p and q momenta in N_{pq} are on-shell, the components of P and ξ are not independent. We have $P^2 = m^2 - \xi^2/2$, so we may consider P on-shell when ξ is small, and $P\xi = 0$, meaning that $\xi^0 = (P/\omega_P)\,\xi^1$. Therefore

$$\left| \int d^4x \, e^{-i\xi x} S(x, P) \right|^2$$
$$= (2\omega_P N)_P^2 \, \exp\left\{ -\left[\left(\frac{R^{00}P^2}{\omega_P^2} + R^{11} \right) \xi_1^2 + R^{22}\xi_2^2 + R^{33}\xi_3^2 \right] \right\} \quad (14.32)$$

We see that the HBT correlations may be parameterized in terms of three "radii," with z corresponding to the "longitudinal" direction, x to the "out" direction, and y to the "side" direction.

$$\left| \int d^4x \, e^{-i\xi x} S(x, P) \right|^2 \sim \exp\left\{ -\left[R_{\text{out}}^2 \xi_1^2 + R_{\text{side}}^2 \xi_2^2 + R_{\text{long}}^2 \xi_3^2 \right] \right\} \quad (14.33)$$

($R_{\text{out}}^2 = R^{00}P^2/\omega_P^2 + R^{11}$, $R_{\text{side}}^2 = R^{22}$, $R_{\text{long}}^2 = R^{33}$). Observe that, in general, we expect $R^{11} \sim R^{22}$, and so the out radius, which is sensitive also to the duration of the emission process (in terms of laboratory time) is predicted to be larger than the side radius. This prediction is not borne out by the data, which show $R_{\text{out}}/R_{\text{side}} \sim 1.25$–$1.5$ [BRAHMS05, PHOBOS05, STAR05, PHENIX05]. This disagreement constitutes the so-called *HBT puzzle*.

This suggests that the emission process occurs early, which reinforces the need for an early onset of the hydrodynamic regime, or else for some new thinking [SoBaDi01, Hum06]. In principle the HBT puzzle is a puzzle only within the framework of hydrodynamical models.

14.1.5 Insights from nonequilibrium quantum field theory

We see from the above analysis that the clue to understanding the physics of RHICs lies in the first fermi/c or so after the collision. This is the point where nonequilibrium field theory methods may have an impact on the theory of RHICs.

The first input for any field theoretic modeling is of course some well-defined initial condition. The basic idea is that each colliding nucleus is not just a bunch of nucleons marching in step, but a rather complex array of gluons and partons. In fact, a naive perturbative calculation yields the result that the number of gluons with a given momentum diverges as the momentum becomes light-like. It is believed that this divergence is cut off at some scale by nonperturbative effects (parton saturation) [Mue01, KhaLev01, KhLeNa01].

A sophisticated model built on this premise is the so-called color glass condensate [IaleMc02, BjoVen01, KrNaVe02, McLVen94a, McLVen94b, McLVen94c].

The basic framework to understand the early evolution of the plasma is the so-called *bottom-up* scenario [BMSS01, MuShWo05]. The hard gluons released from the color glass condensate take part in both elastic and inelastic collisions. Elastic collisions broaden a little the initial gluon distribution (see below) while inelastic collisions contribute to the creation of a soft gluon background. It may be observed that the emission of ultrasoft gluons is suppressed by destructive interference between multiple collision events, the so-called Landau–Pomeranchuk–Migdal effect [BaiKAt03, ArMoYa01a, ArMoYa01b, ArMoYa02, BBGM06]. On the other hand, nearly collinear events are amplified by the small denominators in the transition amplitude [Won04].

The soft gluons thermalize very efficiently. Eventually they become the dominant species, and we have a picture of a few very energetic gluons on a thermalized soft gluon background. The remaining hard gluons decay (through gluon branching, which is a specific form of wave splitting for a non-abelian plasma). The decay of the hard gluons heats up the soft gluons over and above the cooling from the longitudinal expansion of the plasma, and so we may enter the fully hydrodynamic stage at a conveniently high temperature.

The key question in the bottom-up scenario is how fast the soft fields build up from the initial hard quanta. The natural approach would seem to be to write a kinetic equation for those hard gluons [Mue00a, Mue00b], taking into account both elastic and inelastic processes (see also Chapter 11). The result seems to be that the build up of soft fields is too slow to meet the demands of hydrodynamical RHIC models.

At the time of writing, much effort is being spent on elucidating a proposal by S. Mrowczynski which would result on a much faster growth rate [Mro94a, Mro94b] (see [Mro05] for a recent review). Mrowczynski's insight is that the initial gluon distribution must be highly anisotropic. Since gluons with a substantial longitudinal momentum will stream out of the central region, the momentum distribution in the local rest frame is squeezed along the beam. Under these conditions, the so-called filamentation or Weibel instability sets in. Suppose the initial hard gluon distribution results in alternating currents along a transverse direction. These currents create magnetic fields, and the corresponding Lorentz force accelerates particles along the longitudinal direction. Moreover, particles are redistributed in such a way that the initial currents are amplified, thus setting up a positive feedback loop. While the instability lasts, the soft fields are found to increase exponentially. Instabilities do not directly equilibrate the system but rather isotropize it and thus speed up the process of thermalization [Mro07].

Current efforts are aimed at a precise estimate of the growth rates that may be achieved by this mechanism, and to identify possible effects which may knock off the instability. At the time of writing, the most important limiting factor seems to be that the growing soft modes will in turn excite a turbulent ultraviolet cascade [ArLeMo03, ALMY05, Moo05, ArnMoo05, ArMoYa05, ArnMoo06, MuShWo07]. The energy extracted from the hard gluons through the magnetic fields is returned to them through the cascade. The growth of the soft modes turns from exponential to linear, and eventually ceases altogether. It is not clear whether this effect will rule out fast enough thermalization through Weibel instabilities. In principle, it ought to be possible to obtain an answer by coupling the Yang–Mills classical equations for the soft fields to the Wong kinetic equations for the hard fields (see Chapter 11 and [ManMro06, Mro06, RomReb06, DuNaSt07, Str06, RomVen06, Ven07]), but it is hard to carry out numerical simulations within a realistic parameter range.

14.2 Disoriented chiral condensates (DCCs)

Besides deconfinement, other exotic events are thought to lie just above the QCD phase transition. Among these, one of the best researched is the possibility of chiral symmetry restoration. More concretely, the idea is, if it were possible to heat strongly interacting matter above the chiral restoration temperature, and then quenching it again below the critical point, there exists the possibility that the second time around the system will settle into a different vacuum than the one we are familiar with. That would create a new form of matter, the so called "disoriented chiral condensate" (DCC). When brought into contact with the ordinary vacuum, the DCC would decay with a characteristic burst of particles, whose detection would provide a signature of its existence.

Theoretical and experimental interest in DCCs had a strong surge in the early 1990s [KowTay92], further motivated by the unexplained Centauro events seen in cosmic ray experiments [MohSer05]. After several searches both in an *ad hoc* experiment [MINIMAX03] and as a part of larger RHIC program, no clear detection has been reported. However, this null result is actually in agreement with theoretical estimates. New probes are being suggested which could lead to a positive result [AgSoVi06]. We refer the reader to [Bjo97] and [MohSer05] for reviews.

With these experimental perimeters delimited, let us describe in slightly more detail what a DCC is expected to look like. According to the standard models of particle interactions, the fundamental constituents of hadrons are quarks. There are six flavors of quarks, organized into three isospin doublets (u, d), (c, s) and (t, b). The quark masses increase as we go from one doublet to the next; for the (u, d) pair they are of a few MeV, about a GeV for (c, s) and a few GeV for (t, b). In a first approximation, the (u, d) quarks may be taken as massless.

Now, a theory with a massless isospin doublet would be invariant under *independent* global isospin rotations of the left and right quark components. Thus the isospin group should have been $SU(2) \times SU(2)$, rather than the observed $SU(2)$. In particular, for each hadronic state there would be a partner with opposite parity. This is not even approximately observed, and therefore the $SU(2) \times SU(2)$ symmetry must be broken down to the physical isospin $SU(2)$.

The idea is that the quark vacuum is not invariant under $SU(2) \times SU(2)$. Since the algebra of this group is isomorphic to $SO(4)$, it is natural to take the order parameter for this transition (*chiral* symmetry breaking) as a vector in a four-dimensional internal Euclidean space. The symmetric state corresponds to a vanishing order parameter. A nonzero order parameter picks up a definite direction in four-dimensional internal space, therefore breaking the symmetry down to $SO(3)$, with covering group $SU(2)$. From the microscopic point of view, the components of the order parameter express the formation of quark pair condensates, in a mechanism which resembles the formation of Cooper pairs (with a breaking of the $U(1)$ symmetry) in a BCS superconductor.

According to Goldstone's theorem, the breaking of a global symmetry must be followed by the apparition of one massless particle for each broken symmetry. In our case there are three, one for each $SU(2)$ generator, while the Goldstone bosons are the pions. Of course, quarks are not really massless, $SU(2) \times SU(2)$ is not an exact symmetry, and pions are therefore not quite massless, but their masses are small enough, certainly in comparison with the quarks themselves.

In this picture, pions are viewed as the lowest energy excitations of the quark vacuum, and at low energy the standard model is a pion theory. In the broken-symmetry phase, the modulus of the pion vector is fixed by the symmetry-breaking condition, and so pions are represented by a vector living in the unit sphere of Euclidean 4-space. This is the *nonlinear sigma model*. At higher energy, the modulus also becomes dynamical, and we may represent pions as a 4-vector self-interacting via a $SO(4)$-invariant potential. This is the *linear sigma model*, which will be the starting point for our discussion below.

As we have seen, the Bjorken scenario of a RHIC leads to the "baked Alaska" picture of the collision, where the edge of the expanding central region is hotter than its center. The hot plasma layer shields the cool center from interaction with the outer world, and therefore makes it possible for cooling the pion field to develop in a direction (in internal isospin space) different from the (cosmologically chosen) direction outside.

At some point the outer layer will be cool enough that causal contact will be restored, and the "disoriented" pion condensate will register as "ordinary" pions. Suppose that we call z the direction corresponding to neutral pions in isospin space, and that the disoriented pion condensate points in a direction z' at an angle Θ with respect to z. Upon decay into ordinary matter, the ratio of neutral to total number of pions will go roughly as $f = \cos^2 \Theta$. Assuming that all directions in the unit sphere in isospace are equivalent, and recalling that the same f results from angles Θ and $\pi - \Theta$, then the probability to find a ratio between f and $f + df$ would go like $df/\sqrt{f}$. This characteristic distribution is another remarkably simple prediction of the "baked Alaska" scenario. Other signatures of DCC formation involve the nonequilibrium emission of photons [BVHK97, CNLL02].

Going beyond this qualitative picture, we now wish to introduce a microscopic perspective based upon nonequilibrium quantum field theory to provide a more detailed description of the chiral phase transition in the aftermath of the collision. We will largely follow the treatment by Cooper and collaborators [CKMP95, CoKlMo96, LaDaCo96]. To the best of our knowledge, this was also one of the first attempts to apply NEqQFT to a realistic experimental situation. Mean field models have also been investigated [MroMul95, Ran97, AmBjLa97], and there is a proposal to study DCC evolution within a Langevin framework [BeRaSt01].

14.2.1 Self-consistent mean fields in the large N approximation

Adopting the above qualitative picture we now study the evolution of the mean field in a $O(4)$ symmetric theory assumed to describe the low-energy excitations of the QCD vacuum. We shall make one further simplification, namely, instead of $O(4)$ we work with an $O(N)$ theory under the *large N* approximation. We have studied the large N (LN) approximation in Chapter 6. Unlike there, now we have to account for the possibility of symmetry breaking. To avoid misunderstandings, we shall develop the relevant formulae from scratch.

The $O(N)$ invariant action, allowing for spontaneous symmetry breaking, reads

$$S = \int d^4x \left\{ -\frac{1}{2} \partial_\mu \Psi^A \partial^\mu \Psi^A - \frac{\lambda}{8N} \left(\Psi^A \Psi^A - Nv^2 \right)^2 \right\} \qquad (14.34)$$

We scale $\Psi^A = \sqrt{N} \Phi^A$ to get

$$S = N \int d^4x \left\{ -\frac{1}{2} \partial_\mu \Phi^A \partial^\mu \Phi^A - \frac{\lambda}{8} \left(\Phi^A \Phi^A - v^2 \right)^2 \right\} \qquad (14.35)$$

To make the perturbative expansion more manageable, we use the Coleman–Jackiw–Politzer trick of including an auxiliary field $\chi = \lambda \left(\Phi^A \Phi^A - v^2 \right)/2$, by adding a term to the action, which becomes

$$S = N \int d^4x \left\{ -\frac{1}{2} \partial_\mu \Phi^A \partial^\mu \Phi^A - \frac{\lambda}{8} \left(\Phi^A \Phi^A - v^2 \right)^2 \right.$$
$$\left. + \frac{1}{2} \left(\frac{\chi}{\sqrt{\lambda}} - \frac{\sqrt{\lambda}}{2} \left(\Phi^A \Phi^A - v^2 \right) \right)^2 \right\} \qquad (14.36)$$

Expanding out, we get

$$S = N \int d^4x \left\{ \frac{-1}{2} \partial_\mu \Phi^A \partial^\mu \Phi^A + \frac{\chi^2}{2\lambda} + \frac{1}{2} v^2 \chi - \frac{1}{2} \chi \Phi^A \Phi^A \right\} \qquad (14.37)$$

In this new action, strings of fish graphs beyond two loops are no longer 2PI. The next nontrivial graph is the three-pointed star, Fig. 6.10 in Chapter 6, which scales as N^{-1}. Thus, once again, we obtain a closed form for NLO large N.

To obtain this explicit expression, we begin by shifting the field $\Phi^A \to f^A + \varphi^A$, $\chi \to K + \bar{\kappa}$. As usual, we discard linear terms, so

$$S = S\left[f^A, K\right] + N \int d^4x$$
$$\times \left\{ \frac{-1}{2} \partial_\mu \varphi^A \partial^\mu \varphi^A + \frac{\bar{\kappa}^2}{2\lambda} - \frac{1}{2} K \varphi^A \varphi^A - f^A \bar{\kappa} \varphi^A - \frac{1}{2} \bar{\kappa} \varphi^A \varphi^A \right\} \quad (14.38)$$

It is convenient to eliminate the quadratic cross-term, shifting $\bar{\kappa} = \kappa + \lambda f^A \varphi^A$. We get

$$S = S\left[f^A, K\right] + N \int d^4x \left\{ -\frac{1}{2}(\partial\varphi)^2 + \frac{\kappa^2}{2\lambda} - \frac{1}{2} M_{AB}^2 \varphi^A \varphi^B \right.$$
$$\left. - \frac{1}{2} \kappa \varphi^A \varphi^A - \frac{1}{2} \lambda f^A \varphi^A \varphi^B \varphi^B \right\} \quad (14.39)$$

$(M_{\alpha\beta}^2 = K\delta_{AB} + \lambda f_A f_B)$, where the 2PIEA is

$$\Gamma^{\mathrm{NLO}} = S\left[f^A, K\right] + \frac{N}{2} \left\{ \left[\nabla^2 \delta_{AB} - M_{AB}^2\right] G^{AB} + \frac{H}{\lambda} \right\}$$
$$- \frac{i\hbar}{2} \left\{ \mathrm{Tr}\,\ln H + \mathrm{Tr}\,\ln G \right\} + \Gamma_Q^{\mathrm{NLO}} + \mathrm{O}\left(N^{-1}\right) \quad (14.40)$$

$$\Gamma_Q^{\mathrm{NLO}} = \frac{iN^2}{4\hbar} \int d^4x\,d^4x' \left\{ H(x, x')\left[G^{AB}(x, x')\right]^2 + \lambda^2 f^A(x) f^B(x') \Delta^{AB}(x, x') \right\} \quad (14.41)$$

$$\Delta^{AB}(x, x') = G^{AB}(x, x')\left[G^{CD}(x, x')\right]^2 + 2G^{AD}(x, x') G^{CD}(x, x') G^{CB}(x, x') \quad (14.42)$$

Let us write the equations of motion leaving the CTP indices implicit

$$\nabla^2 f^A - K f^A - \lambda G^{AB}(x, x) f_B(x) + \frac{i\lambda^2 N}{2\hbar} \int d^4x'\, f^B(x') \Delta^{AB}(x, x') = 0 \quad (14.43)$$

$$\frac{K}{\lambda} + \frac{1}{2}v^2 - \frac{1}{2} f^A f^A - \frac{1}{2} G^{AA}(x, x) = 0 \quad (14.44)$$

$$1 - \frac{i\lambda\hbar}{N} H^{-1} + \frac{i\lambda N}{2\hbar}\left[G^{AB}(x, x')\right]^2 = 0 \quad (14.45)$$

$$\left[\nabla^2 \delta_{AB} - M_{AB}^2\right] - \frac{i\hbar}{N} G_{AB}^{-1} + \frac{iN}{\hbar} H(x, x') G^{AB}(x, x')$$
$$+ \frac{i\lambda^2 N}{2\hbar} f^A(x) f^B(x')\left[G^{CD}(x, x')\right]^2$$
$$+ \frac{i\lambda^2 N}{\hbar} f^C(x) f^D(x') G^{CD}(x, x') G^{AB}(x, x')$$
$$+ \frac{i\lambda^2 N}{\hbar} f^C(x) f^D(x') G^{CB}(x, x') G^{AD}(x, x')$$
$$+ \frac{i\lambda^2 N}{\hbar} f^A(x) f^D(x') G^{CD}(x, x') G^{CB}(x, x')$$
$$+ \frac{i\lambda^2 N}{\hbar} f^C(x) f^B(x') G^{CD}(x, x') G^{AD}(x, x') = 0 \quad (14.46)$$

It is clear from these equations that the propagators are $O\left(N^{-1}\right)$, and therefore some of the terms we have included are actually of higher order. In removing them, however, we must be careful that we compute factors of N which may arise when summing over internal indices. The resulting equations are

$$\nabla^2 f^A - K f^A = 0 \tag{14.47}$$

$$\frac{K}{\lambda} + \frac{1}{2}v^2 - \frac{1}{2}f^A f^A - \frac{1}{2}G^{AA}(x,x) = 0 \tag{14.48}$$

$$1 - \frac{i\lambda\hbar}{N}H^{-1} + \frac{i\lambda N}{2\hbar}\left[G^{AB}(x,x')\right]^2 = 0 \tag{14.49}$$

$$\left[\nabla^2 \delta_{AB} - M^2_{AB}\right] - \frac{i\hbar}{N}G^{-1}_{AB} + \frac{i\lambda^2 N}{2\hbar}f^A(x)f^B(x')\left[G^{CD}(x,x')\right]^2 = 0 \tag{14.50}$$

We observe that the H propagator does not feed back on the mean fields, so we will not consider its evolution. For the other propagators, it is convenient to discriminate between the "pion" propagators $G^{AB}_\perp$ (which is defined by the property that $G^{AB}_\perp f_B \equiv 0$) and the "sigma" propagator, that is, the propagator for fluctuations along the mean field. Since there are $N-1$ "pions" and only one "sigma," only the former feed back on K. Writing only the equations for the mean field and the pion propagator, we obtain

$$\nabla^2 f^A - K f^A = 0 \tag{14.51}$$

$$\frac{K}{\lambda} + \frac{1}{2}v^2 - \frac{1}{2}f^A f^A - \frac{1}{2}G^{AA}_\perp(x,x) = 0 \tag{14.52}$$

$$\left[\nabla^2 - K\right]\delta_{\perp AB} - \frac{i\hbar}{N}G^{-1}_{\perp AB} = 0 \tag{14.53}$$

where $\delta_{\perp AB}$ is the projector orthogonal to the mean field. These are the equations which determine the mean field evolution. Clearly, they admit a solution where $f^1 = f$, $f^A = 0$ $(A \neq 1)$ and $G^{AB}_\perp = G(x,x')\delta^{AB}_\perp/N$. For such a solution, we obtain

$$\nabla^2 f - K f = 0 \tag{14.54}$$

$$\frac{K}{\lambda} + \frac{1}{2}v^2 - \frac{1}{2}f^2 - \frac{1}{2}G(x,x) = 0 \tag{14.55}$$

$$\left[\nabla^2 - K\right] - i\hbar G^{-1} = 0 \tag{14.56}$$

where we have used the result that $(N-1)/N = 1 - O(1/N)$.

14.2.2 The quantum pion field

The equations for the mean fields and the pion propagator are simplified by the observation that the latter are identical to the equations for the propagators of free fields with a position-dependent mass K. Thus we may introduce a "Heisenberg" pion field Φ, decompose it in modes, and then compute the propagators by summing over modes in the usual way. It is natural to choose a set of modes adapted to the boost symmetry of the baked Alaska scenario. That is, we introduce, instead of the usual Minkowski coordinates t and x, *Rindler* coordinates τ and η, defined by $t = \tau \cosh \eta$ and $x = \tau \sinh \eta$. In these coordinates, the Minkowski metric reads $ds^2 = -d\tau^2 + \tau^2 d\eta^2 + dx_\perp^2$ ($\mathbf{x}_\perp = (y, z)$), and the D'Alembertian $\nabla^2 = -\tau^{-1} \partial_\tau \tau \partial_\tau + \tau^{-2} \partial_\eta^2 + \nabla_\perp^2$. We therefore write

$$\Phi(\tau, \eta, \mathbf{x}_\perp) = \int \frac{d^2 k_\perp dp}{(2\pi)^{3/2}} e^{ik_\perp x_\perp} e^{ip\eta} \left\{ \phi_{pk_\perp}(\tau) a_{pk_\perp} + \phi^*_{pk_\perp}(\tau) a^\dagger_{-p-k_\perp} \right\}$$

(14.57)

where the mode functions obey

$$\left\{ \frac{1}{\tau} \frac{d}{d\tau} \tau \frac{d}{d\tau} + \frac{p^2}{\tau^2} + k_\perp^2 + K \right\} \phi_{pk_\perp}(\tau) = 0$$

(14.58)

and the destruction operators $a_{pk_\perp}$ have the usual commutation relations

$$\left[a_{pk_\perp}, a^\dagger_{p'k'_\perp} \right] = \delta(p - p') \delta^2(k_\perp - k'_\perp)$$

(14.59)

To obtain the usual ETCCRs for the field operators, we must normalize the modes as

$$\phi^*_{pk_\perp}(\tau) \frac{d}{d\tau} \phi_{pk_\perp}(\tau) - \phi_{pk_\perp}(\tau) \frac{d}{d\tau} \phi^*_{pk_\perp}(\tau) = -\frac{i}{\tau}$$

(14.60)

If we make the reasonable assumption that the initial state, defined on some surface $\tau = \tau_0 = $ constant, is an incoherent superposition of states with well-defined occupation numbers as defined from the $a_{pk_\perp}$ particle model, then the coincidence limit in the equation for K reads

$$G = 2 \int \frac{d^2 k_\perp dp}{(2\pi)^3} |\phi_{pk_\perp}(\tau)|^2 \left\{ \frac{1}{2} + n^0_{pk_\perp} \right\}$$

(14.61)

where $n^0_{pk_\perp}$ is the occupation number for the corresponding mode in the initial state.

14.2.3 Adiabatic modes and renormalization

At this point, we have reduced the problem of computing the mean field evolution to a system of $n + 2$ coupled ordinary differential equations, where n is the number of modes we care to include in our numerical solution (already this problem is too complex to attempt a closed analytical solution). Since the number of modes is necessarily finite, in effect we are imposing a momentum cut-off on the theory. This means that the coincidence limit (14.61) is *de facto* finite, but, since it diverges as the cut-off is removed, it is strongly cut-off dependent.

Physics, on the other hand, is supposed to be cut-off *in*dependent, so we should be able to absorb the dependence on the cut-off by renormalizing the parameters in the equation for K, namely λ and v^2, which implies, as a previous necessary condition, that the cut-off dependent part of the coincidence limit depends on the instantaneous value of K, but not on its derivatives.

To analyze the ultraviolet behavior of the mode amplitudes, let us write
$\phi_{pk_\perp}(\tau) = \tau^{-1/2} \varphi_{pk_\perp}(\tau)$, whereby

$$\frac{d^2}{d\tau^2} \varphi_{pk_\perp}(\tau) + \left[\Omega^2_{pk_\perp}(\tau) + \frac{1}{4\tau^2} \right] \varphi_{pk_\perp}(\tau) = 0, \qquad \Omega^2_{pk_\perp}(\tau) = \frac{p^2}{\tau^2} + k_\perp^2 + K \tag{14.62}$$

and $\varphi^*_{pk_\perp} \varphi'_{pk_\perp} - \varphi_{pk_\perp} (\varphi'_{pk_\perp})^* = -i$, where a prime stands for a τ derivative. In this regime $\Omega^2_{pk_\perp}(\tau)$ becomes a slowly varying function of τ. This suggests trying a WKB-type solution

$$F_{pk_\perp}(\tau) = \frac{e^{-iS(\tau)}}{\sqrt{2w_{pk_\perp}(\tau)}}, \qquad S = \int^\tau d\tau' \, w_{pk_\perp}(\tau') \tag{14.63}$$

$F_{pk_\perp}$ is well normalized by construction, and the mode equation becomes

$$w^2_{pk_\perp} = \Omega^2_{pk_\perp}(\tau) + \frac{1}{4\tau^2} - \frac{1}{4} \frac{\left(w^2_{pk_\perp} \right)''}{w^2_{pk_\perp}} + \frac{5}{16} \left[\frac{\left(w^2_{pk_\perp} \right)'}{w^2_{pk_\perp}} \right]^2 \tag{14.64}$$

The hypothesis of slow variation allows us to seek an adiabatic solution, namely, an iterative solution starting from the zeroth order approximation $w^2_{pk_\perp} = \Omega^2_{pk_\perp}(\tau)$. Let us write this solution as a formal series

$$w^2_{pk_\perp} = \sum_{n=0}^\infty W^{(n)} \left[\frac{p^2}{\tau^2}, k_\perp^2, \tau \right] \tag{14.65}$$

where $W^{(n)}$ is a homogeneous function of p^2/τ^2 and $k_\perp^2$ of degree $1 - n$. It follows that

$$\frac{1}{w_{pk_\perp}} = \frac{1}{[W^{(0)}]^{1/2}} - \frac{1}{2} \frac{W^{(1)}}{[W^{(0)}]^{3/2}} + R \tag{14.66}$$

where R vanishes at large momentum as (momentum)$^{-5}$. It is clear that only $W^{(0)}$ and $W^{(1)}$ may contribute to the cut-off dependence. Replacing (14.65) into (14.64), we obtain

$$W^{(0)} = \frac{p^2}{\tau^2} + k_\perp^2 \tag{14.67}$$

$$W^{(1)} = K + \frac{1}{4\tau^2} - \frac{1}{4} \frac{\frac{6p^2}{\tau^4}}{\frac{p^2}{\tau^2} + k_\perp^2} + \frac{5}{16} \left(\frac{\frac{2p^2}{\tau^3}}{\frac{p^2}{\tau^2} + k_\perp^2} \right)^2 \equiv K + \frac{k_\perp^2 \left(\frac{-4p^2}{\tau^2} + k_\perp^2 \right)}{4\tau^2 \left(\frac{p^2}{\tau^2} + k_\perp^2 \right)^2} \tag{14.68}$$

The potentially cut-off dependent terms in the coincidence limit of the propagator are

$$\frac{1}{\left(\frac{p^2}{\tau^2} + k_\perp^2 \right)^{1/2}} - \frac{K}{2 \left(\frac{p^2}{\tau^2} + k_\perp^2 \right)^{3/2}} - \frac{k_\perp^2 \left(\frac{-4p^2}{\tau^2} + k_\perp^2 \right)}{8\tau^2 \left(\frac{p^2}{\tau^2} + k_\perp^2 \right)^{7/2}} \tag{14.69}$$

However, the third term vanishes upon integration (this is easiest to see in polar coordinates). In conclusion, we obtain the same cut-off dependence as from the simple approximation $w^2_{pk_\perp} = \Omega^2_{pk_\perp}(\tau)$, and in passing we have proved that the cut-off dependent terms are functions of the instantaneous value of K, as required.

To complete the renormalization procedure, we write

$$v^2 = v^2_r + \frac{\Lambda^2}{4\pi^2}; \qquad \frac{1}{\lambda} = \frac{1}{\lambda_r} - \frac{1}{8\pi^2} \ln\left(\frac{\Lambda}{\kappa}\right) \qquad (14.70)$$

where κ defines the renormalization point. The finite equations of motion now read

$$\nabla^2 f - K f = 0 \qquad (14.71)$$

$$\frac{K}{\lambda_r} + \frac{1}{2} v^2_r - \frac{1}{2} f^2 - \frac{1}{2} M^2 = 0 \qquad (14.72)$$

$$\frac{d^2}{d\tau^2} \varphi_{pk_\perp}(\tau) + \left[\Omega^2_{pk_\perp}(\tau) + \frac{1}{4\tau^2}\right] \varphi_{pk_\perp}(\tau) = 0 \qquad (14.73)$$

where $\Omega^2_{pk_\perp}(\tau) = \frac{p^2}{\tau^2} + k^2_\perp + K$ and

$$M^2 = \frac{1}{\tau} \int^\Lambda \frac{d^2 k_\perp dp}{(2\pi)^3} \left\{ |\varphi_{pk_\perp}(\tau)|^2 (1 + 2n^0_{pk_\perp}) - \frac{1}{\left(\frac{p^2}{\tau^2} + k^2_\perp\right)^{1/2}} \right.$$
$$\left. + \frac{K\theta\left(\frac{p^2}{\tau^2} + k^2_\perp - \kappa^2\right)}{2\left(\frac{p^2}{\tau^2} + k^2_\perp\right)^{3/2}} \right\} \qquad (14.74)$$

In a typical collision, the initial occupation numbers $n^0_{pk_\perp}$ will be high enough to ensure a large positive M^2, and therefore also K will be positive; in this regime, the symmetric point $f = 0$ is stable. As the system expands and cools, M^2 will go down, and eventually K becomes negative. This event marks the start of the chiral symmetry-breaking transition, and the formation of the disoriented condensate.

For negative K and large enough τ, not only f but also some of the long-wavelength modes will grow exponentially. This will shift the particle spectrum towards the infrared, which becomes the basic signal for DCC formation.

In summary, we have depicted DCC formation as a spinodal decomposition process in an expanding geometry. Since we have already discussed a similar process in Chapter 4, we will not discuss further the evolution of this model. The size and duration of the ordered domains determine the prospective sizes of the DCCs, and therefore the probability of their detection.

15

Nonequilibrium quantum processes in the early universe

As stated in the Preface, we intend the chapters in the last part of the book to illustrate how quantum field theoretical methods can be applied to nonequilibrium statistical processes in several areas of current research, specifically, particle–nuclear processes (in RHIC and DCC), dynamics of cold atoms (BEC) in AMO physics and quantum processes in the early universe (cosmology) and in this endeavor also try to present an introduction to an important subject matter in that area. With this specified emphasis on the applications of techniques of NEqQFT, these accounts are more in the nature of a research topic exercise or extended example than a full review, in that the topics are selected because of the NEqQFT context, and the presentations are illustrations of the methodology. Thus we suggest the reader refer to review articles or monographs to get a more balanced and complete view on different physical approaches to the same subject matter.

In this chapter on cosmology, after a brief introduction to inflationary cosmology, highlighting the stochastic inflation model, we discuss how NEqQFT impacts on some central issues in cosmology. The methodology introduced in Chapters 4–6 covering particle creation mechanisms and the nPI CTP-CGEA/IF functional formalisms for NEq processes can be applied to solve a number of basic problems in cosmology.

Some specific processes have been discussed in earlier parts of this book. In Chapter 5, with the aid of the CGEA and the influence functional [Hu94b] we learned the relationship between the processes of dissipation, fluctuation, noise and decoherence. Then, in Chapter 9, we examined, starting from first principles, under what circumstances the fluctuations of a quantum field transmute into classical, stochastic fluctuations. We used a simple model to illustrate how decoherence comes about in a quantum phase transition. We then used a partitioned interacting scalar field theory in de Sitter spacetime to show how in the stochastic inflation paradigm the long-wavelength sector gets decohered and becomes classical under the influence of the short-wavelength sector acting as noise (more precisely, the rms value of the fluctuations can be treated as classical). Here we continue this investigation in early universe quantum processes, focusing on three major topics: the origin and nature of noise from quantum fields, structure formation from colored noises, and reheating from particle creation after inflation.

15.1 Quantum fluctuations and noise in inflationary cosmology

15.1.1 Inflationary cosmology

In modern cosmology, before the advent of the inflationary universe, the widely accepted model which explains very well the present day observed universe (according to the high-precision experiments of the 1990s and 2000s such as COBE and WMAP) has been the so-called standard model [Pee80] based on the Friedmann–Lemaitre–Robertson–Walker (FLRW) universe. Filled with a classical matter source with equation of state pressure $p = \gamma\rho$ matter density, its scale factor $a(t)$ undergoes a power-law expansion $a(t) = t^\alpha$ in cosmic time t. Thus for a spatially flat FLRW universe, in the matter-dominated era $\gamma = 0$, $\alpha = 2/3$ for a pressureless fluid; and in the radiation-dominated era, $\gamma = 1/3, \alpha = 1/2$ for a relativistic fluid.

Since the 1980s the inflationary cosmology has become a widely accepted paradigm to explain the observed large-scale flatness and homogeneity of the universe [LytRio99, Rio02]. Inflation also provides an efficient mechanism for the magnification of quantum fluctuations to cosmological scales, and the generation of small curvature perturbations which in principle can produce the observed cosmic microwave background (CMB) temperature anisotropies and provide the seeds for the formation of large-scale structures from galaxies to superclusters in today's universe.

Inflationary cosmology can be represented by the same FLRW spacetime, but instead filled with a constant energy density source which drives the universe (again assuming a spatially flat metric) into a phase of exponential expansion, $a(t) = a_0 \exp(Ht)$, where $H = \dot{a}/a$ is the Hubble expansion rate (a dot over a quantity stands for a derivative with respect to cosmic time t). This is the Einstein–de Sitter model obtained by de Sitter in 1917 from a solution of Einstein's equation with a cosmological term. When interpreted as classical matter this constant energy density source corresponds to matter with an unphysical equation of state $p = -\rho$ because it admits acausal propagation. What turned the de Sitter universe into a viable cosmological model was when Guth in 1981 proposed that this constant energy density in the potential energy is associated with the expectation value of a quantum field (the Higgs or the gauge field) which mediates some particle physics symmetry-breaking process in the early universe. Inflation was originally motivated by the removal of monopole overabundance in the GUT epoch, which it does, but turned out to be highly successful in addressing the flatness and horizon issues which are the more significant and immediate problems in cosmology.

The quantum scalar field Φ which drives inflation, known as the inflaton, evolves according to the equation

$$\ddot{\Phi} + 3H\dot{\Phi} + dV[\Phi]/d\Phi = 0, \qquad (15.1)$$

where the potential $V[\Phi]$ can take on a variety of forms, such as the Φ^4 double well potential in Guth's original "old" inflation [Guth81, Sato81]; an almost-flat

Coleman–Weinberg potential (of a massless field with only radiative correction) in the "new" inflation of Albrecht-Steinhardt and Linde [AlbSte82, Lin82]; a $m^2\Phi^2$ potential in Linde's chaotic inflation [Lin85]; an exponential form giving rise to power-law inflation [LucMat85] and many more later models suggested for specific purposes. The main idea is to get the universe into a vacuum energy dominated stage (the *entry problem*), to find ways (or rationale) to sustain the inflation for at least 68 e-folding time so as to produce sufficient entropy content of our present universe, and to get it out of this supercooled stage (the *exit problem*) by reheating it to the radiation-dominated FLRW universe described by the standard model.

Issues in the three stages pertaining to NEqQFT

Much work in the 1980s till now was devoted to the second issue, i.e. finding the right potential for inflation to serve specific purposes (see, e.g. [SteTur84]). Serious work on reheating started in the mid-1990s, but somewhat surprisingly, the very first issue, *the entry problem*, i.e. how did the universe get into a vacuum dominated phase, has not been taken up and pursued in earnest in the inflationary cosmology community, except for a brief period in the early 1990s [SalBon91, Hod90, MMOL91, KBHP91]. In principle one expects this issue can be resolved if we know what had happened in an earlier epoch. In this regard there were studies in quantum cosmology in the 1980s pertaining to this question. There were claims from both the no-boundary wavefunctions proposal of Hartle and Hawking [HarHaw83] and the "birth" by tunneling idea of Vilenkin [Vil83b] that these scenarios admit the de Sitter solution. This is an important issue of principle, related to what metastable states can exist in the pre-inflationary stage, what mechanisms can induce the universe to become vacuum dominated, and the probability it actually did. At the level of ideas there were criticisms of principle and of practice (e.g. [GiHaSt87, HawPag88, HolWal02, KoLiMu02]) and there were many plausibility arguments presented. More quantitative methods involve the derivation and solution of a Fokker–Planck equation for the distribution function constructed out of the universe's quantum state, from which one can examine the likelihood the universe could enter into a metastable state (the false vacuum) and stay there long enough to start inflation. See, e.g. [Sta82].

For the second issue, on *the dynamics of inflation*, in Guth's original model (old inflation) with a double well potential, the universe gets out of the vacuum dominated stage by tunneling. However, the underlying nucleation process happens infrequently and gives rise to a highly inhomogeneous universe. This can be improved upon by invoking a nearly flat potential as in new inflation, or by allowing the inflaton to slowly roll down the quadratic potential as in chaotic inflation. In all these cases, a slow-roll condition is desirable to sustain the inflationary expansion for a reasonable duration.

For the third issue, in conjunction with the so-called "graceful exit" problem, much detailed consideration has been devoted in the last 10 years to the *post-inflation reheating* processes. This epoch after the inflationary expansion contains several stages: preheating, reheating and thermalization. These processes are important because the temperature and entropy generated as the universe reheats after inflation are important parameters which enter into all ensuing cosmological processes.

What we want to point out is that in all three stages, the basic issues can be formulated in the language of NEqQFT, and be addressed with the techniques of NEqQFT we have constructed in earlier chapters. For example, on the "entry" or "get-started" issue, a more productive approach to the investigation on whether any metastable state exists could be by means of the Fokker–Planck equation for the distribution function (or a related master equation for the density matrix) of the universe. The second issue on the energetics of inflation depends strongly on the nature and dynamics of phase transition, whether it is first order via nucleation, as in old inflation or second order via spinodal decomposition as in new inflation. Vital issues in the quantum theory of structure formation, such as when the long-wavelength sector of the inflaton becomes classical, and what kind of noise the short-wavelength sector of quantum fluctuations engender, if any, are fundamentally NEqQFT problems. The third stage of reheating involves particle creation from the rapidly changing inflaton field as it descends a steep potential well, and is reasonably well treated by the CTP 2PI effective action, as we will illustrate in the last part of this chapter.

Stochastic inflation

To address these issues in some detail and to seek solutions, we now specialize and delve into one such theory of inflation known as stochastic inflation which was proposed by Starobinsky [Sta86] (see also earlier work by Vilenkin [Vil83a, Vil83b]) and developed by many [BarBub87, Rey87, PolSta96, GoLiMu87, NaNaSa88, NamSas89, Nam89, LiLiMe94, Hab90, StaYok94, Mat97a, Mat97b, WinVil00]. In this theory the inflation field is divided into two parts at every instant according to their physical wavelengths, i.e.

$$\Phi(x) = \Phi_<(x) + \Phi_>(x) \tag{15.2}$$

The first part $\Phi_<$ (the "system field") consists of field modes whose physical wavelengths associated with physical momenta $p \equiv k/a$ are longer than the de Sitter horizon size, i.e. $p < \sigma H$ where σ is a parameter smaller than unity defining the size of the coarse-graining domain and the shape of the window function. The second part $\Phi_>$ (viewed as the "environment field") consists of field modes whose physical wavelengths are shorter than the horizon size whereby $p > \sigma H$. Inflation continuously shifts additional modes of the environment field into the system, stretching their physical wavelengths beyond the de Sitter horizon size. Technically the system field can be obtained from the total field by introducing a dynamic cut-off in momentum space through a suitable time-dependent window function that filters out the modes whose frequencies are lower than the comoving horizon size.

Due to the exponentially rapid expansion of spacetime, fluctuations of the inflaton field $\Phi(x)$ on super-horizon scales effectively "freeze" in a few Hubble times H^{-1} after they leave the horizon. For this reason, it is often said that after suitable smoothing on the super-horizon scales, the averaged field containing the long-wavelength modes (the system field) $\Phi_<$ can be considered to be classical. The quantum field comprising of shorter wavelength modes (the environment field) can effectively be viewed as a classical noise ξ driving the system field via a Langevin equation of the form

$$\dot{\Phi}_< + \frac{1}{3H}\frac{dV[\Phi_<]}{d\Phi_<} = \xi(\mathbf{x}, t) \tag{15.3}$$

where $V(\Phi)$ is the inflaton potential, and the "noise field" $\xi(\mathbf{x}, t)$ is assumed for simplicity (but not required – this would be true for free fields anyway) to be a Gaussian random field characterized by its two-point function $\langle \xi(\mathbf{x}, t)\xi(\mathbf{x}', t')\rangle$. This noise correlator plays a key role in stochastic inflation.

To examine the form of the noise field, one can first examine a free scalar field in the de Sitter spacetime, wherein the scale factor (assuming a spatially flat FLRW universe) $a(t) \sim e^{Ht}$. (In reality the scalar field can only be approximately massless and the spacetime approximately de Sitter, because otherwise the universe will be forever inflating.) In Starobinsky's original derivation [Sta86] the noise correlator is given by

$$\langle \xi(\mathbf{x}, t)\xi(\mathbf{x}', t')\rangle = \left(\frac{H}{2\pi}\right)^2 \frac{\sin\theta}{\theta}\delta(t - t') \tag{15.4}$$

where $\theta \equiv r/R$; $r \equiv |\mathbf{x} - \mathbf{x}'|$, and R is the spatial averaging scale for the inflaton field:

$$R(t) \equiv [\sigma H a(t)]^{-1} \tag{15.5}$$

with $\sigma \ll 1$.

This equation is the basis for the investigation of structure formation. Two basic issues are: How does the long-wavelength sector become classical, and what is the underlying mechanism? What makes the short-wavelength sector behave like noise, and what kind of noise is it? As will be shown below, the characteristics of the noise field play a pivotal role in determining the spectral function of structures and the decoherence of the system field.

In this model, the partition of the system and environment modes is a crucial element which affects the outcome of structure formation, since the noise generated from it after being amplified in the inflationary dynamics is responsible for the structure of the late universe. Following Starobinsky's proposal [Sta86] many papers have been written using a Langevin equation with a white noise source, but the justification was not so clearly understood. A few authors (e.g. [HuPaZh93b, CalHu95, CalGon97, Mat97a, Mat97b]) took exception to this way of noise generation and suggested that, rather than using a window function for free fields which contains an arbitrary parameter, an interacting quantum field (which the inflaton is assumed to be) when partitioned into two sectors can naturally produce noise which in general is colored and multiplicative. Recently it was pointed out [WinVil00] that the white noise originating from a sharp momentum cut-off (or the window function being a step function in Fourier space) has some pathological behavior, whereas a smooth window function will necessarily lead to a colored noise.

As noticed by Winitzki and Vilenkin (WV) [WinVil00], equation (15.4) shows a surprisingly slow decay of correlations at large distances. For comparison, the two-point function of the time derivatives of the unsmoothed field $\langle \dot{\phi}(\mathbf{x}, t) \dot{\phi}(\mathbf{x}', t') \rangle$ at large separations r behaves as $\propto r^{-4}$ (here the angular brackets denote vacuum expectation value rather than statistical average). One would not expect a smearing of the field operators $\phi(\mathbf{x}, t)$ on scales R to have such an effect on correlations at distances $r \gg R$.

The analysis of WV shows that the origin of the unusual behavior of the correlator found by Starobinsky is the sharp momentum cut-off in his smoothing procedure. With a smooth cut-off, WV recover the r^{-4} behavior independently of the cut-off window function and find that the time dependence of the noise correlator at large times is generically $\propto \exp(-2Ht)$ instead of a sharp δ-function dependence of equation (15.4).

For the correct prediction of the density contrasts in a quantum theory of structure formation in the early universe it is necessary to give a proper treatment of quantum and classical fluctuations and a correct identification of the origin and nature of noise. We have discussed the issue of decoherence in stochastic inflation in Chapter 9. We will discuss the issue of noise and structures in two separate sections below.

15.1.2 Noise in stochastic inflation

Noise from partitioning and smoothing a free field

Consider a free massive (m) scalar field $\Phi(\mathbf{x}, t)$ in a spatially-flat Robertson-Walker (RW) spacetime with metric

$$ds^2 = -dt^2 + a^2(t)d\mathbf{x}^2 = a^2(\eta)(-d\eta^2 + d\mathbf{x}^2) \tag{15.6}$$

where $a(t)$ is the scale factor and η is the conformal time defined by $a(t)d\eta = dt$.

Expanding Φ in normal modes with the basis spatial wavefunctions $e^{i\mathbf{k}\cdot\mathbf{x}}/(2\pi)^{3/2}$ of the spatially flat RW spacetime,

$$\Phi(\mathbf{x}, t) = \int \frac{d^3\mathbf{k}}{(2\pi)^{3/2}} \left[a_{\mathbf{k}} \phi_{\mathbf{k}}(t) e^{i\mathbf{k}\cdot\mathbf{x}} + \text{h.c.} \right] \tag{15.7}$$

the amplitude function $\phi_{\mathbf{k}}(t)$ of the $\mathbf{k}$ mode obeys the equation of motion

$$\ddot{\phi}_{\mathbf{k}}(t) + 3H\dot{\phi}_{\mathbf{k}}(t) + \omega_{\mathbf{k}}^2 \phi_{\mathbf{k}}(t) = 0 \tag{15.8}$$

where $\omega_{\mathbf{k}}^2(t) \equiv p^2 + m^2$, $p \equiv k/a$, $k \equiv |\mathbf{k}|$ and an overdot here denotes derivatives with respect to cosmic time t.

In the conformally related field, the normal mode amplitude $\chi_{\mathbf{k}}(\eta) = \phi_{\mathbf{k}} a(\eta)$ corresponding to $\phi_{\mathbf{k}}$ obeys the equation of motion

$$\chi_{\mathbf{k}}''(\eta) + \left(k^2 + m^2 a^2 - \frac{a''}{a}\right) \chi_{\mathbf{k}}(\eta) = 0 \tag{15.9}$$

where a prime denotes taking the derivative with respect to the conformal time $\partial_\eta = a \partial_t$.

For the de Sitter universe, in a spatially flat RW coordinate representation,

$$a(t) = e^{Ht} \tag{15.10}$$

the expansion rate (Hubble parameter) $H \equiv \dot{a}/a$ is a constant in time and inflation goes on forever. In conformal time (ranging from $-\infty$ to 0)

$$\eta = -\frac{1}{a(t)H} \tag{15.11}$$

the evolution equation for the amplitude function $\chi_{\mathbf{k}}(\eta)$ of the conformally related field becomes

$$\chi_{\mathbf{k}}''(\eta) + \left[k^2 - \frac{1}{\eta^2}\left(\nu^2 - \frac{1}{4}\right)\right] \chi_{\mathbf{k}}(\eta) = 0 \tag{15.12}$$

where the parameter ν is defined as

$$\nu = \sqrt{\frac{9}{4} - \frac{m^2}{H^2}} \equiv \frac{3}{2} - \epsilon_m \tag{15.13}$$

The generic solution to this equation can be expressed in terms of Bessel functions of the first and second kind,

$$c_1 \sqrt{|\eta|} J_\nu(k|\eta|) + c_2 \sqrt{|\eta|} Y_\nu(k|\eta|) \tag{15.14}$$

Requiring each $\chi_{\mathbf{k}}$ to match the plane wave solution $e^{-ik\eta}/\sqrt{2k}$ for $k \gg aH$, when wavelengths are too short to feel any spacetime curvature effects, produces the standard Bunch–Davies solution

$$\chi_{\mathbf{k}}(\eta) = \frac{\sqrt{\pi}}{2} \sqrt{|\eta|} H_\nu^{(1)}(k|\eta|) \tag{15.15}$$

where

$$H_\nu^{(1)}(x) = J_\nu(x) + iY_\nu(x) \tag{15.16}$$

is the Hankel function of the first kind. The amplitude function of the kth normal mode of the original scalar field ϕ is given by

$$\phi_{\mathbf{k}}(\eta) = \frac{\sqrt{\pi}}{2} H|\eta|^{3/2} H_\nu^{(1)}(k|\eta|) \tag{15.17}$$

which in the massless case ($\nu = \frac{3}{2}$) becomes

$$\phi_{\mathbf{k}}(\eta) = H \frac{k\eta - i}{\sqrt{2k^3}} e^{-ik\eta} \tag{15.18}$$

In an expanding universe each mode will successively leave the horizon when its physical wavelength $p^{-1} = a/k$ reaches H^{-1}. Thus for a de Sitter universe, at the horizon crossing, $|k\eta| = 1$.

Spatial averaging and noise

Field fluctuations on super-horizon scales behave effectively as classical fluctuation modes with random amplitudes. This is conventionally described by averaging the field Φ in space over super-horizon scales and treating the resulting field $\Phi_<$ as a classical stochastic field satisfying a Langevin equation with a noise source described by a Gaussian random field of the shorter wavelength modes, given by equation (15.3); see [Sta86, GonLin86, NaNaSa88, NamSas89, Nam89, Mij90, SalBon91].

The averaging of the field Φ is performed by means of a suitable window function $W_s(\mathbf{x}; R)$ with a characteristic smoothing scale R,

$$\bar{\Phi}(\mathbf{x}, t) \equiv \int d^3\mathbf{x}' \phi(\mathbf{x}', t) W_s(\mathbf{x} - \mathbf{x}'; R) \tag{15.19}$$

Here, the physical smoothing scale is taken to be σ^{-1} times larger than the horizon size, with $\sigma \ll 1$. The corresponding comoving scale is

$$R(t) = \frac{1}{\sigma H a(t)} \tag{15.20}$$

The volume-averaged field has a mode expansion

$$\bar{\Phi}(\mathbf{x}, t) = \int \frac{d^3\mathbf{k}}{(2\pi)^{3/2}} \left[w(kR) a_\mathbf{k} \phi_\mathbf{k}(t) e^{i\mathbf{k}\cdot\mathbf{x}} + \text{h.c.} \right] \tag{15.21}$$

where $w(kR)$ is a suitable Fourier transform of the window function W_s. Starobinsky used a sharp step-function cut-off in Fourier space:

$$w(kR) = \theta(1 - kR) \tag{15.22}$$

The volume-averaged inflaton field is treated as a classical field $\Phi_<$ satisfying the Langevin (15.3) under the potential $V(\Phi)$ and an effective "noise field" source $\xi(\mathbf{x}, t)$. In the original proposal the noise source $\xi(\mathbf{x}, t)$ was heuristically defined as a stochastic field that corresponds to the quantum operator of the free field derivative $\dot{\bar{\Phi}}$, in the sense that any average of ξ, such as the correlator $\langle \xi(\mathbf{x}, t) \xi(\mathbf{x}', t') \rangle$, is assumed to be the same as the corresponding quantum expectation values of $\dot{\bar{\Phi}}$ in the vacuum state (which for de Sitter spacetime is the standard Bunch–Davies vacuum). The effective noise field ξ defined in this way is a Gaussian random field with zero mean, so the correlator $\langle \xi(\mathbf{x}, t) \xi(\mathbf{x}', t') \rangle$ completely describes its properties.

We show below the calculation of the noise correlator $\langle \xi(\mathbf{x}, t) \xi(\mathbf{x}', t') \rangle$ from a computation of the corresponding expectation value of the quantum "noise operator" $\dot{\bar{\Phi}}$ following WV [WinVil00]. The noise correlator generally depends on the particular window function $W_s(\mathbf{x}; R)$ and on the parameter σ. These parameters can in principle be related to observational data such as from the WMAP via the standard theory of structure formation, a topic we will come to in a later section.

Correlator of noise

Here we derive the correlators of the effective noise field $\xi(\mathbf{x}, t)$ for an arbitrary smoothing window. In stochastic inflation the noise field $\xi(\mathbf{x}, t)$ is defined through the time derivative of the averaged field $\dot{\bar{\Phi}}$ in mode expansion

$$\dot{\bar{\Phi}}(\mathbf{x}, t) = \int \frac{d^3\mathbf{k}}{(2\pi)^{3/2}} \left[v_k(\eta) a_{\mathbf{k}} e^{i\mathbf{k}\cdot\mathbf{x}} + \text{h.c.} \right] \tag{15.23}$$

where

$$v_k(\eta) \equiv \frac{d}{dt} \left[w(kR) \phi_k(\eta) \right] = \left[-HkRw'(kR) \phi_k(\eta) + w(kR) \dot{\phi}_k(\eta) \right] \tag{15.24}$$

In the limit of $\sigma \ll 1$ we may disregard the second term in the square brackets. The noise correlator then becomes

$$\langle \xi(\mathbf{x}_1 \eta_1) \xi(\mathbf{x}_2, \eta_2) \rangle = \frac{H^4 \eta_1 \eta_2}{4\pi^2 r \sigma^2} \int_0^\infty dk \, \sin kr \, h(k) \tag{15.25}$$

where $h(k)$ is a dimensionless function of two variables η_1, η_2

$$h(k) \equiv (1 + iy_1)(1 - iy_2) e^{ik(\eta_2 - \eta_1)} w'\left(-\frac{y_1}{\sigma}\right) w'\left(-\frac{y_2}{\sigma}\right) \tag{15.26}$$

where $y_1 = k\eta_1$, $y_2 = k\eta_2$. The asymptotic form of equation (15.25) at large r is given by

$$\langle \xi(\mathbf{x}_1, \eta_1) \xi(\mathbf{x}_2, \eta_2) \rangle = -\frac{\left(H^2 \eta_1 \eta_2\right)^2}{2\pi^2 r^4 \sigma^4} |w''(0)|^2 + O\left(r^{-6}\right) \tag{15.27}$$

Now examine the unsmoothed correlator of quantum field derivatives given at arbitrary space and time points by (Appendix C of WV):

$$\left\langle \dot{\Phi}(\mathbf{x}_1, t_1) \dot{\Phi}(\mathbf{x}_2, t_2) \right\rangle = \frac{1}{2\pi^2} \int_0^\infty \dot{\phi}_k(t) \dot{\phi}_k^*(0) \frac{\sin kr}{r} k \, dk$$

$$= \frac{H^4}{2\pi^2} (\eta_1 \eta_2)^2 \frac{3(\eta_1 - \eta_2)^2 + r^2}{\left[(\eta_1 - \eta_2)^2 - r^2\right]^3} \tag{15.28}$$

As expected, it diverges on the lightcone where r becomes $|\eta_1 - \eta_2|$. The asymptotic form of equation (15.28) at large distances r is

$$\left\langle \dot{\Phi}(\mathbf{x}_1, t_1) \dot{\Phi}(\mathbf{x}_2, t_2) \right\rangle = -\frac{H^4 (\eta_1 \eta_2)^2}{2\pi^2 r^4} + O\left(r^{-6}\right) \tag{15.29}$$

We see that the stochastic source correlator (15.27) is very similar to the quantum field correlator (15.29). Note also that the asymptotic (15.27) is essentially independent of the shape of the window function, since the value $|w''(0)|$ as indicated by equation (15.21) has the meaning of the window-averaged squared distance and must be of order 1 because the window profile $W(q)$ starts to decay at $q \sim 1$ by construction.

We can obtain a simpler expression for the correlator in the limit when the smoothing parameter σ is small while the product σHr remains finite. A rescaling $r \to \sigma Hr \equiv \rho$ and the corresponding change of variable $k \equiv \sigma H \kappa$ simplify equation (15.25) because we can omit terms of order σ and smaller; in particular, the product of mode functions is simplified to

$$\phi_k^* (\eta_1) \phi_k (\eta_2) = \frac{1}{2H\kappa^3 \sigma^3} (1 + O(\sigma^2)) \tag{15.30}$$

The leading term in the correlator, expressed through κ and ρ, becomes

$$\langle \xi(\mathbf{x}_1, \eta_1) \xi(\mathbf{x}_2, \eta_2) \rangle = \frac{H^6 \eta_1 \eta_2}{4\pi^2 \rho} \int_0^\infty d\kappa \, \sin \kappa \rho \, w'(-H\eta_1 \kappa) \, w'(-H\eta_2 \kappa) + O(\sigma^2) \tag{15.31}$$

Therefore, in the limit of small σ but finite σHr, the correlator as a function of the "effective distance" ρ and the time difference (expressed by η_2/η_1) becomes independent of σ.

The expression in equation (15.31) allows us to compute the correlator at all distances in the limit of small σ. Under this condition, for a Gaussian smoothing window, $w(p) = \exp(-p^2/2)$, we obtain

$$\langle \xi(\mathbf{x}_1, \eta_1) \xi(\mathbf{x}_2, \eta_2) \rangle$$
$$= \frac{(H^4 \eta_1 \eta_2)^2}{4\pi^2 \rho} \int_0^\infty \exp\left[-H^2 \frac{\eta_1^2 + \eta_2^2}{2} \kappa^2 \right] \kappa^2 \sin \kappa \rho \, d\kappa$$
$$= \frac{(H^4 \eta_1 \eta_2)^2 \mu^4}{4\pi^2 \rho^4} \left[1 - \left(\frac{1}{\mu} - \mu \right) i \sqrt{\frac{\pi}{2}} \mathrm{erf}\left(\frac{i\mu}{\sqrt{2}} \right) \exp\left(-\frac{\mu^2}{2} \right) \right] \tag{15.32}$$

where

$$\mu \equiv \frac{\rho}{\sqrt{H^2 (\eta_1^2 + \eta_2^2)}} \tag{15.33}$$

is a dimensionless quantity. (A plot of this function for $\eta_1 = \eta_2$ can be found in WV.) The leading term of the expression in brackets in equation (15.32) at large μ is $(-2\mu^{-4})$, and since for the Gaussian window $w''(0) = -1$, we recover equation (15.27). The value of the correlator at the coincident points ($\rho = 0$) as a function of time separation is

$$\langle \xi(0, \eta_1) \xi(0, \eta_2) \rangle = \frac{H^4 (\eta_1 \eta_2)^2}{2\pi^2 (\eta_1^2 + \eta_2^2)^2} = \frac{H^4}{8\pi^2} \frac{1}{\cosh^2 H\Delta t} \tag{15.34}$$

We can also obtain the leading asymptotics of the unequal-time correlator at large time separations. Again start with equation (15.25) and assume that the time separation is much greater than the Hubble time, $\eta_2/\eta_1 \equiv a^{-1} \ll 1$. For simplicity we can choose the initial time such that $H\eta_1 = -1$. Using an expansion (see equation (A12) of WV) for $w(a^{-1}k)$ at small $a^{-1}\kappa$ (since the integration is effectively performed over a fixed finite range of k) we obtain

$$\langle \xi(\mathbf{x}_1, \eta_1) \xi(\mathbf{x}_2, \eta_2) \rangle = \frac{H^2 w''(0)}{4\pi^2 \sigma^3 a^2 H r} \int_0^\infty dk\, k \sin kr\, w'\left(\frac{k}{\sigma H}\right)$$

$$\times\, e^{ik/H}\left(1 + i\frac{k}{H}\right) + O\left(a^{-4}\right) \qquad (15.35)$$

The integral in equation (15.35) is time-independent. Therefore the correlator decays as $a^{-2} = \exp(-2Ht)$ with time separation at any fixed distance. This derivation shows how a regular window function produces a colored noise.

Colored noise from coarse graining an interacting field

In addition to using a smoothing window function as illustrated in the above section [WinVil00, Mat93, Rio02] one could make a frequency or wavelength partition, splitting the short- and the long-wavelength sectors. This has been treated in Chapter 5 for a scalar field in Minkowski spacetime and in Chapter 9 for a conformally-related theory in de Sitter spacetime. We now turn to the issue of structure formation from a colored noise.

15.2 Structure formation: Effect of colored noise

A standard mechanism for structure formation is the amplification of primordial density fluctuations by the evolutionary dynamics of spacetime [Sak66, LifKal63, Bar80, Muk05]. In the lowest order approximation the gravitational perturbations (scalar perturbations for matter density and tensor perturbations for gravitational waves) obey linear equations of motion. Their initial values and distributions are stipulated, generally assumed to be a white noise spectrum. In these theories, fashionable in the 1960s and 1970s, the primordial fluctuations are classical in nature. In the standard model of FLRW cosmology, the scale factor of the universe growing in a power law of cosmic time generates a density contrast which turns out to be too small to account for the galaxy masses. The observed nearly scale-invariant spectrum [Har70, Zel72] also does not find any easy explanation in this model [Pee80, ZelNov85].

Inflationary models explain structure formation from amplification of vacuum fluctuations of a scalar field Φ, the inflaton; see [GuthPi82, Sta82, MukChi82, Haw82, BaStTu83, Bra83, MuFeBr92, DeGuLa92, YiViMi91, YiVis92, YiVis93, YiVis93b, GlMaRa82, BoVeHo94, Bur95, Muk05]. Consider the "eternal inflation" stage where the universe has locally a de Sitter geometry, with a constant Hubble radius (de Sitter horizon) $l_h = H^{-1}$. (In reality H cannot strictly be a constant, for otherwise the universe cannot reheat to our present FLRW state.) The physical wavelength l of a mode of the inflaton field is $l = p^{-1} = a/k$, where k is the wavenumber of that mode. As the scale factor increases exponentially, the wavelengths of many modes can grow larger than the horizon size. After the end of the de Sitter phase, the universe begins to reheat, turning into a radiation-dominated Friedmann universe with power law expansion $a(t) \sim t^n$. In this phase, the Hubble radius grows much faster than the physical wavelength, and some inflaton modes will reenter the horizon. The fluctuations of these long-wavelength inflaton modes that went out of the de Sitter horizon and later came back into the FLRW horizon play an important role in determining the large-scale density fluctuations of the early universe, which in time seeded the galaxies.

The stochastic inflation paradigm, after a proper treatment of decoherence of the long wavelength modes[1] and a first-principles derivation of noise (arising from the short wavelength sector), could thus provide a sound rationale for the Langevin equation depicting the dynamics of the inflaton perturbations or the Fokker–Planck equation describing the evolution of their probability distributions.

A key issue in the solution of the Langevin or Fokker–Planck equation is the choice of the initial conditions for the perturbations. Many authors (see [SalBon91, Hod90, MMOL91, KBHP91]) agree that it should be consistent to assume the spatial homogeneity of our observable local patch of the universe, and therefore the vanishing of all fluctuations right before the moment it crosses the horizon size, about 60 e-folds before the end of inflation, since at that time only fluctuations on larger scales could have grown significantly. Therefore, all points inside the present Hubble radius (at that time contained in the same coarsegraining domain) must have the same local value of the scalar field, although this value can be different from the one assumed in other regions of the Universe.

Even if it is generally assumed that inflation started well before the last 60 e-folds, for the white-noise case the evolution of fluctuations is completely insensitive to what happened before that epoch and the constraint really becomes a new initial condition. In contrast, non-Markovian fluctuations generated by colored noises [HuPaZh93b] will retain some memory of the evolution before the constraint.

The linkage of colored noise-generated structure to observations in WMAP was suggested in [MaMuRi04] (MMR), where evidence was found for a blue tilt in the power-spectrum on the largest observable scales as a consequence of the non-Markovian dynamics near the constraint. This is due to the fact that the increased noise correlation time (with respect to the white-noise case) acts as a sort of "inertia" against the growth of the perturbations after the constraint, thereby resulting in a suppression of the power-spectrum on the scales that crossed the horizon in the ensuing few Hubble times. This is an interesting

---

[1] There are different views on how the long-wavelength modes got decohered, including the extreme one that no dynamical explanation needs to be provided. This so-called *decoherence without decoherence* theme first proposed by Polarsky and Starobinsky [KiPoSt98] is attractive more because of its expedience than truth value. The original form has been revised after meeting with criticisms. For a more careful recent study on this proposal, see, e.g. [CamPar05]. A different approach is suggested by Woodard [TsaWoo05, Woo05a, Woo06]. For a recent review, see [Win06].

feature, since the CMB anisotropy measurements made by WMAP [Spe03] give some evidence for a suppression of the low multipoles, consistent with earlier analogous results found by COBE [Ben96], although the statistical significance of such a suppression is not large [TeCoHa03, OTZH04, Efs03, Efs04, BiGoBa04].

A related paper [LMMR04] (LMMR) points out the low multipoles suppression might also be a consequence of the colored noise. Compared to white noise, a smooth choice of the window function will in fact slightly suppress the contribution to the noise given by the field modes whose frequency is immediately higher than the cut-off scale $\sigma(aH)$ (while enhancing the lower frequencies). Right after the time τ_* at which the homogeneity constraint is set on the comoving patch of the universe, fluctuations with $k \lesssim \sigma a_* H_*$ will grow less than in the white-noise case before freezing out, and if σ is not too small this suppression can be effective also on observable scales. Even in the Markovian case, the noise correlation function in configuration space has a dependence of σ beyond the second order which could show up in the power spectrum.

Before getting into the details we want to add a qualifying remark on how this mechanism is placed in relation to other mechanisms so as to avoid a skewed perspective. The colored noise explanation of the suppression of lower multipoles (blue tilt) mode is only one amongst many proposed. As cautioned in the beginning, we select this topic mainly to illustrate some key ideas in NEqQFT, in this case, the effect of quantum noise on structure formation in stochastic inflation. Adopting this perspective we hope that even if at the end the actual physical scenario may not survive over other competing theories, the readers can learn the physics of NEq quantum fields through a detailed analysis of these sample problems.

15.2.1 Colored noise from smooth window functions

Partitioning and smoothing

As discussed earlier, if one uses the cosmological horizon as the partition scale, the environment field $\Phi_>$ consisting of the subhorizon (short-wavelength) modes can be sieved out by the use of a suitable time-dependent high-pass filter in Fourier space. This is achieved by means of a different window function $\tilde{W}_\sigma(y)$, $y \equiv k\eta$ such that $\tilde{W}_\sigma(y) = 0$ for $k|\eta| \ll \sigma$ and $\tilde{W}_\sigma(y) = 1$ for $k|\eta| \gg \sigma$. (Note that this window function used by LMMR is complementary to the one used by WV discussed in the last section, which is a low-pass filter.) The parameter σ defines the size of the coarse-graining domain and an "effective horizon" $\sigma(aH)$:

$$\Phi_> = \int d^3k \frac{\tilde{W}_\sigma(k\eta)}{(2\pi)^{3/2}} \left[a_{\mathbf{k}}\phi_{\mathbf{k}}(\eta)e^{i\mathbf{k}\cdot\mathbf{x}} + \text{h.c.} \right] \tag{15.36}$$

In the stochastic inflation paradigm, the quantum fluctuations on subhorizon scales act as a classical noise source ξ with a given probability distribution $P[\xi]$ in a Langevin equation which drives the super-horizon modes. Our discussions

in the previous section on the origin and nature of noise from quantum fluctuations and on the decoherence of the long-wavelength mode by this noise may
serve as justification for such a proposal. Technically, the quantum problem of
computing the expectation value of the coarse-grained field is thus reduced to
the classical problem of evaluating the mean of the solution to the stochastic
evolution equation averaged over all possible noise configurations.

Following such a prescription, we can split the scalar field $\Phi = \bar{\phi} + \varphi$ into
its statistical mean value $\bar{\phi}$ whose normal mode amplitudes satisfy the classical equation of motion (15.8) and a fluctuation field $\varphi[\xi]$, with zero mean over
the distribution $P[\xi]$.[2] The stochastic equation of motion for the super-horizon
fluctuations was shown before to be

$$\ddot{\varphi}_{\mathbf{k}} + 3H\dot{\varphi}_{\mathbf{k}} - \left(\frac{k^2}{a^2} - m^2\right)\varphi_{\mathbf{k}} = \frac{\xi_{\mathbf{k}}}{a^3} \tag{15.37}$$

or, for the conformally related field normal mode amplitude $\chi_{\mathbf{k}} = a\phi_{\mathbf{k}}$ in con-
formal time η, in a similar decomposition $\chi_{\mathbf{k}} = \bar{\chi}_{\mathbf{k}} + \tilde{\chi}_{\mathbf{k}}$ the mean field satisfies
(15.9) and the fluctuation field modes $\tilde{\chi}_{\mathbf{k}}$ obeys

$$\tilde{\chi}_{\mathbf{k}}'' + \left(k^2 + m^2 a^2 - \frac{a''}{a}\right)\tilde{\chi}_{\mathbf{k}} = \xi_{\mathbf{k}} \tag{15.38}$$

The noise ξ is a Gaussian random field, whose configurations are weighted by
the functional probability distribution

$$P[\xi] = N \exp\left[-\frac{1}{2}\int d^4x\, d^4x'\, \xi(x)\mathcal{N}^{-1}(x, x')\xi(x')\right] \tag{15.39}$$

$$= N \exp\left[-\frac{1}{2}\int d\eta\, d\eta'\, d^3k\, d^3k'\, \xi_{\mathbf{k}}(\eta)\mathcal{N}^{-1}_{\mathbf{k},\mathbf{k}'}(\eta, \eta')\xi_{\mathbf{k}'}(\eta')\right] \tag{15.40}$$

where $\mathcal{N}^{-1}_{\mathbf{k},\mathbf{k}'}(\eta, \eta')$ is the functional inverse of

$$\mathcal{N}_{\mathbf{k},\mathbf{k}'}(\eta, \eta') = \delta(\mathbf{k} + \mathbf{k}')\frac{\mathrm{Re}[f(y)f^*(y')]}{2k^3} \tag{15.41}$$

and, with $y \equiv k\eta$,

$$f(y) = \sqrt{2k^3}(\tilde{W}_\sigma'' \chi_{\mathbf{k}} + 2\tilde{W}_\sigma' \chi_{\mathbf{k}}') \tag{15.42}$$

This probability distribution allows us to calculate the statistical mean value
$\langle\ldots\rangle_\xi$ of any ξ-dependent quantity averaged over the noise field configurations,
defined as

$$\langle\ldots\rangle_\xi = \int \mathcal{D}[\xi]\ldots P[\xi] \tag{15.43}$$

---

[2] This is true at linear order because nonlinear corrections will shift the mean value. Also, if
Φ is the inflaton field then $\bar{\phi}$ should be the homogeneous background, and as such have no
Fourier decomposition. All is well in the case of a test field with no metric fluctuations,
which is what we will assume here.

Then, by definition the mean $\langle \xi(\eta) \rangle_\xi$ of the noise vanishes at all times, while the two-point correlation function is by definition

$$\langle \xi_\mathbf{k}(\eta)\xi_{\mathbf{k}'}(\eta') \rangle_\xi = \mathcal{N}_{\mathbf{k},\mathbf{k}'}(t,t') \tag{15.44}$$

This correlation function, the noise kernel, completely characterizes the statistical properties of the Gaussian noise field. In configuration space it reads

$$\langle \xi(x)\xi(x') \rangle_\xi = \int \frac{d^3k}{(2\pi)^3} e^{i\mathbf{k}\cdot(\mathbf{x}-\mathbf{x}')} \frac{1}{2k^3} \mathrm{Re}[f(y)f^*(y')] \tag{15.45}$$

As we saw before the statistical behavior of the noise depends critically on the shape of the filter. Choosing the special window function $\tilde{W}_\sigma(k\eta) = \theta(k|\eta| - \sigma)$ leads to the standard white-noise two-point correlation function. For $\mathbf{x} = \mathbf{x}'$ it reads

$$\langle \xi(x)\xi(x') \rangle_\xi = \frac{H^3}{4\pi^2}(1 + \mathcal{O}(\sigma^2))\delta(t - t') \tag{15.46}$$

which is highly divergent for $t = t'$ and has a vanishing characteristic correlation time. In contrast a smooth window function yields a correlation function with no divergence and a finite correlation time, therefore producing a colored noise, e.g. with

$$\tilde{W}_\sigma(y) = 1 - e^{-\frac{y^2}{2\sigma^2}} \tag{15.47}$$

the two-point correlation function at $r = 0$ is given by

$$\langle \xi(t)\xi(t') \rangle_\xi = \frac{H^4}{8\pi^2} \frac{1}{\cosh^2(H(t - t'))} + \mathcal{O}(\sigma^2) \tag{15.48}$$

which behaves like $e^{-2H(t-t')}$ asymptotically. This asymptotic behavior is quite general for a wide class of smooth window functions [WinVil00].

Fluctuations and structures

The particular solution of the evolution equation (15.38) for the fluctuations $\tilde{\chi}_\mathbf{k}$ sourced by the noise field ξ can be expressed in terms of the general solutions $\chi_1 = \sqrt{k|\eta|}J_\nu(|y|)$ and $\chi_2 = \sqrt{k|\eta|}Y_\nu(|y|)$ of the homogeneous equation (15.12). This solution reads

$$\tilde{\chi}_\mathbf{k}[\xi](\eta) = \int_{\eta_i}^\eta d\eta'\, g(y,y')\, \xi_\mathbf{k}(y') \tag{15.49}$$

where

$$g(y,y') = \frac{\chi_1(y)\chi_2(y') - \chi_2(y)\chi_1(y')}{\chi_1'(y')\chi_2(y') - \chi_2'(y')\chi_1(y')} \tag{15.50}$$

and η_i is the beginning of inflation, at which we set the initial condition $\tilde{\chi}_\mathbf{k}(\eta_i) = 0$.

Keeping this assumption, LMMR impose the constraint that at a much later time η_* (roughly about 60 e-folds before the end of inflation) there are no fluctuations in that part of the universe corresponding to the present observable sky. This is motivated by the fact that all the points we observe today with substantial homogeneity were included at η_* in the same coarse-grained domain.[3] Physically this amounts to the assumption that at η_* the comoving patch of the universe we observe today has complete homogeneity and all fluctuations on smaller scales were generated later by the stochastic source represented by the noise term.[4] There is no assumption made on the behavior of larger unobservable scales.

We are thus led to consider (for a given noise configuration) a different solution for the subsequent evolution of the fluctuations, obtained as in (15.49) by starting the integration at η_*, when a new (stochastic) initial condition holds. In turn, $\tilde{\chi}_{\mathbf{k}}[\xi](\eta_*)$ is determined again from (15.49) with the usual vanishing initial condition at η_i. However, as long as we are dealing with points inside the present observable universe, we can skip the stochastic initial conditions η_* since their inverse Fourier transform is assumed to vanish. Therefore, in configuration space the subsequent evolution of the fluctuations will only contain noise modes integrated after η_*. Thus, for relevant $\mathbf{x}$'s we may write

$$\varphi(\mathbf{x}, \eta) = \int \frac{d^3\mathbf{k}}{(2\pi)^{3/2}} \frac{e^{i\mathbf{k}\cdot\mathbf{x}}}{a} \int_{\eta_*}^{\eta} dy'\, g(y, y')\, \xi_{\mathbf{k}}(\eta') \tag{15.51}$$

and

$$\varphi(\mathbf{x}, \eta_*) = \int \frac{d^3\mathbf{k}}{(2\pi)^{3/2}} \frac{e^{i\mathbf{k}\cdot\mathbf{x}}}{a_*} \int_{\eta_i}^{\eta_*} d\eta'\, g(y_*, y)\, \xi_{\mathbf{k}}(y') \tag{15.52}$$

where the first equation is only valid for scales inside our observed patch of the universe.

As expected, since the fluctuation $\varphi_{\mathbf{k}}[\xi]$ is linear in ξ, at all times we have that

$$\langle \varphi[\xi](\eta) \rangle_\xi = 0 \tag{15.53}$$

while the two-point correlation function in $\mathbf{x}_1$ and $\mathbf{x}_2 = \mathbf{x}_1 + \mathbf{r}$ can be obtained by integrating the noise correlation function (15.41). LMMR find

$$C(\mathbf{r}, \eta) \equiv \langle \varphi[\xi](\mathbf{x}_1, \eta)\varphi[\xi](\mathbf{x}_2, \eta) \rangle_\xi = \int \frac{d^3\mathbf{k}}{(2\pi)^3} e^{i\mathbf{k}\cdot\mathbf{r}} \frac{|I_1(k)|^2}{2k^3} \tag{15.54}$$

---

[3] Introduced first by Salopek and Bond [SalBon91], such a constraint is necessary if one wants to use the variance of the single-point probability distribution (which has no spatial information encoded) to extract some information on the cosmic microwave background. Without this constraint, the variance will be much larger because fluctuations (specially at the beginning of inflation) add up very rapidly over time. However, this variance can now only be used to model the structure on ultralarge scales (of the order of the wavelength of the first modes crossing the Hubble radius).

[4] In principle, solving the Langevin equation with the full space dependence may not require the imposition of this constraint, because the correlation function is able to distinguish the scales. At any time, as a consequence of the smoothing, fluctuations on scales smaller than the filtering scale will not appear (as in white noise) or will appear only in a finite frequency range around this scale (as with colored noise). In either case imposition of the window function could effectively serve the function of the homogeneity constraint.

where

$$I_1(k) = \frac{\sqrt{2k^3}}{a} \int_{\eta_*}^{\eta} d\eta' \, g(y, y') (\tilde{W}''_\sigma \chi_{\mathbf{k}} + 2\tilde{W}'_\sigma \chi'_{\mathbf{k}}) \qquad (15.55)$$

In the same way one can calculate the correlation function evaluated at η_*, yielding

$$C_*(\mathbf{r}, \eta_*) \equiv \langle \varphi[\xi](\mathbf{x}_1, \eta_*) \varphi[\xi](\mathbf{x}_2, \eta_*) \rangle_\xi = \int \frac{d^3k}{(2\pi)^3} e^{i\mathbf{k}\cdot\mathbf{r}} \frac{|I_2(k)|^2}{2k^3} \qquad (15.56)$$

where $I_2(k)$ has the same form as $I_1(k)$ but it refers to the time interval $[\eta_i, \eta_*]$.

One can also define the mixed correlation function of the scalar field perturbations evaluated at different times:

$$C_\times(\mathbf{r}, \eta, \eta_*) \equiv \langle \varphi[\xi](\mathbf{x}_1, \eta) \varphi[\xi](\mathbf{x}_2, \eta_*) \rangle_\xi = \int \frac{d^3k}{(2\pi)^3} e^{i\mathbf{k}\cdot\mathbf{r}} \frac{\mathrm{Re}[I_1(k) I_2^*(k)]}{2k^3} \qquad (15.57)$$

With these correlation functions $C(\mathbf{r}, \eta)$, $C_*(\mathbf{r}, \eta_*)$, $C_\times(\mathbf{r}, \eta, \eta_*)$ one can proceed to calculate the conditional correlation function of the scalar field perturbations – conditional (subscript c) here referring to the constraint defined in the set-up of the initial conditions described above. In the physically reasonable limit of $\eta_i \ll \eta_*$ LMMR [LMMR04] obtained

$$\langle \varphi(\mathbf{x}_1) \varphi(\mathbf{x}_2) \rangle_{\mathrm{c}} \simeq C(\mathbf{r}) \qquad (15.58)$$

This yields the power spectrum $\mathcal{P}_{\delta\varphi}(k)$ of the fluctuations, defined by

$$\langle \varphi(\mathbf{x}_1) \varphi(\mathbf{x}_2) \rangle_{\mathrm{c}} = \frac{1}{4\pi} \int d^3k \, e^{i\mathbf{k}\cdot\mathbf{r}} \frac{\mathcal{P}_{\delta\varphi}(k)}{k^3} \qquad (15.59)$$

as

$$\mathcal{P}_{\delta\varphi}(k) = \frac{1}{4\pi^2} |I_1(k)|^2 \qquad (15.60)$$

In the small-σ limit the standard scale-invariant result $\mathcal{P}_{\delta\varphi}(k) = H^2/4\pi^2$ is recovered.

15.2.2 Curvature perturbations and blue tilt

So far this treatment has been under the test-field approximation, meaning that the background spacetime where the quantum field propagates is assumed to be fixed, i.e. the de Sitter universe. But in reality the quantum field contributes to the energy–momentum tensor which determines the evolution of the scale factor, via the slow-roll Friedmann equation $H^2 \simeq (8\pi G/3)V(\phi)$, and the field perturbations induce fluctuations in the metric. These metric perturbations need be considered alongside the scalar field perturbations φ. Let ψ be the curvature perturbation, which is gauge dependent. To avoid spurious coordinate effects it is preferable to use the gauge-invariant comoving curvature perturbation $\mathcal{R} = \psi + H(\varphi/\dot{\phi})$ which measures the intrinsic spatial curvature on hypersurfaces of constant time [Rio02] as the physical degrees of freedom.

Defining in conformal time $v = a^2 \phi'/a'$, the variable $u = -v\mathcal{R}$ satisfies the equation of motion

$$u'' - \nabla^2 u - \left(\frac{v''}{v}\right) u = 0 \qquad (15.61)$$

Expanding the last term to first order in the slow-roll parameters $\epsilon_V \equiv -3\dot{H}/H^2$; $\eta_V \equiv V''/3H^2$ formed from the Hubble parameter and the inflaton potential, one finds

$$\frac{v''}{v} \simeq \frac{1}{\eta^2}\left(\nu^2 - \frac{1}{4}\right) \qquad (15.62)$$

where $\nu \simeq \frac{3}{2} + 3\epsilon_V - \eta_V$.

We see that in the slow-roll approximation the gauge-invariant normal modes $u_{\mathbf{k}}$ satisfy the same equation of motion (15.12), the only difference enters in the definition of the parameter ν labeling the solutions:

$$u_{\mathbf{k}}'' + \left[k^2 - \frac{1}{\eta^2}\left(\nu^2 - \frac{1}{4}\right)\right] u_{\mathbf{k}} = \xi_{\mathbf{k}} \qquad (15.63)$$

We can then apply to $\mathcal{R}$ the results derived for the power spectrum of the perturbations of a test scalar field, concluding that for the curvature perturbation we also have $\mathcal{P}_{\mathcal{R}}(k) \propto |I_1(k)|^2$.

In the limit $k|\eta| \ll \sigma \lesssim 1$ which is reasonably satisfied on cosmological scales, the power spectrum simplifies to

$$\mathcal{P}_{\mathcal{R}}(k) = A_{\mathcal{R}}^2 \tilde{W}_\sigma^2(k|\eta_*|)(k|\eta|)^{2\eta_V - 6\epsilon_V} \qquad (15.64)$$

Since $\tilde{W}_\sigma^2 < 1$, this shows a blue tilt on large observable scales with $k \sim \sigma a_* H_*$, corresponding to physical lengths about σ^{-1} times greater than the present Hubble radius. In the limit $\sigma \ll k|\eta_*|$ (since $\tilde{W}_\sigma(k|\eta_*|) \simeq 1$) we recover the ordinary result

$$\mathcal{P}_{\mathcal{R}}(k) = A_{\mathcal{R}}^2 (k|\eta|)^{2\eta_V - 6\epsilon_V} \qquad (15.65)$$

This blue tilt stems from the fact that a smooth window function does not make a sharp separation in Fourier space but it gradually weighs the modes, allowing for a small low-frequency contribution to the short-wavelength part of the field (in terms of which the noise is defined) while depleting modes whose wavelength is immediately smaller than the cut-off scale. The colored noise originated from such a window function is thus able to generate fewer fluctuations than a white noise on scales slightly smaller than the comoving coarse-graining domain.

As a consequence, under the constraint that in our comoving patch of the universe the fluctuations can grow only after η_*, the scales that are leaving the horizon in the following few Hubble times receive fewer "random kicks" before freezing out than in the white-noise case. Therefore, the power spectrum is a function of k smoothly interpolating between the values 0 and 1 it assumes for small and large k, respectively.

This power spectrum can be used to calculate the CMB multipoles predicted by a specific choice of the window function W. Quite generally, we expect to find a suppression of the lowest multipole, which is sensitive to a modification of the power spectrum on this very large scale. However, in order to quantify this suppression one needs to choose the shape of the window function and the precise time η_* at which the constraint is set. As mentioned before the significance of the low multipole suppression varies depending on the choice of the constraint time. Detailed description can be found in LMMR, where our exposition here is adapted from.

15.2.3 Structures from coarse graining an interacting field

As we learned earlier colored noise can also be generated by coarse graining a sector of one partitioned interacting quantum field [CalHu95, CalGon97, Mat97a, Mat97b]. In Chapter 5 we derived the influence functional describing the effect of high-frequency modes on the low-frequency sector. The real part of the influence action contains divergent terms and should be renormalized. The imaginary part is finite and is associated with the decoherence process. From this one can derive the renormalized semiclassical Langevin equation governing the system field (the long-wavelength sector) driven by a noise originating from coarse graining the environment field (the short-wavelength sector). We can use this equation to understand the generation of classical inhomogeneities from quantum fluctuations, obtaining their power spectrum and be able to compare with observational data such as from WMAP.

In the ϕ^4 model used by Lombardo and Nacir [LomNac05] we presented in Chapter 9, there are two such sources ξ_2 and ξ_3, associated with the interaction terms $\phi_<^2 \phi_>^2$ and $\phi_<^3 \phi_>$ respectively. The full influence function is given in (9.112). Reading the noise kernels off that equation, we may now treat the generation of inhomogeneities with noise arising from one interacting quantum field.

We are interested in finding the power spectrum of perturbations to the inflaton field up to $\hbar$ and λ^2 order. To carry this out, we split the system field as $\phi_< = \phi_0(\eta) + \varphi_<$, where we identify $\phi_0(\eta)$ as a classical background field which satisfies the slow-roll conditions. The power spectrum of the field fluctuations $\varphi_<$ may be expressed as $P_\varphi(k) = 2\pi^2 k^{-3}\Delta_\varphi^2(k)$, with $\Delta_\varphi^2(k)$ defined by

$$\langle \varphi_<(\mathbf{x})\varphi_<(\mathbf{x}')\rangle = \int d^3\mathbf{k}\, \frac{\Delta_\varphi^2(k)}{4\pi k^3} \exp(-i\mathbf{k}\cdot\mathbf{r}) \qquad (15.66)$$

where $\mathbf{r} \equiv \mathbf{x} - \mathbf{x}'$.

Expanding the semiclassical Langevin equation up to linear order in the mode amplitude of interest $\varphi_<(\vec{k})$, we obtain

$$\phi_0''(\eta) + 2\tilde{H}\phi_0'(\eta) + 4\lambda a^2 \phi_0^3(\eta) = 0 \qquad (15.67)$$

$$\varphi_<''(\vec{k},\eta) + [k^2 + 12\lambda a^2 \phi_0^2(\eta)]\varphi_<(\vec{k},\eta) + 2\tilde{H}\varphi_<'(\vec{k},\eta) = -\frac{\xi_2(\vec{k},\eta)}{a^2}\phi_0(\eta)$$

$$(15.68)$$

where terms which do not contribute to the power spectrum up to order $\hbar$ have been discarded. The term with the ξ_3 noise source gives a zero contribution due to our approximations and the orthogonality of the Fourier modes. Note the presence of the ξ_2 noise source, which is instrumental to the decoherence process.

A general solution $\varphi_<$ to equation (15.68) is made up of two parts: a part $\varphi_<^q$ which is a solution to the homogeneous equation (i.e. without the source term on the right-hand side) and a particular solution $\varphi_<^\xi$ with vanishing initial conditions. Namely, $\varphi_<(\vec{k}, \eta) = \varphi_<^\xi(\vec{k}, \eta) + \varphi_<^q(\vec{k}, \eta)$. The first part is made up of "intrinsic fluctuations" which coincide with the quantum fluctuations of the free field. The second part is sometimes called "induced fluctuations" referring to the influences from the environment. Under some reasonable approximations the result is analogous to that for the linear quantum Brownian motion (QBM) [Zha90, HuPaZh92, HuPaZh93a, PaHaZu93, Paz94, HalYu96, KUMS97]. Correspondingly the quantity $\Delta_\varphi^2(k)$ has two contributions:

$$\Delta_\varphi^2(k) = \Delta_{\varphi^q}^2(k) + \Delta_{\varphi^\xi}^2(k) \tag{15.69}$$

Because in equation (15.68) the dissipation kernel is assumed to be small, the first part follows an almost unitary evolution of the initial density matrix, yielding the usual result for the case of the free field: $\Delta_{\varphi^q}^2(k) = (H/2\pi)^2 \left(1 + k^2\eta^2\right)$. The second part is due to the ξ_2 noise source and can be expressed as

$$\Delta_{\varphi^\xi}^2(k) = -\lambda^2 \frac{144}{\pi^2} k^3 \int_{\eta_i}^{\eta} d\eta_1 \int_{\eta_i}^{\eta} d\eta_2 \, a^4(\eta_1)a^4(\eta_2)$$
$$\times \phi_0(\eta_1)\phi_0(\eta_2)h(k, \eta, \eta_1)h(k, \eta, \eta_2)$$
$$\times ReG_F^{\Lambda 2}(\eta_1, \eta_2, \vec{k}) \tag{15.70}$$

where

$$h(k, \eta, \eta') \equiv \frac{1}{a(\eta)a(\eta')} \left[\frac{\sin[k(\eta - \eta')]}{k} \left(1 + \frac{1}{k^2\eta\eta'}\right) - \frac{\cos[k(\eta - \eta')]}{k^2\eta\eta'}(\eta - \eta') \right] \tag{15.71}$$

On the other hand, the usual contribution $\Delta_{\varphi^q}^2(k)$ is independent of k for a fixed value of $k\eta$, corresponding to a nearly scale-invariant spectrum, whereas $\Delta_{\varphi^\xi}^2(k)$ depends on k and Λ.

Thus, concerning the influence of the environment on the power spectrum for some modes in the system, the results of Lombardo and Nacir [LomNac05] indicate that the contribution to the spectrum from the unitary evolution of the Bunch–Davies initial condition dominates over the contribution from the system–environment interaction.

15.2.4 Structures from interaction with other fields

In this last subsection we turn to structure formation from colored noise generated from coarse graining some other quantum field(s) the inflaton interacts with, using the two-field model discussed in Chapter 5. We report on the findings of Wu *et al.* [WuNgLeeLeeCha06], who show that the inflaton fluctuations driven by the colored noise are strongly dependent on the onset of inflation and become scale-invariant asymptotically at small scales. These induced fluctuations would grow with time only in a certain intrinsic time-scale. For this proposal to work, one needs to assume that the gravitational perturbations associated with the passive (or induced) quantum field fluctuations can become larger than the active (or intrinsic) fluctuations. Some mechanism should be present to suppress the active fluctuations for this assumption to be valid. Only in the (hitherto not easily explicable) case when the induced fluctuations contribute a significant portion to the density perturbation would they cause a suppression of the density power spectrum on large scales which shows up as a depression of low-l multipoles in CMB. Of special interest to colored-noise induced structure formation is that the observed low CMB quadrupole may open a window on the physics of the first few e-foldings of inflation.

Consider an inflaton field Φ with potential $V(\Phi)$ coupled to a massive scalar quantum field Ψ described by the Lagrangian

$$\mathcal{L} = \frac{-1}{2} g^{\mu\nu} \partial_\mu \Phi \, \partial_\nu \Phi + \frac{-1}{2} g^{\mu\nu} \partial_\mu \Psi \, \partial_\nu \Psi - V(\Phi) - \left(\frac{m_\Psi^2}{2} \Psi^2 + \frac{g^2}{2} \Phi^2 \Psi^2 \right) \tag{15.72}$$

where $V(\Phi)$ is the inflaton potential that complies with the slow-roll conditions and g is a coupling constant between Φ and Ψ. Thus, we can approximate the spacetime during inflation by a de Sitter metric given by equation (15.6). We can rescale a so that at the initial time of the inflation era, $\eta_i = -1/H$. In the influence functional approach [HuPaZh93b, CalHu94, CalHu95, CalGon97, KUMS97, Lee04, LomNac05], the environmental field Ψ is traced out up to the one-loop level. Assuming also that the quantum field has gone through the quantum-to-classical transition, the Langevin equation for Φ is given by:

$$\Phi'' + 2aH\Phi' - \nabla^2 \Phi + a^2 \left[dV(\Phi)/d\Phi + g^2 \langle \Psi^2 \rangle \Phi \right] - g^4 a^2 \Phi$$
$$\times \int d^4 x' a^4(\eta') \theta(\eta - \eta') i \, G_-(x, x') \Phi^2(x') = \frac{\Phi}{a^2} \xi \tag{15.73}$$

where the prime denotes differentiation with respect to η. As we will see later, the quantum fluctuations of Φ will contribute to the mass correction of Ψ at one loop. The dissipation term in this Langevin equation is actually divergent. Wu *et al.* removed the divergence by using the regularization method that sets the ultraviolet cut-off $\Lambda = He^{Ht}$. They found that this term only contributes a mass correction of about $10^{-2} g^4 \bar{\phi}_0^2$ to $m_{\varphi\text{eff}}^2$ (defined after equation (15.76)) as well as a small friction term of order $10^{-2} g^4 \bar{\phi}_0^2 a \dot{\phi}/H$ to equation (15.73). As we have seen before, the environment field Ψ engenders dissipative dynamics in the inflaton field Φ via the kernel G_- and produces a multiplicative colored noise ξ with correlator

$$\langle \xi(x) \xi(x') \rangle = g^4 a^4(\eta) a^4(\eta') G_+(x, x') \tag{15.74}$$

The kernels $G_\pm$ in equations (15.73) and (15.74) can be constructed from the Green's function of Ψ with respect to a particular choice of the initial vacuum state to be specified. They were derived in Chapter 5:

$$G_\pm(x, x') = \langle \Psi(x)\Psi(x') \rangle^2 \pm \langle \Psi(x')\Psi(x) \rangle^2 \tag{15.75}$$

To focus on noise-generated structure, in the solution of equation (15.73), one can first ignore the dissipative term.

Following the stochastic inflation paradigm, after sufficient decoherence, we can decompose $\Phi(\eta, \vec{x}) = \bar{\phi}(\eta) + \varphi(\eta, \mathbf{x})$ into a mean field $\bar{\phi}$ and a fluctuation field φ which obeys the linearized Langevin equation

$$\varphi'' + 2aH\varphi' - \nabla^2\varphi + a^2 m_{\varphi\text{eff}}^2 \varphi = \bar{\phi}\xi/a^2 \tag{15.76}$$

where the effective mass is defined as $m_{\varphi\text{eff}}^2 = d^2 V(\bar{\phi})/d\varphi^2 + g^2 \langle \Psi^2 \rangle$ and the time evolution of $\bar{\phi}$ is governed by $V(\bar{\phi})$. The equation of motion for Ψ from which we construct its Green's function can be read off from its quadratic terms in the Lagrangian (15.72) as

$$\Psi'' + 2aH\Psi' - \nabla^2\Psi + a^2 m_{\Psi\text{eff}}^2 \Psi = 0 \tag{15.77}$$

where $m_{\Psi\text{eff}}^2 = m_\Psi^2 + g^2(\bar{\phi}^2 + \langle \varphi_q^2 \rangle)$. Here $\langle \varphi_q^2 \rangle$ denotes the active or intrinsic quantum fluctuations with a scale-invariant power spectrum given by $\Delta_k^q = H^2/(4\pi^2)$. Let us decompose

$$\Upsilon(x) = \int \frac{d\mathbf{k}}{(2\pi)^{\frac{3}{2}}} Y_\mathbf{k}(\eta) \, e^{i\mathbf{k}\cdot\mathbf{x}} \tag{15.78}$$

where $\Upsilon = \varphi, \xi$, and correspondingly $Y = \varphi_\mathbf{k}, \xi_\mathbf{k}$

$$\Psi(x) = \int \frac{d\mathbf{k}}{(2\pi)^{\frac{3}{2}}} \left[b_\mathbf{k} \psi_k(\eta) \, e^{i\mathbf{k}\cdot\mathbf{x}} + \text{h.c.} \right] \tag{15.79}$$

where $b_\mathbf{k}^\dagger$ and $b_\mathbf{k}$ are creation and annihilation operators satisfying $[b_\mathbf{k}, b_{\mathbf{k}'}^\dagger] = \delta(\mathbf{k} - \mathbf{k}')$.

The solution to equation (15.76) is obtained as

$$\varphi_{\vec{k}} = -i \int_{\eta_i}^{\eta} d\eta' \bar{\phi}(\eta') \xi_{\vec{k}}(\eta') \left[\varphi_k^{(1)}(\eta')\varphi_k^{(2)}(\eta) - \varphi_k^{(2)}(\eta')\varphi_k^{(1)}(\eta) \right] \tag{15.80}$$

where the homogeneous solutions $\varphi_k^{(1),(2)}$ are given by

$$\varphi_k^{(1),(2)} = \frac{1}{2a}(\pi|\eta|)^{\frac{1}{2}} H_\nu^{(1),(2)}(k\eta) \tag{15.81}$$

Here $H_\nu^{(1)}$ and $H_\nu^{(2)}$ are Hankel functions of the first and second kinds respectively and $\nu^2 = 9/4 - m_{\varphi\text{eff}}^2/H^2$. In addition, we have from equation (15.77) that

$$\psi_k = \frac{1}{2a}(\pi|\eta|)^{\frac{1}{2}} \left[c_1 H_\mu^{(1)}(k\eta) + c_2 H_\mu^{(2)}(k\eta) \right] \tag{15.82}$$

where the constants c_1 and c_2 are subject to the normalization condition, $|c_2|^2 - |c_1|^2 = 1$, and $\mu^2 = 9/4 - m_{\Psi\text{eff}}^2/H^2$.

Low ℓ WMAP modes and running spectral index

From this we can calculate the power spectrum of the perturbation $\delta\varphi$. To maintain the slow-roll condition: $m^2_{\phi\text{eff}} = m^2_{\varphi\text{eff}} \ll H^2$ (i.e. $\nu = 3/2$), we require that $g^2 < 1$ and $m^2_\Psi > H^2$. The latter condition limits the growth of $\langle \Psi^2 \rangle$ during inflation to be less than about $10^{-2}H^2$ [BunDav78, VilFor82, EnNgOl88]. Under this condition, $\langle \varphi_q^2 \rangle$ grows linearly as $H^3t/4\pi^2$ [BunDav78, VilFor82, EnNgOl88] and thus $\langle \varphi_q^2 \rangle \simeq H^2$ after about 60 e-foldings (i.e. $Ht \simeq 60$). Therefore, as long as $g^2\bar{\phi}^2 \leq 2H^2$, one can conveniently choose $m^2_{\Psi\text{eff}} = 2H^2$ (i.e. $\mu = 1/2$) for which Ψ takes a very simple form. Also, it was shown that when $\mu = 1/2$ one can select the Bunch–Davies vacuum (i.e. $c_2 = 1$ and $c_1 = 0$) [EnNgOl88]. Hence, using equations (15.74) and (15.80), one obtains

$$\langle \varphi_{\vec{k}}(\eta)\varphi^*_{\vec{k}'}(\eta) \rangle = \frac{2\pi^2}{k^3}\Delta^\xi_k(\eta)\delta(\vec{k} - \vec{k}') \tag{15.83}$$

where the noise-driven power spectrum is given by

$$\Delta^\xi_k(\eta) = \frac{g^4 y^2}{8\pi^4} \int^y_{y_i} dy_1 \int^y_{y_i} dy_2 \bar{\phi}(\eta_1)\bar{\phi}(\eta_2)\frac{\sin y_-}{y_1 y_2 y_-}$$
$$\times \left[\sin(2\Lambda y_-/k)/y_- - 1\right] F(y_1)F(y_2) \tag{15.84}$$

where $y_- = y_2 - y_1$, $y = k\eta$, $y_i = k\eta_i = -k/H$, Λ is the momentum cut-off introduced in the evaluation of the ultraviolet divergent Green's function in equation (15.75), and

$$F(x) = \left(1 + \frac{1}{xy}\right)\sin(x - y) + \left(\frac{1}{x} - \frac{1}{y}\right)\cos(x - y) \tag{15.85}$$

Note that the term $\sin(2\Lambda y_-/k)/y_- \simeq \pi\delta(y_-)$ when $\Lambda \gg k$, so $\Delta^\xi_k(\eta)$ is insensitive to Λ. Both $\bar{\phi}(\eta_1)$ and $\bar{\phi}(\eta_2)$ in equation (15.84) can be approximated as a constant mean field $\bar{\phi}_0$, since we are concerned with large scales at which the rate of change of the mean field at horizon crossing, $d\bar{\phi}/d\ln k \simeq -\sqrt{-2\epsilon\dot{H}}M_{Pl}/H$, where $\epsilon \equiv -\dot{H}/H^2$ is the slow-roll parameter, is consistent with zero up to the scale near the first CMB Doppler peak in WMAP measurements [Spe03]. A plot by Wu et al. of $\Delta^\xi_k(\eta)$ at the horizon-crossing time (defined by $y = -2\pi$) versus k/H shows that the noise-driven fluctuations depend on the onset time of inflation and approach asymptotically to a scale-invariant power spectrum $\Delta^\xi_k \simeq 0.2g^4\bar{\phi}_0^2/(4\pi^2)$ at large k. Within the usual models of inflation, the possible interactions of the inflaton are too restricted for this effect to be observable; however, the fact that interactions do affect the spectrum of primordial fluctuations has some interest on its own.

On the other hand, if the effect of interactions is expected to be important, then a nonperturbative evaluation of the influence functional becomes necessary. We describe below a possible strategy [ZanCal07a].

15.2.5 *Primordial spectrum from nonequilibrium*
renormalization group

The basic idea of RG for systems in equilibrium (where time does not enter in the description) is the coarse graining of the original system, i.e. the change in the resolution with which the system is observed [WilKog74]. Given a system with a range of scales which goes up to wavenumber Λ, if we are only interested in scales up to wavenumber $k < \Lambda$, we can separate the original system in two sectors: a lower wavenumber (soft) sector, with $k' < k$, the relevant system, and a higher wavenumber (hard) sector with $k < k' < \Lambda$, the environment. Once this division is done, the environment modes are eliminated from the description. In equilibrium, this is achieved by computing the coarse-grained "in-out" effective action for the lower sector, complemented with a rescaling of the fields and momenta that restores the cut-off and the coefficient of the q^2 term in the action to their initial values. The elimination of the modes between Λ and k proceeds by infinitesimal steps. In this way, the calculation involves only tree and one-loop diagrams, and the resulting equations form a set of differential equations for the parameters that define the effective action [WegHou73].

Essentially, the same scheme can be used for nonequilibrium systems. We want to compute true expectation values at given times, not transition amplitudes between "in" and "out" asymptotic states, far away in the future and in the past. We want to follow the real and causal evolution of expectation values, for which the usual "in-out" representation is not appropriate. A suitable description of nonequilibrium systems is given within the "closed time path" (CTP) formalism.

It is important to stress two basic differences between the nonequilibrium and equilibrium RG [Lit98]. The IF may be regarded as an action for a theory defined on a "closed time path" (CTP) composed of a first branch (going from the initial time $t = 0$ to a later time $t = T$ when the relevant observations will be performed – that is why we need the density matrix at T) and a second branch returning from T to 0. Thus each physical degree of freedom on the first branch acquires a twin on the second branch – we say the number of degrees of freedom is doubled. The IF is not just a combination of the usual actions for each branch, but also admits direct couplings across the branches. The damping constant κ and the noise constant ν are associated to two of those "mixed" terms. Therefore, the structure of the IF (from now on, CTP action, to emphasize this feature) is much more complex than the usual Euclidean or "in-out" action.

The second fundamental difference is the presence of the parameter T itself. In nonequilibrium evolution, it is important to specify the time-scale over which we shall observe the system. The CTP action contains this physical time-scale T. From the point of view of the RG, this adds one more dimensional parameter to the theory, much as an external field in the Ising model. Physically, because time integrations are restricted to the interval $[0, T]$, energy conservation does not hold at each vertex. This is of paramount importance regarding damping.

The RG for the CTP effective action (obtained by taking the limit $T \to \infty$) was studied by Dalvit and Mazzitelli [DalMaz96, Dal98]; see also [CaHuMa01] and [Pol06, ZanCal07a, ZanCal06b].

In formulating a nonequilibrium RG, we must deal with the fact that the CTP action may have an arbitrary functional dependence on the fields and be nonlocal both in time and space. In principle, one can define an exact RG transformation [DalMaz96], where all three functional dependencies are left open. However, the resulting formalism is too complex to be of practical use. Fortunately, the special properties of the application to thermalization allow for substantial simplifications.

The full RG equations for this theory is given in [ZanCal06b]. Here we shall only highlight those aspects of the calculation most relevant to the application to primordial fluctuation generation.

We shall work with the conformally scaled field $\chi = a\Phi$. For simplicity, we shall treat χ as a field on flat spacetime. This only induces an error of order 1 in the amplitude of the fluctuations at horizon exit.

Let us call $\chi^{1(2)}$ the field variable in the first (resp. second) branch of the CTP. To write down the CTP action, it is best to introduce average and difference variables

$$\chi_- = \chi^1 - \chi^2$$
$$\chi_+ = \left(\chi^1 + \chi^2 \right) / 2 \tag{15.86}$$

In terms of these variables, a generic CTP action may be written as

$$S_{\mathrm{CTP}} = S_0 + S_\lambda + S_{\mathrm{other}} \tag{15.87}$$

where S_0 is the CTP action functional for a free massless field theory, S_λ accounts for a $\lambda \chi^4$-type self-interaction and S_{other} includes all other possible terms. Momentum integrals are bounded by $k = \Lambda$. We shall assume that the initial condition for the RG flow is $S_{\mathrm{other}} = 0$ at the hard scale Λ, so that if it appears at soft scales, it is as a consequence of the RG running itself. Note that this is true, in particular, for the noise and dissipation terms.

To define the nonequilibrium RG we also need to specify the state of the field at the initial time $t = 0$. For simplicity, we shall assume this is the vacuum state for the free action S_0. Observe that this is a nonequilibrium state for the interacting theory.

The value λ_0 of the coupling constant λ at the hard scale Λ may be used as the small parameter in a perturbative expansion of the RG equation. To order λ_0^2, the RG equation for the quartic coupling decouples, and can be solved by itself. The result is that at soft scales k, λ is both scale and T dependent. There is no RG running if $T = 0$, while the usual textbook result is obtained as $T \to \infty$. For all values of T, λ is driven to zero as $k \to 0$ [ZanCal06b]. Thus it is consistent to assume that λ is uniformly small in the relevant scale range.

In particular, in order to compute the RG equations to order λ^2, it will be enough to use in the Feynman graphs the zeroth order propagators, which are those of the massless free theory. The only exception is in computing the effective mass, but this calculation is decoupled from the noise and dissipation terms to order λ^2. Observe that it is at the same time a huge simplification and a strong limitation concerning the range of application of our results, as we expect substantial shifts in the propagators when T approaches the relaxation time of the theory.

Because of the nonzero initial value of λ, other couplings will appear as a result of the RG running. To order λ^2, it is enough to consider quadratic, quartic and six-point terms in the action. All these terms feed back into each other, so they must be taken self-consistently. To compute the amplitude of the fluctuations at horizon exit, however, it is enough to focus on the quadratic terms,

$$S_{\text{other}} \to S_2\left[\chi_-, \chi_+\right] = \int_0^T dt_1 \int_0^T dt_2 \int d^d k \; \left[v_{21}(k; t_1, t_2)\, \chi_-(\mathbf{k}, t_1)\, \chi_+(-\mathbf{k}, t_2)\right.$$
$$\left. + i\, v_{22}(k; t_1, t_2)\, \chi_-(\mathbf{k}, t_1)\, \chi_-(-\mathbf{k}, t_2)\right] \tag{15.88}$$

In principle, the induced quadratic terms will be oscillatory functions of $\Lambda t_{1,2}$. However, by the time a mode reaches the horizon it becomes insensitive to high frequencies. To focus on the slow dynamics, we may project out the mass, dissipation and noise terms on which the oscillations are mounted.

To this end, we introduce two projectors. Given a function of two times $v(k; t_1, t_2)$, we define

$$\mathbf{P}v(k; t_1, t_2) = \mathcal{P}v(k)\, \delta(t_1 - t_2) \tag{15.89}$$

and, if $v(k; t_1, t_2) = 0$ for $t_2 > t_1$,

$$\mathbf{Q}v(k; t_1, t_2) = \mathcal{Q}v(k) \left[2\left(\frac{\partial}{\partial t_2} + \delta(t_2) - \delta(0)\right)\delta(t_1 - t_2)\right] \tag{15.90}$$

where

$$\mathcal{P}v(k) = \frac{1}{T}\int_0^T dt_1 \int_0^T dt_2\, v(k; t_1, t_2) \tag{15.91}$$

and

$$\mathcal{Q}v(k) = \frac{1}{T}\int_0^T dt_1 \int_0^T dt_2\, v(k; t_1, t_2)\, (t_2 - t_1) \tag{15.92}$$

It is easy to verify that $\mathbf{P}^2 = \mathbf{P}$, $\mathbf{Q}^2 = \mathbf{Q}$, and that $\mathbf{Q}\mathbf{P} = \mathbf{P}\mathbf{Q} = 0$. This proves that the decomposition

$$v(k; t_1, t_2) = \mathbf{P}v(k; t_1, t_2) + \mathbf{Q}v(k; t_1, t_2) + \Delta v(k; t_1, t_2) \tag{15.93}$$

is unique. Defining

$$v_0 = \mathcal{P}v_{21} \tag{15.94}$$

and

$$v_1 = Q v_{21} \qquad (15.95)$$

we extract from $v_{21}(k; t_1, t_2)$ two quantities: $-v_0(k)$, which acts as a momentum-dependent mass squared term, and $-v_1(k)/2$, which is equivalent to a damping constant.

If we further expand in powers of wavenumber k

$$v_0(k) = v_0(0) + k \frac{\partial v_0(0)}{\partial k} + \frac{k^2}{2!} \frac{\partial^2 v_0(0)}{\partial k^2} + \dots \qquad (15.96)$$

the linear term vanishes from symmetry, and the appearance of the quadratic term is prevented by performing a field rescaling as part of the RG transformation (thus the field acquires an anomalous dimension). The net effect is then to induce a mass term

$$m^2 = -v_0(0) \qquad (15.97)$$

and a damping constant

$$\kappa = -v_1(0)/2 \qquad (15.98)$$

The noise kernel is obtained in a similar way from the imaginary part of the CTP action, v_{22}.

After these considerations, the relevant CTP action for long-wavelength, slowly varying configurations reduces to

$$S_{\text{CTP}}\left[\chi_-, \chi_+\right] = \int_0^T dt \int d^d k \left[\dot{\chi}_-(\mathbf{k}, t) \, \dot{\chi}_+(-\mathbf{k}, t) - \chi_-(\mathbf{k}, t) \left(k^2 + m^2\right) \chi_+(-\mathbf{k}, t) \right.$$
$$\left. - 2\kappa \, \chi_-(\mathbf{k}, t)\dot{\chi}_+(-\mathbf{k}, t) + \frac{i}{2}\nu \, \chi_-(\mathbf{k}, t)\chi_-(-\mathbf{k}, t) \right] \qquad (15.99)$$

The flow of the RG drives the initial interacting theory towards the free theory (15.99), and allows us to find a relation between expectation values associated with each theory. The relation is

$$G\left(k, t, \mu(\Lambda, T)\right) = (\Lambda/k)^{\alpha(k,T)} \, G\left(\Lambda, (\Lambda/k)^{\beta(k,T)} \, t, \mu(k, T)\right) \qquad (15.100)$$

On the left-hand side, G is the two-field expectation value computed for a mode k at time t, and $\mu(\Lambda, T)$ stands for the set of parameters which define the action at scale Λ. In our case the only parameter is the coupling constant λ. On the right-hand side, G is the expectation value of the theory defined by the set of parameters $\mu(k, T)$, reached after modes between k and Λ have been eliminated. The relevant parameters in $\mu(k, T)$ are $m^2(k, T)$, $\kappa(k, T)$, and $\nu(k, T)$. Finally, the exponents α and β depend on the trajectory followed by the action when it goes from scale Λ to k.

Now we connect to the original problem for the power spectrum of an interacting inflaton field. We must feed the RG group equations with an initial condition at scale Λ and then use the relation (15.100) to obtain the expectation value for the mode k as it exits the horizon. The initial condition, in terms of the conformal field, is given by the CTP action at scale Λ, where t has to be replaced by the conformal time η. The mode k exits the horizon when

$$\eta = -k^{-1} \qquad (15.101)$$

If inflation starts at η^*, the time that the mode k spends inside the horizon is given by

$$\tau_k = -k^{-1} - \eta^* \qquad (15.102)$$

For the physical field (subscript HE stands for horizon exit)

$$\langle \Phi(k,t)\Phi(k,t)\rangle_{\mathrm{HE}} = k^{-2}\, G\,(k,\tau_k,\lambda) \qquad (15.103)$$

From equation (15.100), identifying t and T with τ_k, we get

$$\langle \Phi(k,t)\Phi(k,t)\rangle_{\mathrm{HE}} = k^{-2}\,(\Lambda/k)^{\alpha(k,\tau_k)}$$
$$G\left(\Lambda,\,(\Lambda/k)^{\beta(k,\tau_k)}\,\tau_k, m^2(k,\tau_k), \kappa(k,\tau_k), \nu(k,\tau_k)\right)$$
$$(15.104)$$

Here, the relevant elements of $\mu(k,\tau_k)$ have been shown explicitly. The right-hand side of equation (15.104) can be calculated using the G corresponding to the action (15.99)

$$G\left(k,t,m^2,\kappa,\nu\right) = \left(\frac{2}{k} - \frac{\nu}{\kappa\omega_0^2}\right)\left[\frac{\omega_0^2}{\omega^2} - \frac{\kappa^2}{\omega^2}\cos(2\omega t) + \frac{\kappa}{\omega}\sin(2\omega t)\right]e^{-2\kappa t} + \frac{\nu}{\kappa\omega_0^2}$$
$$(15.105)$$

where $\omega_0^2 = m^2 + k^2$ and $\omega^2 = \omega_0^2 - \kappa^2$ [ZanCal06b].

The expressions for m^2, κ, and ν, and for the exponents α and β, as functions of k and τ_k, are given in [ZanCal06b]. The main effects are introduced by the mass term.

15.3 Reheating in the inflationary universe

We focus here on the so-called reheating regime when the universe began to warm up due to particle creation from excitations of the vacuum fluctuations of the inflaton field and other fields coupled to it. The back-reaction of created particles results in the decay of the inflaton mean field and the turnover of the universe from the inflationary state described by an approximate de Sitter solution to a radiation-dominated FLRW solution depicted in the so-called standard model.

As stated before, the inflationary scenarios can generically be divided into three eras: (1) entrance into a vacuum energy density dominated era, which can be a metastable state of the Higgs field in a GUT era, where the universe begins inflation; (2) a "slow roll" of the inflation field ϕ either from a relatively flat effective potential $V(\phi)$, or from a simple ϕ^2 potential, as in the new or chaotic inflationary cosmology; (3) exiting the inflationary era and entering into an era when the inflaton field undergoes rapid oscillations, where the vacuum energy density is transformed into radiation via particle creation and the universe begins to reheat to a radiation-dominated state.

One can also divide the reheating era roughly into two or even three stages, preheating, heating and thermalization. In the preheating stage the dominant effect is due to parametric particle creation [KoLiSt97]. Brandenberger, Traschen and Shtanov [ShTrBr95], Kofman, Linde and Starobinsky [KoLiSt94] and Boyanovsky *et al.* [BVHS96] first pointed out the importance of parametric resonance at work in this stage. We have explained this mechanism of a rather general nature in Chapter 4, e.g. the narrow and broad resonances. The thermalization process is a difficult and complex one. We discussed some aspects of it in Chapter 12, but the reader should consult representative papers (e.g. [BoVeSa05, Lin90, Muk05, BaTsWa06]) for a better understanding of the specific context of thermalization in post-inflationary reheating.

As explained earlier, since the purpose of these latter chapters is to illustrate how the methods in NEqQFT we have learned can be applied to treat relevant problems in different contexts, the discussions here on reheating are not meant to be of a review nature, where ideally all ideas and approaches ought to be represented. We refer the readers to reviews [BaTsWa06] for a more balanced overall perspective of the physical processes involved. For our more restricted aim here, we shall only describe two examples where we have some first-hand experience in which the full use of the methods of nonequilibrium quantum field theory plays an essential role. These examples concern the back-reaction of the created particles on the inflaton field during preheating, and the generation of primordial magnetic fields as a side-effect of reheating.

15.3.1 Case study I: Back-reaction of Fermi fields during preheating

The earliest analysis of the reheating stage assumed that the decay of the inflaton field could be described perturbatively, by computing the absorption parts of suitable Feynman graphs. That led to an apparent contradiction between the theories of reheating and structure formation, since the latter places very stringent limits on the possible couplings of the inflaton. Moreover, the generation of heavy particles was strongly suppressed, against the expectation that heavy bosons generated during reheating could play a role in baryogenesis.

This seeming difficulty was overcome when it was realized that the decay of the inflaton could proceed very efficiently through the parametric amplification of matter fields, which is an essentially nonperturbative process. As a matter of fact, in these new scenarios enough reheating is obtained even if the inflaton is not coupled to any other field at all, other than the gravitational field.

While the basic mechanism of parametric amplification during reheating are the broad and narrow resonances, they are also strongly affected by the expansion of the universe [RamHu97b]. As we have seen in Chapter 4, the evolution of the field under broad resonance may be described as a series of adiabatic evolutions punctuated by nonadiabatic transitions. The growth factor from one transition to the next depends on the accumulated phase of the field variable. The dynamic geometrical background induces changes in this phase, because the relevant parameters become time dependent, and thus affects the nature of resonance. The general case becomes a sequence of different resonance regimes due to the process of parametric resonance [KoLiSt97, GKLS97, ChNuMi05]. The nonlinearity of the inflaton oscillations also plays an important role. In the general case, the oscillating inflaton field will have a full frequency spectrum, not limited to a few narrow bands, and the amplification of the matter fields may be studied by the methods of particle production from a time-dependent background, which we have discussed in Chapters 4 and 8 [Bas98, ZMCB98, ZMCB99].

Eventually, all relevant matter field modes acquire high occupation numbers and a classical treatment becomes possible [CalGra02]. This observation has been key to progress in the analysis of the fully nonlinear regime, including inflaton fragmentation and the so-called turbulent reheating stage [KhlTka96, KoLiSt97, FelTka00, FeKoLi01, FelKof01, FGGKLT01, MicTka04, PFKP06]. We have discussed similar processes (albeit on nonexpanding spacetimes) in Chapter 12.

Nevertheless, the back-reaction of the created particles has a strong effect on the inflaton even before the classical approximation becomes reliable. The inflaton must be seen as an effectively open system – with all other matter fields providing an environment – and its dynamics is subject to dissipation and noise therefrom [Hu91, SinHu91, LomMaz96, DalMaz96, GreMul97].

It was earlier realized that a description of the inflaton dynamics based on the 1PI effective action (cf. Chapter 6) or similar constructs with a c-number inflaton field as the sole argument is not satisfactory. For one, inflaton fluctuations play a key role in the theory of structure formation and one should to follow their evolution through the reheating stage. Most importantly, the equations of motion as derived from the 1PI effective action are affected by secular terms and become unreliable after several inflaton oscillations. We have found a similar problem in the treatment of Bose–Einstein condensates in Chapter 13.

Although it is possible to extract useful information from these secular terms through dynamical renormalization group analysis, it is best to improve the model, by including the physical processes that cut off the growth of secular terms. The most efficient way of accomplishing this is by going over to a 2PI description (cf. Chapter 6), where the inflaton mean field and fluctuations are treated self-consistently. The 2PI effective action implements the resummation of secular terms, and also incorporates the basic processes that eventually could lead to thermalization, as we have discussed in Chapter 12.

As a concrete example of the application of 2PIEA techniques to the description of preheating, we shall analyze the evolution of the inflaton field coupled to N Fermi fields. Our treatment here follows [RaStHu98]. While Fermi fields are subject to Pauli blocking which, unlike the stimulated emission of Bose fields, opposes particle creation (cf. Chapter 4), the fact that most matter fields in the standard model are fermionic makes them a proper subject of study [KoLiSt97].

We consider a model of a scalar inflaton field Φ (with $\lambda\Phi^4$ self-coupling) interacting with a spinor field via Yukawa coupling. The system consists of the inflaton mean field and variance, and the environment consists of the spinor field(s) Ψ. We construct the CTP-2PI-CGEA, and derive from it the effective dynamical equations for the inflaton field, taking into account its effect on the environment, and back-reaction therefrom in a self-consistent manner.

This problem is a good example of how to apply many of the concepts and techniques presented in earlier chapers. The first step is to derive a set of coupled nonperturbative equations for the inflaton mean field and variance at two loops. Only beginning at two loops will both the inflaton mean field *and the inflaton variance* couple to the spinor degrees of freedom. They are damped by back-reaction from fermion particle production. (Calculations using the 1PI effective action will miss this important effect.) The equations of motion are real and causal, and the gap equation for the two-point function is dissipative due to fermion particle production.

As we emphasized in Chapter 9, there is a subtle yet important distinction between the system–environment division in nonequilibrium statistical mechanics and the system–bath division assumed in thermal field theory. In the latter, one assumes that the propagators for the bath degrees of freedom are *fixed*, finite-temperature equilibrium Green functions, whereas in the case of the CTP-CGEA, the environmental propagators are *slaved* (in the sense of [CalHu95a]) to the dynamics of the system degrees of freedom, and are not fixed a priori to be equilibrium Green functions for all time. This distinction is important for discussions of fermion particle production during reheating, because it is only when the inflaton mean field amplitude is small enough for the use of perturbation theory, that the system–bath split implicit in thermal field theory can be used. Otherwise, one must take into account the effect of the inflaton mean field on the bath (spinor) *propagators*.

As we saw in Chapter 6, the use of the closed time path (CTP) formalism allows formulation of the nonequilibrium dynamics of the inflaton from an appropriately defined initial quantum state. At the onset of the reheating period, the inflaton field's zero mode typically has a large expectation value, whereas all other fields coupled to the inflaton, as well as inflaton modes with momenta greater than the Hubble constant, are to a good approximation in a vacuum state [Bra85].

The model consists of a scalar field Φ (the inflaton field) which is Yukawa-coupled to a spinor field Ψ, in a curved, dynamical, classical background spacetime. The total action

$$S[\Phi, \bar{\Psi}, \Psi, g^{\mu\nu}] = S^{\mathrm{G}}[g^{\mu\nu}] + S^{\mathrm{F}}[\Phi, \bar{\Psi}, \Psi, g^{\mu\nu}] \tag{15.106}$$

consists of a part depicting classical gravity, $S^{\mathrm{G}}[g^{\mu\nu}]$, and a part for the matter fields,

$$S^{\mathrm{F}}[\Phi, \bar{\Psi}, \Psi, g^{\mu\nu}] = S^{\Phi}[\Phi, g^{\mu\nu}] + S^{\Psi}[\bar{\Psi}, \Psi, g^{\mu\nu}] + S^{\mathrm{Y}}[\Phi, \bar{\Psi}, \Psi, g^{\mu\nu}] \tag{15.107}$$

whose scalar (inflaton), spinor (fermion), and Yukawa interaction parts are given by

$$S^{\Phi}[\Phi, g^{\mu\nu}] = -\frac{1}{2}\int d^4x\sqrt{-g}\left[\Phi(\nabla^2 + m^2 + \xi R)\Phi + \frac{\lambda}{12}\Phi^4\right] \tag{15.108}$$

$$S^{\Psi}[\bar{\Psi}, \Psi, g^{\mu\nu}] = \int d^4x\sqrt{-g}\left[\frac{i}{2}\left(\bar{\Psi}\gamma^\mu\nabla_\mu\Psi - (\nabla_\mu\bar{\Psi})\gamma^\mu\Psi\right) - \mu\bar{\Psi}\Psi\right] \tag{15.109}$$

$$S^{\mathrm{Y}}[\Phi, \bar{\Psi}, \Psi, g^{\mu\nu}] = -f\int d^4x\sqrt{-g}\Phi\bar{\Psi}\Psi \tag{15.110}$$

For this theory to be renormalizable in semiclassical gravity, the bare gravity action $S^{\mathrm{G}}[g^{\mu\nu}]$ of equation (15.106) should have the form [DeW75, BirDav82]

$$S^{\mathrm{G}}[g^{\mu\nu}] = \frac{1}{16\pi G}\int d^4x\sqrt{-g}\left[R - 2\Lambda_{\mathrm{c}} + cR^2 + bR^{\alpha\beta}R_{\alpha\beta} + aR^{\alpha\beta\gamma\delta}R_{\alpha\beta\gamma\delta}\right] \tag{15.111}$$

In equations (15.108)–(15.110), m is the scalar field "mass" (with dimensions of inverse length); ξ is the dimensionless coupling to gravity; μ is the spinor field "mass," with dimensions of inverse length; ∇^2 is the Laplace–Beltrami operator in the curved background spacetime with metric tensor $g_{\mu\nu}$; ∇_μ is the covariant derivative compatible with the metric; $\sqrt{-g}$ is the square root of the absolute value of the determinant of the metric; λ is the self-coupling of the inflaton field, with dimensions of $1/\sqrt{\hbar}$; and f is the Yukawa coupling constant, which has dimensions of $1/\sqrt{\hbar}$. In equation (15.111), G is Newton's constant (with dimensions of length divided by mass); R is the scalar curvature; $R_{\mu\nu}$ is the Ricci tensor; $R_{\alpha\beta\gamma\delta}$ is the Riemann tensor; a, b, and c are constants with dimensions of length squared; and Λ_{c} is the cosmological constant, which has dimensions of inverse length squared. The curved spacetime Dirac matrices γ^μ satisfy the anticommutation relation

$$\{\gamma^\mu, \gamma^\nu\}_+ = 2g^{\mu\nu}1_{\mathrm{sp}}, \tag{15.112}$$

in terms of the contravariant metric tensor $g^{\mu\nu}$. The symbol 1_{sp} denotes the identity element in the Dirac algebra.

In four spacetime dimensions the terms with constants a, b, and c are related by a generalized Gauss–Bonnet theorem [Che62], so we have the freedom to choose $a = 0$. It is assumed that there is a definite separation of time-scales between the stage of "preheating" (see, e.g. [RamHu97b]), and fermionic particle production. In addition, the fermion field mass μ is assumed to be much lighter than the inflaton field mass m, i.e. the renormalized parameters m and μ satisfy $m \gg \mu$.

We denote the quantum Heisenberg field operators of the scalar field Φ and the spinor field Ψ by Φ_{H} and Ψ_{H}, respectively, and the quantum state by $|s\rangle$. For consistency with the truncation of the correlation hierarchy at second order, we assume Φ_{H} to have a Gaussian moment expansion in the position basis [MazPaz89], in which case the relevant observables are the scalar mean field

$$\bar{\phi}(x) \equiv \langle s|\Phi_{\mathrm{H}}(x)|s\rangle \tag{15.113}$$

and the mean-squared fluctuations, or variance, of the scalar field

$$\langle s|\Phi_{\mathrm{H}}^2(x)|s\rangle - \langle s|\Phi_{\mathrm{H}}(x)|s\rangle^2 \equiv \langle s|\varphi_{\mathrm{H}}{}^2(x)|s\rangle \tag{15.114}$$

where the last equality follows from the definition of the scalar *fluctuation field*

$$\varphi_{\mathrm{H}}(x) \equiv \Phi_{\mathrm{H}}(x) - \bar{\phi}(x) \tag{15.115}$$

As discussed above, at the end of the preheating period, the inflaton variance can be as large as the square of the amplitude of mean-field oscillations. On the basis of our assumption of separation of time-scales and the conditions which prevail at the onset of reheating, the initial quantum state $|s\rangle$ is assumed to be an appropriately defined vacuum state for the *spinor* field.

The construction of the CTP-2PI-CGEA for the $\Phi\bar{\Psi}\Psi$ theory in a general, curved, background spacetime closely parallels the construction of the CTP-2PI effective action for the $O(N)$ model discussed in Chapter 6 [LomMaz98]. Within the spacetime manifold (whose dynamics must be determined self-consistently through the semiclassical gravitational field equation), let M be defined as the past domain of dependence of a Cauchy hypersurface Σ_*, where Σ_* has been chosen to be far to the future of any dynamics we wish to study. We now define a "CTP" manifold $\mathcal{M}$ as the union of the two copies of M corresponding to the $\{+, -\}$ time branches, with their last Cauchy hypersurfaces Σ_* identified. As in Chapter 6, we define an action functional on the closed time path manifold as the difference of the actions evaluated on each branch. For a function Φ on $\mathcal{M}$, the restrictions of Φ to the $+$ and $-$ time branches are subject to the boundary condition $(\Phi_+)_{|\Sigma_*} = (\Phi_-)_{|\Sigma_*}$ at the hypersurface Σ_*.

Following the general procedure in Chapter 6, we obtain the CTP-2PI-CGEA

$$\Gamma[\bar{\phi}, G, g^{\mu\nu}] = \mathcal{S}^\Phi[\bar{\phi}] - \frac{i\hbar}{2}\ln\det G_{ab} - i\hbar\ln\det F_{ab} + \Gamma_2[\bar{\phi}, G]$$

$$+ \frac{i\hbar}{2}\int_M d^4x\sqrt{-g}\int_M d^4x'\sqrt{-g'}\mathcal{A}^{ab}(x', x)G_{ab}(x, x')$$

$$\tag{15.116}$$

where $\mathcal{A}^{ab}$ is the second functional derivative of the scalar part of the classical action S^Φ, evaluated at $\bar{\phi}$,

$$
\begin{aligned}
i\mathcal{A}^{ab}(x,x') &= \frac{1}{\sqrt{-g}}\left(\frac{\delta^2 S^\Phi}{\delta\Phi_a(x)\delta\Phi_b(x')}[\bar{\phi}]\right)\frac{1}{\sqrt{-g'}} \\
&= -\left[c^{ab}(\nabla_x^2 + m^2 + \xi R(x)) + c^{abcd}\frac{\lambda}{2}\bar{\phi}_c(x)\bar{\phi}_d(x)\right]\frac{\delta(x-x')}{\sqrt{-g'}}
\end{aligned}
$$

(15.117)

The symbol F_{ab} denotes the one-loop CTP spinor propagator, which is defined by

$$
F_{ab}(x,x') \equiv \mathcal{B}_{ab}^{-1}(x,x')
$$

(15.118)

where we are suppressing spinor indices, and the inverse spinor propagator $\mathcal{B}^{ab}$ is defined by

$$
\begin{aligned}
i\mathcal{B}^{ab}(x,x') &= \frac{1}{\sqrt{-g}}\left[\frac{\delta^2(S^\Psi[\bar{\Psi},\Psi] + S^Y[\bar{\Psi},\Psi;\bar{\phi}])}{\delta\Psi_a(x)\delta\bar{\Psi}_b(x')}\right]\frac{1}{\sqrt{-g'}} \\
&= \left(c^{ab}(i\gamma^\mu\nabla'_\mu - \mu) - c^{abc}f\bar{\phi}_c(x')\right)\frac{\delta(x'-x)}{\sqrt{-g}}1_{\text{sp}}
\end{aligned}
$$

(15.119)

It is clear from equation (15.119) that the use of the one-loop spinor propagators in the construction of the CTP-2PI-CGEA represents a nonperturbative resummation in the Yukawa coupling constant, which (as discussed above) goes beyond the standard time-dependent perturbation theory. The boundary conditions which define the inverses of equations (15.117) and (15.119) are the boundary conditions at the initial data surface in the functional integral which in turn define the initial quantum state $|s\rangle$. The one-loop spinor propagators F_{ab}, introduced in Chapter 10, are related to the expectation values of the spinor Heisenberg field operators in the presence of the c-number background field $\bar{\phi}$.

Only diagrams which are two-particle irreducible with respect to cuts of *scalar* lines contribute to Γ_2. The distinction between the CTP-2PI, *coarse-grained* effective action which arises here, and the fully two-particle irreducible effective action (2PI with respect to scalar *and* spinor cuts) is due to the fact that we only Legendre-transformed sources coupled to Φ; i.e. the spinor field is treated as the environment.

We evaluate the functional $\Gamma_2[\bar{\phi}, G, g^{\mu\nu}]$ in a loop expansion, starting with the two-loop term, $\Gamma^{(2)}$. The $\lambda\Phi^4$ self-interaction leads to two terms in the two-loop part of the effective action. They are the "setting sun" diagram, which is $O(\lambda^2)$, and the "double bubble," which is $O(\lambda)$, respectively (cf. Chapter 6). The Yukawa interaction leads to only one diagram in $\Gamma^{(2)}$

$$
\frac{if^2}{2}c^{aa'a''}c^{bb'b''}\int d^4x\sqrt{-g}\int d^4x'\sqrt{-g'}G_{ab}(x,x')\text{tr}_{\text{sp}}\left[F_{a'b'}(x,x')F_{b''a''}(x',x)\right]
$$

(15.120)

where the trace is understood to be over the spinor indices which are not shown, and the three-index symbol c^{abc} is defined as in Chapter 6.

We treat the λ self-interaction using the time-dependent Hartree–Fock approximation [CoJaTo74], which is equivalent to retaining the $O(\lambda)$ (double bubble) graph and dropping the $O(\lambda^2)$ (setting sun) graph. We assume for the present study that the coupling λ is sufficiently small that the $O(\lambda^2)$ diagram is unimportant on the time-scales of interest in the fermion production regime of the inflaton dynamics. The mean-field and gap equations including both the $O(\lambda)$ and the $O(\lambda^2)$ diagrams have been derived in a general curved spacetime in [RamHu97a].

The (bare) semiclassical field equations for the two-point function, mean field, and metric can be obtained from the CTP-2PI-CGEA by functional differentiation with respect to G_{ab}, $\bar{\phi}_a$, and $g^{\mu\nu}$, followed by identifications of $\bar{\phi}$ and $g^{\mu\nu}$ on the two time branches [RamHu97b]. The field equation of semiclassical gravity (with bare parameters) is

$$G_{\mu\nu} + \Lambda_c g_{\mu\nu} + c^{(1)}H_{\mu\nu} + b^{(2)}H_{\mu\nu} = 8\pi G \langle T_{\mu\nu} \rangle \qquad (15.121)$$

where $^{(1,2)}H_{\mu\nu}$ are tensors constructed from the covariant derivatives of the metric and connection forms (e.g. defined in [BirDav82]). The (unrenormalized) quantum energy–momentum tensor is defined by

$$\langle T^{\mu\nu} \rangle = \frac{2}{\sqrt{-g}} \left. \frac{\delta \Gamma[\bar{\phi}, G, g_{\mu\nu}]}{\delta g_{\mu\nu+}} \right|_{\bar{\phi}_+ = \bar{\phi}_- = \bar{\phi}} {}_{g_+^{\mu\nu} = g_-^{\mu\nu} = g^{\mu\nu}} \qquad (15.122)$$

The energy–momentum tensor $\langle T_{\mu\nu} \rangle$ is divergent in four spacetime dimensions, and must be regularized via a covariant procedure [BirDav82, RamHu97b].

Making the two-loop approximation to the CTP-2PI-CGEA, where we take $\Gamma_2 \simeq \hbar^2 \Gamma^{(2)}$, and dropping the $O(\lambda^2)$ diagram from Γ_2, the mean-field equation becomes

$$\left(\nabla^2 + m^2 + \xi R(x) + \frac{\lambda}{6}\bar{\phi}^2(x) + \frac{\lambda\hbar}{2}G(x,x) \right) \bar{\phi}$$
$$+ \hbar f \mathrm{Tr}_{\mathrm{sp}}[F_{ab}(x,x)] - \hbar^2 g^3 \Sigma(x) = 0 \qquad (15.123)$$

where $G(x,x)$ is the coincidence limit of $G_{ab}(x,x')$, and $\Sigma(y)$ is a (self-energy) function defined by

$$\Sigma(y) = -2 \int d^4 x \sqrt{-g} \int d^4 x' \sqrt{-g'} \, \mathrm{Re}\, \mathrm{Tr}_{\mathrm{sp}} \Big[\big(\theta(x,x') G_{(1)}(x',x) F^{(1)}(x,x')$$
$$- G_{\mathrm{R}}(x,x')^\dagger F_{\mathrm{R}}(x,x') \big) F_{\mathrm{R}}(y,x')^\dagger F_{\mathrm{R}}(y,x) \Big] \qquad (15.124)$$

where an index 1 refers to a Hadamard propagator (cf. Chapters 6 and 10), and a subindex R to a retarded propagator. It is clear that the integrand vanishes whenever x or x' is to the future of y. The equation for G_{ab} is given by

$$(G^{-1})^{ba}(x,x') = \mathcal{A}^{ba}(x,x') + \frac{i\lambda\hbar}{4}c^{ba}G_1(x,x)\frac{\delta(x-x')}{\sqrt{-g'}}$$
$$+ \hbar f^2 c^{aa'a''} c^{bb'b''} \mathrm{Tr}_{\mathrm{sp}}[F_{a'b'}(x,x')F_{b''a''}(x',x)] \qquad (15.125)$$

Multiplying equation (15.125) through by G_{ab}, performing a spacetime integration, and taking the 11 component, we obtain

$$\left(\nabla^2 + m^2 + \xi R + \frac{\lambda}{2}\bar{\phi}^2 + \frac{\lambda\hbar}{4}G_1(x,x) \right) G_F(x,x')$$

$$+ \hbar f^2 \int dx'' \sqrt{-g''}\mathcal{K}(x,x'')G_F(x'',x') = -i\frac{\delta(x-x')}{\sqrt{-g'}}$$

$$(15.126)$$

in terms of a kernel $\mathcal{K}(x,x'')$ defined by

$$\mathcal{K}(x,x') = -i\mathrm{Tr}_{\mathrm{sp}}\left[F_F(x,x')^2 - F^+(x,x')^2 \right] = \mathrm{Re}\,\mathrm{Tr}_{\mathrm{sp}}\left[F_R(x,x')F_1(x',x) \right]$$

$$(15.127)$$

which shows that equation (15.126) is manifestly real and causal. The kernel $\mathcal{K}(x,x')$ is dissipative, and it reflects the back-reaction from fermionic particle production induced by the time-dependence of the inflaton *variance*. Equation (15.126) is therefore damped for modes above threshold, and this damping is not accounted for in the 1PI treatments of inflaton dynamics (where only the inflaton mean field is dynamical). As stressed above, the dissipative dynamics of the inflaton two-point function can be important when the inflaton variance is on the order of the square of the inflaton mean-field amplitude; such conditions may exist at the end of preheating.

The set of evolution equations (15.123) for $\bar{\phi}$ and (15.126) for G is formally complete to two loops. Dissipation arises due to the coarse graining of the spinor degrees of freedom. These dynamical equations are valid in a general background spacetime and are useful for reheating studies and more general purposes.

15.3.2 Case study II: Primordial magnetic field generation

Given the difficulties in constructing a suitable model of the reheating stage, to further elucidate its physics it helps to investigate other physical processes coexisting with the reheating of the universe which could have produced an observable imprint either on the CMB or today's large-scale structures.

The two processes most studied are the generation of spin-two and spin-one fields. The former concerns the possible processing of primordial gravitational fluctuations on super-horizon scales, while the latter addresses the feasibility of generating primordial magnetic fields during reheating. Gravitational fluctuations ought to have influenced the spectrum and polarization of the CMB, while a primordial magnetic field could serve as a seed for the magnetic fields observed today in cosmological structures, and should also have affected the CMB [Dod03, Lon98].

Fields with a strength of about a millionth of the Earth's magnetic field are observed both in galaxies and clusters of galaxies. There are at least three good reasons to believe these fields have a cosmological origin. First, the fact that they extend over huge scales. Second, fields are also observed at high red-shift, when dynamo mechanisms have less time to operate. This strongly suggests the field was "already there," though at the time of writing it is unclear exactly how fast dynamo amplification can be [BraSub05]. Third, that in any case "local" mechanisms such as a "galactic dynamo" could amplify an existing seed field, but not create a field from nothing [GraRub01].

The same reasons of scale make it tempting to place the origin of the field in the inflationary era (for primordial but not inflationary mechanisms see [BoVeSi03b, BoVeSi03a, Vac01, VilLea82]). However a large enough magnetic field is not expected to be generated during inflation because of the conformal invariance of the Maxwell field.[5]

We give a semiquantitative discussion here (adopting natural units ($\hbar = c = k_B = 1$)). As in Chapter 7, the field is described by a vector potential A_μ; we rescale the field by the gauge coupling constant, so that the curved space free Lagrangian density reads

$$\mathcal{L} = \frac{-\sqrt{-g}}{4} F^{\mu\nu} F_{\mu\nu} \tag{15.128}$$

with the abelian field tensor equation (7.5) $F_{\mu\nu} = \partial_\mu A_\nu - \partial_\nu A_\mu$. If we assume a conformally flat FLRW metric written in conformal coordinates $(\eta, \vec{x})$ (cf. Section 4.6.2), then in four spacetime dimensions the conformal factor drops out of the free action.

The inhomogeneous Maxwell equations for a field driven by a current

$$J^\mu = \frac{1}{\sqrt{-g}} \frac{\delta S_m}{\delta A_\mu} \tag{15.129}$$

are given by

$$F^{\nu\mu}_{;\nu} = -J^\mu \tag{15.130}$$

During the radiation-dominated era, the current is induced by the Lorentz force acting on the charged plasma, so we have a constitutive relation

$$J^\mu = \sigma F^{\mu\nu} u_\nu \tag{15.131}$$

where

$$u_\mu = a\eta_{0\mu} \tag{15.132}$$

is the 4-velocity of the plasma in conformal coordinates. The conductivity is $\sigma \propto T$ (see below) [GioSha00], so the combination $\bar{\sigma} = a\sigma$ is independent of the scale factor. Since we already noted that the free action is independent of the

---

[5] For the sake of discussion, we gloss over the fact that properly speaking we should not be concerned with a Maxwell field, but rather with a spin-one field which becomes electromagnetic after electroweak symmetry breaking.

scale factor in four spacetime dimensions, the result is that a drops from the Maxwell equations, which read

$$A_{i,i} = 0 \qquad (15.133)$$

$A_0 = 0$ and

$$A_{j,00} + \bar{\sigma} A_{j,0} - A_{j,ii} = 0 \qquad (15.134)$$

For each Fourier mode, the corresponding amplitude behaves as a damped harmonic oscillator. If the comoving wavenumber $k < \bar{\sigma}/2$, the mode is overdamped. There is a fast decaying component

$$f_{\text{fast}} = e^{-\bar{\sigma}\eta} \qquad (15.135)$$

and a slow decaying component

$$f_{\text{slow}} = e^{-k^2\eta/\bar{\sigma}} \qquad (15.136)$$

For long enough wavelengths we may approximate $f_{\text{slow}} = 1$. The boundary conditions are that at the beginning of the radiation-dominated era there are no fields, so $A_j(0) = 0$. From the constitutive relation we get

$$A_{j,0}(0) = -\frac{a^2(0)}{\bar{\sigma}} J_j(0) \qquad (15.137)$$

and so once f_{fast} decays the field settles down to a time-independent value

$$A_j(\infty) = -\frac{J_j(0)}{\sigma^2(0)} \qquad (15.138)$$

Associated with the free action there is an energy–momentum tensor (15.122)

$$T_{\mu\nu} = F_{\lambda\mu} F^{\lambda}_{\ \nu} - \frac{1}{4} g_{\mu\nu} F_{\lambda\sigma} F^{\lambda\sigma} \qquad (15.139)$$

and an energy density

$$\rho = T_{\mu\nu} u^{\mu} u^{\nu} \qquad (15.140)$$

In the asymptotic regime, ρ scales as a^{-4}. Therefore the ratio r between the energy density of the coherent Maxwell field and the total energy density of radiation is constant, provided the cosmic expansion is adiabatic. Disregarding the entropy generated during particle annihilations, we may say r is constant up to our times. A value of $r = 10^{-8}$ is strong enough to originate the galactic fields without further dynamo amplification [TurWid88]. The lowest value of r that could seed the galactic field through dynamo amplification is hard to estimate, as it depends on the details both of the galaxy formation process and of the cosmological model (i.e. the amount of dark energy or the space curvature) [DaLiTo99]. A primordial field should also leave an imprint on the cosmic microwave radiation, but present data only provide upper bounds [YIKM06].

To put these numbers in perspective, we may ask which value of r could be expected for fields coherent over a physical scale L_{phys}, given thermal

equilibrium conditions. Since in the Rayleigh–Jeans part of the spectrum we may assume equipartition, the energy density associated with modes $k < L_{\text{phys}}^{-1}$ is $T_{\text{today}}L_{\text{phys}}^{-3}$, and so $r \approx (L_{\text{phys}}T_{\text{today}})^{-3}$. Using $T_{\text{today}} = 10^{-4}$eV, we get $L_{\text{phys}}T_{\text{today}} \approx 10^{24} \, (L_{\text{phys}}/1 \text{ Mly})$. For a galaxy cluster-size scale, r is way below the interesting range. In this section, we shall use a subindex "today" to indicate that a quantity is evaluated at the present time (we assume $a_{\text{today}} = 1$). Similarly, "reh" will denote the end of reheating, and "equiv" the time of equivalence between matter and radiation.

More generally, r remains constant when both the coherent magnetic field and the thermal cosmic background evolve in a conformally invariant way. So to increase the value of r, we must break conformal invariance. In their seminal work on magnetic field generation [TurWid88], Turner and Widrow have considered a number of possible conformal symmetry-breaking mechanisms.

The hardest way to break the symmetry is to add to the action a direct coupling of the Maxwell field to curvature, such as, for example, $R^{\mu\nu}A_\mu A_\nu$. However, this term breaks gauge along with conformal symmetry, and it is hard to generate in a natural way. Gauge symmetric terms such as $f(R) F^{\mu\nu}F_{\mu\nu}$ are more appealing, partly because they arise naturally from radiative corrections in a curved spacetime. Nevertheless, Mazzitelli and Spedalieri [MazSpe95] have observed that, after proper resummation of the leading quantum corrections, the dependence on curvature is at most logarithmic, and so it is hard to achieve efficient magnetic field generation. A similar conclusion, in a wider set of problems, has been reached recently by Weinberg [Wei05a, Wei06].

Over and above the details of each mechanism, we must consider that the quantity r generated during inflation may well be diluted at reheating. During reheating the density of radiation increases by a factor of at least e^{4N}, where N is the number of e-foldings. Unless the coherent field is also amplified, r decreases by the same amount.

When we consider the generation of magnetic fields during reheating, a new possibility opens up. The abrupt changes in metric during this stage may result in abundant particle creation of charged species. This would generate stochastic currents (recall Chapter 8), which eventually decay onto the Maxwell field [CaKaMa98].

Before we evaluate whether such a mechanism is feasible, let us observe the following. Because the inflaton is a gauge singlet, we do not expect it will decay directly into charged species. Therefore, the model assumes these charged particles are created from the gravitational field, which in turn responds to the changes in the equation of state of the inflaton [PeeVil99].

Spin 1/2 particles such as electrons would be conformally invariant at the high energies prevalent during inflation, so they are not created in large numbers. We must seek a fundamental charged scalar field, of which there is none in the standard model. There are suitable candidates in supersymmetric extensions of the standard model, however [KCMW00].

An alternative which appeals only to known and proven physics is to replace the charged field by the gravitational field itself. Inflation generates tensor gravitational fluctuations, and therefore an inflationary universe is not strictly speaking conformally flat. The evolution of these gravitational fluctuations may result in amplification of the Maxwell field [BaTsWa06, BPTV01, TsaKan05]. However, due to the weakness of the gravitational couplings, it is hard to achieve the desired efficiency.

In the following we shall give an estimate of the field strength to be expected from particle creation of a charged, minimally coupled scalar field ϕ by the end of the reheating period. We decompose the field into its real and imaginary parts $\Phi = (\phi_1 + i\phi_2)/\sqrt{2}$. The current is

$$J^\mu = J_1^\mu + J_2^\mu \tag{15.141}$$

where

$$J_{1\mu} = e\left(\phi_1 \partial_\mu \phi_2 - \phi_2 \partial_\mu \phi_1\right) \tag{15.142}$$

$$J_{2\mu} = -e^2 A_\mu \left(\phi_1^2 + \phi_2^2\right) \tag{15.143}$$

In a linearized analysis we set $J_2 = 0$. Each field is decomposed into modes

$$\phi_i = \int \frac{d^3\mathbf{k}}{(2\pi)^3} e^{i\mathbf{k}\mathbf{x}} \phi_{i\mathbf{k}} \tag{15.144}$$

where

$$\phi_{i\mathbf{k}} = \phi_\mathbf{k} a_{i\mathbf{k}} + \phi_\mathbf{k}^* a_{i-\mathbf{k}}^+ \tag{15.145}$$

leading to a mode decomposition of the current. The spatial components become

$$\mathbf{J}_1 = ie \int \frac{d^3\mathbf{k}}{(2\pi)^3} e^{i\mathbf{k}\mathbf{x}} \int \frac{d^3\mathbf{q}}{(2\pi)^3} \left[2\mathbf{q} - \mathbf{k}\right] \phi_{1\mathbf{k}-\mathbf{q}} \phi_{2\mathbf{q}} \tag{15.146}$$

while the charge density is

$$J_{10} = -ie \int \frac{d^3\mathbf{k}}{(2\pi)^3} e^{i\mathbf{k}\mathbf{x}} \int \frac{d^3\mathbf{q}}{(2\pi)^3} \left[\omega_{2\mathbf{q}} - \omega_{1\mathbf{k}-\mathbf{q}}\right] \phi_{1\mathbf{k}-\mathbf{q}} \phi_{2\mathbf{q}} \tag{15.147}$$

where

$$\omega_{i\mathbf{k}} = \frac{i}{\phi_{i\mathbf{k}}} \frac{d\phi_{i\mathbf{k}}}{dt} \tag{15.148}$$

We are interested in the current averaged over a comoving scale L

$$\mathbf{J}_{1L} = ie \int \frac{d^3\mathbf{k}}{(2\pi)^3} W_L\left[k\right] \int \frac{d^3\mathbf{q}}{(2\pi)^3} \left[2\mathbf{q} - \mathbf{k}\right] \phi_{1\mathbf{k}-\mathbf{q}} \phi_{2\mathbf{q}} \tag{15.149}$$

where W_L is a window function. If the initial state of the field is the vacuum, it is clear that $\langle \mathbf{J}_{1L} \rangle = 0$, but

$$\langle \mathbf{J}_{1L}^2 \rangle = e^2 \int \frac{d^3\mathbf{k}}{(2\pi)^3} W_L\left[k\right]^2 \int \frac{d^3\mathbf{q}}{(2\pi)^3} \left[2\mathbf{q} - \mathbf{k}\right]^2 \left|\phi_\mathbf{q}\right|^2 \left|\phi_{\mathbf{k}-\mathbf{q}}\right|^2 \tag{15.150}$$

To see the meaning of this equation, let us consider (and reject) the case of a conformally coupled field. For conformal coupling, we simply have (cf. Chapter 4)

$$|\phi_{\mathbf{q}}|^2 = \frac{1}{2a^2q} \tag{15.151}$$

It would seem that $\langle \mathbf{J}_{1L}^2 \rangle$ is dominated by very short modes. However, since short modes are supposed to thermalize during reheating, they cannot possibly be described within a linear theory. There must be a comoving cut-off Λ which marks the limit of the linearized approximation. Assuming however $\Lambda \gg L^{-1}$, we see that the dominant contribution to $\langle \mathbf{J}_{1L}^2 \rangle$ comes from modes where $q \approx \Lambda \gg k \approx L^{-1}$. The integrals decouple, and we get

$$\langle \mathbf{J}_{1L}^2 \rangle \approx \frac{e^2 \Lambda^3}{a^4 L^3} \tag{15.152}$$

Under the same approximations, the mean square value of the charge density vanishes.

To transform this into an estimate for the Maxwell field, we need the value of σ at the end of reheating. The usual estimate for the conductivity is $\sigma \approx e^2 n\tau/m$, where n and m are the density and rest mass of the dominant charge carriers, and τ a typical mean free time. If the dominant carriers are just electrons and positrons, then prior to annihilation we have $n \approx T^3$. If we assume that reheating ends as soon as thermal equilibrium is reached, then at that time we may approximate τ by the effective age of the universe $\tau \equiv H^{-1} = m_P T_{\text{reh}}^{-2}$.

To conclude, we evaluate the asymptotic vector potential from (15.138) and the corresponding energy density from (15.140), where we use the result that a space derivative is $\partial \approx L^{-1}$

$$\rho \approx \frac{\langle \mathbf{J}_{1L}^2 \rangle_{\text{reh}}}{a^4 L^2 \sigma_{\text{reh}}^4} \approx \frac{1}{a^4} \left[\frac{m^4 \Lambda^3 H_{\text{reh}}^4}{a_{\text{reh}}^4 L^5 e^6 T_{\text{reh}}^{12}} \right] \tag{15.153}$$

Combining these estimates we get

$$r \approx \left(\frac{m}{m_P e^{3/2}} \right)^4 \left(\frac{T_{\text{reh}}}{m_P} \right)^3 \left(\frac{\Lambda}{a_{\text{reh}} H_{\text{reh}}} \right)^3 \left(\frac{1}{L T_{\text{today}}} \right)^5 \tag{15.154}$$

which is far worse than our previous estimate based on equilibrium conditions.

It is clear from this analysis that to obtain a larger value of r we must amplify the scalar field fluctuations far above the conformal value. During the radiation-dominated era the scalar curvature vanishes and any scalar field is conformally invariant as long as it is effectively massless. But during inflation the behavior is totally different, because while the conformal fields evolve as a^{-1} throughout, the minimal fields freeze upon horizon exit and remain constant until the scalar curvature is suppressed enough during reheating. To see this, let us return to the mode equation (15.8). We write the mode functions as

$$\phi_{\mathbf{k}} = \frac{f_k}{a^{3/2}} \tag{15.155}$$

Using the Friedmann equation $H^2 = \rho/m_P^2$, the continuity equation $\dot\rho = -3H(\rho + p)$ and the equation of state $p = \gamma\rho$, we transform the mode equation into (4.22)

$$\frac{d^2 f_k}{dt^2} + \Omega_k^2(t) f_k(t) = 0 \qquad (15.156)$$

where

$$\Omega_k^2(t) = \frac{k^2}{a^2} + m^2(t) + \frac{9}{4} H^2 \gamma \qquad (15.157)$$

and we allow for the possibility of a time-dependent mass, for example, due to thermal corrections (cf. Chapter 10). During inflation $\gamma = -1$. Ω_k^2 starts positive and becomes negative upon horizon exit. Outside of the horizon there is a growing mode which remains frozen because the growth of the WKB solution just matches the $a^{-3/2}$ suppression, and a decaying mode which soon becomes irrelevant. At some point during reheating γ becomes positive and Ω_k^2 changes sign again; we say the mode "thaws." Neglecting the decaying solution and assuming a long enough wavelength, we have immediately after thawing the q-number amplitudes

$$\phi_{i\mathbf{k}} = A_k F(t) \left[a_{i\mathbf{k}} + a_{i-\mathbf{k}}^+ \right] \qquad (15.158)$$

where $|A_k|$ is of the order of magnitude of the amplitude at horizon exit. Observe that the time-dependent part $F(t)$ is essentially mode-independent: the field is performing "Sakharov" oscillations [Sak66]. This implies the vanishing of the induced charge density.

As a first approximation, we may assume that all modes thaw at the same time at the end of reheating. As compared with the conformal case, the minimally coupled mode amplitudes are amplified by a factor $a_{\rm reh}/a_{\rm exit} = a_{\rm reh} H_{\rm reh}/k$, where we assume that reheating is fast enough that the Hubble rate remains approximately time-independent throughout.

Our estimate for the current at reheating now reads

$$\langle \mathbf{J}_{1L}^2 \rangle_{\rm reh} \approx \frac{e^2 H_{\rm reh}^4}{4} \int \frac{d^3\mathbf{k}}{(2\pi)^3} \, W_L[k]^2 \int \frac{d^3\mathbf{q}}{(2\pi)^3} \, [2\mathbf{q} - \mathbf{k}]^2 \frac{1}{q^3} \frac{1}{(|\mathbf{k} - \mathbf{q}|)^3} \qquad (15.159)$$

The q integral is dominated by peaks at $\mathbf{q} = 0$ and $\mathbf{q} = \mathbf{k}$. They both contribute the same, as they are transformed into each other by the change of variables $\mathbf{q} \to \mathbf{k} - \mathbf{q}$. So it is enough to evaluate the contribution from $q \ll k$.

$$\langle \mathbf{J}_{1L}^2 \rangle_{\rm reh} \approx \frac{e^2 H_{\rm reh}^4}{4} \int \frac{d^3\mathbf{k}}{(2\pi)^3} \frac{W_L[k]^2}{k} \int \frac{dq}{q} \qquad (15.160)$$

We evaluate the logarithmic integral as $\ln[q_{\rm max}/q_{\rm min}]$, where the q's are the longest and shortest modes to leave the horizon during inflation. Therefore

$$\int \frac{dq}{q} \approx N \qquad (15.161)$$

where N is the number of e-foldings. We obtain

$$\langle \mathbf{J}_{1L}^2 \rangle_{\rm reh} \approx \frac{e^2 N H_{\rm reh}^4}{4L^2} \qquad (15.162)$$

The improved estimate for r is

$$r \approx N \left(\frac{m}{m_P e^{3/2}} \right)^4 \left(\frac{T_{\text{reh}}}{m_P} \right)^4 \left(\frac{1}{LT_{\text{today}}} \right)^4 \qquad (15.163)$$

which is still a very small number.

Although prospects are understandably bleak, our argument has a loophole [Fin00]. This is the neglect of the "London" current (15.143). Because of this term, the heavily amplified long-wavelength modes of the scalar field act as a Landau–Ginzburg order parameter in a superconductor [Tin96]. As in the Meissner effect, the photon acquires a (here time-dependent) mass. Kandus *et al.* have shown that an exponential growth of the Maxwell field during reheating as a consequence of parametric amplification is possible [CalKan02]. However, in this case the actual growth factor is sensitive to the details of the reheating scenario, and so it is not possible to obtain generally valid estimates such as the above.

At the end of this discussion, we reach a situation remarkably similar to our description of early thermalization in RHICs in Chapter 14. Both the generation of a primordial magnetic field during reheating and ultrafast equilibration after the collision are demonstrably beyond the possibilities of weakly interacting fields, but could be allowed because of exponential instabilities in strongly nonlinear scenarios. In either problem, we do not have answers yet, but it is clear that finding those answers will require the full application of the methods of nonequilibrium field theory, whose basic principles we have attempted to present in this book.

References

[Aar01] G. Aarts, Spectral function at high temperature in the classical approximation, *Phys. Lett.* **518B**, 315–322 (2001).

[AarBer01] G. Aarts and J. Berges, Nonequilibrium time evolution of the spectral function in quantum field theory, *Phys. Rev. D* **64**, 105010 (2001).

[AarBer02] G. Aarts and J. Berges, Classical aspects of quantum fields far from equilibrium, *Phys. Rev. Lett.* **88**, 041603 (2002).

[AaBoWe00a] G. Aarts, G.F. Bonini and C. Wetterich, Exact and truncated dynamics in nonequilibrium field theory, *Phys. Rev. D* **63**, 025012 (2000).

[AaBoWe00b] G. Aarts, G.F. Bonini and C. Wetterich, On thermalization in classical scalar field theory, *Nucl. Phys.* **B587**, 403 (2000).

[AarMar02] G. Aarts and J. Martinez Resco, Transport coefficients, spectral functions and the lattice, *JHEP* **204**, 54 (2002).

[AarSmit98] G. Aarts and J. Smit, Classical approximation for time-dependent quantum field theory: Diagrammatic analysis for hot scalar fields, *Nucl. Phys.* **B511**, 451 (1998).

[AarSmi99] G. Aarts and J. Smit, Particle production and effective thermalization in inhomogeneous mean field theory, *Phys. Rev. D* **61**, 025002 (1999).

[ABZH96] A. Abada, M. Birse, P. Zhuang and U. Heinz, Remarks on "Relativistic kinetic equations for electromagnetic, scalar and pseudoscalar interactions," *Phys. Rev. D* **54**, 4175 (1996).

[Abb81] L. Abbott, The background field method beyond one loop, *Nucl. Phys.* **B185**, 189 (1981).

[AARS96] Y. Abe, S. Ayik, P.-G. Reinhard and E. Suraud, On stochastic approaches of nuclear dynamics, *Phys. Rep.* **275**, 49 (1996).

[AbrSte72] M. Abramowitz and I. Stegun (eds.), *Handbook of Mathematical Functions* (Dover, New York, 1972).

[AbGoDz75] A.A. Abrikosov, L.P. Gorkov and I.E. Dzyaloshinski, *Methods of Quantum Field Theory in Statistical Physics* (Dover, New York, 1975).

[Adh04] S.K. Adhikari, Mean-field model of jet formation in a collapsing Bose–Einstein condensate, *J. Phys. B At. Mol. Opt. Phys.* **37**, 1185 (2004).

[Adl04] S.L. Adler, *Quantum theory as an emergent phenomenon* (Cambridge University Press, Cambridge, 2004).

[AgSoVi06] M.M. Aggarwal, G. Sood and Y.P. Viyogi, Event-by-event study of DCC-like fluctuation in ultra-relativistic nuclear collisions, *Phys. Lett.* **638B**, 39 (2006).

[AkhPel81] A.I. Akhiezer and S.V. Peletminskii, *Statistical Physics* (Mir, Moscow, 1981).

[AkLeSi01] S. Akkelin, R. Lednicky and Y. Sinyukov, Correlation search for coherent pion emission in heavy ion collisions, *Phys. Rev. C*, **65**, 064904 (2002).

[ABBCFJ99] S. Alamoudi, D. Barci, D. Boyanovsky, A. de Carvalho, E. Fraga, S. Joras and F. Takakura, Dynamical viscosity of nucleating bubbles, *Phys. Rev. D* **60**, 125003 (1999).

[AFJP94] A. Albrecht, P. Ferreira, M. Joyce and T. Prokopec, Inflation and squeezed quantum states, *Phys. Rev. D* **50**, 4807 (1994).

[AlbSte82] A. Albrecht and P.J. Steinhardt, Cosmology for grand unified theories with radiatively induced symmetry breaking, *Phys. Rev. Lett.* **48**, 1220 (1982).

[Ald73] B.J. Alder, Computer dynamics, *An. Rev. Phys. Chem.* **24**, 325–337 (1973).

[Alx99] S. Alexandrov, Effective action and quantum gauge transformations, *Phys. Rev. D* **59**, 125016 (1999).

[AllMis38] J.F. Allen and A.D. Misener, Flow of liquid helium II, *Nature* **141**, 75 (1938).

[AmBjLa97] G. Amelino-Camelia, J.D. Bjorken and S.E. Larsson, Pion production from Baked-Alaska disoriented chiral condensate, *Phys. Rev. D* **56**, 6942 (1997).

[Ana97a] C. Anastopoulos, Coarse grainings and irreversibility in quantum field theory, *Phys. Rev. D* **56**, 1009 (1997).

[Ana97b] C. Anastopoulos, n-particle sector of field theory as a quantum open system, *Phys. Rev. D* **56**, 6702–6705 (1997).

[AnaHal95] C. Anastopoulos and J.J. Halliwell, Generalized uncertainty relations and long-time limits for quantum Brownian motion models, *Phys. Rev. D* **51**, 6870 (1995).

[And04] J. Andersen, Theory of the weakly interacting Bose gas, *Rev. Mod. Phys.* **76**, 599 (2004).

[AndHal93] A. Anderson and J.J. Halliwell, Information-theoretic measure of uncertainty due to quantum and thermal fluctuations, *Phys. Rev. D* **48**, 2753 (1993).

[Ang93] J.R. Anglin, Influence functionals and the accelerating detector, *Phys. Rev. D* **47**, 4525 (1993).

[AngZur96] J.R. Anglin and W.H. Zurek, Decoherence of quantum fields: Pointer states and predictability, *Phys. Rev. D* **53**, 7327 (1996).

[AntBet97] N. Antunes and L. Bettencourt, Out of equilibrium dynamics of quench-induced spontaneous symmetry breaking and topological defect formation, *Phys. Rev. D* **55**, 925 (1997).

[AnBeZu99] N. Antunes, L. Bettencourt and W. Zurek, Vortex string formation in a 3D U(1) temperature quench, *Phys. Rev. Lett.* **82**, 2824 (1999).

[AnGaRi06] N. Antunes, P. Gandra and R. Rivers, Domain formation: decided before, or after the transition?, *Phys. Rev. D* **73**, 125003 (2006).

[AGRS06] N. Antunes, P. Gandra, R. Rivers and A. Swarup, The creation of defects with core condensation, *Phys. Rev. D* **73**, 085012 (2006).

[ALMV06] N. Antunes, F. Lombardo, D. Monteoliva and P. Villar, Decoherence, tunneling and noise-activation in a double-potential well at high and zero temperature, *Phys. Rev. E* **73**, 066105 (2006).

[ACGI00] D. Arbo, M. Castagnino, F. Gaioli and S. Iguri, Minimal irreversible quantum mechanics. The decay of unstable states, *Physica A* **277**, 469 (2000).

[ArDoMo06] P. Arnold, C. Dogan and G. Moore, The bulk viscosity of high-temperature QCD, *Phys. Rev. D* **74**, 085021 (2006).

[ArLeMo03] P. Arnold, J. Lenaghan and G.D. Moore, QCD plasma instabilities and bottom-up thermalization, *JHEP* **0308**, 002 (2003).

[ALMY05] P. Arnold, J. Lenaghan, G.D. Moore and L.G. Yaffe, Apparent thermalization due to plasma instabilities in a quark-gluon plasma, *Phys. Rev. Lett.* **94**, 072302 (2005).

[ArnMoo05] P. Arnold and G.D. Moore, QCD plasma instabilities: the non-abelian cascade, *Phys. Rev. D* **73**, 025006 (2006).

[ArnMoo06] P. Arnold and G.D. Moore, Turbulent spectrum created by non-abelian plasma instabilities, *Phys. Rev. D* **73**, 025013 (2006).

[ArMoYa00] P. Arnold, G.D. Moore and L.G. Yaffe, Transport coefficients in high temperature gauge theories, *JHEP* **0011**, 001 (2000).

[ArMoYa01a] P. Arnold, G.D. Moore and L.G. Yaffe, Photon emission from quark-gluon plasma: complete leading order results, *JHEP* **11**, 57 (2001).

[ArMoYa01b] P. Arnold, G.D. Moore and L.G. Yaffe, Photon emission from ultrarelativistic plasmas, *JHEP* **12**, 9 (2001).

[ArMoYa02] P. Arnold, G.D. Moore and L.G. Yaffe, Photon and glucon emission in relativistic plasmas, *JHEP* **6**, 30 (2002).

[ArMoYa03a] P. Arnold, G.D. Moore and L.G. Yaffe, Effective kinetic theory for high temperature gauge theories, *JHEP* **1**, 30 (2003).

[ArMoYa03b] P. Arnold, G.D. Moore and L.G. Yaffe, Transport coefficients in high temperature gauge theories, 2. Beyond leading log, *JHEP* **5**, 51 (2003).

[ArMoYa05] P. Arnold, G.D. Moore and L.G. Yaffe, Fate of non-Abelian plasma instabilities in $3 + 1$ dimensions, *Phys. Rev. D* **72**, 054003 (2005).

[ArSoYa99a] P. Arnold, D. Son and L.G. Yaffe, Effective dynamics of hot, soft, non-abelian gauge fields, *Phys. Rev. D* **659**, 105020 (1999).

[ArSoYa99b] P. Arnold, D. Son and L. Yaffe, Longitudinal subtleties in diffusive Langevin equations, *Phys. Rev. D* **60**, 025007 (1999).

[ArrSmi02] A. Arrizabalaga and J. Smit, Gauge-fixing dependence of Φ- derivable approximations, *Phys. Rev. D* **66**, 65014 (2002).

[ArSmTr05] A. Arrizabalaga, J. Smit and A. Transberg, Equilibration in φ^4 theory in 3+1 dimensions, *Phys. Rev. D* **72**, 025014 (2005).

[AuGeZa02] P. Aurenche, F. Gelis and H. Zaraket, A simple sum rule for the thermal gluon spectral function and applications, *JHEP* **05**, 43 (2002).

[Baa98] J. Baacke, Choice of initial states in nonequilibrium dynamics, *Phys. Rev. D* **57**, 6398 (1998).

[Baa00a] J. Baacke, Out-of-equilibrium evolution of scalar fields in FRW cosmology: Renormalization and numerical simulations, *Phys. Rev. D* **61**, 024016 (2000).

[Baa00b] J. Baacke, Renormalization of the nonequilibrium dynamics of fermions in a flat FRW universe, *Phys. Rev. D* **62**, 084008 (2000).

[BaBoVe01] J. Baacke, D. Boyanovsky and H.J. de Vega, Initial time singularities in nonequilibrium evolution of condensates and their resolution in the linearized approximation, *Phys. Rev. D* **63**, 45023 (2001).

[BaaHei03b] J. Baacke and A. Heinen, Nonequilibrium evolution of Φ^4 theory in 1+1 dimensions in the two-particle point-irreducible formalism, *Phys. Rev. D* **67**, 105020 (2003).

[BaaHei03a] J. Baacke and A. Heinen, Quantum dynamics of Φ^4 field theory beyond leading order in 1+1 dimensions, *Phys. Rev. D* **68**, 127702 (2003).

[BaiKAt03] V.N. Baier and V.M. Katkov, On theory of Landau–Pomeranchuk–Migdal effect, Plenary talk at the International Conference "I.Ya.Pomeranchuk and the Physics at the Turn of Centuries" Moscow, January 24–28, 2003. To be published in the Proceedings (World Scientific, Singapore) (hep-ph/0305133).

[Bai92] R. Baier, G. Kunstatter and D. Schiff, Gauge dependence of the resummed thermal gluon self-energy, *Nucl. Phys.* **B388**, 287 (1992).

[BMSS01] R. Baier, A. Mueller, D. Schiff and D. Son, Bottom-up thermalization in heavy ion collisions, *Phys. Lett.* **502B**, 51 (2001).

[BaRoWi06a] R. Baier, P. Romatschke and U.A. Wiedemann, Dissipative hydrodynamics and heavy ion collisions, *Phys. Rev. C* **73**, 064903 (2006).

[BaRoWi06b] R. Baier, P. Romatschke and U.A. Wiedemann, Transverse flow in relativistic viscous hydrodynamics, *Nucl. Phys. A* **782**, 313–318 (2007).

[BaiSto04] R. Baier and T. Stockamp, Kinetic equations for Bose–Einstein condensates from the 2PI effective action, hep-ph/0412310.

[BakMah63] P.M. Bakshi and K.T. Mahanthappa, Expectation value formalism in quantum field theory, *J. Math. Phys. (N.Y.)* **4**, 1 (1963); *J. Math. Phys.* **4**, 12 (1963).

[Bal75] R. Balescu, *Equilibrium and Nonequilibrium Statistical Mechanics* (John Wiley, New York, 1975).

[BalVen87] R. Balian and M. Veneroni, Incomplete descriptions, relevant information, and entropy production in collision processes, *Ann. Phys. (N.Y.)* **174**, 229–244 (1987).

[BajaMa04] W. Bao, D. Jaksch and P. Markovich, The dimensional simulation of the formation of collapsing condensates, *J. Phys. B At. Mol. Opt. Phys.* **37**, 329 (2004).

[BaLiVi05] C. Barcelo, S. Liberati and M. Visser, Analogue gravity, *Liv. Rev. Relativity* **8**, 12–113 (2005).

[BFGR01] D. Barci, E. Fraga, M. Gleiser and R. Ramos, Equilibration time scales in homogeneous Bose–Einstein condensate dynamics, *Physica A* **317**, 535–545 (2003).

[BaFrRa02] D.G. Barci, E.S. Fraga, Rudnei O. Ramos, Microscopic evolution of a weakly interacting homogeneous Bose gas, *Laser Phys.* **12**, 43–49 (2002).

[Bar80] J. Bardeen, Gauge-invariant cosmological perturbations, *Phys. Rev. D* **22**, 1882 (1980).

[BarBub87] J.M. Bardeen and G.J. Bublik, Quantum fluctuations and inflation, *Class. Quan. Grav.* **4**, 573 (1987).

[BaStTu83] J.M. Bardeen, P.J. Steinhardt and M.S. Turner, Spontaneous creation of almost scale-free density perturbations in an inflationary universe, *Phys. Rev. D* **28**, 679 (1983).

[Bas98] B.A. Bassett, Inflationary reheating classes via spectral methods, *Phys. Rev. D* **58**, 021303 (1998).

[BPTV01] B.A. Bassett, G. Pollifrone, S. Tsujikawa and F. Viniegra, Preheating – cosmic magnetic dynamo?, *Phys. Rev. D* **63**, 103515 (2001).

[BaTsWa06] B.A. Bassett, S. Tsujikawa and D. Wands, Inflation dynamics and reheating, *Rev. Mod. Phys.* **78**, 537 (2006).

[BasGhi99] A. Bassi and G.-C. Ghirardi, Can the decoherent histories description of reality be considered satisfactory?, *Phys. Lett.* **257A**, 247 (1999).

[Bat59] G.K. Batchelor, *The Theory of Homogeneous Turbulence* (Cambridge University Press, Cambridge, 1959).

[Bay62] G. Baym, Self-consistent approximations in many-body systems, *Phys. Rev.* **127**, 1391 (1962).

[BBGM06] G. Baym, J.-P. Blaizot, F. Gelis and T. Matsui, Landau–Pomeranchuk–Migdal effect in a quark–gluon plasma and the Boltzmann equation, *Phys. Lett. B* **644**, 48–53 (2007).

[BayKad61] G. Baym and L.P. Kadanoff, Conservation laws and correlation functions, *Phys. Rev.* **124**, 287 (1961).

[Bed] P.F. Bedaque, Thermalization and pinch singularities in non-equilibrium quantum field theory, *Phys. Lett.* **344B**, 23 (1995).

[Bek73] J.D. Bekenstein, Black holes and entropy, *Phys. Rev. D* **7**, 2333 (1973).

[Bek83] J.D. Bekenstein, Black hole fluctuations in *Quantum Theory of Gravity* ed. S. Christensen (Adam Hilger, Boston, 1983).

[Bek94] J.D. Bekenstein, Review talk at MG7 (1994), Do we understand black hole entropy?, gr-qc/9409015.

[BekMuk95] J.D. Bekenstein and V. F. Mukhanov, Spectroscopy of the quantum black hole, *Phys. Lett.* **360B**, 7 (1995).

[BelLan56] S. Belen'kji and L. Landau, Hydrodynamic theory of multiple production of particles, *Supplemento al Nuovo* Cimento **3**, 15 (1956).

[Bel58a] S.T. Beliaev, Application of the methods of quantum field theory to a system of bosons, *Zh. Eksp. Teor. Fiz.* **34**, 417 (1958) (*Sov. Phys. JETP* **7**, 289 (1958)).

[Bel58b] S.T. Beliaev, Energy spectrum of a non-ideal Bose gas, *Zh. Eksp. Teor. Fiz.* **34**, 433 (1958) (*Sov. Phys. JETP* **7**, 299 (1958)).

[Ben96] C.L. Bennett *et al.*, 4-Year COBE DMR cosmic microwave background observations: Maps and basic results, *Astrophys. J.* **464**, L1 (1996).

[Ben03] C.L. Bennett *et al.*, First-year Wilkinson Microwave Anisotropy Probe (WMAP) observations: Determination of cosmological parameters, *Astrophys. J. Suppl. Ser.* **148**, 1 (2003).

[BeGlRa98] A. Berera, M. Gleiser and R. Ramos, Strong dissipative behavior in quantum field theory, *Phys. Rev. D* **58**, 123508 (1998).

[BerRam01] A. Berera and R. Ramos, Affinity for scalar fields to dissipate, *Phys. Rev. D* **63**, 103509 (2001).

[Ber66] F.A. Berezin, *The Method of Second Quantization* (Academic Press, New York, 1966).

[Berger74] B. Berger, Quantum graviton creation in a model universe, *Ann. Phys.* (N.Y.) **83**, 458 (1974).

[Berger75a] B. Berger, Quantum cosmology: Exact solution for the Gowdy T^3 model, *Phys. Rev. D* **11**, 2770 (1975).

[Berger75b] B. Berger, Scalar particle creation in an anisotropic universe, *Phys. Rev. D* **12**, 368 (1975).

[Ber02]   J. Berges, Controlled nonperturbative dynamics of quantum fields out of equilibrium, *Nucl. Phys.* **A699**, 847 (2002).

[Ber04a] J. Berges, n-particle irreducible effective action techniques for gauge theories, *Phys. Rev. D* **70**, 105010 (2004).

[Ber04b] J. Berges, Introduction to nonequilibrium quantum field theory, hep-ph/0409233.

[BerBol06] J. Berges and S. Bolsanyi, Range of validity of transport equations, *Phys. Rev. D* **74**, 045022 (2006).

[BerBol06] J. Berges and S. Bolsanyi, Progress in nonequilibrium quantum field theory III, *Nucl. Phys.* **A785**, 58–67 (2007).

[BeBoSe03] J. Berges, S. Bolsanyi and J. Serreau, Thermalization of fermionic quantum fields, *Nucl. Phys.* **B600**, 51 (2003).

[BBSS06] J. Berges, S. Bolsanyi, D. Sexty and I-O. Stamatescu, Lattice simulations of real-time quantum fields, hep-lat/0609058.

[BeBoWe04] J. Berges, S. Bolsanyi and C. Wetterich, Prethermalization, *Phys. Rev. Lett.* **93**, 142002 (2004).

[BeBoWe05] J. Berges, S. Bolsanyi and C. Wetterich, Isotropization far from equilibrium, *Nucl. Phys.* **B727**, 244 (2005).

[BerSer03a] J. Berges and J. Serreau, Parametric resonance in quantum field theory, *Phys. Rev. Lett.* **91**, 111601 (2003).

[BerSer03b] J. Berges and J. Serreau, Progress in nonequilibrium quantum field theory, "Strong and Electroweak Matter" (SEWM 2002), Heidelberg, Germany, 2–5 Oct 2002 (hep-ph/0302210).

[BerSer04] J. Berges and J. Serreau, Progress in nonequilibrium quantum field theory II, "Strong and Electroweak Matter" (SEWM 2004), Helsinki, Finland, 16-19 June (hep-ph/0410330).

[BerSta05] J. Berges and I.O. Stamatescu, Simulating nonequlibrium quantum fields with stochastic quantization techniques, hep-lat/0508030.

[Ber74]   C. Bernard, Feynman rules for gauge theories at finite temperature, *Phys. Rev. D* **9**, 3312 (1974).

[BerDun77] C. Bernard and A. Duncan, Regularization and renormalization of quantum field theory in curved space-time, *Ann. Phys.* (N.Y.) **107**, 201 (1977).

[Bes04]   D.R. Bes, *Quantum Mechanics* (Springer-Verlag, Berlin, 2004).

[BesKur90] D.R. Bes and J. Kurchan, *The Treatment of Collective Coordinates in Many body Systems* (World Scientific, Singapore, 1990).

[Bet01]   L. Bettencourt, Properties of the Langevin and Fokker–Planck equations for scalar fields and their application to the dynamics of second order phase transitions, *Phys. Rev. D* **63**, 45020 (2001).

[BeAnZu00] L. Bettencourt, N. Antunes and W. Zurek, Ginzburg regime and its effect on topological defect formation, *Phys. Rev. D* **62**, 065005 (2000).

[BeCoPa02] L. Bettencourt, F. Cooper and K. Pao, Hydrodynamic scaling from the dynamics of relativistic quantum field theory, *Phys. Rev. Lett.* **89**, 112301 (2002).

[BeHaLy99] L. Bettencourt, S. Habib and G. Lythe, Controlling one-dimensional Langevin dynamics on the lattice, *Phys. Rev. D* **60**, 105039 (1999).

[BeRaSt01] L. Bettencourt, K. Rajagopal and J. Steele, Langevin evolution of disoriented chiral condensate, *Nucl. Phys.* **A693**, 825 (2001).

[BetWet98] L. Bettencourt and C. Wetterich, Time evolution of correlation functions for classical and quantum anharmonic oscillators, hep-ph/9805360.

[BhWaHo02] S.G. Bhongale, R. Walser and M.J. Holland, Memory effects and conservation laws in the quantum kinetic evolution of a dilute Bose gas, *Phys. Rev. A* **66**, 043618 (2002).

[BiGoBa04] P. Bielewicz, K.M. Górski and A.J. Banday, Low order multipole maps of CMB anisotropy derived from WMAP, *MNRAS* **355**, 1283 (2004).

[BirDav82] N.D. Birrell and P.C.W. Davies, *Quantum Fields in Curved Spaces* (Cambridge University Press, Cambridge, 1982).

[BixZwa69] M. Bixon and R. Zwanzig, Boltzmann–Langevin equation and hydrodynamic fluctuations, *Phys. Rev.* **187**, 267 (1969).

[BixZwa71] M. Bixon and R. Zwanzig, Brownian motion of a nonlinear oscillator, *J. Stat. Phys.* **3**, 245 (1971).

[BjoVen01] J. Bjoraker and R. Venugopalan, From a colored glass condensate to the gluon plasma: equilibration in high energy heavy ion collisions, *Phys. Rev. C* **63**, 024609 (2001).

[Bjo83] J.D. Bjorken, Highly relativistic nucleus-nucleus collisions: the central rapidity region, *Phys. Rev. D* **27**, 140 (1983).

[Bjo97] J.D. Bjorken, Disoriented chiral condensate: theory and phenomenology, *Acta Phys. Polon. B* **28**, 2773 (1997).

[BjoDre64] J. Bjorken and S. Drell, *Relativistic Quantum Mechanics* (McGraw-Hill, New York, 1964).

[BjoDre65] J. Bjorken and S. Drell, *Relativistic Quantum Fields* (McGraw-Hill, New York, 1965).

[BlaIan99] J.-P. Blaizot and E. Iancu, A Boltzmann equation for the QCD plasma, *Nucl. Phys.* **B557**, 183 (1999).

[BlaIan02] J.-P. Blaizot and E. Iancu, The quark–gluon plasma: collective dynamics and hard thermal loops, *Phys. Rep.* **359**, 355 (2002).

[BllaRe03] J.-P. Blaizot, E. Iancu and U. Reinosa, Renormalizability of Φ-derivable approximations in scalar ϕ^4 theory, *Phys. Lett.* **568B**, 160 (2003).

[Bod98] D. Bodeker, Effective dynamics of soft non-abelian gauge fields at finite temperature, *Phys. Lett.* **426B**, 351 (1998).

[Bod99] D. Bodeker, From hard thermal loops to Langevin dynamics, *Nucl. Phys.* **B559**, 502 (1999).

[Bol64] L. Boltzmann, *Lectures on Gas Theory* (Dover, New York, 1964).

[BKLS86] L. Bombelli, R.K. Koul, J. Lee, and R.D. Sorkin, Quantum source of entropy for black holes, *Phys. Rev. D* **34**, 373 (1986).

[BonSen04] K. Bongs and K. Sengstock, Physics with coherent matter waves, cond-mat/0403128.

[BooYip91] J.P. Boon and S. Yip, *Molecular Hydrodynamics* (Dover, New York, 1991).

[Bordag] M. Bordag (ed.) Proceedings of the Fifth Workshop on Quantum Field Theory under the Influence of External Conditions, Leipzig, Germany, Sep. 10-14, 2001. World Scientific, Singapore (2002). Appeared as Special Issue of *Int. J. Mod. Phys. A*, **17** (2002).

[Bordag96] M. Bordag (ed.) Quantum Field Theory under the Influence of external condition. Proceedings of the Third Workshop, Leipzig, Germany, Sept. 18-22 (1995). Stuttgart, Germany, Teubner (1996). (Teubner-Texte zur Physik), 1996.

[Bordag99] M. Bordag (ed.) The Casimir Effect 50 Years Later. Proceedings of the Fourth Workshop on Quantum Field Theory under the Influence of External Conditions, Leipzig, Germany, Sep. 14-18 (1998). World Scientific, Singapore (1999).

[BotMal90] W. Botermans and R. Malfliet, Quantum transport theory of nuclear matter, *Phys. Rep.* **198**, 115 (1990).

[BowRue75] R. Bowen and D. Ruelle, The ergodic theory of Axiom-A flows, *Inventiones Mathematicae*, **29**, 181 (1975).

[BowMom98] M.J. Bowick and A. Momen, Domain formation in finite-time quenches, *Phys. Rev. D* **58**, 085014 (1998).

[Boy02] D. Boyanovsky, S.-Y. Wang, D.-S. Lee, H.-L. Yu, and S. M. Alamoudi, Nonequilibrium relaxation of Bose–Einstein condensates: Real-time equations of motion and Ward identities, *Annals Phys.* **300**, 1–31 (2002).

[BoCaVe02] D. Boyanovsky, F. Cao and H. de Vega, Inflation from tsunami-waves, *Nucl. Phys.* **B632**, 121–154 (2002).

[BoyVeg93] D. Boyanovsky and H. de Vega, Quantum rolling down out of equilibrium, *Phys. Rev. D* **47**, 2343 (1993).

[BoVeHo94] D. Boyanovsky, H. de Vega and R. Holman, Nonequilibrium evolution of scalar fields in FRW cosmologies, *Phys. Rev. D* **49**, 2769 (1994).

[BoVeHo99] D. Boyanovsky, H. de Vega, and R. Holman, "Non-equilibrium phase transitions in condensed matter and cosmology: spinodal decomposition, condensates and defects." Lectures delivered at the NATO Advanced Study Institute: Topological Defects and the Non-Equilibrium Dynamics of Symmetry Breaking Phase Transitions, hep-ph/9903534.

[BVHK97] D. Boyanovsky, H. de Vega, R. Holman and S. Prem Kumar, Nonequilibrium production of photons via $\pi^0 \to 2\gamma$ in disoriented chiral condensates, *Phys. Rev. D* **56**, 3929 (1997).

[BBHKP98] D. Boyanovsky, H. de Vega, R. Holman, S. Kumar and R. Pisarski, Nonequilibrium evolution of a "tsunami," a high multiplicity initial quantum state: Dynamical symmetry breaking, *Phys. Rev. D* **57**, 3653 (1998).

[BVHLS95] D. Boyanovsky, H. de Vega, R. Holman, D.S. Lee and A. Singh, Dissipation via particle production in scalar field theories, *Phys. Rev. D* **51**, 4419 (1995).

[BVHS96] D. Boyanovsky, H. de Vega, R. Holman and J.F.J. Salgado, Analytic and numerical study of reheating dynamics, *Phys. Rev. D* **54**, 7570 (1996).

[BVHS99a] D. Boyanovsky, H. de Vega, R. Holman and J.F.J. Salgado, Nonequilibrium Bose–Einstein condensates, dynamical scaling, and symmetric evolution in the large N Φ^4 theory, *Phys. Rev. D* **59**, 125009 (1999).

[BVHS99b] D. Boyanovsky, H. de Vega, R. Holman and M. Simionato, Dynamical renormalization group resummation of finite temperature infrared divergences, *Phys. Rev. D* **60**, 065003 (1999).

[BoVeSa05] D. Boyanovsky, H. de Vega and N.G. Sanchez, The classical and quantum inflaton: the precise inflationary potential and quantum inflaton decay after WMAP, Lecture given at "The Density Perturbation in the Universe," Demokritos Center, Athens, Greece, June 2004 (astro-ph/0503128).

[BoVeSi03a] D. Boyanovsky, H. de Vega and M. Simionato, Magnetic field generation from non-equilibrium phase transitions, *Phys. Rev. D* **67**, 023502 (2003).

[BoVeSi03b] D. Boyanovsky, H. de Vega and M. Simionato, Large scale magnetogenesis from a non-equilibrium phase transition in the radiation dominated era, *Phys. Rev. D* **67**, 123505 (2003).

[BoVeWa00] D. Boyanovsky, H. de Vega and S.Y. Wang, Dynamical RG approach to quantum kinetics, *Phys. Rev. D* **61**, 65006 (2000).

[BoDeVe04] D. Boyanovsky, C. Destri and H. de Vega, The approach to thermalization in the classical ϕ^4 theory in 1+1 dimensions: energy cascades and universal scaling, *Phys. Rev. D* **69** 045003 (2004).

[BoLeeSi93] D. Boyanovsky, D.S. Lee, and A. Singh, Phase transitions out of equilibrium: Domain formation and growth, *Phys. Rev. D* **48**, 800 (1993).

[BraPis90a] E. Braaten and R. Pisarski, Soft amplitudes in hot gauge theories: a general analysis, *Nucl. Phys.* **B337**, 569 (1990).

[BraPis90b] E. Braaten and R. Pisarski, Deducing hard thermal loops from Ward identities, *Nucl. Phys.* **B339**, 310 (1990).

[BraPis92] E. Braaten and R.D. Pisarski, Simple effective Lagrangian for hard thermal loops, *Phys. Rev. D* **45**, 1827–1830 (1992).

[BrBlGa05] A.S. Bradley, P.B. Blakie and C.W. Gardiner, Properties of the stochastic Gross–Pitaievskii equation: projected Ehrenfest relations and the optimal plane wave basis, *J. Phys. B, At. Mol. Opt. Phys.* **38**, 4259 (2005).

[BRAHMS05] BRAHMS Collaboration, Quark–gluon plasma and color glass condensate at RHIC? The perspective from the BRAHMS experiment, *Nucl. Phys.* **A757**, 1 (2005).

[Bra85] R.H. Brandenberger, Quantum field theory methods and inflationary universe models, *Rev. Mod. Phys.* **57**, 1 (1985).

[Bra83] R. Brandenberger, R. Kahn and W. Press, Cosmological perturbations in the early universe, *Phys. Rev. D* **28**, 1809 (1983).

[BraMag99] R.H. Brandenberger and J. Maguiejo, Cosmic defects and cosmology, astro-ph/0002030. Lecture notes of the International School on Cosmology, Kish Island, Ireland, Jan. 22 – Feb. 4, 1999.

[BrMuPr92] R.H. Brandenberger, V. Mukhanov and T. Prokopec, Entropy of a classical stochastic field and cosmological perturbations, *Phys. Rev. Lett.* **69**, 3606 (1992).

[BrMuPr93] R.H. Brandenberger, V. Mukhanov and T. Prokopec, Entropy of the gravitational field, *Phys. Rev. D* **48**, 2443 (1993).

[BraSub05] A. Brandenburg and K. Subramanian, Astrophysical magnetic fields and nonlinear dynamo theory, *Phys. Rep.* **417**, 1 (2005).

[Bre05] G.K. Brennen *et al.*, Scalable register initialization for quantum computing in an optical lattice, *J. Phys. B* **38**, 1687 (2005).

[BroCar79] L.S. Brown and L.J. Carson, Quantum-mechanical parametric amplification, *Phys. Rev. A* **20**, 2486 (1979).

[Bru93] T. Brun, Quasiclassical equations of motion for nonlinear Brownian systems, *Phys. Rev. D* **47**, 3383 (1993).

[Bru96] T. Brun and J.J. Halliwell, Decoherence of hydrodynamic histories: A simple spin model, *Phys. Rev. D* **54**, 2899 (1996).

[Bug03] K.A. Bugaev, Relativistic kinetic equations for finite domains and the freeze-out problem, *Phys. Rev. Lett.* **90**, 252301 (2003).

[BunDav78] T.S. Bunch and P.C.W. Davies, Quantum field theory in de Sitter space; renormalization by point splitting, *Proc. Roy. Soc. A* **360**, 117 (1978).

[BunPar79] T.S. Bunch and L. Parker, Feynman propagator in curved spacetime: A momentum-space representation, *Phys. Rev. D* **20**, 2499 (1979).

[Bur95] Olga Buryak, Stochastic dynamics of large-scale inflation in De Sitter space, *Phys. Rev. D* **53**, 1763 (1996).

[BuLiRi05] C. Bustamante, J. Liphardt, and F. Ritort, The nonequilibrium thermodynamics of small systems. *Phys. Today* **58**, 43 (2005).

[CalLeg83a] A.O. Caldeira and A.J. Leggett, Path integral approach to quantum brownian motion, *Physica* **121A**, 587 (1983).

[CalLeg83b] A.O. Caldeira and A.J. Leggett, Quantum tunnelling in a dissipative system, *Ann. Phys. (N.Y.)* **149**, 374 (1983).

[CalLeg85] A.O. Caldeira and A.J. Leggett, Influence of damping on quantum interference: An exactly soluble model, *Phys. Rev. A* **31**, 1059 (1985).

[CaCoJa70] C.G. Callen, S. Coleman and R. Jackiw, A new improved energy-momentum tensor, *Ann. Phys. (N.Y.)* **59**, 42 (1970).

[CalWel51] H. Callen and T. Welton, Irreversibility and generalized noise, *Phys. Rev.* **83**, 34 (1951).

[Cal89] E. Calzetta, Spinodal decomposition in quantum field theory, *Ann. Phys. (N.Y.)* **190**, 32 (1989).

[Cal98] E. Calzetta, Relativistic fluctuating hydrodynamics, *Class. Q. Grav.* **15**, 653 (1998).

[CalGon97] E. Calzetta and S. Gonorazky, Primordial fluctuations from nonlinear couplings, *Phys. Rev. D* **55**, 1812 (1997).

[CalGra02] E. Calzetta and M. Graña, Reheating and turbulence, *Phys. Rev. D* **65**, 063522 (2002).

[CaHaHu88] E. Calzetta, S. Habib and B.L. Hu, Quantum kinetic field theory in curved spacetime: covariant Wigner function and Liouville–Vlasov equation, *Phys. Rev. D* **37**, 2901 (1988).

[CalHu87] E. Calzetta and B.L. Hu, Closed time-path functional formalism in curved spacetime: Application to cosmological back-reaction problems, *Phys. Rev. D* **35**, 495 (1987).

[CalHu88] E. Calzetta and B.L. Hu, Nonequilibrium quantum fields: Closed-time-path effective action, Wigner function, and Boltzmann equation, *Phys. Rev. D* **37**, 2878 (1988).

[CalHu89] E. Calzetta and B.L. Hu, Dissipation of quantum fields from particle creation, *Phys. Rev. D* **40**, 656 (1989).

[CalHu93] E. Calzetta and B.L. Hu, Decoherence of correlation histories, in *Directions in General Relativity*, vol. II: *Brill Festschrift*, ed. B. L. Hu and T. A. Jacobson (Cambridge University Press, Cambridge, 1993).

[CalHu94] E. Calzetta and B.L. Hu, Noise and fluctuations in semiclassical gravity, *Phys. Rev. D* **49**, 6636 (1994).

[CalHu95] E. Calzetta and B.L. Hu, Quantum fluctuations, decoherence of the mean field, and structure formation in the early universe, *Phys. Rev. D* **52**, 6770 (1995).

[CalHu95a] E. Calzetta and B.L. Hu, Correlations, decoherence, dissipation and noise in quantum field theory, in *Heat Kernel Techniques and Quantum Gravity,* vol. 4 of *Discourses in Mathematics and Its Applications,* ed. S. A. Fulling (A&M University Press, College Station, TX, 1995), hep-th/9501040.

[CalHu97] E. Calzetta and B.L. Hu, Stochastic behavior of effective field theories across the threshold, *Phys. Rev. D* **55**, 3536 (1997).

[CalHu00] E. Calzetta and B.L. Hu, Stochastic dynamics of correlations in quantum field theory: from Schwinger–Dyson to Boltzmann–Langevin equations, *Phys. Rev. D* **61**, 025012 (2000).

[CalHu03] E. Calzetta and B.L. Hu, Bose–Einstein condensate collapse and dynamical squeezing of vacuum fluctuations, *Phys. Rev. A* **68**, 043625 (2003).

[CaHuMa01] E. Calzetta, B.L. Hu, Francisco D. Mazzitelli, Coarse-grained effective action and renormalization group theory in semiclassical gravity and cosmology, *Phys. Rep.* **352**, 459–520 (2001).

[CaHuRa00] E. Calzetta, B.L. Hu, S.A. Ramsey, Hydrodynamic transport functions from quantum kinetic theory, *Phys. Rev. D* **61** 125013 (2000).

[CaHuRe06] E. Calzetta, B.L. Hu and A.M. Rey, Bose–Einstein condensate superfluid-Mott-insulator transition in an optical lattice, *Phys. Rev. A* **73**, 023610 (2006).

[CaHuVe07] E. Calzetta, B.L. Hu and E. Verdaguer, Stochastic Gross–Pitaevskii equation for BEC via coarse-grained effective action, *Int. J. Mod. Phys, B* **21**, 4239 (2007).

[CaJaPA86] E. Calzetta, I. Jack and L. Parker, Quantum gauge fields at high curvature, *Phys. Rev. D* **33**, 953 (1986).

[CalKan02] E. Calzetta and A. Kandus, Self-consistent estimates of magnetic fields from reheating, *Phys. Rev. D* **65**, 063004 (2002).

[CaKaMa98] E. Calzetta, A. Kandus and F.D. Mazzitelli, Primordial magnetic fields induced by cosmological particle creation, *Phys. Rev. D* **57**, 7139 (1998).

[CalMaz90] E. Calzetta and D. Mazzitelli, Decoherence and particle creation, *Phys. Rev. D* **42**, 4066 (1990).

[CaRoVo01] E. Calzetta, A. Roura and E. Verdaguer, Vacuum decay in quantum field theory, *Phys. Rev. D* **64**, 105008 (2001).

[CaRoVe03] E. Calzetta, A. Roura and E. Verdaguer, Stochastic description for open quantum systems, *Physica A* **319**, 188 (2003).

[CalSak92] E. Calzetta and M. Sakellariadou, Inflation in inhomogeneous cosmology, *Phys. Rev. D* **45**, 2802 (1992).

[CalThi00] E. Calzetta and M. Thibeault, Relativistic theories of interacting fields and fluids, *Phys. Rev. D* **63**, 103507 (2001).

[CalThi03] E. Calzetta and M. Thibeault, Macroscopic description of preheating, hep-ph/0309135.

[CamPar04] D. Campo and R. Parentani, Space-time correlations in inflationary spectra: A wave-packet analysis, *Phys. Rev. D* **70**, 105020 (2004).

[CamPar05] David Campo and Renaud Parentani, Inflationary spectra and partially decohered distributions, *Phys. Rev. D* **72**, 045015 (2005).

[CamVer96] A. Campos and E. Verdaguer, Stochastic semiclassical equations for weakly inhomogeneous cosmologies, *Phys. Rev. D* **53**, 1927 (1996).

[CanSci77] P. Candelas and D.W. Sciama, Irreversible thermodynamics of black holes, *Phys. Rev. Lett.* **38**, 1372 (1977).

[Car93] H. Carmichael, *An Open Systems Approach to Quantum Optics* (Springer-Verlag, Berlin, 1993).

[Car99] H. Carmichael, *Statistical Methods in Quantum Optics*, vol. I (Springer-Verlag, Berlin, 1999, corrected second printing 2001).

[CaDeKo00] M. Carrington, H. Defu and R. Kobes, Shear viscosity in phi-fourth theory from an extended ladder resummation, *Phys. Rev. D* **62**, 25010 (2000).

[CaDeKo01] M. Carrington, H. Defu and R. Kobes, Non-linear response from transport theory and QFT at finite temperature, *Phys. Rev. D* **64**, 25001 (2001).

[CarKob98] M. Carrington and R. Kobes, General cancellation of ladder graphs at finite temperature, *Phys. Rev. D* **57**, 6372 (1998).

[CaKoPe98] M. Carrington, R. Kobes and E. Petitgirard, Cancellation of ladder graphs in an effective expansion, *Phys. Rev. D* **57**, 2631 (1998).

[CaKuZa03] M. Carrington, G. Kunstatter and H. Zaraket, 2PI effective action and gauge invariance problems, *Eur. Phys. J.* **C42**, 253–259 (2005).

[CarDuo73] P. Carruthers and M. Duong-van, Rapidity and angular distributions of charged secondaries, *Phys. Rev. D* **8**, 859 (1973).

[CarZac83] P. Carruthers and F. Zachariasen, Quantum collision theory with phase-space distributions, *Rev. Mod. Phys.* **55**, 245 (1983).

[Cas04] Y. Castin, Simple theoretical tools for low dimension Bose gases, quantum gases in low dimensions, *J. Phys. IV France,* **116**, 89 (2004).

[CasDum97] Y. Castin and R. Dum, Instability and depletion of an excited Bose–Einstein condensate in a trap, *Phys. Rev. Lett.* **79**, 3553 (1997).

[CasDum98] Y. Castin and R. Dum, Low-temperature Bose–Einstein condensates in time-dependent traps: Beyond the U(1) symmetry breaking approach, *Phys. Rev. A* **57**, 3008 (1998).

[CavSch85] C.M. Caves and B.L. Schumacher, New formalism for two-photon quantum optics. I. Quadrature phases and squeezed states, *Phys. Rev. A* **31**, 3068, 3093 (1985).

[Cer69] C. Cercignani, *Mathematical Methods in Kinetic Theory* (Macmillan, New York, 1969).

[Cha43] S. Chandrasekhar, Stochastic problems in physics and astronomy, *Rev. Mod. Phys.* **15**, 1 (1943).

[ChaCow39] S. Chapman and T. Cowling, *The Mathematical Theory of Non-uniform Gases* (Cambridge University Press, Cambridge, 1939) (reissued 1990).

[CNLL02] Y.Y. Charng, K.W. Ng, C.Y. Lin and D.S. Lee, Photon production from non-equilibrium disoriented chiral condensates in a spherical expansion, *Phys. Lett.* **548B**, 175 (2002).

[ChNuMi05] T. Charters, A. Nunes and J.P. Mimoso, Phase dynamics and particle production in preheating, *Phys. Rev. D* **71**, 083515 (2005).

[Che62] S.S. Chern, Geometry of a quadratic differential form *J. Soc. Ind. Appl. Math.* **10**, 751 (1962).

[ChWiDi77] Y. Choquet-Bruhat, C. deWitt-Morette and M. Dillard-Bleick, *Analysis, Manifolds and Physics* (North-Holland, Amsterdam, 1977).

[ChoSuHa80] K. Chou, Z. Su and B. Hao, Closed time path Green's functions and critical dynamics, *Phys. Rev. B* **22**, 3385 (1980).

[CSHY85] K. Chou and Z. Su and B. Hao and L. Yu, Equilibrium and nonequilibrium formalisms made unified, *Phys. Rep.* **118**, 1 (1985).

[ChCoPh95] Nobel Lectures: S. Chu. The manipulation of neutral particles, *Rev. Mod. Phys.* **70**, 685 (1998); C.N. Cohen-Tannoudji, Manipulating atoms with photons, *ibid.*, 707; W.D. Phillips, Laser cooling and trapping of neutral atoms, *ibid*, 721.

[Cla03a] N. Claussen, Dynamics of Bose–Einstein condensates near a Feshbach resonance in ^{85}Rb. Ph.D. Thesis, University of Colorado (2003).

[Cla03b] N. Claussen *et al.*, Very-high-precision bound-state spectroscopy near a ^{85}Rb Feshbach resonance, *Phys. Rev. A* **67**, 060701(R) (2003).

[Coh07] T.D. Cohen, Is there a "most perfect fluid" consistent with quantum field theory?, hep-th/0702136.

[Col85] S. Coleman, *Aspects of Symmetry* (Cambridge University Press, Cambridge, 1985).

[CoJaPo74] S. Coleman, R. Jackiw and H.D. Politzer, Spontaneous symmetry breaking in the O(N) model for large N, *Phys. Rev. D* **10**, 2491 (1974).

[CoDuJo77] J.C. Collins, A. Duncan and S.D. Joglekar, Trace and dilatation anomalies in gauge theories, *Phys. Rev. D* **16**, 438 (1977).

[CoPaPe95] G. Compagno, R. Passante and F. Persico, *Atom-Field Interactions and Dressed Atom* (Cambridge University Press, Cambridge, 1995).

[CDHR98] F. Cooper, J. Dawson, S. Habib and R. Ryne, Chaos in time dependent variational approximations to quantum dynamics, *Phys. Rev. E* **57**, 1489 (1998).

[CEKMS93] F. Cooper, J. Eisenberg, Y. Kluger, E. Mottola and B. Svetisky, Particle production in the central rapidity region, *Phys. Rev. D* **48**, 190 (1993).

[CooFry74] F. Cooper and G. Frye, Single-particle distribution in the hydrodynamic and statistical models of multiparticle production, *Phys. Rev. D* **10**, 186 (1974).

[CoFrSc74] F. Cooper, G. Frye and E. Schonberg, Landau's hydrodynamic model of particle production and electron-positron annihilation into hadrons, *Phys. Rev. D* **11**, 192 (1974).

[CHKM97] F. Cooper, S. Habib, Y. Kluger and E. Mottola, Nonequilibrium dynamics of symmetry breaking in $\lambda\Phi^4$ theory, *Phys. Rev. D* **55**, 6471 (1997).

[CHKMPA94] F. Cooper, S. Habib, Y. Kluger, E. Mottola, J.P. Paz and P. Anderson, Nonequilibrium quantum fields in the large-N expansion, *Phys. Rev. D* **50**, 2848 (1994).

[CoKlMo96] F. Cooper, Y. Kluger and E. Mottola, Anomalous transverse distribution of pions as a signal for the production of disoriented chiral condensates, *Phys. Rev. C* **54** 3298 (1996).

[CKMP95] F. Cooper, Y. Kluger, E. Mottola and J.P. Paz, Quantum evolution of disoriented chiral condensates, *Phys. Rev. D* **51**, 2377 (1995).

[CooMot87] F. Cooper and E. Mottola, Initial-value problems in quantum field theory in the large-N approximation, *Phys. Rev. D* **36**, 3114 (1987).

[CoPiSt86] F. Cooper, S.Y. Pi, and P. Stancioff, Quantum dynamics in a time dependent variational approximation, *Phys. Rev. D* **34**, 3831 (1986).

[CoEnWi99] E.A. Cornell, J.R. Ensher and C.E. Wieman, Experiments in dilute atomic Bose–Einstein condensation, Varenna conference on Bose–Einstein condensation, July 1998 (cond-mat/9903109).

[CorWie02] E.A. Cornell and C.E. Wieman, Nobel Lecture: Bose–Einstein condensation in a dilute gas, the first 70 years and some recent experiments, *Rev. Mod. Phys.* **74**, 875–893 (2002).

[CoThWi06] S.L. Cornish, S.T. Thompson and C.E. Wieman, Formation of bright matter-wave solitons during the collapse of attractive Bose–Einstein condensates, *Phys. Rev. Lett.* **96**, 170401 (2006).

[CoJaTo74] J.M. Cornwall, R. Jackiw and E. Tomboulis, Effective action for composite operators, *Phys. Rev. D* **10**, 2428 (1974).

[Cro80] C. Cronstrom, A simple and complete Lorentz – covariant gauge condition, *Phys. Lett.* **90B**, 267 (1980).

[Cse94] L.P. Csernai, *Introduction to Relativistic Heavy Ion Collisions* (John Wiley, Chichester, 1994).

[Dad99] I. Dadic, Two mechanisms for the elimination of pinch singularities in out of equilibrium thermal field theories, *Phys. Rev. D* **59**, 125012 (1999).

[DahLas67] H.D. Dahmen and G. Jona Lasino, Variational formulation of quantum field theory II Nuovo Cimiento, **LIIA**, 807 (1967).

[Dal98] D.A.R. Dalvit, Quantum corrections and effective action in field theory, doctoral thesis, University of Buenos Aires.

[DaDzOn02] D.A.R. Dalvit, J. Dziarmaga and R. Onofrio, Continuous quantum measurement of a Bose–Einstein condensate: a stochastic Gross–Pitaievskii equation, *Phys. Rev. A* **65**, 053604 (2002).

[DalMaz96] D.A.R. Dalvit and F.D. Mazzitelli, Exact CTP renormalization group equation for the coarse-grained effective action, *Phys. Rev. D* **54**, 6338 (1996).

[Dan84a] P. Danielewicz, Quantum theory of nonequilibrium processes I, *Ann. Phys.* (N.Y.) **152**, 239 (1984).

[Dan84b] P. Danielewicz, Quantum theory of nonequilibrium processes II. Application to nuclear collisions, *Ann. Phys.* (N.Y.) **152**, 305 (1984).

[Davies76] E.B. Davies, *The Quantum Theory of Open Systems* (Academic Press, London, 1976).

[DavFul77] P.C.W. Davies and S.A. Fulling, Radiation from moving mirrors and from black holes, *Proc. Roy. Soc. Lond. A* **356**, 237 (1977).

[DavPaz97] L. Dávila Romero and J. Pablo Paz, Decoherence and initial correlations in quantum Brownian motion, *Phys. Rev. A* **55**, 4070–4083 (1997).

[DaLiTo99] A.-C. Davis, M. Lilley and O. Tornkvist, Relaxing the bounds on primordial magnetic seed fields, *Phys. Rev. D* **60**, 021301 (1999).

[DeGuLa92] N. Deruelle, C. Gundlach and D. Langlois, Vacuum density fluctuations in extended chaotic inflation, *Phys. Rev. D* **46**, 5337 (1992).

[DeuDru02] P. Deuar and P. Drummond, Gauge P representations for quantum-dynamical problems, *Phys. Rev. A* **62**, 033812 (2002).

[DeW64] B. DeWitt, Dynamical theory of groups and fields, in *Relativity, Groups and Topology*, ed. C. and B. DeWitt (Gordon and Breach, New York, 1964).

[DeW67] B. DeWitt, Quantum theory of gravity I: the canonical theory, *Phys. Rev.* **160**, 1113 (1967).

[DeW75] B. DeWitt, Quantum field theory in curved spacetime, *Phys. Rep. C* **19**, 295 (1975).

[DeW79] B. DeWitt, The formal structure of quantum gravity, in *Recent Developments in Gravitation*, ed. M. Levy and S. Deser (Plenum, New York, 1979).

[DeW81] B. DeWitt, A gauge invariant effective action, in *quantum gravity 2*, ed. C. Isham, R. Penrose and D. Sciama (Clarendon Press, Oxford, 1981).

[DeW84] B. DeWitt, The spacetime approach to quantum field theory, in *Relativity, Groups and Topology II*, ed. B. DeWitt and R. Stora (Elsevier, Amsterdam, 1984).

[DeW86] B. DeWitt, Effective action for expectation values in *Quantum Concepts in Space and Time*, ed. R. Penrose and C.J. Isham (Clarendon Press, Oxford, Clarendon, 1986).

[DeW87] B. DeWitt, The effective action, in *Quantum Field Theory and Quantum Statistics*, ed. I. Batalin, C. Isham and G. Vilkovisky (A. Hilger, Bristol, 1987).

[DeWMol98] B. DeWitt, C. Molina-Paris, Quantum gravity without ghosts, *Mod. Phys. Lett. A* **13**, 2475–2478 (1998).

[Dir50] P.A.M. Dirac, Generalized Hamiltonian dynamics, *Can. J. Math.* **2**, 129 (1950).

[Dir58] P.A.M. Dirac, *The Principles of Quantum Mechanics* (Oxford University Press, Oxford, 1958).

[Dir58b] P.A.M. Dirac, Generalized Hamiltonian dynamics, *Proc. R. Soc. London A* **246**, 326 (1958).

[Dir65] P.A.M. Dirac, *Lectures on Quantum Field Theory* (Yeshiva University, New York, 1965).

[Dod03] S. Dodelson, *Modern Cosmology* (Academic Press, Amsterdam, 2003).

[DolJac74] L. Dolan and R. Jackiw, Symmetry behavior at finite temperature, *Phys. Rev. D* **9**, 3320 (1974).

[DomMar64a] C. de Dominicis and P.C. Martin, Stationary entropy principle and renormalization in normal and superfluid systems I, *J. Math. Phys.* **5**, 14 (1964).

[DomMar64b] C. de Dominicis and P.C. Martin, Stationary entropy principle and renormalization in normal and superfluid systems II, *J. Math. Phys.* **5**, 31 (1964).

[DomRit02] P. Domokos and H. Ritsch, Collective cooling and self-organization of atoms in a cavity, *Phys. Rev. Lett.* **89**, 253003 (2002).

[Don01] E. Donley *et al.*, Dynamics of collapsing and exploding Bose–Einstein condensates, *Nature* **412**, 295 (2001).

[Dor81] R. Dorfman, Some recent development in the kinetic theory, in H.J. Raveché (editor) *Perspectives in Statistical Physics: M.S. Green Memorial Volume* (North Holland, Amsterdam, 1981).

[Dor99] R. Dorfman, *An Introduction to Chaos in Nonequilibrium Statistical Mechanics* (Cambridge University Press, Cambridge, 1999).

[DowKen96] F. Dowker and A. Kent, On the consistent histories approach to quantum mechanics, *J. Stat. Phys.* **82**, 1575 (1996).

[DrDeKh04] P. Drummond, P. Deuar and K. Kheruntsyan, Canonical Bose gas simulations with stochastic gauges, *Phys. Rev. Lett.* **92**, 040405 (2004).

[Dub67] D.F. DuBois, in *Lectures in Theoretical Physics*, vol. IX C, *Kinetic Theory*, ed. W.E. Brittin (Gordon and Breach, New York, 1967).

[DuSto01] R. Duine and H. Stoof, Stochastic dynamics of a trapped Bose–Einstein condensate, *Phys. Rev. A* **65**, 13603 (2001).

[DuNaSt07] A. Dumitru, Y. Nara and M. Strickland, Ultraviolet avalanche in anisotropic non-abelian plasmas, *Phys. Rev. D* **75**, 025016 (2007).

[Dzi05b] J. Dziarmaga, Images of a Bose-Einstein condensate at finite temperature, "Quantum Optics VI," June 13–18, 2005, Krynica, Poland (cond-mat/0506723).

[DzLaZu99] J. Dziarmaga, P. Laguna and W.H. Zurek, Symmetry breaking with a slant: topological defects after an inhomogeneous quench, *Phys. Rev. Lett.* **82**, 4749 (1999).

[DSZB04] J. Dziarmaga, A. Smerzi, W. Zurek and A. Bishop, Non-equilibrium Mott transition in a lattice of Bose–Einstein condensates, Proceedings of NATO ASI Patterns of symmetry breaking, Krakow, Poland, Sept. 2002, cond-mat/0403607.

[EbJaPi88] O. Eboli, R. Jackiw and S.Y. Pi, Quantum fields out of thermal equilibrium, *Phys. Rev. D* **37**, 3557 (1988).

[Efs03] G. Efstathiou, The statistical significance of the low CMB mulitipoles, *MNRAS* **346**, L26 (2003).

[Efs04] G. Efstathiou, A maximum likelihood analysis of the low CMB multipoles from WMAP, *MNRAS* **348**, 885 (2004).

[EiOCSt95] M. Van Eijck, D. O'Connor and C.R. Stephens, Critical temperature and amplitude ratios from a finite-temperature renormalization group, *Int. J. Mod. Phys. A* **10**, 3343 (1995).

[Ein05] A. Einstein, On the movement of small particles suspended in a stationary liquid demanded by the molecular kinetic theory of heat, *Ann. der Phys.* **17**, 549 (1905), reprinted in [Fur56].

[Ein17] A. Einstein, Zur Quantentheorie der Strahlung, *Physik Z.* **18**, 121 (1917) (Engl. tr. On the quantum theory of radiation, reprinted in B.L. van der Waerden, *Sources of Quantum Mechanics* (Dover, New York, 1968)).

[Ein24] A. Einstein, Quantentheorie des einatomigen idealen Gases (Quantum theory of monatomic ideal gases), Sitzungsberichte der Preussichen Akademie der Wissenschaften Physikalisch-Mathematische Klasse: 261–267 (1924).

[Ein25] A. Einstein, Quantentheorie des einatomigen idealen Gases, 2. Abhandlung. Sitzungsberichte der Preussischen Akademie der Wissenschaften (Berlin), Physikalisch-mathematische Klasse: 3–14 (1925).

[EiPoRo35] A. Einstein, B. Podolsky and N. Rosen, Can quantum-mechanical description of physical reality be considered complete?, *Phys. Rev.* **47**, 777 (1935); reprinted in J.A. Wheeler and W.H. Zurek (editors), *Quantum Theory and Measurement* (Princeton University Press, Princeton, 1983).

[ElKrVo06] V.B. Eltsov, M. Krusius and G.E. Volovik, Transition to superfluid turbulence, *J. of Low Temp. Phys.* **145**, 89–106 (2006).

[Elz02] H.-Th. Elze, Fluid dynamics of relativistic quantum dust, *J. Phys. G* **28**, 2235 (2002).

[ElGyVa86] H.-Th. Elze, M. Gyulassy and D. Vasak, Transport equations for the QCD quark Wigner operator, *Nucl. Phys.* **B276**, 706 (1986).

[EnNgOl88] K. Enqvist, K.-W. Ng and K.A. Olive, Cosmic abundances of very heavy neutrinos, *Nucl. Phys.* **B303**, 713 (1988).

[EvCoMo93] D.J. Evans, E.G.D. Cohen and G.P. Morris, Probability of second law violations in shearing steady states, *Phys. Rev. Lett.* **71**, 2401 (1993).

[EvaSea94] D.J. Evans and D.J. Searles, Equilibrium microstates which generate second law violating steady states, *Phys. Rev. E* **50**, 1645 (1994).

[Far64] I. Farquhar, *Ergodic Theory in Statistical Mechanics* (John Wiley, New York, 1964).

[FGGKLT01] G. Felder, J. Garcia-Bellido, P.B. Greene, L. Kofman, A. Linde and I. Tkachev, Dynamics of symmetry breaking and tachyonic preheating, *Phys. Rev. Lett.* **87**, 011601 (2001).

[FelKof01] G. Felder and L. Kofman, The development of equilibrium after preheating, *Phys. Rev. D* **63**, 103503 (2001).

[FeKoLi01] G. Felder, L. Kofman and A. Linde, Tachyonic instability and dynamics of spontaneous symmetry breaking, *Phys. Rev. D* **64**, 123517 (2001).

[FelTka00] G. Felder and I. Tkachev, LATTICEEASY: A program for lattice simulations of scalar fields in an expanding universe, hep-ph/0011159.

[FeyHib65] R. Feynman and A. Hibbs, *Quantum Mechanics and Path Integrals* (McGraw-Hill, New York, 1965).

[FeyVer63] R. Feynman and F. Vernon, The theory of a general quantum system interacting with a linear dissipative system, *Ann. Phys.* (N.Y.) **24**, 118 (1963).

[FiGaJe06] F. Filion-Gourdeau, J.-S. Gagnon and S. Jeon, All orders Boltzmann collision term from the multiple scattering expansion of the self-energy, *Phys. Rev. D* **74**, 025010 (2006).

[Fin00] F. Finelli, Cosmological magnetic fields by parametric resonance, astro-ph/0007290. COSMO 99: 3rd International Conference on Particle Physics and the Early Universe, Trieste, Italy, 27 Sep.–3 Oct. 1999.

[FiHaHu79] M.V. Fischetti, J.B. Hartle and B.L. Hu, Quantum effects in the early universe I: Influence of trace anomalies on homogeneous, isotropic, classical geometries, *Phys. Rev. D* **20**, 1757 (1979).

[Fis74] M. Fisher, The renormalization group in the theory of critical behavior, *Rev. Mod. Phys.* **46**, 597 (1974).

[Fis83] M. Fisher, in *Critical Phenomena*, ed. F.J.W. Hahne (Springer-Verlag, Berlin 1983).

[FWGF89] M. Fisher, P. Weichman, G. Grinstein and D. Fisher, Boson localization and the superfluid-insulator transition, *Phys. Rev. B* **40**, 546 (89).

[Foc37] V.A. Fock, Die Eigenzeit in der klassichten und in der Quantenmechanik, *Phys. Zeit. Sowjetunion* **12**, 404 (1937).

[Fon94] O. Fonarev, Wigner function and QKT in curved space-time and external fields, *J. Math Phys.* **35**, 2105 (1994).

[FoKaMa65] G.W. Ford, M. Kac and P. Mazur, Statistical mechanics of assemblies of coupled oscillators, *J. Math. Phys.* **6**, 504 (1965).

[ForPar77] L. Ford and L. Parker, Quantized gravitational wave perturbations in Robertson–Walker universes, *Phys. Rev. D* **16**, 1601 (1977).

[FoxUhl70a] R. Fox and G. Uhlembeck, Contributions to non-equilibrium thermodynamics. I. Theory of hydrodynamical fluctuations, *Phys. Fluids* **13**, 1893 (1970).

[FoxUhl70b] R. Fox and G. Uhlembeck, Contributions to nonequilibrium thermodynamics. II. Fluctuation theory for the Boltzmann equation, *Phys. Fluids* **13**, 2881 (1970).

[FrGiSh91] E.S. Fradkin, D.M. Gitman, and S.M. Shvartsman, *Quantum Electrodynamics with Unstable Vacuum* (Springer-Verlag, Berlin, 1991).

[FASA05] P. Freire, N. Antunes, P. Salmi and A. Achúcarro, The role of dissipation in biasing the vacuum selection in quantum field theory at finite temperature, *Phys. Rev. D* **72**, 045017 (2005).

[FEOS96] F. Freire, M. Van Eijck, D. O'Connor and C.R. Stephens, Finite-temperature renormalization group predictions: The critical temperature exponents, and amplitude ratios, preprint hep-th/9601165.

[FreTay90] J. Frenkel and J. Taylor, High temperature limit of thermal QCD, *Nucl. Phys.* **B334**, 199 (1990).

[Fri95] U. Frisch, *Turbulence, the Legacy of A. N. Kolmogorov* (Cambridge University Press, Cambridge, 1995).

[FroNov93] V. Frolov and I.D. Novikov, Dynamical origin of the entropy of a black hole, *Phys. Rev. D* **48**, 4545 (1993).

[Fuj80] K. Fujikawa, Comment on chiral and conformal anomalies, *Phys. Rev. Lett.* **44**, 1733 (1980).

[FKYY96] H. Fujisaki, K. Kumekawa, M. Yamaguchi and M. Yoshimura, Particle production and dissipative cosmic field, *Phys. Rev. D* **53**, 6805 (1996).

[Ful72] S.A. Fulling, Scalar quantum field theory in a closed universe of constant curvature. Ph.D. Thesis, Princeton University (1972).

[Ful89] S.A. Fulling, *Aspects of Quantum Field Theory in Curved Spacetime* (Cambridge University Press, Cambridge, 1989).

[Ful95] S.A. Fulling (ed.) *Heat Kernel Techniques and Quantum Gravity* (Texas A&M Press, College Station, 1995).

[FulDav76] S.A. Fulling and P.C.W. Davies, Radiation from a moving mirror in two dimensional space-time: conformal anomaly, *Proc. R. Soc. Lond. A* **348**, 393 (1976).

[FulPar74] S.A. Fulling and L. Parker, Renormalization in the theory of a quantized scalar field interacting with a Robertson–Walker spacetime, *Ann. Phys.* (N.Y.) **87**, 176 (1974).

[FuPaHu74] S.A. Fulling, L. Parker and B.L. Hu, Conformal energy-momentum tensor in curved spacetime: Adiabatic regularization and renormalization, *Phys. Rev. D* **10**, 3905 (1974); Erratum: *Phys. Rev. D.* **11**, 1714 (1975).

[Fur56] R. Furth (ed.), *Investigations on the Theory of the Brownian Movement* (Dover, New York, 1956).

[Gal98] G. Gallavotti, Chaotic dynamics, fluctuations, nonequilibrium ensembles, *Chaos* **8**, 384 (1998).

[GalCoh95] G. Gallavotti and E. Cohen, Dynamical ensembles in nonequilibrium statistical mechanics, *Phys. Rev. Lett.* **74**, 2694 (1995).

[GaFrTo01] A. Gammal, T. Frederico and L. Tomio, Critical number of atoms for attractive Bose–Einstein condensates with cylindrically symmetric traps, *Phys. Rev. A* **64**, 055602 (2001).

[Gar90] C.W. Gardiner, *Handbook of Stochastic Methods* (Springer-Verlag, Berlin, 1990).

[Gar97] C. Gardiner, Particle-number-conserving Bogoliubov method which demonstrates the validity of the time-dependent Gross–Pitaevskii equation for a highly condensed Bose gas, *Phys. Rev. A* **56**, 1414 (1997).

[GaAnFu01] C. Gardiner, J. Anglin and T. Fudge, The stochastic Gross–Pitaevskii equation, *J. Phys. B At. Mol. Opt. Phys.* **35**, 1555 (2002).

[GarDav03] C. Gardiner and M. Davis, The stochastic Gross–Pitaevskii equation II, cond-mat/0308044.

[GarZol97] C. Gardiner and P. Zoller, Quantum kinetic theory, *Phys. Rev. A* **55**, 2902 (1997).

[GarZol98] C. Gardiner and P. Zoller, Quantum kinetic theory III, *Phys. Rev. A* **58**, 536 (1998).

[GarZol00a] C. Gardiner and P. Zoller, Quantum kinetic theory V: quantum kinetic master equation for mutual interaction of condensate and noncondensate, *Phys. Rev. A* **61**, 033601 (2000).

[GarZol00b] C.W. Gardiner and P. Zoller, *Quantum Noise* (Springer-Verlag, Berlin, 2000).

[GJDCZ00] S. Gardiner, D. Jaksch, R. Dum, J. Cirac and P. Zoller, Nonlinear matter wave dynamics with a chaotic potential, *Phys. Rev. A* **62**, 23612 (2000).

[GarMor07] S. Gardiner and S.A. Morgan, A number-conserving approach to a minimal self-consistent treatment of condensate and non-condensate dynamics in a degenerate Bose gas, *Phys. Rev. A* 75, 043621 (2007).

[GBSS05] T. Gasenzer, J. Berges, M. Schmidt and M. Seco, Nonperturbative dynamical many-body theory of a Bose–Einstein condensate, *Phys. Rev. A* **72**, 063604 (2005).

[GBSS07] T. Gasenzer, J. Berges, M. Schmidt and M. Seco, Ultracold quantum gases far from equilibrium, *Nucl. Phys. A* **785**, 214–217 (2007).

[Gas98] P. Gaspard, *Chaos, Scattering and Statistical Mechanics* (Cambridge University Press, Cambridge, 1998).

[Gas04] P. Gaspard, Time-reversed dynamical entropy and irreversibility in Markovian random processes, *J. Stat. Phys.* **117**, 599 (2004).

[Gas05] P. Gaspard, Dynamical systems theory of irreversibility in P. Collet *et al.* (eds.), *Chaotic Dynamics and Transport in Classical and Quantum Systems* (Kluwer, Dordrecht, 2005).

[Gas06] P. Gaspard, Hamiltonian dynamics, nanosystems, and nonequilibrium statistical mechanics, Lecture Notes for the International Summer School, *Fundamental Problems in Statistical Physics XI* (Leuven, Belgium, September 4–17, 2005).

[GasGio93] M. Gasperini and M. Giovanni, Entropy production in the cosmological amplification of the vacuum fluctuations, *Phys. Lett.* **301B**, 334 (1993).

[GasGioVen93] M. Gasperini, M. Giovanni and G. Veneziano, Squeezed thermal vacuum and the maximum scale for inflation, *Phys. Rev. D* **48**, R439 (1993).

[Gei96] K. Geiger, Quantum field kinetics of QCD: Quark–gluon transport theory for light-cone dominated processes, *Phys. Rev. D* **54**, 949 (1996).

[Gei97] K. Geiger, Nonequilibrium QCD: Interplay of hard and soft dynamics in high-energy multigluon beams, *Phys. Rev. D* **56**, 2665 (1997).

[Gei99] K. Geiger, Propagation of gluons from a nonperturbative evolution equation in axial gauges, *Phys. Rev. D* **60**, 034012 (1999).

[GeScSe01] F. Gelis, D. Schiff and J. Serreau, Simple out-of-equilibrium field theory formalism?, *Phys. Rev. D* **64**, 56006 (2001).

[GelHar90] M. Gell-Mann and J.B. Hartle, Quantum mechanics in the light of quantum cosmology in *Complexity, Entropy and the Physics of Information*, ed. W.H. Zurek (Addison-Wesley, Reading, MA, 1990).

[GelHar] M. Gell-Mann and J.B. Hartle, Quantum mechanics in the light of quantum cosmology, in *Proceedings of the Third International Symposium on the Foundations of Quantum Mechanics in the Light of New Technology*, ed. S. Kobayashi. S. Kobayashi, H. Ezawa, Y. Murayama and S. Nomura (Physical Society of Japan, Tokyo, 1990).

[GelHar06] M. Gell-Mann and J.B. Hartle, Quasiclasical coarse graining and thermodynamic entropy, quant-ph/0609190. Dedicated to Rafael Sorkin on his 60th birthday.

[GerReb03] A. Gerhold and A. Rebhan, Gauge dependence identities for color superconducting QCD, *Phys. Rev. D* **68**, 011502 (2003).

[Ger95] R. Geroch, Relativistic theories of dissipative fluids, *J. Math. Phys.* **36**, 4226 (1995).

[Ger01] R. Geroch, On hyperbolic "theories" of relativistic dissipative fluids, gr-qc/0103112.

[GerLin90] R. Geroch and L. Lindblom, Dissipative relativistic fluid theories of divergence type, *Phys. Rev. D* **41**, 1855 (1990).

[GiHaSt87] G.W. Gibbons, S.W. Hawking and J.M. Stewart, A natural measure on the set of all universes, *Nucl. Phys.* **B281**, 736 (1987).

[Ginz87] V.L. Ginzburg (ed.) *Issues in Intense Field Quantum Electrodynamics*, Proc. Lebedev Phys. Inst. Sci. USSR, vol. **168** (Nova Science, Commack, 1987).

[Ginz95] V.L. Ginzburg (ed.) *Quantum Electrodynamics with Unstable Vacuum*, Proc. Lebedev Phys. Inst. Sci. Rus., vol. **220** (Nova Science, Commack, 1995).

[GioSha00] M. Giovannini and M. Shaposhnikov, Primordial magnetic fields from inflation?, *Phys. Rev. D* **62**, 103512 (2000).

[Gir60] M. Girardeau, Relationship between systems of impenetrable bosons and fermions in one dimension, *J. Math. Phys.* **1**, 516 (1960).

[GirArn98] M. Girardeau, Comment on Particle-number-conserving Bogoliubov method which demonstrates the validity of the time-dependent Gross–Pitaevskii equation for a highly condensed Bose gas, *Phys. Rev. A* **58**, 775 (1998).

[GirArn59] M. Girardeau and R. Arnowitt, Theory of many-boson systems: Pair theory, *Phys. Rev.* **113**, 755 (1959).

[GKJKSZ96] D. Giulini, C. Kiefer, E. Joos, J. Kupsch, I.O. Stamatescu and H.D. Zeh, *Decoherence and the Appearance of a Classical World in Quantum Theory* (Springer-Verlag, Berlin, 1996).

[Gla05] J.R. Glauber, 100 Years of Light Quanta, Nobel Lecture (2005).

[GlMaRa82] M. Gleiser, G.C. Marques and R.O. Ramos, Evaluation of thermal corrections to false vacuum decay rates, *Phys. Rev. D* **48**, 1571 (1993).

[GleRam94] M. Gleiser and R. Ramos, Microphysical approach to nonequilibrium dynamics of quantum fields, *Phys. Rev. D* **50**, 2441 (1994).

[Gold61] J. Goldstone, Field theories with superconductor solutions, *Nuovo Cimento* **19**, 154 (1961).

[GoPaSa95] J. Gomis, J. Paris and S. Samuel, Antibracket, antifields and gauge theory quantization, *Phys. Rep.* **259**, 1 (1995).

[GonLin86] A.S. Goncharov and A.D. Linde, Tunneling in an expanding universe: Euclidean and Hamiltonian approaches, *Sov. J. Part. Nucl.* **17**, 369 (1986).

[GoLiMu87] A.S. Goncharov, A.D. Linde, and V. F. Mukhanov, The global structure of the inflationary universe, *Int. J. Mod. Phys. A* **2**, 561 (1987).

[GKGB04] K. Goral, T. Kohler, T. Gasenzer and K. Burnett, Dynamics of correlations in atomic Bose–Einstein condensates, *J. Mod. Opt.* **51**, 1731 (2004).

[GorKar04] A. Gorban and I. Karlin, Uniqueness of thermodynamic projector and kinetic basis of molecular individualism, *Physica A* **336**, 391 (2004).

[GoKaZi04] A. Gorban, I. Karlin and A. Zinovyev, Constructive methods of invariant manifolds for kinetic problems, *Phys. Rep.* **396**, 197 (2004).

[Gra82] H. Grabert, *Projection Operator Techniques in Nonequilibrium Statistical Mechanics* (Springer-Verlag, Berlin, 1982).

[GrScIn88] H. Grabert, P. Schramm and G.L. Ingold, Quantum Brownian motion, the functional integral approach, *Phys. Rep.* **168**, 115 (1988).

[GraRub01] D. Grasso and H. Rubinstein, Magnetic fields in the early universe, *Phys. Rep.* **348**, 163 (2001).

[GKLS97] P. Greene, L. Kofman, A. Linde and A. Starobinsky, Structure of resonance in preheating after inflation, *Phys. Rev. D* **56**, 6175 (1997).

[GreLeu98] C. Greiner and S. Leupold, Stochastic interpretation of Kadanoff–Baym equations and their relation to Langevin processes, *Ann. Phys.* (N.Y.) **270**, 328–390 (1998).

[GreMul97] C. Greiner and B. Müller, Classical fields near thermal equilibrium, *Phys. Rev. D* **55**, 1026 (1997).

[GMEHB02] C. Greiner *et al.*, Quantum phase transition from a superfluid to a Mott insulator in a gas of ultracold atoms, *Nature* **415**, 39 (2002).

[Greiner] W. Greiner, B. Müller and J. Rafelski, *Quantum Electrodynamics of Strong Fields* (Springer-Verlag, Berlin, 1985).

[GrReBr96] W. Greiner, J. Reinhardt and D.A. Bromley, *Field Quantization* (Springer-Verlag, Berlin, 1996).

[GrMaMo88] A.A. Grib, S.G. Mamaev and V.M. Mostepanenko, *Vacuum Quantum Effects in Strong Fields* (Atomizdat, Moscow, 1988; Friedmann Laboratory Publishing, St. Petersburg, 1994).

[Gri96] A. Griffin, Conserving and gapless approximations for an inhomogeneous Bose gas at finite temperatures, *Phys. Rev. B* **53**, 9341 (1996).

[Gri84] R.B. Griffiths, Consistent histories and the interpretation of quantum mechanics, *J. Stat. Phys.* **36**, 219 (1984).

[Gri93] R.B. Griffiths, Consistent interpretation of quantum mechanics using quantum trajectories, *Phys. Rev. Lett.* **70**, 2201 (1993).

[Gri74] L. Grishchuk, Amplification of gravitational waves in an istropic universe, *Zh. Eksp. Teor. Fiz.* **67**, 825 (1974) (*Sov. Phys. JETP* **40**, 409 (1975)).

[GriSid90] L. Grishchuk and Y.V. Sidorov, Squeezed quantum states of relic gravitons and primordial density fluctuations, *Phys. Rev. D* **42**, 3413 (1990).

[GrLeWe80] S.R. de Groot, W.A. van Leeuwen and Ch.G. van Weert, *Relativistic Kinetic Theory* (North-Holland, Amsterdam, 1980).

[Guth81] A.H. Guth, Inflationary universe: A possible solution to the horizon and flatness problems, *Phys. Rev. D* **23**, 347 (1981).

[GuthPi82] A.H. Guth and S.Y. Pi, Fluctuations in the new inflationary universe, *Phys. Rev. Lett.* **49**, 1110 (1982).

[GutPi85] A.H. Guth and S.Y. Pi, Quantum mechanics of the scalar field in the new inflationary universe, *Phys. Rev. D* **32**, 1899 (1985).

[Gyu01] M. Gyulassy, Lectures on the theory of high energy A+A at RHIC, *Lect. Notes Phys.* **583**, 37–79 (2002) (nucl-th/0106072).

[GyKaWi79] M. Gyulassy, S. Kauffmann and L. Wilson, Pion interferometry of nuclear collisions. I. Theory, *Phys. Rev. C* **20**, 2267 (1979).

[GyRiZh96a] M. Gyulassy, D. Rischke and B. Zhang, Transverse shocks in the turbulent gluon plasma produced in ultra-relativistic A+A, nucl-th/9606045. Proc. of the Int. Conf. on Nucl. Phys. "Structure of Vacuum and Elementary Matter," March 10-16, Wilderness, South Africa.

[GyRiZh97] M. Gyulassy, D. Rischke and B. Zhang, Hot spots and turbulent initial conditions of quark-gluon plasmas in nuclear collisions, *Nucl. Phys.* **A613**, 397–434 (1997).

[Hab90] S. Habib, Exact Langevin equation in a cosmological setting, *Phys. Rev. D* **39**, 2871 (1990).

[Hab92] S. Habib, Stochastic inflation: Quantum phase-space approach, *Phys. Rev. D* **46**, 2408 (1992).

[Hab04] S. Habib, Gaussian dynamics is classical dynamics, quant-ph/0406011.

[HKMP96] S. Habib, Y. Kluger, E. Mottola and J. Paz, Dissipation and decoherence in mean field theory, *Phys. Rev. Lett.* **76**, 4660 (1996).

[HabLaf90] S. Habib and R. Laflamme, Wigner function and decoherence in quantum cosmology, *Phys. Rev. D* **42**, 4056 (1990).

[HabLyt00] S. Habib and G. Lythe, Dynamics of kinks: nucleation, diffusion and annihilation, *Phys. Rev. Lett.* **84**, 1070 (2000).

[HaMoMo99] S. Habib, C. Molina-Paris and E. Mottola, Energy-momentum tensor of particles created in an expanding universe, *Phys. Rev. D* **61**, 024010 (1999).

[HakAmb85] V. Hakim and V. Ambegaokar, Quantum theory of a free particle interacting with a linearly dissipative environment, *Phys. Rev. A* **32**, 423 (1985).

[Hal81] F. Haldane, Effective harmonic-fluid approach to low-energy properties of one-dimensional quantum fluids, *Phys. Rev. Lett.* **47**, 1840 (1981).

[Hal89] J.J. Halliwell, Decoherence in quantum cosmology, *Phys. Rev. D* **39**, 2912 (1989).

[Hal93] J.J. Halliwell, Quantum-mechanical histories and the uncertainty principle: Information-theoretic inequalities, *Phys. Rev. D* **48**, 2739, 4785 (1993).

[Hal98] J.J. Halliwell, Decoherent histories and hydrodynamic equations, *Phys. Rev. D* **58**, 105015 (1998).

[HaPeZu94] J. Halliwell, J. Perez Mercader and W. Zurek (eds.), *The Physical Origin of Time Asymmetry* (Cambridge University Press, Cambridge, 1994).

[HalYu96] J.J. Halliwell and T. Yu, Alternative derivation of the Hu–Paz–Zhang master equation of quantum Brownian motion, *Phys. Rev. D* **53**, 2012 (1996).

[Har70] E. Harrison, Fluctuations at the threshold of classical cosmology, *Phys. Rev. D* **1**, 2726 (1970).

[Hart93] C.F. Hart, Theory and renormalization of the gauge-invariant effective action, *Phys. Rev. D* **28**, 1993 (1983).

[Har80] J.B. Hartle, Quantum effects in the early universe IV: nonlocal effects in particle production in anisotropic models, *Phys. Rev. D* **22**, 2091 (1980).

[Har81] J.B. Hartle, Quantum effects in the early universe V: finite particle production without trace anomalies, *Phys. Rev. D* **23**, 2121 (1981).

[Har92] J.B. Hartle, Spacetime quantum mechanics and the quantum mechanics of spacetime, Lectures given at Summer School, NATO ASI, Session, LVII: Gravitation and Quantifications, B. Juliz & J. Zinn-Justin (eds). Les Houches, France, 1992 (Elsevier, Amsterdam, 1995).

[Har93] J.B. Hartle, Quantum mechanics of closed systems, in *Directions in General Relativity*, vol. 1, ed. B.L. Hu, M.P. Ryan and C.V. Vishveswara (Cambridge University Press, Cambridge, 1993).

[HarGel93] J.B. Hartle and M. Gell-Mann, Classical equations for quantum systems, *Phys. Rev. D* **47**, 3345 (1993).

[HarHaw83] J.B. Hartle and S.W. Hawking, Wave function of the universe, *Phys. Rev. D* **28**, 2960 (1983).

[HarHor81] J.B. Hartle and G. Horowitz, Ground-state expectation value of the metric in the 1 / N or semiclassical approximation to quantum gravity, *Phys. Rev. D* **24**, 257 (1981).

508

References

[HarHu79] J.B. Hartle and B.L. Hu, Quantum effects in the early universe II: effective action for scalar fields in homogeneous cosmologies with small anisotropy, *Phys. Rev. D* **20**, 1772 (1979).

[HarHu80] J.B. Hartle and B.L. Hu, Quantum effects in the early universe III: dissipation of anisotropy by scalar particle production, *Phys. Rev. D* **21**, 2756 (1980).

[HaLaMa95] J.B. Hartle, R. Laflamme, and D. Marolf, Conservation laws in the quantum mechanics of closed systems, *Phys. Rev. D* **51**, 7007 (1995).

[HatKug80] H. Hata and T. Kugo, Operator formalism of statistical mechanics of gauge theory in covariant gauges, *Phys. Rev. D* **21**, 3333 (1980).

[Haw75] S.W. Hawking, Particle creation by black holes, *Commun. Math. Phys.* **43**, 199 (1975).

[Haw82] S.W. Hawking, The development of irregularities in a single bubble inflationary universe, *Phys. Lett.* **115B**, 295 (1982).

[HawPag88] S.W. Hawking and Don N. Page, How probable is inflation?, *Nucl. Phys.* **B298**, 789 (1988).

[HeeKno02a] H. van Hees and J. Knoll, Renormalization in self-consistent approximation schemes at finite temperature: Theory, *Phys. Rev. D* **65**, 25010 (2002).

[HeeKno02b] H. van Hees and J. Knoll, Renormalization of self-consistent approximation schemes at finite temperature. II. Applications to the sunset diagram, *Phys. Rev. D* **65**, 105005 (2002).

[HeeKno02c] H. van Hees and J. Knoll, Renormalization in self-consistent approximation schemes at finite temperature. III. Global symmetries, *Phys. Rev. D* **66**, 25028 (2002).

[Hei83] U. Heinz, Kinetic theory for plasmas with non-Abelian interactions, *Phys. Rev. Lett.* **51**, 351 (1983).

[Hei96] U. Heinz, Hanbury-Brown/Twiss interferometry for relativistic heavy ion collisions: theoretical aspects, nucl-th/9609029. Lectures given at the NATO ASI on "Correlations and Clustering Phenomena in Subatomic Physics" at Dronten, Netherlands, Aug. 4–18, 1996, to be published in the NATO ASI Proceedings Series by Plenum Publ. Corp. (M.N. Harakeh, O. Scholten, and J.K. Koch, eds.)

[HeiKol02a] U. Heinz and P. Kolb, Early thermalization at RHIC, *Nucl. Phys.* **A702**, 269–280 (2002).

[HciKol02b] U.W. Heinz and P.F. Kolb, Two RHIC puzzles: Early thermalization and the HBT problem, in *Proc. 18th Winter Workshop on Nuclear Dynamics*, ed. R. Bellwied, J. Harris, and W. Bauer, pp. 205–216. (EP Systema, Debrecen, Hungary, 2002) (hep-ph/0204061).

[HeiWon02] U. Heinz and S. Wong, Elliptic flow from a transversally thermalized fireball, *Phys. Rev. C* **66**, 014907 (2002).

[Hei94] H. Heiselberg, Viscosities of quark-gluon plasma, *Phys. Rev. D* **49**, 4739 (1994).

[HeiEul36] W. Heisenberg and H. Euler, Consequences of the Dirac theory of the positron, *Z. Physik* **98**, 14 (1936).

[HelJan07] M.P. Heller and R.A. Janik, Viscous hydrodynamics relaxation time from AdS/CFT, hep-th/0703243.

[HenTei92] M. Henneaux and C. Teitelboim, *Quantization of Gauge Systems* (Princeton University Press, Princeton, NJ, 1992).

[Hen95] P. Henning, A transport equation for quantum fields with continuous mass spectrum, *Nucl. Phys.* **A582**, 633 (1995).

[Her02] C.P. Herzog, The hydrodynamics of M-theory, *JHEP* **0212**, 026 (2002).

[Her03] C.P. Herzog, Sound of M theory, *Phys. Rev. D* **68**, 024013 (2003).

[HerSon03] C.P. Herzog and D.T. Son, Schwinger-Keldysh propagators from AdS/CFT correspondence, *JHEP* **03**, 46 (2003).

[HOSW84] M. Hillery, R.F. O'Connell, M.O. Scully and E.P. Wigner, Distribution functions in physics: Fundamentals, *Phys. Rep.* **106**, 121 (1984).

[Hin07] G. Hinshaw *et al.*, Three-year Wilkinson Microwave Anisotropy Probe (WMAP) Observations: Temperature Analysis, *ApJS*, 170, 288 (2007).

[Hin75] J.O. Hinze, *Turbulence* (McGraw-Hill, New York, 1975).

[Hir01] T. Hirano, Is early thermalization achieved only near midrapidity in Au+Au collisions at $\sqrt{s_{NN}} = 130$ GeV?, *Phys. Rev. C* **65**, 11901(R) (2001).

[HirNar04] T. Hirano and Y. Nara, Hydrodynamic afterburner for the color glass condensate and the parton energy loss, *Nucl. Phys.* **A743**, 305 (2004).

[HirTsu02] T. Hirano and K. Tsuda, Collective flow and two-pion correlations from a relativistic hydrodynamic model with early chemical freeze-out, *Phys. Rev. C* **66**, 054905 (2002).

[HisLin83] W. Hiscock and L. Lindblom, Stability and causality in dissipative relativistic fluids, *Ann. Phys.* (N.Y.) **151**, 466 (1983).

[HisLin85] W. Hiscock and L. Lindblom, Generic instabilities in first-order dissipative relativistic fluid theories, *Phys. Rev. D* **31**, 725 (1985).

[Hod90] H.M. Hodges, Is double inflation likely?, *Phys. Rev. Lett.* **64**, 1080 (1990).

[HACCES06] M.A. Hoefer *et al.*, On dispersive and classical shock waves in Bose–Einstein condensates and gas dynamics, *Phys. Rev. A* **74**, 023623 (2006).

[HohMar65] P. Hohenberg and P. Martin, Microscopic theory of superfluid helium, *Ann. Phys.* (N.Y.) **34**, 291 (1965).

[HolWal02] S. Hollands and R.M. Wald, Comment on inflation and alternative cosmology, hep-th/0210001.

[HorSch87] R. Horsley and W. Schoenmaker, Quantum field theories out of thermal equilibrium I, *Nucl. Phys.* **B280**, 716 (1987); II: *Nucl. Phys.*, 735 (1987).

[HosSak84] A. Hosoya and M. Sakagami, Time development of Higgs field at finite temperature, *Phys. Rev. D* **29**, 2228 (1984).

[HoSata84] A. Hosoya, M. Sakagami and M. Tasaki, Nonequilibrium thermodynamics in field theory: transport coefficients, *Ann. Phys.* **154**, 229 (1984).

[Hu72] B.L. Hu, Scalar waves and tensor perturbation, in the mixmaster universe Ph.D. Thesis, Princeton University, 1972.

[Hu74] B.L. Hu, Scalar waves in the mixmaster universe II: Particle creation, *Phys. Rev. D* **9**, 3263 (1974).

[Hu81] B.L. Hu, Effect of finite temperature quantum fields on the early universe, *Phys. Lett.* **B103**, 331 (1981).

[Hu82] B.L. Hu, Vacuum viscosity description of quantum processes in the early universe, *Phys. Lett.* **90A**, 375 (1982).

[Hu83] B.L. Hu, Quantum dissipative processes and gravitational entropy of the universe, *Phys. Lett.* **97A**, 368 (1983).

[Hu84] B.L. Hu, Vacuum viscosity and entropy generation quantum gravitational processes in the early universe, in *Cosmology of the Early Universe* ed. L.Z. Fang and R. Ruffini (World Scientific, Singapore, 1984).

[Hu89] B.L. Hu, Dissipation in quantum fields and semiclassical gravity, *Physica A* **158**, 399–424 (1989).

[Hu91] B.L. Hu, Coarse graining and back reaction in inflationary and minisuperspace cosmology in *Relativity and Gravitation: Classical and Quantum, Proceedings of the SILARG VII Symposium*, Cocoyoc, Mexico, 1990, ed. J.C. D'Olivo *et al.* (World Scientific, Singapore, 1991), p. 33.

[Hu94a] B.L. Hu, Fluctuation, dissipation and irreversibility in cosmology, in *The Physical Origin of Time-Asymmetry*, Huelva, Spain, 1991 ed. J.J. Halliwell, J. Perez-Mercader and W.H. Zurek (Cambridge University Press, Cambridge, 1994).

[Hu94b] B.L. Hu, Quantum statistical fields in gravitation and cosmology, in *Proc. Third International Workshop on Thermal Field Theory and Applications*, ed. R. Kobes and G. Kunstatter (World Scientific, Singapore, 1994), gr-qc/9403061.

[Hu99] B.L. Hu, Stochastic gravity, *Int. J. Theor. Phys.* **38**, 2987 (1999).

[Hu02] B.L. Hu, A kinetic theory approach to quantum gravity, *Int. J. Theor. Phys.* **41**, 2111–2138 (2002).

[Hu05] B.L. Hu, Can Spacetime be a Condensate? *Int. J. Theor. Phys.* **44**, 1785-1806 (2005).

[HuKan87] B.L. Hu and H.E. Kandrup, Entropy generation in cosmological particle creation and interactions: A statistical subdynamics analysis, *Phys. Rev. D* **35**, 1776 (1987).

[HuKaMa94] B.L. Hu, G.W. Kang and A. Matacz, Squeezed vacua and the quantum statistics of cosmological particle creation, *Int. J. Mod. Phys. A* **9**, 991 (1994).

[HuMat94] B.L. Hu and A. Matacz, Quantum Brownian motion in a bath of parametric oscillators, *Phys. Rev. D* **49**, 6612 (1994).

[HuMat95] B.L. Hu and A. Matacz, Einstein–Langevin equation for backreactions in semiclassical cosmology, *Phys. Rev. D* **51**, 1577 (1995).

[HuMat96] B.L. Hu and A. Matacz, Quantum noise in gravitation and cosmology, invited talk at the Workshop on *Fluctuations and Order*, ed. M. Millonas (Springer-Verlag, Berlin, 1996). Univ. Maryland preprint 94-44 (1994).

[HuPar77] B.L. Hu and L. Parker, Effect of gravitation creation in isotropically expanding universes, *Phys. Lett.* **63A**, 217 (1977).

[HuPar78] B.L. Hu and L. Parker, Anisotropy damping through quantum effects in the early universe, *Phys. Rev. D* **17**, 933 (1978).

[HuPav86] B.L. Hu and D. Pavón, Intrinsic measures of field entropy in cosmological particle creation, *Phys. Lett.* **180B**, 329 (1986).

[HuPaZh92] B.L. Hu, J.P. Paz and Y. Zhang, Quantum Brownian motion in a general environment: Exact master equation with nonlocal dissipation and colored noise, *Phys. Rev. D* **45**, 2843 (1992).

[HuPaZh93a] B.L. Hu, J.P. Paz and Y. Zhang, Quantum Brownian motion in a general environment. II. Nonlinear coupling and perturbative approach, *Phys. Rev. D* **47**, 1576 (1993).

[HuPaZh93b] B.L. Hu, J.P. Paz, and Y. Zhang, in *The Origin of Structure in the Universe*, ed. E. Gunzig and P. Nardone (NATO ASI Series) (Plenum Press, New York, 1993).

[HuRav96] B.L. Hu and A. Raval, Thermal radiation from black holes and cosmological spacetimes, *Mod. Phys. Lett. Af* **32/33**, 2625–2638 (1996).

[HuSin95] B.L. Hu and S. Sinha, Fluctuation-dissipation relation in cosmology, *Phys. Rev. D* **51**, 1587 (1995).

[HuVer02] B.L. Hu and E. Verdaguer, Recent advances in stochastic gravity: Theory and issues, in P. Bergmann and V. de Sabata (eds.), *Advances in the Interplay between Quantum and Gravity Physics* (Kluwer, Amsterdam, 2002).

[HuVer03] B.L. Hu and E. Verdaguer, Stochastic gravity: a primer with applications, *Class. Q. Grav.* **20**, R1 (2003).

[HuVer04] B.L. Hu and E. Verdaguer, Stochastic gravity: theory and applications, living reviews in relativity, **7**, 3 (2004). [update in arXiv:0802.0658]

[HuZha93b] B.L. Hu and Yuhong Zhang, Squeezed states and uncertainty relation at finite temperature, *Mod. Phys. Lett. A* **8**, 3575 (1993).

[HuZha95] B.L. Hu and Yuhong Zhang, Uncertainty relation for a quantum open system, *Int. J. Mod. Phys. A* **10**, 4537 (1995).

[Hua87] K. Huang, *Statistical Physics* (John Wiley, New York, 1987).

[Hua98] K. Huang, *Quantum Field Theory* (Wiley-Interscience, New York, 1998).

[HugPin59] N.M. Hugenholtz and D. Pines, Ground-state energy and excitation spectrum of a system of interacting bosons, *Phys. Rev.* **116**, 489 (1959).

[Hum06] T. Humanic, Can hadronic rescattering explain the large elliptic flow and small HBT radii seen at RHIC?, *Acta Phys. Hung. A* **25**, 1–9 (2006).

[HKHRV01] P. Huovinen, P.F. Kolb, U.W. Heinz, P.V. Ruuskanen and S.A. Voloshin, Radial and elliptic flow at RHIC: further predictions, *Phys. Lett.* **503B**, 58 (2001).

[IaleMc02] E. Iancu, A. Leonidov and L. McLerran, The colour glass condensate: An introduction, hep-ph/0202270.

[IbaCal99] D. Ibaceta and E. Calzetta, Counting defects in an instantaneous quench, *Phys. Rev. E* **60**, 2999 (1999).

[Idz05a] Z. Idziaszek, Microcanonical fluctuations of the condensate in a weakly interacting Bose gas, *Phys. Rev. A* **71**, 053604 (2005).

[Ike04] T. Ikeda, Effect of memory on relaxation in a scalar field theory, *Phys. Rev. D* **69**, 105018 (2004).

[IlltMa75] J. Iliopoulos, C. Itzykson and A. Martin, Functional methods and perturbation theory, *Rev. Mod. Phys.* **47**, 165 (1975).

[Inc56] E. Ince, *Ordinary Differential Equations* (Dover, New York, 1956).

[Isr72] W. Israel, The relativistic Boltzmann equation, in L. O'Raifeartaigh (ed.), *General Relativity: Papers in Honour of J. L. Synge* (Clarendon Press, Oxford, 1972).

[Isr88] W. Israel, Covariant fluid mechanics and thermodynamics: an introduction, in A. Anile and Y. Choquet-Bruhat (eds.), *Relativistic Fluid Dynamics* (Springer, New York, 1988).

[ItzZub80] C. Itzykson and J.-B. Zuber, *Quantum Field Theory* (McGraw-Hill, New York, 1980).

[IvKnVo99] Yu.B. Ivanov, J. Knoll and D.N. Voskresensky, Self-consistent approximations to non-equilibrium many-body theory, *Nucl. Phys.* **A657**, 413 (1999).

[IvKnVo00] Yu.B. Ivanov, J. Knoll, D. Voskresensky, Resonance transport and kinetic entropy, *Nucl. Phys.* **A672**, 313 (2000).

[IvKnVo03] Yu.B. Ivanov, J. Knoll and D.N. Voskresensky, Self-consistent approach to off-shell transport, nucl-th/0303006. Submitted to issue of Phys. Atom. Nucl. dedicated to S.T. Belyaev on the occasion of his 80th birthday.

[Jac74] R. Jackiw, Functional evaluation of the effective potential, *Phys. Rev. D* **9**, 1686 (1974).

[JacKer79] R. Jackiw and A. Kerman, Time-dependent variational principle and the effective action, *Phys. Lett.* **71A**, 158 (1979).

[Jac02] J.D. Jackson, From Lorentz to Coulomb and other explicit gauge transformations, *Am. J. Phys.* **70**, 917 (2002).

[Jaigar04] P. Jain and C. Gardiner, A phase space method for the Bose–Hubbard model, cond-mat/0404642.

[Jak02] A. Jakovac, Generalized BE for on-shell particle production in a hot plasma, *Phys. Rev. D* **66**, 125003 (2002).

[JGGZ98] D. Jaksch, C.W. Gardiner, K.M. Gheri and P. Zoller, Quantum kinetic theory IV, *Phys. Rev. A* **58**, 1450 (1998).

[JaGaZo97] D. Jaksch, C.W. Gardiner and P. Zoller, Quantum kinetic theory II, *Phys. Rev. A* **56**, 575 (1997).

[JakZol04] D. Jaksch and P. Zoller, The cold atom Hubbard toolbox, *Annals of Physics*, **315**, 52 (2005). (cond-mat/0410614).

[Jar97] C. Jarzynski, Nonequilibrium equality for free energy differences, *Phys. Rev. Lett.* **78**, 2690 (1997).

[Jeo93] S. Jeon, Computing spectral densities in finite temperature field theory, *Phys. Rev. D* **47**, 4586 (1993).

[Jeo95] S. Jeon, Hydrodynamic transport coefficients in relativistic scalar field theory, *Phys. Rev. D* **52**, 3591 (1995).

[JeoYaf96] S. Jeon and L. Yaffe, From quantum field theory to hydrodynamics, *Phys. Rev. D* **53**, 5799 (1996).

[JooZeh85] E. Joos and H.D. Zeh, *Z. Phys. B* **59**, 223 (1985).

[Jor86] R.D. Jordan, Effective field equations for expectation values, *Phys. Rev. D* **33**, 444 (1986).

[JuCaGr04] S. Juchem, W. Cassing and C. Greiner, Quantum dynamics and thermalization for out-of equilibrium ϕ^4 theory, *Phys. Rev. D* **69**, 025006 (2004).

[KacLog76] M. Kac and J. Logan, Fluctuations and the Boltzmann equation. I, *Phys. Rev. A* **13**, 458 (1976).

[KacLog79] M. Kac and J. Logan, Fluctuations, in *Fluctuation Phenomena*, ed. E. Montroll and J. Lebowitz (Elsevier, New York, 1979).

[Kad76] L. Kadanoff, Notes on Migdal's recursion formulas, *Ann. Phys.* (N.Y.) **100**, 359 (1976).

[Kad77] L. Kadanoff, The application of renormalization group techniques to quarks and strings, *Rev. Mod. Phys.* **49**, 267 (1977).

References

[KadBay62] L. Kadanoff and G. Baym, *Quantum Statistical Mechanics* (Benjamin, New York, 1962).

[KadMar63] L. Kadanoff and P. Martin, Hydrodynamic equations and correlation functions, *Ann. Phys.* (N.Y.) **24**, 419 (1963).

[KaSuSh96] Yu. Kagan, E. Surkov and G. Shlyapnikov, Evolution of a Bose-condensed gas under variations of the confining potential, *Phys. Rev. A* **54**, R1753 (1996).

[KaSuSh97] Yu. Kagan, E. Surkov and G. Shlyapnikov, Evolution and global collapse of trapped Bose condensates under variations of the scattering length, *Phys. Rev. Lett.* **79**, 2604 (1997).

[KamUme64] S. Kamefuchi and H. Umezawa, Bose fields and inequivalent representations, *Nuovo Cimento* **31**, 429 (1964) Appendix A.

[Kam81] N.G. van Kampen, *Stochastic Processes in Physics and Chemistry* (North-Holland, Amsterdam, 1985).

[Kam85] N.G. van Kampen, Elimination of fast variables, *Phys. Rep.* **124**, 70 (1985).

[Kan88a] H.E. Kandrup, Entropy generation, particle creation, and quantum field theory in a cosmological spacetime: When do number and entropy increase?, *Phys. Rev. D* **37**, 3505 (1988).

[Kan88b] H.E. Kandrup, Particle number and random phases, *Phys. Rev. D* **38**, 1773 (1988).

[KCMW00] A. Kandus, E. Calzetta, F. Mazzitelli and C. Wagner, Cosmological magnetic fields from gauge mediated supersymmetry breaking models, *Phys. Lett.* **472B**, 287 (2000).

[KanKad65] J.W. Kane and L. Kadanoff, Green's functions and superfluid hydrodynamics, *J. Math. Phys.* **6**, 1902 (1965).

[Kap38] P. Kapitza, Viscosity of liquid helium below the λ-point, *Nature* **141**, 74 (1938).

[Kap89] J. Kapusta, *Finite Temperature Field Theory* (Cambridge University Press, Cambridge, 1989).

[KarGor02] I. Karlin and A. Gorban, Hydrodynamics from Grad's equations, *Ann. Phys* (Leipzig) **V11**, 783 (2002) (cond-mat/0209560).

[KarGor03] I. Karlin and A. Gorban, Methods of nonlinear kinetics, cond-mat/0306062.

[KarRiv97] G. Karra and R.J. Rivers, Initial vortex densities after a temperature quench, *Phys. Lett.* **414B**, 28 (1997).

[Kel64] L.V. Keldysh, Diagram technique for nonequilbrium processes, *Zh. Eksp. Teor. Fiz.* **47**, 1515 (1964) (*Sov. Phys. JETP* **20**, 1018 (1965)).

[Kes94] E. Keski-Vakkuri, Coarse-grained entropy and stimulated emission in curved space-time, *Phys. Rev. D* **49**, 2122 (1994).

[Ket02] W. Ketterle, Nobel lecture: When atoms behave as waves: Bose–Einstein condensation and the atom laser, *Rev. Mod. Phys.* **74**, 1131–1151 (2002).

[KhaLev01] D. Kharzeev and E. Levin, Manifestations of high density QCD in the first RHIC data, *Phys. Lett.* **B523**, 79–87 (2001).

[KhLeNa01] D. Kharzeev, E. Levin and M. Nardi, The onset of classical QCD dynamics in relativistic heavy ion collisions, hep-ph/0111315.

[KhlTka96] S.Yu. Khlebnikov and I. Tkachev, Classical decay of the inflaton, *Phys. Rev. Lett.* **77**, 219 (1996).

[Kib76] T.W.B. Kibble, Topology of cosmic domains and strings, *J. Phys. A* **9**, 1387 (1976).

[Kib80] T.W.B. Kibble, Some implications of a cosmological phase transition, *Phys. Rep.* **67**, 183 (1980).

[Kib88] T.W.B. Kibble, Topology of cosmic domains and strings, *J. Phys. A* **9**, 1387 (1988).

[Kib02] T.W.B. Kibble, Symmetry breaking and defects, Lectures at NATO ASI "Patterns of symmetry breaking," Cracow, Sept. 2002, cond-mat/0211110.

[KiPoSt98] C. Kiefer, D. Polarski, and A.A. Starobinsky, Quantum to classical transition for fluctuations in the early universe, *Int. J. Mod. Phys. D* **7**, 455 (1998).

[KiPoSt00] C. Kiefer, D. Polarski and A.A. Starobinsky, Entropy of gravitons produced in the early universe, *Phys. Rev. D* **62**, 043518 (2000).

[Kle29] O. Klein, Die Reflexion von Elektronen an einem Potentialsprung nach der relativistischen Dynamic von Dirac, *Z. Phys.* **53**, 157 (1929).

[Kle90] H. Kleinert. *Path Integrals in Quantum Mechanics, Statistics, and Polymer Physics* (World Scientific, Singapore, 1990).

[KESCM92] Y. Kluger, J.M. Eisenberg, B. Svetisky, F. Cooper and E. Mottola, Fermion pair production in a strong electric field, *Phys. Rev. D* **45**, 4659 (1992).

[KlMoEi98] Y. Kluger, E. Mottola and J.M. Eisenberg, Quantum Vlasov equation and its Markov limit, *Phys. Rev. D* **58**, 125015 (1998).

[KnIvVo01] J. Knoll, Yu. Ivanov, D.N. Voskresensky, Exact conservation laws of the gradient expanded KBEs, *Ann. Phys. (N.Y.)* **293**, 126 (2001).

[KoEzMuNo90] S. Kobayashi, H. Ezawa, Y. Murayama and S. Nomura, *Proceedings of the 3rd International Symposium on the Foundations of Quantum Mechanics in the Light of New Technology* (Physical Society of Japan, Tokyo, 1989).

[KobKun89] R. Kobes and G. Kunstatter, The gauge dependence of the gluon damping constant, *Physica A* **158**, 192 (1989).

[KoKuRe91] R. Kobes, G. Kunstatter and A. Rebhan, Gauge dependence identities and their application at finite temperature, *Nucl. Phys.* **B355**, 1 (1991).

[KKHOSK06] V.V. Kocharovsky *et al.*, Fluctuations in ideal and interacting Bose–Einstein condensates, *Adv. Atom., Mol. Opt. Phys.* **53**, 291 (2006).

[KBHP91] L. Kofman, G.R. Blumenthal, H. Hodges and J.R. Primack, Generation of nonflat and non-Gaussian perturbations from inflation, *ASP Conf. Ser.* **15**, 339 (1991).

[KoLiMu02] Lev Kofman, Andrei Linde and V. Mukhanov, Inflationary theory and alternative cosmology, *JHEP* **0210**, 057 (2002).

[KoLiSt94] L. Kofman, A. Linde and A.A. Starobinsky, Reheating after inflation, *Phys. Rev. Lett.* **73**, 3195 (1994).

[KoLiSt97] L. Kovman, A. Linde and A.A. Starobinsky, Towards the theory of reheating after inflation, *Phys. Rev. D* **56**, 3258 (1997).

[Koh05] M. Köhl *et al.*, Superfluid to Mott insulator transition in one, two and three dimensions, *J. Low T. Phys.* **138**, 635 (2005).

[KohBur02] T. Kohler and K. Burnett, Microscopic quantum dynamics approach to the dilute condensed Bose gas, *Phys. Rev. A* **65**, 033601 (2002).

[Koi02] T. Koide, Derivation of transport equations using the time-dependent projection operator method, Prog. Theor. Phys. **107**, 525 (2002).

[Koi07] T. Koide, Microscopic formula of transport coefficients for causal hydrodynamics, *Phys. Rev. E* **75**, 060103(R) (2007).

[Kok96] D. Koks, Decoherence, entropy and thermal radiance using influence functionals. Ph.D. Thesis, University of Adelaide (1996).

[KoMaHu97] D. Koks, Andrew Matacz and B.L. Hu, Entropy and uncertainty of squeezed quantum open systems, *Phys. Rev. D* **55**, 5917 (1997).

[KolRap03] P.F. Kolb and R. Rapp, Transverse flow and hadrochemistry in Au+Au collisions at $\sqrt{sNN}$=200 GeV, *Phys. Rev. C* **67**, 044903 (2003).

[KoSoHe00] P. Kolb, J. Sollfrank and U. Heinz, Anisotropic transverse flow and the quark-hadron phase transition, *Phys. Rev. C* **62**, 54909 (2000).

[KoSoSt05] P.K. Kovtun, D.T. Son and A.O. Starinets, Viscosity in strongly interacting quantum field theories from black hole physics, *Phys. Rev. Lett.* **94**, 111601 (2005).

[KowTay92] K.L. Kowalski and C.C. Taylor, Disoriented chiral condensate: A white paper for the full acceptance detector, hep-ph/9211282.

[KraReb04] U. Kraemmer and A. Rebhan, Advances in perturbative thermal field theory, *Rep. Prog. Phys.* **67**, 351 (2004).

[Kra60] R. Kraichnan, Dynamics of nonlinear stochastic systems, *J. Math. Phys.* **2**, 124 (1960).

[Kra40] H.A. Kramers, Brownian motion in a field of force and the diffusion model of chemical reactions, *Physica* **7**, 284 (1940).

[KrNaVe02] A. Krasnitz, Y. Nara and R. Venugopalan, Elliptic flow of colored glass in high energy heavy ion collisions, hep-ph/0204361.

[KrNaOrRe97] H.-O. Kreiss, G. Nagy, O. Ortiz and O. Reula, Global existence and exponential decay for hyperbolic dissipative relativistic fluid theories, *J. Math. Phys.* **38**, 5272 (1997).

[KBKS97] D. Kremp, M. Bonitz, W.D. Kraeft and M. Schlanges, Non-Markovian Boltzmann equation, *Ann. Phys.* (NY) **258**, 320 (1997).

[Kre81] H.J. Kreuzer, *Nonequilibrium Thermodynamics and its Statistical Foundations* (Oxford University Press, Oxford, 1981).

[KrOxZa94] M. Kruczenski, L.E. Oxman and M. Zaldarriaga, Large squeezing behavior of cosmological entropy generation, *Class. Quant. Grav.* **11**, 2317 (1994).

[Kry44] N.N. Krylov, Relaxation processes in statistical physics, *Nature* **153**, (1944) 709.

[Kry79] N.N. Krylov, *Works on the Foundations of Statistical Mechanics* (Princeton University Press, Princeton, NJ, 1979).

[Kub57] R. Kubo, Statistical-mechanical theory of irreversible processes I, *J. Phys. Soc. Japan* **12**, 570 (1957).

[KuToHa91] R. Kubo, M. Toda and N. Hashitsume, *Statistical Physics II, Nonequilibrium Statistical Mechanics* (Springer-Verlag, Berlin, 1991).

[ToKuSa92] R. Kubo, M. Toda and N. Saito, *Statistical Physics I, Equilibrium Statistical Mechanics* (Springer-Verlag, Berlin, 1992).

[KuYoNa57] R. Kubo, M. Yokota and S. Nakajima, Statistical-mechanical theory of irreversible processes. II, *J. Phys. Soc. Japan* **12**, 1203 (1957).

[KUMS97] Hiroto Kubotani, Tomoko Uesugi, Masahiro Morikawa and Akio Sugamoto, Classicalization of quantum fluctuations in an inflationary universe, *Prog. of Theor. Phys.*, **98**, 1063 (1997).

[KugOji79] T. Kugo and I. Ojima, Local covariant operator formalism of non-abelian gauge theories and quark confinement problem, *Supp. Prog. Theo. Phys.* **66**, 1 (1979).

[KunTsu06] T. Kunihiro and K. Tsumura, Application of the renormalization group method to the reduction of transport equations, *J. Phys. A* **39**, 8089–8104 (2006).

[Kur98] J. Kurchan, Fluctuation theorem for stochastic dynamics, *J. Phys. A* **31**, 3719 (1998).

[Kur05] J. Kurchan, Non-equilibrium work relations, cond-mat/0511073.

[LafMat93] R. Laflamme and A. Matacz, Decoherence functional and inhomogeneities in the early universe, *Int. J. Mod. Phys. D* **2**, 171–182 (1993).

[LagZur97] P. Laguna and W. Zurek, Density of kinks after a quench: When symmetry breaks, how big are the pieces?, *Phys. Rev. Lett.* **78**, 2519 (1997).

[LagZur98] P. Laguna and W. Zurek, Critical dynamics of symmetry breaking: quenches, dissipation and cosmology, *Phys. Rev. D* **58**, 085021 (1998).

[LaDaCo96] M. Lampert, J. Dawson and F. Cooper, Time evolution of the chiral phase transition during a spherical expansion, *Phys. Rev. D* **54**, 2213 (1996).

[LanLif57] L. Landau and E. Lifshitz, *J. Exptl. Theor. Phys.* (USSR) **32**, 618 (1957) (Engl. trans. *Sov. Phys. JETP* **5**, 512 (1957)).

[LanLif59] L. Landau and E. Lifshitz, *Fluid Mechanics* (Pergamon Press, Oxford, 1959).

[LanLif69] L. Landau and E. Lifshitz, *Mécanique* (Mir, Moscow, 1969). Mechanics, Pergamon Press, Oxford, 1960.

[LanLif76] L. Landau and E. Lifshitz, *Meccanica Quantistica* (Mir, Moscow, 1976). Quantum Mechanics, The non-relativistic theory, Pergamon Press, Oxford, 1958.

[LanLif72] L. Landau and E. Lifchitz, *Relativistic Quantum Theory* (Pergamon Press, Oxford, 1972).

[LaLiPi80a] L. Landau, E. Lifshitz and L. Pitaevskii, *Statistical Physics*, vol. I (Pergamon Press, London, 1980).

[LaLiPi80b] L. Landau, E. Lifshitz and L. Pitaevskii, *Statistical Physics*, vol. II (Pergamon Press, London, 1980).

[LanPom53a] L. Landau and I. Pomeranchuk Limit on the applicability of electron Bremsstrahlung theory and pair creation at high energies. *Doklady Akad. Nauk. SSSR* **92**, 535 (1953).

[LanPom53b] L. Landau and I. Pomeranchuk Avalanche processes of electrons at high energies. *Doklady Akad. Nauk. SSSR* **92**, 735 (1953).

[LanReb92] P. Landshoff and A. Rebhan, Covariant gauges at finite temperature, *Nucl. Phys.* **B383**, 607 (1992).

[LanReb93] P. Landshoff and A. Rebhan, Thermalization of longitudinal gluons, *Nucl. Phys.* **B410**, 23 (1993).

[LanWer87] N.P. Landsman and Ch.G. van Weert, Real and imaginary time field theory at finite temperature and density, *Phys. Rep.* **145**, 141 (1987).

[Lan75] O.E. Lanford, Time evolution of large classical systems, in J. Moser (ed.), *Dynamical Systems, Theory and Applications* (Springer, New York, 1975).

[LaCaId99] R. Laura, M. Castagnino and R. Id Betan, Perturbative method for generalized spectral decompositions, *Physica A* **271**, 357 (1999).

[Law89] I. Lawrie, Perturbative description of dissipation in nonequilibrium field theory, *Phys. Rev. D* **40**, 3330 (1989).

[Law92] I. Lawrie, Feynman rules for nonequilibrium field theory, *J. Phys. A* **25**, 6493 (1992).

[Law99] I. Lawrie, Perturbative nonequilibrium dynamics of phase transitions in an expanding universe, *Phys. Rev. D* **60**, 63510 (1999).

[LawKer00] I. Lawrie and D. McKernan, Nonequilibrium perturbation theory for spin-1/2 fields, *Phys. Rev. D* **62**, 105032 (2000).

[Lea01] S.M. Leach *et al.*, Enhancement of superhorizon scale inflationary curvature perturbations, *Phys. Rev. D* **64**, 023512 (2001).

[LeB91] M. Le Bellac, *Quantum and Statistical Field Theory* (Oxford University Press, Oxford, 1991).

[LeB96] M. Le Bellac, *Thermal Field Theory* (Cambridge University Press, Cambridge, 1996).

[Leb93] J.L. Lebowitz, Macroscopic laws, microscopic dynamics, time's arrow and Botlzmann's entropy, *Physica A* **194**, 1 (1993).

[Lee81] T.D. Lee, *Particle Physics and Introduction to Field Theory* (Harwood, Zurich, 1981).

[LeeBoy93] D.S. Lee and D. Boyanovsky, Dynamics of phase transitions induced by a heat bath, *Nucl. Phys.* **B406**, 631 (1993).

[Lee04] W. Lee, Y.Y. Charng, D.S. Lee, and L.Z. Fang, Off-equilibrium dynamics of the primordial perturbations in the inflationary universe: The O(N) model, *Phys. Rev. D* **69**, 123522 (2004).

[Leg06] A. J. Leggett, *Quantum Liquids* (Oxford University Press, Oxford, 2006).

[LePoSt97] J. Lesgourgues, D. Polarski and A.A. Starobinsky, Quantum-to-classical transition of cosmological perturbations for non-vacuum initial states, *Nucl. Phys.* **B497**, 479 (1997).

[Lib98] R. Liboff, *Kinetic Theory* (John Wiley, New York, 1998).

[Lif46] E.M. Lifshitz, On the gravitational stability of the expanding universe, *Zh. Eksp. Teor. Phys.* **16**, 587 (1946) (*J. Phys. USSR* **10**, 116 (1946)).

[LifKal63] E.M. Lifshitz and I. Khalatnikov, Investigations in relativistic cosmology, *Adv. Phys.* **12**, 185 (1963).

[LifPit81] E. Lifshitz and L.P. Pitaievskii, *Physical Kinetics* (Pergamon Press, Oxford, 1981).

[LMMR04] M. Liguori, S. Matarrese, M.A. Musso and A. Riotto, Stochastic inflation and the lower multipoles in the CMB anisotropies, *JCAP* **0408**, 011 (2004).

[Lin82] A.D. Linde, Coleman–Weinberg theory and the new inflationary universe scenario, *Phys. Lett.* **114B**, 431 (1982).

[Lin85] A. Linde, Initial conditions for inflation, *Phys. Lett.* **162B**, 281 (1985).

[Lin90] A.D. Linde, *Particle Physics and Inflationary Cosmology* (Harwood, Chur, Switzerland, 1990).

[LiLiMe94] A.D. Linde, D.A. Linde and A. Mezhlumian, From the Big Bang theory to the theory of a stationary universe, *Phys. Rev. D* **49**, 1783 (1994).

[LinWes90] K. Lindenberg and B.J. West, *The Nonequilibrium Statistical Mechanics of Open and Closed Systems* (VCH Press, New York, 1990).

[Lin87] J. Lindig, Not all adiabatic vacua are physical states, *Phys. Rev. D* **59**, 064011 (1999).

[Lit98] D. Litim, Wilsonian flow equations and thermal field theory, hep-ph/9811272 (1998).

[LitMan02] D. Litim and C. Manuel, Semiclassical transport theory for non-abelian plasmas, *Phys. Rep.* **364**, 451 (2002).

[LomMaz96] F.C. Lombardo and F.D. Mazzitelli, Coarse graining and decoherence in quantum field theory, *Phys. Rev. D* **53**, 2001 (1996).

[LomMaz97] F.C. Lombardo and F.D. Mazzitelli, Einstein–Langevin equations from running coupling constants, *Phys. Rev. D* **55**, 3889 (1997).

[LomMaz98] F.C. Lombardo and F.D. Mazzitelli, Influence functional in two-dimensional dilaton gravity, *Phys. Rev. D* **58**, 024009 (1998).

[LoMaMo00] F.C. Lombardo, F.D. Mazzitelli and D. Monteoliva, Classicality of the order parameter during a phase transition, *Phys. Rev. D* **62**, 045016 (2000).

[LoMaRi01] F.C. Lombardo, F.D. Mazzitelli and R.J. Rivers, Classical behaviour after a phase transition, *Phys. Lett.* **B523**, 317 (2001).

[LoMaRi03] F.C. Lombardo, F.D. Mazzitelli and R.J. Rivers, Decoherence in field theory: general couplings and slow quenches, *Nucl. Phys.* **B672**, 462 (2003).

[LomNac05] F.C. Lombardo and D. Nacir, Decoherence during inflation: The generation of classical inhomogeneities, *Phys. Rev. D* **72**, 063506 (2005).

[LoRiVi07] F.C. Lombardo, R.J. Rivers and P. Villar, Decoherence of domains and defects at phase transitions, *Phys. Lett.* **648B**, 64 (2007).

[LomVil05] F.C. Lombardo and P. Villar, Decoherence induced by zero-point fluctuations in quantum Brownian motion, *Phys. Lett.* **336A**, 16 (2005).

[Lon38] F. London, The λ-phenomenon of liquid helium and the Bose–Einstein degeneracy, *Nature* **141**, 643.

[Lon98] M.S. Longair, *Galaxy Formation* (Springer-Verlag, Berlin, 1998).

[LucMat85] F. Lucchin and S. Matarrese, Power-law inflation, *Phys. Rev. D* **32**, 1316 (1985).

[LunRam02] E. Lundh and J. Rammer, Effective action approach to a trapped Bose gas, *Phys. Rev. A* **66**, 33607 (2002).

[LutWar60] J. Luttinger and J. Ward, Ground-state energy of many-fermion systems, *Phys. Rev.* **118**, 1417 (1960).

[LynBel77] D. Lynden-Bell and R.M. Lynden-Bell, On the negative specific heat paradox, *MNRAS* **181**, 405 (1977).

[LytRio99] D.H. Lyth and A. Riotto, Particle physics models of inflation and the cosmological density perturbation, *Phys. Rep.* **314**, 1 (1999).

[Ma76] S.K. Ma, *Modern Theory of Critical Phenomena* (Benjamin, London, 1976).

[Ma85] S.K. Ma, *Statistical Mechanics* (World Scientific, Singapore, 1985).

[Mac74] G.W. Mackey, Ergodic theory and its significance for statistical mechanics and probability theory, *Adv. Maths.* **12**, 178–268 (1974).

[Mac92] M.C. Mackey, *Time's Arrow: The Origin of Thermodynamic Behavior* (Springer-Verlag, New York, 1992).

[Mad27] E. Madelung, Quantentheorie in hydrodynamischer Form, *Zeitschrift fur Physik* **XL**, 322 (1927).

[Mal99] J. Maldacena, The Large-N limit of superconformal field theories and supergravity, *Int. J. Theor. Phys.* **38**, 1113 (1999).

[Man95] L. Mandel and E. Wolf, *Optical Coherence and Quantum Optics* (Cambridge University Press, Cambridge, 1995).

[ManMro06] C. Manuel and S. Mrowcynski, Chromo-hydrodynamic approach to the unstable quark-gluon plasma, hep-ph/0606276.

[MarSch59] P. Martin and J. Schwinger, Theory of many-particle systems I, *Phys. Rev.* **115**, 1342 (1959).

[MarVer99a] R. Martin and E. Verdaguer, On the semiclassical Einstein–Langevin equation, *Phys. Lett.* **465B**, 113 (1999).

[MarVer99b] R. Martin and E. Verdaguer, Stochastic semiclassical gravity, *Phys. Rev. D* **60**, 084008 (1999).

[MarVer99c] R. Martin and E. Verdaguer, An effective stochastic semiclassical theory for the gravitational field, *Int. J. Theor. Phys.* **38**, 3049 (1999).

[MarVer00] R. Martin and E. Verdaguer, Stochastic semiclassical fluctuations in Minkowski spacetime, *Phys. Rev. D* **61**, 124024 (2000).

[Mat93] A.L. Matacz, The emergence of classical behaviour in the quantum fluctuations of a scalar field in an expanding universe, *Class. Quant. Grav.* **10**, 509 (1993).

[Mat94] A.L. Matacz, The coherent state representation of quantum fluctuations in the early universe, *Phys. Rev. D* **49**, 788–798 (1994).

[Mat97a] A. Matacz, A new theory of stochastic inflation, *Phys. Rev. D* **55**, 1860 (1997).

[Mat97b] A. Matacz, Inflation and the fine-tuning problem, *Phys. Rev. D* **56**, 1836 (1997).

[Mat90] S. Matarrese, Statistical properties of curvature perturbations generated during inflation, in *6th Moriond Astrophysics Meeting: The Early Universe and Cosmic Structures*, ed. J.M. Alimi *et al.* Proceedings of the 25th Rencont re de Moriond (Editions Frontieres, Paris, 1990), p. 21.

[MaMuRi04] S. Matarrese, M. A. Musso, and A. Riotto, The influence of super-horizon scales on cosmological observables generated during inflation, *J. Cosmol. Astropart. Phys.* **5**, 8 (2004).

[MazMon70] P. Mazur and E. Montroll, Poincaré cycles, ergodicity, and irreversibility in assemblies of coupled harmonic oscillators, *J. Math. Phys.* **1**, 70–84 (1970).

[MazPaz89] F.D. Mazzitelli and J.P. Paz, Gaussian and $1/N$ approximations in semiclassical cosmology, *Phys. Rev. D* **39**, 2234 (1989).

[MaPaHa89] F.D. Mazzitelli, J.P. Paz and C. El Hasi, Reheating of the Universe and evolution of the inflaton, *Phys. Rev. D* **40**, 955 (1989).

[MazSpe95] F.D. Mazzitelli and F. Spedalieri, Scalar electrodynamics and primordial magnetic fields, *Phys. Rev. D* **52**, 6694 (1995).

[McC94] W.D. McComb, *The Physics of Fluid Turbulence* (Clarendon Press, Oxford, 1994).

[McL72a] D.W. McLaughlin, Path integrals, asymptotics and singular perturbations, *J. Math. Phys.* **13**, 734 (1972).

[McL72b] D.W. McLaughlin, Complex time, contour independent path integrals, and barrier penetration, *J. Math. Phys.* **13**, 1099 (1972).

[McLVen94a] L.D. McLerran and R. Venugopalan, Computing quark and gluon distribution functions for very large nuclei, *Phys. Rev. D* **49**, 2233 (1994).

[McLVen94b] L.D. McLerran and R. Venugopalan, Gluon distribution functions for very large nuclei at small transverse momentum, *Phys. Rev. D* **49**, 3352 (1994).

[McLVen94c] L.D. McLerran and R. Venugopalan, Green's function in the color field of a large nucleus, *Phys. Rev. D* **50**, 2225 (1994).

[MicTka04] R. Micha and I. Tkachev, Turbulent thermalization, *Phys. Rev. D* **70**, 043538 (2004).

[Mig56] A.B. Migdal, Bremsstrahlung and pair production in condensed media at high energies, *Phys. Rev.* **103**, 1811 (1956).

[MACDH00] B. Mihaila, T. Athan, F. Cooper, J. Dawson and S. Habib, Exact and approximated dynamics of the quantum mechanical $O(N)$ model, *Phys. Rev. D* **62**, 125015 (2000).

[MiDaCo01] B. Mihaila, J. Dawson and F. Cooper, Resumming the large-N approximation for time evolving quantum systems, *Phys. Rev. D* **63**, 096003 (2001).

[Mij90] M. Mijic, Random walk after the big bang, *Phys. Rev. D* **42**, 2469 (1990).

[MilWya83] K. Milfeld and R. Wyatt, Study, extension, and application of Floquet theory for quantum molecular systems in an oscillating field, *Phys. Rev. A* **27**, 72 (1983).

[Mil69] R. Mills, *Propagators for Many-particle Systems* (Gordon and Breach, New York, 1969).

[MINIMAX03] MINIMAX Collaboration, Search for disoriented chiral condensate at the Fermilab Tevatron, *Phys. Rev. D* **61**, 032003 (2000).

[MiThWh72] C. Misner, K. Thorne and J.A. Wheeler, *Gravitation* (Freeman, San Francisco, 1972).

[MohSer05] B. Mohanty and J. Serreau, Disoriented chiral condensate: Theory and experiment, *Phys. Rep.* **414**, 263 (2005).

[MMOL91] S. Mollerach, S. Matarrese, A. Ortolan and F. Lucchin, Stochastic inflation in a simple two field model, *Phys. Rev. D* **44**, 1670 (1991).

[Mol67] B.R. Mollow, Quantum statistics of coupled oscillator systems, *Phys. Rev.* **162**, 1256 (1967).

[MolGyu00] D. Molnar and M. Gyulassy, New solutions to covariant nonequilibrium dynamics, *Phys. Rev. C* **62**, 054907 (2000).

[MonPaz01] D. Monteoliva and J.P. Paz, Decoherence in a classically chaotic quantum system: Entropy production and quantum–classical correspondence, quant-ph/0106090.

[Moo05] G.D. Moore, Numerical studies of QGP plasma instabilities and implications, hep-ph/0511203.

[MorCas03] C. Mora and Y. Castin, Extension of Bogoliubov theory to quasicondensates, *Phys. Rev. A* **67**, 053615 (2003).

[MorCal87] C. Morais Smith, A.O. Caldeira, Generalized Feynman–Vernon approach to dissipative quantum systems, *Phys. Rev. A* **36**, 3509 (1987).

[Mor51] C. Morette, On the definition and approximation of Feynman's path integrals, *Phys. Rev.* **81**, 848 (1951).

[Mor99] S. Morgan, A gapless theory of Bose–Einstein condensation in dilute gases at finite temperature, D.Phil. Thesis, Oxford University.

[Mor00] S. Morgan, A gapless theory of Bose–Einstein condensation in dilute gases at finite temperature, *J. Phys. B: At. Mol. Opt. Phys.* **33**, 3847 (2000).

[Mor04] S. Morgan, The response of Bose–Einstein condensates to external perturbations at finite temperature, *Phys. Rev. A* **69**, 023609 (2004).

[MCBE98] S.A. Morgan, S. Choi, K. Burnett and M. Edwards, Nonlinear mixing of quasiparticles in an inhomogeneous Bose condensate, *Phys. Rev. A* **57**, 3818 (1988).

[Mor58] H. Mori, Statistical-mechanical theory of transport in fluids, *Phys. Rev.* **112**, 1829 (1958).

[Mor65] H. Mori, Transport, collective motion, and Brownian motion, *Prog. Theor. Phys.* **423**, 1338 (1965).

[Mor86] M. Morikawa, Classical fluctuations in dissipative quantum systems, *Phys. Rev. D* **33**, 3607 (1986).

[Mor90] M. Morikawa, Dissipation and fluctuation of quantum fields in expanding universes, *Phys. Rev. D* **42**, 1027 (1990).

[MorSas84] M. Morikawa and M. Sasaki, Entropy production in the inflationary universe, *Prog. Theor. Phys.* **72**, 782 (1984).

[MMNH02] K. Morita, S. Muroya, C. Nonaka and T. Hirano, Comparison of space-time evolutions of hot/dense matter in $\sqrt{s_{NN}}=17$ and 130 GeV relativistic heavy ion collisions based on a hydrodynamical model, *Phys. Rev. C* **66**, 054904 (2002).

[MorRop99] V. Morozov and G. Röpke, The mixed Green's function approach to QK with initial correlations, *Ann. Phys.* **278**, 127 (1999).

[Mos02] I. Moss, Derivative expansions of the non-equilibrium effective action, *Nucl. Phys.* **B631**, 500 (2002).

[Mot86] E. Mottola, Quantum fluctuation-dissipation theorem for general relativity, *Phys. Rev. D* **33**, 2136 (1986).

[Mot03] E. Mottola, Gauge invariance in 2PI effective actions, Proceedings of SEWM 2002, hep-ph/0304279.

[Moy49] J.E. Moyal, Stochastic processes and statistical physics, *J. R. Stat. Soc. London, Ser. B* **11**, 150 (1949).

[Mro89] S. Mrowczynski, Kinetic-theory approach to quark-gluon plasma oscillations, *Phys. Rev. D* **39**, 1940 (1989).

[Mro94a] S. Mrowczynski, Plasma instability at the initial stage of ultrarelativistic heavy-ion collisions, *Phys. Lett.* **314B**, 118 (1994).

[Mro94b] S. Mrowczynski, Color collective effects at the early stage of ultrarelativistic heavy-ion collisions, *Phys. Rev. C* **49**, 2191 (1994).

[Mro97] S. Mrowczynski, Transport theory of massless fields, *Phys. Rev. D* **56**, 2265 (1997).

[Mro05] S. Mrowczynski, Instabilities driven equilibration of the quark-gluon plasma, hep-ph/0511052.

[Mro06] S. Mrowcynski, Early stage thermalization via instabilities, hep-ph/0611067.

[Mro07] S. Mrowczynski, private communication.

[MroDan90] S. Mrowczynski and P. Danielewicz, Green function approach to transport theory of scalar fields, *Nucl. Phys.* **B342**, 345 (1990).

[MroHei94] St. Mrowczynski and U. Heinz, Towards relativistic transport-theory of nuclear matter, *Ann. Phys.* **229**, 1 (1994).

[MroMul95] S. Mrowczynski and B. Muller, Reheating after supercooling in the chiral phase transition, *Phys. Lett.* **363B**, 1 (1995).

[Mue00a] A.H. Mueller, The Boltzmann equation for gluons at early times after a heavy ion collision, *Phys. Lett.* **475B**, 220 (2000).

[Mue00b] A.H. Mueller, Toward equilibration in the early stages after a high energy heavy ion collision, *Nucl. Phys.* **B572**, 227 (2000).

[Mue01] A.H. Mueller, Parton saturation – An overview, hep-ph/0111244.

[MuShWo05] A.H. Mueller, A.I. Shoshi and S.M.H. Wong, A modified bottom-up thermalization in heavy ion collisions, *Eur. Phys. J.* **A29**, 49 (2006).

[MuShWo07] A.H. Mueller, A.I. Shoshi and S.M.H. Wong, On Kolmogorov wave turbulence in QCD, *Nucl. Phys.* **B760**, 145 (2007).

[MueSon04] A.H. Mueller and D.T. Son, On the equivalence between the Boltzmann equation and classical field theory at large occupation numbers, *Phys. Lett.* **582B**, 279 (2004).

[Muk05] V. Mukhanov, *Physical Foundations of Cosmology* (Cambridge University Press, Cambridge, 2005).

[MukChi82] V.F. Mukhanov and V.F. Chibisov, Vacuum energy and large-scale structure of the universe, *JETP* **83**, 475–487 (1982).

[MuFeBr92] V. Mukhanov, H. Feldman and R. Brandenberger, Theory of cosmological perturbations, *Phys. Rep.* **215**, 203 (1992).

[MulNag06] B. Muller and J. Nagle, Results from the Relativistic Heavy Ion Collider, nucl-th/0602029 (Ann. Rev. Nucl. and Part. Phys. 2006).

[NaOrRe94] G. Nagy, O. Ortiz and O. Reula, The behavior of hyperbolic heat equations' solutions near their parabolic limits, *J. Math. Phys.* **35**, 4334 (1994).

[Nak58] S. Nakajima, On quantum theory of transport phenomena – steady diffusion, *Progr. Theor. Phys.* **20**, 948 (1958).

[Nak66] N. Nakanishi, Covariant quantization of the electromagnetic field in the Landau gauge, *Prog. Theor. Phys.* **35**, 1111 (1966).

[NaNaSa88] K. Nakao, Y. Nambu and M. Sasaki, Stochastic dynamics of new inflation, *Prog. Theor. Phys.* **80**, 1041 (1988).

[Nam89] Y. Nambu, Stochastic dynamics of an inflationary model and initial distribution of universes, *Prog. Theor. Phys.* **81**, 1037 (1989).

[NamSas89] Y. Nambu and M. Sasaki, Stochastic approach to chaotic inflation and the distribution of universes, *Phys. Lett.* **219B**, 240 (1989).

[NegOrl98] J. Negele and H. Orland, *Quantum Many-particle Systems* (Perseus, New York, 1998).

[NeuHil27] J. von Neumann and D. Hilbert, Uber die Grundlagen der Quantenmechanik, Mathematische Annalen (1927), in von Neumann's *Collected Works* (Pergamon, Oxford, 1961), **1**, 111.

[NiCsKi98] S. Nickerson, T. Csorgo and D. Kiang, Testing the core-halo model on Bose–Einstein correlation functions, *Phys. Rev. C* **57**, 3251 (1998).

[Nic95] G. Nicolis, *Introduction to Nonlinear Science* (Cambridge University Press, Cambridge, 1995).

[Nie02] A. Niegawa, Fermion propagator in out of equilibrium quantum-field system and the Boltzmann equation, *Phys. Rev. D* **65**, 56009 (2002).

[Nie75] N.K. Nielsen, On the gauge dependence of spontaneous symmetry breaking in gauge theories, *Nucl. Phys.* **B101**, 173 (1975).

[NieSem84a] A.J. Niemi and G.W. Semenoff, Finite-temperature quantum field theory in Minkowski space, *Ann. Phys.* (N.Y.) **152**, 105 (1984).

[NieSem84b] A.J. Niemi and G.W. Semenoff, Thermodynamic calculations in relativistic finite-temperature quantum field theories, *Nucl. Phys.* **B230**, 181 (1984).

[NobaGa05] A.A. Norrie, R.J. Ballagh and C.W. Gardiner, Quantum turbulence in condensate collisions: an application of the classical field method, *Phys. Rev. Lett.* **94**, 040401 (2005).

[NobaGa06] A.A. Norrie, R.J. Ballagh and C.W. Gardiner, Quantum turbulence and correlations in Bose–Einstein condensate collisions, *Phys. Rev. A* **73**, 043617 (2006).

520 *References*

[NorCor75] R. Norton and J. Cornwall, On the formalism of relativistic many body theory, *Ann. Phys.* (N.Y.) **91**, 106 (1975).

[Nov65] E.A. Novikov, Functionals and the random-force method in turbulence theory, *Sov. Phys. JETP* **20**, 1290 (1965).

[Nyq28] H. Nyquist, Thermal agitation of electric charge in conductors, *Phys. Rev.* **32**, 110 (1928).

[OCoSte94a] D.J. O'Connor and C.R. Stephens, "Environmentally friendly" renormalization, *Int. J. Mod. Phys. A* **9**, 2805 (1994).

[OCoSte94b] D.J. O'Connor and C.R. Stephens, Effective critical exponents for dimensional crossover and quantum systems from an environmentally friendly renormalization group, *Phys. Rev. Lett.* **72**, 506 (1994).

[Oji81] I. Ojima, Gauge fields at finite temperature: Thermo field dynamics and the KMS condition and their extension to gauge theories, *Ann. Phys.* (N.Y.) **137**, 1 (1981).

[OTZH04] A. de Oliveira-Costa, M. Tegmark, M. Zaldarriaga and A. Hamilton, The significance of the largest scale CMB fluctuations in WMAP, *Phys. Rev. D* **69**, 063516 (2004).

[Omn88] R. Omnès, Logical reformulation of quantum mechanics I. Foundations, *J. Stat. Phys.* **53**, 893, 933, 957 (1988).

[Omn90] R. Omnès, From Hilbert space to common sense: A synthesis of recent progress in the interpretation of quantum mechanics, *Ann. Phys.* (N.Y.) **201**, 354 (1990).

[Omn92] R. Omnès, Consistent interpretations of quantum mechanics, *Rev. Mod. Phys.* **64**, 339 (1992).

[Omn94] R. Omnès, *The Interpretation of Quantum Mechanics* (Princeton University Press, Princeton, NJ, 1994).

[Oos02] D. van Oosten, P. van der Straten and H.T.C. Stoof, Mott insulators in an optical lattice with high filling factors, cond-mat/0205066.

[OrbBel67] J. Orban and A. Belleman, Velocity-inversion and irreversibility in a dilute gas of hard disks, *Phys. Lett.* **24A**, 620–1 (1967).

[Pag93] D.N. Page, Average entropy of a subsystem, *Phys. Rev. Lett.* **71**, 1291 (1993).

[Pap84] A. Papooulis, *Probability of Random Variables and Stochastic Processes* (McGraw-Hill, New York, 1984).

[Par04] B. Paredes *et al.*, Tonks–Girardeau gas of ultracold atoms in an optical lattice, *Nature* **429**, 277 (2004).

[Par88] G. Parisi, *Statistical Field Theory* (Addison-Wesley, Reading, Massachusetts, 1988).

[Park66] L. Parker, The creation of particles in an expanding universe. Ph.D. Thesis, Harvard University (1966).

[Park68] L. Parker, Particle creation in expanding universes, *Phys. Rev. Lett.* **21**, 562 (1968).

[Park69] L. Parker, Quantized fields and particle creation in expanding universes. I, *Phys. Rev.* **183**, 1057 (1969).

[Park71] L. Parker, Quantized fields and particle creation in expanding universes. II, *Phys. Rev. D* **3**, 346 (1971).

[Park76] L. Parker, Thermal radiation produced by the expansion of the universe, *Nature* **261**, 20 (1976).

[Park77] L. Parker, The production of elementary particles in strong gravitational fields, in *Asymptotic Structure of Space-Time*, ed. F.P. Esposito and L. Witten (Plenum Press, New York, 1977).

[ParFul73] L. Parker and S.A. Fulling, Quantized matter fields and the avoidance of singularities in general relativity, *Phys. Rev. D* **7**, 2357 (1973).

[ParFul74] L. Parker and S.A. Fulling, Adiabatic regularization of the energy-momentum tensor of a quantized field in homogeneous spaces, *Phys. Rev. D* **9**, 341 (1974).

[ParSim93] L. Parker and J. Simon, Einstein equations with quantum corrections reduced to second order, *Phys. Rev. D* **47**, 1339 (1993).

[ParZha93] L. Parker and Y. Zhang, Relativistic condensate as a source for inflation, *Phys. Rev. D* **47**, 416 (1993).

[ParZha95] L. Parker and Y. Zhang, Cosmological perturbations of a relativistic condensate, *Phys. Rev. D* **51**, 2703 (1995).

[PLURH00] H. Pastawski, P. Levstein, G. Usaj, J. Raya and J. Hirschinger, A nuclear magnetic resonance answer to the Boltzmann–Loschmidt controversy?, *Physica A* **283**, 166 (2000).

[Paz90a] J.P. Paz, Anisotropy dissipation in the early universe: Finite-temperature effects reexamined, *Phys. Rev. D* **41**, 1054 (1990).

[Paz90b] J.P. Paz, Dissipative effects during the oscillations around a true vacuum, *Phys. Rev. D* **42**, 529 (1990).

[Paz94] J.P. Paz, Decoherence in quantum Brownian motion, in *The Physical Origin of Time-Asymmetry*, Huelva, Spain, 1991 ed. J.J. Halliwell, J. Perez-Mercader and W.H. Zurek (Cambridge University Press, Cambridge, 1994).

[PaHaZu93] J.P. Paz, S. Habib and W.H. Zurek, Reduction of the wave packet: Preferred observable and decoherence time scale, *Phys. Rev. D* **47**, 488 (1993).

[Paz00] J.P. Paz and W. Zurek, Environment-induced decoherence and the transition from quantum to classical, lectures at the 72nd Les Houches Summer School "Coherent Matter Waves" (1999) (quant-ph/0010011).

[Pee80] P.J.E. Peebles, *Large Scale Structure of the Universe* (Princeton University Press, Princeton, NJ, 1980).

[PeeVil99] P.J.E. Peebles and A. Vilenkin, Quintessential inflation, *Phys. Rev. D* **59**, 063505 (1999).

[Pen70] O. Penrose, *Foundations of Statistical Mechanics* (Pergamon Press, Oxford, 1970).

[Pen79] O. Penrose, Foundations of statistical mechanics, *Rep. Prog. Theor. Phys.* **42**, 1937–2006 (1979).

[PenOns56] O. Penrose and L. Onsager, Bose–Einstein condensation and liquid helium, *Phys. Rev.* **104**, 576 (1956).

[Per98] I. Percival, *Quantum State Diffusion* (Cambridge University Press, Cambridge, 1998).

[Per93] A. Peres, *Quantum Theory: Concepts and Methods* (Kluwer, Dordrecht, 1993).

[PesSch95] M. Peskin and D. Schroeder, *An Introduction to Quantum Field Theory* (Addison-Wesley, New York, 1995).

[PetSmi02] C. Pethick and H. Smith, *Bose–Einstein Condensation in Dilute Gases* (Cambridge University Press, Cambridge, 2002).

[Pet69] A.Z. Petrov, *Einstein Spaces* (Pergamon Press, Oxford, 1969).

[PHENIX05] PHENIX Collaboration, Formation of dense partonic matter in relativistic nucleus–nucleus collisions at RHIC: Experimental evaluation by the PHENIX collaboration, *Nucl. Phys.* **A757**, 184 (2005).

[PHOBOS05] PHOBOS Collaboration, The PHOBOS perspective on discoveries at RHIC, *Nucl. Phys.* **A757**, 28 (2005).

[PinRam06] M. Pinto and R. Ramos, Inverse symmetry breaking in multi-scalar field theories, *J. Phys. A* **39**, 6649–6655 (2006).

[PRSC02] M. Pinto, R. Ramos and F. de Souza Cruz, Unusual transition patterns in Bose–Einstein condensation, cond-mat/0204416.

[Pis89a] R. Pisarski, Renormalized gauge propagator in hot gauge theories, *Physica A* **158**, 146 (1989).

[Pis89b] R. Pisarski, How to compute scattering amplitudes in hot gauge theories, *Physica A* **158**, 246 (1989).

[Pla59] M. Planck, *The Theory of Heat Radiation* (Dover, New York, 1959).

[PFKP06] D. Podolsky, G. Felder, L. Kovman and M. Peloso, Equation of state and beginning of thermalization after preheating, *Phys. Rev. D* **73**, 023501 (2006).

[PolSta96] D. Polarski and A.A. Starobinsky, Semiclassicality and decoherence of cosmological perturbations, *Class. Quantum Grav.* **13**, 377 (1996).

[PoSoSt02a] G. Policastro, D.T. Son and A.O. Starinets, From AdS/CFT correspondence to hydrodynamics, *JHEP* **0209**, 43 (2002).

[PoSoSt02b] G. Policastro, D.T. Son and A.O. Starinets, From AdS/CFT correspondence to hydrodynamics, II. Sound waves, *JHEP* **0212**, 54 (2002).

[Pol06] J. Polonyi, Quantum-classical crossover in electrodynamics, *Phys. Rev. D* **74**, 065014 (2006).

[Pri62] I. Prigogine, *Nonequilibrium Statistical Mechanics* (John Wiley, New York, 1962).

[Pri73] I. Prigogine, Irreversibility as a symmetry-breaking process, *Nature* **246**, 67–71 (1973).

[PrBuSt98] N.P. Proukakis, K. Burnett and H.T.C. Stoof, Microscopic treatment of binary interactions in the nonequilibrium dynamics of partially Bose-condensed trapped gases, *Phys. Rev. A* **57**, 1230 (1998).

[Pup04] G. Pupillo *et al.*, Scalable quantum computation in systems with Bose–Hubbard dynamics, *J. Mod. Opt.* **51**, 16 (2004).

[PuWiPr06] G. Pupillo, C. Williams and N. Prokof'ev, Effects of finite temperature on the Mott insulator state, *Phys. Rev. A* **73**, 013408 (2006).

[RajMar82] A.K. Rajagopal and J.T. Marshall, New coherent states with applications to time-dependent systems, *Phys. Rev. A* **26**, 2977 (1982).

[Raj87] R. Rajaraman, *Solitons and Instantons* (Elsevier, Amsterdam, 1987).

[Ram98] J. Rammer, *Quantum Transport Theory* (Perseus, MA, 1998).

[Ram80] P. Ramond, *Field Theory, a Modern Primer* (Addison-Wesley, New York, 1980).

[RamNav00] R. Ramos and F. Navarro, Chaotic symmetry breaking and dissipative two-field dynamics, *Phys. Rev. D* **62**, 85016 (2000).

[RamHu97a] S. Ramsey and B.L. Hu, O(N) quantum fields in curved spacetime, *Phys. Rev. D* **56**, 661 (1997).

[RamHu97b] S. Ramsey and B.L. Hu, Nonequilibrium inflaton dynamics and reheating: Back reaction of parametric particle creation and curved spacetime effects, *Phys. Rev. D* **56**, 678 (1997).

[RaStHu98] S. Ramsey, A. Stylianopoulos and B.L. Hu, Nonequilibrium inflaton dynamics and reheating. II. Fermion production, noise and stochasticity, *Phys. Rev.* **D57**, 6003 (1998).

[Ran97] J. Randrup, Mean-field treatment of the linear σ model in dynamical calculation of DCC observables, *Nucl. Phys. A* **616**, 531 (1997).

[Rau94] J. Rau, Pair production in the quantum Boltzmann equation, *Phys. Rev. D* **50**, 6911 (1994).

[RauMue96] J. Rau and B. Mueller, From reversible quantum microdynamics to irreversible quantum transport, *Phys. Rep.* **272**, 1 (1996).

[RaHuAn96] A. Raval, B.L. Hu and J. Anglin, Stochastic theory of accelerated detectors in quantum fields, *Phys. Rev. D* **53**, 7003–7019 (1996).

[RaHuKo97] A. Raval, B.L. Hu and D. Koks, Near-thermal radiation in detectors, mirrors, and black holes: A stochastic approach, *Phys. Rev. D* **55**, 4795 (1997).

[Reb87] A. Rebhan, The Vilkovisky–DeWitt effective action and its application to Yang–Mills theories, *Nucl. Phys.* **B288**, 832 (1987).

[Reg04] C.A. Regal, M. Greiner and D.S. Jin, Observation of resonance condensation of fermionic atom pairs, *Phys. Rev. Lett.* **92**, 040403 (2004).

[Rei98] L. Reichl, *A Modern Course in Statistical Physics* (John Wiley, New York, 1998).

[Rei65] F. Reif, *Fundamentals of Statistical and Thermal Physics* (McGraw-Hill, New York, 1965).

[Rei67] F. Reif, *Statistical Physics – Berkeley Physics Course* (McGraw-Hill, New York, 1967).

[ReiToe94] P.-G. Reinhard and C. Toeffer, Correlations in nuclei and nuclear dynamics, *Int. J. Mod. Phys. E* **3**, 436 (1994).

[ReiSer06] U. Reinosa and J. Serreau, 2PI effective action for gauge theories: renormalization, *JHEP* **0607**, 028 (2006).

[Rei05] A. Reischl, K. Schmidt and G. Uhrig, Temperature in one-dimensional bosonic Mott insulators, *Phys. Rev. A* **72**, 063609 (2005).

[Rey04] A.M. Rey, Ultra cold bosonic atoms in optical lattices, D.Phil. Thesis, Maryland University at College Park (2004).

[ReBlCl03] A.M. Rey, P.B. Blakie and C.W. Clark, Dynamics of a period 3 pattern-loaded BEC in an optical lattice, *Phys. Rev. A* **67**, 053610 (2003).

[RHCC05] A.M. Rey, B.L. Hu, E. Calzetta and C. Clark, Quantum kinetic theory of a Bose–Einstein gas confined in a lattice, *Phys. Rev. A* **72**, 023604 (2005).

[RHCRC04] A.M. Rey, B.L. Hu, E. Calzetta, A. Roura and C. Clark, Nonequilibrium dynamics of optical lattice-loaded BEC atoms: Beyond HFB approximation, *Phys. Rev. A* **69**, 033610 (2004).

[Rey87] S.J. Rey, Dynamics of inflationary phase transition, *Nucl. Phys.* **B284**, 706 (1987).

[RhPiWa71] W.-K. Rhim, A. Pines and J. Waugh, Violation of the spin-temperature hypothesis, *Phys. Rev. Lett.* **25**, 218–220 (1971).

[Rio02] A. Riotto, "Inflation and the theory of cosmological perturbations," Lecture given at ICTP Summer School on Astroparticle Physics and Cosmology, Trieste, Italy, 17 Jun – 5 Jul 2002, published in Trieste 2002, *Astroparticle physics and cosmology*, 317–413.

[Ris98] D. Rischke, Fluid dynamics for relativistic nuclear collisions, Proceedings of the 11th Chris Engelbrecht Summer School in Theoretical Physics, Cape Town, Feb. 4–13, 1998 (nucl-th/9809044).

[Ris89] H. Risken, *The Fokker–Planck Equation* (Springer-Verlag, Berlin, 1989).

[Riv01] R.J. Rivers, Zurek–Kibble causality bounds in time-dependent Ginzburg–Landau theory and quantum field theory, *J. Low T. Phys.* **124**, 41 (2001).

[RiKaKa00] R.J. Rivers, E. Kavoussanaki, and G. Karra, The onset of phase transitions in condensed matter and relativistic QFT, *Cond. Matt. Phys.* **3**, 133 (2000).

[RivLom05] R.J. Rivers and F.C. Lombardo, Onset of classical behaviour after a phase transition, *Braz. J. Phys.* **35**, 397–402 (2005).

[RiLoMa02] R.J. Rivers, F.C. Lombardo, and F.D. Mazzitelli, The formation of classical defects after a slow quantum phase transition, *Phys. Lett.* **B539**, 1 (2002).

[Rob05] D.C. Roberts, Probing Landau and Beliav damping rates in Bose–Einstein condensates using ultraslow light experiments, *Phys. Rev. A* **72**, 065602 (2005).

[Rom69] P. Roman, *Introduction to Quantum Field Theory* (John Wiley, New York, 1969).

[RomReb06] P. Romatschke and A. Rebhan, Plasma instabilities in an anisotropically expanding geometry, hep-ph/0605064.

[RomVen06] P. Romatschke and R. Venugopalan, The unstable glasma, hep-ph/0605045.

[Roo74] R. Root, Effective potential for O(N) model to order 1/N, *Phys. Rev. D* **10**, 3322 (1974).

[Rou02] A. Roura, Descripció estocàstica de sistemes quàntics oberts i pertorbacions cosmològiques, Ph.D. Thesis, University of Barcelona (2001).

[RouVer99] A. Roura and E. Verdaguer, Mode decomposition and renormalization in semiclassical gravity, *Phys. Rev. D* **60**, 107503 (1999).

[Rub60] R.J. Rubin, Statistical dynamics of simple cubic lattices. Model for the study of Brownian motion, *J. Math. Phys.* **1**, 309 (1960).

[Rub61] R.J. Rubin, Statistical dynamics of simple cubic lattices. Model for the study of Brownian motion. II, *J. Math. Phys.* **2**, 373 (1961).

[Rue76] D. Ruelle, A measure associated with Axiom-A attractors, *Am. J. Math*, **98**, 619 (1976).

[RueEck85] D. Ruelle and J.P. Eckmann, Ergodic theory of chaos and strange attractors, *Rev. Mod. Phys.* **57**, 617 (1985).

[RueSin86] D. Ruelle and Ya.G. Sinai, From dynamical systems to statistical mechanics and back, *Physica A* **140**, 1 (1986).

[Sac99] S. Sachdev, *Quantum Phase Transitions* (Cambridge University Press, Cambridge, 1999).

[SaiUed03] H. Saito and M. Ueda, A consistent picture of a collapsing Bose–Einstein condensate, *J. Phys. Soc. Japan* **72** (Suppl. C), 127 (2003).

[Sak66] A.D. Sakharov, The initial stage of an expanding universe and the appearance of a nonuniform distribution of matter, *Sov. Phys. JETP* **22**, 241 (1966).

[SalBon91] D.S. Salopek and J.R. Bond, Stochastic inflation and nonlinear gravity, *Phys. Rev. D* **43**, 1005 (1991).

[SanMig89] J.M. Sancho and M. San Miguel, Langevin equations with colored noise, in *Noise in Nonlinear Dynamical Systems: Theory Experiment, Simulation*, vol 1, ed. F. Moss and P.V.E. McClintock (Cambridge University Press, Cambridge, 1989), pp. 110–160.

[SanShl02] L. Santos and G. Shlyapnikov, Collapse dynamics of trapped Bose–Einstein condensates, *Phys. Rev. A* **66**, 011602(R) (2002).

[Sato81] K. Sato, Cosmological baryon-number domain structure and the first order phase transition of a vacuum, *Phys. Lett.* **99B**, 66 (1981).

[Sau31] F. Sauter, Über das Verhalten eines Elektrons in homogen elektrischen Feld nach der relativistischen Theorie Diracs, *Z. Phys.* **69**, 742 (1931).

[Sau32] F. Sauter, Zum 'kleinschen Paradoxon', *Z. Phys.* **73**, 547 (1932).

[SaRoHo03] C. Savage, N. Robins and J. Hope, Bose–Einstein condensate collapse: A comparison between theory and experiment, *Phys. Rev. A* **67**, 014304 (2003).

[SaOHTh97] T. Savard, K. O'Hara and J. Thomas, Laser-noise-induced heating in far-off resonance optical traps, *Phys. Rev. A* **56**, R1095 (1997).

[SSMKE04] C. Schori *et al.*, Excitations of a superfluid in a 3D optical lattice, *Phys. Rev. Lett.* **93**, 240402 (2004).

[Sch81] L.S. Schulman, *Techniques and Applications of Path Integration* (John Wiley, New York, 1981).

[Sch97] L.S. Schulman, *Time's Arrow and Quantum Measurement* (Cambridge University Press, Cambridge, 1997).

[Sch86] B.L. Schumacher, Quantum mechanical pure states with Gaussian wave functions, *Phys. Rep.* **135**, 317 (1986).

[Sch51] J. Schwinger, On gauge invariance and vacuum polarization, *Phys. Rev.* **82**, 664 (1951).

[Sch60] J. Schwinger, Field theory methods in non-field-theory contexts, Brandeis 1960 Summer Institute in Theoretical Physics.

[Sch61] J. Schwinger, Brownian motion of a quantum oscillator, *J. Math. Phys.* **2**, 407 (1961).

[Sch70] J. Schwinger, *Particles, Sources and Fields*, vol. 1 (Addison-Wesley, Reading, MA, 1970).

[Sci79] D.W. Sciama, Thermal and quantum fluctuations in special and general relativity: An Einstein synthesis, in *Relativity, Quanta and Cosmology – Centenario di Einstein* ed. F. DeFinis (Editrici Giunti Barbera Universitaria, Florence, Italy, 1979).

[ScHuGa06] R.G. Scott, D.A.W. Hutchinson and C.W. Gardiner, Disruption of reflecting Bose–Einstein condensates due to inter-atomic interactions and quantum noise, cond-mat/0608135.

[SelCre06] U. Seljak and P. Creminelli, ICTP Workshop on non-Gaussianity in Cosmology, July 2006.

[SelZal96] U. Seljak and M. Zaldarriaga, A line of sight integration approach to cosmic microwave background anisotropies, *Astrophys. J.* **469**, 437 (1996).

[SeKrBo00] D. Semkat, D. Kremp and M. Bonitz, Kadanoff–Baym equations and non-Markovian BE in generalized T-matrix approximation, *J. Math. Phys.* **41**, 7458 (2000).

[SexUrb69] R.U. Sexl and H.K. Urbantke, Production of particles by gravitational fields, *Phys. Rev.* **179**, 1247 (1969).

[ShiGri98] H. Shi and A. Griffin, Finite-temperature excitations in a dilute Bose-condensed gas, *Phys. Rep.* **304**, 1 (1998).

[Shi65] J. Shirley, Solution of the Schrodinger equation with a Hamiltonian periodic in time, *Phys. Rev.* **138**, B979 (1965).

[ShTrBr95] Y. Shtanov, J. Traschen and R. Brandenberger, Universe reheating after inflation, *Phys. Rev. D* **51**, 5438–5455 (1995).

[Shu88] E.V. Shuryak, *The QCD Vacuum, Hadrons and the Superdense Matter* (World Scientific, Singapore, 1988).

[Sinai72] Ya.G. Sinai, Gibbs measures in ergodic theory, *Russ. Math. Surveys* **27**, 21 (1972).

[SiCaWi06] A. Sinatra, Y. Castin and E. Witkowska, Non-diffusive phase spreading of a Bose–Einstein condensate at finite temperature, *Phys. Rev. A* **75**, 0033616 (2007).

[Sin97] S. Sinha, Decoherence at absolute zero, *Phys. Lett. A* **228** (1997).

[SinHu89] S. Sinha and B.L. Hu, Symmetry behavior in cosmological spacetimes: Effect of slowly varying background fields, *Phys. Rev. D* **38**, 2423 (1989).

[SinHu91] S. Sinha and B.L. Hu, Validity of the minisuperspace approximation: An example from interacting quantum field theory, *Phys. Rev. D* **44**, 1028 (1991).

[Sin99] Y. Sinyukov, Boson spectra and correlations for thermally locally equilibrium systems, *Heavy Ion Phys.* **10**, 113–136 (1999).

[SiAkHa02] Y. Sinyukov, S. Akkelin and Y. Hama, On freeze-out problem in hydro-kinetic approach to A+A collisions, *Phys. Rev. Lett.* **89**, 052301 (2002).

[SRSBTP97] S.A. Smolyansky, G. Roepke, S. Schmidt, D. Blaschke, V.D. Toneev and A.V. Prozorkevich, Dynamical derivation of a quantum kinetic equation for particle production in the Schwinger mechanism, hep-ph/9712377 (unpublished).

[SoBaDi01] S. Soff, S. Bass and A. Dumitru, Pion Interferometry at RHIC: Probing a thermalized quark-gluon plasma?, *Phys. Rev. Lett.* **86**, 3981 (2001).

[Son96] D.T. Son, Reheating and thermalization in a simple scalar model, *Phys. Rev. D* **54**, 3745 (1996).

[Son97] D.T. Son, Effective non-perturbative real-time dynamics of soft modes in hot gauge theories, hep-ph/9707351.

[Sou02] K. Southwell (ed.), Nature insight: Ultracold matter, *Nature* **416**, 205 (2002).

[SPRS02] F. de Souza Cruz, M. Pinto, R. Ramos and P. Sena, Higher order evaluation of the critical temperature for interacting homogeneous dilute Bose gases, *Phys. Rev. A* **65**, 053613 (2002).

[Spe03] D.N. Spergel *et al.*, First year Wilkinson Microwave Anisotropy Probe (WMAP) observations: Determination of cosmological parameters, *Astrophys. J. Suppl.* **148**, 175 (2003).

[Spo91] H. Spohn, *Large Scale Dynamics of Interacting Particles* (Springer-Verlag, Berlin, 1991).

[STAR05] STAR Collaboration, Experimental and theoretical challenges in the search for the quark-gluon plasma: The STAR Collaboration's critical assesment of the evidence from RHIC collisions, *Nucl. Phys.* **A757**, 102 (2005).

[Sta82] A.A. Starobinsky, Dynamics of phase transition in the new inflationary universe scenario and generation of perturbations, *Phys. Lett.* **117B**, 175 (1982).

[Sta86] A.A. Starobinsky, Stochastic De Sitter inflationary stage in the early universe in *Field Theory, Quantum Gravity and Strings*, ed. H. J. de Vega and N. Sanchez (Springer-Verlag, Berlin, 1986).

[StaYok94] A.A. Starobinsky and J. Yokoyama, Equilibrium state of a self-interacting scalar field in the de Sitter background, *Phys. Rev. D* **50**, 6357–6368 (1994).

[Ste98] M.J. Steel *et al.*, Dynamical quantum noise in trapped Bose–Einstein condensates, *Phys. Rev. A* **58**, 4824 (1998).

[SteTur84] P.J. Steinhardt and M.S. Turner, Prescription for successful new inflation, *Phys. Rev. D* **29**, 2162–2171 (1984).

[Ste98] C.R. Stephens, Why two renormalization groups are better than one, *Int. J. Mod. Phys.* **B12**, 1379–1396 (1998).

[Ste00] G.J. Stephens, Unraveling critical dynamics: The formation and evolution of textures, *Phys. Rev. D* **61**, 085002 (2000).

[StBeZu02] G. Stephens, L. Bettencourt and W. Zurek, Critical dynamics of gauge systems: spontaneous vortex formation in 2D superconductors, *Phys. Rev. Lett.* **88**, 137004 (2002).

[SCHR99] G. Stephens, E. Calzetta, B.L. Hu and S. Ramsey, Defect formation and critical dynamics in the early Universe, *Phys. Rev. D* **59**, 045009 (1999).

[Ste02] E.D. Stewart, Spectrum of density perturbations produced during inflation to leading order in a general slow-roll approximation, *Phys. Rev. D* **65**, 103508 (2002).

[SMSKE04] T. Stoferle *et al.*, Transition from a strongly interacting 1D superfluid to a Mott insulator, *Phys. Rev. Lett.* **92**, 130403 (2004).

[Sto99] H. Stoof, Field theory for trapped atomic gases, *J. Low Temp. Phys.* **114**, 11 (1999) (cond-mat/9910441).

[StrWig80] R.F. Streater and A.S. Wightman, *PCT, Spin and Statistics, and All That* (Benjamin/Cummings, New York, 1980).

[Str06] M. Strickland, Thermalization and plasma instabilities, hep-ph/0608173.

[StrVaf96] A. Strominger and C. Vafa, Mirror symmetry is T duality, spectroscopy of the quantum black hole, *Nucl. Phys.* **B479**, 243 (1996).

[SCYC88] Z. Su, L.Y. Chen, X. Yu and K. Chou, Influence functional and closed-time-path Green's function, *Phys. Rev. B* **37**, 9810 (1988).

[Sun82] K. Sundermeyer, *Constrained Dynamics* (Springer-Verlag, Berlin, 1982).

[TayWon90] J.C. Taylor and S.M. Wong, The effective action of hard thermal loops in QCD, *Nucl. Phys.* **B346**, 115 (1990).

[Tea03] D. Teaney, The effect of shear viscosity on spectra, elliptic flow, and HBT radii, *Phys. Rev. C* **68**, 034913 (2003).

[TeLaSh01] D. Teaney, J. Lauret and E.V. Shuryak, A hydrodynamic description of heavy ion collisions at the SPS and RHIC, nucl-th/0110037.

[TeCoHa03] M. Tegmark, A. de Oliveira-Costa and A. Hamilton, A high resolution foreground cleaned CMB map from WMAP, *Phys. Rev. D* **68**, 123523 (2003).

[TemGas06] K. Temme and T. Gasenzer, Non-equilibrium dynamics of a Bose Einstein condensate in an optical lattice in the 2PI $1/N$ approach, cond-mat/0607116.

[Tin96] M. Tinkham, *Introduction to Superconductivity* (McGraw-Hill, New York, 1996).

[TsaKan05] C. Tsagas and A. Kandus, Superadiabatic-type magnetic amplification in conventional cosmology, *Phys. Rev. D* **71**, 123506 (2005).

[TsaWoo05] N.C. Tsamis and R.P. Woodard, Stochastic quantum gravitational inflation, *Nucl. Phys. B* **724**, 295–328 (2005).

[TsuGri03] S. Tsuchiya and A. Griffin, Damping of Bogoliubov excitations in optical lattices, cond-matt/0311321.

[TsuGri05] S. Tsuchiya and A. Griffin, Landau damping of Bogoliubov excitations in optical lattices at finite temperature, cond-mat/0506016.

[TurWid88] M.S. Turner and L.M. Widrow, Inflation-produced, large-scale magnetic fields, *Phys. Rev. D* **37**, 2743 (1988).

[UedSai03] M. Ueda and H. Saito, A consistent picture of a collapsing Bose–Einstein condensate, *J. Phys. Soc. Jpn.* **72** (Suppl. C), 127 (2003).

[UmMaTa82] H. Umezawa, H. Matsumoto and M. Tachiki, *Thermo Field Dynamics and Condensed States* (North-Holland, Amsterdam, 1982).

[UMDdemo] University of Maryland Lecture Demonstration F4-12. http://www.physics.umd.edu/deptinfo/facilities/lecdem/f4-12.html.

[Unr76] W.G. Unruh, Notes on black-hole evaporation, *Phys. Rev. D* **14**, 870 (1976).

[UnrSch07] W.G. Unruh and R. Schutzhold (eds.), *Quantum Analogues from Phase Transitions to Black holes and Cosmology* (LNP718, Springer, New York 2007).

[UnrZur89] W.G. Unruh and W.H. Zurek, Reduction of a wave packet in quantum Brownian motion, *Phys. Rev. D* **40**, 1071 (1989).

[Vac01] T. Vachaspati, Estimate of the primordial magnetic field helicity, *Phys. Rev. Lett.* **87**, 251302 (2001).

[Ven07] R. Venugopalan, Multiparticle production in the Glasma at NLO and plasma instabilities, hep-ph/0702019.

[Vil83a] A. Vilenkin, Quantum fluctuations in the new inflationary universe, *Nucl. Phys.* **B226**, 527 (1983).

[Vil83b] A. Vilenkin, Birth of inflationary universes, *Phys. Rev. D* **27**, 2848 (1983).

[VilFor82] A. Vilenkin and L.H. Ford, Gravitational effects upon cosmological phase transitions, *Phys. Rev. D* **26**, 1231 (1982).

[VilLea82] A. Vilenkin and D.A. Leahy, Parity nonconservation and the origin of cosmic magnetic fields, *Ap. J.* **254**, 77 (1982).

[VilShe00] A. Vilenkin and E.P.S. Shellard, *Cosmic Strings and Other Topological Defects* (Cambridge University Press, Cambridge, 2000).

[Vil84] G.A. Vilkovisky, The gospel according to DeWitt, in *Quantum Theory of Gravity*, ed. S.M. Christensen (Hilger, Bristol, 1984).

[Vle28] J.H. Van Vleck, The correspondence principle in the statistical interpretation of quantum mechanics, Proc. N.A.S. **14**, 178 (1928).

[VolZha96] S. Voloshin and Y. Zhang, Flow study in relativistic nuclear collisions by Fourier expansion of azimuthal particle distributions, *Z. Phys. C* **70**, 665 (1996). (hep-ph/9407282).

[WWCH01] J. Wachter, R. Walser, J. Cooper and M. Holland, Equivalence of kinetic theories of Bose–Einstein condensation, *Phys. Rev. A* **64**, 053612 (2001).

[WWCH02a] J. Wachter, R. Walser, J. Cooper and M. Holland, Erratum: Equivalence of kinetic theories of Bose–Einstein condensation, *Phys. Rev. A* **65**, 039904 (2002).

[WWCH02b] J. Wachter, R. Walser, J. Cooper and M. Holland, Gapless kinetic theory beyond the Popov approximation, cond-mat/0212432.

[WaCoHo00] R. Walser, J. Cooper and M. Holland, Reversible and irreversible evolution of a condensed bosonic gas, *Phys. Rev. A* **63**, 013607 (2000).

[WWCH99] R. Walser, J. Williams, J. Cooper, M.J. Holland *et al.*, Quantum kinetic theory for a condensed bosonic gas, *Phys. Rev. A* **59**, 3878 (1999).

[Wan97] X.N. Wang, A pQCD-based approach to parton production and equilibration in high energy nuclear collisions, *Phys. Rep.* **280**, 287 (1997).

[WBVS00] S.Y. Wang, D. Boyanovsky, H. de Vega and D.S. Lee, Real time nonequilibrium dynamics in hot QED plasmas: dynamical renormalization group approach *Phys. Rev. D* **62**, 105026 (2000).

[WanHei99] E. Wang and U. Heinz, Non-perturbative calculation of shear viscosity, *Phys. Lett.* **471B**, 208 (1999).

[WanHei02] E. Wang and U. Heinz, A generalized fluctuation-dissipation theorem for nonlinear response functions, *Phys. Rev. D* **66**, 025008 (2002).

[WaHeZh96] E. Wang, U. Heinz and X. Zhuang, Viscosity in hot scalar field theory, *Phys. Rev. D* **53**, 5978 (1996).

[WRSG02] Q. Wang, K. Redlich, H. Stocker and W. Greiner, Kinetic equation for gluons in the background gauge of QCD, *Phys. Rev. Lett.* **88**, 132303 (2002).

[WRSG03] Q. Wang, K. Redlich, H. Stöcker and W. Greiner, From the Dyson–Schwinger to the transport equation in the background field gauge of QCD, *Nucl. Phys.* **A714**, 293–334 (2003).

[WegHou73] F. Wegner and A. Houghton, Renormalization group equation for critical phenomena, *Phys. Rev. A* **8**, 401 (1973).

[Wei72] S. Weinberg, *Gravitation and Cosmology: Principles and Applications of the General Theory of Relativity* (John Wiley, New York, 1972).

[Wei74] S. Weinberg, Gauge and global symmetries at high temperatures, *Phys. Rev. D* **9**, 3357 (1974).

[Wei95] S. Weinberg, *The Quantum Theory of Fields*, vol. I: *Foundations* (Cambridge University Press, Cambridge, 1995).

[Wei96] S. Weinberg, *The Quantum Theory of Fields*, vol. II: *Modern Applications* (Cambridge University Press, Cambridge, 1996).

[Wei00] S. Weinberg, *The Quantum Theory of Fields*, vol. III: *Supersymmetry* (Cambridge University Press, Cambridge, 2000).

[Wei05a] S. Weinberg, Quantum contributions to cosmological correlations, *Phys. Rev. D* **72**, 043514 (2005).

[Wei06] S. Weinberg, Quantum contributions to cosmological correlations. II. Can these corections become large?, *Phys. Rev. D* **74**, 023508 (2006).

[Wei05b] S. Weinstock, Boltzmann collision term, *Phys. Rev. D* **73**, 025005 (2006).

[Wei93] U. Weiss, *Quantum Dissipative Systems* (World Scientific, Singapore, 1993).

[Wei88] M. Weissbluth, *Photon-Atom Interactions* (Academic Press, San Diego, 1988), Chapter 1.

[WhiWat40] E. Whittaker and G. Watson, *A Course of Modern Analysis* (Cambridge University Press, Cambridge, 1940).

[WieHei99] U. Wiedemann and U. Heinz, Particle interferometry for relativistic heavy-ion collisions, *Phys. Rept.* **319**, 145–230 (1999).

[Wig32] E. Wigner, On the quantum correction for thermodynamic equilibrium, *Phys. Rev.* **40**, 749 (1932).

[Wig68] E.P. Wigner, Application of group theory to the special functions of mathematical physics (Unpublished lecture notes, Princeton University, Princeton NJ, 1955); see also J. Talman, Special functions: A group theoretical approach (Benjamin, New York, 1968), based on Wigner's Lectures.

[WilPic74] C.R. Willis and R.H. Picard, Time-dependent projection-operator approach to master equations for coupled systems, *Phys. Rev. A* **9**, 1343 (1974).

[Wil82] K. Wilson, The renormalization group and critical phenomena, *Rev. Mod. Phys.* **55**, 583 (1982).

[WilKog74] K. Wilson and J. Kogut, The renormalization group and the ϵ expansion, *Phys. Rep.* **12**, 75 (1974).

[Win06] S. Winitzki, Predictions in eternal inflation, gr-qc/0612164.

[WinVil00] S. Winitzki and A. Vilenkin, Effective noise in stochastic description of inflation, *Phys. Rev. D* **61**, 084008 (2000).

[Win84] J. Winter, Covariant extension of the Wigner transformation to non-Abelian Yang–Mills symmetries for a Vlasov equation approach to the quark-gluon plasma, *J. Phys. (Paris)* **45**, C6 (1984).

[Win85] J. Winter, Wigner transformation in curved space-time and the curvature correction of the Vlasov equation for semiclassical gravitating systems, *Phys. Rev. D* **32**, 1871 (1985).

[Won94] C.-Y. Wong, *Introduction to High-Energy Heavy Ion Collisions* (World Scientific, Singapore, 1994).

[Won70] S.K. Wong, Field and particle equations for the classical Yang–Mills field and particles with isotopic spin, *Nuovo Cim.* **65A**, 689 (1970).

[Won04] S.M.H. Wong, Out-of-equilibrium collinear enhanced equilibration in the bottom-up thermalization scenario in heavy ion collisions, hep-ph/0404222.

[Woo05a] R.P. Woodard, A leading logarithm approximation for inflationary quantum field theory, *Nucl. Phys. B* – Proceedings Supplements **148**, 108–119 (2005).

[Woo06] R.P. Woodard, Generalizing Starobinskii's formalism to Yukawa theory and to scalar QED, Proceedings of NEB XII, Nauplion, Greece, June 29 – July 2, 2006 (gr-qc/0608037).

[WuNgLeeLeeCha06] Chun-Hsien Wu, Kin-Wang Ng, Wolung Lee, Da-Shin Lee and Yeo-Yie Charng, Quantum noise and large-scale cosmic microwave background anisotropy, *JCAP* **0702**, 006 (2007).

[WuHoSa05] S. Wüster, J. Hope and C. Savage, Collapsing Bose–Einstein condensates beyond the Gross–Pitaevskii approximation, *Phys. Rev. A* **71**, 033604 (2005).

[WDBDBH07] S. Wurster et al., Quantum depletion of collapsing Bose–Einstein condensates, *Phys. Rev. A* **75**, 043611 (2007).

[YIKM06] D.G. Yamazaki, K. Ichiki, T. Kajino and G.J. Mathews, Constraints on the evolution of the primordial magnetic field from the small-scale cosmic microwave background angular anisotropy, *Astrophys. J.* **646**, 719–729 (2006).

[YatZur98] A. Yates and W. Zurek, Vortex formation in two dimensions: when symmetry breaks, how big are the pieces?, *Phys. Rev. Lett.* **80**, 5477 (1998).

[YiVis92] I. Yi and E.T. Vishniac, Scaling solution for cosmological fluctuations and large scale structure, *Phys. Rev. D* **45**, 3441 (1992).

[YiVis93] I. Yi and E.T. Vishniac, Stochastic analysis of the initial condition constraints on chaotic inflation, *Phys. Rev. D* **47**, 5280, 5295 (1993).

[YiVis93b] I. Yi and E.T. Vishniac, Inflationary stochastic dynamics and the statistics of large-scale structure, Astrophys. *J. Suppl.* **86**, 333 (1993).

[YiViMi91] I. Yi, E.T. Vishniac and S. Mineshige, Generation of non-Gaussian fluctuations during chaotic inflation, *Phys. Rev. D* **43**, 362 (1991).

[Yur02] V.A. Yurovsky, Quantum effects on dynamics of instabilities in Bose–Einstein condensates, *Phys. Rev. A* **65**, 33605 (2002).

[ZaLvFa92] V.E. Zakharov, V.S. L'vov and G. Falkovich, *Kolmogorov Spectra of Turbulence I: Wave Turbulence* (Springer-Verlag, Berlin, 1992).

[ZMCB98] V. Zanchin, A. Maia Jr., W. Craig and R. Brandenberger, Reheating in the presence of noise, *Phys. Rev. D* **57**, 4651 (1998).

[ZMCB99] V. Zanchin, A. Maia Jr., W. Craig and R. Brandenberger, Reheating in the presence of inhomogeneous noise, *Phys. Rev. D* **60**, 023505 (1999).

[ZanCal06b] J. Zanella and E. Calzetta, Renormalization group study of damping in nonequilibrium field theory, hep-th/0611222 (2006).

[ZanCal07a] J. Zanella and E. Calzetta, Inflation and nonequilibrium renormalization group, *J. Phys. A: Math. Theor.* **40**, 7037–7041 (2007).

[Zel70] Ya.B. Zel'dovich, Particle production in cosmology, *Pis'ma Zh. Eksp. Teor. Fiz*, **12**, 443 (1970) (*JETP Lett.* **12**, 307 (1970)).

[Zel72] Ya.B. Zel'dovich, A hypothesis, unifying the structure and the entropy of the Universe, *MNRAS* **160**, 1P (1972).

[ZelNov85] Ya.B. Zel'dovich and I.D. Novikov, *Relativistic Cosmology*, vol. 2 (University of Chicago Press, Chicago, 1985).

[ZelSta71] Ya.B. Zel'dovich and A.A. Starobinsky, Particle production and vacuum polarization in an anisotropic gravitational field, *Zh. Teor. Eksp. Fiz.* **61**, 2161 (1971) (*Sov. Phys. JETP* **34**, 1159 (1972)).

[Zha90] Yuhong Zhang, Stochastic properties of interacting quantum fields, Ph.D. Thesis, University of Maryland, 1990.

[ZhaHei96a] P. Zhuang and U. Heinz, Relativistic kinetic equations for electromagnetic, scalar and pseudoscalar interactions, *Phys. Rev. D* **53**, 2096 (1996).

[ZhaHei96b] P. Zhuang and U. Heinz, Relativistic quantum transport theory for electrodynamics, *Ann. Phys.* (N.Y.) **245**, 311 (1996).

[ZhuHei98] P. Zhuang and U. Heinz, Equal-time hierarchies for quantum transport theory, *Phys. Rev. D* **57**, 6525 (1998).

[ZiBrSc01] J.P. Zibin, R. Brandenberger and D. Scott, Backreaction and the parametric resonance of cosmological fluctuations, *Phys. Rev. D* **63**, 043511 (2001).

[Zin93] J. Zinn-Justin, *Quantum Field Theory and Critical Phenomena* (Clarendon Press, Oxford, 1993).

[Zub74] D.N. Zubarev, *Nonequilibrium Statistical Thermodynamics* (Plenum, New York, 1974).

[Zur81] W.H. Zurek, Pointer basis of quantum apparatus: Into what mixture does the wave packet collapse?, *Phys. Rev. D* **24**, 1516 (1981).

[Zur82] W.H. Zurek, Environment-induced superselection rules, *Phys. Rev. D* **26**, 1862 (1982); and in *Frontiers of Nonequilibrium Statistical Physics*, ed. G.T. Moore and M.O. Scully (Plenum, New York, 1986).

[Zur85] W.H. Zurek, Cosmological experiments in superfluid helium?, *Nature* **317**, 505 (1985).

[Zur91] W.H. Zurek, Decoherence and the transition from quantum to classical, *Physics Today* **44**, 36 (1991).

[Zur93] W.H. Zurek, Preferred states, predictability, classicality and the environment-induced decoherence, *Prog. Theor. Phys.* **89**, 281 (1993).

[Zur96] W.H. Zurek, Cosmological experiments in condensed matter systems, *Phys. Rep.* **276**, 177 (1996).

[Zur03] W.H. Zurek, Decoherence, einselection, and the quantum origins of the classical, *Rev. Mod. Phys.* **75**, 715–775 (2003).

[ZuHaPa03] W.H. Zurek, S. Habib and J.P. Paz, Coherent states via decoherence, *Phys. Rev. Lett.* **70**, 1187 (1993).

[ZurTho85] W.H. Zurek and K.S. Thorne, Statistical mechanical origin of the entropy of a rotating, charged black hole, *Phys. Rev. Lett.* **54**, 2171 (1985).

[Zwa60] R. Zwanzig, Ensemble method in the theory of irreversibility, *J. Chem. Phys.* **33**, 1338 (1960).

[Zwa61] R. Zwanzig, Statistical mechanics of irreversibility, in *Lectures in Theoretical Physics* III, ed. W.E. Britten, B.W. Downes and J. Downs (Interscience, New York 1961) pp. 106–141.

[Zwa73] R. Zwanzig, Nonlinear generalized Langevin equations, *J. Stat. Phys.* **9**, 215 (1973).

[Zwa01] R. Zwanzig, *Nonequilibrium Statistical Mechanics* (Oxford University Press, Oxford, 2001).

Index

Printed in the United States
by Baker & Taylor Publisher Services